GEOLOGICAL EVOLUTION OF NORTH AMERICA

GEOLOGICAL EVOLUTION OF NORTH AMERICA

THIRD EDITION

Colin W. Stearn
Robert L. Carroll
Thomas H. Clark
McGill University

JOHN WILEY & SONS
NEW YORK · SANTA BARBARA · CHICHESTER · BRISBANE · TORONTO

Designed By Fern Logan
Cover Photo by Eliot Porter

Library of Congress Cataloging in Publication Data:

Stearn, Colin William, 1928–
Geological evolution of North America.

First–2nd ed. by T. H. Clark and C. W. Stearn.
Includes bibliographical references and indexes.
1. Geology—North America. I. Carroll, Robert Lynn, 1938– joint author. II. Clark, Thomas Henry, 1893– joint author. III. Title.
QE71.S88 1978 551.7′0097 78-8124
ISBN 0-471-07252-4

Printed in the United States of America

10 9 8 7 6 5 4 3 2 1

PREFACE

Rapid changes in the basic concepts of historical geology within the last decade have outdated any but the most recently published textbooks. When the second edition of the *Geological Evolution of North America* was published, the impact of studies of the ocean basins on the interpretation of earth history was just beginning to be felt. By the middle of the 1970s, the basic tenets of the "new global tectonics"—continental motion, the nature of the boundaries of the lithosphere plates, the apparent movement of the poles—had been widely accepted by historical geologists; they now form the basis of what might be called the "new" historical geology. Although difficulties in the interpretation of the history of the earth on the basis of plate motion and relative changes in polar positions remain, the basic principles of the new global tectonics are widely accepted. We attempt to summarize the geological history of North America in these terms.

The second edition has been completely rewritten and reorganized. We have retained the original concept, that the history of the continent is best considered from a regional approach and that the discipline of historical geology is better introduced by studying the history of a single continent instead of surveying the history of the whole world to gain a superficial knowledge. Part One of this edition starts with two chapters in which the basis of global tectonics is explained and the major tectonic units of North America are introduced. Part Two contains chapters on the interpretation of the stratigraphic record and the construction of the relative and absolute time scales. In these chapters the discussion of paleontology and biostratigraphy has been integrated with that of sedimentation and stratigraphy. The integration of the sections on paleontology with those on the physical evolution of the continent is continued throughout the book. Part Two ends with a chapter on the evolution and origin of life.

Part Three starts with a discussion of the Canadian Shield. The chapters on the platform rocks are organized on the basis of the sequences proposed by L. L. Sloss, as in the second edition, but the integration of the paleontological sections into these chapters has increased their size, so that each sequence of the Paleozoic is treated in a separate chapter.

The descriptions of the marginal mountain chains of the continent—the Cordilleran, Appalachian, and Innuitian belts—have been completely rewritten in the light of new information and the interpretation of this information in terms of plate tectonics. Cordilleran geology offers an opportunity to discuss the interaction between oceanic and continental lithosphere plates over long periods of time. The history of the Appalachian chain is used to illustrate the results of continental rifting and plate collision.

Throughout the book, we have tried to give students a knowledge of the principles by which the earth's history is interpreted as they learn the geology of the North American continent. For this reason certain typical patterns of sedimentation, or steps of evolution, have been explained in detail. Like its predecessors, the book has been designed to be used in a course for students who have already taken a semester of physical or general geology. It could be used by students without a previous course in geology who have had some instruction in the meaning of rock names.

For this edition the original authors have been joined by Robert Carroll, who is responsible for the preparation of the sections of the text on vertebrate paleontology, paleobotany, and evolution. Colin Stearn is primarily responsible for the sections on the physical evolution of the continent and invertebrate paleontology. All the authors have critically read and contributed to the whole text.

The diagrams for this edition have been almost entirely redrafted by Mary Joan Stearn, whom we thank for her meticulous and talented work. Although the figures have been taken from many sources, they have all been redrawn to a consistent level of detail. We are grateful to the many scientists who have allowed us to use illustrative material from their research; their contributions are acknowledged in the figure captions.

We also thank the many colleagues who read parts of the manuscript and commented critically on them. We are particularly grateful to Eric Mountjoy, who read many of the chapters. Other colleagues who have helped us in this way include Ronald Doig, John Elson, Reinhard Hesse, T. E. Morris, Andrew Hynes, Rennick Spence, and Hans Trettin.

Finally, we offer our thanks to all the field geologists and paleontologists whose original work on the outcrop and subsurface section have made a compilation of the evolution of North America possible.

Colin W. Stearn
Robert L. Carroll
Thomas H. Clark

CONTENTS

CREDITS FOR PART OPENING PAGES

PART ONE The Gulf of Aden and the mouth of the Red Sea looking northeast from a Gemini spacecraft. The spreading of the crust in the ocean sectors is separating the Arabian plate on the top from the African plate at the bottom. The recent nature of the rifting is shown by the matching of the coastlines (Photograph courtesy of NASA)

PART TWO Beds of the Upper Cambrian Potsdam Sandstone. Ausable Chasm, New York. (Courtesy of the Ausable Chasm Co.)

PART THREE Unconformable contact between the Pre-Paleozoic metamorphic rocks of the Shield in the bed of the Churchill River and flatlying Upper Ordovician limestone of the platform, northern Manitoba. (Courtesy of the Geological Survey of Canada, Ottawa, 20841-B)

PART FOUR Folded and faulted Devonian rocks of the Rocky Mountains at Mount Strange, about 30 km north of Jasper, Alberta. (Courtesy of Chrevon Standard Limited.)

PART FIVE Topography of the Valley and Ridge Province in southern Pennsylvania. North is at the bottom. Image produced by side-scanning radar. (Courtesy of the Remote Sensing Laboratory, Center for Research Inc. University of Kansas.)

PART SIX Glaciers flow outward from the Viking Ice Cap, Ellesmere Island. (Courtesy of the Department of Energy, Mines and Resources, Ottawa.)

PART ONE

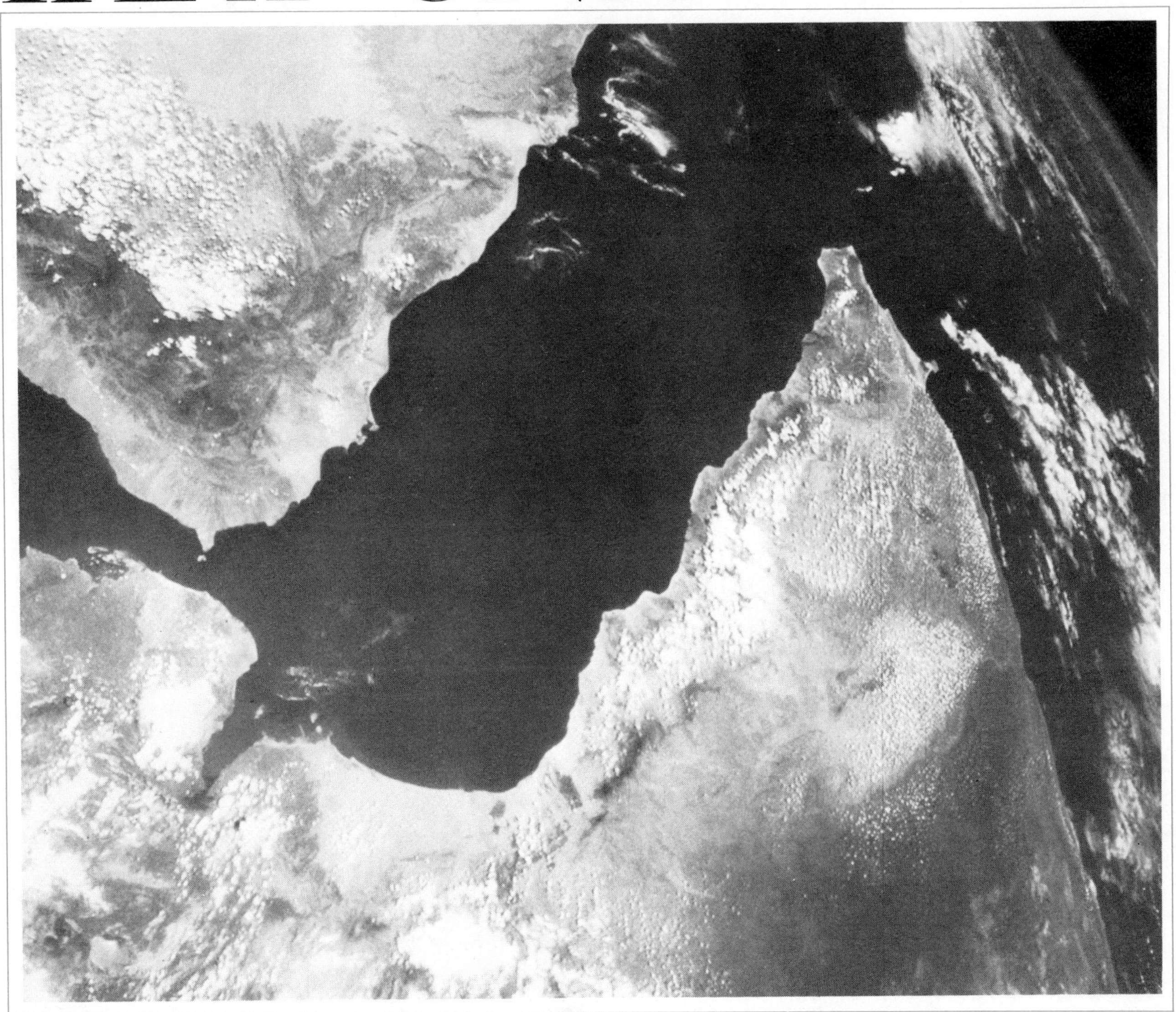

framework of the earth

CHAPTER ONE

The earth's crust today is the product of millions of years of change, shaped by external and internal forces and bearing the scars of these forces for us to interpret.

Although much of our knowledge of the earth's history is restricted to changes in its surface, these changes are the result of the interaction of surface processes with uplift, subsidence, deformation, and metamorphism controlled by forces arising deep within the earth's interior. Geologists can investigate surficial processes such as weathering, erosion, and sedimentation directly, but processes of the interior must be deduced from studies of seismic waves, magnetic fields, variations in gravity, and the products of volcanoes. Although some of the physical properties of the interior of the earth are established, geologists do not generally agree on the processes at work beneath the crust, and reconstruction of the nature and rates of subcrustal processes in the geological past is highly speculative. Geologists know that the earth beneath the crust is not static; it generates heat, ejects material through volcanoes, moves in earthquakes, generates the earth's magnetic field, and controls the rise and fall of continents and oceans. From greatly accelerated studies of the oceans within the last decade, geologists have produced a synthetic theory of how the earth works.

THE EARTH'S INTERIOR

The **core** of the earth is about 7000 km in diameter (Fig. 1–1). Earthquake waves penetrating the core reveal a solid, inner core 2500 km in diameter, and a liquid, outer core. The entire core is believed to be an alloy of nickel and iron at temperature of

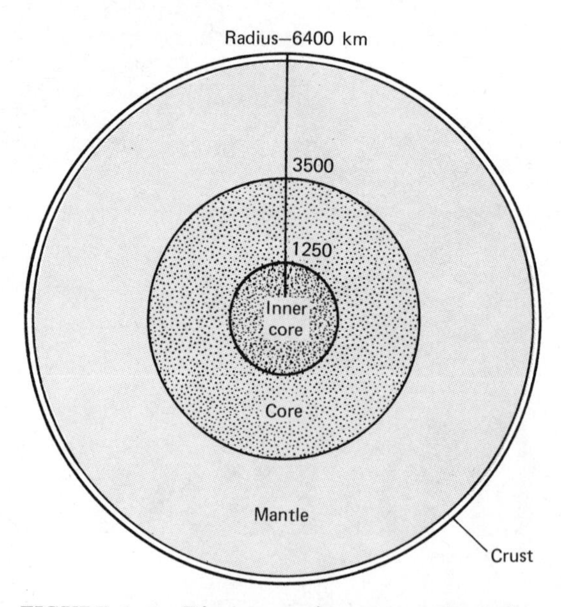

FIGURE 1–1 **Diagrammatic cross-section of the earth showing the interior zones.**

CONTINENTS AND OCEANS

around 4000°C. As the planet rotates, the motion of the liquid outer core is thought to be responsible for the magnetic field of the earth.

The **mantle** forms a shell almost 3000 km thick that surrounds the core. The upper part of the mantle is probably peridotite, a rock composed of olivine and plagioclase. Peridotite is found rarely on the earth's surface, only where deformation or igneous intrusion has brought up samples from deep within the earth. The properties of the lower mantle, inferred from the transmission of earthquake waves and vertical temperature and pressure gradients, are unlike those of any surface rock. Under extreme pressure in the earth, compounds with which we are familiar at the surface would take forms in which their atomic particles were closely packed to save space. Such high-density silicates and oxides of magnesium and iron are probably the main constituents of the lower mantle.

The **crust** of the earth is an irregular layer 12 to 75 km thick, completely enclosing the mantle. Its lower boundary is marked by a change in velocity of earthquake waves first recognized by a Yugoslav seismologist, Andrija Mohorovicic, and therefore called the **Mohorovicic discontinuity** (Fig. 1–2). The crust is thickest under large mountain chains like the Himalayas, and thinnest under ocean basins. The crust is divided into a lower part composed of basaltic rocks called **sima**, and an upper

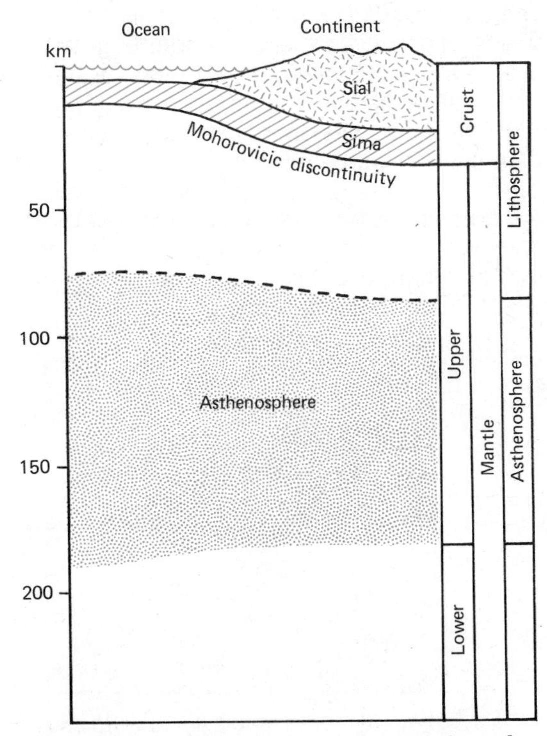

FIGURE 1–2 **Diagrammatic cross-section of the upper zones of the earth at the edge of a continent.**

part composed of granitic rocks called **sial.** The sialic layer is confined to the continents and thins and disappears at the borders of the ocean basins, which have floors entirely of basalt, the simatic layer.

Studies of the gravity field of the earth show that the major mountain chains are underlain by segments of the crust that are less dense than those under areas that are topographically lower. The greater mass of the mountains is compensated by the lesser density of the crust below, so that at a certain level in the earth (the **level of compensation,** at about 60 km) each segment of the crust exerts the same weight. The various parts of the earth's crust appear to be in a state of flotational equilibrium that is called **isostasy,** like blocks of wood of different lengths floating in a body of water and rising to different heights above its surface (Fig. 1–3). Like such blocks, the parts of the crust respond to loading at the surface by sinking, and to removal of load by rising. The sialic upper layer of the crust is composed of granitic rocks of a specific gravity of 2.6–2.7. The simatic layer below, of basaltic rocks, has a specific gravity of 2.8–2.9. The continents, therefore, stand higher than the oceans because this section of the earth's crust and subjacent mantle above the compensation level contains a greater proportion of lighter material than the crust and mantle above this level under the oceans (Fig. 1–4). In areas of large mountain ranges, such as the Himalayas, study of earthquake wave paths shows that the lighter sialic crust is greatly thickened, forming a root whose buoyancy supports the mass of the mountains. Not all compensation for topography can be accounted for by variations in the thickness of the crust. Variations in the density both of the crust and the upper mantle probably play a role in isostatic compensation.

Seismologists have outlined a zone 75 to 175 km below the top of the mantle in which the velocity of seismic waves drops by about 6 per cent. In this low-velocity layer, the temperature–pressure relationship is postulated to be such that the mantle is partially melted and therefore has little strength. As little as 1 per cent of the rock need be melted to explain its properties. It is much like wax in that it is weak when under long-term stress. This layer is called the **asthenosphere;** the solid uppermost mantle and crust together are called the **lithosphere** (Fig. 1–2).

Most of the mountain ranges of the world are associated with boundaries between oceans and continents. The major exception to this generalization is the Alpine–Himalaya chain between Africa and Europe and India and the rest of Asia. Even this moun-

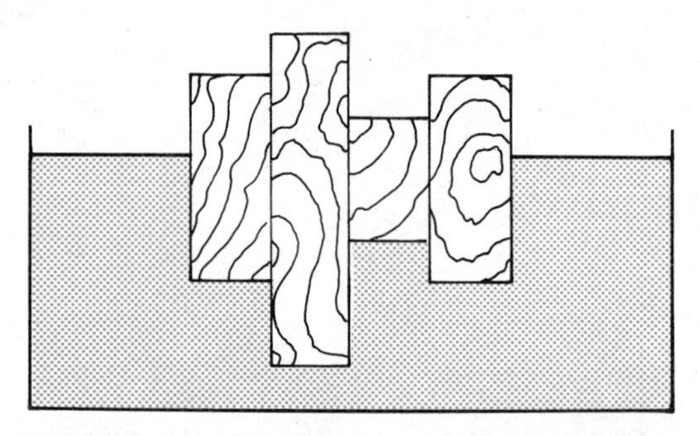

FIGURE 1–3 **Blocks of wood floating in water illustrate the principle of isostasy. The surface topography is determined by the shape of the blocks and their density.**

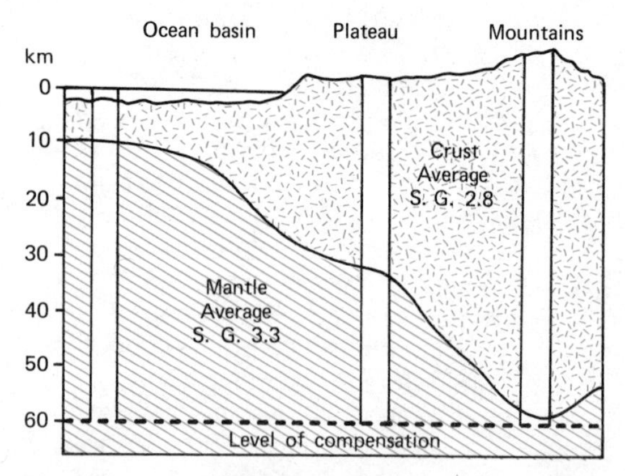

FIGURE 1–4 **Isostatic compensation by the "roots of mountains" hypothesis. The three columns have the same mass at the level of compensation.**

tain system was born of the events taking place at an ocean–continent junction, but the oceanic area has since been eliminated by the collision of continental blocks. To understand this interaction of continents and oceans we must examine the structure of the oceans.

MID-OCEAN RIDGES

Modern depth-sounding instruments operating on the sonar principle developed during the Second World War have revolutionized our concept of the topography of the ocean floor in the last 20 years. With sounding lines the oceanographers of the 19th century were able to trace only the vague outlines of vast ridges traversing the axes of the ocean basins. Continuously recording depth-sounders have mapped the details of this **mid-ocean ridge** system, and it is now recognized as the major mountain system of the world, stretching for 40,000 km with a relief of 3000 m. The Mid-Atlantic Ridge is the best known segment of the ridge system (Fig. 1–5). It follows a sinuous course in the Atlantic Ocean, approximately mid-way between the bordering continents. The relief of the ridge increases from the adjacent abyssal plains, through a foothills belt, to the axis, which is marked by jagged fault scarps and a central valley similar in form to the rift valley of Africa. Photographs taken by deep-diving submersibles show the axial zone of the ridge to be a chaotic landscape dominated by evidence of recent underwater volcanic eruptions (Fig. 1–6). The ridge breaks sea level in Iceland, the Azores, and a few volcanic islands of the South Atlantic, but for most of its length it is about 300 m below the surface.

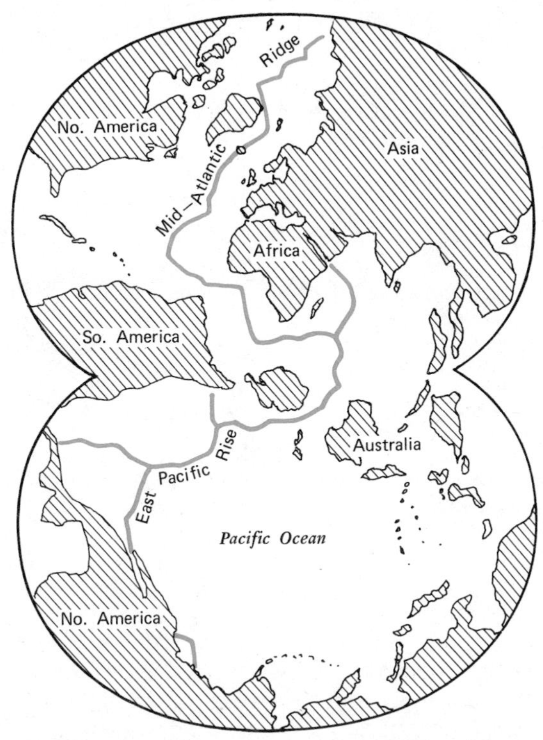

FIGURE 1–5 **The system of mid-ocean ridges. Antarctica is in the center of the projection; the continents surrounding the Atlantic are in the upper hemisphere, and those surrounding the Pacific are in the lower hemisphere. (After J. T. Wilson; courtesy of the *American Scientist*.)**

The mid-ocean ridge system can be followed eastward from the Mid-Atlantic Ridge around the Cape of Good Hope into the Indian Ocean, where it divides into two branches, a branch continuing eastward between Australia and Antarctica and a branch trending northward toward the Red Sea (Fig. 1–5). The southern branch crosses the South Pacific Ocean and then swings northward to form the East Pacific Rise. The north end of the Mid-Atlantic Ridge crosses the Arctic Ocean and appears to end at the Siberian coastline.

The nature of the mid-ocean ridge system is revealed by several lines of evidence. Areas of the ridge above water are composed of very young or active volcanoes. The flow of heat from the interior of the earth at the axes of the ridges is high, and evidence of subsea volcanism and shallow-focus earthquakes indicate the highly unstable nature of the crust. The age of volcanic islands in the oceans increases regularly away from the central ridges. Where the ridges intersect continental crust, as in the Red Sea and the Gulf of California, the matching of the opposing coasts shows

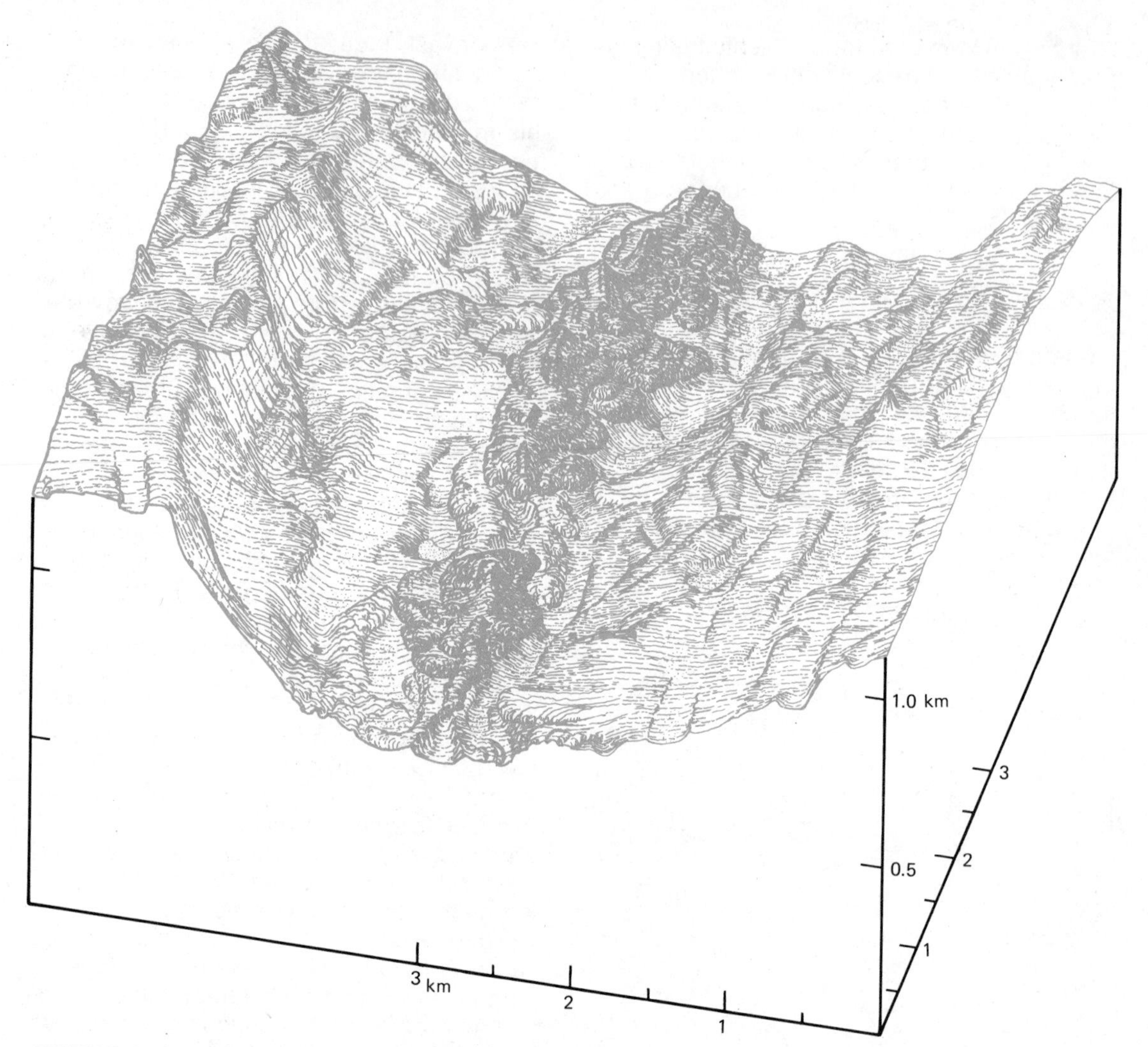

FIGURE 1–6 **Orthographic drawing of the central part of the rift valley of the mid-Atlantic Ridge. The youngest lava in the center is darkest. Vertical exaggeration × 2. (After J.G. Moore, H.S. Fleming, and J.D. Phillips; Courtesy of the Geological Society of America.)**

that the crust is pulling apart. Although this sort of evidence suggests spreading of the crust and upwelling of volcanic material along the ridges, the evidence that is most convincing for this explanation comes from study of magnetic anomalies of the sea floor.

As explained more fully in Chapter 7, the polarity of the earth's magnetic field periodically reverses, that is, the North Pole becomes the South Pole, and vice versa. Within the last 2 million years this reversal has occurred nine times. The polarity of the field is recorded in the magnetic minerals that crystallize from molten igneous rocks. A thick succession of lava flows rich in iron minerals may record several reversals that took place during the interval of repeated extrusion. Magnetic surveys of the oceans reveal that the sea floor shows anomalies from the normal magnetic field arranged as stripes parallel to the axes of the mid-ocean ridge system (Fig. 1–7). Following an idea proposed by Harry Hess that the ridges

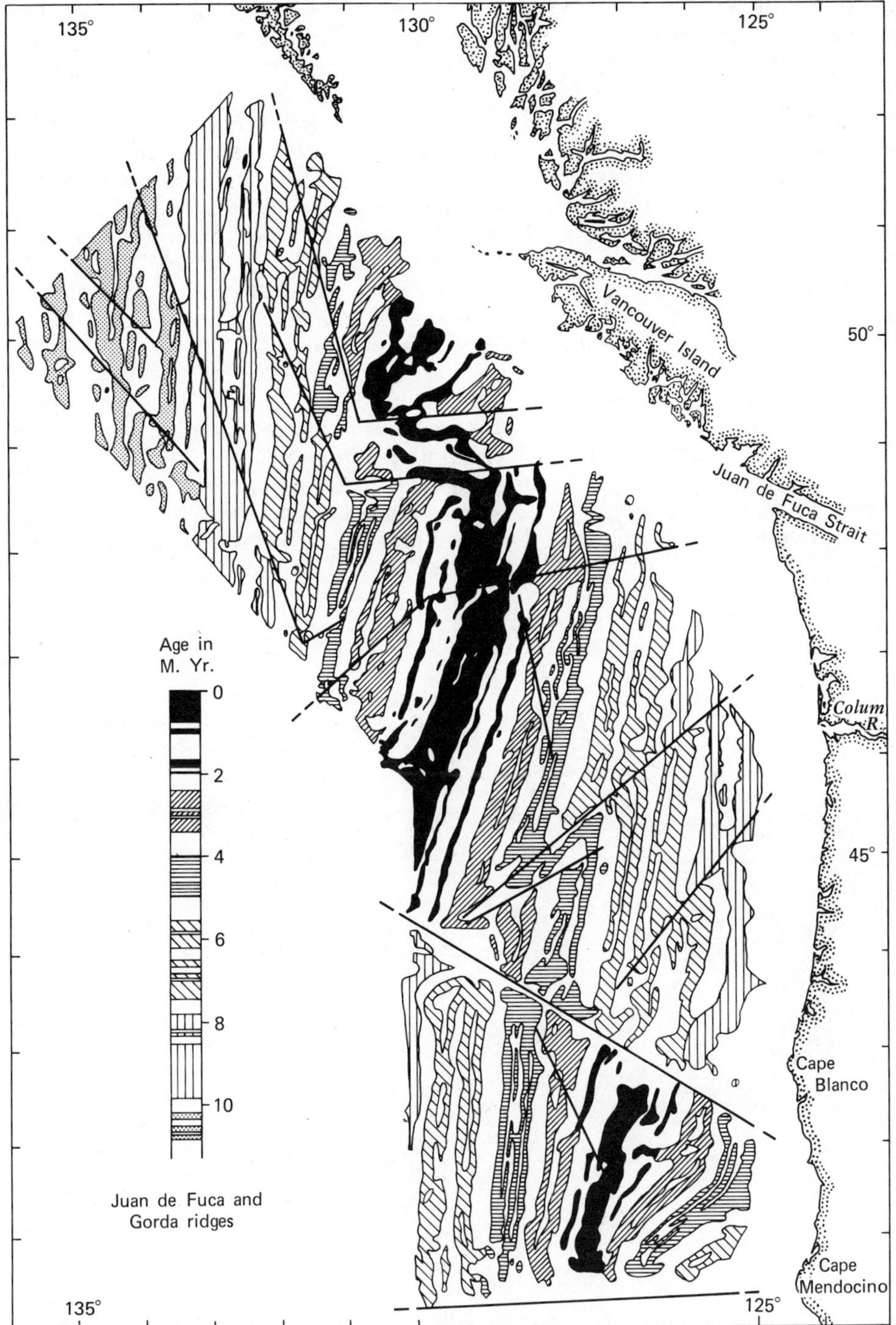

FIGURE 1–7 **Pattern of magnetic anomalies in the northwest Pacific. The scale for the ages of the anomalies, to the left of the diagram, indicates the reversal of the magnetic polarity of the earth in the last 10 million years. Patterned segments of the column and the magnetic stripes on the ocean floor indicate times of normal polarity. The Juan da Fuca and Gorda spreading centers—displaced by transform faults—are indicated by solid pattern. (After E.A. Silver, F.J. Vine, A.D. Raff, and R. G. Mason; courtesy of the Geological Society of America.)**

might be centers from which the ocean floor was spreading outward, Frederick Vine and Drummond Matthews concluded that the magnetic anomaly stripes were caused by changes in the earth's polarity and preserved in the extruded lavas as new crust formed at the ridge crest. Each band represents the lava extruded during a period of stable polarity. The stripes are more or less symmetrical on either side of the mid-ocean ridges, as would be expected on the basis of this hypothesis (Fig. 1–8). Most of the known facts about mid-ocean ridges are explained by the theory that they are areas where material from the mantle is welling up as dike intrusions and volcanic extrusions, and increasing the area of the ocean floor, which moves outward in response to this axial growth. Growth is not uniform along the whole length of the ridge, but the anomaly pattern suggests that the rate of spreading averages between 1 and 7 cm per year.

The oceanic areas on either side of the mid-ocean ridges are partial spherical shells, moving away from each other over the weak layer in the mantle. The geometry of movement of such shells on a sphere was investigated in the 18th century by the mathematician Leonard Euler. Euler showed that any movement of a shell with respect to another stationary shell can be resolved into simple rotation around a pole (Fig. 1–9). Such rotation of one shell away from another would result in the opening between them of a pie-shaped area, wide at the equator of the sphere and narrow toward the pole of spreading. Irregularities in the earth's lithosphere preclude the opening of a straight rift and make the course irregularly zigzag. The solidification of new lithosphere in the

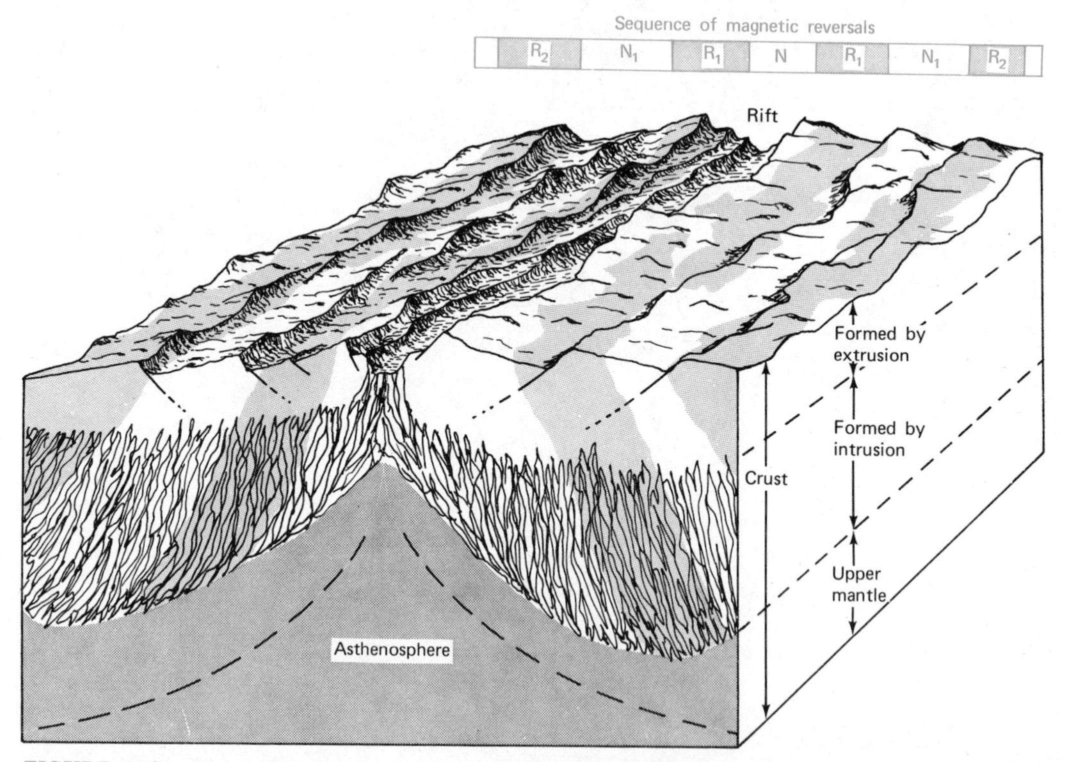

FIGURE 1–8 **Block diagram of a mid-ocean ridge spreading zone, showing how the upwelling of the mantle forms the lithosphere by the intrusion of masses of dikes (irregular line pattern) and the extrusion of submarine lavas. The reversing polarity of the earth's magnetic field produces successive sectors of the lithosphere with alternating remanent magnetic polarities. N = normal, R = reversed.**

spreading zone tends to reduce these irregularities to straight rift segments offset by a series of faults that parallel the direction of spreading and are concentric with the spreading pole. These faults were called **transform faults** by Tuzo Wilson. The horizontal shearing motion along them is confined to the segments joining the spreading axis, but traces of formerly active segments may stretch far into the adjacent shells (Fig. 1–10). The presence of such transform faults crossing the mid-ocean ridge systems in all parts of the world is additional confirmation of the nature of the ridge system.

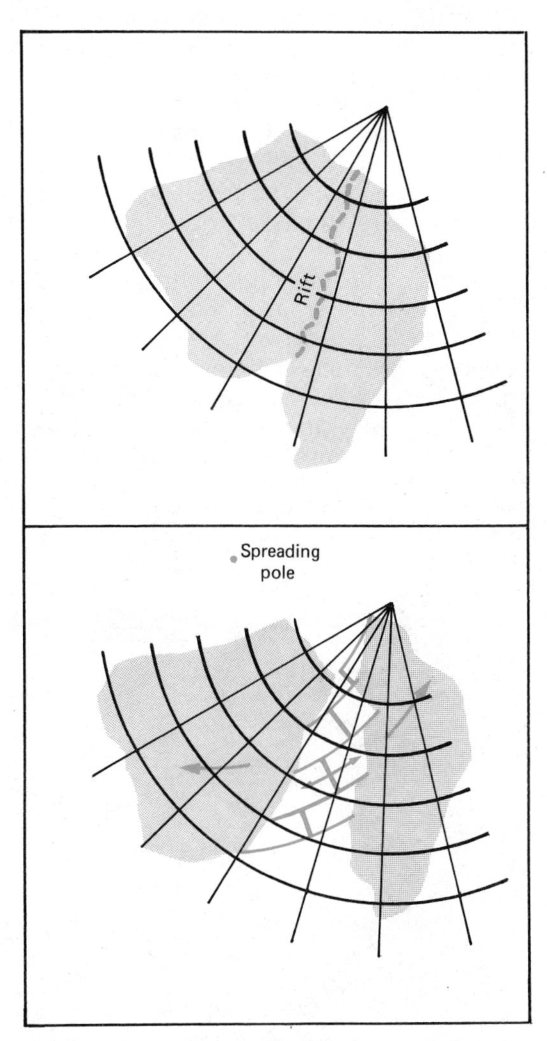

FIGURE 1–9 **The separation of two continents along an irregular rift zone and the formation of transform faults. The radial and concentric lines provide a reference frame for the movement of the plates. The continents move away from each other by rotation of the plates about a spreading pole to which the transform faults are concentric.**

OCEANIC TRENCHES

The deepest part of the ocean basins is the Vityaz Deep (10,860 m) east of the Philippines. It is one of a series of trenches that border the volcanic island arcs of the western Pacific on the oceanward side. From the Philippines the trench is essentially contin-

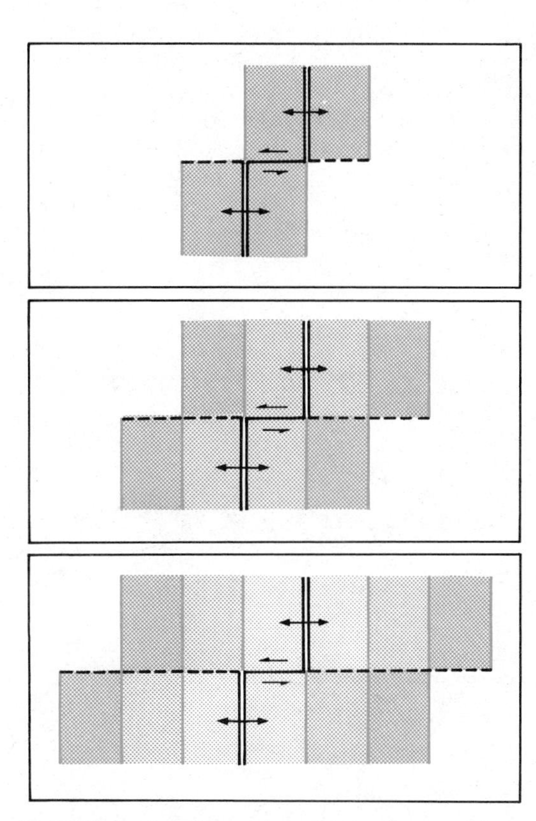

FIGURE 1–10 **Geometry of transform faulting. The successive segments of oceanic lithosphere added at the spreading centers are indicated by shading. Movement along the transform faults is confined to the segment between the spreading centers; the inactive parts of the faults are shown by dashed lines.**

uous northward along the Ryukyu Islands, through the Japanese and Kurile Trenches to the Aleutian Trench that forms the northern border of the Pacific Basin. The eastern border of the Pacific Ocean along the Central and South American coast is marked by a trench bordered on the east by continental crust, not volcanic islands.

Earthquake foci associated with the trench–island arc systems define zones dipping away from the ocean basins beneath the island arcs. These zones are known as Benioff zones, after the American seismologist, Hugo Benioff, who delineated them. The topography, volcanism, and earthquake distribution, along with other geophysical evidence, suggest that along these trenches the lithosphere, composed of crust and subjacent mantle, is being pushed deep into the mantle where it is melting and being assimilated (Fig. 1–11). Such areas have been called **subduction zones** because the lithosphere is being led, or pulled down, into the lower mantle. The assimilation of the lithosphere plate is believed to give rise to the characteristic volcanism of the island arc or continental border above the descending slab. Evidence of subduction at the trenches produces a consistent picture of the oceanic lithosphere, created at the spreading centers of the mid-ocean ridges and destroyed at the subduction zones of the trenches.

PLATE TECTONICS

If two world maps, one showing the location of earthquake foci, the other the position of mid-ocean spreading axes and the peripheral subduction zones, are superposed, the parts of the earth's surface between the latter features are shown to be stable and almost without quakes. The only zones of the crust generating earthquakes are those that are being formed in spreading centers, consumed in subduction zones, or are shearing past each other on transform faults. In the late 1960's several geologists recognized that earthquake belts define the boundaries of rigid and stable lithosphere plates (Fig. 1–12). Most plates include both continental and oceanic crustal segments. Most of the earth's crust can be divided into six great plates: Americas, Eurasia, Africa, India, Antarctica, and Pacific. As further plate boun-

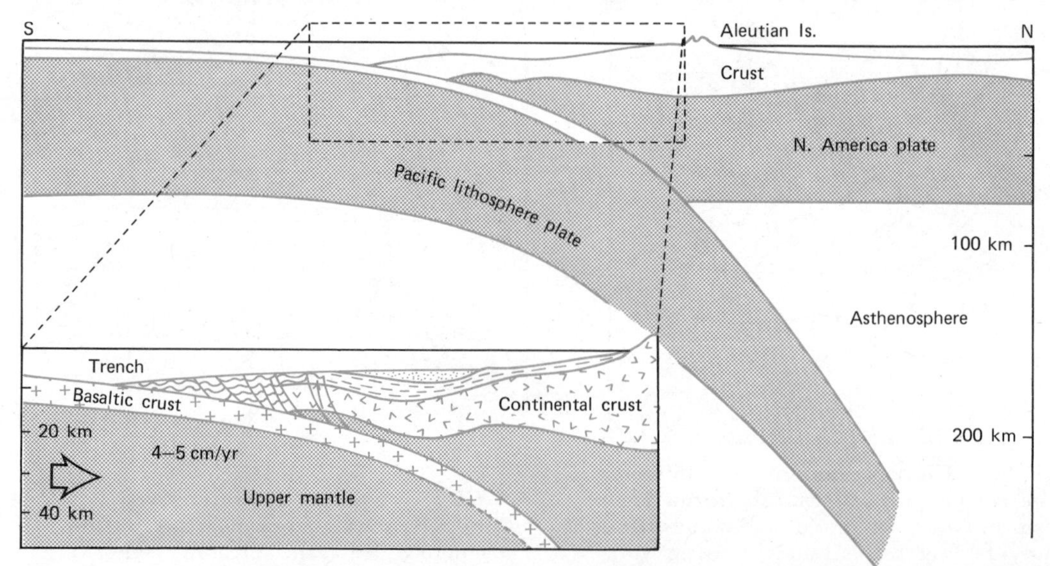

FIGURE 1–11 **Interpretive cross-section of the Aleutian trench from seismic and gravity data. The region of the trench is enlarged at the lower left to show the accretionary prism of sediments at the subduction zone. No vertical exaggeration. (After J.T. Grow)**

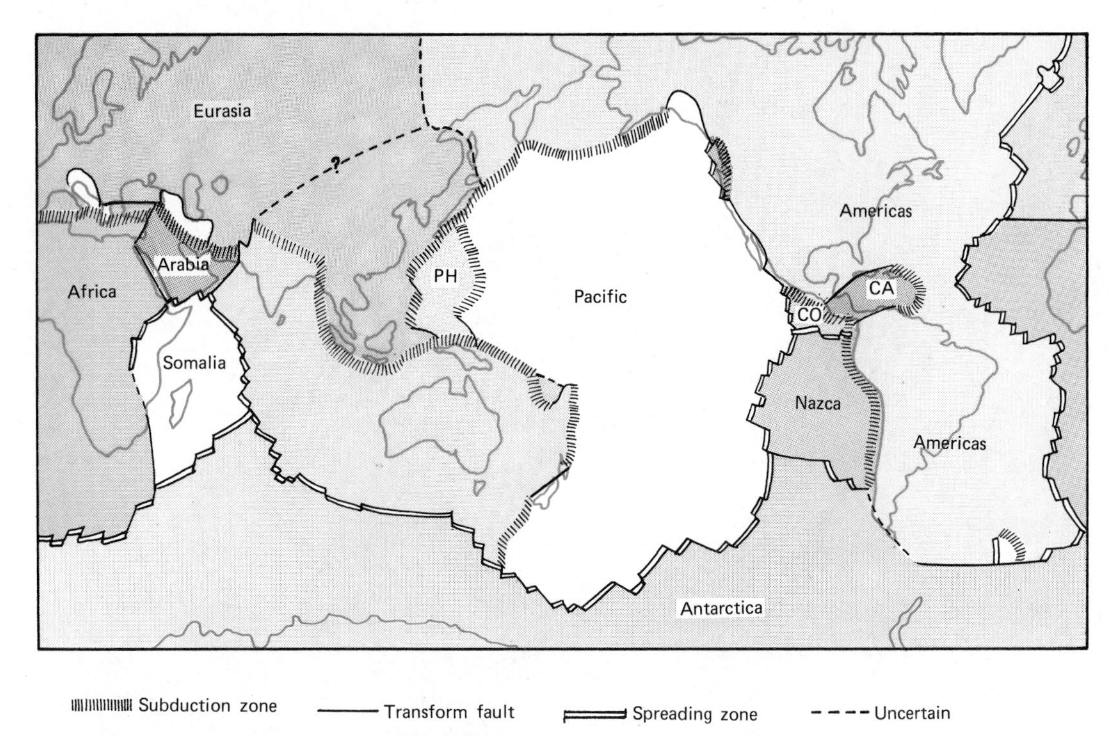

FIGURE 1–12 **The major lithosphere plates of the world showing the nature of the plate boundaries. PH = Philippines, CO = Cocos, CA = Caribbea. (Modified from several sources.)**

daries were recognized, six additional plates were named: Nazca, Somalia, Philippines, Arabia, Caribbea, and Cocos. Several smaller plates are defined by the complex relationships of spreading centers, subduction zones, and transform faults in the southwestern Pacific and the Mediterranean region.

Plate boundaries may be of three types: spreading centers, subduction zones, and transform faults. The boundaries of the northern section of the Americas plate will illustrate the complex relationships between these types (Fig. 1–13). The Americas plate is bounded on the east by the Mid-Atlantic spreading center, which extends from the Siberian coast across the Arctic Ocean, through Iceland, and into the axis of the Atlantic Ocean. The southern boundary against the Caribbean plate is marked by a complex of trench subduction zones and transform faults. The Puerto Rico Trench appears to mark a subduction zone, but the topographic expression is masked southward toward Barbados, probably by filling with sediments. On either end the Puerto Rico subduction zone joins large transform faults, the northern one of which crosses Guatemala and is expressed on the Caribbean sea floor as the Cayman Trench. These transform faults join the Middle America Trench, which follows the Pacific coastline of Central America. The spreading center of the East Pacific Rise meets North America at the Gulf of California, and is displaced northwestward along transform faults in southern California to the sea floor off the Oregon–Washington coast. The largest of these transform faults is the San Andreas fault. The western edge of the Americas plate off British Columbia is a transform fault along which the northward-moving Pacific plate is shearing past the northwestward-moving Americas plate and being

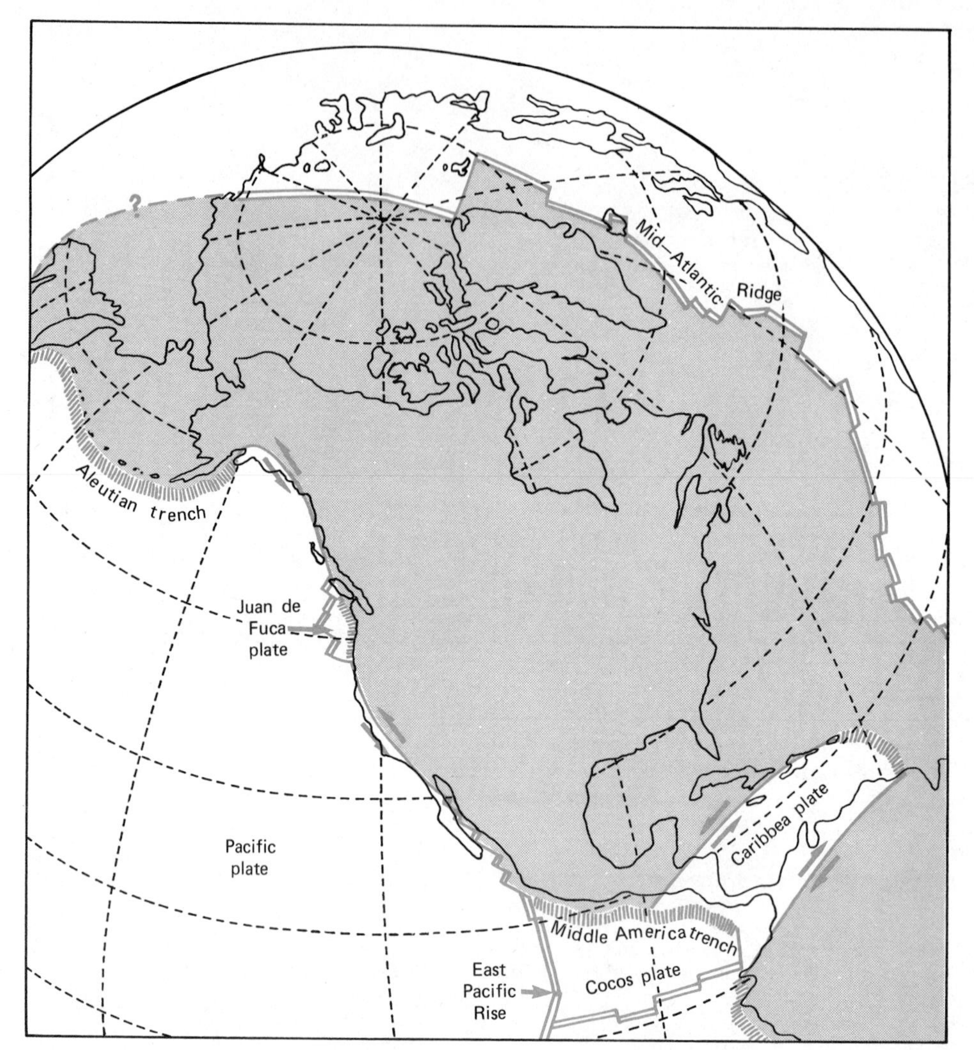

FIGURE 1–13 **Nature of the boundaries of the north segment of the Americas plate.**

subducted along the trench that marks the southern margin of the Aleutian arc. The tip of Siberia east of the Verkhoyansk Mountains is thought to be part of the Americas plate, but the nature of its boundary with the Eurasian plate is doubtful.

On flat maps the directions of plate motion within a single plate may appear to diverge due to distortions caused by projection. The geometry of the motions is best illustrated on a globe with spherical shells that can be fitted to the surface.

The direction of movement of the plates can be determined by the movements of the earth in quakes along their boundaries, by the geometrical relationships of spreading centers and subduction zones, and by the displacement of geological markers in the crust along transform faults. The motion of the lithosphere plates is made possible by the low strength of the asthenosphere. The rate of movement can be measured by the spacing of the magnetic anomalies on either side of the spreading centers. Another method of determining the movement of a plate is illustrated by the Hawaiian Islands. These islands are the eastern end of a long line of seamounts that stretches across the

Pacific toward the Aleutians. The lavas of the islands and seamounts increase in age westward, away from the presently active volcanic center near the easternmost island, Hawaii (Fig. 1–14). The distribution of ages along the seamount chain can be explained logically if the present site of Hawaii is a "hot spot" or "plume" in the mantle below the lithosphere over which the Pacific plate moves northwestward, leaving behind a line of volcanoes. This explanation is confirmed by evidence from other sources, that the movement of the Pacific plate for the past tens of millions of years has been northwestward, parallel to the line of seamounts. About 3500 km west of Hawaii the seamount chain turns northward at a conspicuous angle, and beyond this point is called the Emperor seamount chain. The angle records a change in the direction of motion of the plate 40 million years ago. The line of seamounts indicates only relative motion between the plume and the overlying plate, and does not require that the plume itself be stationary with respect to the mantle or other plates. In fact, geophysicists believe that such plumes are unlikely to remain stationary as the mantle itself is mobile. The study of these seamount chains also indicates that small areas of the mantle may preserve a concentration of thermal energy and upwelling of mantle material for many millions of years. Another volcanic center that may mark a mantle plume is Iceland, on the Mid-Atlantic Ridge.

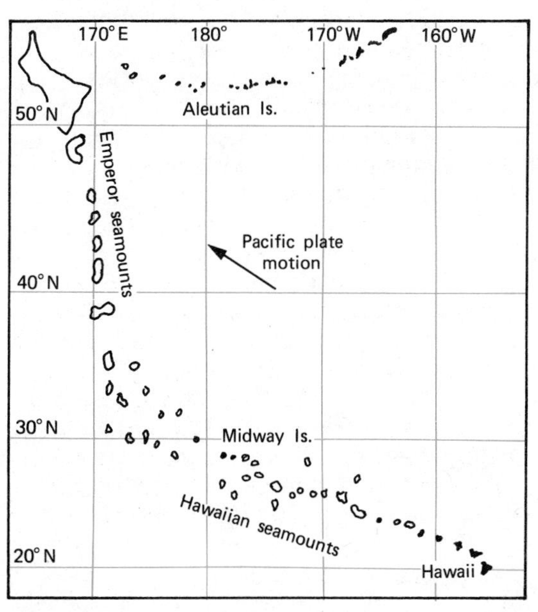

FIGURE 1–14 **Hawaiian and Emperor seamount chain mark the track of the mantle plume presently below the island of Hawaii.**

Since the recognition in the late 1960's that the lithosphere is composed of a series of interacting plates, geologists and geophysicists have been concerned with tracing the positions of the plate boundaries, not only in their present positions but also in past configurations. The task of reconstructing the relationships of the mobile lithosphere plates in the past is a formidable one that is just starting. Throughout much of this book we shall be exploring the possible relationships between the events of earth history recorded in the stratigraphic column and the position of the lithosphere plate boundaries of the past.

CONTINENTAL DRIFT

Long before the nature of the mid-ocean ridge system was recognized, a small group of geologists speculated that Africa and South America were parts of one continent that had split apart. The theory of continental drift was first set forth in detail by Alfred Wegener in 1912. He postulated a continent called **Pangaea** whose fragmentation gave rise to the present distribution of the continents (Fig. 1–15). During the break-up India separated from southeast Africa and drifted northward to impinge on the Asiatic continent. Australia, at first attached to Antarctica, drifted eastward into the Pacific. Where the leading edges of these continental rafts encountered opposition from the sea floor, Wegener believed they were compressed into folded mountain ranges, such as the Cordillera of the Americas. Although the coastlines of Africa and South America match with hardly a gap or overlap (Fig. 1–16), geologists generally did not accept the hypothesis because they could not comprehend how the relatively thin sialic rafts of the continents could maintain their integrity

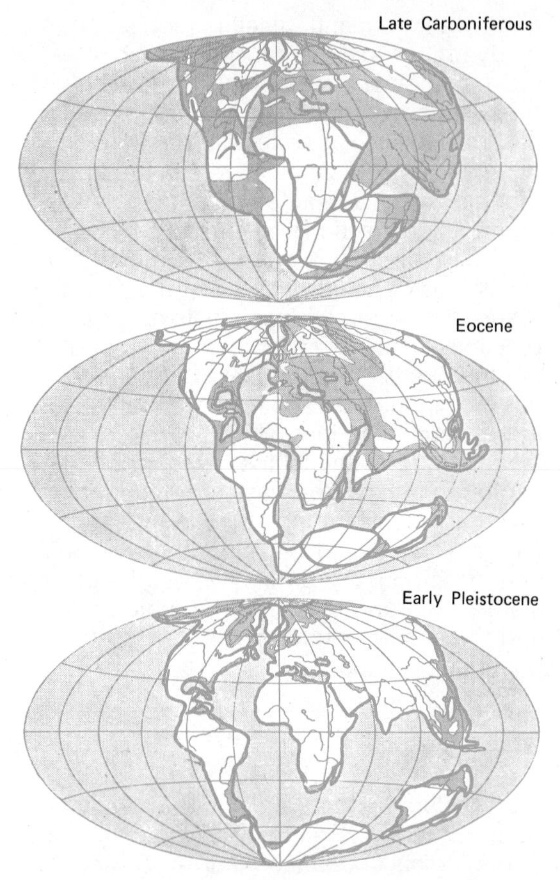

FIGURE 1–15 **Wegener's reconstruction of the break-up of Pangaca. (After A. Wegener, 1912; from Arthur Holmes, *Principles of Physical Geology,* 2nd ed. New York: The Ronald Press Co.).**

as they plowed through the simatic ocean basins, nor could they imagine a force that would impel them.

The correspondence between the two southern continents is geological as well as topographical. The history, as recorded in sedimentary rocks of these southern continents before their separation, is strikingly similar, and parallels between the sequences of beds in continents now widely separated are easily drawn. The Cape System of folded mountains that crosses the southern tip of Africa is believed to be continued in the Sierra de la Ventana of Argentina. Two hundred and fifty million years ago much of the southern hemisphere was covered by an ice sheet, but the distribution of the glacial deposits from this ice age only make sense if the southern continents were united, as Wegener suggested. In South Africa the Dwyka tillite covers wide areas and overlies a polished and striated pavement of basement rocks. From these striations and other indications the direction of movement of the ice can be determined. Similar glacial deposits of the same age occur in India, South America, and Australia (Fig. 1–17). Reconstruction of Pangaea by assembling the continents not only allows interpretation of these now scattered deposits as the product of a single continental ice sheet, but also integrates the present disparate directions of ice flow into an approximately radial pattern.

Similarities between the Atlantic coastline of North America and the western coastlines of Europe and northwestern Africa are not as evident as those between South America and southern Africa. However, when the continents are fitted together at the continental shelf edge, and the Grand Banks off Newfoundland are placed into the notch of Gibraltar, the fit becomes more obvious (Fig. 1–16). On both sides of the Atlantic, chains of folded mountains strike out into the ocean. Although the structure of these ranges cannot be easily followed where they are buried beneath the sediments of the wide continental shelves, the closing of the Atlantic restores continuity to these mountain chains and leaves little doubt as to the validity of the reconstruction. Similarity of the geology on either side of the North Atlantic Ocean will be discussed further in Chapters 17 and 18.

Areas where incipient continental rifting may be taking place today include the Red Sea where the Indian Ocean spreading center ends, and the Gulf of California where the East Pacific Rise meets North America. Possibly these areas resemble the Mid-Atlantic region when the rift that was to become an ocean was just beginning.

AGE OF THE OCEAN BASINS

Much of the new information available about the oceans has resulted from the develop-

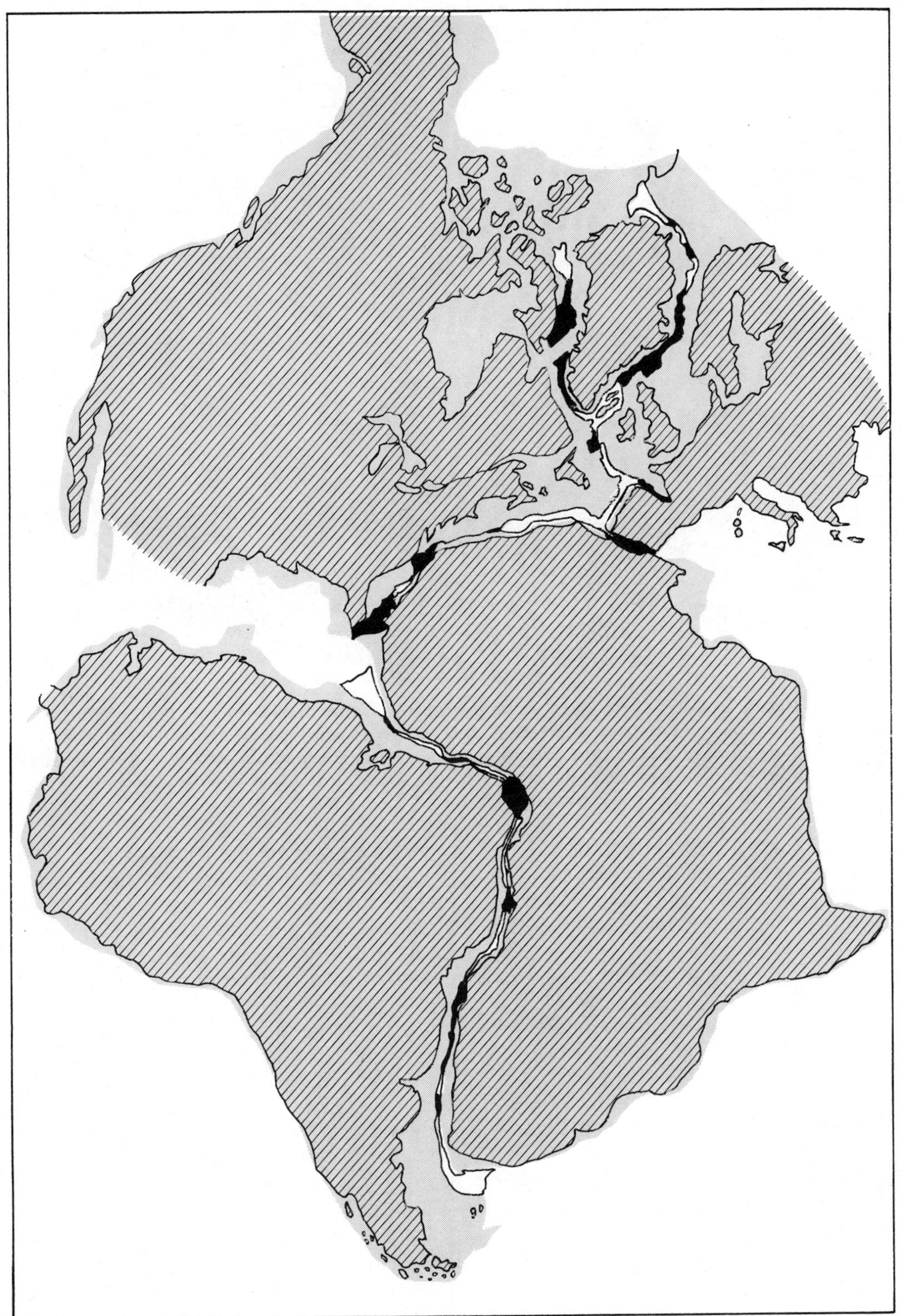

FIGURE 1–16 **Fit of the Atlantic Ocean optimized by computor at the 500-fathom contour on the continental slope. Areas where the continental plates overlap are colored dark; gaps are white. (After E. W. Bullard; courtesy of the Royal Society, London.)**

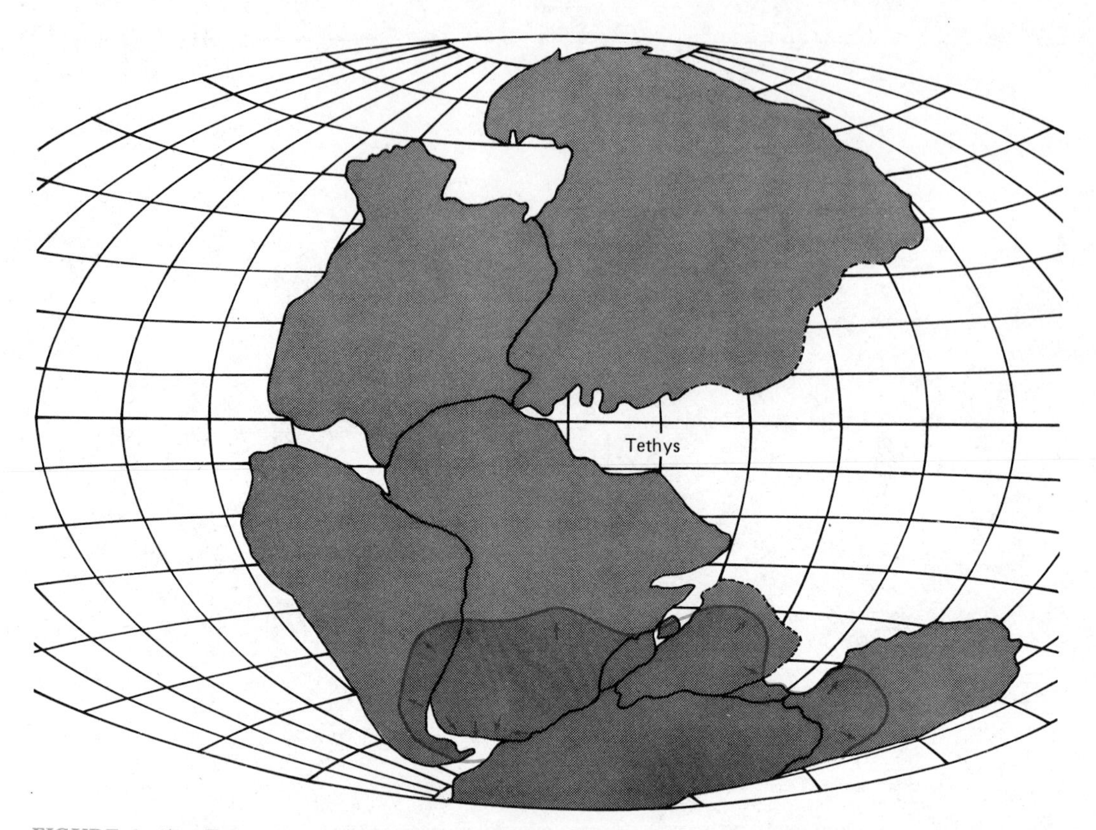

FIGURE 1–17 **Reconstruction of Pangaea at the beginning of the Mesozoic Era, about 200 million years ago. The extent of late Paleozoic glaciation, and the arrows indicating the direction of movement of the ice, have been plotted on the reconstruction. Notice the position of the Tethys seaway, which developed into a mountain belt that spans half the world. (Modified from R. Dietz and J.C. Holden; courtesy of the American Geophysical Union.)**

ment of a technology for drilling in deep water. The first vessel designed for this work, the *Glomar Challenger,* has an oil-well drilling rig mounted over a large hole in the bottom of the ship. Laterally thrusting engines are used in conjunction with the propeller to keep the ship in position over the hole being drilled in the ocean floor hundreds of meters below. The *Glomar Challenger* has drilled over 100 holes in the oceans since its first cruise in 1968 in water of depths over 6000 m, yielding cores with 1300-m penetration of the ocean floor. These holes penetrate sediments composed of tiny calcareous shells of floating micro-organisms and terrigenous particles. Some holes have bottomed in basaltic rocks considered to be the top of the oceanic or simatic layer of the crust. The evolution of the planktonic micro-organisms in these sediments is now well known and their relative ages can be determined readily (see also Chapter 5). The basalts can be dated in years by isotopic methods explained in Chapter 5.

As might be expected, the rocks along the mid-ocean ridges are very young and the age of the basaltic ocean basement increases away from the ridges just as do the ages of the volcanic islands. No sediment covers the recently extruded ridge crest areas, but a wedge of sediments thickens regularly away from their axes (Fig. 1–18). In holes drilled successively farther away from the axes, older beds appear at the base of the sedimentary wedge because the

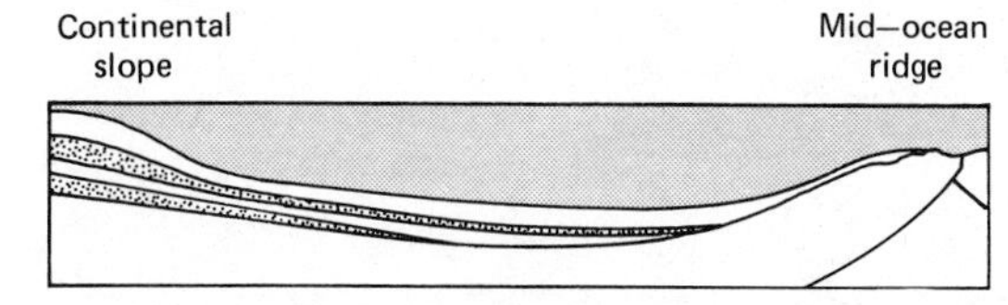

FIGURE 1–18 **Thickening of the sedimentary layers on the oceanic lithosphere away from a spreading center.**

floor of the ocean is older in these peripheral areas, having been formed in the early stages of spreading.

The oldest sedimentary rocks in the Pacific Basin, 150 to 180 million years old, are found at the periphery where oceanic crust is being subducted along the trenches. Older sediments that might tell us about the early history of the Pacific have been subducted along these western trenches, assimilated into the mantle, and regurgitated as volcanic material in the island arcs. The Pacific Basin, as we know it, is therefore a relatively young feature of the earth; possibly it existed before 180 million years ago, but record of that existence and the arrangement of continents around that basin has been swallowed by the mantle and is unlikely ever to be available to us. Reconstruction of the history of the Atlantic side of the globe during the last 500 million years now seems possible on the basis of its oceanic sediments and the geology of the continental borders.

BREAK-UP OF PANGAEA

Geologists can now revise Wegener's reconstruction of the origin of the continents, using new information on the nature of the ocean basins. The distribution of the continents prior to 200 million years ago is largely undetermined, but we can trace the history of parts of the globe, such as the North Atlantic Basin, before this time. Most geologists are agreed that about 200 million years ago the continents together formed a single mass called Pangaea, as shown in Fig. 1–17. A conspicuous feature of Pangaea was an equatorial sea, the ancestral Mediterranean, which is called **Tethys.** As shown in the figure, the group of continents was considerably southeast of their present position and the continents moved northwestward with respect to the geographic poles during the fragmentation. The area covered by glacial ice late in the Paleozoic Era, about 250 million years ago, is also shown to illustrate the harmony achieved by the continental reconstruction.

The modern Atlantic spreading center appeared about 180 million years ago (Fig. 1–19) and began the separation of North America from North Africa and South America. Meanwhile, the persistent subduction zone that followed the trace of Tethys began consuming oceanic crust as India started its long northward course toward collision with Asia. The giant transform fault that then separated India from the Australian and Pacific blocks is now inactive and marked by the Ninety East Ridge in the Indian Ocean. By 135 million years ago (Fig. 1–20) the Atlantic spreading center had penetrated between northern Europe and North America, and a southern segment was forming between South America and Africa. The westward drift of the Americas was compensated by a long north–south subduction zone along their Pacific margins. The Atlantic spreading center for a short time separated Greenland from Labrador, but soon shifted to the east side of Greenland to complete the formation of the North Atlantic Ocean as we now know it. Sixty-five million years ago (Fig. 1–21) major mountain building events were taking place along the Alpine–Himalaya trend as the old Tethys Ocean closed along its subduction zones. Australia was still attached to Antarctica but about to start its northward movement across the South Pacific to its present position. Along the west side of the Americas plate mountains were being formed by its interaction with the Pacific plate along the coastal subduction zones.

The processes of mountain making are called **orogeny.** They are intimately related to the movement of the continents and the nature of the plate boundaries. In Chapter 2 this interaction of continent and ocean in orogeny will be examined further.

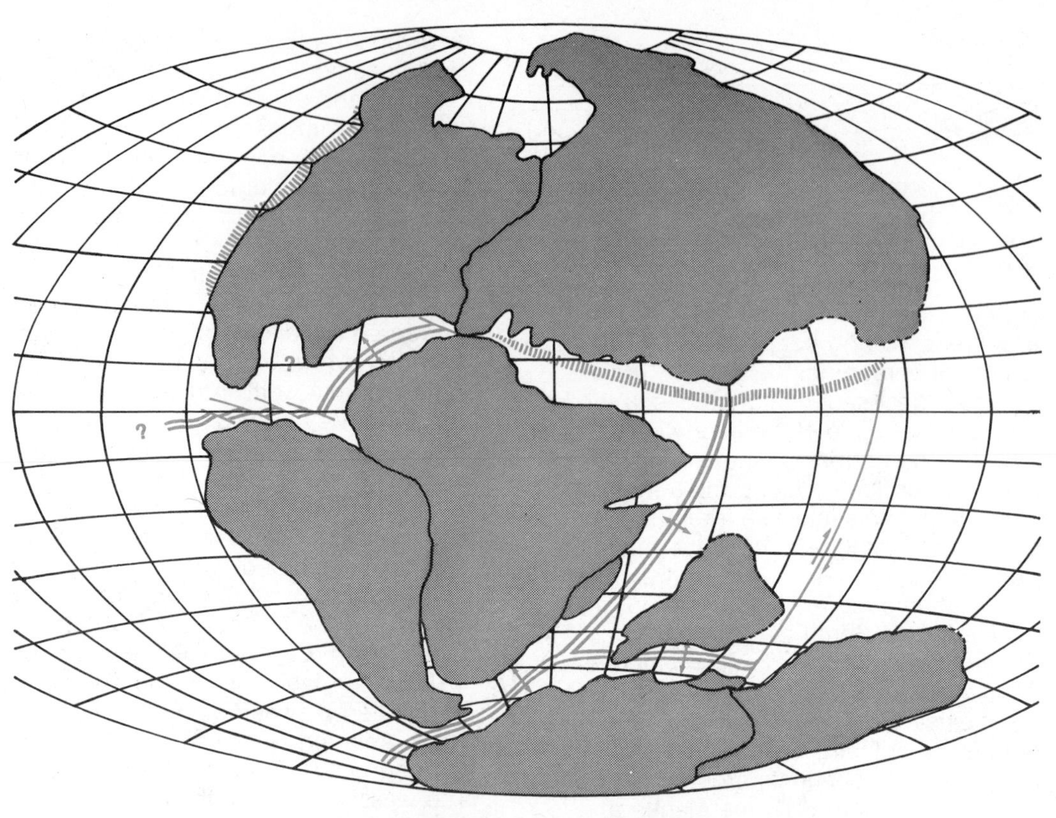

FIGURE 1–19
The break-up of Pangaea at the end of the Triassic Period. The northern group of continents, known as Laurasia, has split away from the southern group, Gondwana. North America has begun to separate from South America by the formation of the Mid-Atlantic spreading center. (Modified from R. Dietz and J. C. Holden; courtesy of the American Geophysical Union.)

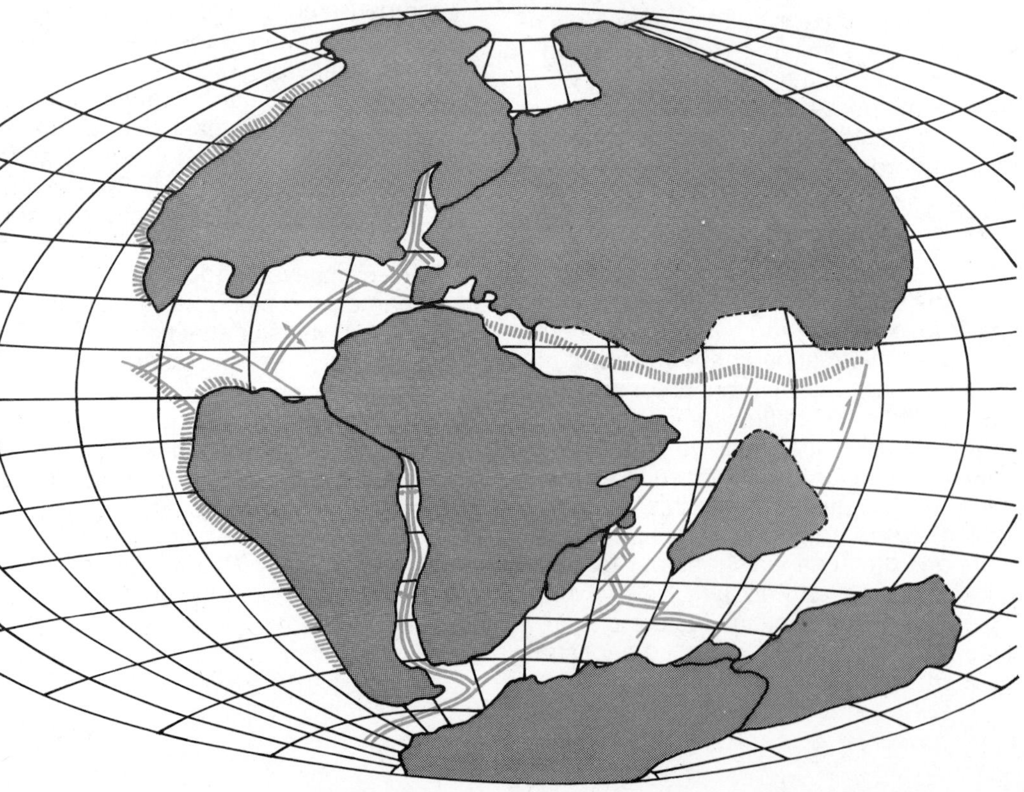

FIGURE 1–20
The break-up of Pangaea at the end of the Jurassic Period, about 135 million years ago. Europe and North America have begun to separate, and India is moving toward the Tethyan subduction zone. (Modified from R. Dietz and J. C. Holden; courtesy of the American Geophysical Union.)

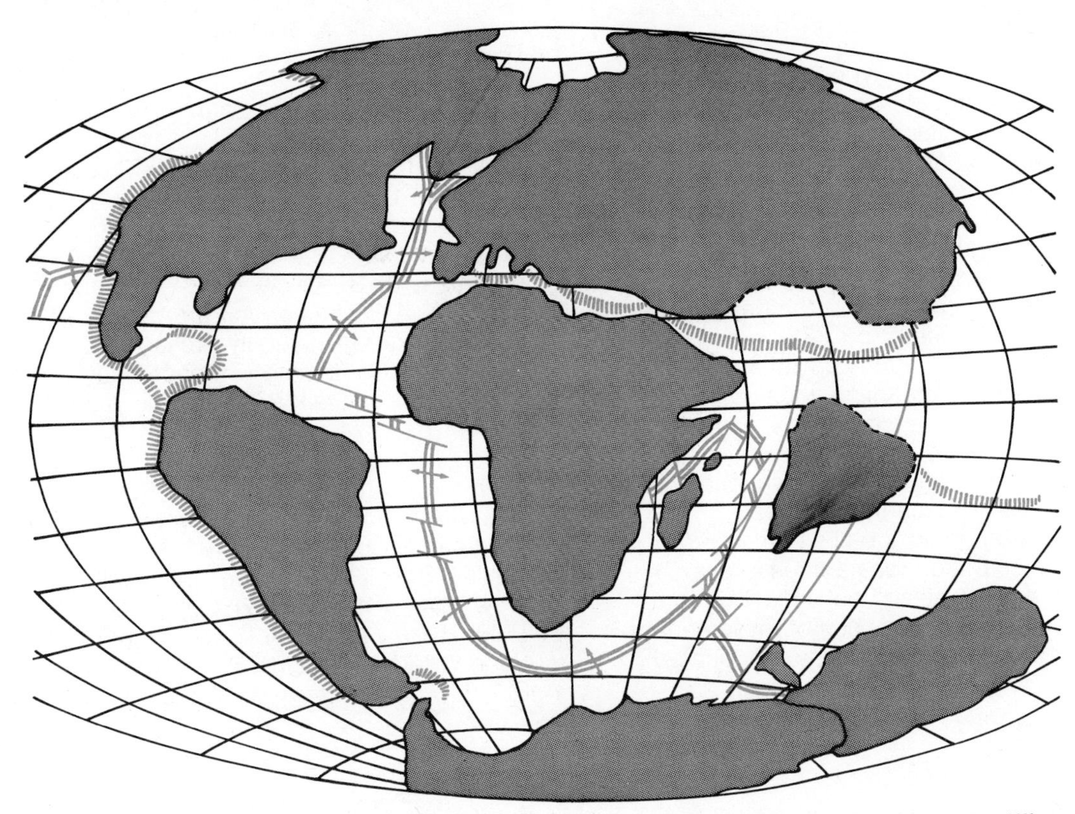

FIGURE 1–21 **The breakup of Pangaea at the end of the Cretaceous Period, about 65 million years ago. The South Atlantic Ocean is forming and the North Atlantic spreading zone is shifting from one side of Greenland to the other; India is approaching its collision with Asia. (Modified from R. Dietz and J. C. Holden; courtesy of the American Geophysical Union.)**

SUMMARY

The crust of the earth and the upper mantle comprising the lithosphere rest on a layer of partially melted mantle of little strength called the asthenosphere.

The mid-ocean ridge system marks places where the mantle is being converted to oceanic crust through injection and extrusion.

Spreading movements on a global scale from the mid-ocean ridges are accompanied by lateral displacement of the spreading axis along transform faults.

Oceanic crust is consumed at trenches in subduction zones where a descending lithosphere plate is assimilated in the mantle.

The crust can be divided into six major and many minor plates bordered by spreading centers, transform faults, and subduction zones.

Motion of the plates over relatively stable plumes in the mantle may leave behind chains of islands or other evidence of volcanism.

The movement of the continents away from each other and the Mid-Atlantic Ridge was deduced by the correspondence of the coastlines and geology on either side of the ocean, long before the nature of spreading centers was realized.

The break-up of the continents that were once welded together in the supercontinent, Pangaea, can be explained on the basis of our new understanding of the processes at work in the ocean basins.

QUESTIONS

1. Describe three types of plate boundaries and the geological processes that take place at each.
2. Famous earthquakes have taken place in San Francisco in 1906, Tokyo, Japan, in 1923, and Managua, Nicaragua, in 1972. What type of lithosphere plate motion is likely to have been responsible for the earthquake in each case?
3. By using an atlas and reference books on volcanism, locate five active volcanoes on island arcs, five on mid-ocean ridges, and five that are located on the edges of continents bordered by oceanic trenches.
4. By examining a bathymetric map of the sea floor, locate lines of seamounts "trailing" behind volcanic islands. What direction of movement of the plate on which they are located do they indicate?

SUGGESTIONS FOR FURTHER READING

CALDER, N. *The Restless Earth.* New York: Viking Press, 1972.

CONDIE, K. C. *Plate Tectonics and Crustal Evolution.* London Pergamon Press, 1976.

COX, A., ed. *Plate Tectonics and Geomagnetic Reversals.* San Francisco: W. H. Freeman, 1973.

HALLAM, A. *A Revolution in Earth Science.* Oxford: Oxford University Press, 1973.

SULLIVAN, W. *Continents in Motion: the New Earth Debate.* New York: McGraw-Hill & Co., 1974.

WILSON, J. T., ed. *Continents Adrift and Continents Aground.* Readings from Scientific American, San Francisco: W. H. Freeman & Co., 1976.

WINDLEY, B. F. *The Evolving Continents.* New York: John Wiley & Sons, 1977.

WYLLIE, P. *How the Earth Works.* New York: John Wiley & Sons, 1976.

YORK, D. *Planet Earth.* New York: McGraw-Hill & Co., 1975.

CHAPTER TWO

There are mountains and mountains. To people who live in the flatlands a mountain may be any change in relief of a few hundred meters. Volcanoes, formed by the outpouring and explosive ejection of molten rocks, would be considered mountains by most people, yet their structure and origin are relatively simple compared to those of the Rockies or the Himalayas. To geologists, mountains are belts of deformed rocks of continental dimensions that form the major elevations on the earth's surface. Whereas a volcano is built on the earth's surface and can be eroded away with hardly a trace, the structures produced in the deformation of a mountain chain may affect the entire thickness of the crust and remain after the relief of the chain has been eliminated by erosion. Through a century of study of the complex stratigraphy and structure of mountain chains, geologists are now beginning to reach a consensus on their origin.

ANATOMY OF A MOUNTAIN CHAIN

Later in this book the history of the three major mountain chains of North America will be examined in some detail. In this chapter we shall review features that mountain chains the world over have in common and theories that have been advanced to explain the way they were formed.

Sedimentary, igneous, and metamorphic rocks within mountain chains show signs of lateral compression and burial deep in the earth's crust. Since sedimentary rocks, which are formed at the earth's surface, react more obviously to increases in temperature and pressure, they show these effects more clearly than the others. Taken together, the structures of mountain chains strongly indicate that the crust of the earth has been shortened by lateral compression across the chain. In most chains a zone of metamorphic and igneous rocks on the oceanward side can be distinguished from a zone of deformed sedimentary rocks on the continentward side. The terms "continentward" and "oceanward" can, of course, be applied strictly only to mountain chains that are now marginal to continents, such as those of North America, but even where mountain chains now exist between continental plates, remnants of oceanic crust in the chain suggest that they were once marginal to a segment of oceanic crust that has since been subducted.

A traverse through a typical marginal mountain chain, from the continent to ocean, would first cross a zone of slightly metamorphosed sedimentary strata that are folded

OROGENESIS

and cut by thrust faults in such a way as to suggest that they were shoved by a mighty force away from the ocean and piled up over the continental platform (Fig. 2–1). The thrust faults of this belt dip away from the continent and the anticlinal folds are overturned toward the continent. On the continental edge of the belt of deformation the faults dip at low angles, but toward the central part of the chain they become steeper and are evidence of more prominent vertical forces.

Igneous rocks are common in the oceanward zone. **Batholiths** are large bodies, usually composed of rocks like granite and granodiorite. The crystallinity of the batholithic rocks shows that they have cooled slowly at depth beneath an insulating layer of crustal rocks and have been exposed at the surface through erosion of their cover. The borders of some of these great masses of igneous rock are sharply defined, and along them the molten magma has penetrated the country rock as dikes and sills. The heat and gases of the molten rock causes mineralogical and textural changes in the country rock into which it is intruded, in a zone surrounding the batholith called a **contact metamorphic aureole.** Because the igneous rocks cut across the bedding and schistosity in the country rock, such batholiths are said to be **discordant.**

The borders of other batholiths are gradational, and their internal structure is gneissic (Fig. 2–2). They occur in regionally metamorphosed terranes of schists and

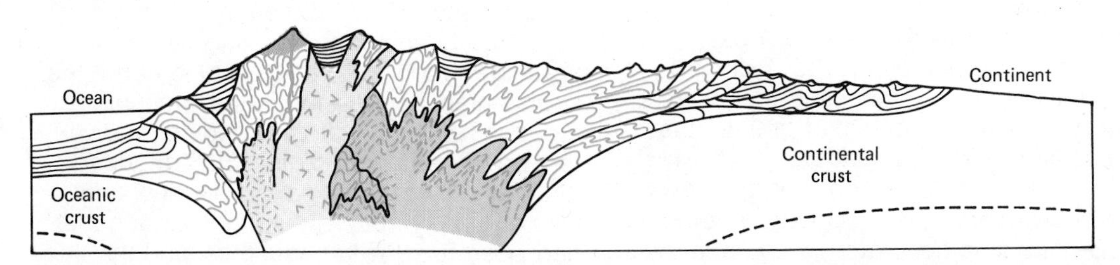

FIGURE 2–1 **Cross-section through a typical mountain chain marginal to a continent. Metamorphic and igneous rocks are indicated by color, sedimentary rocks are in black. Batholiths, some composed of foliated gneiss and some massive, are common on the oceanward side of the chain.**

FIGURE 2–2 **Migmatitic gneiss of the Grenville Group, Ontario. (C. W. Stearn, photograph.)**

gneisses in which the zone of thermal metamorphism is less marked. A zone of **migmatites** (literally, mixed rocks) is common around the borders of such batholiths. Migmatites are composed of thinly alternating layers of igneous and metamorphic rock, the product of selective melting of the more fusible components. The gradation from metamorphic to igneous rock near such batholiths suggests that heat and solutions beneath these mountain chains so alter the sediment that, layer by layer, they pass through metamorphic stages to become igneous rock. As most batholithic rocks have compositions similar to that of granite, the process is called **granitization.**

Batholiths with gradational borders are called **concordant.** They appear to be huge lenses of magma generated in place in the depths of the crust by the effects of heat, pressure, and solutions. When such a molten mass migrates upward by breaking away piecemeal the unmelted rock above it and penetrates the higher zones in the crust, it cools as a discordant batholith.

In addition to the metamorphism induced by contact with batholiths, the oceanward side of mountain chains is commonly affected by a pervasive regional metamorphism in which sedimentary rocks become slates, phyllites, schists, and gneisses. Minerals found in such rocks can be interpreted in terms of various conditions of temperature, hydrostatic pressure, and shearing stress that occur when surface rocks are moved by deforming forces into deeper and hotter zones of the earth's crust.

In mountainous areas volcanic rocks hundreds of meters thick are intercalated between the sedimentary rocks and metamorphosed with them. Some of the volcanic rocks are composed of ash ejected by gas pressure from volcanoes, and some are pillow lavas extruded under water. Recently formed mountain chains remain the sites of volcanic action long after the lateral forces that have caused the deformation have ceased, and great volcanic edifices are built on the deformed and metamorphosed sedimentary, volcanic, and intrusive rocks (Fig. 2–3).

Thickness Relationships

In 1857, James Hall, a paleontologist from the state of New York, expounded a theory for the origin of mountains before the American Association for the Advancement of Science. Hall had noted that the folded strata of the Appalachian Mountains were much thicker than those in the plains to the west where the rocks were not folded. This observation, based on his experience in New York, has been verified by geologists for all the folded mountain ranges of the earth: the sedimentary succession in such mountain areas is much thicker than beds of equivalent age in adjacent plains. Hall deduced that the site of the Appalachian Mountains was originally a trough-like depression that had been filled with sediments, and that the folding of the rocks was due to the collapse of the sediments into the center of the trough. He offered no explanation for the up-

FIGURE 2–3 **Mount Garibaldi, southern British Columbia—a volcano active until recently—one of a chain of volcanoes along the central Pacific coast, located above the presumed subduction area of the Juan da Fuca plate. (Photograph courtesy of Austin Post, U. of Washington.)**

lift of the mountains, but believed that the amount of uplift was dependent on the thickness of the sediments. As James Dana later caustically remarked, Hall had ". . . promoted a theory for the origin of mountains with the origin of mountains left out." However, Dana accepted the theory in large part and developed it in classic papers published later in the 19th century. Dana proposed the name "geosynclinal" for the trough, but this was soon modified to "**geosyncline.**" He believed that the formation and subsequent deformation of the geosyncline were caused by compressive forces acting horizontally rather than the collapse of the sediments toward the center of the geosyncline.

In 1923 Charles Schuchert proposed a classification of geosynclines, naming the classes by adding prefixes to obtain such names as monogeosyncline, polygeosyncline, etc. None of Schuchert's terms is in common use today, but a similar, widely used set was introduced by Hans Stille in 1940. Stille noted that most geosynclines have an oceanward side containing volcanics interbedded with the sedimentary rocks, and a continentward side without volcanics. These parts he named the **eugeosyncline** (true geosyncline) and **miogeosyn-**

cline (lesser, or less of a geosyncline), respectively. Because the depositional site of Stille's miogeosyncline is a seaward slope rather than a trough (syncline), Robert Dietz suggested that it be referred to as the **miogeocline.**

Sedimentary Rocks in Mountain Ranges

Not only are sedimentary rocks thicker in mountainous areas than in the plains, but their lithology is also different. The rocks of the plains are shales, quartz sandstones, limestones, and dolomites; those of the mountains are mudstones, graywackes, litharenites, and tuffs. **Graywacke** is a greenish-gray, poorly sorted, impure sandstone containing more than 15 per cent fine-grained matrix of clay. **Litharenites** are sandstones composed of grains of rock fragments rather than minerals. The rock fragments in these sediments are metamorphic, sedimentary, and volcanic and must have been derived from the erosion of areas of complex geology with volcanoes and deformed sediments. The continuing coarseness of graywackes through successions thousands of meters thick indicates that the source areas were continuously being uplifted as they were eroded, or else replaced by source areas of equal relief.

FIGURE 2–4 **Graded beds of graywacke in a flysch sequence—Cloridorme Formation, Ordovician, Quebec. (Courtesy of R. Hesse.)**

In many mountain areas graywacke beds are regularly interbedded with mudstones in units less than a meter thick, (Fig. 2–4). This association is commonly referred to by the German term "**flysch.**" In the graywacke beds in flysch sequences the size of the sedimentary particles grades from coarse at the bottom to fine at the top. Such graded beds have been interpreted as the deposits of turbidity currents spreading out over the ocean floor in deep water. A **turbidity current** is a flow of water that contains sediment suspended by the turbulent motion, within a body of clearer water. The turbidity flow may be triggered by the slumping of sediments on a slope, and once set in motion, it is capable of maintaining its integrity and flow for distances of tens, hundreds, and even thousands of kilometers over low slopes. When the current of the flow decreases as the level sea floor is reached, its competence falls and the particles in suspension are deposited from the base of the flow in order of size to produce a graded bed. Scour markings on the base of the graded bed reflect the direction from which the turbidity current flowed (Fig. 2–5). Another structure common in **turbidites,** as the deposits of turbidity currents are called, is **convolute bedding.** The bedding is said to be convolute when the layers are complexly contorted and disrupted. Plotting of the direction of scour marks shows that some turbidity currents flowed along the length of the geosynclinal trough, others across it. The currents are believed to start by the slumping of sediments on the flanks of the trough and later to turn along its length in the axis of the depression.

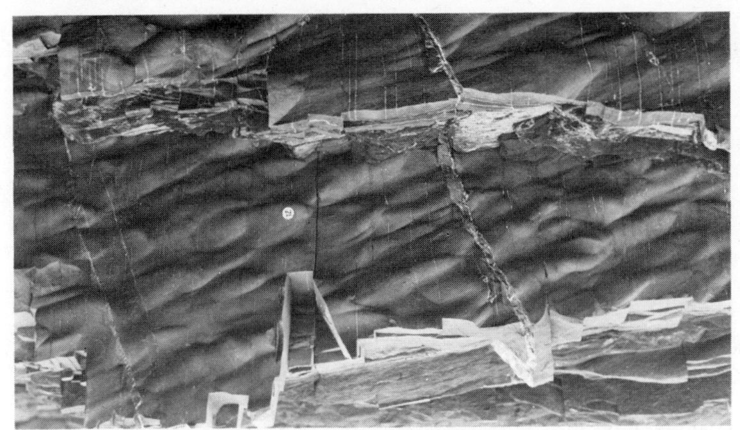

FIGURE 2–5 **Bottom view of several beds of graywacke in a flysch sequence, showing scour marks known as flute casts. The shape of the marks indicates that the turbidity currents depositing the beds flowed to the upper right. St. Roch beds, Quebec. (Courtesy of E.W. Mountjoy.)**

Sedimentary structures, mineralogical studies, and changes in thickness of beds and the grain size of particles give information on the direction from which the geosynclinal trough was supplied with sediments. In the oldest sediments in mountain chains, deposited when the geosyncline first formed, the direction of supply appears to have been from the continent. The source of supply changes to the oceanward side of the trough later in the history of many geosynclines. The nature of such oceanside source areas has been a subject of discussion among geologists for years, for when the succeeding mountains have been formed little trace of the source remains.

The Sequence of Deformation

The sedimentary rocks of mountain chains preserve the record of events in the geosyncline from which the mountains were formed. This record shows that the deformation of the rocks, their metamorphism, and the formation of batholiths did not take place in one event following the interval of geosynclinal deposition. Instead, geosynclines must have been sections of the crust with great mobility, scenes of episodic deformation, metamorphism, extrusion, and intrusion. At times mountainous ridges were uplifted in the troughs only to be eroded, submerged, and covered with younger sediments.

The whole process of mountain building is called **orogenesis** (Greek: *oros* = mountain); the events in which mountains are formed are called **orogenies.** Episodes of intense orogenesis in the history of mountain chains are generally named from some geographic feature of the area, such as the Taconic Orogeny in the Appalachians (Taconic Range of Massachusetts and New York) or the Antler Orogeny in the Cordillera (Antler Peak of Nevada). The orogeny in which the chain reaches its present form is commonly referred to by the name of the chain itself, as in Alpine Orogeny or Cordilleran Orogeny, but such usage does not imply that these events were single, short intervals of deformation; commonly they include many episodes of folding, faulting, uplift, intrusion, and volcanism.

Elevation of Mountains

The topography of mountain chains is a product of the interaction of forces within the earth that lift them and forces of weathering and erosion that wear them away. Little evidence can be found to show that the deformation of geosynclinal rocks directly causes or is immediately followed by their elevation into a mountain range. The elevation of mountains may be delayed after their structure has been formed, or elevation may appear to accompany deformation. The elevation of the chain after deformation is a response to disturbances in isostasy caused by subsidence of the geosyncline and orogeny. The accumulation of a thick prism of sediments in a geosynclinal trough and its lateral compression during orogeny may create a "root" of lighter rock protruding into the heavier asthenosphere. When lighter rocks take the place of heavier rocks in the earth they form a negative gravity anomaly that expresses a deficiency of mass. Isostatic equilibrium can be restored by the raising of the mountain area and by the underflow of heavier material until the balance pictured in Fig. 1–4 is attained. The persistence of surface relief so formed can be appreciated by returning to the analogy of the blocks of wood floating in water (Fig. 1–3). If one of the blocks of wood is 10 cm thick and the rest 4 cm, and the blocks float with half their thickness above water, the 10-cm block, representing the part of the crust thickened in orogeny, will float 3 cm above its neighbors. The effect of erosion on the mountain range can be simulated by planing 3 cm off the 10-cm block. Although the removed portion equals the height by which the block stood above its neighbors, the block still floats $1\frac{1}{2}$ cm above them. Not until it has been reduced to 4 cm will the surfaces of the blocks be at the same elevation.

The uplifted remnants of erosion surfaces preserved in the topography of mountainous areas and the volume of sediments supplied by the erosion of such areas to adjacent basins of deposition are evidence of the long sustained, yet episodic, nature of the uplift of folded mountain chains. The crust of the earth is much stronger than the analogy suggests—it is strong enough so that the isostatic rebound does not immediately compensate for erosion, and during this lag forces of erosion may reduce local areas or even the whole chain to a surface of low relief. Uplift is commonly accompanied by faulting at high angles, as the crust is not uniform enough to respond evenly to the uplift without breaking into segments.

GEOSYNCLINES AND TRENCHES

Geologists working in mountain chains have been able to decipher their past history without reaching a consensus on the causes of orogenesis. Geologists working from ships have recently assembled a model of the behavior of the crust beneath the sea. Land-based geologists, realizing that the key events in the history of geosynclines took place beneath the sea, have for years been hampered in their search for modern geosynclinal counterparts by ignorance of the nature of the oceanic crust. New knowledge of the oceans offers hope that the origin of mountains may be understood not only in terms of the geologic record of what took place in the past, but also, by analogy, with what is taking place today beneath the sea. The suggestion that geosynclines are the ancient equivalents of modern oceanic trenches was made at the beginning of this century by Emile Haug, but so little has been known of them until recently that comparisons could not be made in detail.

Middle America Trench

To determine how far the model of the structure of oceanic trenches set forth by geophysicists in recent years can throw light on the history of mountain chains, we must examine such trenches in more detail than was done in Chapter 1. The various forms of the Pacific trenches reflect different stages of development and different environments of development. The best studied trench on the borders of the North American plate is the Middle America trench along the southwest coast of central Mexico, where the Cocos plate is being subducted (Fig. 2–6).

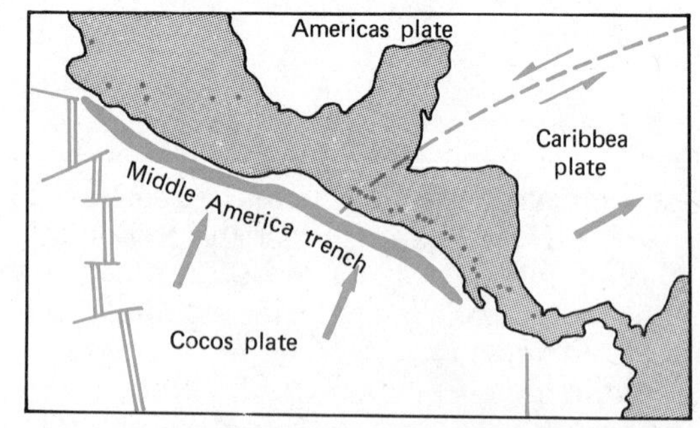

FIGURE 2–6 **Structural setting of the Middle America trench. The large arrows show the direction of movement of the plates. The dots on the Central American mainland designate volcanoes active in historical times.**

The Middle America trench (Fig. 2–7) is bordered by high volcanic mountains reaching 4200 m at the intersection of the central Mexico volcanic belt and the Sierra Madre Occidental. The axis of the trench is about 4800 m below sea level, and the relief between mountains and trench about 9000 m in 300 km. The landside slope of the trench is steeper (6° to 15°) than the seaside slope (3° to 6°). Geophysical measurements suggest that the oceanic crustal plate slopes eastward toward a fault in the subduction zone. The reflection of sound waves from bedding horizons in the sediments accumulating in the trench allows the geologist to determine their structure, and sampling by dredge and piston corer allow him to examine their composition. The Middle America trench contains about 700 m of sands, silts, and clays and these appear to be undeformed. The deposits in the trench are graded and show other sedimentary structures typical of turbidites. Sediment eroded from the mountains is brought to the landside of the trench and periodically slumps down to the depths where turbidity currents turn along the length of the trough.

In the Middle America trench little sediment has accumulated so far, but in other trenches a thick prism of sediments may be present. The motion of subduction may extensively deform these sediments as they accumulate, elevating them into islands to form a second, non-volcanic island arc in front of the volcanic one. South of the Aleutian arc (Fig. 1–11) a large mass of deformed sediments now separates the islands from the trench axis. The extent of the **accretionary prism,** as this mass of sediments is called in modern trenches, is related to the rate of supply of sediments to the trough and the rate of subduction. Between the Aleutian volcanic arc and the Alaskan and Siberian mainland, the Bering Sea occupies a much shallower basin than the trench. In the lithology of their sediments and their association with volcanic rocks and deformation, the oceanic trenches and their neighboring island arcs closely reproduce the geosynclinal environment reconstructed by geologists from their study of mountain belts. Basins like the Bering Sea between the volcanic areas and the continents have much in common with the miogeoclines of continental margins.

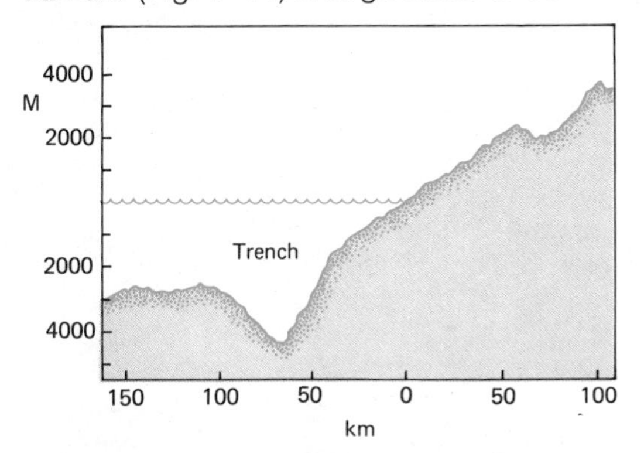

FIGURE 2–7 **Cross-section of the Middle American trench and adjacent Mexico. (From information of R.L. Fisher.)**

Igneous Rocks and Trenches. Volcanism at spreading centers, such as the Mid-Atlantic Ridge, is characterized by a type of basalt called **theoleitic** (from Thule = Iceland). It is relatively deficient in the alkali elements, sodium, and potassium.

The descending slab of oceanic lithosphere subducted in trenches undergoes changes in composition and physical properties as it is heated and assimilated in the asthenosphere. The descending plate is thrust into the hotter zones of the mantle and melting takes place. Deep in the subduction zone trench sediments, water, oceanic crust, and upper mantle are combined to produce magmas rich in sodium and potassium. The volcanoes fed from such subduction zones commonly erupt andesitic lavas, and most of the giant volcanoes that rim the Pacific are of this composition. Some of the lavas in island arcs are alkali basalts, dacites, and rhyolites, but tholeitic basalts are not formed in this environment. At depth, complex processes that are not well understood are believed to produce granitic rocks that are intruded as batholiths. Granite batholiths are not typical of island arc environments, but are found associated with some of the larger arcs, such as the Japanese, and where subduction zones dip beneath continental masses.

In some mountain chains the presence of ultramafic rocks, such as peridotites, dunites, and eclogites suggests that deformation of the descending plate in the subduction zone has sheared off parts of the upper mantle and carried it into the accretionary prism. This process has been called **obduction.** These mantle rocks, which are sometimes referred to as the ophiolite suite, have been regarded by geologists for many years as particularly characteristic of the axis of geosynclinal troughs and deep-water sedimentation.

Sediment Supply. At certain times in their development geosynclines were supplied with sediment from the erosion of adjacent continental nuclei. At other times the volcanic highlands associated with the formation of deep water troughs have supplied sediment, and at still other times supply has come from sources of doubtful nature located in areas now occupied by oceanic crust.

The geographic relationships of modern plate boundaries (Fig. 1–12) show that not only oceanic crust is brought to subduction zones by the motion of the plates. Where two subduction zones come together, as south of Japan, an island arc system, in this case the Mariannas arc, many be consumed in a trench. A spreading center may also move into a subduction zone as is now happening at the north end of the Middle America trench, and has happened recently as the Americas plate encountered the East Pacific Rise. As India moved northward toward Asia and oceanic crust was consumed along the Tethyan subduction zone (Fig. 1–21), the two continental blocks were eventually brought into collision. Continental crust is too light to be easily subducted. Where India and Asia have met some continental crust has been thrust into the subduction zone, plugging it up by doubling the thickness of the crust in this area and forming a massive, buoyant root for the highest mountains in the world.

Against the background of the various possible interactions of subduction zones and other crustal structures, the variety of source areas that supplied sediment to geosynclines can be compared to modern counterparts. Geosynclines may have been supplied from the ocean side by islands from another arc–trench system, by oceanic islands associated with spreading centers or mantle plumes, or by the approach of a section of continental lithosphere as an oceanic segment of the lithosphere was closed by consumption in a subduction zone. Each of these structures will have its distinct lithology and will supply sediments that may allow it to be identified in the sedimentary record of the geosyncline. Other sediments may come from volcanic island arcs formed as a result of the subduction or from the continental mass to which the geosyncline is marginal. Finally, the episodic deformation of the geosyncline may raise welts or ridges of the sediment accumulating there, which are eroded and redeposited in the adjacent basins. As the geosyncline might appear to be devouring itself, this type of sediment supply has been called "cannibalistic."

A General Orogenic Model

Despite the fact that ancient geosynclines and their resultant mountain chains, as well as modern trench systems, show a wide range of histories and geometries, a general model of the events that take place when a geosyncline becomes a mountain chain can be attempted.

At the edge of a continent bordered by a trench two suites of sediments are deposited (Fig. 2–8A). On the depressed edge of the sialic continental crust a sequence of limestones, sandstones, and shales is deposited as a seaward thickening wedge (*a* in Fig. 2–8A) up to several thousand meters thick. Because the sediments making up most of this prism were deposited in shallow water, the edge of the continent must have subsided, keeping pace with deposition. This is the miogeoclinal environment. Farther offshore in the trench, clastic sediments escaping from the miogeocline and slumping into deeper water mix with oceanic sediment (*b* in Fig. 2–8A). As subduction carries the oceanic lithosphere downward, a prism of sediment is built up

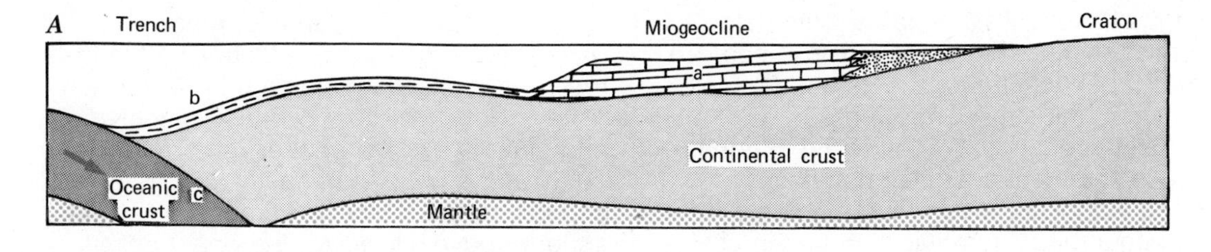

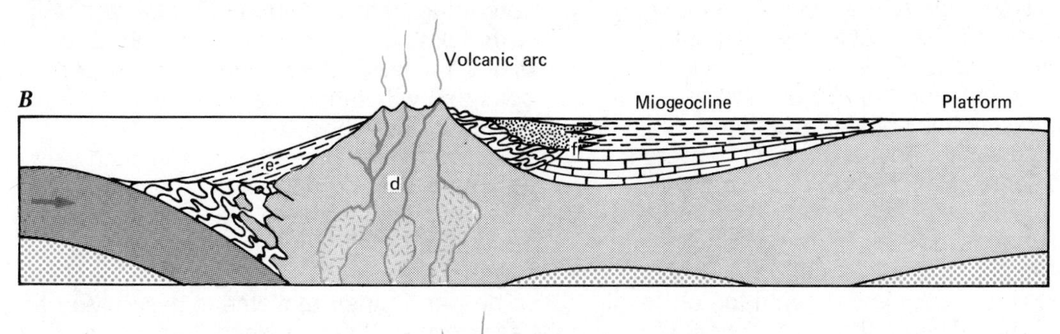

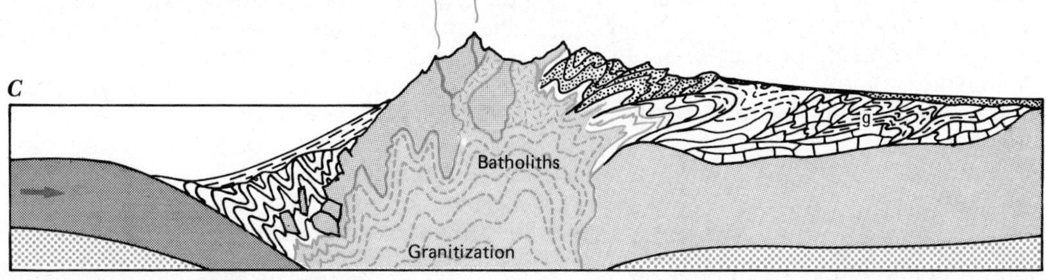

FIGURE 2–8 **A general orogenic model. a = platform sediments accumulating on the miogeocline, b = deep water sediments beginning to accumulate at the trench, c = subduction of oceanic crust, d = volcanic island arc, e = flysch facies sediments accumulating in the trench, f = sediments derived from the volcanic arc, g = thrust sheets of miogeoclinal sediments pushed against the edge of the craton.**

against the continent. A volcanic arc forms above the subduction zone (*d* in Fig. 2–8B), fed by the assimilation of the oceanic plate. The erosion products from the ridge and oceanic source areas feed turbidity currents, depositing turbidities in the trough forming the suite of sediments called flysch (*e* in Fig. 2–8B). Coarse clastics, lava flows, and volcanic ash are also supplied by the volcanic arc to basins between the islands and to the continentward side of the arc, where they interfinger with the miogeoclinal sediments deposited on the continental margin. Continued subduction periodically deforms the sediments and lavas accumulating in the trench and around the island arc (*f* in Fig. 2–8A). The environment controlled by the volcanic arc and trench has been defined as the eugeosynclinal realm.

Continued compression of the continental edge and assimilation of the descending slab generate large volumes of magma beneath the continental margin. Granitization of the sediments in the accreting prism takes place and granite magmas rise toward the surface. Increase in temperature and pressure causes widespread metamorphism of the sediments and volcanics at the base of the accreting prism and miogeocline. The rocks of the trench and accretionary prism are thrust toward the continent, overriding and pushing before them the wedge of sediment that has accumulated in the miogeocline (Fig. 2–8C). These sediments break into thrust sheets and flex into recumbent folds that are piled against the edge of the continent. The crustal shortening and sediment accumulation have formed a thicker crust, which rises into isostatic equilibrium forming a mountain chain.

This sequence of events is greatly generalized and does not fit any particular mountain chain, as will become evident when we consider the histories of the various geosynclines of North America.

If geosynclines are the trench–arc environments of deposition of the past, of what value is the retention of the name and concept of geosyncline? Perhaps none, but at present geology is in a period of reconciling the new oceanic data with the old continental data, and until such reconciliation is complete, the term "geosyncline" is useful for referring to the basin, or basins, of deposition from which a mountain chain has arisen without implication as to modern equivalents.

Mechanisms

The compressional forces whose effects we see in the structure of mountain chains can be explained by the movement of the lithosphere plates. But what causes the movement of the plates? Alfred Wegener set forth the concept of continental drift over 60 years ago, but his theory did not receive widespread acceptance because he could not offer an acceptable mechanism. Are geologists in similar difficulties to explain the movement of lithosphere plates?

At the surface a transfer of material is taking place from hot spreading centers to cool subduction zones. Crust is being made in the spreading centers and lost in the subduction zones. If, as we believe, this has been going on for millions of years and the size of the earth has not changed, then material must be returned from the trenches to the spreading centers below the lithosphere plates. Calculations from the study of seismic waves showing that the mantle increases greatly in strength below about 300 km place a limit to the depth at which this returning movement can take place. The pattern is best explained as basically the result of convection: material is heated and rises under spreading centers, is cooled and sinks beneath trenches, and travels laterally at the base of the asthenosphere counter to the plate movement in the lithosphere above. However, the pattern of spreading centers and subduction zones is complex and the upper mantle cannot be divided simply into a series of convection cells that will satisfy the plate motions observed in the lithosphere.

Mechanisms other than convection probably also operate. The plates generally move downhill from the highs of the spreading centers to the topographically lower subduction zones, and gravity may play some part in their movement. Jason Morgan has suggested that plumes of rising material that originate in the lower mantle, by spreading out in the asthenosphere, drive the plates above them. Most geophysicists are skeptical that the properties of the lower mantle would allow large-scale convective movements there.

Although most geophysicists agree that the driving mechanism of plate motion must be associated with convection, and its basic cause is unequal distribution of heat sources in the mantle, the arrangement of the convective cells and the mechanisms of plate motion are still subjects of much research and controversy.

CONTINENTAL STRUCTURE

The structures of the continental parts of the lithosphere plates have been investigated for 200 years and are much better known than the structures of the oceanic parts. Most continents can be divided into three major areas according to general structure and behavior of the crust.

1. The **craton** (Greek: strength or power) or nucleus of the continents, a stable, largely rigid area of relatively low relief;
2. The **geosynclinal mountain systems** that border the craton;

3. The **coastal plains** and **continental shelves** that separate the older mountain chains from the oceanic crust.

As the subject of this book is the North American continent, we shall use it as an example to illustrate these divisions (Fig. 2–9).

The Craton

The stable central area of the continent can be divided into two sectors that differ in surface geology. In North America the area in the middle of the continent where ancient rocks are exposed is called the Canadian Shield. The term "**shield**" is used as a crude description of its form—a low broad dome. The crescent-shaped area of the craton that surrounds the shield and is underlain by essentially flat-lying sedimentary strata of younger age is called the **platform.** The old pre-Paleozoic rocks of the shield extend out under the whole of the craton and project through the younger rocks of the platform as isolated outcrops along its edge in Texas, Arizona, Wyoming, South Dakota, etc. Deep wells in the plains of the United States and Canada penetrate these older, generally metamorphosed rocks beneath the younger

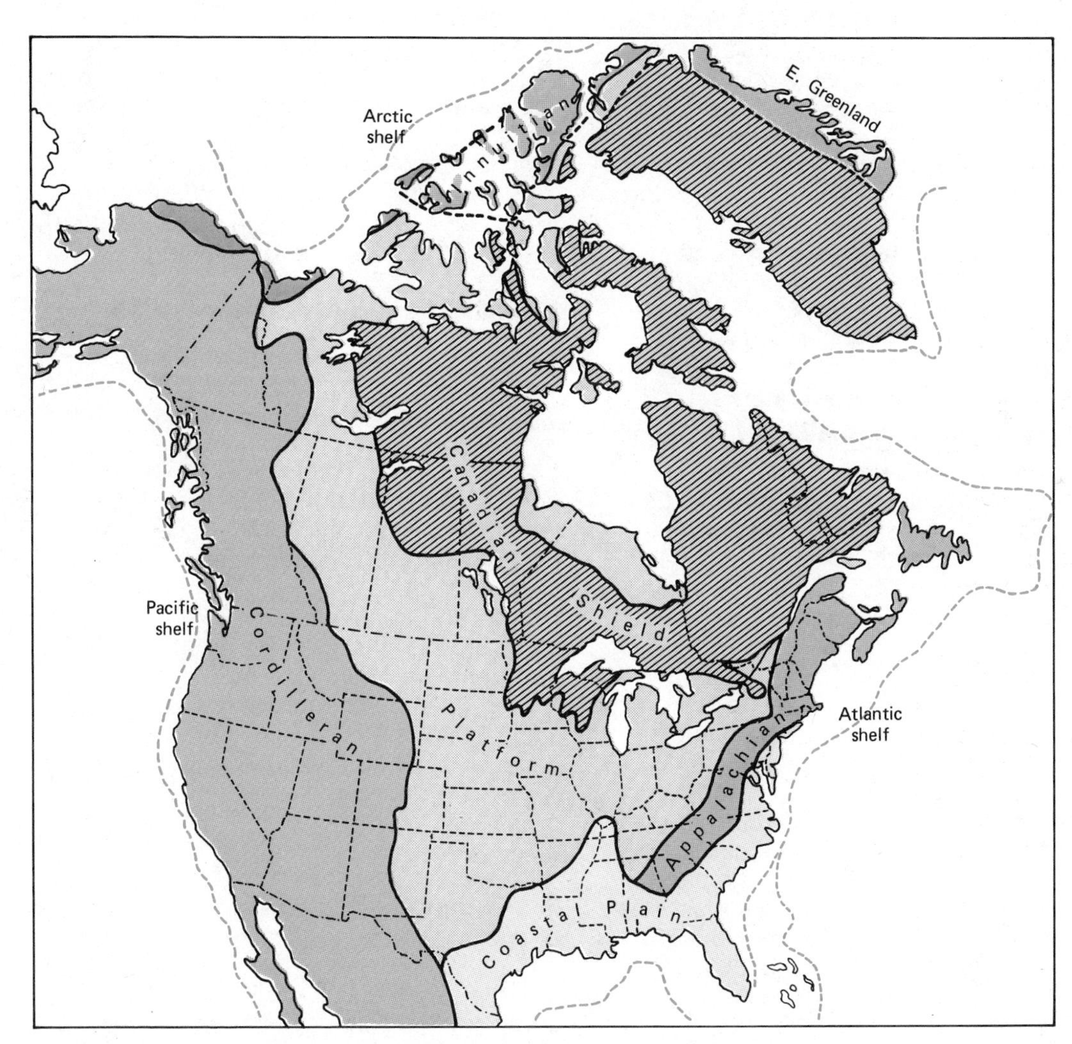

FIGURE 2–9 **Major structural divisions of the North American continent.**

sedimentary beds. The boundary between the shield and the platform is marked by the erosional edge of the younger sedimentary strata, and is constantly moving outward as erosion strips the platform covering from the shield. The platform sediments are a mere veneer compared to the thickness of the crust, but they reach thicknesses of several thousand meters in Colorado and Alberta.

Most of the rocks of the Canadian Shield are schists, gneisses, and granites, the kinds of rocks that are characteristic of younger mountain chains. In Chapter 9 we shall look more closely at the history of this area, which, during the early stages of earth history, was mobile and subject to repeated orogenesis, but has since been stable and for the last 600 million years subject to only broad movements of uplift and subsidence.

During this interval the platform has been only slightly more mobile, locally subsiding into basins and uplifted along arches that will be studied in Chapters 10 and 11. At the edge of the platform the covering sediments thicken rapidly toward the miogeocline and are caught up in the deformation of the geosynclinal mountains.

Geosynclinal Mountain Systems

The North American craton is surrounded by four chains of mountains formed from geosynclines. Generally, these mountains are separated from the shield by thousands of kilometers of platform rocks, but in Greenland and southeastern Canada they impinge on the shield. The youngest and highest of the mountain chains follows the Pacific coastline from Alaska to Central America. This great backbone of the Americas is made up of many mountain ranges that are collectively called the North American Cordillera (Cor-di-yérà).

The eastern margin of the continent is bordered by the older and more subdued Appalachian Mountain System. The chain can be followed from Newfoundland to Alabama, where it is covered by the younger sediments of the Coastal Plains. Its continuation north of the Gulf of Mexico toward the Cordillera is obscured by these coastal plains sediments, but in the Ouachita Mountains of Oklahoma and Arkansas and the Marathon Mountains of western Texas the deformed belt emerges for short distances.

In the Canadian Arctic Islands the craton is bounded on the north by the folds of the Innuitian Mountain System. The junction of this system with the Cordillera near Alaska is hidden by younger sediments and the waters of the Beaufort Sea. Its eastern extension crosses the northern tip of Greenland and meets a belt of deformed rocks that strikes north along the eastern coast of Greenland. The East Greenland Mountain chain is the edge of a much larger belt of deformed rocks that was divided between Greenland, Norway, and the British Isles when the North Atlantic opened. If the continents are restored to their pre-rifting positions (Fig. 1–16), the East Greenland Mountains strike southward through the British Isles and join with the north end of the Appalachian chain of Newfoundland, completing the ring of the Canadian Shield.

Coastal Plains and Continental Shelves

Along the Atlantic and Gulf of Mexico coasts of the United States sandstones and shales form a wedge of sediments that underlies the coastal plain and extend seaward to form a broad continental shelf. At the edge of the shelf in the Gulf the wedge is tens of kilometers thick. Only a narrow shelf has formed in the Pacific Ocean beside the Cordillera, but along the coast of the Arctic Islands and in Alaska a narrow strip of rocks on shore and a more extensive shelf constitute the Arctic Coastal Plain.

The geology of each of these major divisions of North America is studied in further detail in succeeding sections of this book.

SUMMARY

Zones of mountain building show evidence of crustal shortening.

In a typical mountain chain marginal to a continent, the continental side is formed of sedimentary rocks thrust and folded against the continental platform; the oceanward side is characterized by large masses of granitic rock called batholiths and extensive metamorphism.

The greater thickness of the sedimentary succession in mountains compared to that of equivalent rocks beneath adjacent continental interiors indicates that the site of the mountains was previously a trough. The sedimentary rocks that accumulated in these troughs, called geosynclines, are commonly composed of graded beds of graywacke deposited by turbidity currents.

The sedimentary succession of geosynclines preserves the record of many periods of crustal mobility, compression, and elevation.

The elevation of mountains is probably related to the thickening of the light crust of the earth at the site of the chain and its rise into isostatic equilibrium.

Research on ocean basins suggests that environments of ancient geosynclines are duplicated in modern oceanic trenches, island arcs, and basins associated with these features. The environments of the shelf at the continental margin where quartz sandstones and carbonate rocks accumulate is called the miogeocline. Sediments and lavas accumulating in the trench and basins associated with island arcs are called eugeosynclinal or, simply, geosynclinal.

Interaction of lithosphere plates causes compression of continental margins and widespread deformation of geosynclinal and miogeoclinal environments in which the sediments and volcanic rocks are thrust toward the continent, metamorphosed, and intruded in a major mountain building episode called an orogeny.

Movement of the lithosphere plates may be caused by convection in the mantle owing to unequal distribution of heat sources in the earth, but its cause is not strictly established.

Continents consist of central stable cores called cratons, surrounded by geosynclinal mountain ranges that may be, in turn, bordered by coastal plains and continental shelves.

QUESTIONS

1. Discuss the similarities and differences between oceanic trenches and geosynclines.
2. How are intrusive and extrusive igneous rocks related to the subduction process?
3. Why are geosynclinal mountain chains such persistent features of the earth's crust despite the levelling processes of weathering and erosion?
4. If the ocean basins are no more than 180 million years old, how can we have a fossil record of marine life extending back 3.5 billion years?

SUGGESTIONS FOR FURTHER READING

DEWEY, J. F. "Plate Tectonics." *Scientific American,* 228:56–68, 1972.

KAHLE, C. F., ed. "Plate Tectonics, Assessments and Reassessments." *American Association of Petroleum Geologists Memoir No. 23,* 1974.

LEPICHON, X., FRANCHETEAU, J., and BONNIN, J. *Plate Tectonics.* Amsterdam: Elsevier Scientific Publishing Co., 1973.

MARVIN, U. B. *Continental Drift, the Evolution of a Concept.* Washington: Smithsonian Institution Press, 1973.

MCELHINNY, M. W. *Paleomagnetism and Plate Tectonics.* Cambridge: Cambridge University Press, 1973.

MCKENZIE, D. P. and RICHTER, F. "Convection Currents in the Earth's Mantle." *Scientific American,* 235:72–89, 1976.

SMITH, A. G. "Plate Tectonics and Orogeny: A Review." *Tectonophysics,* 33:215–285, 1976.

WILSON, J. T., ed. *Continents Adrift and Continents Aground, Readings from Scientific American.* San Francisco: W. H. Freeman & Co., 1976.

PART TWO

sedimentary record

CHAPTER THREE

The concepts of plate tectonics provide a basis for the understanding of the major physical events in the history of the earth. The record of these events in the past must be interpreted from the igneous, metamorphic, and sedimentary rocks that were formed and reformed in reponse to the changing conditions within the crust and on it. The history of the earth's surface is best recorded in the sedimentary rocks that were deposited on it in response to changing conditions—submergence and emergence of the land, mountain building and erosion of highlands, deposition and weathering. Each sedimentary rock is a product of an environment where it was deposited, an agent which transported the loose sediment, and a source area from which its constituents were derived and, if properly interpreted, can give evidence about the nature of each. The interpretation of the succession of sedimentary units (or strata) is the discipline called **stratigraphy.**

Major tectonic processes and orogenic events are best studied on a global scale; sedimentary processes and patterns of deposition are better studied in local areas. The local section, a succession of beds exposed in a quarry, road cut, or stream bed, provides the information on which most stratigraphic studies are based. Most sedimentary rocks are layered or divided into beds. Each layer represents an episode of deposition. The order of depositional events is read in a stratigraphic section, from the base to the top, in the order in which the beds were laid down.

A local succession of sedimentary beds provides information about past conditions in a limited geographic area. Information from many such sections must be compiled to reconstruct the past condition of the whole continent. In the following chapter we shall consider the methods of making such compilations; in this chapter we are concerned with the information that can be found in a single stratigraphic section. The section of rocks at Niagara Falls is well exposed, contains a variety of rock types, and has been studied in detail by many geologists. It will provide an example of this type of stratigraphic analysis (Fig. 3–1).

INTERPRETATION OF SEDIMENTARY ROCKS

The Local Stratigraphic Section

At Niagara Falls the drainage of the Great Lakes plunges over an escarpment of sedimentary rocks that can be traced from northern New York State into southern Ontario. The rocks exposed near Niagara are illus-

STRATIGRAPHY

trated in Figure 3–2. At the base of the escarpment is a red shale. It is overlain by 4 m of white sandstone, 15 m of gray-green shale, 12 m of red and green shale, and so on. Each set of beds, consisting of a single rock type such as sandstone, shale, or limestone, or some uniform combination of these types, is called a **formation.** Each formation is named after a geographical locality designated by the geologist who first described the rocks and designated the **type locality.** If the formation consists of a single rock type, that rock name may follow the geographic name in the designation of the formation, for example, Queenston Shale. If the formation is composed of two or more rock types that are interbedded, the term "Formation" follows the geographic name, as in Grimsby Formation. Subdivisions of formations are called **members.**

The interpretation of the environments that produced the sequence of sedimentary

FIGURE 3–1 **Niagara Falls. The drainage of the upper Great Lakes pours over an escarpment formed by resistant strata of Silurian age. (Courtesy of Ontario Hydro.)**

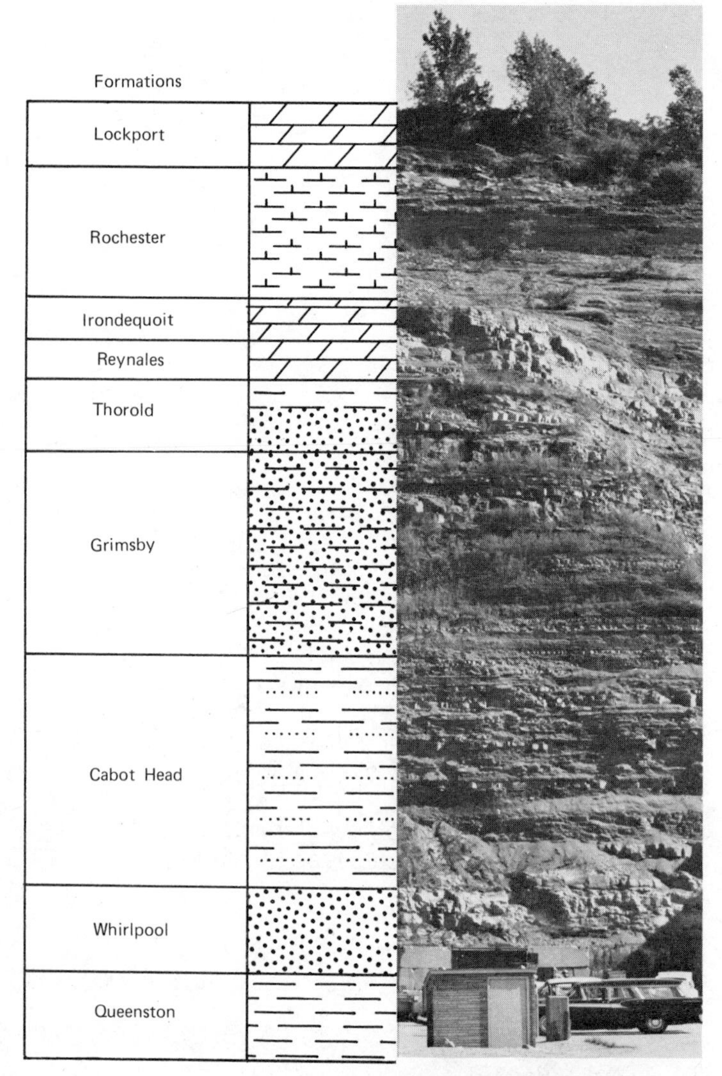

FIGURE 3–2 **Formations of the Niagara Escarpment as exposed at Decew Falls, near Niagara Falls, Ontario (M.J. Copeland; courtesy of the Geological Survey of Canada.)**

rocks at Niagara is based on a principle first stated clearly two centuries ago by James Hutton. Hutton believed that the history of the earth preserved in sedimentary rocks could be explained by the slow workings of the processes we can observe around us, acting over a great length of time. His doctrine has long been known as *uniformity* or *uniformitarianism* and is summed up in the phrase: the present is the key to the past. To illustrate the application of this principle we will look more closely at the Grimsby Formation at the base of the Niagara section.

The Grimsby Formation consists of thin beds of red shale and sandstone. The sandstones show fine internal layers that lie oblique to the planes that delimit the larger beds. Such a structure is called **cross bedding.** The surfaces of some of the beds are marked by holes, fine ridges, and grooves and irregular sausages of sediment penetrate the beds (Fig. 3–3). The grains of the sandstone are fine, well-rounded, and coated with a thin film of iron oxide that gives them their red color. To reconstruct the environment in which the Grimsby Formation was deposited we must find an area where similar sediment is being deposited today.

FIGURE 3–3 **One-meter bed of the Grimsby Sandstone at Rochester, New York; completely burrowed by unknown organisms, probably crustaceans. (Courtesy of Michael Risk.)**

Similar red sands and muds are accumulating today on the shores of the Bay of Fundy on the North Atlantic Coast. The ebb and flow of the tidal currents produces cross bedding in the fine sands. The sand grains, derived from the erosion of neighboring red sandstone cliffs, are covered with a film of iron oxide. When the tide is out the sediment is marked by the castings and feeding traces of marine organisms that live in the mud (Fig. 3–4). Similar conditions of deposition must have produced the Grimsby Formation.

FOSSILS

In the Grimsby Formation at Niagara grooves, trails, and casts are evidence of animals that inhabited the mud flats of an ancient shoreline. Some of the formations in this section contain no organic remains, but most are characterized by a particular suite of fossils and different types of preservation.

Preservation of Fossils

Fossils are the remains of organisms, or traces of their existence, preserved in rocks. The study of fossils is the science of **paleontology.** Most organisms are composed of soft and hard tissue. The soft tissue is composed of compounds of carbon that are readily broken down by bacterial action when the organisms die. The hard tissue forming a skeleton, shell, or covering, may also be composed of organic compounds but, more commonly, is reinforced by mineral matter, such as calcium carbonate (Fig. 3–5), calcium phosphate, silica, etc., that resists decomposition. One type of fossil includes organisms that have been preserved whole and unchanged, soft and hard tissue together. These are the rarest of fossils, exemplified by the carcasses of the woolly mammoth from the frozen gravels of Siberia, preserved by the cold so perfectly that in one specimen the last meal remained undigested in the stomach and the flesh remained edible after several thousand years.

Preservation of only the hard parts in their original state is far more common.

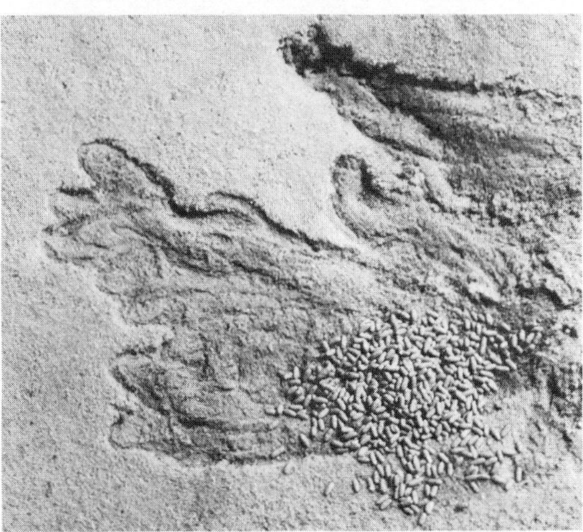

FIGURE 3–4 **Ancient and modern feeding traces. *Top:* The bedding plane of the Grimsby Formation with the trace of a feeding worm. *Bottom:* The trace of a living echiuroid worm, with the fecal pellets produced by the animal. (Courtesy of Michael Risk and the American Association for the Advancement of Science.)**

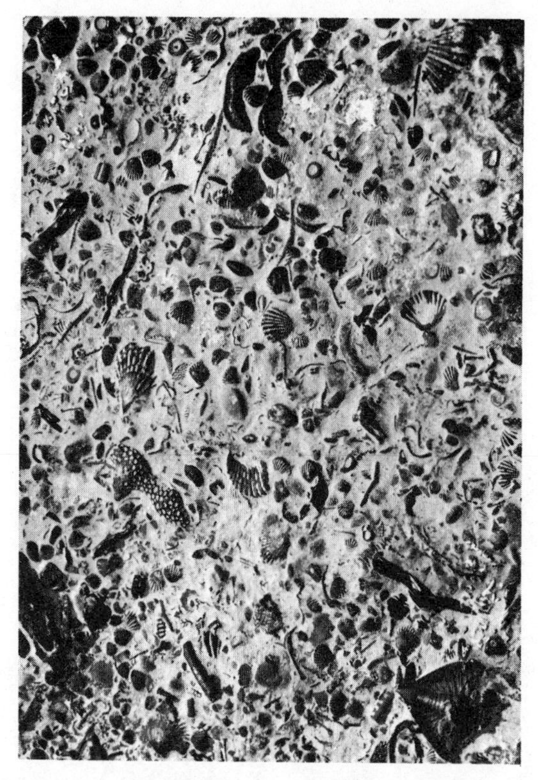

FIGURE 3–5 **Middle Silurian limestone containing many brachiopods and bryozoans reinforced by calcium carbonate. The weathering of the matrix has left the fossils in relief. (Courtesy of the Redpath Museum.)**

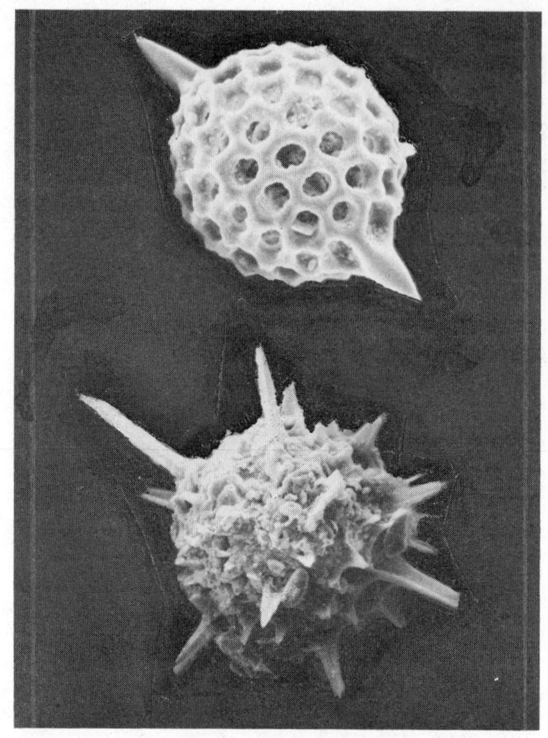

FIGURE 3–6 **Radiolarians build minute shells of silica, which have been preserved without change in a deep-sea core from the Labrador Sea. The tests are about 0.05 mm across. (Scanning Electron micrograph, C.W. Stearn; courtesy of S. Chough.)**

Bones, woody matter, and the exoskeletons of insects, composed of a protein-like substance called **chitin**, are all relatively enduring compared to the soft tissue of organisms, yet all are subject to decay by bacteria. Shells that are composed of inorganic compounds such as calcium carbonate and silica (Fig. 3–6) resist decay and are the most abundant types of fossils in relatively young sedimentary rocks. Most shells and bones contain organic matter distributed as layers and tubes within their substance. Upon death of the organisms this organic matter decomposes and its place may be taken by minerals, of much the same chemical composition as the shell, deposited by water seeping through the rock. A shell preserved in this way is said to be **permineralized.** It contains all the calcium carbonate in the original shell plus enough newly deposited calcium carbonate to fill the spaces vacated by the organic matter. Most of the fossils in older rocks are permineralized.

The fossils of plants and the protective shells of insects and other arthropods commonly appear as black coatings on bedding planes of sedimentary rocks. The fossils were originally composed of such organic compounds as chitin, cutase, and cellulose that have reached high enough temperatures to lose by distillation all their volatile materials—oxygen, nitrogen, hydrogen, and hydrocarbons—leaving a residue of almost pure carbon. Thus, a fossil leaf consisting of a thin film of carbon on the surface of shale (Fig. 3–7) is, in effect, a thin film of coal. Coal is, after all, the remains of plant matter

FIGURE 3–7 **Fern-like leaves (*Pecopteris*) of Pennsylvanian age, preserved by carbonization of the original plant material. (Courtesy of the Redpath Museum.)**

FIGURE 3–8 **This net-like specimen of the dendroid graptolite *Dictyonema* is a carbonized film on the surface of a Silurian dolomite. (Courtesy of the Redpath Museum.)**

that has lost most of its gaseous constituents largely as the result of increased temperature. Trilobites, insects, graptolites (Fig. 3–8), and other groups of animals whose shells were composed of such organic compounds as chitin may be preserved in this way. The process is referred to as **distillation** or **carbon concentration.**

In the Niagara section the beds of the Reynales Dolomite contain crowded masses of regular cavities. If these cavities are filled with a setting rubber compound and the rock is dissolved away, the cavities can be shown to have the shapes of fossil shells. After the shells were enclosed within the accumulating sediment, water seeping through it dissolved out the shell substance, leaving a mold of the shell (Fig. 3–9). When the rock is broken open the sediment surfaces against which the shell once rested will preserve its size, shape, and ornamentation. When the paleontologist introduces rubber into the cavity of the mold, he makes a cast which is a replica of the original shell. Mineral matter deposited by solutions in a mold may make a natural cast. Where the interior of a fossil shell is filled with sediment before the shell is dissolved away, a mold of the interior surface of the shell is preserved as a fossil (Fig. 3–10). Insects preserved in the gum of coniferous trees, that turns to amber in time, rarely possess any organic substance and are, therefore, molds of the exterior of the organism (Fig. 3–11).

FIGURE 3–9 **Cluster of the brachiopod *Pentamerus oblongus,* seen from the base of the bed in which specimens are preserved in life position as interior molds. Although this specimen is from the Middle Silurian of Alabama, similar specimens occur in the Reynales Dolomite. (Courtesy of A.M. Zeigler, A.J. Boucot, and R.P. Sheldon and the *Journal of Paleontology.*)**

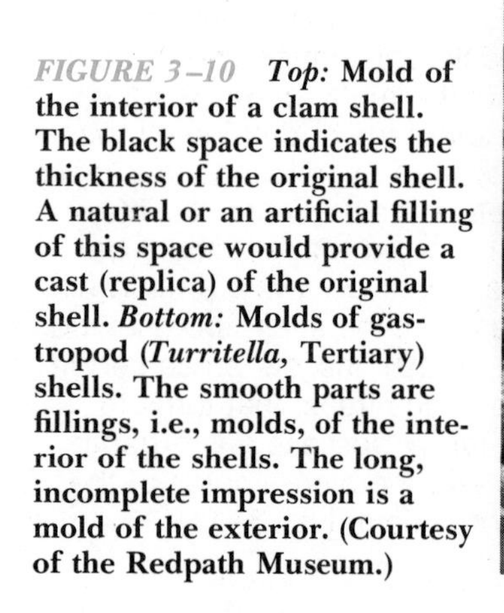

FIGURE 3–10 ***Top:* Mold of the interior of a clam shell. The black space indicates the thickness of the original shell. A natural or an artificial filling of this space would provide a cast (replica) of the original shell. *Bottom:* Molds of gastropod (*Turritella,* Tertiary) shells. The smooth parts are fillings, i.e., molds, of the interior of the shells. The long, incomplete impression is a mold of the exterior. (Courtesy of the Redpath Museum.)**

FIGURE 3–11 **Mold of caddis fly preserved in Tertiary amber from the Baltic region. (Courtesy of the Smithsonian Institution.)**

The tracks and trails of animals feeding, crawling, or running on soft sediment could be considered to be molds of the exterior surfaces of the animals making them, but are usually classified as a separate type of fossil called **ichnofossils** or **trace fossils.** Such fossils may give valuable evidence about the way in which the animal making them lived and moved (Fig. 3–12). Other types of ichnofossils consist of burrows made by animals in soft sediment, and borings of parasites and predators in the hard tissue of their hosts and prey.

Here and there in sedimentary rocks are found brachiopods composed of the mineral pyrite, and corals composed of silica in the form of chert. Both these organisms, when living, secreted hard parts that were originally calcium carbonate, and their substance must have been replaced in the preservation process by entirely different materials. The process of **replacement** is not fully understood, but once enclosed in the sediment, the shell must encounter conditions in which the original material is dissolved at the same time as the replacing mineral is deposited. Silicified fossils are more resistant to weathering than either the original shell material or the limestone sediment surrounding them (Fig. 3–13). When limestone containing silicified fossils is dissolved away with acid or weathers away naturally, the fossils released may show fine spines and delicate ornamentation that would be destroyed if they were collected by breaking them from the rock with a hammer (Figs. 3–14, 3–15).

Many marine invertebrates pass sediment through their digestive systems to feed on the interstitial organic matter, and excrete the mineral particles commonly in the form of small, ovoid aggregates called **pellets.** Some limestones are composed entirely of such pelletted material and, in a sense, the pellets are molds of the gut of the animal, and hence fossils. The term **coprolite** is used for the fossilized excrement of all animals, but it is commonly reserved for that of terrestrial vertebrates. This material may be useful in reconstructing the diet of fossil animals from the undigested food that may be identified in it.

Most sedimentary rocks contain markings, cracks, and indistinct shapes that could have been formed by organisms or could be the result of inorganic processes involved in the deposition, deformation, or weathering of the rock. Such markings are commonly referred to as **pseudofossils** or **problematica.** Current marks, raindrop prints, cone-in-cone structure, and distorted bedding are commonly mistaken for fossils by those unfamiliar with sedimentary rocks. Some complex pseudofossils have been the subject of debate over their nature for years.

In summary, an organism of the past may have entered the paleontological record by a variety of routes; permineralization, carbon concentration, replacement, as a mold or cast, as an ichnofossil, etc. Examination of the Rochester Shale at Niagara shows that many of the organisms living at the time of its deposition achieved this form of immortality, but a similar study of the Whirlpool Sandstone reveals no fossils. Either no organisms lived in the sea where the Whirlpool Sandstone was being deposited

FIGURE 3–12 **Trackways. *Left: Laoporus,* an amphibian. *Right: Paleohelcura,* an invertebrate, probably an arthropod. Both from the Permian beds of the Grand Canyon. (Courtesy of the Smithsonian Institution.)**

FIGURE 3–13 **Silicified coral (*Halysites*, Silurian). The coral colony, originally composed of calcium carbonate, has been completely replaced by silica, the superior hardness and resistance to weathering of which allow it to stand out in prominent relief to the limestone in which it is embedded. (Courtesy of the Redpath Museum.)**

FIGURE 3–14 **Polished surface of a silicified tree. Note the faithful reproduction of radial and concentric structures in the original wood. Clover Creek, Idaho.**

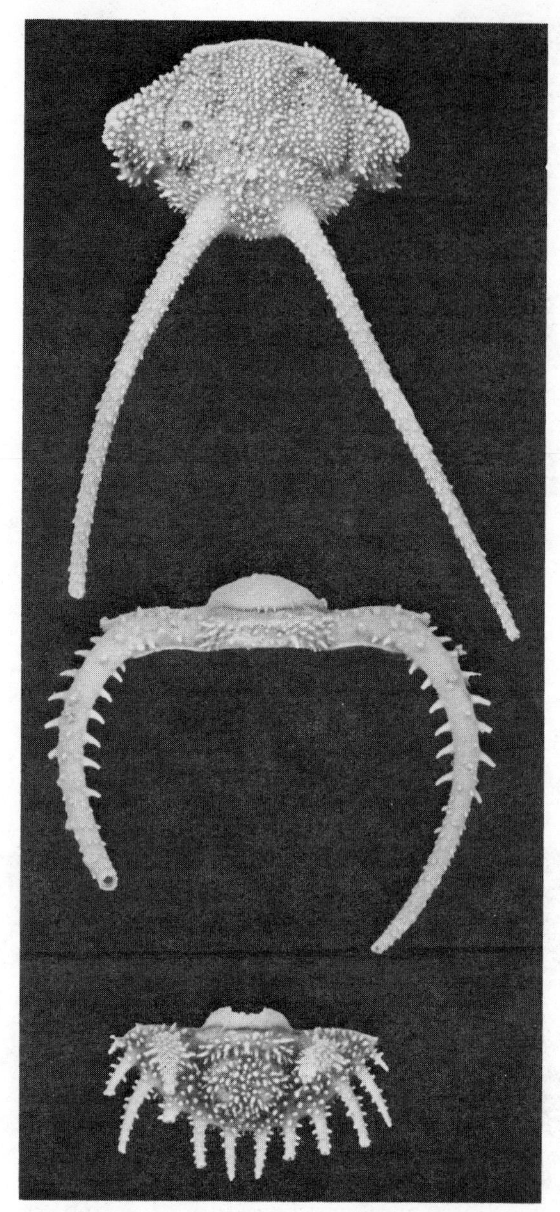

FIGURE 3–15 **Detailed preservation of spiny trilobites by silicification. Cephalon, a thoracic segment, and pygidium of *Apianusus barbatus*, from the Middle Ordovician limestones of Virginia, dissolved out of the limestone with acid. (Photograph by Harry B. Whittington of specimens in the U.S. National Museum.)**

or else none of them left behind preservable remains. What chance has an organism to form a fossil and contribute to our understanding of the life of the past?

Nature of the Paleontological Record

In the open forest and around coral reefs in shallow seas thousands of animals die each year, yet these environments are not littered with their bodies, skeletons, or shells. The cleanliness of these environments is a measure of the efficiency of bacterial decay, scavengers, and predators. The soft tissues of dying organisms will almost certainly be destroyed by these agencies without leaving a trace; even their hard tissue is attacked by boring worms, sponges, bivalves, fungi, and algae in the sea and by scavenger rodents and insects on land. Observations of modern environments suggest that fossilization was a rare accident in the geological past, and in all probability not one organism in a million in ancient communities left a trace behind. To become a fossil an organism must be buried quickly in sediment to remove it from the local garbage collectors and from the destructive effects of inorganic agencies such as waves, currents, abrasion, and dissolution.

Despite these obstacles to the preservation of the individual organism, we must remember that many modern species, and certainly many extinct ones also, are represented by millions of individuals. Many species had ranges of millions of years in time and must have been represented during this span by astronomical numbers of individuals. The chances that such long-lived, abundant species will be preserved in the fossil record are very good.

Much has been written about the inadequacy or imperfection of the fossil record. Paleontologists prefer not to use these terms; they prefer to discuss the chances of preservation of the various kinds of life as the "bias of the record." In many ways the paleontological record is not a representative record of the life of the past; that is, the sample is biased in many directions. Most important to the paleontologist is the bias against soft-bodied organisms. The representation of modern invertebrate phyla in the fossil record is directly proportional to their ability to secrete a hard shell or skeleton. Of some groups, such as the jellyfish, worms, and many protozoans, we have practically no fossil record except vague impressions on bedding planes. Possibly, orders or classes of soft-bodied organisms have existed in the past and left no paleontological record.

The record is biased in several other ways that can be mentioned only briefly here. Most fossil-bearing rocks were deposited in shallow marine waters along continental shelves; few deep-sea sediments are available to the paleontologist for collecting because the record of this environment remains at the bottom of the ocean basins where it has been sampled by only a few drill holes. Unlike the sediments of the continental borders, oceanic sediments are rarely lifted by forces in the crust to dry land where they can be studied in detail. The record of terrestrial animals and plants is biased in favor of lowland species. Organisms that lived on higher ground lived in areas of erosion where no sediments were being laid down in which they could become buried. To interpret the record of life correctly the paleontologist must keep these biases in mind.

Various paleontologists have tried to estimate the number of species that have existed since life began and the proportion that have been preserved and observed by man. George Simpson estimated that about 300 million species of animals have lived on earth. Wyatt Durham suggested that about 4 million species of marine organisms with preservable hard parts have existed since complex life began. James Valentine believes that life in its early stages of development was far less diverse than it is now, and estimates that between 1.5 and 0.3 million species of marine invertebrates capable of preservation may have lived. The large spread of these figures reflects the inadequacy of data and the range of assumptions made by the estimators. Paleontologists have collected and described a small proportion of the species that estimates imply lie in the rocks, waiting to be discovered. So far, only about 100,000 species of marine invertebrates have been described.

In biological classifications (Appendix) species are grouped into genera, genera into families, families into orders, and so on. Although only a small proportion of the species that have ever lived have been described, a much greater proportion of the number of genera is known, and for more abundant groups of fossils secreting hard parts, we probably have a record of at least one representative from all the families.

The fossils preserved in any succession of rocks, like that at Niagara, comprise a small part of the many organisms that lived in the sea and by the shore while these sediments were being deposited. In addition, the proportions of the various organisms in the fossils collected by the paleontologist from these beds will be different from those in the living community because each organism does not have the same chance of being preserved.

Each formation of the Niagara section has a characteristic fauna. The Rochester Shale is full of bryozoans and brachiopods; the Reynales Dolomite has large casts of the brachiopod *Pentamerus oblongus;* the Grimsby Formation has ichnofossils. Some of the reasons for their differences are easy to understand, others pose problems that paleontologists have struggled with for nearly 200 years.

PALEOECOLOGY

Different environments or habitats on land or in the sea support different animals and plants, and these differences are a familiar part of the modern world. The plants that grow in marshes are not found on dry hillsides, nor do the birds that live in the pinewoods frequent the meadows. The mussels that cling to rocky, wave-washed shores are not found in the shifting sands of beaches. The ecologist is concerned not only with the interaction between various members of plant and animal communities, but with the influence of physical factors in the environment of the community. Each environment is characterized by animals and plants that are adapted to their physical and biological surroundings and that react to changes in these conditions. The classification of environments is the beginning of the study of ecological relationships.

The ecologist classifies the environments of the sea largely on the basis of depth: the **littoral** zone between high and low tides, the **neritic** or **sub-littoral** zone of the continental shelves, the **bathyal** zone of the continental slopes, and the **abyssal** zone of the ocean deeps (Fig. 3–16). These broad environments can be further divided many times into habitats, each of which is characterized by its own community of animals and plants. The organisms that live on the ocean floor are said to be **benthonic,** those that swim in the water above are **nectonic,** and those that float passively are **planktonic.** The sum totals of each of these assemblages are called, respectively, the **benthos,** the **necton,** and the **plankton.**

The roles that organisms play in the marine environment may be classified on the basis of the methods they use to obtain food and their position in the food chain. This is a **trophic** classification. Micro-organisms that are capable of using inorganic nutrients from sea water to make complex organic compounds using the energy of sunlight are called **primary producers.** They are food for many marine invertebrates that pass water through a variety of filtering apparatuses, and are called **filter** or **suspension feeders.** Other animals pass sediment through their bodies, digesting the micro-organisms and waste organic material

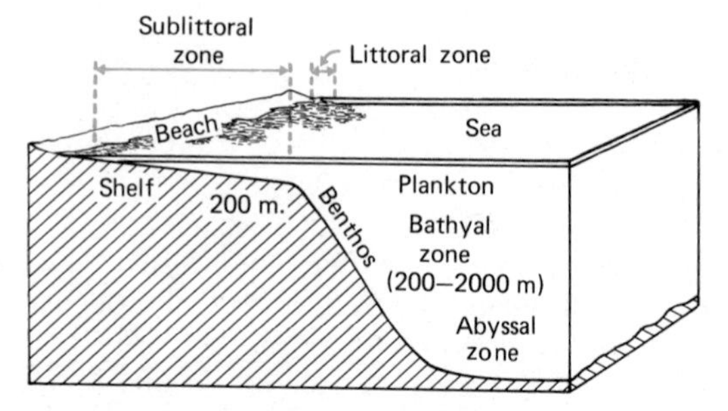

FIGURE 3–16 **Division of the waters and floor of the sea into ecological zones.**

in its interstices. These are called **infaunal deposit feeders** if they live burrowing within the sediments, or **epifaunal deposit feeders** if they live on the sediment surface. Scavengers eat dead organisms on the sea floor: predators eat other animals that are alive. Parasites depend on the life processes of their hosts. In each marine community different organisms play these interdependent roles.

Terrestrial communities have the same trophic structure. The primary producers are rooted plants rather than micro-organisms. Herbivores eat the plants and are, in turn, eaten by carnivores, which may also eat each other.

In the past, as at present, each organism occupied a particular environment, and this environment may be reflected in the sediment laid down where the animal lived. Animals of the coastal deltas were preserved in cross-bedded red sands and shales. Brachiopods and bryozoans of the sublittoral zone were enclosed in shaly limestones. The study of the relationship of fossil animals to the environments in which they lived and the sediments in which they were buried is the science of **paleoecology.**

From the physical characteristics of sedimentary rocks much can be deduced about the environment in which they were deposited by applying the principle of uniformity. The fossils contained in the rocks can be used in a similar way to aid the reconstruction of past conditions. By comparing animal communities found in modern seas with fossil communities, the paleontologist can deduce the physical factors that controlled ancient environments. The division of a modern reef into ecological zones on the basis of animals adapted to various environments (fore-reef, reef-crest, reef-flat, lagoon) can be duplicated in the distribution of fossils preserved in ancient reefs. Thus, coral faunas found in reefs millions of years old can be interpreted in terms of conditions in modern reefs on the basis of comparison of the fossils with modern communities.

By studying the form of fossils, and comparing them with modern species whose adaptations are known, the paleoecologist can reconstruct the ancient community. Fig.

17–12 is a reconstruction of three communities of clams living in deltaic marine sands of 450 million years ago. On the basis of the form of their shells some clams are identified as infaunal deposit feeders, others are epifaunal suspension feeders. Other elements of the fauna are located in trophic positions that seem logical to the paleoecologist as the fossils are compared with animals of modern seas.

Because both the lithology (rock nature) of a sedimentary rock and its fossils are partly dependent on the environment in which they were deposited, certain rocks are closely associated with distinctive fossil communities. Quartz sandstones formed on a beach or in very shallow water commonly contain worm burrows. Gray siltstones are characterized by clams that lived in estuarine conditions. All fossils were, to a greater or lesser degree, associated with a particular environment, but some are found in a greater range of lithologies than others. The fossils of animals that floated or swam in marine environments are found in many different kinds of bottom deposits, for their existence at the sea surface was independent of the material deposited on the bottom. Graptolites (see Appendix for description), extinct floating animals that lived 500 million to 300 million years ago, are particularly abundant in black shales, but they are occasionally found in limestones, siltstones, and even sandstones into which they were deposited from the water above when they died.

The fossil faunas in a section such as that at Niagara reflect the environments of deposition at that site. Yet the parade of life we witness in the sedimentary succession is more than just the comings and goings of organisms adapted to a succession of environments. If a long enough time is recorded in a sedimentary succession, we will see that the fossil species are not identical in recurring formations of similar lithology and presumably representative of similar environments. The succession of life from such thick sequences as the sediments exposed in the Grand Canyon (Fig. 3–17) shows a progression from simple to more complex forms. Superimposed on the differences caused by environmental change is a progression of life which allows the geologist to divide the succession of beds into units defined by fossils.

BIOSTRATIGRAPHY

The interval of existence on earth of each species that left its fossils in the paleontological record is recorded in sedimentary beds that accumulated during the time interval and define its range. Fossils of the species appear in beds laid down during that interval where the environment of deposition was such that the animal could live there successfully or be carried there by natural agencies after death elsewhere. That each formation in a succession is characterized by a unique assemblage of fossils was first recognized late in the 18th century.

An English civil engineer, William Smith (1769–1837), spent much of his life surveying for and supervising the digging of canals. The sedimentary rocks of England had already been divided into units and named, and some work had been done on British fossils. In studying the formations that were cut through in his canals, Smith noticed that the beds all dipped toward the southeast, and that, as one traveled in that direction one passed over the edges of successively younger beds. Smith was able to develop a stratigraphic column of formations and for the first time to predict what kind of rock would be found in a water well or in the next village to the northwest. This stratigraphic and structural synthesis would have been enough to place Smith among the founders of stratigraphic geology (he is commonly referred to as "Strata Smith"), but his second discovery was of even greater importance. He noticed that wherever he dug into a certain formation, the fossils that turned up were always the same, and that those fossils did not occur in any other formation. This, he discovered, was true for the entire country. Here, then, was a way of identifying the presence of a certain formation from fossils picked up and brought to him by his assistants, or from the study of fossils in the cabinets of local collectors.

FIGURE 3–17 **Sedimentary strata exposed in the Grand Canyon of the Colorado River record a span of a billion years of earth history. The lithology and ages of the formations are shown on the left side in a columnar section. (Courtesy of the U.S. Geological Survey.)**

For a dozen years Smith accumulated a tremendous amount of information, undeniably linking certain fossils to certain formations, and discussed his findings freely with interested persons. In 1815, he published a hand-colored map of England and Wales, showing with surprising accuracy of detail the geological formations then known (Fig. 3–18). There followed the next year an explanatory pamphlet entitled "Strata Identified by Organized Fossils." Fortunately, Smith lived to see his revolutionary innovation accepted and respected and his map made the prototype of geological maps of other localities. To these intangibles were added a Royal Society gold medal and a modest pension from the Crown.

Cuvier and Brogniart

While William Smith was laying the foundations of biostratigraphy, two scientists in France were coming to the same conclusions concerning stratigraphic succession and diagnostic fossils. Both were well established in Paris—Georges Cuvier (1769–1832) first as professor of comparative anatomy at the University of Paris and later as Perpetual Secretary of the Institute of France, and Alexandre Brogniart (1770–1847) as professor of mineralogy at the Museum of Natural History in Paris. Together these two men, zoologist and geologist, scoured the countryside around Paris and accumulated enough information so that by 1808 they were able to publish their conclusions: that the strata of the Paris basin were arranged in a definite succession, and that each subdivision contained its own peculiar assemblage of fossils. Although this publication antedated Smith's map and report, it must be remembered that for a decade before, Smith had distributed to anyone in-

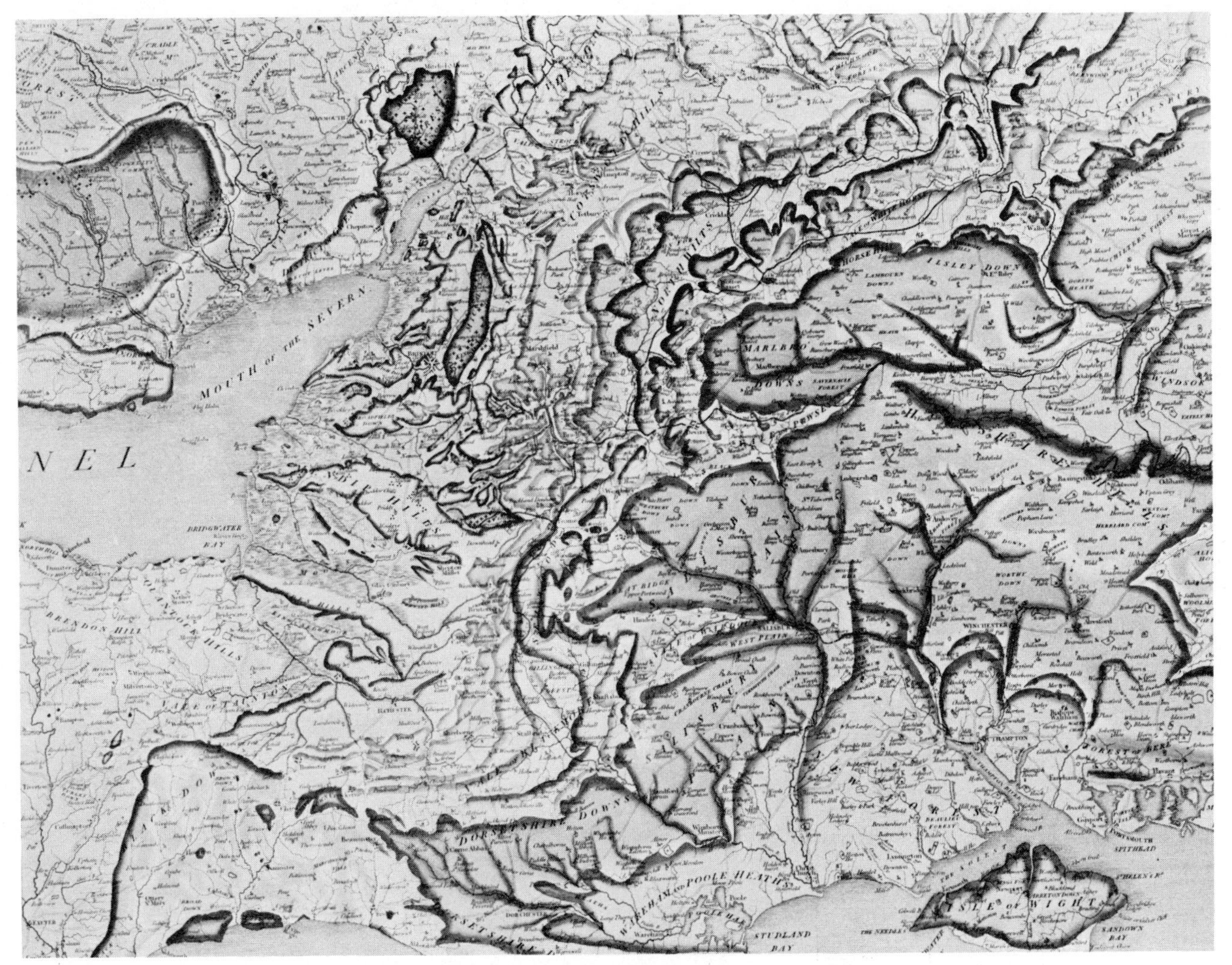

FIGURE 3–18 **Part of William Smith's map of England, published in 1815, showing area east of the Bristol Channel. The formations are distinguished by bands that are darker on the lower side. (Courtesy of the McGill University Libraries.)**

terested an account of his labors and results.

In the days of Smith, Cuvier, and Brogniart the progression of life that we now speak of as evolution was appreciated by only a few scientists whose views were widely regarded as unorthodox. None of these three supported the evolutionists, and Cuvier formulated a theory of repeated creations to explain changes in the fossil record. They did discover that each life form had a span of existence that is expressed in a thickness of sedimentary strata within which this organism is found as a fossil. The fundamental unit of biostratigraphy, the science of the distribution of fossils in time, is called the **zone.** A zone is a thickness of strata defined by the range of a single fossil or the ranges of a group of fossils. In a sense the range of each fossil defines a zone that could be used in the division of strata, but some fossils, because of their abundance or limited occurence in time, are more valuable than others for comparing successions from place to place or in dividing successions into smaller units. In practice, most zones are defined on the basis of assemblages of fossils, but for the sake of uniformity the name of one of them is chosen to designate the zone.

The paleontologist collecting fossils from a stratigraphic section notes their stratigraphic position and compiles the information on a chart showing their overlapping ranges (Fig. 3–19). Within the overlapping ranges he selects horizons defined by the beginnings and endings of ranges that he believes may have regional significance, and uses them to define sequences of zones named from a prominent fossil within the zone. These zones may be united into stages and series to form a hierarchy of biostratigraphic divisions of rock strata. The **stage** is a succession of beds characterized by one or more zones, and a **series** is a succession of beds which includes several stages. Generally, a scheme of stages and zones cannot be set up by the study of a

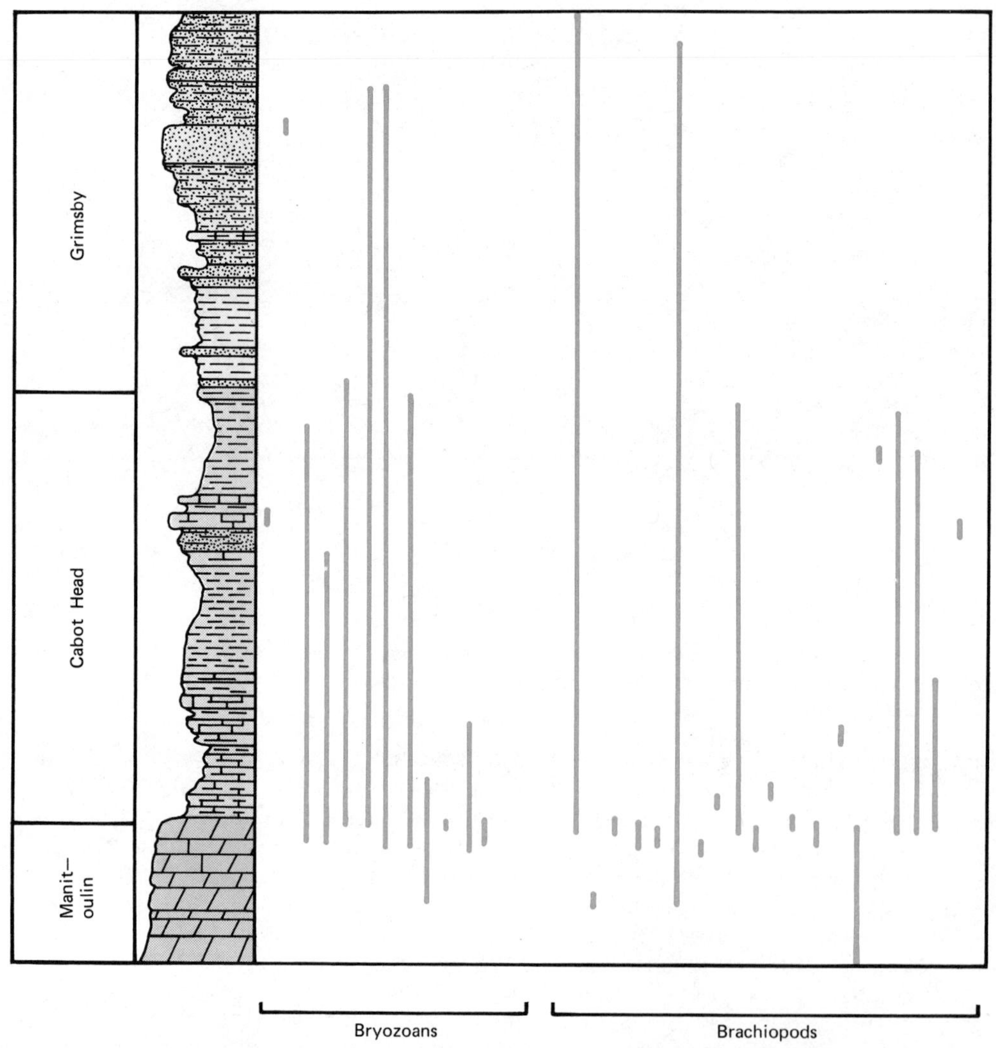

FIGURE 3–19 **Ranges of brachiopod and bryozoan fossils in the lower part of the Silurian section along the Niagara Escarpment at Hamilton, Ontario. (Courtesy of T.E. Bolton.)**

single exposure, but is the result of regional study of many exposures (Fig. 3–20).

The succession of fossil life revealed in a set of sedimentary beds is the product of the interaction of environment and life. Not only does the environment determine what organisms can live and be preserved in a particular place, but organisms adapt and change with the changing environment, modifying their abilities to live under particular environmental conditions. We call this adaptation, and modification **evolution**; we shall examine it in some detail in Chapter 8. In Chapter 5 we shall see how the parade of life produced by this complex interaction of life and environment allows stratigraphers to determine the relative ages of sedimentary rocks.

MARINE AND NON-MARINE SEDIMENTS

Many of the formations of the Niagara section contain fossils of such animals as crinoids and brachiopods that are now found only in the waters of the oceans, and other groups, such as the bryozoans, that only reach abundance there. These beds must have been deposited in the sea. We are forced to conclude that for a time in the past the sea inundated this section of the continent. More extensive observations show that large areas of the continents of the world are covered with layers of sedimentary rock whose structures, textures, and fossils show that they are of marine origin. To understand the history of a continent the

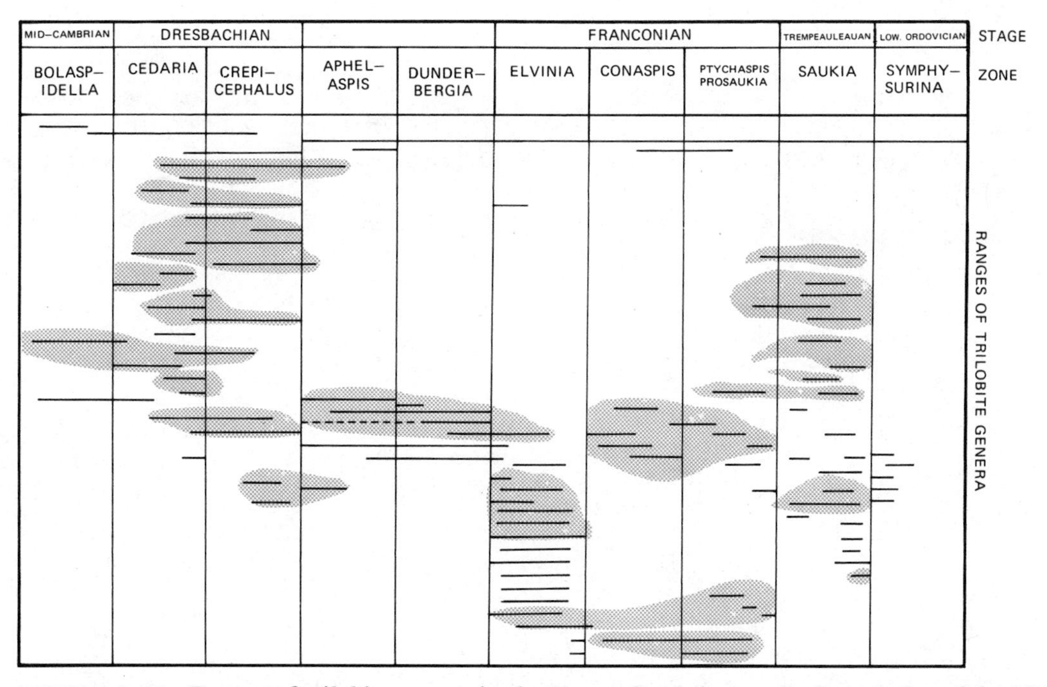

FIGURE 3–20 **Ranges of trilobite genera in the Upper Cambrian, or Croixan Series of the Llano region, Texas. The vertical lines represent the ranges of genera, the shaded areas families. The horizontal lines not enclosed within an area of shading represent genera which are not grouped into families on this chart. A zone is named after a prominent member of the fossil fauna, and its lower boundary is defined by the first appearance of the fossils concerned. (Courtesy of C. Lochman-Balk and the Paleontological Society.)**

historical geologist must be able to distinguish sedimentary rocks laid down in the ocean from those deposited in lakes, swamps, rivers, and deserts.

Unfortunately the structures and textures found in non-marine or terrestrial sedimentary rocks can also be formed on the sea floor. Geologists may rely on color to distinguish marine from non-marine sediments, but color may be misleading. Many non-marine sediments are colored red, brown, or yellow by a small content of iron oxide (hematite) or hydrous iron oxides (limonite). The iron in these oxides, formed when iron-rich minerals are weathered in the presence of oxygen in the air, is in the ferric state. When the sediments are brought to the sea, they lose their color because the iron oxides are reduced on the sea floor from the ferric state to the colorless, ferrous state. The reducing environment is formed by bacteria that use up all the available oxygen as they decompose waste organic matter in the sediments. Because of this reducing action developed on the sea floor, most marine sediments are gray; but some ferric oxides will slip through the reducing barrier and color marine sediments red. The reducing action of bacteria is not limited to the ocean; in lakes and swamps iron compounds are reduced and lose their reddish colors. Though most red-colored sediments are non-marine, some are marine; most marine sediments are gray, but many non-marine sediments are also gray.

The most reliable indicators of the environment of deposition are fossils contained in sedimentary rocks. Certain groups of invertebrate animals today cannot live in fresh water or out of water. Among such animals are the corals, sea stars, sea urchins, brachiopods, barnacles, and squids. Because we have no reason to believe the habitat requirements of these animals have changed, we conclude that sedimentary rocks containing them as fossils are marine. The marine or non-marine origin of some sedimentary rocks that lack both fossils and distinctive sedimentary structures must remain undetermined until new methods are developed. Throughout the geologic column, most sedimentary rocks investigated by geologists are of marine origin.

The succession of formations of the Niagara Escarpment can be interpreted as a succession of environments of deposition by applying the principle of uniformity and a knowledge of modern environments of deposition. The sequence of changes in time must be read from bottom to top as the products of later events were superposed on those of earlier ones. The Queenston Shale is the result of rivers washing red muds into an area slightly above sea level. The Whirlpool Sandstone reflects the submergence of the area beneath a shallow sea where waves sifted, rounded, and cleaned the quartz grains. The water deepened during the deposition of the overlying Cabot Head Shale and many marine animals lived along the sea floor, leaving their shells as fossils in the mud. The Grimsby Formation records the shallowing of the sea as red sandstones and muds built a delta near to, or above, sea level. The white Thorold Sandstone marks the renewed advance of the sea bringing back conditions like those in which the Whirlpool Sandstone was deposited. The rest of the formations, composed of limestones and shales with marine fossils, are evidence of the persistence of the sea until the end of the interval of time represented by this succession. By interpreting the conditions of deposition of each formation the geologist reconstructs a history of the rise and fall of the sea.

SUMMARY

Sedimentary successions are divided into formations on the basis of their lithology. Formations are named after their type localities.

Environments in which formations were deposited can be determined by comparing their lithology with that of the sediments now accumulating under known conditions (i.e., applying the principle of uniformity, that the present is the key to the past).

Marine sedimentary rocks widely distributed on the continents indicate that the oceans have inundated the continents several times in the geological past.

Although red colors may be indicators of terrestrial origin of sedimentary rocks, fossils are the most reliable criteria for determining whether a sediment is marine or non-marine. A stratigraphic section records a sequence of environmental changes in an area.

Organisms may be preserved in sedimentary rocks: (1) unchanged, (2) as unchanged hard tissue only, (3) permineralized, (4) by carbon concentration or distillation, (5) as molds and casts, (6) as ichnofossils (burrows, feeding traces, trails, etc.), or (7) as replacements by silica, pyrite, or other minerals.

Bacterial decay, scavengers, predators, and forces that mechanically break up organisms after death make the likelihood of an organism being preserved in the paleontological record very small. The paleontological record does not preserve faithfully the life of the past, but is biased in several ways. Only a small proportion of the animals that have ever lived have been preserved in rocks, and still fewer have been discovered and described by paleontologists.

Paleoecology is the study of the relationship between the fossil organism and the environment in which it lived. Paleontologists reconstruct ancient communities by comparing them with modern ones and assigning the organisms a position in the ancient trophic structure.

Each fossil species is preserved in the fossil record in a set of beds deposited during its time range on earth. Smith, Cuvier, and Brogniart found that each formation contains a characteristic set of fossils, and they laid the foundation for biostratigraphic division of rock successions into zones and stages.

QUESTIONS

1. What features along a modern shoreline might be preserved which would help the geologist find an ancient shoreline?
2. What obstacles stand in the way of an organism with a hard shell "finding a place" in the paleontological record?
3. How do the following types of organisms find their food and where do they live: infaunal deposit feeder, benthonic predator, epifaunal filter feeder, nectonic suspension feeder. Give an example of each.
4. Why are the fossils of some marine animals always associated with a particular lithology and others are found in many different kinds of sediments?

SUGGESTIONS FOR FURTHER READING

BEERBOWER, J. R. *Search for the Past,* Englewood Cliffs, N.J.:Prentice-Hall Inc., 2nd ed., 1968.

BLACK, R. M. *Elements of Palaeontology.* London: MacMillan & Co., 1970.

DURHAM, J. W. "The incompleteness of our knowledge of the Fossil Record." *Journal of Paleontology,* 41: 559–565, 1967.

LANE, N. G. *Life of the Past,* Columbus, Charles E. Merrill Publishing Co., 1978

LAPORTE, L. F. *Ancient Environments.* Englewood Cliffs, N.J.: Prentice-Hall Inc., 1968.

MCALESTER, A. L. *The History of Life.* Englewood Cliffs, N.J.: Prentice Hall Inc., 2nd ed., 1976.

RUDWICK, M. J. S. *The Meaning of Fossils: Episodes in the History of Palaeontology.* New York: Neale Watson Academic Publishers, 2nd ed., 1977.

CHAPTER FOUR

The stratigraphic section at Niagara Falls yields information about the sequence of events during an interval of geological time at a single locality. To understand the events of the whole continent of North America the geologist must take many such local observations and compile them into a regional picture. If the stratigraphic sections are analogous to books that are read by the stratigrapher and paleontologist, then the reconstruction of the conditions of a continent is like extracting and reconciling the essential arguments from many books in a library to write a review of a subject.

A further analogy may help to explain the processes of compilation. Each stratigraphic unit is like a pile of undated, local newspapers accumulated from day to day; each formation is an event recorded in them. If read from the bottom of the pile upward, the newspapers record the sequence of events in a particular city. The task of historians trying to understand the history of the continent is to place the events recorded in the different cities in their relative positions in a time frame. If the undated newspapers from Montreal, Chicago, and Los Angeles record outbreaks of flu, historians will be uncertain whether the disease broke out simultaneously in the three cities or an epidemic swept across the country in one direction or the other. Without dates, relying solely on events, they must search the papers for an event that they know must have affected the cities at the same time, such as an eclipse of the sun or the tremor of a catastrophic earthquake. Then they can establish the time equivalence of events around this one.

The process of matching formations of separate stratigraphic sections in order to establish their time equivalence is known as **correlation.** This chapter is concerned with the methods used by the stratigrapher and paleontologist in correlating stratigraphic sections and the methods they use to produce from them a regional picture of conditions in the past.

FACIES CHANGES

Establishing Lithologic Geometry

The Niagara Escarpment extends for 250 km from New York into Ontario. Figure 4–1 illustrates graphically three successions, or sections, of sedimentary rocks that a geologist has measured at intervals along its length. The information given by the section at Niagara Falls was discussed in the last chapter. To establish the lateral extent of the formations the geologist must join their boundaries in order to produce a logical

CORRELATION

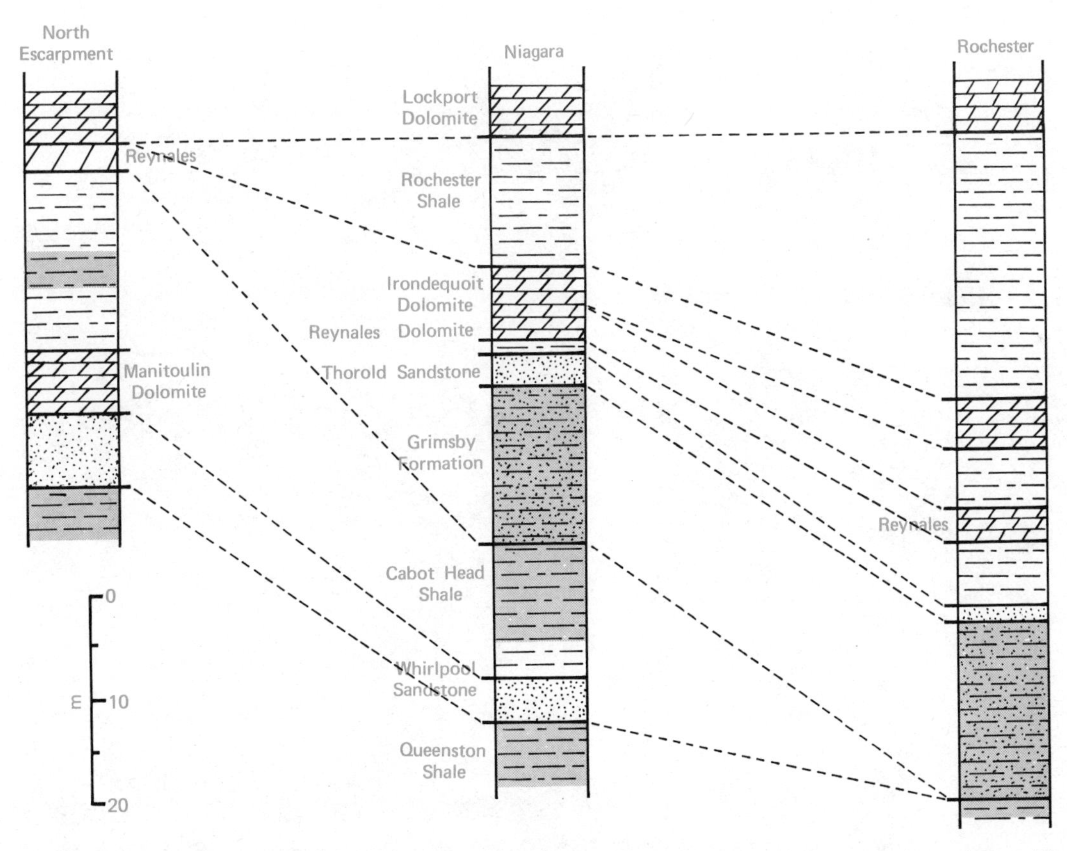

FIGURE 4–1 **Three stratigraphic sections along the Niagara Escarpment at Rochester, New York, at Niagara Falls, and at the north end of the escarpment.**

interpretation of the conditions of deposition in the area.

At the crest of the escarpment is the bed over which the water pours at Niagara, the Lockport Dolomite. Beds similar in lithology and of the same thickness can be found at the top of the other two sections. The dolomite beds can reasonably be assumed to represent the same formation in each section, but if any doubt remains, the geologists can trace the outcrop of the dolomite almost continuously by walking along the top of the escarpment from one section to another. More commonly, the sections being compared by the stratigrapher are isolated from one another by later deposition, subsequent erosion, or deformation of the crust, and continuity of formations cannot be demonstrated.

The Reynales Dolomite does not change in lithology between the three sections and may be identified easily by a thin layer packed with the shells of the brachiopod *Pentamerus oblongus* (Fig. 3–9) at its base. However, the beds above and below the Reynales Dolomite are not the same in the three sections. Below it at Niagara are, in descending order, a white sandstone (Thorold), a red sandstone and shale (Grimsby), a gray shale (Cabot Head), and another white sandstone (Whirlpool). At Rochester the Reynales rests on the Maplewood Shale, the Thorold Sandstone, and it, in turn, on the red sandstones and shales of the Grimsby Formation (Fig. 4–2). At the north end of the escarpment the Reynales is underlain by a reddish, silty shale (Cabot Head), gray dolomite (Manitoulin), and

FIGURE 4–2 **The Silurian section at Rochester, New York. Some of the formations are: (1) Grimsby Formation, (2) Thorold Sandstone, (3) Maplewood Shale, (4) Reynales Limestone, (5) Sodus and Williamson shales, (6) Irondequoit Limestone. (Courtesy of Z. P. Bowen.)**

white sandstone (Whirlpool). As the stratigraphic interval between the Whirlpool and Reynales does not seem to include significant breaks in sedimentation, the sequences Maplewood–Thorold–Grimsby, Thorold–Grimsby–Cabot Head, and Cabot Head–Manitoulin must grade laterally into each other. Measurement of a section intermediate between that at Niagara and the north escarpment shows the progressive thinning northward of the Grimsby and Thorold formations and the appearance and thickening northward of the Manitoulin Dolomite. The zone of contact of the Cabot Head and Grimsby formations in all these sections can be seen to be marked by a gradational sequence in which beds of shale alternate with beds of sandstone. Sandstone beds in this contact zone traced northward along the escarpment die out in the shale. Shale beds traced southward die out in the sandstone (Fig. 4–3). The beds of the two formations interpenetrate like fingers of one hand placed between those of the other. Along the escarpment northward the basal beds of the Grimsby Formation pass laterally into the Cabot Head Shale until at the north end none of the Grimsby Formation is left. This interfingering of lithologies shows that the interfingering parts of the two formations were deposited at the same time and is a result of the lateral oscillation of the boundary of the two environments of deposition. Such oscillation is likely to be caused by changes in the supply of sand to the area of deposition, but could also be caused by changes in sea level or rate of subsidence of the area. When the sand supply from the land to the southeast was great, a bed of sandstone was built out by streams over the mud environment (Figs. 4–3) and the sand–mud boundary moved northward. When the supply waned the area occupied by sand shrank southward and the mud was again deposited over the sand layer. In time the sand environment of the Grimsby Formation extended slowly northward over the mud environment of the Cabot Head until this phase of sedimentation was ended by the deposition of the Thorold Sandstone.

An interfingering relationship can be shown to exist also between the Cabot Head Shale and the Manitoulin Dolomite at its base. Toward Niagara the beds of dolomite interfinger with, and die out in, the basal part of the shale until none of the formation is left.

Using these and other interfingering relationships, the geometry of the formations along the escarpment can now be filled in (Fig. 4–4). The interfingering is shown by a zigzag line. The gradation of the Grimsby Formation into the Cabot Head Shale is called a **facies change**, and the figure is

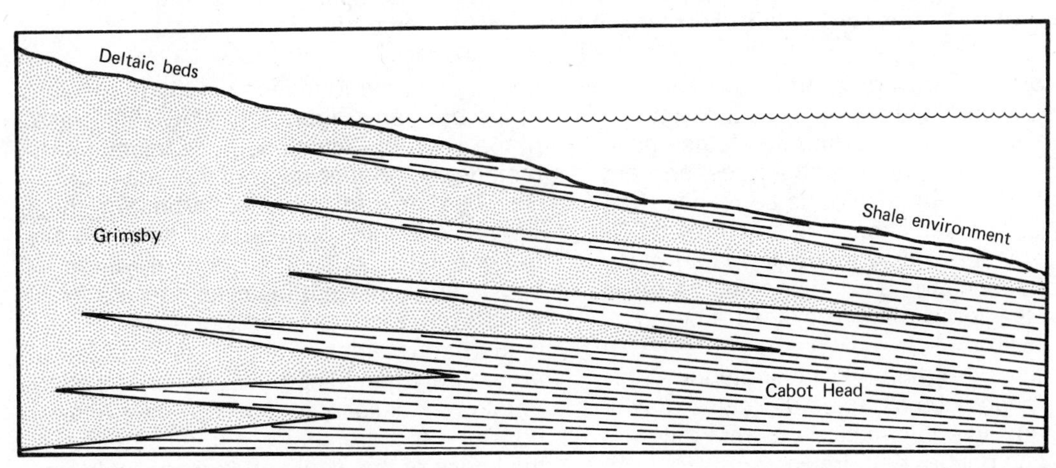

FIGURE 4–3 **The interfingering relationship between the Grimsby Formation and the Cabot Head Shale along the Niagara Escarpment.**

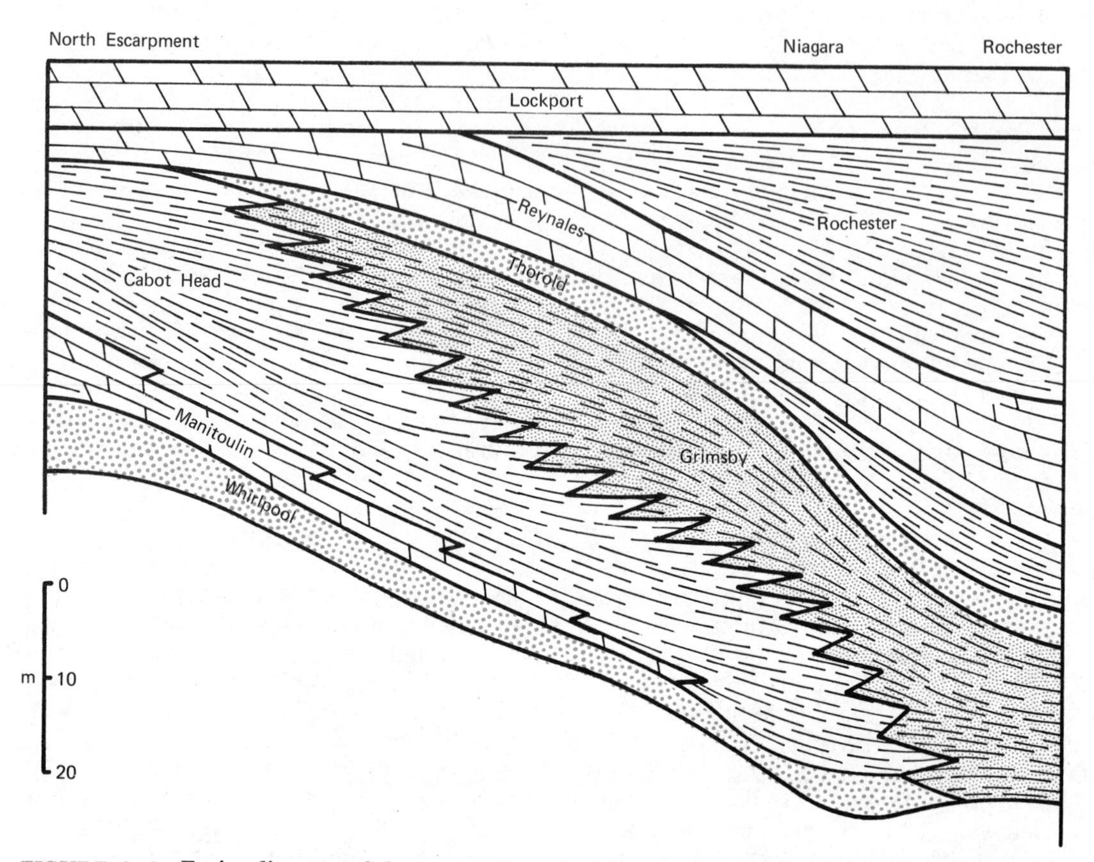

FIGURE 4–4 **Facies diagram of the strata along the Niagara Escarpment showing the interpretation of the facies changes determined by correlation. The base of the Lockport Dolomite is the datum plane, that is, it is considered to be horizontal for the construction of the diagram.**

said to be a **facies diagram.** The term "facies" refers to the total aspect of a sequence of strata, including lithologic characters, bedding, fossils, structures, and any other features that make the sequence distinctive. A facies change is any regional change in these characters and is a reflection of the changing conditions in the basin where the sedimentary rocks were being laid down.

Figures 4–4 and 4–5 illustrate the different appearances of facies diagrams drawn from the same information but with different formation boundaries drawn as horizontal. Figure 4–4 shows the base of the Lockport Dolomite as horizontal and distorts the relationships of the formations in the lower part of the diagram. Figure 4–5 shows the base of the Thorold Sandstone as horizontal and distorts the relationships of the upper part. Because the earth's crust tilted during the deposition of this sequence of beds, no single orientation of such a diagram will be satisfactory unless it involves only a short period of deposition. The slopes of the formation boundaries in such facies diagrams are entirely unrealistic, for the vertical exaggeration necessary to illustrate their relationships is extreme. On a true scale the thickness of the whole sequence would be represented by the finest line on the diagram.

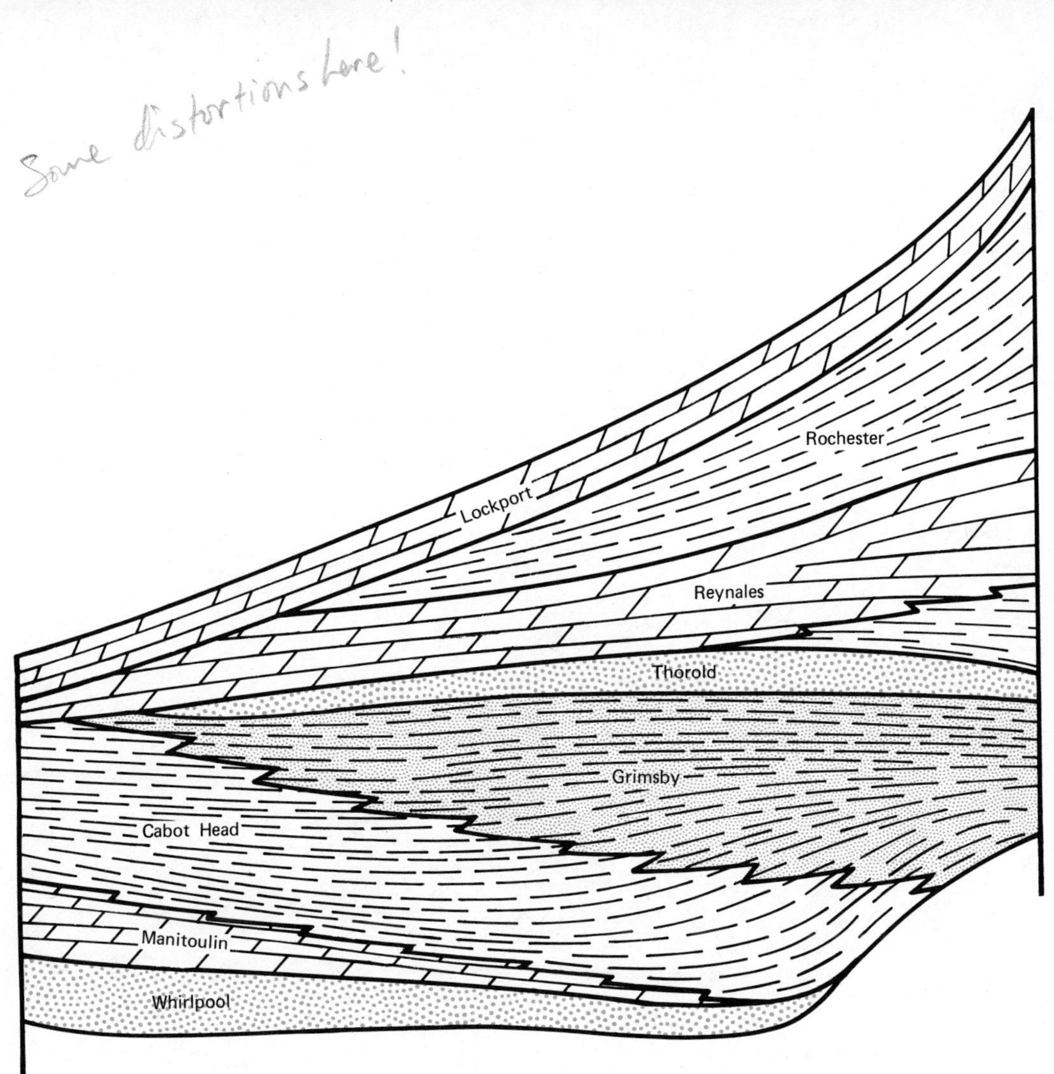

FIGURE 4–5 **Facies diagram of the same strata as in Fig. 4–4, constructed to show the difference in appearance if the base of the Thorold Formation is used as a datum plane.**

Key Beds

The tracing of interfingering relationship and bedding planes provides a local means of determining time equivalence, but since the bedding cannot be traced for more than a few hundred feet, they are not useful for establishing regional correlation. We can see the sequence of events in various places and relate them from place to place, but unless we can find an event that must have been recorded in each place at the same time, we cannot correlate the sequences, that is, establish a regional time framework. Along the Niagara Escarpment the Reynales Dolomite may record such a widespread depositional event, but the sedimentary sequence in eastern Tennessee provides a better example.

Volcanic eruptions may discharge great masses of ash into the atmosphere. The ash is distributed by the wind and settles out to form a thin bed of sediment over a large area. Such a bed is a useful time marker for it records a single event over the area concerned. Figure 17–10 shows the facies change between shale and limestone in eastern Tennessee and the horizons of several ash beds that cross the boundary between the two facies. The ash beds are **key**

FIGURE 4–6 **Radioactivity logs of two wells drilled through the Silurian rocks of southwestern Ontario. The graph on the right side of the lithologic symbols represents the measure of the response of the rocks to bombardment by a source of neutrons. The graph on the left is a measure of the natural radioactivity of the various formations. Where the graphs go off scale on the right, they are continued in shaded form on the left of the scale. Note the similar graphs of similar lithologies, the accuracy with which the depths of the formational contacts can be determined, and the correlations that can be drawn between the formations.**

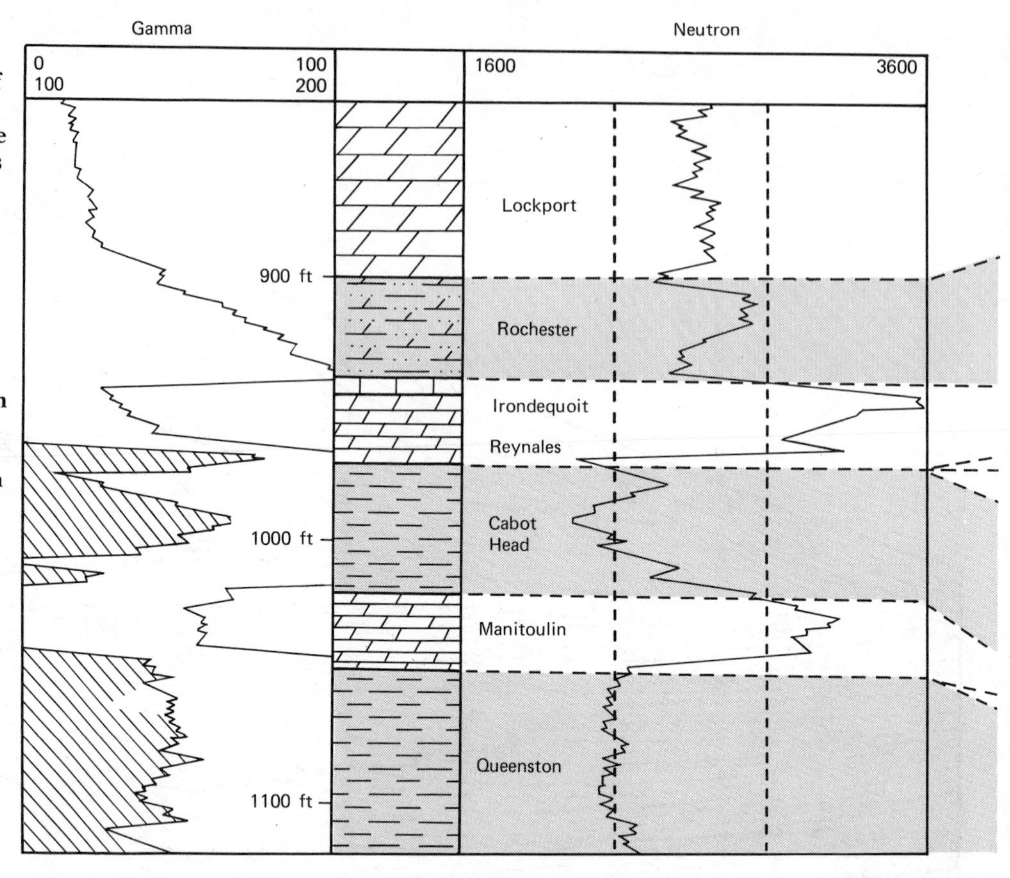

beds in that they establish the contemporaneity of the rocks in their stratigraphic vicinity. Other key beds appear to be caused by sudden changes of sea level that affect whole basins, initiating the deposition of layers of distinctive lithology (see also Chapter 18).

These examples from Niagara and Tennessee illustrate an important general principle in stratigraphy: that boundaries between formations are rarely parallel to time planes. Two formations that grade vertically into one another through a zone of interfingering, like the Grimsby Formation and the Cabot Head Shale, also grade laterally into one another. This observation was first clearly stated by Johannes Walther and has become to be known as **Walther's Law**: sedimentary facies that replace each other vertically, also replace each other laterally.

In summary, the geometric relationships between separate sections of sedimentary formations can be worked out by noting the relationship of units that are persistent to those that change laterally, and by studying interfingering relationships. Unless the sections contain key beds whose lateral synchroneity can be demonstrated, this type of reconstruction does not prove the regional time equivalence of the formations. Fortunately, other methods discussed later in this chapter help the stratigrapher to establish a time framework for the pattern of sedimentation that emerges from his comparison of sections.

Subsurface Information

Sedimentary formations are not two-dimensional bodies as they appear on cross sections, but are three-dimensional and basically tabular. The rocks of the Niagara Escarpment dip southward at less than 1° toward the Michigan Basin (see Fig. 10–2).

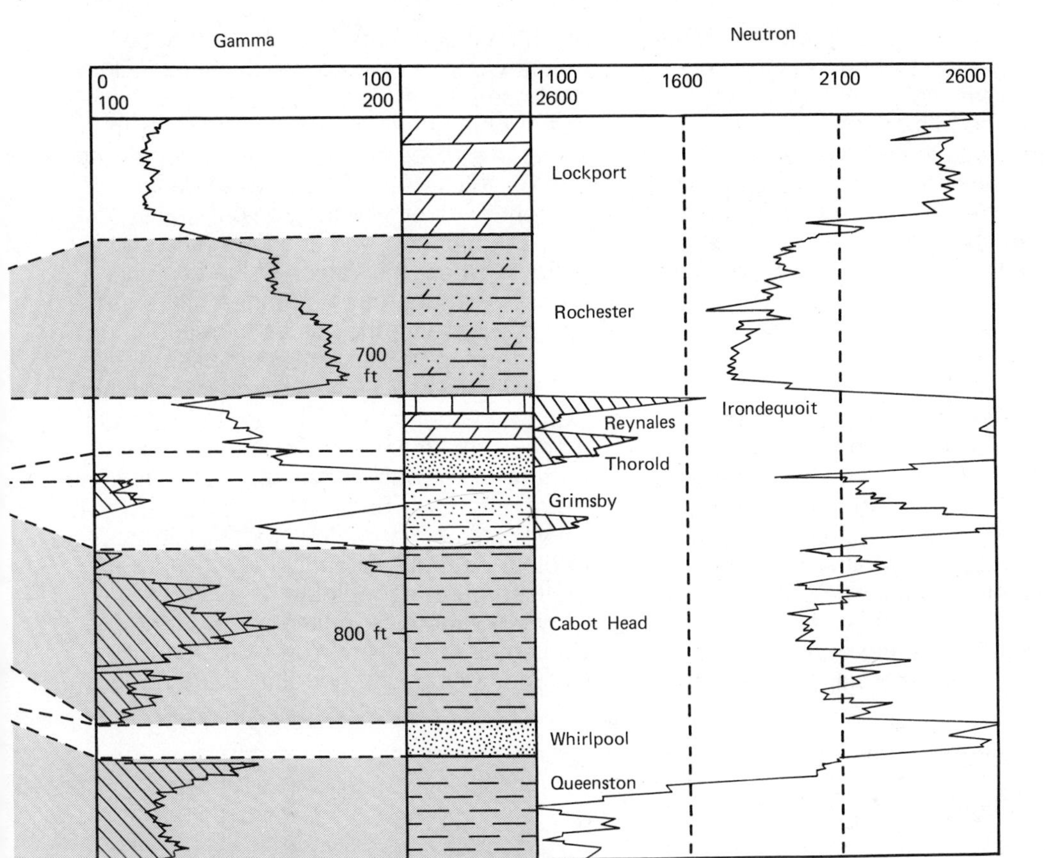

As their dip carries them deeper these rocks are overlain by younger and younger layers so that 100 km south of the escarpment they are hundreds of meters below the surface. A far greater volume of this tabular body of rocks is buried than is exposed along the escarpment. These buried rocks can, however, be studied in the cores and cuttings of holes drilled down to them through the overlying sedimentary rocks in a search for oil, gas, or water. Drilling is expensive but if the chance of finding resources is good, it is justified by the prospect of profit. The sandstones of the Cataract Group trap gas in many fields a few tens of kilometers south of the escarpment, and limestones of the Niagara Group contain both oil and gas 100 km south of it. A three-dimensional picture of the rocks, whose edges are exposed at the escarpment, can be constructed with information gained by drilling for this gas and oil.

The geologist gets stratigraphic information from an oil well in several ways. The drilling bit shaves and chips fragments of rock from the sedimentary layers, and the drilling mud that is pumped through the bit brings these samples to the surface where they can be recovered. By identifying the samples the geologist can make a log, or section, of the rock through which the bit has passed. After the hole has been drilled the rock that forms its walls may be investigated by a variety of instruments lowered into the hole. The records of instruments that measure the resistivity of the rock and the natural electric currents flowing in them are called **electric logs.** Other instruments measure the radioactivity of the sedimentary layers, their reaction to neutron bombardment, the velocity of sound in them, and other properties. Each of these can be interpreted by the expert in terms of the lithology, porosity, or fluid content of the rocks tested (Fig. 4–6). From these mechanical logs the geologist interprets the succession of sedimentary rocks in the well and matches the formations from well to well and from well to outcrop section. The amount of information on the sedimentary rocks of North America revealed by such subsurface methods now far exceeds that derived from the study of surface exposures.

STRATIGRAPHIC MAPS

A geological map records the distribution of stratigraphic formations and other rock bodies on the surface of the land. The three dimensions of sedimentary units are best illustrated by other maps. Regional variations in the thickness of a formation or other stratigraphic unit are shown on an **isopach** map. The word "isopach" is derived from the Greek words meaning equal thickness. The lines on an isopach map connect points at which the formation, bed, or other sedimentary unit is of the same thickness. On his base map the geologist plots the thickness of the unit intersected in wells or measured in surface sections. The regional pattern of "thicks" and "thins" is emphasized by contour lines like the contour lines on a topographic map, which shows hills and valleys. Where the values of the contours increase in a topographic map, a hill is indicated; but where the values increase in an isopach map, the sediments thicken (Fig. 4–7). Figure 4–8 is an isopach map of the Cataract Group of the Niagara Escarpment. It shows that this set of beds thins along a broad arch trending northeast between Lake Huron and Lake Erie.

Variations in the lithology of a formation, or other stratigraphic unit, may also be expressed on a map. Figure 4–9 is a map showing variations in lithology within the Clinton Group of New York, Michigan, and Ontario. Methods of correlation are not pre-

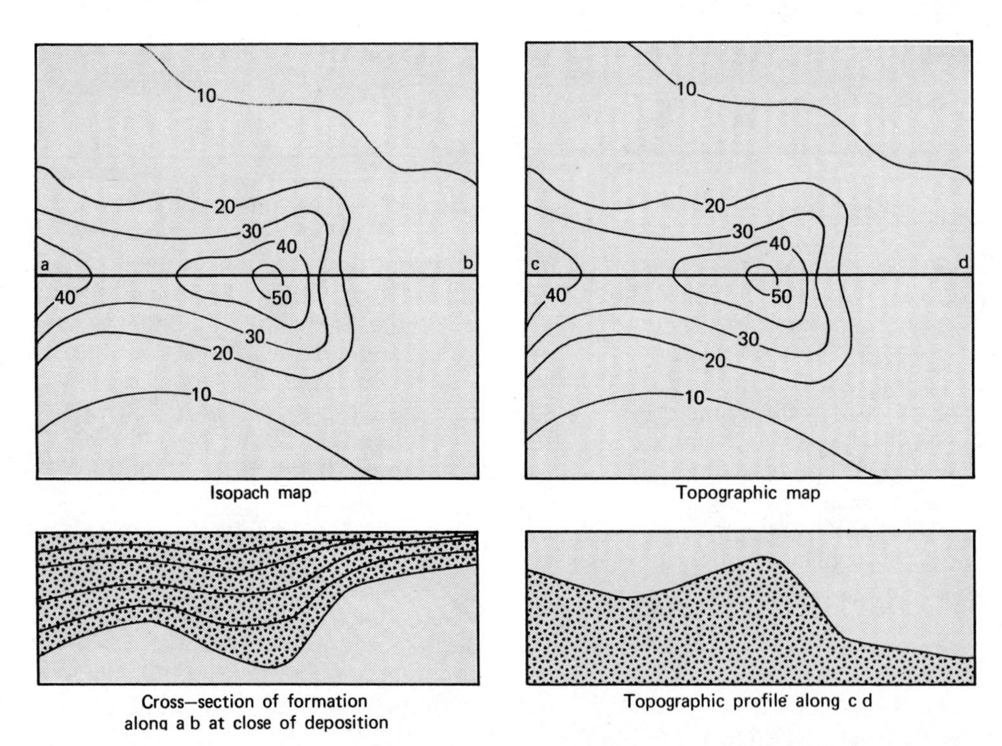

FIGURE 4–7 **Comparison of an isopach and a topographic map. In an isopach map the datum from which the measurements of thickness are made is the upper surface of the formation. In a topographic map the altitudes are measured from the datum of sea level.**

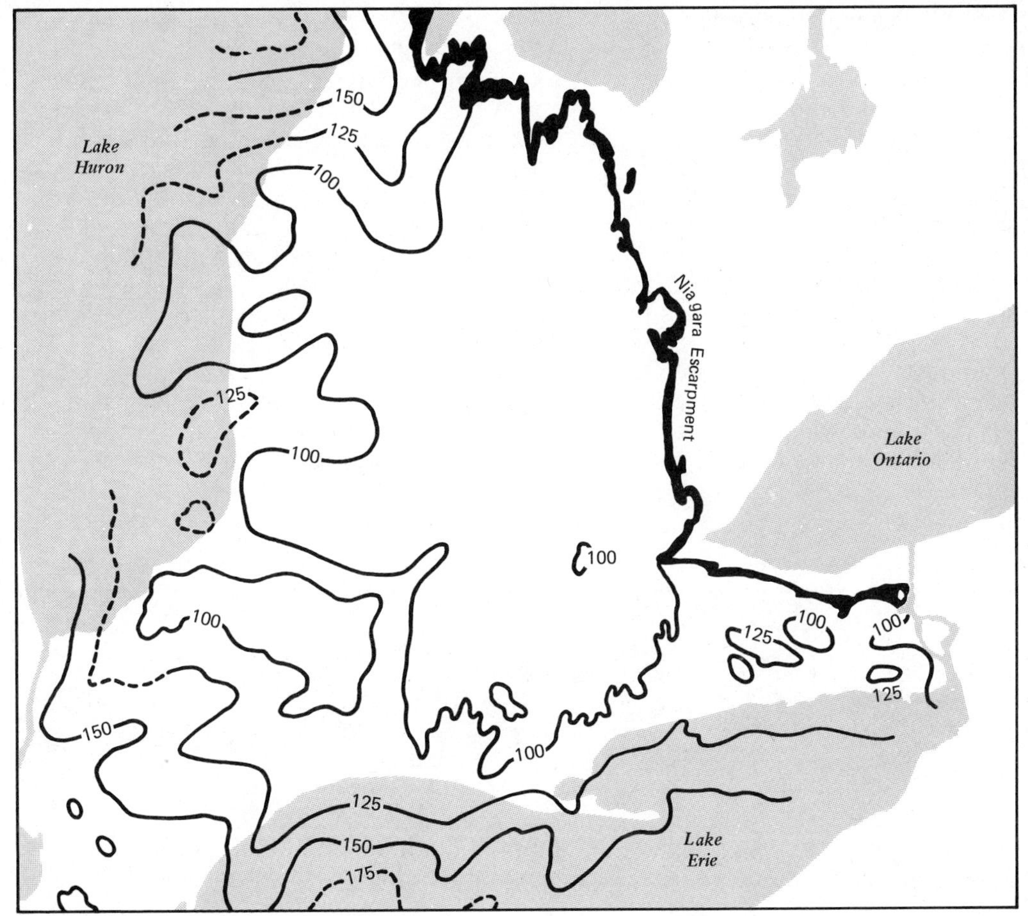

FIGURE 4–8
Isopach map of the Cataract Group in southwestern Ontario. Note how the group thins over a broad arch in the center of the map. (After B.F. Sanford; courtesy of the Geological Survey of Canada.)

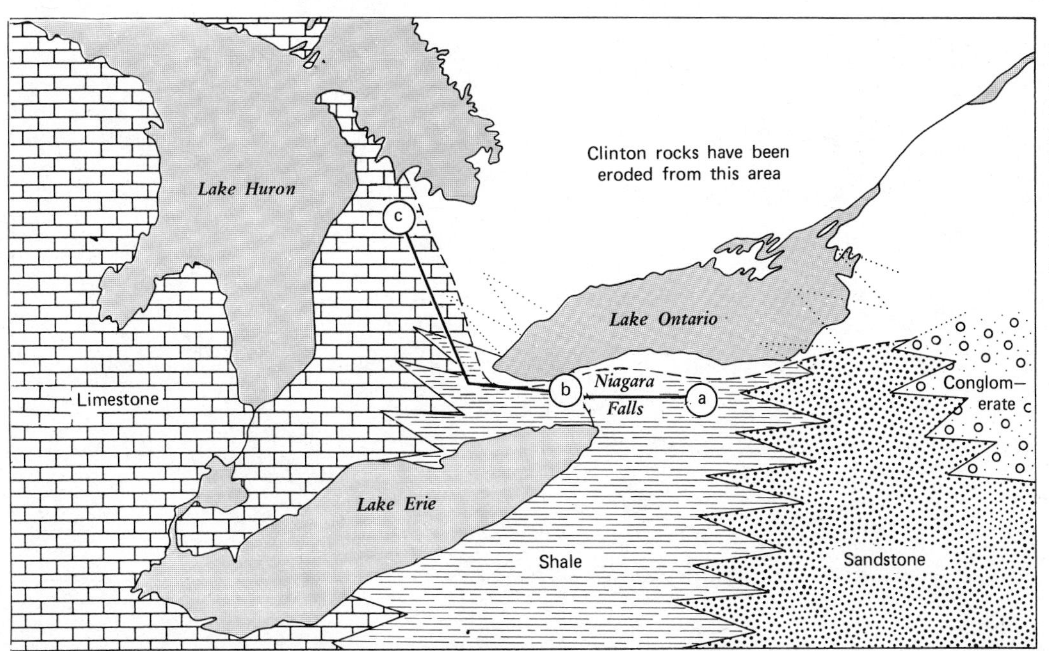

FIGURE 4–9
Qualitative facies map of the Clinton Group in New York, Michigan, and Ontario. The lines of the facies diagrams of Figs. 4–4 and 4–5 is indicated by *a–b–c*.

cise enough to allow the construction of a map that shows the distribution of the various lithologies at some particular instant in the past. However, the average position of the areas of sandstone, limestone, and shale deposition can be reconstructed for the interval of time during which a set of beds, such as the Clinton Group, was deposited. On this map the boundaries between facies are plotted as zigzag lines embracing the belt in which two facies interfinger. The map (Fig. 4–9) shows the influence of an eastern source of clastics on sedimentation in New York and Ontario and the position of the limestone sea in the clear water, beyond the influence of the clastics. Other such facies maps will be found throughout this book.

Such a map gives only a qualitative impression of the lithologic variation in a stratigraphic interval; other types of maps give quantitative information. From the interval to be mapped the geologist measures the aggregate thickness of beds of a given lithology, such as sandstone, in all the wells and surface sections of a given area. These values are then plotted on a map at the location of each section and contours joining points of equal thickness of the lithology are drawn. Instead of the thickness of the lithology, the map may show the proportion of the interval that is represented by that lithology. The contour lines on these maps are called **isoliths** because they join points of equal lithology. Figure 4–10 is an isolith map showing the percentage of sandstone

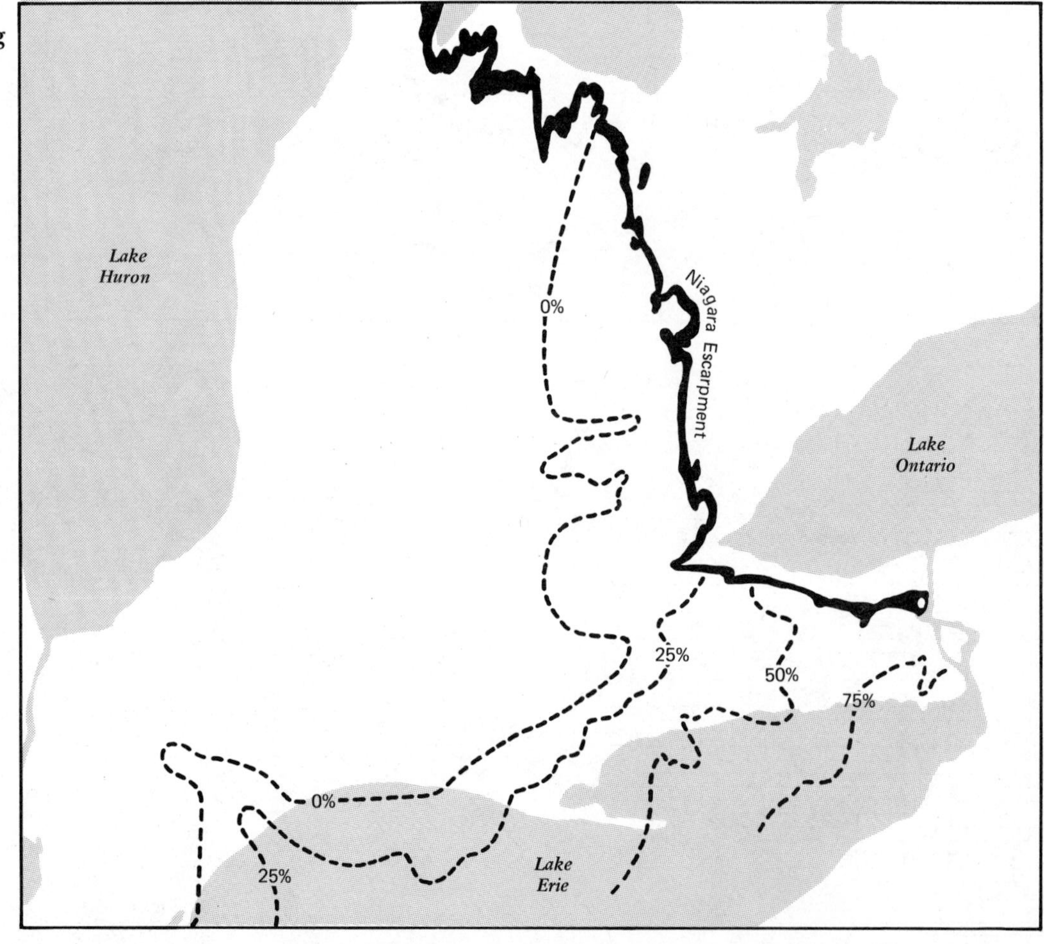

FIGURE 4–10
Isolith map showing the percentage of sandstone in the Cataract Group of southwestern Ontario. (After B.F. Sanford; courtesy of the Geological Survey of Canada.)

beds in the Cataract Group in southwestern Ontario.

Aggregate thicknesses measured from surface and subsurface stratigraphic sections may be used to make maps showing the proportion of two or three types of rock in an interval. If geologists want to illustrate the areal variation of the proportion of sand to shale, they plot and contour the ratio of the aggregate thicknesses of the two lithologies:

$$\text{Sand–shale ratio} = \frac{\text{Aggregate thickness of sandstone beds}}{\text{Aggregate thickness of shale beds}}$$

Two ratios may be contoured on the same map to show the variation of three rock types in an area.

The facies of a stratigraphic unit can be shown quantitatively in many ways on a map by the use of ratios and patterns. Figure 4–11 is a map of the facies of the Cataract Group south and west of the Niagara Escarpment, showing the proportions of sandstone, shale, and limestone encountered in wells and surface sections of this group. The patterns illustrate the occurrence of the limestone and dolomite facies along the arch defined by the isopachs (compare Fig. 4–8) and the greater proportion of shale on its flanks. The increase in the proportion of sandstone toward the east is also shown, as in Fig. 4–9, but this map includes much more information about other lithologies in the section than the simpler isolith map. Figure 4–11 is called a

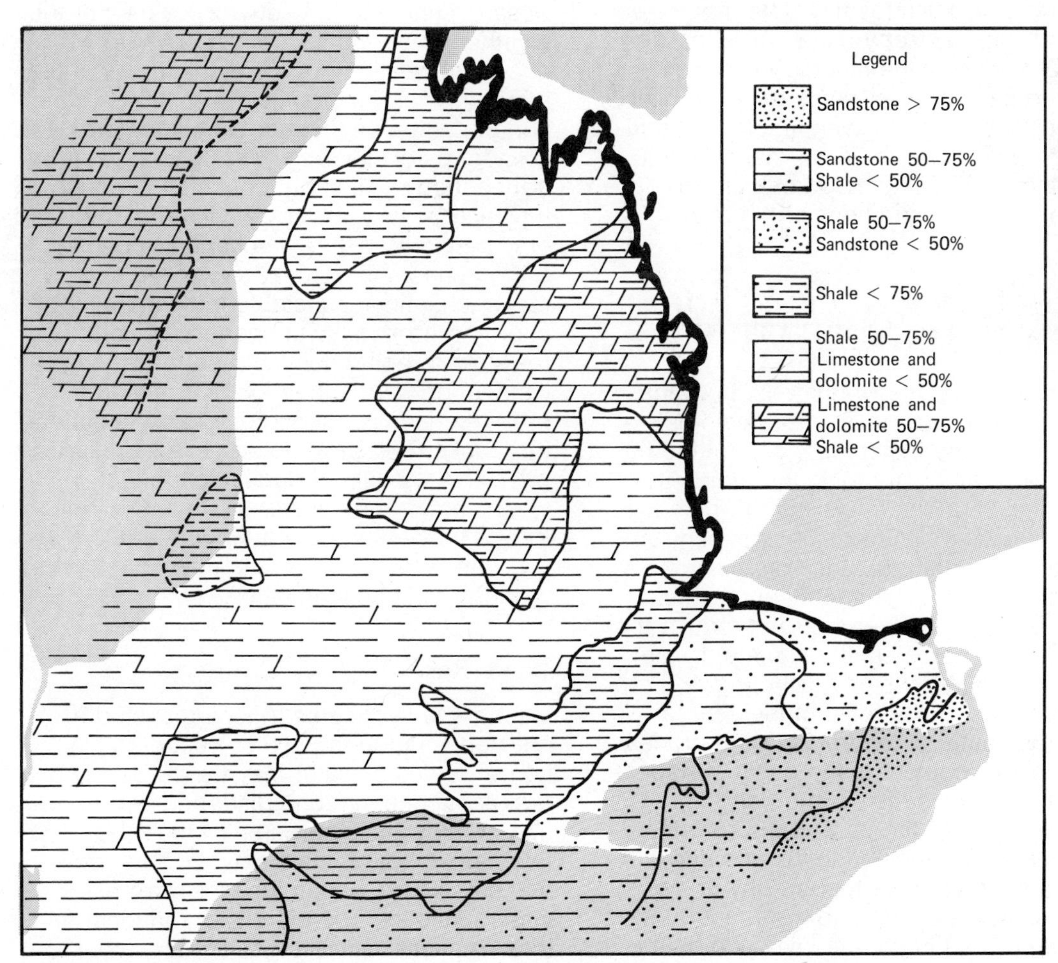

FIGURE 4–11
Lithofacies map of the Cataract Group in southwestern Ontario. The patterns show the proportions of sandstone, shale, and carbonate rocks in various regions. (After B.F. Sanford; courtesy of the Geological Survey of Canada).

lithofacies map and may be constructed for any set of lithologic components. A lithofacies map shows quantitatively the areal variation in the proportion of two or more sedimentary rocks in the stratigraphic interval being mapped. If the lithology of the sections is recorded on punched cards and introduced into a computer and plotter, the machine can be programmed to draw a lithofacies map. Many oil companies and government agencies now record the stratigraphic information from wells in a form that can be used by computers, and some geologists even make their field notes by punching cards instead of giving written descriptions.

Petroleum geologists have invented a wide variety of stratigraphic maps such as these to serve their individual needs. **Facies departure maps** show the degree by which sections at a given locality differ from a particular facies; **entropy maps** show the degree of mixing of the lithologies in sections; and **vertical variation maps** may show the number of beds of a given lithology within a stratigraphic interval. These maps are the tools the geologist uses, wherever appropriate, to interpret conditions of the past.

NATURE AND THICKNESS OF SEDIMENTARY FORMATIONS

To reconstruct the environments in which sedimentation took place, the geologist must assess the factors that control the nature of a sedimentary rock, the rate of its deposition, and the thickness to which it accumulates. The nature of the sediment deposited in a basin depends on many factors, among the most important of which are: the type of rock eroded to produce the sedimentary particles, the climate under which the processes of weathering and erosion took place, the relief of the source area, the nature of the transporting process, and the environment of the basin of deposition. If the source area is high and of rugged topography, the particles supplied by weathering and erosion to the streams will be coarser than if the topography of the same source area were low and subdued. Climate has an important influence on the size of particles supplied by weathering and erosion. For example, high ranges, if located in a hot humid climate, will be weathered largely by chemical decomposition and will supply only the finest detrital particles; low hills, if located in an arid region where mechanical weathering is dominant and rainfall is infrequent but violent, will supply boulder- and cobble-sized particles. The historical geologist must recognize that a change in the sedimentary record from coarse to fine sedimentary particles can be interpreted in various ways—as a change in the topography of the source area, for example, or as a change in climate.

The mineralogy of the detrital particles supplied to the basin of deposition will depend on the minerals present in the source rock being weathered. The weathering of basalts cannot supply quartz grains, nor the weathering of granites, olivine. Where chemical weathering is dominant, few of the minerals that are chemically unstable at the surface of the earth, such as mica, hornblende, pyroxene, and the feldspars, will be transported to sites of deposition because they will be changed by weathering into clay minerals. Where mechanical weathering controls the breakdown of the source rocks, chemically unstable minerals may be supplied essentially unchanged. Quartz is resistant to both chemical and mechanical weathering and is therefore a common mineral in igneous, metamorphic, and sedimentary rocks. The other detrital minerals supplied to the basin of deposition reflect weathering conditions at their source, but before they enter the stratigraphic record they are also subject to the agents at work in the area of accumulation.

Wave Base

Most of the sediments a geologist studies are deposited in the sea or in large lakes. Large bodies of water may be divided into a turbulent zone, or zone disturbed by the action of the waves, and a still water zone that lies below their influence. No simple, permanent surface separates these two environments; the boundary is gradational and impermanent and changes from moment to moment with the character of the waves. By far the greatest energy for the erosion and modification of sedimentary particles is available near the shoreline where surf is active. The indefinite plane down to which wave action extends is called **wave base**; the plane to which surf action is effective is much shallower.

The difference between sediment deposited above or below wave base is important to the stratigrapher. Sedimentary rocks deposited in the zone of wave motion (sometimes referred to as a high-energy environment) are much different from those deposited in deeper, quieter water (low-energy environment). Mechanically or chemically stable mineral grains deposited in the zone of wave motion are apt to be well rounded and sorted according to size by the constant grinding and shifting action of waves. If mechanically or chemically unstable minerals, such as feldspar and mica, are brought into such an environment, they are shifted about by the waves on the sea floor until they are broken in fragments of clay size or decompose to clay minerals and soluble compounds. Only the most stable constituents, such as quartz and clay minerals are left. The waves then distribute the particles into areas of differing wave competency, where they are deposited. The clay is deposited in the quietest water; the sand in more agitated water.

Sedimentary particles introduced into a basin whose surface of deposition is below wave base are incorporated in the accumulating material with little modification. If coarse, unsorted particles of unstable minerals are delivered to the basin, they will be deposited without being sorted, abraded, or decomposed. The mud, silt, and sand of such sediments are mixed together to form "dirty" or "impure" sandstones, such as graywackes, rather than sorted into quartz sandstone and clay shale (Fig. 4–12). Graywacke is a greenish-gray, poorly sorted, impure sandstone containing a high proportion of rock fragments, feldspar, mica, and other chemically and mechanically unstable grains. Such sediment is often referred to as "poured in" sediment. Other poorly sorted sediments may be formed in lakes or streams where the volume of material supplied by weathering and erosion far exceeds the capacity of the agents at the depositional site to sort and modify it. **Arkose**, a generally brown sandstone containing at least 25 per cent feldspar, is commonly formed in this environment.

Subsidence

The earth's crust is unstable—rising, falling, shifting. The presence of marine sedimentary rocks thousands of meters thick on the continents is evidence of relative vertical movements between land and sea of the order of thousands of meters. Episodes of deposition on continental areas are commonly associated with subsidence of the earth's crust to form a basin in which the sediments accumulate.

In areas that are subsiding rapidly, the surface of the sediment will continuously be carried below the zone of the waves as more and more detritus is brought into the basin, and the strata deposited will be of the "poured in" type. In areas of slow subsidence, the surface of deposition will stay in the zone of the waves for a long time, and the motion of the waves will grind and sort the sedimentary grains. If the supply of sedimentary detritus is small, the slowly subsiding area will remain a shallow water environment, but, if the supply is too great for waves and currents to distribute, the surface of deposition will rise above sea level and a subaerial delta will form. Such deltaic sediments, because they are not wave-washed, tend to be similar to those deposited in rapidly subsiding basins. The relationship between subsidence and supply at the depositional site, therefore, determines to a great extent the character of the sediment deposited, and, conversely, the rate of subsidence and its relationship to wave base can be interpreted from the sediment.

The thickness to which sediments can accumulate, as well as their character, is also determined by the subsidence of the basin of deposition. The limit to which sediments can accumulate in a large, stable basin is the upper surface of a delta built into the basin. Additional sediment brought

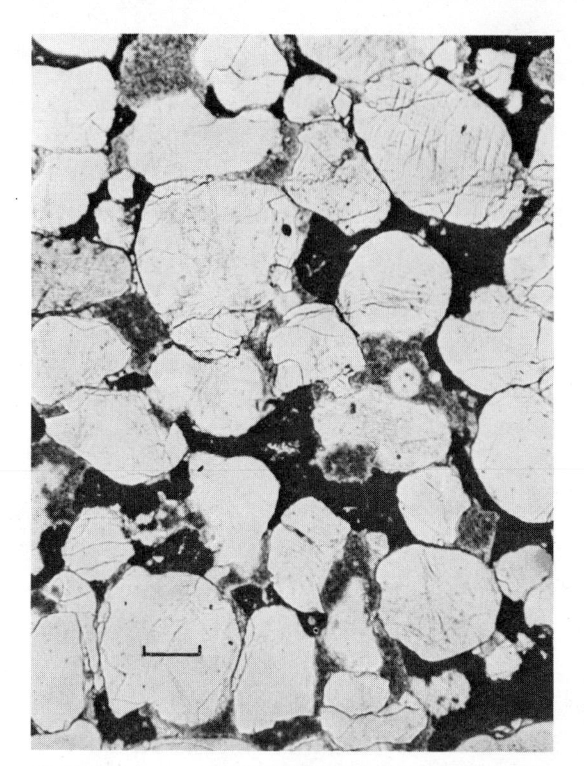
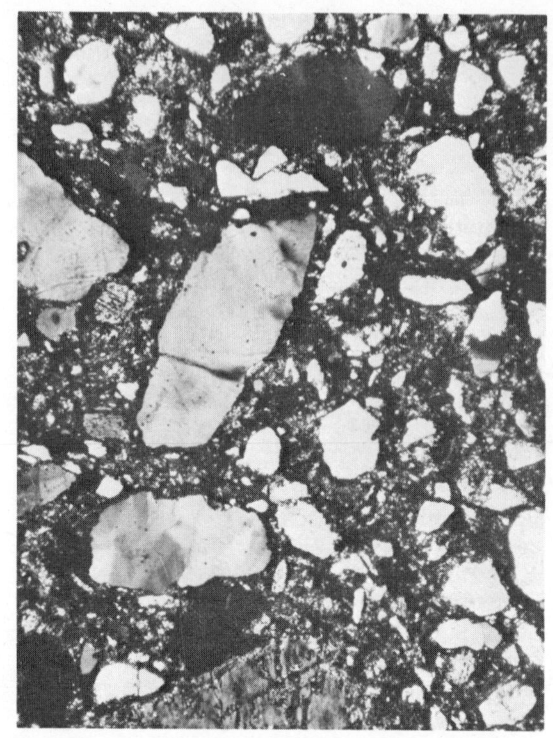

FIGURE 4–12 **Photomicrographs of two Lower Cambrian sandstones to show the contrasting textures of the wave-worked and "poured-in" types of sandstones. *Left:* A quartz sandstone composed of rounded, well-sorted grains of quartz cemented by iron oxide. The black line is 1/10 mm long. *Right:* Angular, poorly sorted grains in a clayey matrix, characterizing the texture of graywackes.**

to the top of a delta that is not subsiding does not add to the thickness of the accumulating sediments, but is carried onward by the distributary streams and dumped over the edge of the delta as foreset beds to extend it seaward. A deltaic deposit can grow in thickness only by the subsidence of the floor on which it is being built. If a large quantity of sediment is entering a basin so that it is silted up near the distributary streams, the thickness of the accumulating sediments will depend on the subsidence; if a small quantity is entering a basin in which subsidence is rapid, the thickness will be dependent on the supply, for the sedimentary surface will never approach wave base.

Sediments deposited in small lakes, intermontane basins, and river flood plains are subject to no such controls. Compared to those in the sea these deposits are temporary and slight changes in the regimen of erosion will remove them from the sedimentary record.

These controls apply to clastic sediments formed from detrital particles derived by weathering and erosion. Limestone is formed on the ocean floor from the accumulating shells of marine animals that extract calcium carbonate from sea water or by the inorganic precipitation of calcium carbonate. Calcium ions derived from chemical weathering are carried by streams to the sea in solution; carbonate ions reach the sea by dissolution of atmospheric carbon dioxide or as the metabolic wastes of organisms. Since the time when animals evolved calcium carbonate shells 500 million years ago, most limestones have been formed in shallow water where marine organisms thrive. Their discarded shells and skeletons are sedimentary particles that are moved and broken by the wave motion like detrital grains. Unless they form wave-resistant structures like reefs, shallow-water limestones can only accumulate to the level of wave base and their thickness is controlled by the subsidence of the depositional site, like that of detrital sediments. Limestones of abyssal environments, formed from the accumulating shells of planktonic organisms, are far below wave base control and are analogous to the graywackes of the detrital sediment suite.

Where supply of sediment is adequate, the thickness to which it may accumulate is a function of the subsidence of the basin of deposition. In such areas an isopach map of a formation may be a valuable indication of the subsidence of the basin during the deposition of the unit. A series of isopach maps of successive formations provides a history of the basin's structural behavior. The term "tectonics" is used to describe the structural behavior of an area as a whole, whether stable, subsiding or rising gently, or being thrust into mountain ranges. The study of the influence of such structural behavior on the nature and thickness of sedimentary rocks is called **sedimentary tectonics.**

CONTINUITY OF THE SEDIMENTARY RECORD

In the depths of the sea where sedimentary particles slowly settle from above, the sedimentary record may be uninterrupted for millions of years. In shallow, inshore waters where the surface of deposition is near wave base, small changes in sea level expose or inundate large areas of depositional basins, and rates of subsidence and uplift are variable; sediment does not accumulate continuously, and periodically that which has accumulated is removed by being raised above wave base or above sea level into the zone of subaerial erosion. If the area has been raised above sea level during such a break in deposition, a bedding surface will be produced marked by irregularities formed by differential weathering and erosion (Fig. 4–13). An ancient soil may be preserved along such surfaces. Such surfaces of erosion are called **disconformities**; their recognition is essential to the understanding of geological history. The beds above and below a disconformity are parallel, but if deformation took place before sedimentation resumed, the beds below the erosion surface will be tilted or folded. Such a surface is called an **angular unconformity** (Fig. 4–14) and indicates a sequence of events involving uplift, deformation, and erosion before submergence and renewed deposition.

The stratigraphic section at Niagara studied in Chapter 4 is about 76 m thick. Using methods which will be described in the next chapter, the geologists can estimate the length of geological time during which this thickness of rock accumulated to be 20 million years. If deposition had been continuous, the average rate of accumulation of this section then must have been about 0.004 mm per year. Yet, individual beds in the section show well-preserved fossil shells, cross bedding, corals buried in place, worm burrows filled with sand by the advancing tide, all evidence of rapid deposition. As no major disconformities are evident in the section, we must conclude that there are many minor ones that we cannot detect. Each bedding plane must represent a gap in the sedimentary record of greater duration than the episodes of deposition represented by the adjacent sedimentary beds. Such minor breaks in the record were called **diastems** by Joseph Barrell, but they are not qualitatively different from disconformities. The complex interaction of wave base, sea level, subsidence, uplift, and sediment supply in shallow-water areas results in the rapid deposition of some beds and the repeated erosion, reworking, and redeposition of others.

Such low average rates of sedimentation are characteristic of shallow, slowly subsiding seas covering continental platforms. In rapidly subsiding troughs at the edges of the continents, gaps in sedimentation are rare and the average rate of accumulation may be thousands of times higher.

FIGURE 4–13 **Disconformity along the surface, marked by the ruler, between sandy limestones strata below and a brecciated calcareous crust developed during a period of weathering. Pleistocene beds of Bermuda. (Courtesy of E.W. Mountjoy.)**

PALEONTOLOGICAL CORRELATION

Smith, Cuvier, and Brogniart recognized that each unit of the stratigraphic column is characterized by a suite of fossils that can help the geologist recognize the unit. They used fossils to identify the rock units they were mapping in areas where the nature of the rock was not distinctive enough for recognition. We now realize that the differences they detected are evidences of a progression of life through time that we call **organic evolution.** These progressive changes provide a scale, dependent on time and independent of changes in lithology, that can be used in the correlation of formations. The matching of lithologies can only determine the geometry of the stratigraphic units; to reconstruct the history of an area the stratigrapher must have a method of establishing the time at which certain events recorded in the rocks occurred. Such a measure is provided by the paleontological record.

Guide Fossils

After the stratigrapher has collected fossils from the sections, bed by bed, and determined the local ranges of the fossils, he can assemble these ranges into a zonal scheme and match them from section to section to establish a time framework. In this matching process not all fossils are of equal value, nor can quantitative measures of the similarity between the suites of fossils be used

FIGURE 4–14 **Angular unconformity between nearly horizontal Upper Devonian strata above, and vertical beds of Proterozoic sandstone below, near Ouray, Colorado. (Courtesy of the Pan American Petroleum Corporation.)**

to estimate their separation in time. As discussed in the last chapter, the occurrence of a fossil in a particular bed depends on two controls; first, the environment of deposition must be one in which the organism could have lived and, second, the moment of deposition must be within the time range of the organism. The most useful fossils for correlation are those whose way of life frees them from dependence on the sedimentary environment in which they become entombed. In the marine environment such organisms are the swimmers and floaters which, after wide dispersal by currents and their own efforts, fall dead on to any one of a variety of sedimentary environments.

The relative value of fossils can be illustrated by an example (Fig. 4–15). The stratigrapher measures five sections through a sequence of beds in the walls of canyons cut into a high plateau. The positions in which he finds various fossils are indicated by letters on the figure. Some fossils, such as *a*, range throughout the section and appear in every limestone unit. These fossils represent slowly evolving or stable organisms whose occurrence is controlled by the sedimentary facies. They are, therefore, of little value in establishing time relationships. Fossil *b* has a small vertical range in the sandstone formation at the base of the section and could be useful for correlating the section if it can be shown not to be a facies fossil, that is, dependent on that par-

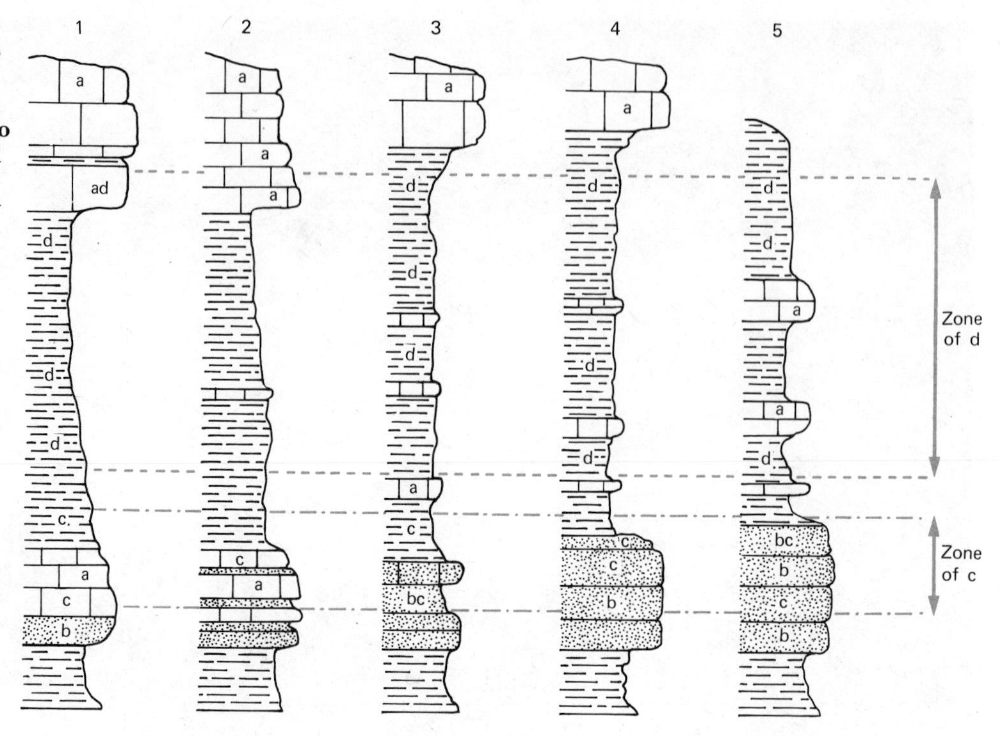

FIGURE 4–15 **Distribution of four fossil species (*a,b,c,d*) in five hypothetical sections. Species *a* is long-ranging but appears to be a facies fossil confined to sandstones; species *c* and *d* are found in narrow stratigraphic ranges within various lithologies and are, therefore, valuable guide fossils and define biostratigraphic units useful for correlation.**

ticular lithology for its presence. Fossil *c* has a narrow range near the base of the sections and occurs in the limestone, shale, and sandstone, defining a zone that is useful for correlating these varied lithologies. The upper boundary of the zone of *c* crosses from the shale facies into the sandstone facies between Sections 1 and 3, proving that the upper boundary of the sandstone unit rises in time in this direction (probably through interfingering with the limestone and shale) and that sandstone deposition persisted in Section 5 longer than in Section 1. In the same way the top of the zone of *d* shows that the base of the limestone unit at the plateau top is not parallel to time and that limestone deposition at Section 5 started later than at Section 1. Boundaries of formations that can be shown to cross time planes are said to be **diachronous.**

Fossils *c* and *d* are called guide fossils because they are more useful to the stratigrapher in correlation than the facies fossils whose occurrence is largely dependent on the environment of deposition. During different periods of the evolution of life different organisms have been useful as guide fossils, but most have been swimmers and floaters. The graptolites of the early Paleozoic Era (Appendix) were probably planktonic and form excellent guide fossils. The ammonoid cephalopods (Appendix) later became important as correlation tools, as they were dispersed widely in the sea through their own swimming activity and by the drifting of their buoyant shells in the sea after their death. In the Mesozoic Era the single-celled foraminifers developed planktonic representatives that are the most useful guide fossils of the Cenozoic Era (Fig. 4–16). In continental sediments pollen grains and spores of plants, blown by the wind, make excellent guide fossils. These very small fossils (microfossils) are particularly useful to paleontologists correlating subsurface formations by means of cuttings brought up in oil wells, for they are small enough to escape destruction by the grinding action of the drilling bit.

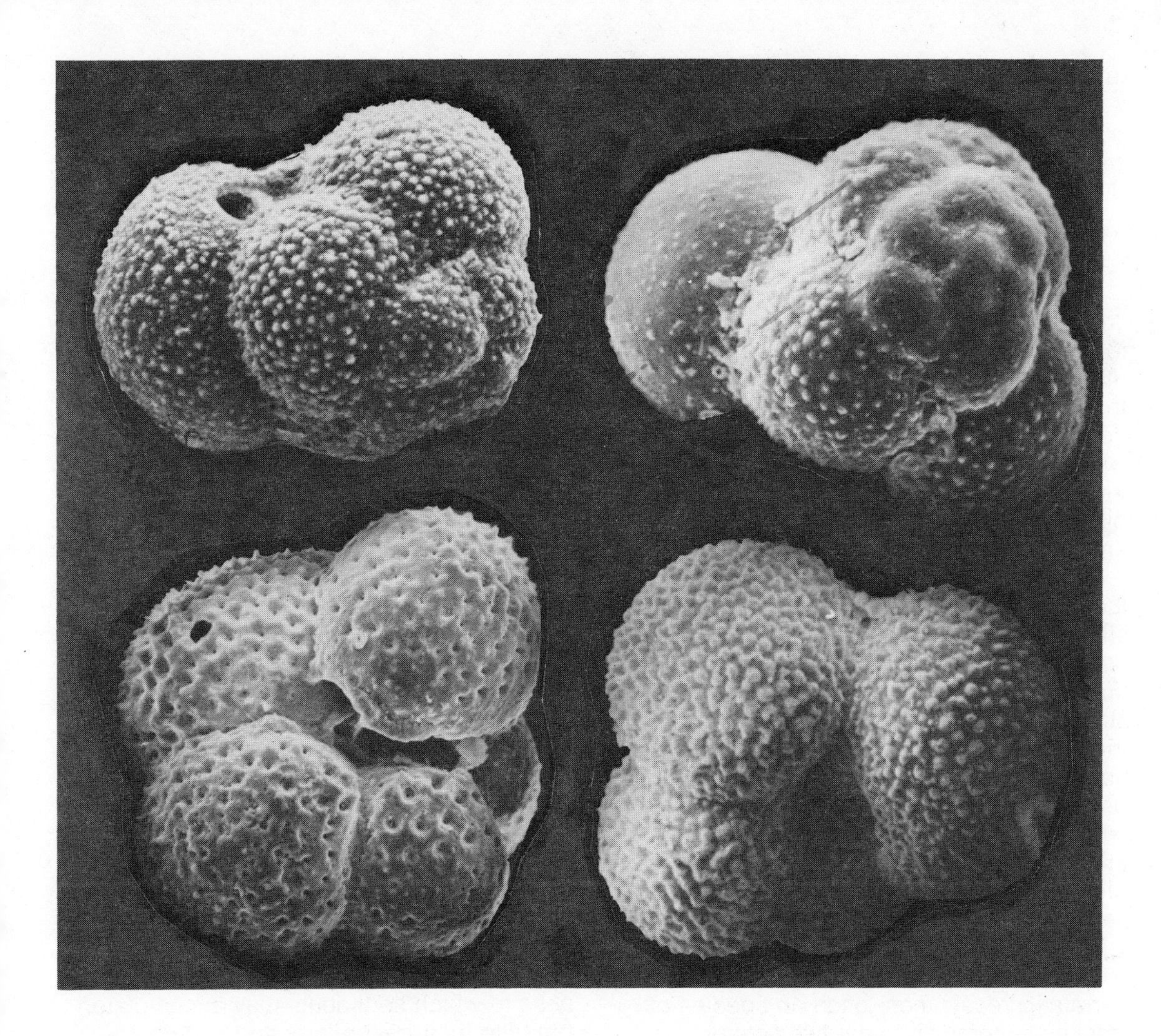

FIGURE 4–16 **Planktonic Foraminifera of Cenozoic beds used for worldwide correlation. From a core from the Labrador Sea (Scanning electron micrographs by C.W. Stearn; courtesy of S. Chough.)**

Migration

One of the assumptions of paleontological correlation is that the dispersal of organisms is, when considered in relation to the vast length of geological time, instantaneous. If an organism were to take several million years to migrate from point X to point Y, then it would appear in the stratigraphic section at Y at a later time, higher in the succession. The rate at which fossil organisms may have spread in favorable environment can be estimated from studies of the dispersal of modern organisms similar to those used in correlation. A good example is the movement of the slipper-shell *Crepidula* along the south coast of England after it was introduced from North America by shipping. During a 25-year period this snail species spread from the Thames to the Isle of Wight, a distance of 320 km. The adult snail hardly moves at all and this dispersal must be attributed to movement during the free-swimming larval stages. At this rate of migration, *Crepidula* could be carried around the world in 3000 years—geologically speaking, an instant. Most marine invertebrates used in paleontological correlation have free-swimming or floating larval stages that assure that the animal is widely dispersed to find favorable environments in which to settle.

INTERCONTINENTAL CORRELATION

Just as most organisms today are restricted to certain parts of the earth, so in the past most organisms were not of worldwide distribution. Most guide fossils can be identified only within a basin of deposition or within areas of continental dimensions. A few, however, are worldwide in their distribution and define zones useful for intercontinental correlation.

The zones and stages that depend for their definition on the ranges of guide fossils may also be of local or intercontinental usefulness. For most parts of geological time each continent has a sequence of stages that constitute a time framework for interpreting the history of that continent. The northern European sections were the first to be studied and have received the most attention. They have been divided into stages that provide a worldwide standard to which biostratigraphic schemes of other continents can be related. Recognition of the guide fossils of these standard stages in rocks of distant continents makes intercontinental correlation easy; more often no species of fossils are common to the two sets of stages, and comparisons must be made on the bases of genera, stage of evolution, or through an intermediate section. If the problem is to correlate Australian and English stages that have no species in common, an intermediate section might be found in North America that shows their contemporaneity by the occurrence of both the English and Australian species in the same beds.

SUMMARY

Correlation is the process of matching formations in order to determine time equivalence.

From a group of isolated sections the stratigrapher establishes the geometry of the depositional environments through tracing continuity of outcrop, determining persistent beds, and analyzing interfingering relationships.

The term facies refers to the total aspect of a sedimentary unit (formation, group, bed), including lithology, bedding, fossils, and other primary structures.

Key beds, such as volcanic ash layers, can establish time levels in a sedimentary sequence.

The subsurface extent of sedimentary units is revealed by drilling. Quantitative aspects of the variation of sedimentary formations can be expressed in isolith, isopach, lithofacies, and other types of stratigraphic maps.

The mineralogy of sedimentary rocks is determined by the climate and relief of the source area, the agent of transport to the basin of deposition, and the processes at work at the site of deposition. Wave base separates the zone of turbulence in the sea from the zone of quiet water. The relationship between sediment supply, wave base, and subsidence controls not only the thickness to which sediments may accumulate, but also their mineralogy and sorting.

Periods of erosion in a stratigraphic sequence form disconformities if the beds on either side of the erosional surface are parallel, angular unconformities if the beds below are tilted or folded, and diastems if the breaks are small and difficult to detect.

The paleontological record of the progression of life can provide a scale independent of lithology to establish time relationships between formations. The most useful fossils for this process of correlation are those of rapidly evolving swimmers and floaters that become preserved in a variety of sedimentary environments.

QUESTIONS

1. Summarize the evidence that supports Walther's Law of sedimentary facies.
2. What information is plotted on the following types of maps: contour, isopach, lithofacies, isolith, limestone-shale ratio?
3. Contrast conditions of source, weathering, transport, and deposition that result in a quartz sandstone and a graywacke composed largely of basalt fragments.
4. What properties are characteristic of an ideal guide fossil?

SUGGESTIONS FOR FURTHER READING

AGER, D. *The Nature of the Stratigraphical Record.* London: MacMillan Ltd., 1973.

EICHER, D. L. *Geologic Time.* Englewood Cliffs, N.J.: Prentice-Hall Inc., 2nd ed., 1976.

HARBAUGH, J. W. *Stratigraphy and Geologic Time.* Dubuque, Iowa: W. C. Brown Co., 1968.

MATTHEWS, R. K. *Dynamic Stratigraphy,* Englewood Cliffs, N.J.: Prentice-Hall Inc., 1974.

RAUP, D. M., and STANLEY, S. M. *Principles of Paleontology.* San Francisco: W. H. Freeman Inc., 2nd ed., 1978.

SHAW, A. B. *Time in Stratigraphy.* New York: McGraw-Hill Inc., 1964.

CHAPTER FIVE

LENGTH OF GEOLOGICAL TIME

The concept of the great length of geological time is the most important contribution geology has made to the other sciences and to culture. The theories of the astronomer and the cosmologist are limited by the time scale of the earth's history as determined by the geologist. The time of origin of the earth, of meteorites, and of the elements themselves can now be suggested by methods based on radioactivity. Through the joint studies of paleontologists and geochronologists, the rates at which animals have evolved have been estimated and the major events in the procession of life have been dated. The immense length of time through which our planet has developed must shape our viewpoint of the place of man in the system of nature.

Until the beginning of the 19th century most scientists believed that all geological events had to be crowded into the history of an earth a few thousand years old. The age of the earth was calculated by Bishop Ussher in 1659 on the basis of historical and traditional events recorded in the Old Testament. His calculation, that creation occurred in 4004 B.C., molded the next two hundred years of thinking about the age of our planet. To compress the sequence of events within such a short time, many geologists adopted a viewpoint that earth history consisted of closely spaced catastrophic events—eruptions, floods, earthquakes, mass extinctions, and instantaneous re-creations. An anti-catastrophist position was clearly set forth by James Hutton at the end of the 18th century. He believed that the processes that shaped earth history operated in the past no faster than they do today, and this concept of uniformity led him to interpret the geological record as representing a vast interval of time.

By the latter half of the 19th century uniformity had become widely accepted and catastrophism, in its extreme form, was dead. The physicists and astronomers then attempted to estimate the age of the earth. The most influential of these was Lord Kelvin. Assuming the sun's heat to come from gravitational contraction, he made several estimates of the age of the solar system and of life on earth. These estimates, in the range of a few tens of millions of years, were widely accepted toward the end of the century.

Just before the turn of the century Antoine Becquerel discovered the phenomenon of radioactivity. Certain atoms, said to be radioactive, are unstable; they decay, either directly or through a series of intermediate products, to stable atoms. The rate of this decay can be accurately measured and is not affected by changes in temperature, pressure, or any other known environmental factor (Fig. 5–1). The rate of decay of one radioactive element into its daughter products can be expressed in a figure called its **half-life.** Thorium, as an example, has a half-life of approximately 13,000,000,000 years (1.38×10^{10} years). After the passing of this immense period of time only half the thorium in a given sample will remain, and in another like period half of that half will have decayed, and so on.

The possibility of dating radioactive minerals and rocks by measuring the accumulated products of the decay was first perceived by Ernest Rutherford and Bertram Boltwood about 1905. Boltwood speculated that lead was the end product of the decay

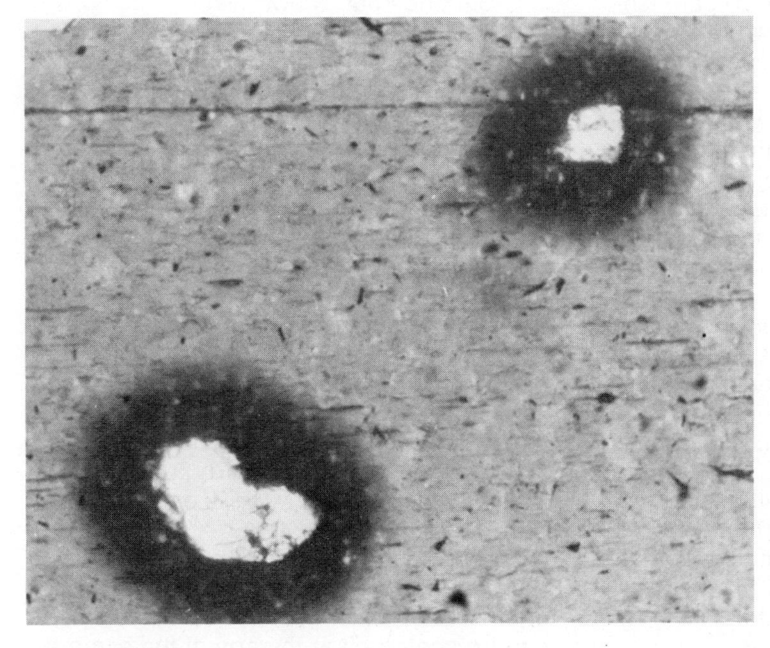

FIGURE 5–1 **Tiny crystals of zircon embedded in a biotite crystal contain enough uranium that the alpha particles from its disintegration have produced a halo or disrupted structure around each inclusion. Zircon crystals are about 0.05 mm across. (Photomicrograph by C.W. Stearn.)**

of uranium and applied his theory to analyses of uranium minerals to determine their age. He came to the amazing conclusion that one of the samples was over 2 billion years old and several were about 500 million years old. Although his results were crude, he had shown that the scale of geological time was far beyond Kelvin's estimate.

LEAD METHODS

Boltwood assumed that the lead and uranium in his samples of uraninite were each composed of only one kind of atom. Since then physicists have found that most elements consist of several kinds of atoms, or **isotopes.** The nuclei of atoms are composed of protons and neutrons. A proton is a nuclear particle of unit mass and positive charge; the neutron has the same mass as the proton but carries no charge. The number of protons in the nucleus, designated by the atomic number, determines its chemical properties. The number of neutrons and protons together determines its atomic weight. The atomic weight is a number which represents the weight of one atom of an element compared with the number 16, which represents the weight of an atom of oxygen. Isotopes of an element have the same number of protons in the nucleus, the same atomic number, and the same chemical properties, but they have different atomic weights because they have different numbers of neutrons in the nucleus. The atomic number of an isotope is properly written as a subscript after the chemical symbol of the element, and the atomic weight as a superscript before it. For example, the important isotopes of uranium for dating rocks are $^{238}U_{92}$ and $^{235}U_{92}$. As the atomic number is implicit in the chemical symbol, it is commonly omitted and the atomic weight is written immediately after the name, or symbol, as uranium 238, or U 238.

These two isotopes of uranium and one of thorium ($^{232}Th_{90}$) all decay to isotopes of lead through several intermediate steps at the following rates:

U 238 to Pb 206 with a half life of
4.5×10^9 years

U 235 to Pb 207 with a half life of
0.7×10^9 years

Th 232 to Pb 208 with a half life of
13.8×10^9 years

The isotopes in a sample can be separated and measured in an instrument called a **mass spectrometer.** As charged atoms travel through a strong magnetic field this instrument measures the curvature of their paths, which is proportional to their masses.

The ideal sample for uranium—lead dating is one which has crystallized from a melt or has been precipitated from a solution. Any decay products that had formed before the crystallization or precipitation have been dispersed in the melt or solution. As soon as it crystallizes the ideal sample begins to accumulate decay products at a constant rate and none of these products leaves the mineral until it is analyzed. If all uranium minerals were deposited under such conditions, most would allow the calculation of three independent ages from the three decay series listed above. Unfortunately, most samples have lost some of the decay products from the system during the decay period, and therefore produce different ages for the time of crystallization from calculations of the different series.

Experience has shown that the loss of lead produced by uranium decay is much less in some minerals than in others. For instance, the mineral zircon, which commonly contains measurable quantities of lead and uranium and is widespread in igneous and metamorphic rocks, retains most of the decay products (Fig. 5–1). Consequently, uranium–lead methods of dating are now applied largely to zircons and the decay series used are the two uranium series, because thorium 232 has such a long half-life that within the span of earth history little of the decay products accumulate. Despite the loss of some of the decay products, accurate age determinations can be made by plotting the isotopic ratios Pb 207/U 235 and Pb 206/U 238 on what is called a **Concordia diagram.**

Because the two uranium series have different decay rates, a plot of the two ratios against each other produces a curve that can be scaled off in the age of the sample in years (Fig. 5–2). If no lead loss has occurred in the decay process, that is, the system has remained closed, the isotopic ratios of samples should fall on the Concordia curve. If a group of samples from which lead loss has occurred are taken from the same area, their isotopic ratios would plot below the Concordia curve and will define a straight line, i.e., a chord (Fig. 5–2). The upper intersection of this straight line with the curve is interpreted as the true age of the sample and the lower intercept as the time at which leakage of decay products occurred. By using such techniques the geochronologist can take account of the lead losses and produce reliable, consistent rock ages by the uranium–lead method.

RUBIDIUM–STRONTIUM METHOD

Rubidium is one of the rarer elements but substitutes for potassium in the crystal structures of potassium minerals. Twenty-eight per cent of common rubidium is composed of the radioactive isotope rubidium 87, which changes to strontium 87 through electron emission. Both isotopes are measured in the mass spectrometer and a calculation of the age of the sample is made with the knowledge that the half-life of the transition is 4.99×10^{10} years. The proportion of strontium that was originally present in the rock and not derived from rubidium decay can be calculated by measuring the amount of strontium 86, an isotope that is not of radioactive origin. The half-life of this transition is so long that the rubidium–strontium method is not useful for dating rocks less than 100 million years old, because in such rocks too little of the decay products have accumulated to be reliably measured.

In practice, a group of rocks of different strontium content are collected from the rock body to be dated. Plotting of the ratio of isotopes Rb 87/Sr 86 against Sr 87/Sr 86 defines a straight line called an **isochron** (Fig. 5–3). With time, the rubidium 87—strontium 86 ratio decreases through the decay of rubidium 87, and the strontium 87—strontium 86 ratio increases as strontium

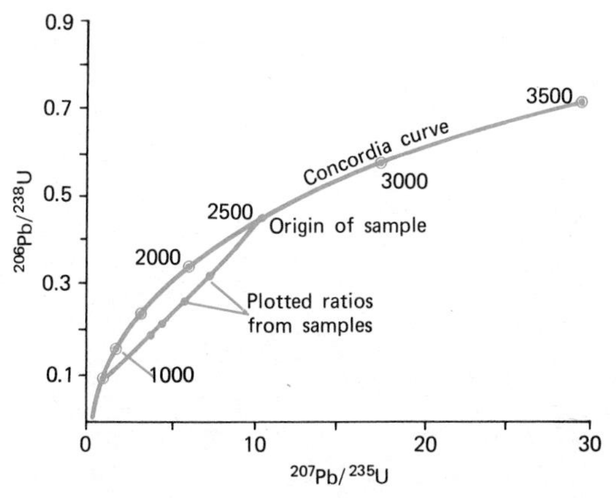

FIGURE 5–2 **Concordia diagram for determining the age of samples by the uranium–lead method. The curve shows the change in the ratios of the isotopes with time and can be scaled off in millions of years as in this plot. Actual samples generally plot below the Concordia curve along a line. The intersection of this line with the Concordia curve indicates the age of the sample.**

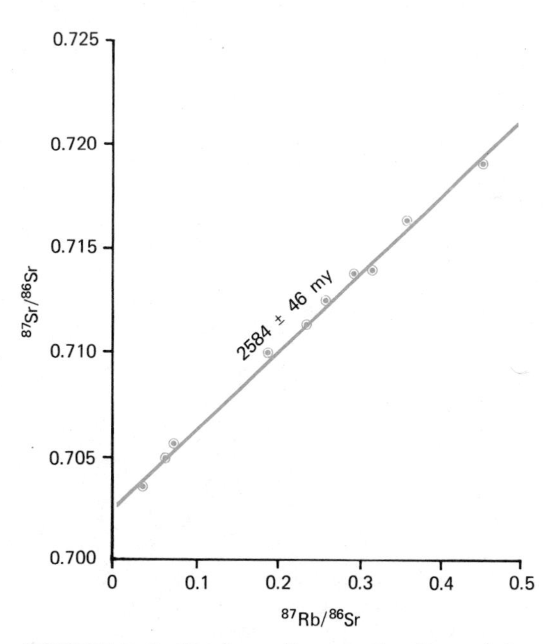

FIGURE 5–3 **Isochron diagram for determining the age of a sample by the rubidium–strontium method. Isotopic ratios of several samples from the same body plot along a straight line whose slope is proportional to the age of the sample. (Courtesy of R. Doig.)**

87 is produced by the decay process. The rocks that originally had the most rubidium 87 (with respect to the reference isotope Sr 86) generate proportionally the greatest amount of strontium 87. The older the rocks are, the more the original isotope ratios will have changed, and the greater the increase of slope of the isochron line.

POTASSIUM–ARGON METHOD

The isotope of potassium that is used in radioactive dating is potassium 40. It comprises about 0.01 per cent of all potassium. Although this is a small proportion, potassium is a common element in the minerals of the earth's crust, such as orthoclase, microcline, and muscovite, and the radioactivity of this isotope accounts for a large part of the normal background radiation that can be detected at the earth's surface. The radioactive decay of potassium 40 yields an isotope of argon (A 40) and an isotope of calcium (Ca 40) (Fig. 5–4). Although only about 10 per cent of the potassium 40 changes to argon 40, it is easier to measure this branch of the disintegration than the one that gives rise to Ca 40 because ordinary, that is nonradiogenic, calcium 40 is abundant and difficult to separate from that derived from potassium 40. Measurements of the potassium and argon isotopes are made in a mass spectrometer, and the age is calculated from the knowledge that the half-life of this transition is 1.31×10^9 years. Measurements and calculations must be made to eliminate the effects of the more abundant isotopes of both potassium and argon which do not take part in the decay series.

Argon is a gas and will escape from a mineral if it is extensively fractured or heated above 300°C. Even though samples are selected carefully for their freshness and lack of mechanical damage, potassium–argon ages of older rocks are generally lower than those obtained by other methods, owing to argon leakage. The potassium–argon method is, therefore, used largely on younger rocks which are difficult to date with the decay series that have long half-lives and have not had as much time to lose their argon. The method can also be useful in dating the last metamorphic event in a mountain-built area. When rocks are heated during metamorphism above the temperature at which argon dissipates rapidly (300°C) the decay products of K40 are released, and when the temperature falls begin to accumulate again. The amount of argon that accumulates in such rocks is a measure of the time of the metamorphism.

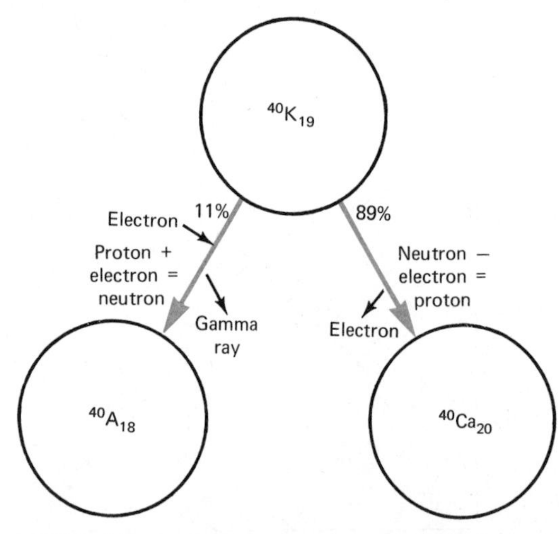

FIGURE 5–4 **Diagrammatic representation of the decay of potassium 40. Eleven per cent decays through electron capture, the union of a proton and an electron to produce a neutron and a gamma ray and the formation of argon 40. Eighty nine per cent decays by electron emission and the formation of a proton from a neutron to produce calcium 40.**

RADIOCARBON DATING

Carbon 14 is a radioactive isotope formed in the atmosphere by the cosmic ray bombardment of nitrogen 14. It is present in a small but constant percentage of the carbon dioxide of the air and sea, in plants that extract this carbon dioxide from the air and sea to build tissues, in animals that eat plants, and in animals that eat animals that eat plants. Carbon 14 begins to decay as soon as it is formed and has a half-life of

5750 years. The constant proportion of carbon 14 to ordinary carbon 12 in the modern atmosphere is the result of an equilibrium between its rate of creation and its rate of decay. Living organisms maintain this equilibrium proportion by constantly interchanging their carbon with that of the air through respiration. When the organism dies this interchange ceases and the content of carbon 14 steadily decreases as the isotope decays. After about 50,000 years so little carbon 14 is left that even our most sensitive counting equipment can barely detect the radioactivity of the isotope. Recently proposed new methods that use cyclotrons or linear accelerators to separate and measure the isotopes of carbon in a sample give promise of extending the usefulness of the radiocarbon method to samples up to 100,000 years old.

Any objects containing organic carbon can be dated, but wood, charcoal, peat, shells, and carbonates precipitated in lakes are most commonly used. The method is of more use to the archeologist than the geologist for its range in time is so limited. One of the basic assumptions of the method is that the cosmic ray flux has been constant since the animal or plant that supplied the sample died. Paleomagnetic evidence discussed in Chapter 7 suggests that the intensity of cosmic ray bombardment may have varied considerably during this period and may have produced minor errors in the dates produced by the radiocarbon method. These errors have been recognized by dating of objects of known historical age and are now routinely corrected.

GROWTH OF THE RELATIVE TIME SCALE

Ideal materials for the geochronologist are fresh, unweathered igneous rocks rich in the radioactive isotopes discussed above. As sedimentary rocks are composed largely of minerals derived from other rocks, they are not easily dated by any of the methods discussed above, although some attempts have been made to date potassium-rich minerals that form in sediments on the sea floor at the time of deposition. Yet of all rocks sedimentary rocks are the most sensitive to changing conditions on the earth's crust, and in them the most complete and easily understood record of geological history has been preserved. How can the record of these rocks be dated in years?

The first divisions of the sedimentary rock succession in northern Europe were made in the last part of the 18th century. The units were separated on the basis of their superposition, degree of consolidation and metamorphism, and extent of deformation, and not purely on lithologic grounds. The Primary rocks were highly deformed, crystalline metamorphic and igneous rocks at the base of the section. The Secondary rocks were generally flat-lying, little metamorphosed sedimentary rocks lying on top of them. The Alluvium, composed of sands and gravels, overlay everything. When additional strata were found between the Primary and Secondary, they were called the Transition series; beds between the Secondary and the Alluvium became the Tertiary. As refinements to this crude scheme of division were made and the fossil content of the beds was revealed, the names of major phases in the progression of life came to be applied to the major divisions of the sedimentary succession. Thus, the Transition beds characterized by the dominance of invertebrates were said to be deposited during the Paleozoic Era (ancient life). The Secondary rocks dominated by the reptiles and ammonites came to be called Mesozoic (middle life), and the Tertiary and Alluvium (also called Quaternary), in which a fauna of mammals and modern-looking invertebrates appeared, were called Cenozoic (recent life).

On the continent and in England attention was focused early on the beds that contained the seams of coal, then being exploited for the rapid industrialization of Europe. In 1832 the English geologist William Conybeare proposed to call them the Carboniferous order. Below these beds in England lay a thick succession of continental red sandstones that came to be known as the Old Red Sandstone.

Beneath the Old Red Sandstone (Fig. 5–5) were slates, quartzites, and limestones, occurring mostly in Wales and southwestern England. Two friends, Roderick Murchison and Adam Sedgwick, the first a retired officer of Wellington's army and the second a Cambridge don, undertook the task of unraveling the complex geology of that region. They were successful in reducing the strata to some sort of order, and each was convinced he had recognized a sequence of rocks older than anything then known. Murchison, who had been working in southern Wales, a region inhabited in the early days by a tribe known as the Silures, called his rocks the Silurian System. Sedgwick had discovered what seemed to him to be a still older assemblage which he named the Cambrian series, taking the name from the ancient name for Wales. Unfortunately, on further investigation, a thickness of rocks appeared to be common to both. These Murchison claimed for his Silurian, and Sedgwick for his Cambrian. Neither would yield and the partnership in geological research that had proved so fruitful was dissolved. For forty years the controversy concerning the overlapping systems divided geologists over the world into two camps, even after Charles Lapworth of Edinburgh proposed as a compromise that the controversial strata should be made into a separate system, which he designated the Ordovician from the name of an ancient British tribe. The Ordovician System was not recognized immediately in Britain nor on the continent, but the name gained wide acceptance in North America.

While still friends, Murchison and Sedgwick spent one season in Devon, where they found folded and faulted limestones and slates containing fossils that were recognized to be older than those of the Carboniferous System and younger than those in the Silurian System. They recognized that the marine limestones and shales of Devon must occupy the stratigraphic position of the continental Old Red Sandstone of central and northern Britain. They jointly proposed the euphonius name Devonian for this new system.

In 1840 and 1841, Murchison, who had become internationally famous for his book on the Silurian System, spent two seasons investigating the coal resources of Russia and found there, in the province of Perm, a thick succession that he called the Permian System. He recognized that this succession

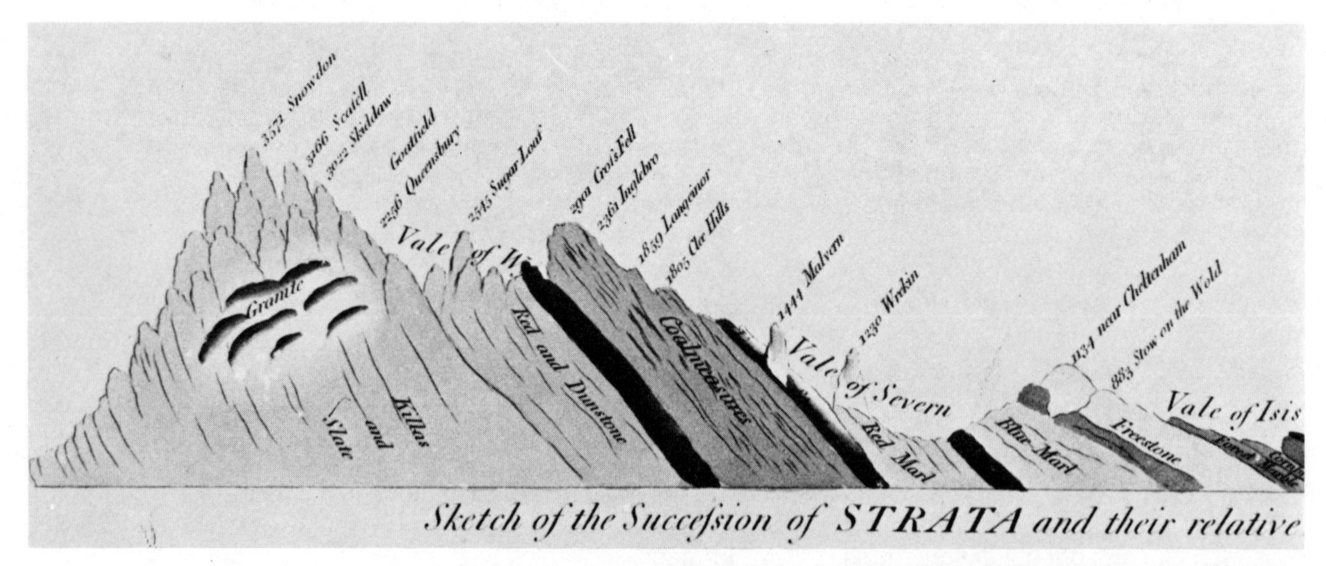

FIGURE 5–5 **William Smith's cross-section of the rocks of southern England (1815), showing the slates of Wales on the left overlain by red sandstone and coal measures, and the valley of the Thames to the right. (Courtesy of the McGill University Libraries.)**

was in the same stratigraphic position as the unfossiliferous red sandstones and magnesian limestones that overlie the Carboniferous System in Britain, and decided the unit should be given a new name.

By 1850 the old Transition series, now called the Paleozoic, was divided into the following systems:

Permian
Carboniferous
Devonian
Silurian
Cambrian

In North America a striking unconformity in the sedimentary rocks of the Carboniferous System had allowed geologists to distinguish an upper Pennsylvanian and a lower Mississippian System. This, with the adoption of the Ordovician System, completed the subdivision of the Paleozoic rocks into a scheme that has resisted alteration for the last century.

The Mesozoic rocks were studied in great detail in Europe in the early part of the 19th century. There they had long been divided into three parts, the youngest of which contained the chalk of the English Channel and was therefore conveniently called the Cretaceous System. The lowest part, because of the ease with which it could be divided into three units, chiefly on the basis of color, came to be called the Triassic; the middle part, well exposed in the Jura Mountains, was called the Jurassic System.

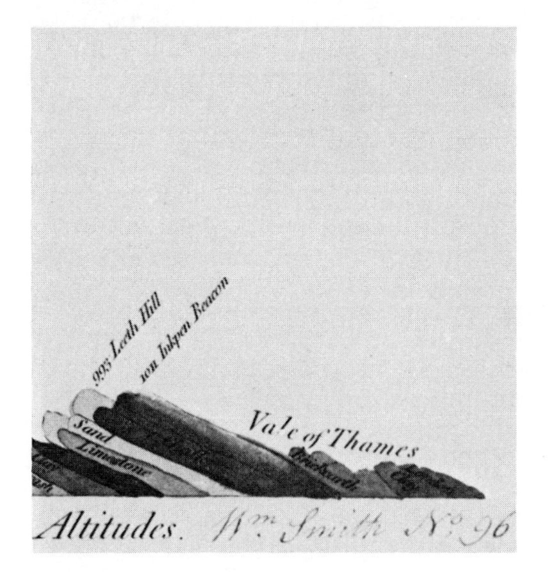

By reference to the sections in the Paris basin, southwestern France, and Italy, the Tertiary rocks were divided into groups largely on the basis of the percentage of living mollusks contained in them. Charles Lyell coined the terms Pliocene (more recent), Miocene (less recent), and Eocene (dawn of recent) for Tertiary beds that contain more than 35 per cent, 17–18 per cent, and $3\frac{1}{2}$ per cent modern mollusks, respectively. Lyell later added the term Pleistocene (most recent), and others suggested Oligocene (few recent) and Paleocene (ancient recent). The terms Tertiary and Quaternary are part of the ordinal system long out of use for older rocks and should be discarded. However, they are both in use today and refer to pre-Pleistocene and post-Pliocene Cenozoic rocks, respectively. Little need has been felt by geologists for a division of the Tertiary and Quaternary rocks into systems. If systems are recognized they are usually the Paleogene for the first three series and the Neogene for the last three.

In this way a composite section of sedimentary rocks was built up for Europe, the component parts of which were taken from local sequences of rocks. Where the boundaries between systems are gradational and not lithologically distinct, disputes arose concerning their positions that had to be settled by arbitration. By paleontological correlation between the sections geologists found that some of the systems as originally established overlapped in time on others. Because the systems were coming to be used as divisions of a time scale, their overlap in time was undesirable.

The base of the Cambrian System is the horizon at which abundant fossils of multicellular animals appear in the stratigraphic record. The rocks below the Cambrian have been referred to as Precambrian rocks since early in the 19th century, when the term Primary became obsolete. However, in many parts of the world Cambrian rocks are conformably underlain by a thick succession of sparsely fossiliferous rocks that have received many different names, such as Eocambrian, Infracambrian, Sparagmite, Vendian, etc. More recently this sequence of rocks has been shown to be characterized by a fauna of early, complex soft-bodied animals distinct from the Cambrian fauna, and to contain at its base in many parts of the world sediments indicative of extensive glaciation. For these rocks the name Ediacaran, first proposed by H. and G. Termier, is used in this book. The Ediacaran is considered to be the lowest system of the Paleozoic Erathem as proposed by Preston Cloud. The rocks below the Ediacaran can be referred to the Proterozoic and Archean Erathems, or less formally, as pre-Paleozoic rocks. The classification of these pre-Paleozoic rocks is considered further in Chapter 9.

Not all the systems proposed in the early days of stratigraphy were founded in northern Europe. In 1841 Ebenezer Emmons of New York State proposed the Taconic System for the slates and quartzites of the eastern part of the state, and a New York System for the limestones and shales that intervened between it and the "Old Red Sandstone." As late as 1911 E. O. Ulrich of the United States Geological Survey proposed a revision of Paleozoic rocks that would have introduced four new systems. These proposals were not widely accepted, even in North America, because paleontologists found that they could recognize the European systems by using paleontological correlation and another set of names was superfluous. In areas where the recognition of the standard systems is difficult, use of local systems has persisted. The Karoo System is used in central Africa for late Paleozoic and early Mesozoic continental sediments that contain only vertebrate fossils and are difficult to correlate with the standard marine sections. Local systems are also still in use in pre-Paleozoic rocks because they are not, at present, adequately correlated from continent to continent.

NATURE OF THE RELATIVE TIME SCALE

While the stratigraphy of northern Europe was being organized into a succession of mutually exclusive systems and these were being used for standards of correlation, the divisions came to have time connotation. The term "period" was introduced to refer to the time taken to deposit a system at the locality where it was first established. Rocks in North America that could be correlated paleontologically with the limestones and shales in Devon were said to have been deposited during the Devonian Period. The succession of rock systems had given birth to a succession of intervals of time, the standard, relative time scale (Fig. 5–6). In the 19th century geologists had no way of knowing the relative lengths of the periods or the total time taken to deposit the succession. The scale was, therefore, a relative one. It could be used to organize earth history, to place events in their chronological order, and establish synchrony of events around the world; but, until the techniques of isotope dating were introduced, it could not be used to date sedimentary strata in years or to give information on the rate of geological processes. Although they are based on sections of rock at the type or original sections of the systems, the periods that form the relative time scale are abstract, intangible divisions of time like the hours of the day. The periods of the time scale are united into eras (Paleozoic, Mesozoic, etc.) and subdivided into epochs (early, middle, and late for most periods).

As paleontological correlation was used to identify systems around the world, the term "system" took on a biostratigraphic meaning. The rocks of the Niagara Escarpment were correlated with the typical Silurian rocks of Britain and were, therefore, said to represent the Silurian System in New York and Ontario. Beyond its type section, the system became a division of rocks defined by the range of fossils encompassing many zones and stages.

With the development of other methods of determining time equivalence, the term

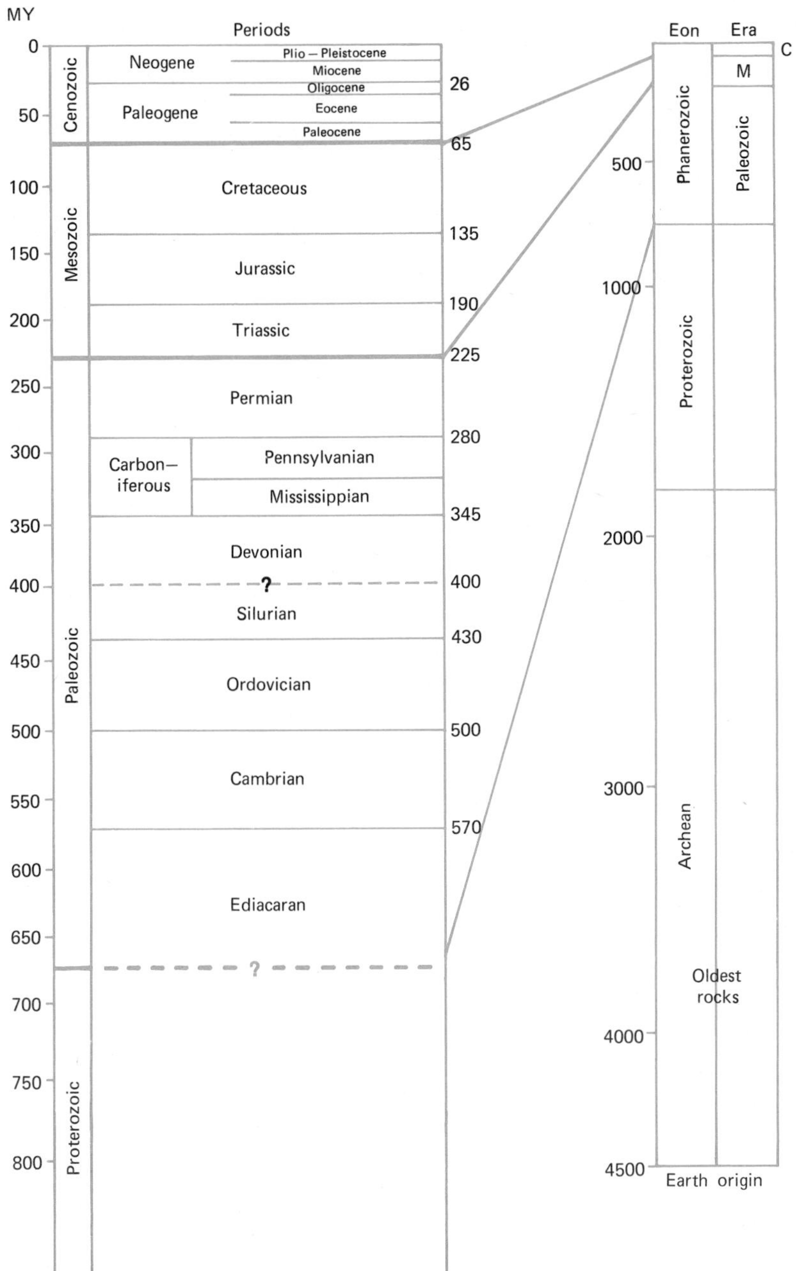

FIGURE 5–6 **The Geological Time Scale. The height of the boxes representing the periods is proportional to their length. The Pleistocene is too short to be represented separately on this scale. On the right the division of earth history into eras and eons shows the large proportion occupied by the Archean and Proterozoic eons.**

"system" has come to have a slightly different meaning. It is said to be a time-stratigraphic unit, that is a set of strata defined by the time during which it was deposited (rather than the fossil ranges). The distinction between time-stratigraphic and biostratigraphic units is a subtle one. While nearly all correlation today is in practice paleontological, methods based on radioactive dating, key beds, and paleomagnetism (see

Chapter 7) are coming into common use. The latter are strictly dependent on time, but paleontological comparisons measure time only indirectly and may be misleading when organisms have migrated slowly or are controlled by facies changes.

At present then, the stratigrapher recognizes four sets of terms for the classification of strata and events:

1. LITHOLOGIC TERMS—divisions of rock strata based on lithology. *Group, Formation, Member*
2. BIOSTRATIGRAPHIC TERMS—divisions of rock strata based on the ranges of fossils. *Series* (of some geologists), *Stage, Zone*
3. TIME TERMS—divisions of time based on the interval of deposition of a stratigraphic unit at a standard section. *Eon, Era, Period, Epoch, Age*
4. TIME-STRATIGRAPHIC TERMS—divisions of rock strata based on time of deposition. *Erathem, System, Series*

An erathem comprises the rock laid down during an era; a system, the rocks laid down during a period; and a series, the rocks laid down during an epoch. An age is the time taken to deposit a stage.

Type Sections

Type sections, or standard sections, form the basis for the definition of all these terms. **Type sections** are the stratigraphic sections where the units were first described, but where these have proved unsatisfactory a section described later may be substituted and is then called the **standard section.** The rocks of the Devonian type section are highly deformed and difficult to place in their original order, and therefore, unsatisfactory as a standard for correlation. The succession of Devonian rocks in Belgium and adjacent Germany is less deformed, easily accessible, thick, complete, and very fossiliferous, and is therefore used as a standard for worldwide correlation. Stratigraphers search for similarly well exposed, complete, and fossiliferous sections of all parts of the stratigraphic record to act as standards for correlation, or **stratotypes.** International commissions decide where in these stratotypes important boundaries should be drawn. Stratigraphers may then use various methods of correlation to identify the boundaries in their own sections. This method has been figuratively called "driving the golden spike" as the commission set, for all time, the boundary of the stratigraphic unit at a definite horizon in a definite section. International agreement on the place and horizon may be difficult but recently, the Silurian–Devonian boundary was fixed by a commission at Klonk in Czechoslovakia.

From its type section a formation is recognized laterally as far as its lithologic unity extends. Biostratigraphic units are extended from the type section of a stage or zone by paleontological criteria: where the fossils that characterize the unit cannot be found, the stage or zone does not exist. Periods of the standard relative time-scale are recognized away from the type sections of the systems on the basis of correlation criteria that should be strictly time-dependent. The boundaries of the periods can now be approximated in years (Fig. 5–6), and events falling within these boundaries can be identified as occurring during the periods. Unfortunately, few sedimentary rocks can be dated in years at present. For most sedimentary rocks paleontological correlation is all that is available, and systems recognized paleontologically away from type sections are not strictly time-stratigraphic in nature but biostratigraphic.

Where time criteria for correlation are available, the time-stratigraphic nature of systems and series is recognized; where only paleontological criteria are available, stratigraphers tend to regard these units as biostratigraphic in nature. This confusing situation will gradually be resolved as criteria for correlation are refined.

DATING THE GEOLOGICAL TIME SCALE

If sedimentary rocks on which the relative time scale is based are difficult to date in terms of years, how are the dates for the beginnings and endings of the periods obtained? These figures are continually being refined as more rocks are dated, and the time scales in older reference works will therefore differ slightly from those in newer ones. Only rock samples that can be accurately placed in the relative time scale and have an appropriate composition for isotopic dating are useful for determining the length in years of the periods and epochs.

Most of the minerals that can be accurately dated by isotopic methods occur in igneous rocks, either in large batholiths or in the dikes and sills that emanate from them (Fig. 5–7). Intrusions are younger than the rocks they penetrate and older than the rocks that overlie them unconformably. If the ages of these bounding strata are close together, such as late Ordovician and early Silurian, then the time of intrusion is precisely dated in the relative time scale. The date in years of such an intrusion would give an accurate estimate of the Ordovician–Silurian boundary. More commonly, the ages of the bounding strata are widely separated because igneous intrusion is associated with mountain building, and before strata over the unconformity can be deposited, the mountains must be eroded to a level where permanent sediment can accumulate. In addition, the sedimentary rocks of mountainous areas are commonly metamorphosed during deformation, and fossils that might indicate their relative age are destroyed. Igneous rocks in New England that intrude such early Paleozoic rocks of indefinite age and are overlain by Triassic sandstones, although they can be determined as 230 million years old, do not tell us anything about the absolute age of any part of the relative time scale. From other intrusions more precisely dated in the relative time scale, we know that 230 million years ago was late in the Permian Period and therefore can determine the relative age of these New England rocks by their absolute age.

Lavas may also be dated by the potassium–argon and rubidium–strontium methods. If they are enclosed in fossiliferous sedimentary rocks which give their relative ages, such occurrences may be valuable for dating the relative time scale. Volcanic ash from an eruption may settle in the sea or in a lake, forming a bed rich in potassium feldspar. Such beds make excellent points for calibration of the relative time scale, for they can be dated by the potassium–argon method and are commonly closely associated with fossil-bearing sedimentary rocks, or contain fossils themselves. The latest Cretaceous rocks of Al-

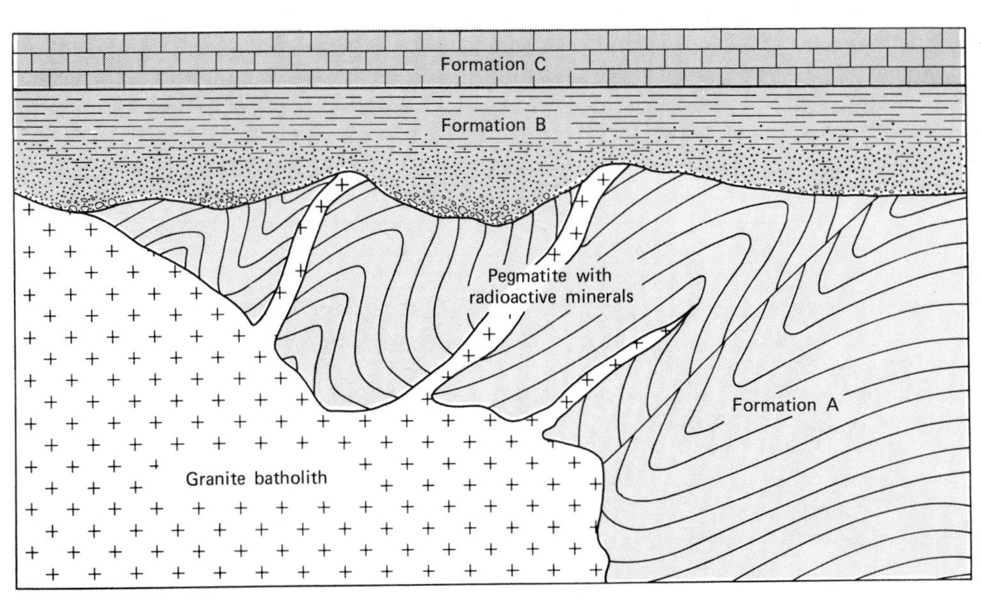

FIGURE 5–7 **Relative dating of radioactive deposits. Pegmatite dikes intruding formation A carry radioactive minerals that can be dated in years. These minerals are younger than formation A, in which they are emplaced, and older than formations B and C, which overlie the intrusives unconformably.**

berta, dated by volcanic ash associated with coal as 63 million years old, give us an estimate of the time of the Cenozoic–Mesozoic boundary. The potassium–argon method has been extensively used in the correlation and dating of the volcanic ash beds in the mammal-bearing Cenozoic beds of the western United States.

Attempts have been made to find methods of isotope dating that could be applied directly to sedimentary rocks. Lead methods give a date of 350 million years for the latest Devonian Chattanooga Shale, which is relatively rich in uranium. Such uranium-rich shales are so rare in the stratigraphic record that the technique of using them to calibrate the scale is not of general application. Attempts have also been made to use minerals like glauconite, that form on the ocean floor during the deposition of sediment, for potassium–argon dating, but these minerals tend to lose argon relatively quickly and give minimum ages only.

THE GEOLOGICAL PERSPECTIVE

The oldest rocks so far dated are gneisses from the western coast of central Greenland and the adjacent coast of Labrador and are 3.8 billion years old. Recently dated gneisses from Minnesota are only slightly younger than these. The earth as a planet appears, from considerations explained in Chapter 6, to be about 4.5 billion years old. What happened during the first hundreds of millions of years we can only guess as no record has been discovered in the rocks of that time. The time between the origin of the earth and the accumulation of a permanent record in the rocks can be called Pregeological time and the time after, Geological time. These two terms are analogous to the terms "historic" and "prehistoric."

The length of geological time indicated by the geochronologist is so great that, like astronomical distances, it taxes the imagination. The figures are given more realism by means of an analogy. If we condense time by a factor of a million so that one year represents a million years, we can compare the 4 billion years of geological time to the 4000 years since 2000 B.C. (Fig. 5–8). The oldest rocks of the earth's crust would be formed at a time equivalent on our condensed scale to the period of the Middle Kingdom of Egyptian civilization. Following this analogy, pre-Paleozoic time would then include the Egyptian, Greek, and Roman empires, and the Dark and Middle Ages up to the beginning of the Renaissance, or about 1370. The time of the Hundred Years' War between Britain and France and the heroic deeds of Joan of Arc would correspond roughly to the beginning of the Ordovician Period. The end of the Paleozoic would correspond to the mid-1700's, when the English of the American and Canadian colonies were struggling with the French for the domination of the New World. The end of the Mesozoic would correspond to 1914, when the First World War began in Europe. The Cenozoic Era would correspond to the 64 years since then. On our scaled-down calendar, humans would have arrived on the scene just last year. In all the 4000 years representing geological time on the one-millionth scale the human species, *Homo sapiens,* occupies less than a single year.

SUMMARY

Discoveries concerning the length of geological time have had a profound effect on humans' understanding of their place in the universe.

Dating methods depend basically on measuring the accumulation of decay products of radioactive isotopes and determination of the rates of decay.

Analysis of uranium-bearing minerals allows the geochronologist to use the two decay series U 238–Pb 206 and U 235–Pb 207 to obtain an accurate age in years, to estimate the amount of lead loss from the system, and to estimate the time of the loss.

The decays of Rb 87 to Sr 87 and of K 40 to A 40 allow many different igneous rocks to be dated. The decay of C 14 can only be

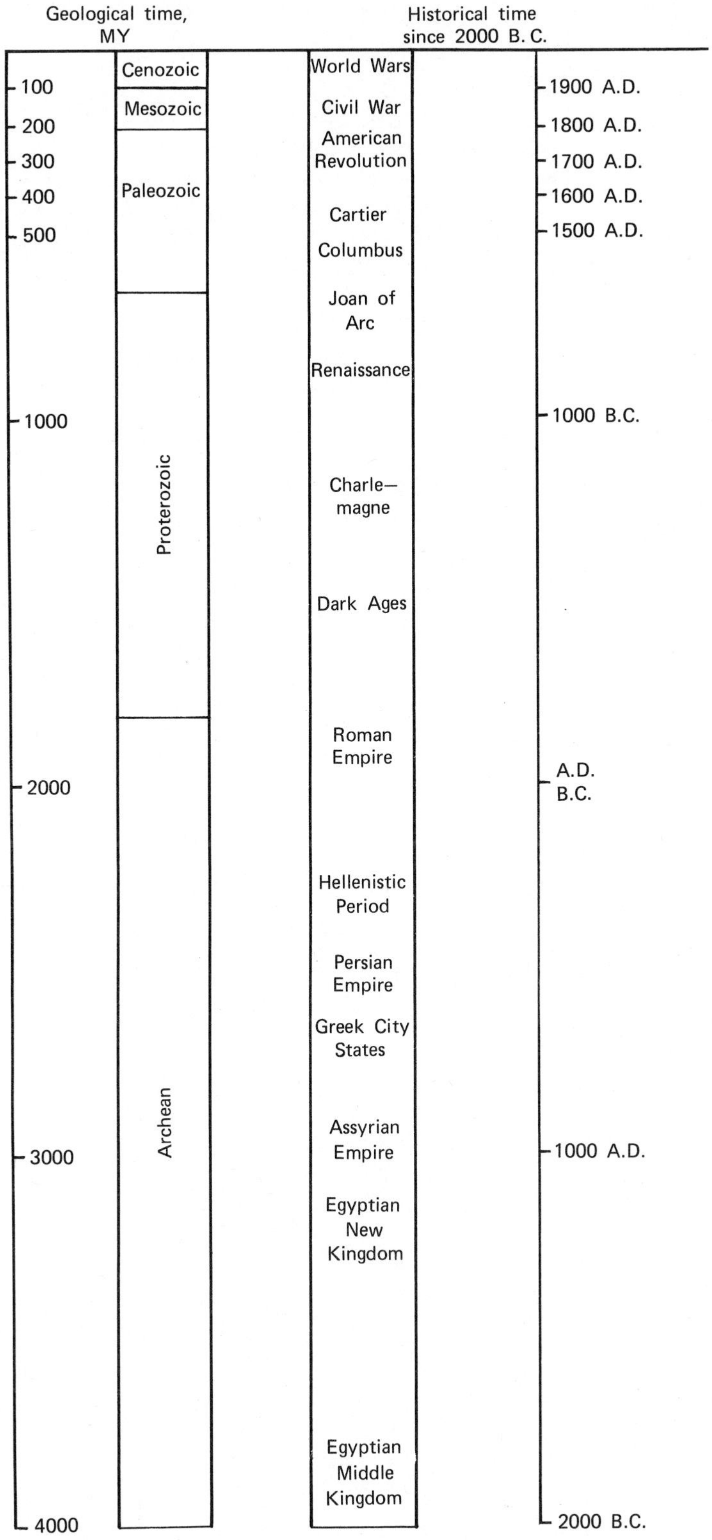

Geological time, MY

Historical time since 2000 B.C.

Cenozoic
Mesozoic
Paleozoic
Proterozoic
Archean

100
200
300
400
500
1000
2000
3000
4000

World Wars
Civil War
American Revolution
Cartier
Columbus
Joan of Arc
Renaissance
Charle–magne
Dark Ages
Roman Empire
Hellenistic Period
Persian Empire
Greek City States
Assyrian Empire
Egyptian New Kingdom
Egyptian Middle Kingdom

1900 A.D.
1800 A.D.
1700 A.D.
1600 A.D.
1500 A.D.
1000 B.C.
A.D.
B.C.
1000 A.D.
2000 B.C.

FIGURE 5–8 **Comparison of the length of geological time with the last 4000 years of the history of civilization.**

used for dating events of the last 10,000 years.

The standard relative time scale was derived from lithologic units established in northern Europe and called systems. These came to have time significance as geologists compared other sequences with them by means of fossils.

Four sets of terms are used in classifying and correlating sedimentary rocks: lithologic terms for divisions of rock strata based on lithology, biostratigraphic terms for divisions of rock strata based on the ranges of fossils, time terms for divisions of time based on intervals of deposition of strata at standard sections, and time-stratigraphic terms for divisions of rock strata based on time of deposition.

Only by isotopic dating of accurately, stratigraphically placed samples can the dates of the periods and epochs be determined.

QUESTIONS

1. What factors or events may produce discrepancies from the true age in the measurement of ages by radiocarbon, lead, and potassium-argon methods?
2. Lyell based his division of the Cenozoic epochs partly on the percentage of living species in the faunas of the rocks. In the light of the nature of the fossil record (Chapter 3), discuss why such a system should not be extended to the whole of the time scale and why it was not successful even in the Cenozoic.
3. Compare time-stratigraphic and biostratigraphic units in terms of their definition and extension from type sections.
4. Invent and describe the geology of an ideal quarry in which as many as possible of the methods described in this chapter could be used to make a correlation between the relative and absolute time scales.

SUGGESTIONS FOR FURTHER READING

BERRY, W. B. N. *Growth of a Prehistoric Time Scale.* San Francisco: W. H. Freeman & Co., 1968.

FAUL, H. A., "A History of Geologic Time.", *American Scientist,* 66: 159–165, 1978.

MOORBATH, S. *"The Oldest Rocks and the Growth of the Continents." Scientific American,* 234: 32–43, 1977.

YORK, D., and FARQUHAR, R. M. *Earth's Age and Geochronology.* London: Pergamon Press, 1972.

CHAPTER SIX

Most of this book is about the history of the planet Earth, and particularly of North America, as it is recorded in the rocks of the crust. Because the major features of the earth, such as the core, mantle, crust, ocean basins, and continents were formed long before any of the rocks we have so far found in the crust, we must step back in time and inquire into the origin of the earth as a planet to understand its framework. Such a search leads to the investigation of other similar planets of the sun's family and to the origin of the sun itself. Logically, our inquiry might involve the origin of the universe and of matter, but then we would have strayed far from the field of geology into relativistic physics. In the review that follows consideration of the earth's place in the universe will be restricted to those aspects that relate to the origin of its internal structure and its early history.

EARTH'S PLACE IN THE UNIVERSE

The earth is one of nine planets that circle the sun. The planets have a common direction and zone of motion around the sun, and all but two spin on their axes in a common direction. Uranus rotates in the opposite direction on an axis inclined about 6° to its orbital plane and appears to have tipped over from an orientation like that of the other planets, and Venus rotates very slowly in a retrograde sense. All the planets, except the one farthest from, and the two nearest the sun, have satellites (or "moons"), most of which circle their planets in the same sense that the planets circle the sun. The planets can be divided into an inner set of "terrestrial planets" (Mercury, Venus, Mars, Earth) with high specific gravities about 4 to $5\frac{1}{2}$, and an outer set (Jupiter, Saturn, Uranus, Neptune) with much lower specific gravities of 0.7 to 1.7. Pluto, usually the outermost planet but at present occupying the part of its orbit that lies within that of Neptune, does not fit several of these generalities and may be an escaped moon of Neptune rather than one of the original planets. The sun is an average star with a diameter of 1.4 million kilometers (just over 100 times the diameter of Earth), a specific gravity of 1.4, an exterior temperature of 6000°C, and a central temperature estimated at 1.5×10^6°C. If there are planets circling around some other stars in the universe, they would be too small for us to see even with our most powerful telescopes (Fig. 6–1). However, careful analysis of the motions of some neighboring stars suggests that they do have invisible companions of planetary size. The stars are so widely separated in space that light trav-

THE EARTH IN SPACE

FIGURE 6–1 **Spiral nebula in Pisces, a system of stars like our own galaxy, the Milky Way. (Photograph courtesy of the Mount Wilson and Palomar observatories.)**

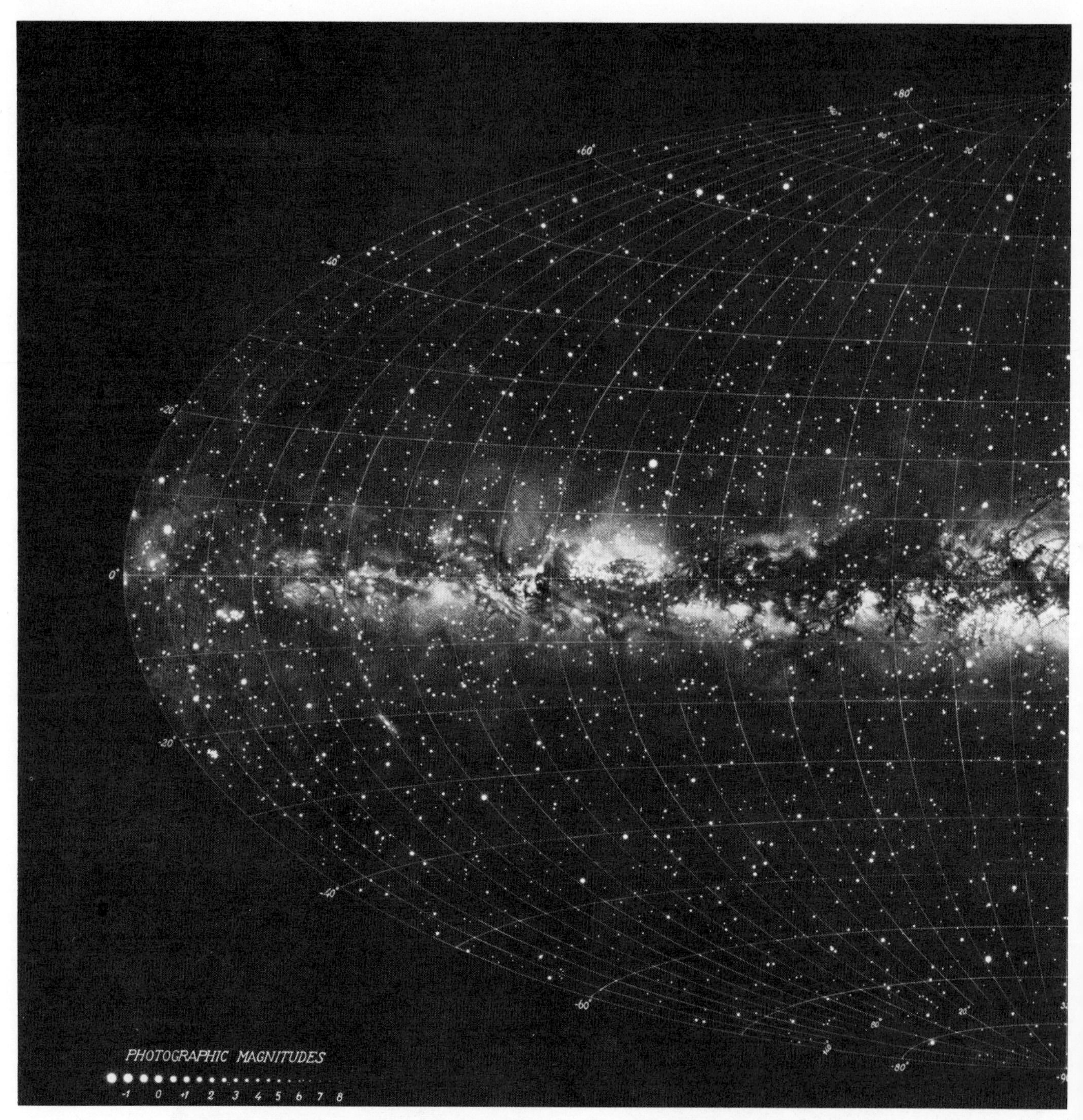

eling at 300,000 km per second from the nearest one takes more than four years to reach us. The distance to remote astronomical objects can be expressed in **light years**, the distance traveled by light in a year or about 10^{13} km.

The 5000 stars visible to the naked eye all belong to our galaxy, the Milky Way (Fig.

FIGURE 6–2 **Diagram of the Milky Way. (Courtesy of the Royal University Observatory, Lund, Sweden.)**

6–2). This is a vast disc containing 100,000 million (10^{11}) stars, that bulges at the center and thins outward toward the edges. The galaxy is 100,000 light years in diameter and its rotation carries the solar system around in a period of about 250 million years. Not all the matter in the Milky Way galaxy consists of stars and their possible planets; between the stars telescopes reveal great masses of dust and gases, such as hydrogen and helium, that are uncondensed even at space temperatures near absolute zero.

The gas can be detected by optical telescopes where it fluoresces under ultraviolet radiation from neighboring stars; the dust can be detected by the way it obscures and absorbs the light of the stars behind it (Figs. 6–3, 6–4). Because characteristic radio waves are emitted by the gases, the radiotelescope is a far more useful instrument for the study of this interstellar material than an optical one. By far, the most abundant of the interstellar gases is hydrogen. Radiotelescopes have also detected a steadily growing number of molecules in the interstellar material. Some of these are relatively simple, like molecular hydrogen (H_2), hydroxyl (OH), and ammonia (NH_3), but some, such as formic acid ($HCOOH$), methylacetylene (CH_3C_2H), and methylamine (CH_3NH_2), are complex organic compounds.

The stars appear to pile up into dense masses along the Milky Way and appear sparsely scattered at right angles to this starry zone at the galactic poles (Fig. 6–2). The stars appear more densely spaced in the Milky Way because the observer is looking through 25,000 to 75,000 light years of stars along the flat diameter of the disc. In other directions our line of sight passes through a shorter distance and fewer stars before encountering the starless space outside the galaxy. By mapping zones in which hydrogen is dense, radioastronomers have defined five concentric bands that constitute the "spiral arms" of our galaxy (Fig. 6–5). The solar system is near the inner edge of

FIGURE 6–3 **Luminous interstellar matter distributed thinly in the filamentary nebula in Cygnus. (Photograph from the Mount Wilson and Palomar observatories.)**

the Orion arm, about 30,000 light years from the dense core of the galaxy which obscures the spiral arms on the side opposite the sun.

Although the center of the galaxy is obscured by clouds of dust, it properties have been surveyed recently by a variety of detectors and telescopes placed on rockets, orbiting satellites, and space stations, outside the filtering effects of the earth's atmosphere. In the core the stars are much closer together than in the periphery of the galaxy; in the densest part they are 10 million times closer than they are in the vicinity of the sun. The core is a powerful source of radio waves, X-rays, and infrared radiation. Hydrogen streams away from it to the outer parts of the galaxy.

FIGURE 6–4 **Dark, obscuring clouds of interstellar matter block out the light from behind in the "Horsehead" nebula in Orion. Taken in red light by the 200-inch telescope. (Photograph from the Mount Wilson and Palomar observatories.)**

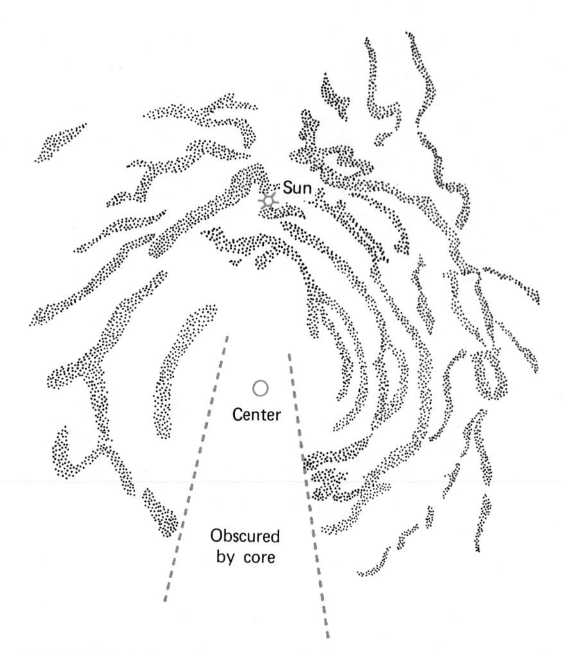

FIGURE 6–5 **Structure of the spiral arms of our galaxy as revealed by radioastronomy. The shaded areas mark bands in which hydrogen is concentrated. The dense matter in the core of the galaxy obscures the structure of the side away from our sun. (Courtesy of Oort, Kerr, and Westerhout.)**

Other Galaxies

Early astronomers, equipped with low-power telescopes, identified many diffuse light sources in the sky as **nebulae** (clouds). Giant telescopes of high resolution showed that many of these nebulae are great spiral structures of stars and are located outside our Milky Way Galaxy. The great spiral nebula in Andromeda (Fig. 6–6), the only one of these that can be seen easily with the naked eye, is similar to the Milky Way in shape but larger. The average distance between galaxies in the vicinity of the Milky Way is about 1 million light years. If the galaxies were the size of a one-cent coin, on this scale they would be separated by about 20 cm, and would stretch away for about 1.2 km to the limit of our telescopic observation.

When light received by telescope from these distant star systems is passed through a prism, it is broken up into a spectrum which reveals something about the composition of the light sources. V. M. Slipher found that the spectra originating in these galaxies are displaced toward the red end relative to corresponding spectra originating on earth, and Edwin Hubble established that the amount of the displacement is proportional to the distance of the systems from us. The farther these galaxies are from us, the greater is the extent of the red shift. The shift can be explained in terms of the "Doppler Effect," that is, the wavelength of radiation received from an emitting body that is approaching or receding from the observer is proportional to the speed of the emitter. The spectral shift to longer wavelengths implies that the galaxies are receding from our own and that the velocity of recession is proportional to their distance. The farthest detected so far is traveling at a speed of 120,000 km per second, almost 40 per cent of the velocity of light. This galaxy is at present about 6 billion light years away, at the outer limits of our observation with the 200-in Hale telescope, at Mount Palomar, California, the largest in operation in the world.

The interpretation of the red shift by the Doppler Effect has been re-examined in the light of the discovery of **quasars.** These are powerful radio sources of apparent low luminosity and extreme red shift, suggesting recession at 90 per cent of the speed of light. If their red shift is caused by their recession, and they follow the recession-distance relationship established for galaxies, then they are immensely far away and the amount of energy they release in radio waves is almost unbelievable. Fred Hoyle and William Fowler have postulated that the red shift of quasars may be unrelated to their recession and may be caused by the gigantic gravity field of a rapidly spinning star 100 million times the mass of the sun. The effect of such a powerful gravity field, they suggest, would be to decrease the energy of the escaping radiation, causing a shift to redder wavelengths. Astronomers

FIGURE 6–6 **Spiral nebula in Andromeda, a galaxy of the local group believed to be much like our own galaxy. (Photograph from the Mount Wilson and Palomar observatories.)**

are not agreed that such an explanation will explain the many enigmatic properties of the quasars nor that it is applicable to other bodies such as the galaxies.

From the recession of galaxies some scientists have concluded that initially, all matter was in a condensed state from which it exploded and is still expanding. The rate at which the galaxies are receding indicates that the cosmic explosion took place about 10 billion years ago.

Stars and the Sun

Spectrographic surveys of our galaxy that astronomers believe are applicable to the universe as a whole indicate that 95 per cent of all the atoms (88 per cent by weight) are hydrogen and 3 per cent (12 per cent by weight) are helium. Some of the hydrogen in the galaxy is a diffuse interstellar gas, but by far the greatest amount is contained in the stars and is the "fuel" they burn to shine. The stars, including our sun, derive their energy from thermonuclear reactions between hydrogen, carbon, nitrogen, and helium at temperatures of several millions of degrees absolute. The most important of these changes involves the conversion of hydrogen, through several intermediate steps, to helium. During the conversion there is a loss of mass of about 0.07 per cent. This loss of mass is converted into energy in the relationship that should be familiar to those living in the atomic age, $E = mc^2$. This equation, proposed by Albert Einstein, expresses the relationship between the energy released (E) by conversion of a mass of material (m) and the speed of light (c). This source of energy can be tapped only at the expense of hydrogen and at temperatures that have been reached on the earth's surface only in thermonuclear explosions of the hydrogen bomb. At present about three-quarters of the sun's mass is hydrogen, and it is being converted to helium at a rate of 660 million metric tons per second.

The major force active in the formation of stars from interstellar material appears to be gravity. Scattered in the galaxy are globular dust clouds of approximately the same mass as stars (Fig. 6–7). Gravity will draw together gas and particles of dust with ever increasing force as they approach each other, and the contraction will produce heat. Eventually, if the cloud had been of appropriate size, enough heat will be generated by the gravitational contraction to start the fusion reaction of hydrogen "burning" that characterizes all stars. Young stars start cool and red but soon reach an equilibrium temperature at which the rate of hydrogen consumption is balanced by the radiation of energy. A star of mass comparable to our sun can be expected to remain in this equilibrium condition for more than 10 billion years before it becomes unstable. In the later stages of evolution stars expand and contract, lose much of their matter to space in stellar explosions called **novae**, and become feebly glowing, dense, contracting masses.

To the historical geologist, the early stages in the evolution of stars when first they condense from the diffuse mass of dust and gases is most important, for at this critical stage current theory suggests that a planetary system may emerge from the nebular chaos.

ORIGIN OF THE SOLAR SYSTEM

The solar system is believed to have formed from a contracting cloud of dust and gas with a dense central zone that was to become the sun. Because the gas and dust were already embedded in a rotating galaxy, different parts had different velocities and as gravitation pulled the particles together, the mass was given a rotational motion. Continuing gravitational contraction increased the speed of rotation and pulled the mass into a disc-like form with a central globular core. The primeval solar cloud is

FIGURE 6–7 **Dark clots of dust against a luminous background in Sagittarius. The masses of dust may be embryonic stars. (Courtesy of the Lick Observatory.)**

postulated to have contained, besides an overwhelming proportion of hydrogen, the molecular constituents of interstellar dust clouds and a small proportion of elements of higher atomic weights.

Since Immanuel Kant (1766) originated the hypothesis, several mechanisms have been proposed to explain how the planets and their satellites condensed from the solar cloud. Many of the proposed models for this process cannot account for the fact that the sun now rotates quite slowly (once every 25 days at the equator), whereas in a spinning, contracting mass most of the angular momentum should end up in the central body.

The dust particles and gas of the solar cloud separated into belts in the present orbits of the planets. Gravitational attraction between adjacent particles pulled them into larger masses. This process of accretion produced swarms of particles and gas, each rotating about a center and orbiting the core of the cloud. These so-called **protoplanets** consisted mostly of hydrogen, with simple compounds of hydrogen, such as ammonia and methane in solid form, and a relatively small amount of the heavier elements. Recent models of the accretion process suggest that it may have been accomplished in a few tens of thousands of years. When the central mass of the solar cloud had been heated by contraction to a critical temperature, it began to convert hydrogen to helium and to shine as the sun does today. The outer planets of low specific gravity are composed largely of hydrogen and its simple compounds and are representative of the composition of the solar cloud. The inner or terrestrial planets of high specific gravity have relatively little hydrogen and are concentrations of the elements heavier than hydrogen, in particular oxygen, silicon, aluminum, sodium, calcium, and iron that are extremely rare in the universe as a whole. According to a widely held theory, heavier elements are produced at extremely high temperatures by the fusion of additional neutrons to the nuclei of the lighter elements and the combination of lighter nuclei. The temperatures required to build heavy elements from light ones are reached only in the interior of exploding stars, the novae. When such stellar explosions occur these heavier elements are sprayed out into space where they may be caught up in another condensing cloud and participate in the formation of another star.

These heavier elements could have been distributed evenly through the solar cloud and in the protoplanets, but some mechanisms concentrated them in the protoplanets. As the sun reached the stage of stardom, the great release of radiation could have heated the protoplanets near the center, releasing their hydrogen and helium into space and vaporizing the solid hydrogen compounds such as ammonia and methane. The protoplanets in the outer part of the disc were not heated to the temperature of those between Mars and the sun, and therefore retained most of their original light gases and solid compounds. When the sun was born the earth's protoplanet may have lost hundreds of times its present mass in lighter matter. On the other hand, the solar nebula was possibly not of uniform composition at the time of protoplanet formation, and differences in the composition of the inner and outer planets may reflect this original zonation.

Each of the protoplanets circling the sun gave rise to satellites in the same way that the solar disc differentiated into protoplanets. The number of moons separating from the protoplanet swirls seems to have depended on the distance of the latter from the disturbing gravitational attraction of the sun. Eventually the particles of the protoplanet were drawn into the solid masses of the planets by gravitational attraction.

The features of this model of the origin of the solar system of particular significance to the historical geologist are: (1) the earth originated by the collection of solid particles from a cloud of interstellar matter; (2) this aggregate originally included hydrogen, helium, methane, and ammonia, as well as elements now common in the earth in relatively small amounts; (3) most of these gases and "ices" were driven off the protoplanet by solar radiation.

Meteorites that occasionally streak through the earth's atmosphere and impact its surface appear to be remnants of the phase of accretion. Possibly they are particles that failed to be absorbed in the planets, or more likely they are products of the disintegration of one or more of the larger masses in the cloud. Most of those hitting the earth are composed of silicates like the crust or mantle, but a few are composed of an iron-nickel alloy like that believed to form the core of the earth. The meteorites can be dated by isotopic techniques discussed in Chapter 5, and all so far studied give an age of nearly 4.6 billion years. This has been generally accepted as the approximate time of the origin of the solar system.

Further information on the late stages of the formation of the solar system has recently been obtained from the exploration of the earth's satellite, the moon.

THE MOON

The most spectacular aspects of lunar exploration are the manned Apollo space flights, but at least an equal amount of information has been gained from about 40 unmanned probes, landers, and orbiters. Through the remote sensing instruments borne by these vehicles, the instruments deployed in the six manned missions, and the observations of the astronauts, the topography, geology, and history of the moon have been outlined. Still, six visits to an area whose surface is 20 per cent of the land area of the earth is sufficient for only a very superficial knowledge of its geology, and much remains to be learned about this most accessible of the extraterrestrial bodies.

The face of the "man in the moon" is formed by three large dark areas that reflect less of the sun's light to earth than the rest of the surface (Fig. 6–8). These roughly circular areas, on closer inspection, appear to be smooth dark plains that, because of their fanciful resemblance to the oceans of the earth, were called **maria** (Latin: seas; singular, mare). The surface of the moon, even in the maria, is pockmarked with innumerable craters ranging in diameter from several hundreds of kilometers to several micrometers (Fig. 6–9). The largest of the craters are not on the tops of mountains but are great walled plains whose floors are below the level of the moon's surface outside the crater wall. Nearly all these craters are formed by the impact of meteorites on the moon's surface. Because the moon has no atmosphere, meteorites are not vaporized before they reach the surface as on earth, but, approaching at speeds greater than the speed of sound, on impact cause an explosive ejection of material that forms the crater rampart. The infall of meteorites over a period of 4 billion years has so shattered the moon's surface that it is everywhere covered with a soil, or regolith, 2 to 20 m thick, composed of broken rock and dust locally compacted and partially fused into a solid breccia. The roughly circular shape of the maria suggests that they are very large impact scars formed by meteorites several kilometers in diameter during the moon's early history. The side of the moon that faces continually away from the earth is scarred with many craters, but the maria are confined to the near side for reasons not understood at present.

The rocks brought back for analysis by the astronauts are all fragments from the lunar soil, material blasted from the crust of the moon by meteorite impact. By far, the largest number of samples are composed of basalt, not unlike the basalts that erupt from oceanic islands on earth, but without any evidence of the influences of water that pervades most earthly volcanic rocks. The basaltic rocks underlie the maria and range in age from about 3.1 to 3.9 billion years. In the Apollo missions a few samples of anorthosite, a rock composed almost entirely of plagioclase feldspar that is also found on earth, were collected. The anorthosite samples give ages of 4.1 to 4.4 billion years and are believed to have been blasted by impact from the highland areas between the maria.

The moon's diameter is about one-quarter that of the earth. Its density is about 60 per cent of the earth's. Seismic experiments left by the astronauts and monitored from

FIGURE 6–8 **Composite photograph of the face of the moon, combining photographs of the first and third quarters. (Lick Observatory Photograph.)**

the earth have shown that the moon is much more rigid than the earth and that moonquakes reverberate for much longer. These waves and calculations of the moment of inertia reveal that the moon may have a core that is much smaller than the earth's (radius 500 km), a mantle (1200 km thick), and a thin crust (60 to 100 km thick).

Not all the surface features of the moon are of impact origin. Hills and escarpments that are like volcanic landforms on earth indicate that volcanic processes have been, and perhaps still are, active there. The flat floors of the maria are believed to have resulted from the flooding by basalt of these great impact scars between 3.9 and 3.2 billion years ago (Fig. 6–10). Still, the topography of the moon is an impressive testimony of the time when there was much unaccreted meteoritic material left over from

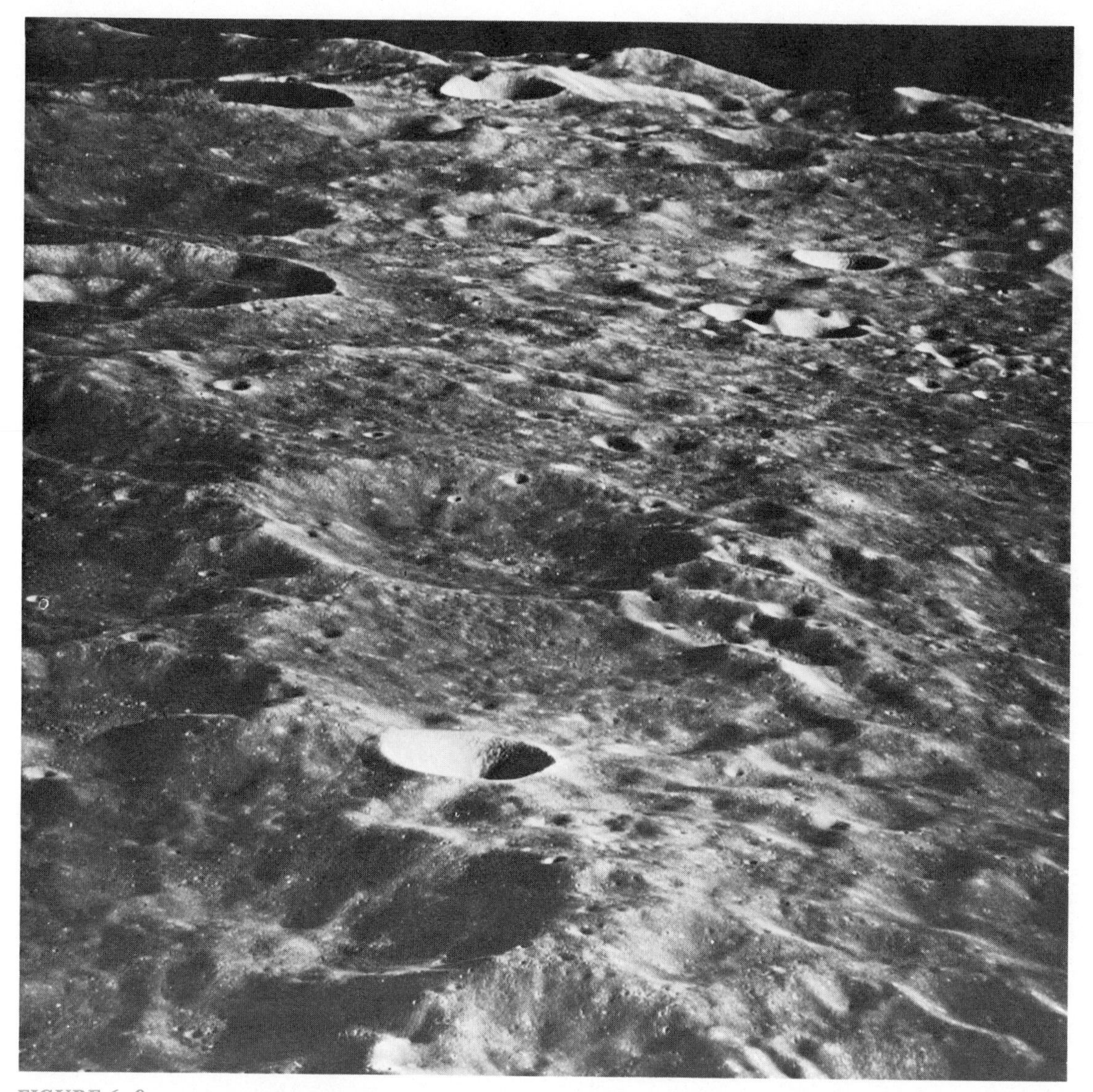

FIGURE 6–9 **Cratered topography of the moon as seen from the Apollo 10 Mission. Note the differing sharpness of the crater outlines indicating their relative ages. (Courtesy of NASA.)**

the formation of the protoplanets. The surface of the moon has not changed much in the last 3 billion years since the maria lakes solidified, for without an atmosphere the moon's surface is not subject to the processes of weathering and erosion that degrade the earth's topography. Only the incessant infall of small meteorites from space and the bombardment by particles of atomic size from the sun slowly soften the contours of the surface and "age" the craters' outlines.

Those who study the moon have not reached a consensus on the mechanism of its formation. Theories that require it to have been torn from the earth during an early stage in the formation of the solar system no longer seem tenable; nevertheless, many astronomers believe that the moon and earth may have been closer at some time in

FIGURE 6–10 **Margin of the Mare Imbrium showing the basalt of the mare flowing into irregularities in the margin, and the "young" craters in the mare basalts. Note also the straight-walled valley with a sinous rille in the foreground. (Lunar Orbiter photograph; courtesy of NASA.)**

the past. Probably the moon represents an aggregation of matter from the solar cloud that was captured by the earth during the period when the protoplanets were consolidating into planets. By 4.4 billion years ago the highland areas of the moon had been formed and were extensively cratered by meteorite infall. About 4.2 billion years ago several major impact events excavated the large maria characteristic of the moon's nearside. Soon after, the internal temperature must have risen, mobilizing the materials that had accreted in an essentially cold state and flooding the maria with basalt. For the last 3 billion years relatively little has happened on the moon's surface, except for the infall of meteorites at a steadily decreasing rate.

MARS AND VENUS

Spacecraft launched by the United States and the Soviet Union have orbited and landed on Mars and Venus. These missions to the earth's neighboring planets have already thrown some light on the origin of these planets and of the whole solar system. Telescopic observations are incapable of resolving into meaningful topography the surfaces of either planet, but spacecraft passing by, orbiting, and landing have sent back pictures of the surfaces and recorded geophysical properties.

Pictures from Mariner and Viking spacecraft show that much of the surface of Mars is pockmarked with craters like that of the moon (Fig. 6–11). However, the topogra-

FIGURE 6–11 **Impact scars on the surface of Mars in the Rasena region traversed by a channel, 25 km at greatest width, that appears to be cut by water. (Mariner photograph; courtesy of the Jet Propulsion Laboratory.)**

phy of Mars shows many features not related to impact. A large valley 6000 m deep, sometimes referred to as the Valles Marineris, shows branching tributaries characteristic of excavation by water (Fig. 6–12). Braided channels and gullies look much like those carved on earth by running water. Other topographic features may have been sculptured by what seem to be dust storms observed occasionally to sweep the planet's surface. Smooth contours of parts of the surface suggest that a blanket of dust

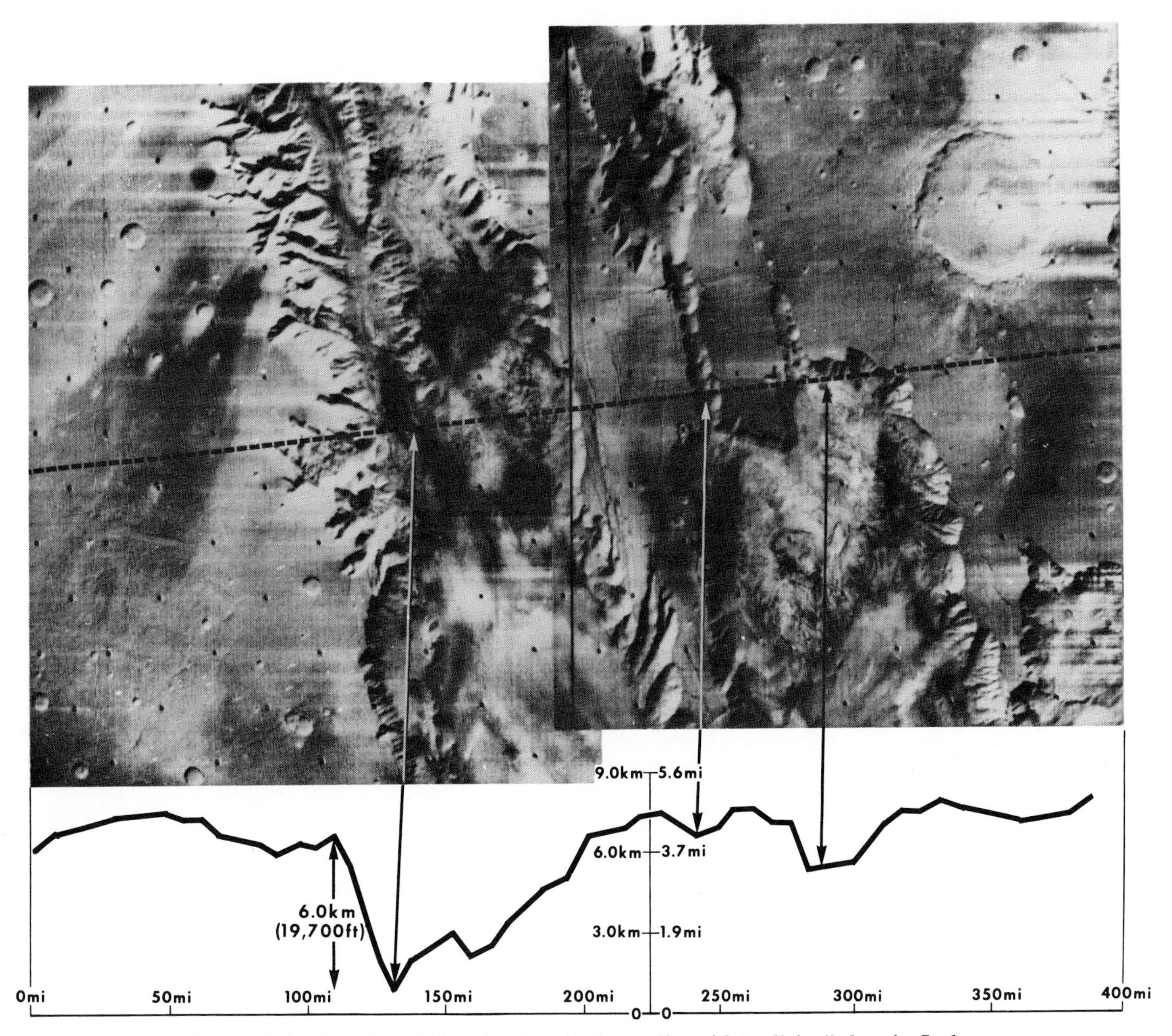

FIGURE 6–12 **Valles Marineris region of Mars showing the deep valley with "gullying" along its flanks and a topographic profile to indicate its proportions. (Mariner photograph; courtesy of the Jet Propulsion Laboratory.)**

has settled, obscuring the topography below. Spectacular shield volcanoes are revealed; the largest of these, Olympus Mons, is about 400 km across at the base, much larger than any earthly volcano (Fig. 6–13).

The mystery of Mars's apparently water-sculptured landscape is that the atmosphere and surface of the planet contain no liquid water. Robert Sharp and Michael Malin suggest that the channels are 3 billion years old and date from a time when water was present on the planet. What little atmosphere Mars has now (one two-hundredth the density of Earth's) is composed mainly of carbon dioxide. At its poles the carbon dioxide freezes into "dry ice," forming ice caps that wax and wane with the seasons. Recent evidence suggests that some of this "ice" is frozen water.

Little is known about the internal struc-

FIGURE 6–13 **The giant shield volcano Olympus Mons on Mars. The cone is 400 km across at the base. (Mariner photograph; courtesy of the Jet Propulsion Laboratory.)**

ture of Mars because seismic stations landed by the Viking spacecraft have not operated satisfactorily. It has a specific gravity of about 3.9, considerably lower than that of the earth. As Mars is smaller and of much less mass than Earth, its internal zones are less compacted gravitationally and it could be of similar composition despite this difference in density.

The experiments on Martian soil performed by the Viking landers in 1976 are open to various interpretations but have not convinced most scientists that life is present on the planet.

Mars is about one-half the diameter of Earth, but Venus is almost identical in size to Earth. Unlike Mars, its surface topography is obscured by a thick atmosphere composed almost entirely of carbon dioxide. Russian space probes passing through the atmosphere and impacting the surface reveal that the pressure there is 100 times higher than on the surface of the earth and the temperature is 475°C. Radar pulses bounced off the surface suggest a rough topography below the thick clouds and suggest that large lava-filled basins and volcanoes may be present on Venus. The overall density of Venus (5.2) suggests that it may have a zoned, internal structure like Earth, and analyses made by the Russian space probe before it succumbed to the extreme conditions on the surface indicate that the surficial rocks are granitic and not unlike crustal rocks of our planet.

Recent pictures sent to Earth by a Mariner spacecraft show that the surface of Mercury is highly cratered, much like that of the moon.

THE EARLY EARTH

What can we learn about the early history of the earth from the moon, Mars, and Venus?

1. On the observable surfaces of these bodies impact scars are prominent features of the topography.
2. The surfaces of neighboring planets also show prominent volcanic features but folded mountain chains, like those of the earth, are apparently lacking.
3. On Mars and Venus carbon dioxide is the most important constituent of the atmosphere and water is rare (the opposite of Earth). Mars may have had water on its surface at some stage in its history, and may still contain small amounts beneath its surface.
4. The moon is internally zoned into core, mantle, and crust; Venus and Mars may also be so zoned, but present evidence is not conclusive.
5. Under conditions so far detected, life as we know it is impossible on Venus and the moon, and unlikely on Mars.

If the cratered surfaces of the moon and the terrestrial planets are the remnants of an early stage in their history when the solar system was still full of particles being gathered into the planets by gravity, why is the surface of the earth so different? Without an atmosphere (moon) or with little atmosphere (Mars) the surface of a planet will not be attacked by the agents of weathering and erosion, and impact scars formed hundreds of millions of years ago will still be recognizable. The earth about 4 billion years ago probably looked much like the moon, an intensely cratered sphere, but the processes controlled by our atmosphere have planed off and filled in most of these craters. Those that have survived from the accretionary phase or been formed by relatively recent impact are called **astroblemes** and can be recognized as circular depressions (Fig. 6–14). None appears to be as large as the lunar maria, unless such features as Hudson Bay are ascribed to impact origin.

The concentric zonation of the earth into a heavy core, in intermediate mantle, and a light crust must have taken place during an early phase. Possibly, but not probably, the particles assembling to make the earth's protoplanet accumulated in order of their density, the nickel-iron first, next the heavy silicates of the mantle, and finally, the light silicates of the crust. Most geophysicists believe that the fragments in the solar cloud accumulated in the protoplanet at random, forming an unzoned mass. If so, then the density zonation of the earth developed during a hot stage that mobilized the metallic iron and nickel so that they migrated to the

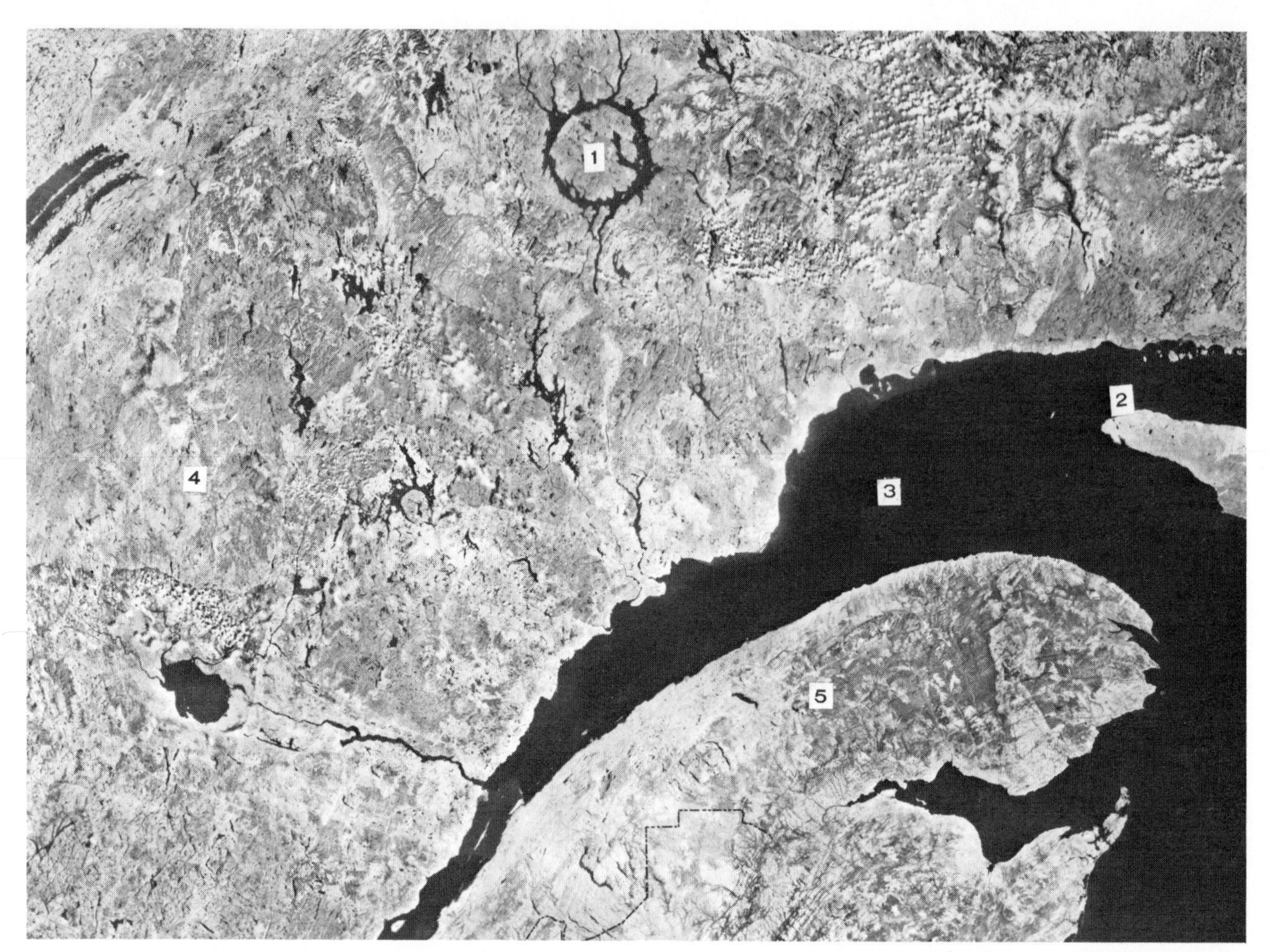

FIGURE 6–14 **Mosaic of Landsat images of the estuary of the St. Lawrence in eastern Canada. 1. The Manicouagan astrobleme. The circular depression around a central uplift is accentuated by the lake produced by the power dam on the Manicouagan River. The structure is about 65 km across. 2. Anticosti Island, underlain by almost flat-lying platform limestones, 3. Estuary of the St. Lawrence River. 4. Metamorphic rocks of the Grenville province of the Canadian Shield. 5. Folded Paleozoic rocks in the Gaspé Peninsula of the Appalachian Mountain System. (Portion of original photograph supplied by the Surveys and Mapping Branch, Department of Energy, Mines and Resources, Canada.)**

core. The heat to raise the planet to the 2000°C necessary for this migration could have come partly from gravitational compaction of the protoplanet, but is more likely to have been derived from the disintegration of many short-lived radioactive isotopes that were present at that time but which have now disappeared, and others that were present then in quantity but have since become rare as they converted to their daughter products. The presence of volatile elements, such as mercury and arsenic, in the crust suggests that the whole earth was never completely molten and that possibly the differentiation of the planet took place by solid diffusion of the constituents at a temperature about half of that required for complete fusion. The density zonation of the moon must have taken place about the same time, 4.1 billion years ago.

The separation of continental crust from oceanic crust may have started during this period about 4 billion years ago. Many geologists believe that this process did not take place once and for all at some time in the past, but has been progressive through much of geological time and is still taking place. The discussion of the early crust of the earth in Chapter 10 throws further light on this problem.

SUMMARY

The planet Earth is one of a family of nine planets revolving around an average star, our sun. Myriads of other stars, together with interstellar gas and dust, make up our galaxy.

Other galaxies are receding from ours and from their neighbors at rates proportional to their distances from us.

Hydrogen and helium, by far the most abundant elements in the universe, form most of the stars and interstellar material.

Gravitational contraction of masses of dust and gas heats them to temperatures of millions of degrees celsius when nuclear transmutation of hydrogen to helium takes place, and they then become luminous stars.

The planets of the solar system are believed to have formed by cold accretion of particles in a contracting and rotating dust cloud that surrounded the sun as it formed from interstellar matter. The early concentrations of matter in the solar cloud contained much hydrogen and lighter gases, but these were volatilized from the inner planets by the heat from the birth of the sun.

The surface of the moon is a remnant of the early stages of the solar system when the infall of meteorites from the solar cloud was much faster than now. Like Earth, the moon appears to have passed through a stage when the temperature rose to near the melting point of iron and the heavier metals found their way to the center to form a core.

Mars and Mercury also show the effects of many impact events. The atmosphere of Venus, composed largely of carbon dioxide, obscures its surface completely.

In its early stages of accretion the surface of the earth must have looked much like that of the moon, but forces of weathering and erosion (absent on the moon) have removed the evidence of this accretionary stage.

QUESTIONS

1. How do modern theories of the origin of the solar system account for the differences between the inner and outer planets?
2. Draw a series of diagrams in progressively larger scales showing the relationship between (a) the earth and the sun, (b) the sun and our galaxy, and (c) our galaxy and the Andromeda galaxy, which is located 2.6 million km away. What would be the diameter of the dot representing the earth in diagram c?
3. What are the main differences and similarities between the rocks and soils of the moon and those of the earth?
4. Compare in tabular form the major features of the earth, moon, Mars, and Venus under headings such as size, surface features, atmosphere, internal structure, rocks, and soils.
5. With binoculars and a star map locate on a clear November night some of the major stars of our galaxy, the luminous nebulosity in Orion, the spiral nebula in Andromeda, Algol (the variable star in Perseus), the dark cloud of the Coal Sack against the Milky Way in Cygnus, and the great cluster of stars in Perseus.

SUGGESTIONS FOR FURTHER READING

BOVA, B. *The New Astronomies.* New York: New American Library, 1972.

CARR, M. H. "The Volcanoes of Mars," *Scientific American,* 234: 32–43. 1976.

COLE, F. W. *Fundamental Astronomy: The Solar System and Beyond.* New York: John Wiley & Sons, 1974.

JURA, M. "Interstellar Clouds and Molecular Hydrogen." *American Scientist,* 65: 446–454, 1977.

HEAD, J. W., WOOD, C. A., and MUTCH, T. A. "Geologic Evolution of the Terrestrial Planets." *American Scientist,* 65: 21–29, 1976.

LEWIS, J. S. "The Chemistry of the Solar System." *Scientific American,* 230: 51–65, 1974.

MURRAY, B. C. "Mars from Mariner 9." *Scientific American,* 229: 48–69, 1973.

REEVES, H. ed. "Symposium on the Origin of the Solar System, Nice 1972." *Colloques internationaux du Centre national de la Recherche scientifique, no. 207,* France, 1974.

TAYLOR, S. R. *Lunar Science: A Post-Apollo View.* New York: Pergamon Press, 1975.

YOUNG, R. S. "Viking on Mars: A Preliminary Survey." *American Scientist,* 64: 620–627, 1976.

CHAPTER SEVEN

UNIFORMITY

The orthodox position of scientists at the end of the 18th century was that the world had been created a few thousand years ago and its subsequent history had been characterized by violent catastrophes, the latest of which was the Biblical flood. In response to these beliefs James Hutton (1726–1797) advanced the concept of **uniformity** or **uniformitarianism.** Basically, Huttonian uniformity postulated that the record of the past could be interpreted on the assumption that processes at work today operated in the same way and at similar rates in the past. An immense length of geological time was required for such a "steady state" world, and Hutton saw this implication when he wrote that his world model implied "no vestige of a beginning, no prospect of an end." Hutton's uniformity was advocated by Charles Lyell (1797–1875) as the key to the interpretation of geological history, and was the theme of his textbook, *Principles of Geology,* the influence of which was immense throughout the 19th century.

Although uniformity proved to be a powerful tool to combat excesses of the catastrophists of the early 19th century, exceptions to its universal applicability soon became apparent. Charles Darwin's (1809–1882) demonstration that life progressed through a long evolutionary history was at first unattractive to Lyell because it was basically anti-uniformity; that is, it postulated that the world was not always the same but a basic part of it progressively changed through a long interval of geological time. Lord Kelvin (1824–1907), the father of thermodynamics, pointed out that the concept of uniformity implied that the earth was a sort of perpetual motion machine, never expending energy or running down, and this was contrary to the second law of thermodynamics.

King Hubbert has recently amplified Lord Kelvin's critique of uniformity and has demonstrated that the planet Earth must be changing from an initial state of high energy to one of lower energy. The downward increase in temperature measured in mines and boreholes (the geothermal gradient) shows that an outward flow of heat from the earth's interior exists. Most geologists believe that this heat is responsible for the formation of mountains and other topographic relief, probably acting by creating convection currents which move lithosphere plates (see Chapter 2). As erosion of mountains proceeds, the potential energy stored in the mountain building process is dissipated as heat, which is radiated into space. Thus, each orogeny and its following erosion cycle diminish the total amount of energy available on earth.

Radioactive isotopes are likely to pro-

UNIFORMITY AND PROGRESS

vide a substantial proportion of the heat that causes convection and plate motion and leads to the processes of orogeny. Since the world began 4.5 billion years ago, radioactive isotopes have been decaying to stable isotopes. The most abundant of these isotopes have very long half-lives (see Chapter 5), but calculations show that the heat generated by them is now only about two-thirds of that produced in the early history of the earth. The laws of radioactive decay are such that one-half of the isotope decays during the first half-life, a quarter in the second, an eighth in the third, and so on. Rapid changes in the heat-generating power, therefore, must have taken place within the first 2 billion years, and since the beginning of the Paleozoic Era little change has taken place. These calculations demonstrate the danger of rigidly holding to a presumption that the erly history of the earth can be interpreted in terms of present conditions.

The first assumption of Hutton's uniformity, that past events must be interpreted in terms of present processes, is just another way of stating a basic presumption of all science: that the laws of nature are invariate in time, that matter does not behave one way in an experiment today and another way tomorrow, that nature is not capricious. However, the second part, that rates of processes have not changed, cannot be correct if applied to the whole of earth history. In its early stages the earth must have been much different from today (Fig. 7–1) and since that early stage changes in life, atmosphere, oceans, temperature, and rotation have taken place progressively.

ATMOSPHERE AND OCEANS

The present **atmosphere** of the earth is essentially four-fifths nitrogen and one-fifth oxygen. All other gases make up less than 1 per cent. This mixture of gases is not found on the other planets of the solar system. Our nearest neighbors, Mars and Venus, have atmospheres of carbon dioxide; Mercury, apparently, has no atmosphere; and the outer planets are largely gaseous and solid ammonia, methane, and hydrogen. In composition the outer planets are like the gases and "ices" detected in interstellar space and postulated to be major constituents of the solar dust cloud. The atmospheres of all the protoplanets in the dust cloud are believed to have been composed of hydrogen, helium, and the hydrides—methane (CH_4) and ammonia (NH_3) (see Chapter 6). When the sun was formed in the center of the solar cloud, these gases are believed to have been volatilized and swept off the

FIGURE 7–1 **Among the oldest rocks in the world so far dated are the Amitsoq gneisses, about 3750 million years old. The gneiss is cut by a mafic dike that has since been metamorphosed to an amphibolite. Godthabsfjord, southwest Greenland. (Courtesy of the Geological Survey of Greenland.)**

inner planets. How then did the earth's unique atmosphere develop and why are the atmospheres of the neighboring planets so different?

The atmosphere and the ocean are closely coupled systems. The gases of the atmosphere dissolve in, or exsolve from, the sea in accordance with changes in temperature and pressure. The waters of the sea are constantly being evaporated into water vapor of the atmosphere and returning as rain and runoff. Questions about the origin of the earth's atmosphere cannot, therefore, be separated from those concerning the origin of the **hydrosphere**, as the sea and fresh water bodies of the earth are called collectively.

Excess Volatiles

Rivers bring dissolved salts to the sea along with sedimentary particles. The dissolved ions accumulate there, except where isolated areas of the sea evaporate and the salts precipitate forming solid deposits. If the dissolved ions of the sea have been solely derived from the weathering products of rocks brought to it by rivers, multiplication of the annual dissolved load of rivers by the age of the oceans should give values comparable to the amount of these ions in the sea water now. For some ions, such as sodium, the product of rate of annual supply and age in years is close to the total mass of that ion in the sea now, but other materials, notably the volatile elements and compounds such as chlorine, carbon dioxide, and sulfur are present in the oceans in far larger quantities than would be expected from the analysis of weathering products transported to the sea in rivers. The sea contains 60 times the chlorine expected from these calculations.

The excess volatiles might have been present originally in the oceans or they might have been added through geological time by some agency other than weathering. If all the chloride ions present in the oceans had been present originally without the accompanying ions that now neutralize them, the sea would have been a one-molar solution of hydrochloric acid and would have reacted violently with the crustal rocks. Because the products of such reactions are not found in the early deposits of the oceans, the hypothesis that the early oceans contained all the chloride ions from the start is not likely to be correct.

In searching for alternate sources for the excess volatiles, William Rubey found that the gases needed were just those that are known to emanate from volcanoes today. Crude estimates of the rate at which such gases as water vapor, carbon dioxide, hydrogen chloride, nitrogen, and others are now released from volcanoes suggest that over the great length of geological time volcanic action is capable of supplying gases for the atmosphere and water and ions for the hydrosphere.

Primitive Atmosphere

The escape of gases from the earth's crust and mantle is referred to as the degassing of the planet. If the primitive atmosphere was supplied by degassing, it may have resembled the gases that now emanate from volcanoes, lava flows, and fumaroles, such as water vapor, carbon dioxide, and nitrogen. The absence of oxygen in this primitive atmosphere is confirmed by considerations related to the origin of life discussed in Chapter 8. The change from the primitive atmosphere to the modern one involved the conversion of some of the carbon dioxide to oxygen and the removal from the atmosphere of most of the rest of it.

Plants use the sun's energy in photosynthesis to combine carbon dioxide and water to produce starches and oxygen. Analyses of sedimentary rocks for minute traces or the degradation products of chlorophyll indicate that photosynthetic organisms were in existence about 3.5 billion years ago, and from that time onward their activity converted carbon dioxide to oxygen.

Carbon dioxide was steadily removed from the atmosphere-hydrosphere system by the deposition of carbonate (limestone, dolomite, siderite) in the sea. In pre-Paleozoic time precipitation was largely inorganic or associated with the activity of algae, but about 500 million years ago animals began using calcium carbonate for their shells and since then have formed large quantities of limestone. The total carbon dioxide locked up as carbonate ions in limestone has been estimated to be almost 1000 times that now present in the atmosphere and hydrosphere and in living organisms. If all the carbon dioxide in limestones were converted into a gas and organisms had not developed photosynthesis, we would have an atmosphere like that of Venus.

Oceans

Degassing of the earth was certainly much faster in the first half-billion years when it passed through a hot stage and gravity differentiation was taking place. In this period water vapor emanating from the crust and mantle condensed to form the first oceans. The oldest metamorphic rocks, believed to be derived from sedimentary rocks deposited in water, are about 3.8 billion years old and are evidence that oceans existed at this time. If the hypothesis of the derivation of the atmosphere and hydrosphere by degassing is correct, then the total volume of the seas must have increased, but the salinity could have been much the same throughout geological time. The basic similarity of marine invertebrates of the early Paleozoic to those of the present seas suggests that the salinity has varied little in the past 500 million years. So little is known about the configuration of the ocean basins in pre-Mesozoic times that the consequences of a steadily increasing hydrosphere upon sedimentation and paleogeography cannot be calculated at present.

MAGNETIC FIELD

The major component of the present magnetic field of the earth can be represented as a bar magnet, or dipole, whose axis is inclined at 11.5° to the axis of rotation. The dipole field is distorted by minor irregularities that drift westward at 0.2° of longitude per year. The geomagnetic poles, which represent the emergence of the axis of the field at the earth's surface, have moved 5° of longitude and a fraction of a degree of latitude in historic times. Geophysicists can show that over the past 1000 or 2000 years the geomagnetic pole has corresponded closely with the geographic or rotational pole and the present departure is unusual. They deduce that this association has always existed.

Remanent Magnetism

Sensitive instruments have determined that many sedimentary and igneous rocks retain a record of the magnetic field in which they were deposited or crystallized. All minerals originating in or on the earth are subject to the earth's magnetic field; those with magnetic susceptibility retain a record of that field. Magnetic mineral grains settling through water or air tend to be aligned and record the earth's field. Mineral grains crystallizing from igneous rocks adopt the magnetic field within which they form. The magnetism that remains in a susceptible substance after the field that induced it changed or disappeared is called **remanent magnetism.** This property of rocks allows the geologist to study changes in the earth's magnetic field through geological time, a study called **paleomagnetism.**

The direction (an azimuth with respect to the modern geographic pole) of the remanent magnetism in the rock shows the direction of the ancient magnetic pole. The inclination of the remanent magnetic field with respect to the horizontal shows the magnetic latitude of the specimen, or its distance from the ancient poles. The inclination, like that of a dip needle, is horizontal at the geomagnetic equator, vertical at the geomagnetic poles, and of intermediate value in between.

Reversals of Field

Early studies in remanent magnetism showed that the direction of magnetization of rocks of certain ages is reversed with respect to the present field of the earth, that is, it differed by 180° from the present polarity, and compass needles that now point north would then have pointed south. Further determinations delineated epochs characterized by this reversal and other epochs when it was normal. (The term epoch commonly used to refer to a time interval characterized by a particular polarity should not be confused with the time stratigraphic term discussed in Chapter 4). Within normal or reversed polarity epochs short intervals of opposite polarity, lasting probably less than 100,000 years, are called **events.** The epochs are named after scientists who have made contributions to the study of the earth's magnetic field, and the events after geographic localities where they were recognized.

The chronology of reversals for the Cenozoic and much of the Mesozoic Era is now well known and Fig. 7–2 shows that changes were common. At other times, such as during the Permian Period, the polarity of the earth was stable. Transition zones preserving the change from one polarity to the other are rarely preserved and difficult to interpret. At present geophysicists are uncer-

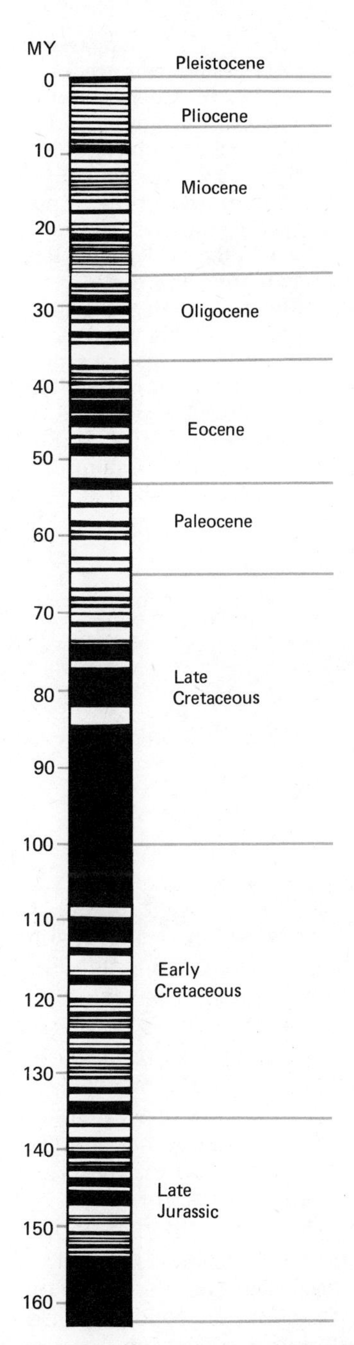

FIGURE 7–2 **Reversals in the earth's polarity during the past 160 million years. The dark intervals represent normal polarity; the light ones, reversed. (Courtesy of R.L. Larsen, Geological Society of America.)**

tain as to whether the field collapses and builds in the opposite direction or whether the dipole tilts over through 180°.

Field reversals have been used for correlation in closely spaced sections where identification of a polarity epoch by some independent method is available. In sediment cores from deep sea drilling sites, the uppermost layers of normal polarity that have just been deposited must correspond to the modern Brunhes Normal Epoch, the reversed layers below them, to the Matuyama Reversed Epoch, and so on. Polarity events and epochs in the sediment can be matched from section to section by counting downward. Isotopic dating of lava flows may also serve to identify a particular polarity epoch for correlation purposes. The reversal of the earth's field is also recorded in the oceanic crust rocks formed at oceanic spreading centers as discussed in Chapter 1.

Geophysicists believe that the magnetic field is maintained by currents in the liquid core related to the earth's rotation. The mechanism by which such currents are reversed periodically is uncertain, but the results are clearly recorded in the rocks.

Polar Wandering

When the positions of the north magnetic pole are plotted from remanent magnetism of North American rocks, the pole is determined to have been in the position of the present Pacific Ocean during the early Paleozoic. From there it moved northward in late Paleozoic time, and during the Mesozoic Era it moved in an arc westward and then northward through the present position of Siberia toward its modern location (Fig. 7–3). Another way of illustrating the changes in the earth's field through the ages is to plot the geomagnetic equator and **isoclines** (lines of equal magnetic inclination) on either side of it (Fig. 7–4). Such maps show the equator crossing from southwest to northeast through the central part of the continent during much of the early Paleozoic Era and forming an arc from northern California to Newfoundland as late as the Permian Period. Although these results

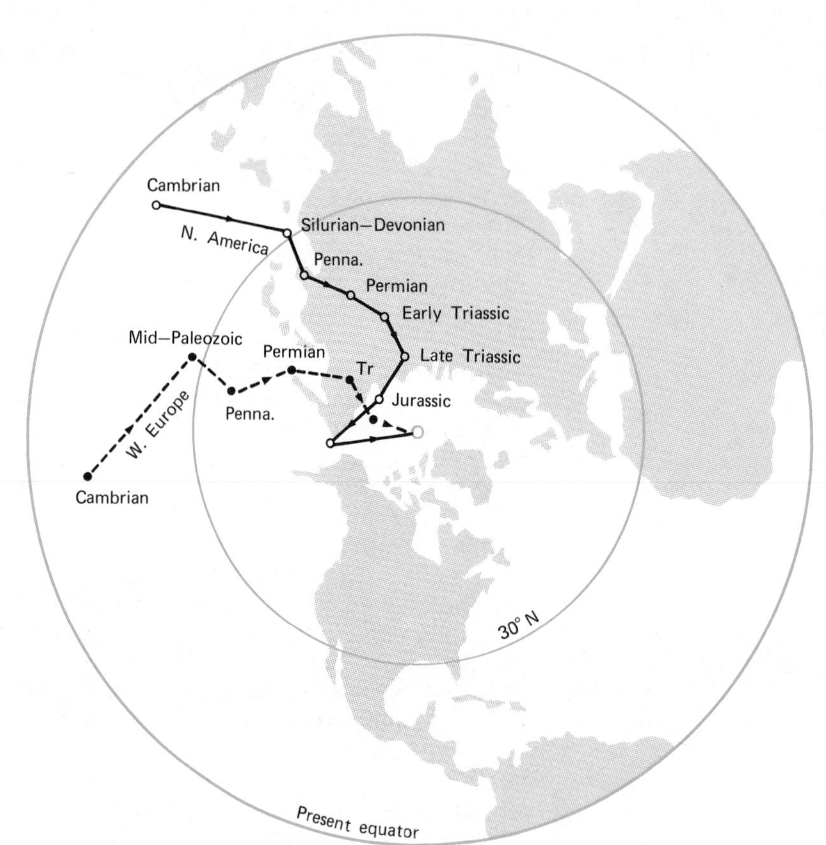

FIGURE 7–3 **Apparent polar wandering tracks for North America and western Europe plotted on a polar projection of the world with the continents in their present position. Note the parallelism of the tracks in the late Paleozoic and early Mesozoic, indicating that the continents acted as one during this interval, and their lack of parallelism after Triassic time. (Courtesy of M.W. McElhinny and the Cambridge University Press.)**

are startling, they provide answers to many paleoclimatological problems, such as the distribution of Paleozoic reefs, evaporites, and coal deposits, which will be discussed in detail later.

As long as paleomagnetic investigation is confined to North America, a single unequivocal wandering curve can be plotted. If the investigation is extended to Europe, the polar wandering curve for that continent is found to be distinctly different. As Fig. 7–3 shows, it is displaced to the east of the curve for North America by about 30° longitude in Permian time and follows a parallel course to it until Triassic time, when the two curves begin to converge toward their present meeting at the north magnetic pole. Because the remanent magnetism of rocks from the two continents must have been oriented to the same geomagnetic pole throughout geological time, the present divergence can be explained only as a result of the relative motion of the continents. The direction of this movement can be determined by the relationship of the two curves. The initial parallelism of the two polar curves shows that the continents were welded together and acted as one in late Paleozoic time, but the distance between the curves is a reflection of the relative counterclockwise rotation of North America necessary to close the North Atlantic gap. The drift of the continents to their present positions is reflected in the convergence of the curves to the present magnetic pole. Because polar wandering curves have been determined for many of the continents for the last 2 billion years, the location of the magnetic pole preserved in remanent magnetism may provide a method of dating and correlating rocks where paleontological methods are not applicable.

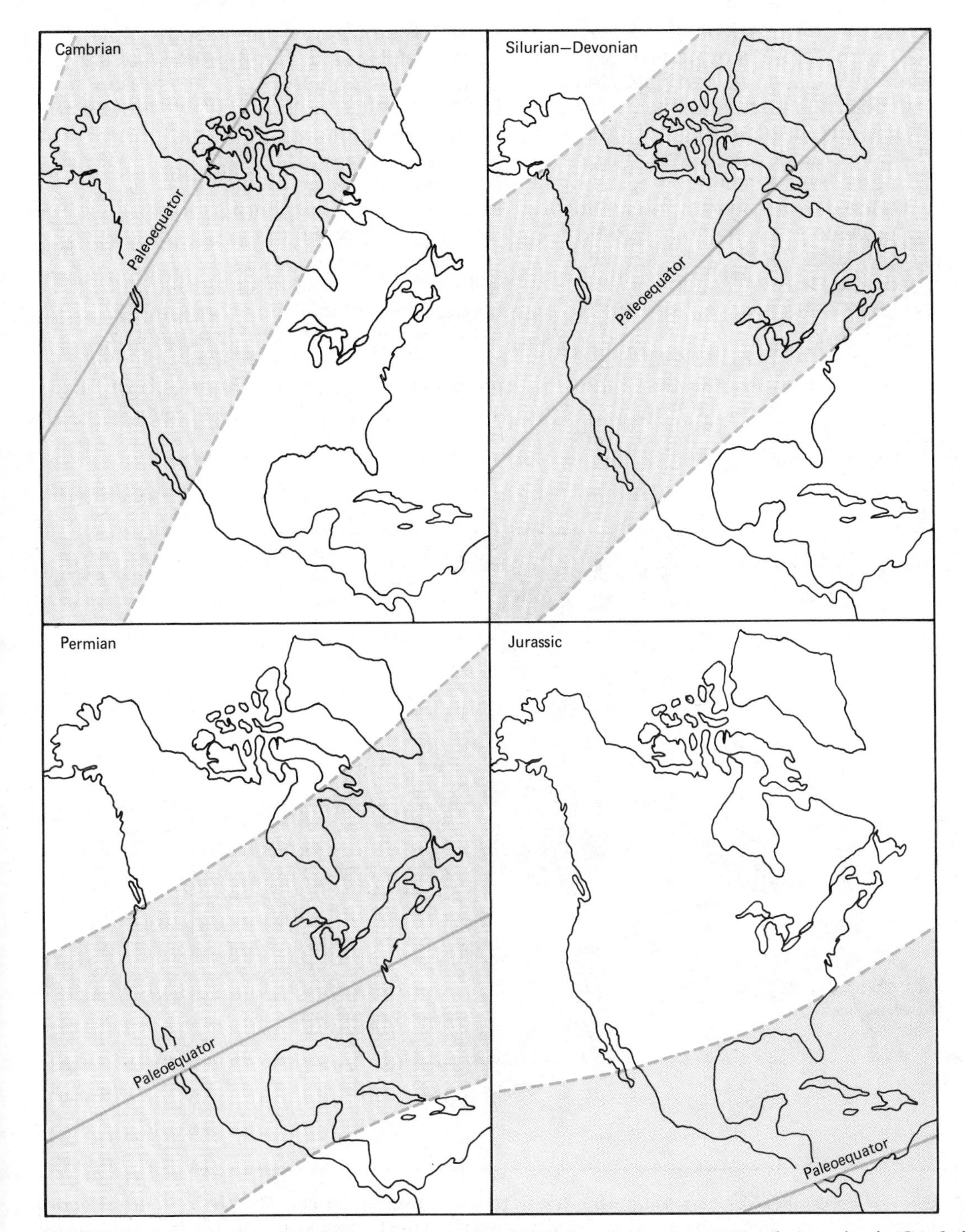

FIGURE 7–4 **Position of the paleoequator and zone of the tropics across North America in Cambrian, Silurian–Devonian, Permian, and Jurassic time, determined by paleomagnetic studies. Constructed from polar positions shown in Fig. 7–3.**

If the successive positions of the magnetic pole during pre-Paleozoic time are plotted from information gained from North American rocks, the path traced includes a number of so-called "hairpin" curves (Fig. 7–5). The correspondence in time of these sharp changes in the direction of polar wandering with the major orogenic periods that are used to divide the Archean and Proterozoic time suggests that the apparent movement of the poles may be accompanied by major changes in the relative positions of lithosphere plates.

Because the magnetic and rotational poles of the earth are closely linked, and we know of no mechanism that is likely to change the axis of rotation of the earth in space, the term "polar wandering" is a misnomer. The pole is stable in space; it is the surface of the earth that moves. What the polar wandering curves express is the mobility of the whole lithosphere with respect to the pole of rotation. Paleomagnetism not only supports evidence that the continents have moved relative to each other, but shows that the lithosphere as a whole is moving with respect to the pole of rotation. The motions of Europe and North America with respect to each other are shown by the convergence of their polar wandering tracks, but the northward movement and counterclockwise rotation of these continents with respect to the poles is best illustrated by successive positions of the paleoequator (Fig. 7–4).

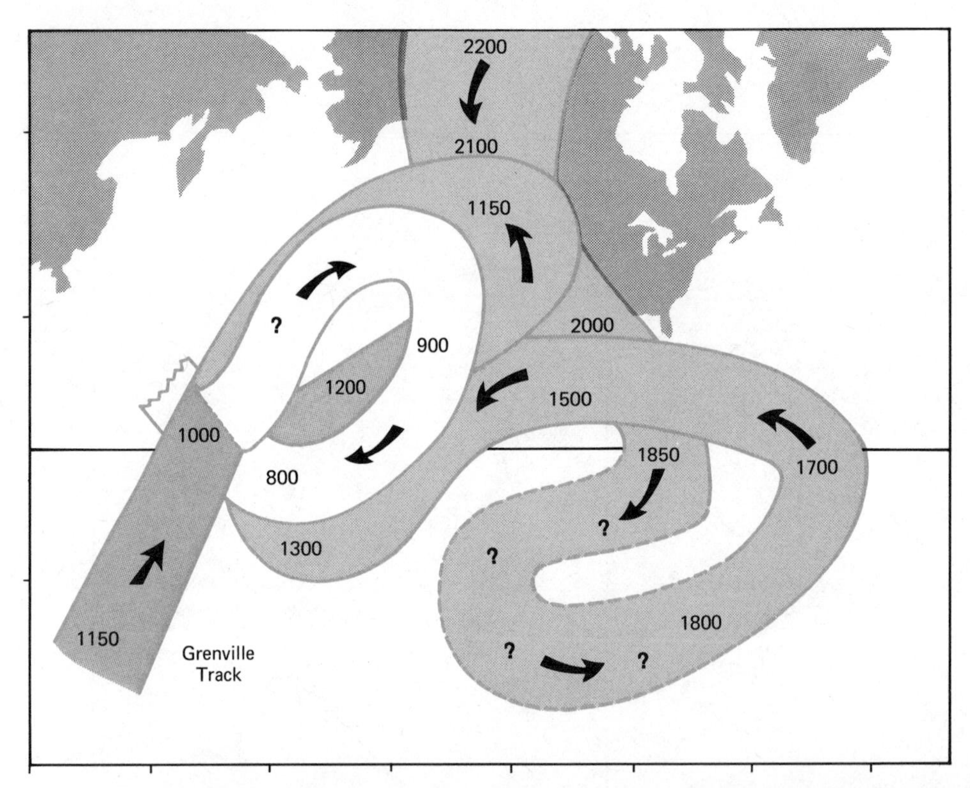

FIGURE 7–5 **Pre-Paleozoic polar wandering tracks for the Canadian Shield. The "hairpin turns" in the track appear to be correlated with episodes of major orogeny in the Canadian shield. The figures along the track indicate the time in millions of years when the pole occupied that position. The fact that poles from the Grenville province of the shield form a separate track joining that of the other provinces about 1000 million years ago has suggested that that province was separated from the rest of the shield until that time. (Courtesy of E. Irving and the Royal Society.)**

The earth now rotates about 365¼ times during the completion of its orbit of the sun, but evidence is accumulating to show that it did not always spin at this rate and that the number of days in the year has, therefore, changed in geological time. Lengthening of the day at the rate of approximately two seconds in 100,000 years has been calculated by astronomers using such data as the historical records of solar eclipses.

Paleontological evidence suggests that similar changes have taken place for much of geological time. The rate at which some invertebrates secrete calcium carbonate for their skeletons and shells varies cyclically with the seasons, phases of the moon, days, and tides. Longitudinal sections through the shells of clams (Fig. 7–6) show a complex banding that is caused by the interaction of these cycles and by biological rhythms. Only a few invertebrate shells show all these cycles and many show none at all. The outer covering of some modern corals shows fine ridges of daily growth and coarser banding formed annually. On the average there are about 360 bands in each annual increment because most, but not all, the days in the year are recorded by a band of calcification. When the banding on fossil corals is examined, they show greater numbers of presumedly daily bands between the presumedly yearly increments. Corals from the Carboniferous Period show 385 to 390 growth lines per annum, and some from the Devonian Period range between 385 and 410 lines per annum (Fig. 7–7). Specimens that preserve annual and daily increments are rare and the reliability of many as recorders of the cycles is open to question. Much work must be done before precise determinations of the length of the day can be made for all the periods. However, preliminary results suggest that the days of Cambrian time may have been only 21½ hours long and that there were about 410 of them in a year. Unfortunately, criteria for extending the curve into pre-Palezoic time are missing, due to the absence of carbonate secreting fossils. Attempts to use stromatolites for this purpose have met with little success.

The change in the period of rotation of the earth has been ascribed to the influence of the moon in producing the braking effect of tides acting in shallow seas, or to internal changes within the earth, such as the growth of the core or the expansion of its radius, both of which would change its moment of inertia. An expansion of the earth's radius by about 0.6 mm per year has been claimed to be sufficient to explain the separation of the continental plates (if the expansion took place only in oceanic sectors) and also changes in the rate of rotation. However, the geometry of the movement of the continents is such that the expansion hypothesis is not successful in explaining the movement of crustal plates and is unlikely to be the main cause of the lengthening of the day. If the tidal braking effect is an important one, then it follows that the moon must be receding from the earth and in the past occupied a position closer to it. The consequences of all these interactions have not yet been resolved, but the possibility that the moon may have been considerably closer to the earth and, therefore, raised

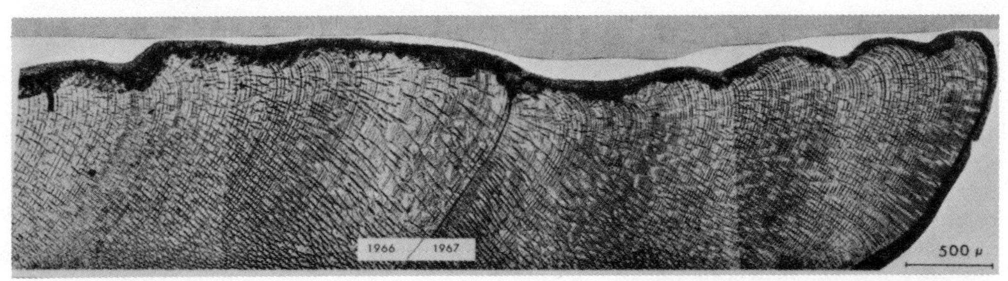

FIGURE 7–6 **Bi-daily rhythmic growth banding in the modern clam *Mercenaria*, × 56. (Courtesy of G. Panella and C. MacClintock and the *Journal of Paleontology*.)**

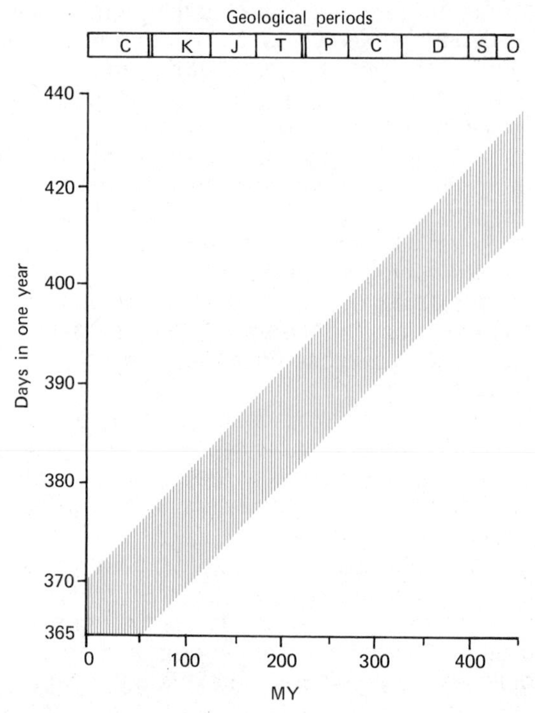

FIGURE 7–7 **General trend of the lengthening of the day in geological time determined by counting daily and annual banding on hard parts of fossils. K = Cretaceous.**

greater tides in the past must be considered by the historical geologist. The stratigraphic record shows no evidence that within the last billion years the tides have been more extreme than at present.

TEMPERATURE

Heat is received at the earth's surface by radiation from the sun and from the interior of the earth. The insulating and filtering effects of the atmospheric water vapor, ozone, and carbon dioxide are important in determining the surface temperature of the earth. Unfortunately, knowledge of the long-term variations in geothermal and solar radiation is sketchy and reconstruction of both the early atmosphere and surface temperature of the earth is speculative. However, geological evidence can be examined for indications of major systematic changes in the earth's surface temperatures.

In the introductory part of this chapter the probability was discussed that the heat reaching the surface from the interior in the earth's early history was at least half again as great as today. That water was present on the earth's surface 3.8 billion years ago, and, therefore, its temperature was between 0° and 100° Celsius, is established by sedimentary rocks in early Archean successions. Since that time the existence of water-laid sedimentary rocks in the geological record is testimony that the earth's surface temperature has always been in this range, except locally where glaciation has taken place. The major decrease in the geothermal heat flow is likely to have taken place between 4.5 and 4 billion years ago, and since that time the rate at which radioactive isotopes became extinct has been greatly slowed because only isotopes with very slow decay rates remain.

At present the amount of energy received by the earth's outer atmosphere from the sun varies very little from year to year. The sunspot cycle has a very small effect on solar output. Whether such constancy of solar emission can be extrapolated backward in time for hundreds of millions of years depends on the model for the sun that is adopted by the extrapolator. A recent model of the sun's interior implies that its radiation was only 70 per cent of its present value 4.5 billion years ago and has increased steadily since. The low value of solar emission postulated for this model is not acceptable in the light of the evidence for liquid water on the early earth unless surface temperatures were maintained by increased geothermal heat flow. Astronomers have recently re-examined this model of the sun's interior because it fails to explain why so few **neutrinos** (a fundamental subatomic particle) from the sun reach the earth. Geological evidence supports a solar model giving essentially constant radiation for the last 4 billion years.

The effect of solar radiation on climate and temperature is influenced by the "greenhouse effect." Shorter wavelengths of the solar emission penetrate the water vapor in the atmosphere and heat the surface. The heated surface radiates the energy at longer wavelengths to which water vapor is not transparent. The re-radiated heat is "trapped" near the surface where it heats the lower atmosphere. Lower solar emission would evaporate less water from the ocean, decreasing the water vapor in the atmosphere, lessening the greenhouse effect, and lowering the surface temperature by a factor far greater than that caused by the decrease in radiation alone. In the same way, anything that increases the amount of water vapor in the atmosphere, such as an increase in the area of the oceans, should increase the greenhouse effect.

The distribution of solar heat to various parts of the earth depends on the tilt of the axis of rotation to the plane of the earth's orbit. The tilt causes the seasons. The axis wobbles, or precesses, regularly in space, describing a cone with a 3° apex once every 21,000 years. M. Milankovitch suggested that the cyclic variations caused by the precession and the eccentricity of the earth's orbit in the radiation received by the polar regions might be sufficient to start and stop periods of glaciation. The hypothesis is still one of considerable controversy.

The character of some sedimentary rocks is climatically controlled; the deposition of all sedimentary rocks is influenced by climate (see Chapter 4). Desert sandstones and saline deposits are evidence of aridity; coal implies the presence of humid swamps in a warm climate. At certain times in earth history such deposits have been more abundant than at others, for example, coal in the Carboniferous and evaporites in the Permian periods. However, the existence on earth now of a range of climates that produces ice caps, hot deserts, and humid swamps suggests that these deposits do not reflect changes in the total temperature of the earth but merely in the distribution of solar and geothermal energy at its surface.

The widespread occurence of tillites and associated sediments produced by glaciation has been considered to be evidence of systematic change in the solar radiation. Four periods of glaciations within the last 1 billion years have been recognized:

1. The present one which began about 10 million years ago;
2. The late Paleozoic glaciation in the Southern Hemisphere (Chapter 2) about 250 million years ago;
3. The Ordovician–Silurian glaciation in northern Africa (Chapter 17) about 450 million years ago;
4. The late Proterozoic glaciation of about 850 million years ago.

Although glacial deposits are present in still older rocks, they are of isolated occurrence, difficult to correlate, and, therefore, of doubtful worldwide significance. Some have claimed to see within this sequence a cycle of 200 to 250 million years by recognizing an Ediacaran glaciation about 600 million years ago. The correspondence in frequency of this cycle and the period of rotation of the Milky Way galaxy has suggested that periodic changes in the sun's luminosity, or of the path between the sun and the earth, are correlated with the passage of the solar system around the galaxy. The galactic cycle theory is highly speculative and glaciations are more likely to have been caused by a combination of geographic and climatological conditions that depend on the relative positions of the continental masses and the geographic poles as the lithosphere plates move. The causes of the Pleistocene ice age are more fully discussed in Chapter 21.

The temperature of the seas of the past has been measured by determining the proportions of two oxygen isotopes in fossil shells. The ratio of oxygen 18 to oxygen 16 in shells is sensitive to the temperature of the water in which the shell was secreted, being higher in colder water. Providing there has been no mineralogical change in the shell in preservation, the information about the temperature of the water during the time of secretion of the shell can be extracted from a fossil shell by measuring the isotope ratio. Unfortunately, most fossil shells have been changed by replacement, infiltration, encrustation, or selective solution and are not suitable for this kind of analysis. Figure 7–8 is an example of results obtained by the analysis of Cenozoic planktonic foraminifers. The changes in water temperature of the sea of this era are clearly shown.

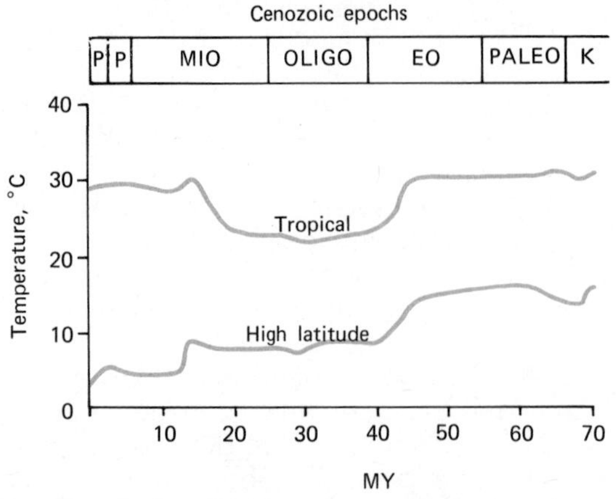

FIGURE 7–8 **Changes in the temperature of the oceans determined by oxygen isotope analyses of fossil foraminifers from deep-sea cores. The upper line represents analyses of planktonic species believed to be representative of surface tropical waters. The lower line represents analyses of benthonic species believed to be representative of cold, high latitude waters. (Courtesy of S.M. Savin and the Geological Society of America.)**

In summary, most geological evidence favors relative stability in solar and geothermal heat sources after an initial interval when the crust was thin and cooling from the hot stage that followed the birth of the planet. Climatic change since Archean time appears to have been caused by modification in the distribution of this energy by atmospheric conditions and the arrangement of the continents and oceans.

SUMMARY

Although the concept of uniformity proposed nearly 200 years ago by James Hutton is useful in the interpretation of the geological past, it cannot be strictly valid. Some of the conditions and processes of the early part of earth history have no counterparts today and did not recur in the history of the planet.

The present composition of the atmosphere and oceans is best explained by contributions from volcanic sources, which were much more active in the early stages of earth history, as the planet was degassing, than today. The early atmosphere was oxygen-free and rich in carbon dioxide. Carbon dioxide was converted to oxygen by photosynthesis and removed from the atmosphere in the formation of limestone.

The magnetic field of the earth has reversed many times in earth history. Certain time intervals were characterized by frequent reversals, others by relative stability. The position of the continents relative to the poles has also changed in geological time, resulting in progressive movement since early Paleozoic time of the equator from a position across Alaska from the Gulf of Alaska to the Arctic Ocean in a clockwise sweep across North America to its present position. Differences in the apparent polar wandering paths of lithosphere plates reflect the relative motion of these plates.

Daily and yearly growth bands on fossils indicate that in early Paleozoic time the rotation of the earth was faster, the day was considerably shorter, and the year included over 400 days.

Prior to 4 billion years ago geothermal heat from short-lived radio-isotopes made the earth's surface much hotter than at present, but sedimentary rocks 3.8 billion years old indicate that the surface temperature has been between 0° and 100°C (the liquid range of water) since that time. Changes in atmospheric composition may have affected the heat contributed by the sun to the earth's surface. Recurrent glaciations have been postulated to reflect an event in earth history with a period of 200 to 250 million years, but are more likely to be caused by a combination of topography and atmospheric factors.

QUESTIONS

1. Write a short essay on the origin, history, and significance of carbon dioxide in the earth's atmosphere.
2. Trace the stages in the evolution of the atmosphere from the protoplanet stage to the present day.
3. Summarize the various methods discussed in this chapter and in Chapters 1 and 2 by which the continental positions of the past and the subsequent movement of the lithosphere plates can be reconstructed.
4. If the changes in the positions of the continents in geological time is explained by expansion of the earth in the oceanic sectors so that the radius increased by 0.6 mm per year, calculate the value of gravity in mid-Paleozoic time relative to that at present. What evidence of a different value of gravity would you expect to find in the rocks and fossils of that time?

SUGGESTIONS FOR FURTHER READING

HUBBERT, M. K. "Critique of the principle of Uniformity." *Geological Society of America Special Paper 89,* 3–33, 1967.

MCELHINNY, M. W., *Paleomagnetism and Plate Tectonics.* Cambridge: Cambridge University Press, 1973.

NEWMAN, M. J. and ROOD, R. T. "Implications of solar evolution for the earth's early atmosphere." *Science,* 198: 1035–1037, 1977.

ROSENBERG, G. D. and RUNCORN, S. K. eds. *Growth Rhythms and the History of the Earth's Rotation.* New York, John Wiley & Sons, 1975.

SCRUTON, C. and HIPKIN, R. G. "Long term changes in the rate of rotation of the earth." *Earth Science Reviews,* 9: 259–274, 1973.

STRANGWAY, D. W. *The History of the Earth's Magnetic Field.* New York: McGraw Hill Inc., 1970.

TARLING, D. H. *Principles and Applications of Palaeomagnetism.* Cambridge: Cambridge University Press, 1971.

VERHOOGAN, J., *et al., The Earth, Chapter 14* (*Some Aspects of the Chemical Evolution of the Earth*). New York: Holt, Rinehart & Winston, 1971.

CHAPTER EIGHT

Progressive changes in the environment of the earth's surface have been caused by the appearance of life and its evolution toward greater complexity. Throughout the remainder of this book, repeated reference will be made to the interaction of life and environment through geological time. In this chapter, we are concerned with the beginning and early history of life and the mechanism of evolution.

ORIGIN OF LIFE

Discussion of the details of the origin of life on earth has become primarily a topic of biochemistry rather than of geology. The geologist, however, may contribute to the understanding of this event by supplying geochemical information about the nature of the environment in which it took place, a framework of time within which it must have taken place, and evidence for early life supplied by the fossil record. The subject of the origin of life is a highly speculative one, and involves the construction of models or scenarios of what might have happened on the basis of very limited evidence. In these circumstances, theories change rapidly as new information becomes available. Although the multiplicity of theories may suggest that most of the ideas have little permanent value, science progresses in areas of rapidly expanding knowledge by progressively limiting the possible explanations until an explanation that is generally acceptable is established.

The Environment

For the first one-half billion years of the earth's history the crust was apparently too hot for the accumulation of water in its liquid phase. Sediments as much as 3.8 billion years old indicate the minimum date for initial formation of the oceans, with a water temperature within the range where living organisms could survive. The first evidence of primitive life is found in sediments only slightly younger.

On the basis of present-day forms, biologists recognize a number of attributes which are common to all living things and are considered necessary for life: (1) organization, (2) constancy of form, (3) motion, (4) response to stimulus, (5) metabolism, (6) growth, (7) reproduction, (8) adaptation. These attributes are all interrelated, and while growth and movement can never be observed in fossils, they may be assumed from the pattern of organization and the presence of particular chemical substances in the fossils and surrounding sediments.

Hydrogen, oxygen, carbon, nitrogen,

LIFE AND EVOLUTION

phosphorus, and sulfur are among the basic constituents of living systems. These elements are combined in molecules that can be classified into six groups: (1) water acts as a medium in which chemical reactions take place, as well as a source of oxygen and hydrogen; (2) carbohydrates are the source of the energy used by organisms; (3) fats, oils, and lipids are used for storage of energy within organisms; (4) proteins (in turn composed of amino acids) serve as structural elements and as enzymes to facilitate chemical reactions; (5) nucleic acids direct the synthesis of proteins and the reproductive processes; (6) adenosine phosphates provide means of energy storage and transfer within cells.

The initial steps in the evolution of life can be reconstructed on the basis of the chemical and physical environment on the surface of the earth and the lower atmosphere in the early pre-Paleozoic. In addition to water, in both liquid and gaseous phases, the early atmosphere is thought to have consisted of hydrogen, carbon dioxide, nitrogen, and lesser amounts of ammonia and methane. The chemistry of the most primitive rocks indicates that gaseous oxygen was not present in either the early atmosphere or in the water on the surface of the earth at this stage.

The lack of molecular oxygen and ozone in the atmosphere would have allowed a heavy influx of ultraviolet radiation onto the surface of the earth. Locally, additional energy sources may have been available in the form of heat from lava or hot springs, and periodical electrical discharges from atmospheric storms. Experiments using these forms of energy in association with the gases expected in the atmosphere of the primitive earth have produced significant quantities of many of the basic molecules of which living organisms are composed, including a wide range of sugars and the essential amino acids.

The next stage, the integration of the complex molecules into systems capable of controlled metabolism and reproduction, was a giant step, the causes of which are not yet understood. Presumably, nonbiologically produced carbon compounds in considerable variety accumulated to fairly high concentrations in what has been termed a primordial "organic soup." More complex combinations of molecules would develop as a result of polymerization and other factors controlled simply by the relative concentration of the various constituents and their chemical affinities. Some large molecules would tend to assimilate small organic components, exhibiting growth of a very crude manner, controlled only by external factors such as the availability of particular smaller molecules. The molecules that could grow most rapidly and assimilate the largest range of smaller molecules would survive, while other, smaller molecules would be consumed. Utilization of particular molecules, such as lipids and proteins, might enable these crude metabolic aggregates to form thin walls or coats, partially separating them from the environment, and so allowing them to control the nature and concentration of the substances with which they could interact chemically.

Despite the complexity of these events, recent discoveries indicate that they occurred relatively quickly. Fossil evidence demonstrates the appearance of organisms at the level of bacteria and blue-green algae within 500 million years after the initial solidification of the earth's crust. There is some indirect evidence that a major division between typical bacteria and blue-green algae on one hand, and a strikingly more primitive group of anaerobic forms, the methanogenes, had already occurred prior to 3.4 billion years ago.

The Earliest Fossils

Sediments approximately 3.5 billion years old, from Swaziland, contain simple carbon compounds in abundance, as do other sediments of this age, and some consistently organized structures that have been attributed to living organisms. Further evidence of early life can be seen in deposits 3.4 billion years old. Very fine-grained cherts of this age, from the Figtree Group in South Africa, contain remains attributed to bacteria (Fig. 8–1). The remains, described by Elso Barghoorn, are tiny rods, 0.5 to 0.7 micrometers in length and from 0.2 to 0.3 micrometers in diameter. In cross-section the cell wall, in some cases, can be seen to consist of an inner and outer layer, totalling approximately 0.015 micrometers in thickness. This size and configuration is so close to that seen in primitive living bacteria that it is reasonable to conclude that the form over 3 billion years old (named *Eobacterium*) had similar metabolic and reproductive attributes as well. Associated with these cherts are trace quantities of complex organic compounds that could only have been formed by living systems. Despite this evidence several paleontologists have expressed skepticism that these objects are fossils, and maintain that the first recognizable cells are preserved in the Gunflint Chert, which is approximately 2 billion years old.

The living bacteria with the general structure of *Eobacterium* obtain energy from simple organic compounds present in the water and soil. They do not make use of atmospheric oxygen, and, in fact, can only live in environments in which oxygen is absent. The inability to utilize atmospheric oxygen strictly limits the metabolic pathways of synthesis and energy transfer available to these bacteria, and presumably reflects their retention of patterns that were first established under the primitive, oxygen-free atmosphere.

Reproduction and Variation

In addition to the metabolic limitation of living bacteria, and almost certainly the early Archean forms as well, are inherent limitations of their genetic apparatus that were presumably major factors in the slow rate of evolutionary progress in the first 2 billion years of life.

All organisms must have a means of governing all aspects of their metabolism and structure. This regulation is provided by **deoxyribonucleic acid,** a complex molecule composed of a sequence of subunits called **nucleotides.** A nucleotide contains a phosphoric acid molecule joined to a five-carbon sugar (deoxyribose) which, in turn, is joined to either adenine, thymine, guanine, or cytosine (Fig. 8–2). The linear sequence of these nucleotide subunits in deoxyribonucleic acid controls the synthesis of proteins, and thereby establishes the structure of the cell and determines its metabolic activity. In the simplest organisms, deoxyribonucleic acid serves both to control the structure of the cell and as a basis for its reproduction. The sequence of nucleotide subunits has,

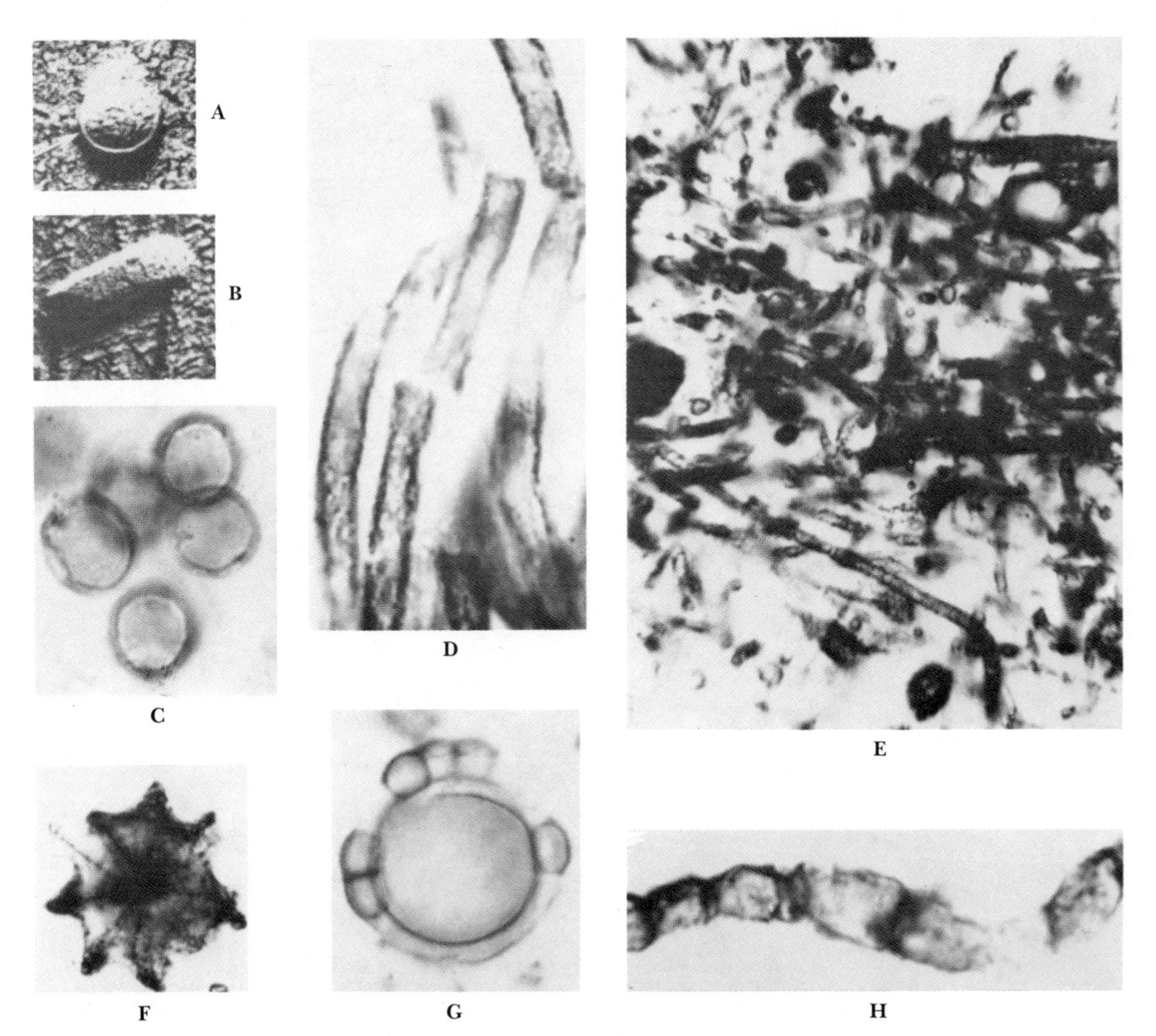

FIGURE 8–1 **Archean and Proterozoic fossils. A, B. *Eobacterium* from the Figtree Group, South Africa, 3.4 billion years old (approximate magnification ×5000). C, D. Late Proterozoic algal filaments and single algal cells from Bitter Springs Limestone, Northern Territory, Australia, 700–900 million years old (approximate magnification ×2500). E. Tangled algal filaments and spores (approximate magnification ×800). F, G. Organisms of unknown affinity (approximate magnification ×2400). H. Single algal strand (approximate magnification ×2400). E–H from Gunflint Chert, Ontario, 1.9 billion years old. (Courtesy of E. S. Barghoorn.)**

therefore, been called the "genetic code."

One of the functions of the cell is the orderly duplication of the nucleotide sequence in the deoxyribonucleic acid so that the genetic code can be transmitted to new cells. The four types of nucleotides readily form complementary pairs: those including cytosine paired with those containing guanine; those with adenine paired with thymine. In deoxyribonucleic acid, the paired nucleotides are arranged in a double helix, one side of which is complementary to the other (Fig. 8–3); for every guanine nucleotide in one strand, there is a cytosine in the

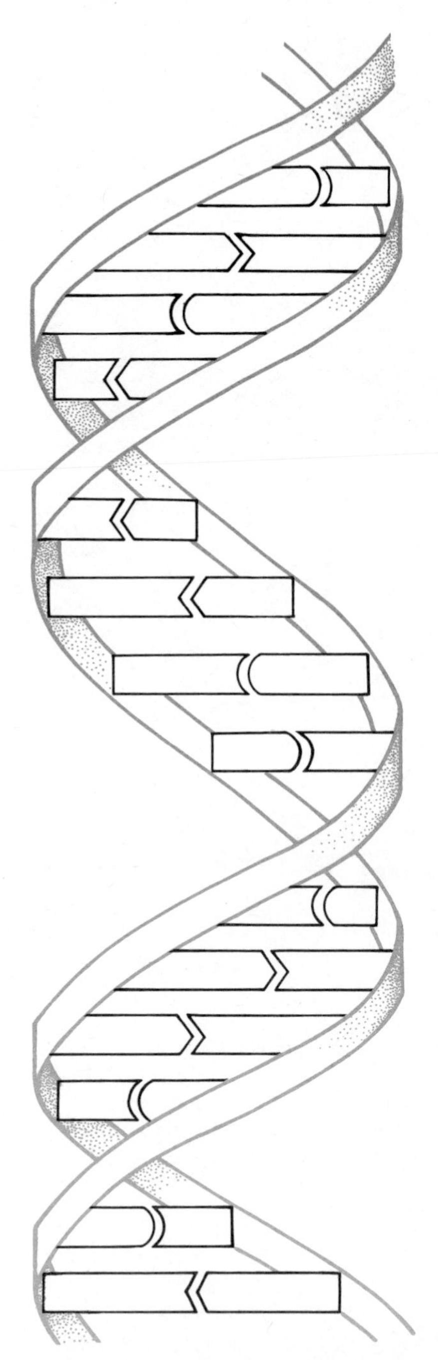

FIGURE 8–2 **Schematic drawing showing the double-stranded helix of DNA (deoxyribonucleic acid). The ribbons represent the external backbone of phosphate and sugars. The horizontal bars represent the base pairs: adenine–thymine; guanine–cytosine.**

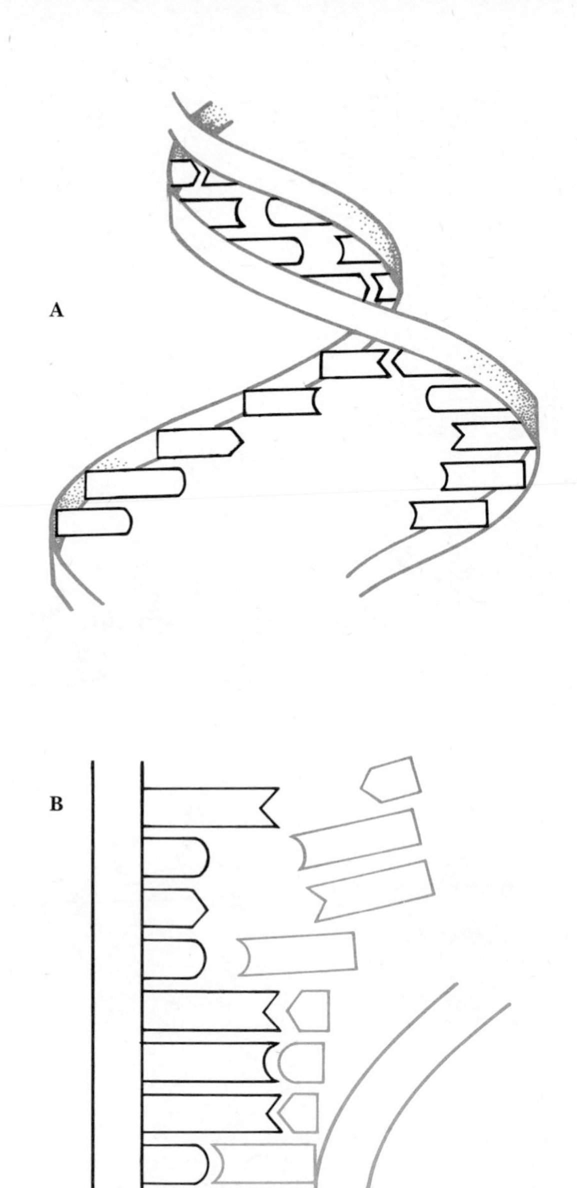

FIGURE 8–3 **Replication of DNA. A. The two strands of the double helix unwind. B. Both strands (only one of which is illustrated) act as templates for the assembly of new, complimentary strands. Newly synthesized bases become associated according to their appropriate pairs: adenine to thymine; thymine to adenine; guanine to cytocine; and cytocine to guanine. C. Newly synthesized phosphate**

and sugar form the backbone for the new strand of DNA. Such a path of replication insures genetic constancy from generation to generation. Anything that interferes with this orderly replication can be termed a mutation.

other, and for every thymine, an adenine. At the time of cell division, each strand separates like a zipper being opened and each, by attracting nucleotides, forms a replica so that the double helix is duplicated exactly. The two complementary strands in the double helix of deoxyribonucleic acid provide only a single copy of the genetic information. This system, if without flaws, would generate identical organisms as long as the environment was such that they could survive. The measure of its perfection can be seen in the persistence of primitive bacteria for over 3 billion years.

Mutation

Despite its permanence the bacterial way of life allows for only a limited utilization of the environment. One of the most important aspects of life is the ability to change. This change need not be immediately advantageous; indeed, the basis for hereditary variability is the perpetuation of errors (or mutations) that are largely random. During the duplication of the deoxyribonucleic acid one or more of the nucleotides may be lost, doubled, or its position changed. These and other changes may, depending on their specific position in the nucleotide sequence, result in greater or lesser change in the proteins produced within the cell, and so will affect the potential for survival of the organism (Fig. 8–4).

As long as the environment remains stable, almost any change in the nucleotide sequence that results in a modification of the proteins will be detrimental to the individual, lessening its ability to utilize the available food source or reducing its reproductive efficiency. Organisms with such errors in the genetic material will either die or have reduced potential for growth and reproduction. In either case, the altered genetic material will usually not be reproduced for more than a few generations. Very rarely, the haphazard changes will be beneficial rather than deleterious, enabling the individual to utilize the environment more efficiently. As a result, more and more individuals in each generation perpetuate the new genetic configuration. The potential for change in the genetic material, even if usually deleterious, is a necessary characteristic of all organisms since it allows them to adjust to changes in the environment.

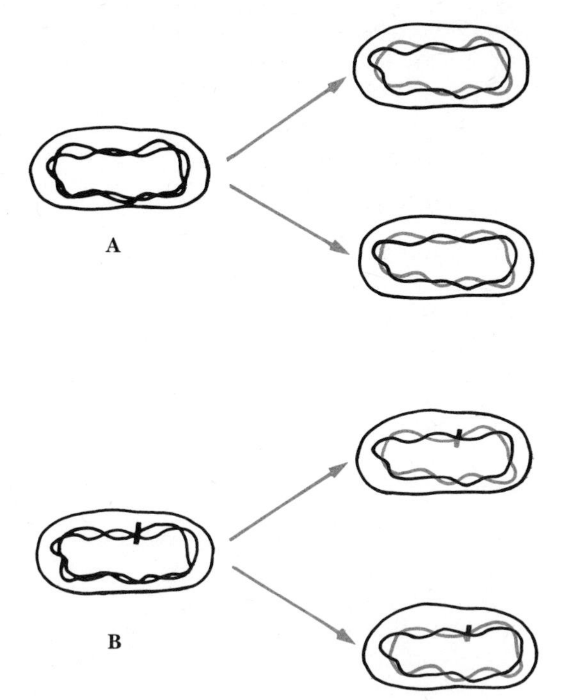

FIGURE 8–4 **A. Reproduction by binary fission. No regular device for genetic interchange between individuals. Without mutations, progeny resemble parents in all traits. Genetic material is not organized into chromosomes, and is circular rather than linear in form. B. Mutation (indicated by the cross bar) affecting both strands of DNA affect all progeny. If the trait is disadvantageous, all will be selected against. Nearly all new traits are eliminated in a single generation. Colored strand is new DNA, synthesized at time of reproduction.**

Such changes were primarily of a physical or chemical nature at first, but as various habitats came to be occupied by living forms, they themselves changed the characteristics of the environment.

Photosynthesis

Modification of such primitive organisms remained very slow, however. One of the major factors limiting the evolution of bacteria and related organisms is that they possess only a single, functional copy of the genetic material. Rarely can it depart from the standard pattern without leading to the death of the organism. Because of this, only the organisms that evolved methods of keeping changes at a minimum, or repairing damage or mistakes, survived. Nevertheless, very important changes did occur at this level of organization. The most far-reaching of these was **photosynthesis**, the ability to bond hydrogen with carbon dioxide to produce simple sugars. With the perfection of this reaction, these primitive forms of life were capable of manufacturing their own food rather than depending on the inorganic production of carbon compounds. Photosynthesis is exhibited by some primitive living bacteria, as well as by algae and higher plants (Fig. 8–5). Presumably, the very early photosynthetic organisms used hydrogen either in the form of hydrogen gas (H_2) or in hydrogen sulfide (H_2S), and were restricted to parts of the environment totally lacking oxygen. Water, the most plentiful source of hydrogen, was almost certainly not an available source initially, because the use of hydrogen would have freed extremely active oxygen ions. The great chemical activity of the oxygen could destroy or render inactive most of the chemical compounds of the primitive organisms. The scarcity of gaseous hydrogen, hydrogen sulfide, and similar compounds placed a premium on organisms that could utilize the ubiquitous water molecule as a source of hydrogen. Some organisms may have evolved the ability to react the released oxygen with iron at the same time that they incorporated the hydrogen into sugars. Ferrous salts of iron, presumably, were

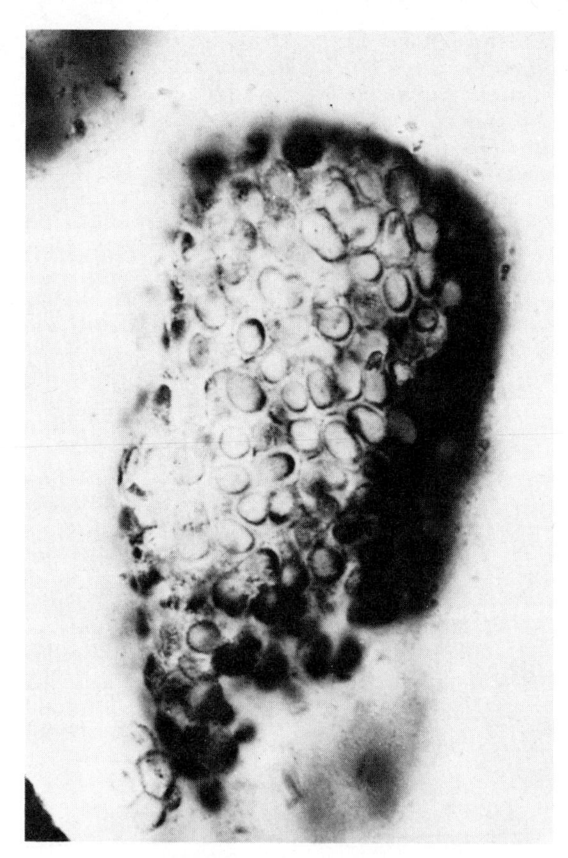

FIGURE 8–5 **Cells of blue-green algae of Aphebian age from the Belcher Islands, Hudson Bay. These cells are practically identical to those found in modern pustulose algal mats. They are preserved in chert. The mass of cells is about 0.05 mm across. (Courtesy of H. Hofmann and the *American Journal of Science*.)**

important components in water in the early stages of the earth's history. At approximately the time at which fossil evidence suggests that photosynthesis, utilizing water as a hydrogen source, was evolving, the geological record shows wide-scale deposits of a particular type of rock—banded iron formation (Fig. 8–6). Today, such rock serves as an important source of iron ore mined extensively in the United States and Canada (Chapter 9). The banding is produced by thin, alternating layers of chert and iron silicate or carbonate, and ferric and ferrous iron, the composition of which

FIGURE 8–6 **Polished specimen of iron formation. The dark bands are hematite of the specular variety; the lighter bands are jasper, a form of chert rich in iron oxides. (Photograph by C. W. Stearn.)**

may have been produced by the following chemical reactions:

$4FeO + O_2 \rightarrow 2Fe_2O_3$ and

$$6FeO + O_2 \rightarrow 2Fe_3O_4$$

For both reactions, the free oxygen may have been produced by primitive, photosynthetic organisms. Periodic variation in the availability of iron in a particular area controlled which of the reactions was realized. Local depletion of iron would have resulted in the reduction of the rate of photosynthesis, or even the death of the organism due to the build-up of toxic oxygen. Alternating population cycles could thus have been responsible for the alternation observed in the banded formation. With some exceptions, the world's banded ironstone appears to have been deposited between 2.8 and 1.8 billion years ago.

The most common fossils in pre-Paleozoic rocks, the **stromatolites**, may have evolved oxygen-mediating enzymes in order to protect their cells from molecular oxygen. Stromatolites consist of thin layers of sedimentary calcium carbonate grains built up concentrically into masses that may reach tens of meters in dimensions. Stromatolites are uncommon in modern seas, but can be observed forming now in very saline environments. Filaments of blue-green algae form a gelatinous mat on the surface of the sediment. Fine particles of calcium carbonate settle out of the water and are trapped on the filaments of the mat, forming a lamina; the algae then grow through the lamina, forming another slimy layer ready to receive more sediment. In some stromatolites, the algae actually appear to secrete the calcium carbonate within the cells, but most stromatolites show only a lamination and no trace of the organism that controlled their formation. The oldest stromatolites in the Pongola Group of South Africa are about 3 billion years old, but may not have been formed by photosynthetic organisms. Degradation products of chlorophyll found in ancient sediments suggest that the production of oxygen by photosynthesis is likely to have begun 3.5 billion years ago, but for many hundreds of millions of years it was probably a rare gas.

Banded iron formation is rare in rocks younger than about 1.8 billion years. Iron deposits in more recent rocks are in the form of hematite and appear to indicate the presence of free atmospheric oxygen. This change can be interpreted as signaling the evolution of a biota resistant to atmospheric oxygen. The presence of oxygen in the atmosphere would have enabled organisms to develop more efficient metabolic processes by which they could derive energy from the complete oxidation of sugar to carbon dioxide and water:

$$C_6H_{12}O_6 + 6O_2 \rightarrow 6CO_2 + 6H_2O$$

This reaction produces approximately twelve times the energy for the equivalent amount of sugar compared with the fermentation breakdown of sugar practiced by anaerobic organisms. Complete oxidation of sugar provides much more varied metabolic pathways for synthesis within the cell. This reaction characterizes all forms of life above the level of the bacteria and blue-green algae. These higher forms are collectively termed **eukaryotes**, in contrast to the more primitive **prokaryotes** (bacteria and blue-green algae).

Only where the level of oxygen in the water is at least 1 per cent of the present atmospheric level is this process of respiration, as the oxidation of sugar can be called, efficient. This level is sometimes spoken of as the **Pasteur Point.** Such a concentration of oxygen was probably reached in the seas first in "oases" around densely spaced areas of photosynthesizing plants, such as in masses of stromatolites (Fig. 8–7). The time at which it reached the 1 per cent level, generally, in the oceans is a matter of discussion among paleontologists. L. V. Berkner and L. C. Marshall suggested that the Pasteur Point was not reached until the beginning of the Cambrian Period, when abundant fossils first appear; Preston Cloud believes that it was reached between 1.2 and 0.6 billion years ago; and William Schopf as early as 1.9 billion years ago. Schopf bases his conclusions on the appearance of eukaryotic organisms and on the oxidized nature of rocks of this age. At present, we can conclude that the oxygen level and efficacy of the ozone shield increased during Archean and Proterozoic

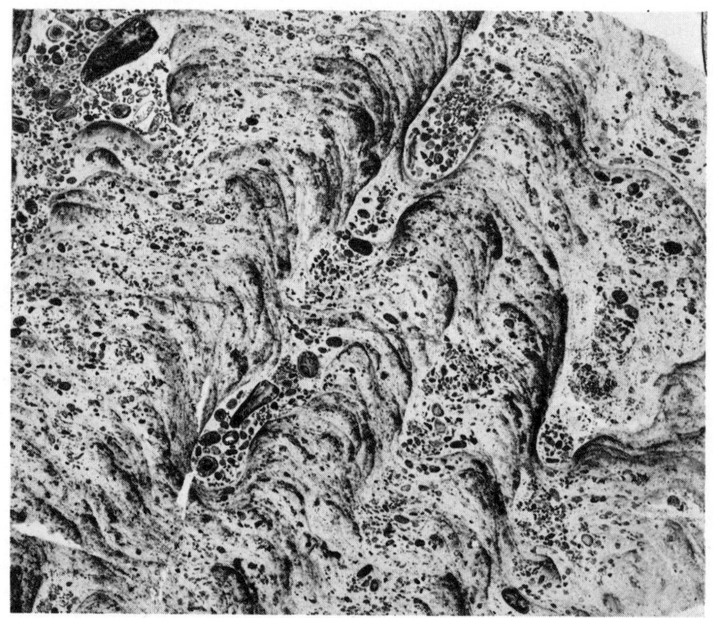

FIGURE 8–7 **Thin-section of a stromatolite bed from the Gunflint Formation, Whitefish River, Ontario. The dark layers are crowded with microfossils that may have grown in an oxygen oasis. (Courtesy of H. Hofmann, Geological Survey of Canada.)**

time through the activities of plants, but we cannot place a strict time scale on the attainment of various levels. We shall return to the discussion of atmospheric evolution in our consideration of the environment of the pre-Paleozoic eras and the origin of the Cambrian fauna.

Evolutionary Significance of the Eukaryotes

The cellular organization of the eukaryote cell is more sophisticated than that of the prokaryotes. It includes a true nucleus distinguished from the rest of the cell contents by a membrane and the localization of particular metabolic activities in structures called **organelles.** The eukaryotic cell is further characterized by the presence of two copies of the genetic code rather than the single functional copy of the prokaryotes.

In the prokaryotes the genetic material is dispersed throughout the cell and divides when the cell divides in reproduction. In the eukaryotes it is organized into linear structures called **chromosomes** (Fig. 8–8). The series of nucleic acid units that determines a specific inheritable feature of an organism is called a **gene.** The genes occupy a specific position along the chromosome in the nucleus of the cell. At some stage in the life history, each chromosome is paired, and the genetic material is represented by two homologous copies. If the genes that determine a particular character or trait, such as whether the human blood group will be of type A or B, are the same in these two homologous chromosomes, the organism is said to be **homozygous** for that trait. If the homologous position on the two chromosomes is occupied by different expressions of the same trait, that is, one for A type and the other for B, the organism is **heterozygous.**

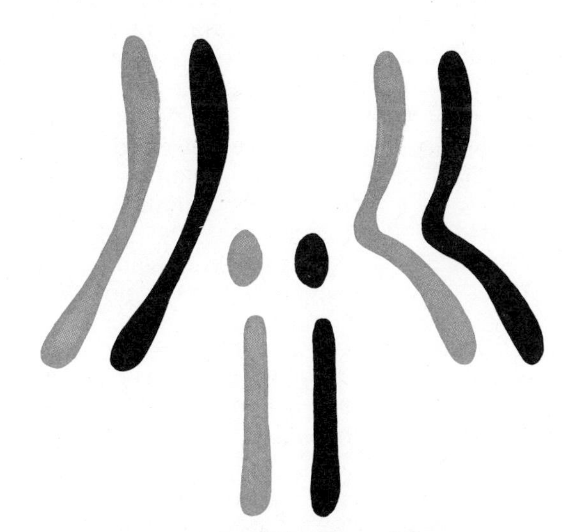

FIGURE 8–8 **Stylized drawing of the chromosomes of the fruit fly *Drosophila melanogaster*. Drawing emphasizes the paired nature of the chromosomes in the body cells of eukaryotic organisms. Each chromosome carries a double helix of DNA. Deleterious mutations in one chromosome may not affect the organism if the other chromosome is not similarly damaged.**

In sexual reproduction, which is characteristic of eukaryotes, the chromosomes are first duplicated and then separate in the formation of gametes (Fig. 8–9). The paired number is reestablished at the time of fertilization. Thus, a new combination of genes is produced and the new organism may be homozygous for a trait for which its ancestor was heterozygous, or vice versa. In the prokaryotes true sexual reproduction is unknown. Bacteria may come together to transfer genetic material, but blue-green algae are known only to split in reproducing.

The presence of two copies forms the basis for the pattern of evolution we see in all higher organisms. With two copies of the genetic information, the potential for evolutionary change is almost infinitely increased. In contrast to the condition in prokaryotes, the two copies of the genetic information permit a large degree of variability without jeopardizing the survival of the organism. Change can occur in one copy of the genetic material without altering the functioning of the mature organism. Frequently, expression of a trait will occur only as a result of mating between two organisms with the same mutation. If, as is usually the case, the newly generated pattern is less viable than the parental condition, the individual that expresses it will die

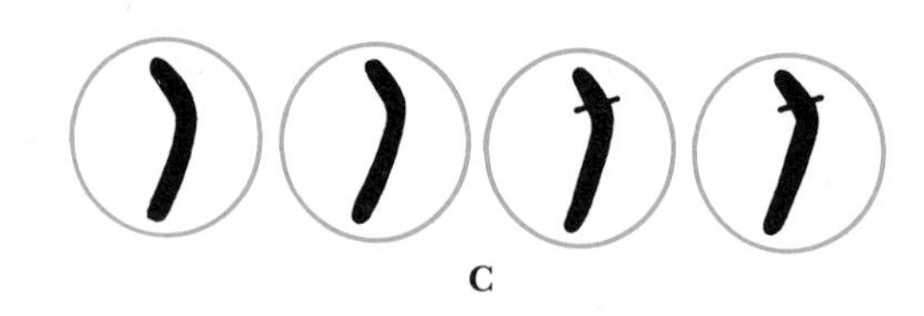

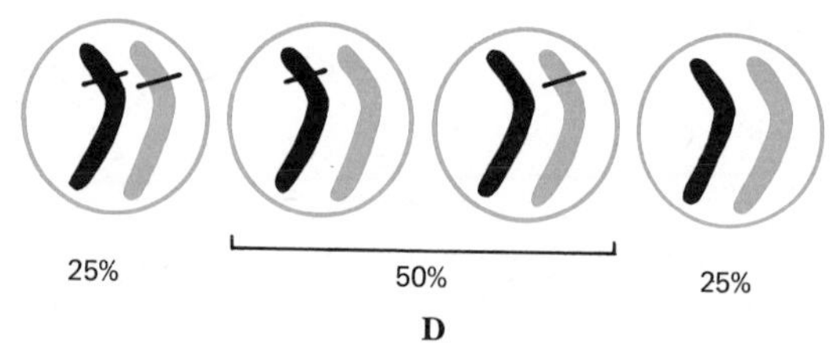

FIGURE 8–9 **Eukaryotic reproduction. A. Nucleus of a cell with only a single pair of chromosomes. Chromosomes become closely associated prior to replication. Cross bar indicates position of mutation affecting one chromosome. B. Duplication of chromosomes occurs as a preliminary step in the formation of gametes. C. Gametes are formed, each with a single chromosome. If the adult was heterozygous for a particular trait, half the gametes will have one allele, and half the other. D. At the time of mating, the diploid chromosome number is re-established. If both parents are heterozygous for the same trait, the zygotes (fertilized eggs) will appear in the ratio of 25 per cent homozygous for one allele, 25 per cent homozygous for the other allele, and 50 per cent heterozygous. In eukaryotes, a trait may remain in the population even if lethal in the homozygous condition, as long as it is not deleterious in the heterozygous condition. Colors distinguish the chromosomes of the two parents.**

or will reproduce less successfully. Unlike the condition in prokaryotes, the trait will not be lost, but will be retained in individuals which only have a single copy. The changes can be retained until such time as they confer an advantage on the organism as the result of further genetic change, or when changes in the environment favor an altered organism.

In addition to the advantage of retaining a large number of genetic variations that can be utilized by the organism in meeting future challenges of a changing environment, the presence of a double copy of the genetic material makes possible interchanges between the two chromosomes at the time when the two copies are formed, prior to sexual reproduction. In higher organisms with thousands of bits of genetic information, the potential variation that can be provided by this interchange, or recombination, is practically limitless. In theory, the potential for variation in the offspring of a single mating of advanced eukaryotes is greater than the number of atoms in the universe. Such a potential for variation provides enormous evolutionary flexibility. Fossil evidence (Chapter 9) suggests that cells with eukaryotic structure had developed about 1.5 billion years ago.

Despite the early development of a modern genetic constitution, the next 800 million years of the earth's history provide almost no evidence of evolutionary progression above the level of unicellular organisms. Not until Ediacaran time does the fossil record give any evidence of complicated organisms heralding the arthropods, molluscs, and annelids of the Cambrian Period. The quantity of oxygen available in the water and atmosphere may have restricted the evolutionary potential of these forms. The paucity of soft-bodied organisms in the fossil record of the Paleozoic suggests that such forms may have been quite diverse and highly evolved in the Proterozoic, but the particular conditions of deposition that would have enabled their preservation were lacking.

EVOLUTION

The term "evolution" has been used in a general way to refer to progressive changes in any object, from the evolution of stars or the landscape of the earth to changes in organisms that are responsible for the complex array of plants and animals now living on the earth. In biology, evolution refers to the gradual modification of all organisms, from generation to generation.

Although the fossil record provides the most definite proof of the evolutionary modification of organisms, the mechanics of evolution are much more specifically evaluated on the basis of living forms that can be manipulated and studied in detail.

Adaptation

Evolutionary changes in organisms have long been recognized as adaptations to changing environments. Early theories that attempted to explain the mechanism of evolution assumed that the specific nature of the environment directly controlled the course of evolution. The lack of acceptance of various early theories of evolution, such as that of Lamarck, was in part due to the lack of any physical explanation of how the organism could modify itself in response to environmental requirements, and particularly how it could pass these changes on to subsequent generations. A more general reason for the rejection by the scientific community of various evolutionary theories of the 18th and early 19th centuries was adherence to religious and philosophical concepts that stressed the fixity of organisms and their creation, in the recent past, in essentially modern form. The scientific, if not the religious, objections to the evidence of evolution were soon swept away following the publication of Darwin's *The Origin of Species* in 1859.

Selection Theory

Darwin's greatest contribution was the marshalling of the evidence of comparative anatomy, zoogeography, and the breeding of plants and animals to demonstrate the evolutionary origin of all organisms. In addition, Darwin outlined a mechanism whereby organisms could respond to changes in the environment. Darwin recognized that variability was a characteristic of all adequately known organisms. He reasoned that the variability was related to the relative ability of an organism to cope with the environment. This could be most clearly seen in traits that directly affected survival and the ability to leave numerous progeny. Darwin then assumed that the variably expressed traits were passed on to the next generation, and that those that were most useful to the species were passed on to succeeding generations in greatest numbers. The possession of advantageous traits has been termed "fitness," and so the expression "survival of the fittest" has been applied to this process.

From observable changes in characteristics seen through the human agency of plant and animal breeding, Darwin saw that organisms could be progressively modified to a startling degree. He extrapolated these observations to the much greater time scale available during geological history, and reasoned that all plant and animal diversity could be accounted for by selection of naturally occurring variants in populations. It is this concept that forms the basis of the Darwinian selection theory.

Mutations and Variability

This theory of evolution was widely accepted by most biologists in the late 19th and early 20th centuries. Several factors must be considered in evaluating selection theory: (1) the nature and origin of variability; (2) the means of inheriting and maintaining variability; (3) the manner in which natural selection acts on the variability. At the time of Darwin's writing, none of these factors was adequately understood.

The question of the maintenance and inheritance of variability through many generations was one of the most troubling problems for Darwin. Ironically, the basis of the inheritance had been established by 1866, from Gregor Mendel's study of sweet peas. This work may have been known to Darwin, but its great significance was not appreciated until the early years of the 20th century. The most important concept of Mendel's work was that of particulate inheritance. All characteristics are controlled by factors, later termed "genes," that are maintained as discrete entities for generation after generation. For Mendel and the early geneticists, the genes remained largely theoretical, but they have since been more specifically characterized from studies in both genetics and biochemistry.

Within the last 20 years, through the development of molecular biology and biochemistry, the chemical nature of the genetic material has been determined. There are still relatively few cases in which the precise chemical basis for particular traits is known in its entirety, but the principle is well understood for at least some characteristics in a considerable range of organisms. The occurrence of variability in organisms is now understood to involve irregularities in the replication of the genetic information encoded in the nucleic acids. Such irregularities are called **mutations.**

Mutations, the source of variation, have already been considered in the discussion of the genetic apparatus of prokaryotes. A factor which should again be emphasized is their essentially random nature. Mutation rate has been found to be controlled by environmental factors, but no mechanism is known whereby mutations of greater selective value can be produced preferentially in any environment.

Many studies of both laboratory and wild populations have been carried out to determine the effectiveness of selection in controlling the frequency of particular genes. In order to consider the relative value of these forces, mutation rates and selection coefficients in various species must be recorded. Detailed studies of various organisms indicate a great range in mutation rates, not only from group to group, but among different traits within a particular species. Each gene in each type of organism has its own mutation rate, and even this is variable, depending on the remainder of the genetic composition and, within a limited extent, environmental factors as well. Nevertheless, average mutation rates for a wide range of organisms can be cited, and range from 1/50,000 to 1/200,000 per site per generation in higher (eukaryotic) organisms. Rates are considerably lower among prokaryotes. Such mutation rates must be assessed against the total number of mutations expected in higher organisms and the total number of sites capable of mutation (approximately 100,000). One or two of these sites might be expected to show new mutation in any one generation. On the basis of new mutations, it is clear that natural selection has little to act upon. New mutations, however, are only a very small proportion of the total variability residual in normal populations. It is now estimated that up to half of the genes in higher organisms normally exist in two or more forms in any population.

Higher organisms may have up to 50,000 alternative genes at any one time on which selection can operate. As a result of recombination (Fig. 8–10), these alternatives can be arranged in an almost infinite variety of combinations—enough to account for the genetic distinction of all individual organisms practicing sexual reproduction. The vast store of variability present in a population is the result of millions of years of accumulating mutations.

Given the known mutation rates, new mutations usually play an extremely small role in changes of a population, even over hundreds of generations. Changes in mutation rates are, therefore, highly unlikely to alter the genetic composition of species and so direct the rate or course of evolution. In order to increase the mutation rate sufficiently to alter appreciably the number of mutations in a population, such exposure to radioactivity or other mutagenic agents would be required that destruction of the non-genetic portion of the cells would occur faster than mutations could accumulate.

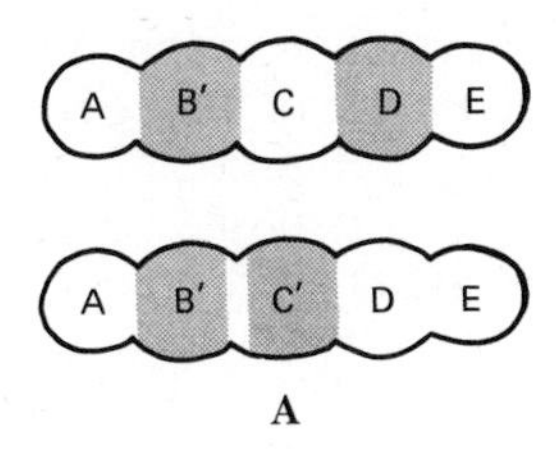

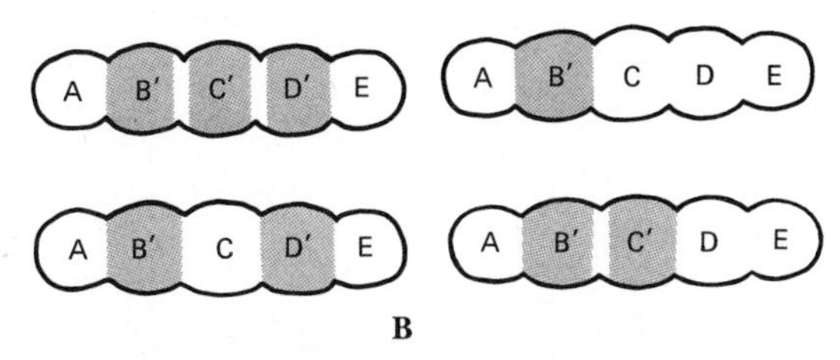

FIGURE 8–10 **Recombination. Homologous chromosomes. Letters A–E indicate normal genes; A′–E′ are mutant alleles. The chromosomes are homozygous for traits A, B, and E, and heterozygous for traits C and D. During replication (as in the maturation of the sex cells), segments of the chromosomes may be exchanged so that the alleles are recombined. The chromosome in each of four gametes may have the configuration shown in B.**

Selection Pressure

Measurements of the force of natural selection are more difficult than determinations of mutation rates, and can never be more than approximate. The problem is that each alternate of each trait must be considered as if it alone were being acted on by selection, while the other 100,000-odd traits are assumed to be neutral. This is clearly an unwarranted assumption, but only equally artificial laboratory experiments with organisms of very closely controlled genetic variability can provide a basis for the study of selection acting on single genes. Nevertheless, these studies, together with field observations, provide evidence of a range of selection rates. At one extreme, certain traits lead to the death of organism (complete lethality) as soon as the mutation is expressed. Such traits may be referred to as having a selection coefficient of 1, or an adaptive value of zero. Other characteristics, while recognizably different from the normal or wild condition, seem to have no effect on the survival or reproductive success of the organisms that possess them. These traits are said to have a selection coefficient of zero (selection does not affect their frequency), or an adaptive value of 1. Some traits may also be assigned an adaptive value greater than 1 if they are more successful than their normal alternative. Selection coefficients as low as about 1 per cent have been calculated. Below this level, measurement becomes very difficult and slight differences in the environment and the remainder of the genes make consistent measurement difficult. Even at this rate, the slightly more advantageous trait would reach a frequency of 100 per cent, although it might require hundreds of generations.

Comparison of observed mutation rates and measured selection coefficients indicates that selection commonly operates much more effectively in controlling gene frequencies. The ultimate source of variability is the occurrence of new mutations, but the frequency of different alternative traits is controlled by selection. Selection is the final arbitrator of gene frequency, and so controls the course of evolution.

Selection has been considered the controlling factor even among the randomly "reproducing," pre-biological, chemical aggregates that preceded the first living forms. The molecules that were able to "grow" by the aggregation or assimilation of the widest range of nutrients would selectively survive other, less efficient molecules. At the prokaryotic level, each mutation would be affected immediately by extremely stringent selection. The very long time period between the first appearance of prokaryotes and the emergence of eukaryotes suggests that selection served as a strong stabilizing force, quickly eliminating most genetic variants. Once the eukaryotic pattern, with a double complement of the genetic material, evolved, numbers of variants could be retained without prejudice to the survival of the individual. Evolution could then proceed at a more rapid pace. Unfortunately, the fossil record of metazoans in the 1000 million years or more between the appearance of the first eukaryotes and the Ediacaran biota is extremely incomplete. The multiplicity of complex organisms within the Cambrian indicates that most of the basic metazoan patterns had evolved long before the appearance of these groups in the fossil record.

With the establishment of the eukaryotic pattern, there is no evidence of any other major changes in the basic scheme within which evolution acts: selection of naturally occurring variants within the population, with new variants continually being produced, at a slow rate, by mutations.

SUMMARY

Life began more than 3 billion years ago through interaction of molecules of the early atmosphere of the earth: water, hydrogen, carbon dioxide, nitrogen, ammonia, and methane. Free oxygen is thought to have been absent at this stage of the earth's development. These gases and water combined under the influence of heat and ultraviolet radiation to form more complex molecules: sugars, amino acids, fats, and lipids. These, in turn, must have been integrated into self-replicating units—the earliest forms of life—but the specific ways in which this might have come about have not yet been established.

The oldest remains thought to represent living organisms have the shape of primitive living bacteria and probably depended on pre-existing organic molecules for food. Maintenance of their constant structure and consistent metabolism required complex molecules such as the nucleic acids of living plants and animals. The nucleic acids also provide a basis for controlled reproduction. They are made up of phosphoric acid, sugar, and a series of four different bases: adenine, thymine, guanine, and cytosine. The nucleic acids are in the form of two strands, wound together as a helix. Together, the two strands serve as a single functional copy of the information which controls the metabolism and reproduction of the organism. Evolutionary change occurs through the alteration of the sequence of bases that make up the nucleic acids. Because there is only a single functional copy of this genetic material in primitive organisms (termed prokaryotes: the living bacteria and blue-green algae), evolutionary change was very slow during the first 2 billion years of life on earth.

The ability to use the energy of sunlight to manufacture sugar from hydrogen and carbon dioxide is thought to have evolved as long ago as 3 billion years among primitive prokaryotic organisms. The presence of large deposits of banded iron formation in rocks 2.8 to 1.8 million years old may have resulted from the trapping of oxygen liberated by primitive photosynthetic organisms, not able to tolerate the molecular oxygen that would have been released by the removal of hydrogen from water. Hematite, common in more recent rocks, indicates the presence of atmospheric oxygen. Once organisms evolved the ability to cope with atmospheric oxygen, they might have made use of it in the oxidative breakdown of sugars, a process that produces much more energy than the anaerobic metabolism of primitive forms of life. Oxidative metabolism is a feature of all organisms above the level of blue-green algae and bacteria. These are termed eukaryotes and are further characterized by the presence of two functional copies of the genetic material, arranged in chromosomes. In eukaryotes, each gene (the basis of a single inheritable trait) is normally represented by homologous copies on each of the pair of chromosomes. The genes may be identical, in which case the individual is said to be homozygous for the particular trait, or different (the heterozygous condition). Two copies of the genetic information give eukaryotes much more evolutionary flexibility than prokaryotes. Change may occur in one copy without jeopardizing survival. Considerable genetic variability can be accommodated, allowing species the potential for rapid adaptation to changing environments.

Evolution occurs through the differential survival and reproduction of individuals with different genetic compositions. With continual new variants produced by mutations, there is a possibility for progressive change, which has led to the diversity of organisms, living and extinct.

QUESTIONS

1. What was the probable composition of the earth's atmosphere at the time that life originated?
2. What effect did the evolution of photosynthesis, dependent on water for a hydrogen source, have on the composition of the atmosphere?
3. What was a possible cause of the deposition of banded iron formation early in the history of life?
4. What chemical substances are necessary for life?
5. What physical conditions are necessary for life as we know it?
6. What is the genetic code?
7. Discuss the relative importance of mutation and selection in determining the course of evolution in eukaryotes.
8. Outline the tenets of Darwin's selection theory.
9. What might be the effects of a great increase in mutation rate on the evolution of a particular species?

SUGGESTIONS FOR FURTHER READING

BARGHOORN, E. S. "The Oldest Fossils," *Scientific American,* 30–42, May 1971.

CLOUD, P., and GIVOR, A. "The Oxygen Cycle," *Scientific American, 111*–123, September 1970.

DOBZHANSKY, T., AYALA, F. J., STEBBINS, G. L., and VALENTINE, J.W. *Evolution.* New York: W. H. Freeman, 1977.

HALLAM, A., ed. *Patterns of Evolution.* New York: Elsevier, 1977.

KNOLL, A. H., and BARGHOORN, E.S. "Archean Microfossils Showing Cell Division from the Swaziland System of South Africa," *Science,* 198: 396–398, 1977.

SCHWARTZ, R.M., and DAYHOFF, M. O., "Origins of Prokaryotes, Eukaryotes, Mitochondria, and Chloroplasts," *Science,* 199: 395–403, 1978.

SIMPSON, G. G. *The Major Features of Evolution.* New York: Columbia University Press, 1953.

STEBBINS, G. L. "Processes of Organic Evolution," *Concepts of Modern Biology Series.* Englewood Cliffs, N.J.: Prentice-Hall, Inc., 1971.

PART THREE

the craton

CHAPTER NINE

Each of the continents of the world has a nucleus or core of metamorphosed pre-Paleozoic rocks that are either exposed as a shield area or covered thinly with younger platform sediments. The Canadian Shield, which forms the nucleus of North America, is an irregularly elliptical area that includes Greenland and most of eastern and central Canada, and extends into the United States in New York, Minnesota, Wisconsin, and Michigan. The term "shield" refers to the subdued topography and low domal form of continental nuclear areas. However, it is not appropriate for the Canadian Shield, the center of which is depressed. The central depression is occupied by the waters of Hudson Bay and by Paleozoic rocks that geophysical surveys show may reach thicknesses of 3000 m.

A large part of the Canadian Shield is a low, flat, glaciated peneplain dotted with innumerable lakes that were formed either in depressions scoured out of the rock or in stream courses damned by glacial drift (Fig. 9–1). Along its eastern and southeastern margin the shield has been elevated and dissected into mountains over 1000 m high, but relief of more than 100 m is uncommon in the rest of the shield. The topography reflects the structure and lithology of the bedrock. Valleys follow belts of rock less resistant to erosion, and high ground is underlain by resistant rocks, a perfection of adjustment attainable only after a long period of erosion. The similarity of topography emerging from beneath the surrounding Paleozoic strata as they are stripped off by erosion, to that of the central zones of the shield, suggests that the subdued shield surface was developed in pre-Paleozoic time and has been little changed by later erosional and depositional cycles.

The oldest rocks of the Canadian Shield are the oldest rocks so far recognized in the world. In the Godthabsfjord region of western Greenland gneisses have been dated at 3.8 billion years. In the shield a record of five-sixths of geological time is preserved. The older shield rocks are the only tangible evidence of the earth's infancy and provide a testing ground for our theories about its origin and early stages. During their deposition the continents grew, the first mountains were thrust up, and the first living things stirred in the "organic soup."

The interpretation of the rocks of the shield is made difficult by their lack of guide fossils, the incompleteness of the record they preserve, and the deformation and metamorphism that they have undergone. Without guide fossils the geologist must rely on lithologic methods of doubtful validity or on isotopic dates to correlate sediments from one basin to another. Compared to the

THE SHIELD

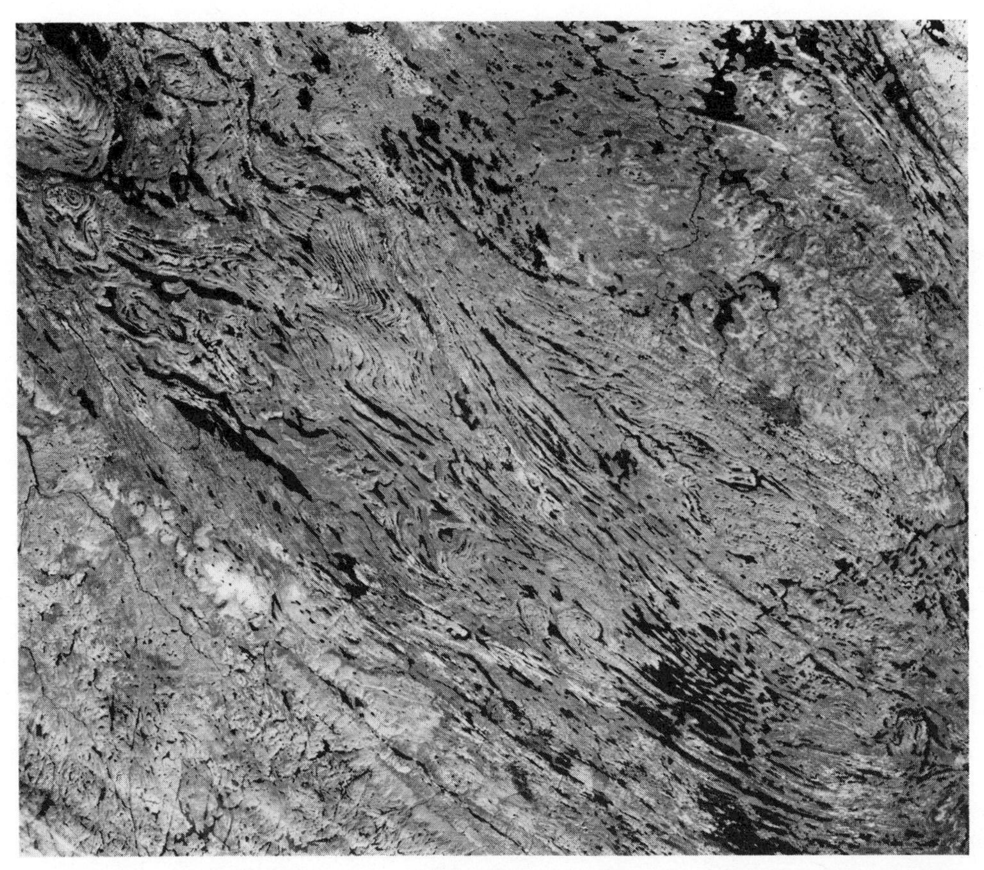

FIGURE 9–1 **Folded Proterozoic rocks of the Labrador trough, northern Quebec. Note the abundance of lakes (black) which reflect the structure in the folded strata. Landsat image, taken from a height of about 900 km above the earth. (NASA photograph.)**

younger record, less of the sedimentary rocks of the shield is available for study because the shield has been exposed longer to destruction by erosion, and because the older the sediments are, the more likely they are to be hidden by later sedimentary layers. Much of the shield has passed through several episodes of orogeny and its rocks have been metamorphosed to schists and gneisses. The grade of metamorphism of these rocks is not generally higher than that of the cores of modern mountain chains, such as the Cordillera. However, in studying such mountain cores the geologist can compare the metamorphic rocks with their unmetamorphosed equivalents in other parts of the range, but in the shield such equivalents are either unavailable or unidentifiable due to lack of criteria for establishing correlation. So far isotopic dating has proved to be the only method of placing shield rocks in a regional time framework.

The rocks of the shield contain some of the world's richest deposits of iron ore, copper, nickel, gold, silver, uranium, and many other valuable materials. The search for these mineral riches has stimulated scientific investigation of the shield and has yielded not only new mineral wealth but new insights into the earth's early history.

DIVISION OF PRE-PALEOZOIC TIME

For many decades the term "Precambrian" has been used to refer to all rocks older than the base of the Cambrian System. The uppermost Precambrian rocks are locally conformable with Cambrian rocks and are more closely allied in degree of metamorphism and fossil content with the older Paleozoic systems than with older Precambrian rocks. Unlike these older rocks, the uppermost Precambrian sediments and volcanics occur only on the periphery of the shield and in the surrounding mountain systems. The name used for these youngest Precambrian rocks differs from continent to continent (Ediacaran, Infracambrian, Eocambrian are examples). In this book the term "Ediacaran," derived from South Australia, is used to refer to this sub-Cambrian unit and it is considered to be the oldest division of the Paleozoic Erathem. The base of the Paleozoic can be defined by the appearance of evidence of metazoan life usually in the form of trace fossils of burrowing organisms. If the Cambrian System is no longer considered to be the base of the Paleozoic Erathem, then the term "Precambrian" is not generally useful and the term "pre-Paleozoic" is more appropriate.

The division of the Paleozoic, Mesozoic, and Cenozoic eras is based on type sections and these divisions are recognized around the world on the basis of paleontological correlation. In the absence of guide fossils geologists investigating older rocks have not been able to use this method. Division of pre-Paleozoic time is based on isotopic dating of orogenic episodes. Locally, geologists have established sequences of depositional and orogenic intervals. Those orogenic episodes that have been recognized through isotopic dating as having a wide extent have been used as marker events to subdivide pre-Paleozoic time into eras. Although some of these time terms, such as Archean and Proterozoic, have been used outside North America, the orogenic episodes have not been demonstrated to be of worldwide occurrence and, therefore, no pre-Paleozoic time divisions are common to all continents.

Since the beginning of the investigation of the shield, geologists have recognized a major break within pre-Paleozoic rocks, marked not only by a widespread unconformity but also by a basic change in lithology and style of deformation. The rocks below the unconformity are called Archean (ancient) and those above, Proterozoic (primitive life). Unfortunately, only Proterozoic is analogous to terms like Paleozoic and Mesozoic, but Archean is so widely used that a change to Archeozoic now seems unlikely. The orogenic episode that brought the Archean to a close is called the Kenoran, and occurred about 2500 million years ago. Proterozoic rocks of the Canadian Shield have been divided by C. H. Stockwell into three eras on the basis of an earlier Hudsonian orogeny (about 1800 million years ago) and a later Grenville orogeny (about 1000 million years ago) (Fig. 9–2). The three eras are called the Aphebian (oldest), Helikian, and Hadrynian (youngest). These names are not derived from geographic localities; they refer to stages in the growth of humans and are derived from the Greek (aphebos = past the prime of life; helikia = maturity; hadrynia = coming to maturity). The terms "Archean" and "Proterozoic," which refer to divisions of time larger than these eras, have been called eons. The term "Phanerozoic" (apparent or evident life) is useful in referring to the eon comprising the Paleozoic, Mesozoic, and Cenozoic eras. The United States Geological Survey has also adopted a system of division of Precambrian time based on orogenies but designating the time divisions by the last letters of the alphabet as shown below:

Division	Boundary
Paleozoic Era	
Precambrian Z	
	800 million years
Precambrian Y	
	1600 million years
Precambrian X	
	2500 million years
Precambrian W	

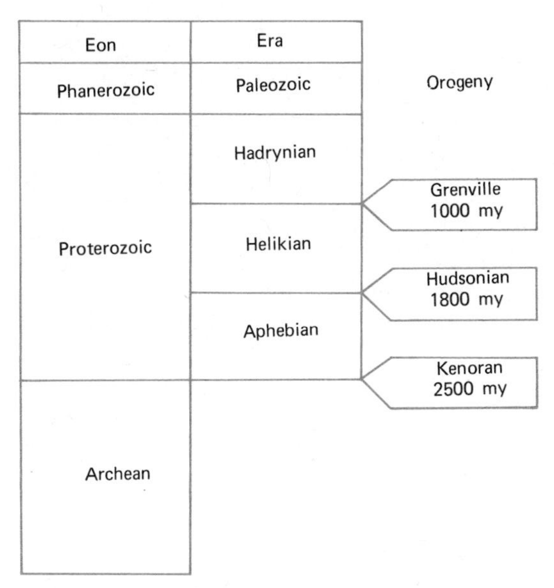

FIGURE 9–2 **Divisions of Proterozoic and Archean time.**

GEOLOGY OF A TYPICAL AREA

Study of a single area shows the nature of the evidence from which geologists reconstruct Archean and Proterozoic events in the shield. The region from Timmins, Ontario, to Val d'Or, Quebec, is geologically well known because of its wealth of mineral deposits, and will serve as a model.

Gneiss, Graywacke, and Greenstone

The geological map (Fig. 9–3) shows a trough, or "pod," of metamorphosed volcanic and sedimentary rocks surrounded by granite gneiss. Similar gneiss covers about 80 per cent of the Canadian Shield. The contacts of the volcanic and sedimentary rocks with the granite gneiss appear to be intrusive or faulted; the pod seems to float within a "sea" of granitic rock. The gneissic granite that underlies so much of the shield is not a homogeneous intrusion; rather, it consists of the coalescence of thousands of bodies of various sizes. Much of it may represent highly metamorphosed sediments or volcanics.

The volcanic and sedimentary rocks of the pod are folded into a complex syncline and metamorphosed. Graywacke is the most common sedimentary rock. Many of the graywackes show repeated graded bedding like the sediments found in trenches of the present day, a feature usually indicative of deposition by turbidity currents. Not all of the sediments in the pod are graywackes; a few can be identified as quartzites and arkoses. Conglomerates are common and contain as pebbles a great variety of volcanic, sedimentary, and intrusive rocks. Particular importance has been attached to conglomerates that contain granite pebbles, for granites are typical of mountain building episodes and such pebbles must be evidence of a sequence of orogenesis, uplift, and erosion.

The lavas in the pods are called greenstones because they have been changed

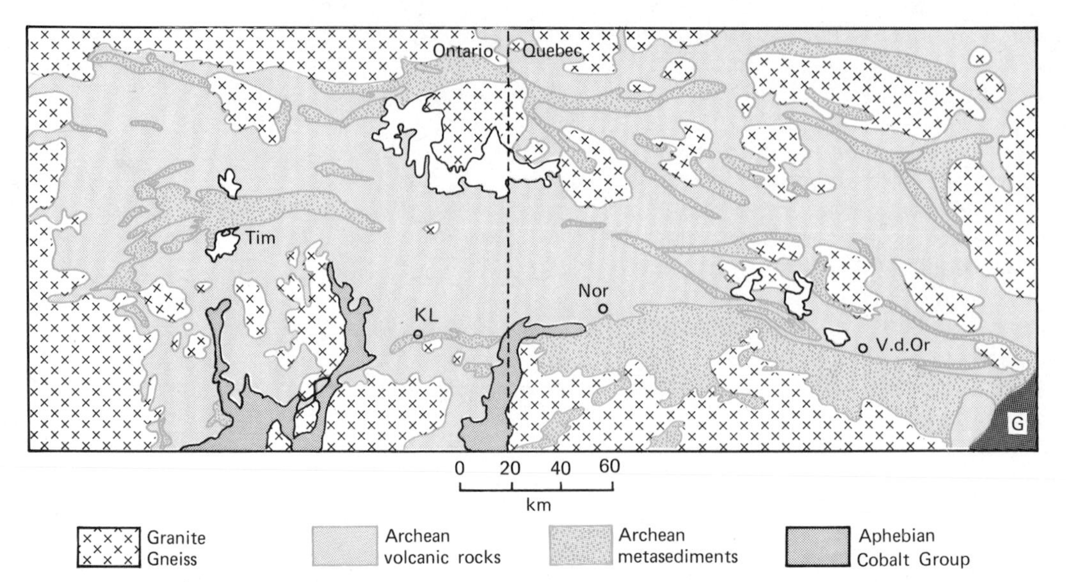

FIGURE 9–3 **Simplified geological map of the Timiskaming Greenstone belt of Ontario and Quebec whose boundary runs through the center of the map. V.d'Or = Val d'Or, KL = Kirkland Lake, Tim = Timmins, Nor = Noranda, G = edge of the Grenville province. (From the Geological Maps of the provinces.)**

FIGURE 9–4 **Pillow structure in lavas, Northwest Territories. Shapes indicate that the tops of the flows face the upper left-hand corner. (Courtesy of the Geological Survey of Canada.)**

largely to secondary green minerals (epidote, chlorite, etc.) during metamorphism. Pillow structures present in many of these beds show that they were intruded under water (Fig. 9–4). Chemical analyses of these much altered lavas indicate that extrusive rocks, ranging from basalts through andesites to rhyolites and their pyroclastic equivalents such as tuff and agglomerate, were important constituents of the pod. A wide variety of igneous bodies intrude the sedimentary–volcanic rocks of the pods.

The metamorphosed sequence of volcanic and sedimentary rocks downfolded into the granite gneiss is of Archean age. Its deformation and metamorphism was accomplished in the Kenoran orogeny about 2500 million years ago. The thickness of the sedimentary–volcanic sequence is difficult to measure due to the intensity of the deformation that has produced vertical dips in most of the belt, but estimates of about 12,000 m have been made.

In this area two Archean groups separated over a considerable area by an angular unconformity can be distinguished. The older Keewatin Group consists largely of volcanic rocks; the younger group, called the Timiskaming, consists largely of sedimentary rocks (Fig. 9–5). A granite pebble conglomerate at the base of the Timiskaming Group reaches thicknesses of 1000 m in places and suggests that a major orogeny occurred near the basin of deposition. However, no intrusive bodies can be identified with this orogeny and possibly the pebbles are derived from a granitic terrane outside the greenstone belt.

Proterozoic Rocks

In the southern part of the Timmins–Val d'Or region the Archean rocks are overlain unconformably by less deformed sedimentary rocks of the Cobalt Group. These rocks are of Aphebian age. They consist largely of quartzites with some limestones, conglomerates and dolomites. The basal formation of the group, the Gowganda Conglomerate, consists of about 100 m of unsorted and unbedded conglomerate overlain by finely graded graywacke (Fig. 9–6). Some of the pebbles in the conglomerate have striated facets similar to those abraded in a modern glacier, and locally the conglomerate rests on a smooth surface of older rocks that resembles those cut and polished by the movements of modern glaciers. This evidence shows that the Gowganda Conglomerate is, in part, an early Proterozoic tillite, and its continuation for over 300 km in one direction suggests that a large area may have been glaciated at this time. The banded graywackes that overlie this ancient tillite may represent varved clays that were deposited in glacial lakes where pebbles were dropped from floating ice (Fig. 9–7).

Along the north shore of Lake Huron a sequence of older sediments and volcanics comprising the Elliot Lake, Hough Lake, and Quirke Lake groups intervenes between the Cobalt Group and Archean rocks. Within this lower succession two other conglomerates occur that are believed to be of glacial origin, showing that the episode of Aphebian glaciation was not only widespread but also recurrent. The Aphebian sequence north of Lake Huron, comprising the Cobalt and these three lower groups, reaches a total thickness of 12,000 m and is referred to as the Huronian Supergroup. The basal conglomerate of the supergroup is notable for the concentrations of uranium minerals that are mined from it. There is as yet no consensus as to whether the minerals were concentrated in the gravels as a placer deposit or introduced by hot solutions later.

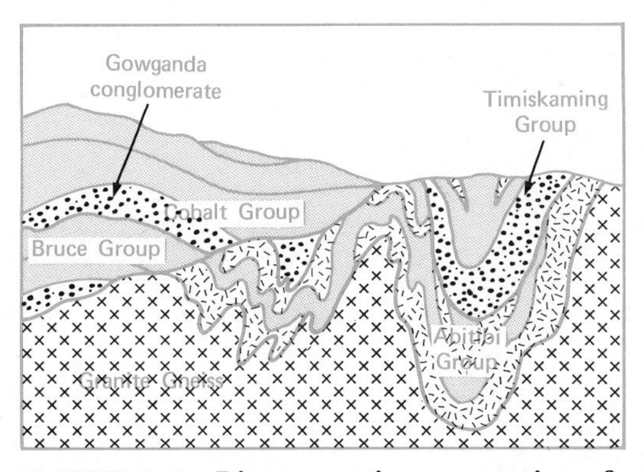

FIGURE 9–5 **Diagrammatic cross-section of the structural relationships of the stratigraphic units in the Timiskaming Greenstone belt.**

STRUCTURAL TRENDS AND PROVINCES

The strike of the belts folded into the gneiss of the Timmins–Val d'Or area is east–west and this structural trend can be followed throughout a large area of the shield south and east of Hudson Bay. However, west of Hudson Bay the folded structures strike northeast–southwest and in the Adirondacks the axes of the folds make swirling patterns. The different trends in the metamorphosed rocks of the areas of the shield reflect both the different directions from which the deforming forces acted in the rocks and the nature of their response to the deforming

FIGURE 9–6 **Gowganda conglomerate, Cobalt District, Ontario. The lack of sorting and variety of boulders suggest that it is an ancient tillite. (Courtesy of the Ontario Department of Mines.)**

FIGURE 9–7 **Finely laminated sediments of the Cobalt Group which may be ancient varved clays. Note the "dropstone" pebble that may have melted out of an iceberg in a glacial lake. (Courtesy of M. J. Frarey and the Geological Survey of Canada.)**

forces. These trends define structural provinces, which are illustrated in Fig. 9–8.

In each of these structural provinces the metamorphic rocks are of approximately the same age; each province was affected by one major orogeny. The Superior and Slave provinces, and part of the Nain province, were metamorphosed in the Kenoran orogeny about 2500 million years ago. The Churchill, Southern, and Bear provinces were subjected to the Hudsonian orogeny (called the Penokean in the Lake Superior area) about 1800 million years ago. The rocks of the Nain province record the Elsonian orogeny of 1300 million years ago, and those of the Grenville province were last deformed in the Grenville orogeny of about 1000 million years ago. The fact that these orogenies are also used to define the major time divisions of the Archean and Protero-

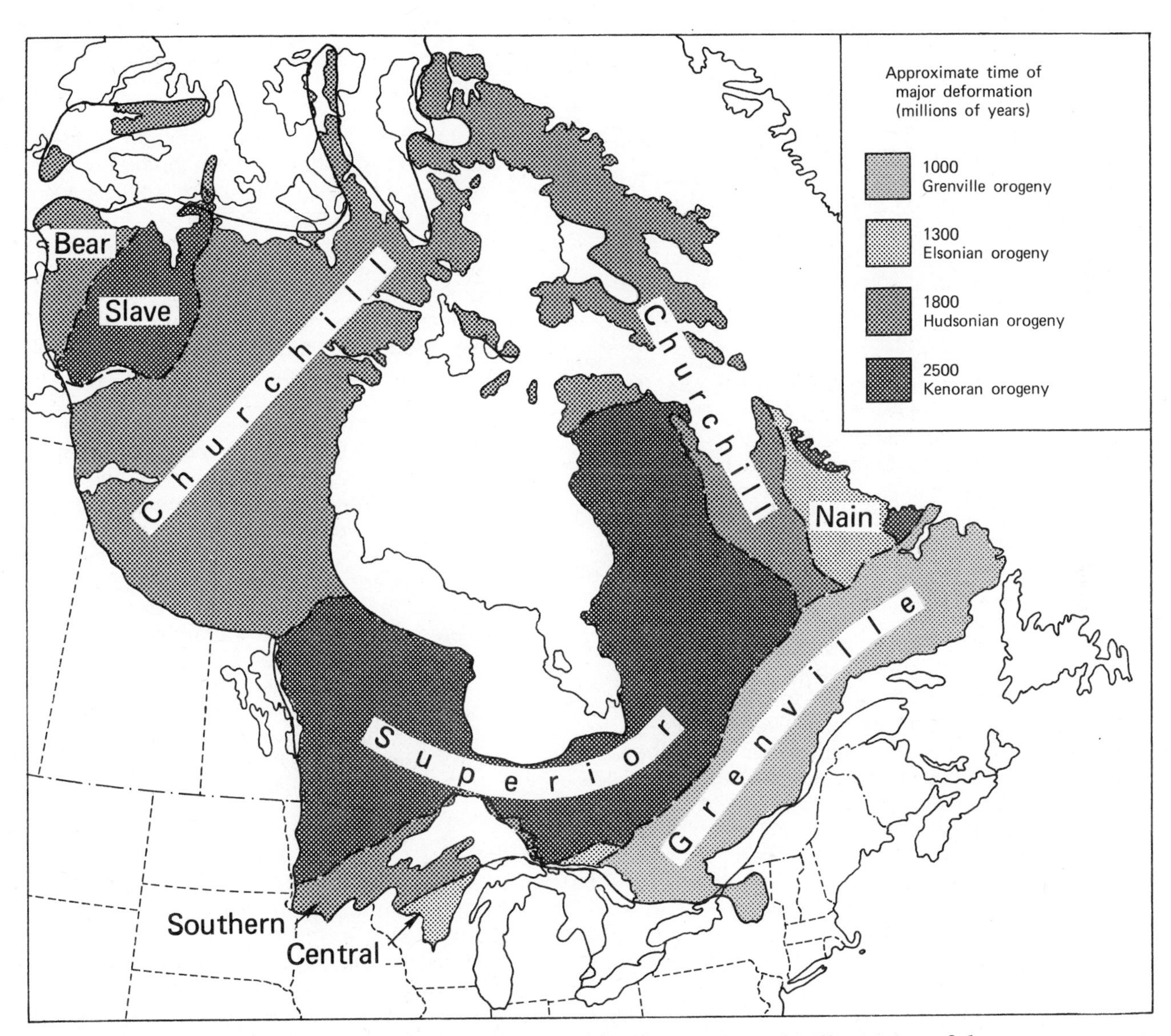

FIGURE 9–8 **Structural provinces of the Canadian Shield. (After C. H. Stockwell; courtesy of the Geological Survey of Canada.)**

zoic does not imply that the provinces are occupied by rocks of one age only. In the Grenville province rocks of Archean and Aphebian ages, in addition to rocks of Helikian age, can be shown to have been affected by the Grenville orogeny.

ARCHEAN ROCKS

Archean rocks, not only in the Superior province, but also in the Slave province and in other shield areas of the world, are characterized by greenstone belts like the one described above in the Timmins–Val d'Or area. These belts are unlike any basins of deposition active at present or in the post-Archean record. In the greenstone belts a succession of volcanic and sedimentary rocks is common in many areas. It starts with basaltic flows interpreted as fissure eruptions; continues with intermediate and felsic rocks, such as andesites and rhyolites, erupted from volcanic centers with abundant pyroclastic rocks; and terminates with blankets of sediment derived from the erosion of the volcanic pile. These belts do not show miogeoclinal and eugeosynclinal realms as do modern trenches and younger geosynclines. Their relationship to the marginal granites and highly metamorphosed gneisses that surround them has been a problem for many years. Since the last part of the 19th century the granites and gneisses have been described as intrusive and, therefore, younger than the greenstone belts. The belts were thought to be remnants, the bases of deeply downfolded synclines, of more extensive volcanic–sedimentary folded belts engulfed at the base by intrusive granites and then deeply eroded to remove all but the deepest folds. Presently, the early suggestion of William Logan, who first described the shield rocks in Canada, that the granite gneiss is older than the greenstones and part of an ancient crust, has been revived. Some of the sediments in the greenstone belts are derived from the weathering of areas of granitic rocks which must have existed as continental nuclei before the belts were formed. The thickness and wedge-shape geometry of the granite pebble conglomerates also suggest that granitic source areas were close to the margins of the greenstone basins.

In the absence of modern or post-Archean analogues of the greenstone belts, reconstruction of Archean crustal conditions is highly speculative. The crust in Archean time was probably thinner than now; the crust and mantle were certainly hotter and more mobile; and continental sialic areas were smaller and more abundant. The first continental nuclei of sialic materials may have separated from the mantle about 4 billion years ago, or may have fallen in from space in the accretion process of the planet. At the beginning of Archean time volcanic rock welled up from the highly active mantle between such thin, small protocontinents and cooled in marginal troughs, interbedded with the erosion products of the primitive continents and earlier lavas. The highly mobile crust was frequently deformed but whether the type of deformation was that associated with plate tectonics and subduction zones is doubtful.

The Kenoran orogeny at the close of the Archean Eon marks a major change in the development of the earth's crust. Geochemical and geophysical measurements indicate that during late Archean time the crust achieved its present thickness of about 40 km under the continents. The Kenoran event that deformed and metamorphosed the rocks of the Superior and Slave provinces appears to have affected a far wider area than subsequent orogenies. The mechanism by which the crust was thickened in this orogeny is open to speculation and further research. When the Aphebian Era opened, after a considerable interval of erosion, crustal conditions were much different.

APHEBIAN ROCKS

In Aphebian rocks the oldest geosynclines can be recognized. The oldest major sequences of quartzites, dolomites, and limestones were deposited in elongate troughs at the margins of the Archean continents. Such sediments are characteristic of miogeoclines, the transition zone between cratons and their bordering trenches.

The Circum-Ungava geosyncline (Fig. 9–9) is a good example of an Aphebian geosyncline. It follows an arc that almost encloses the northeastern half of the Superior province. Its eastern segment, the Labrador trough, is most extensively studied because it is the site of large iron deposits around Shefferville, Quebec. The western side of the trough contains a miogeoclinal suite, but eastward, away from the Superior craton, the trough filling is largely made up of volcanics and graywackes, suggestive of deposition in an oceanic trench. This volcanic realm is not evident in the northern or western segments of the geosyncline exposed in the Belcher Islands (Fig. 9–10). The limestones of the miogeoclinal realm contain fine examples of stromatolites (Fig. 9–11). The southern end of the Labrador trough passes from the Churchill province into the Grenville province and has been caught up in the Grenvillian orogeny. The metamorphic grade of its sediments and iron ores has been considerably raised over that of the rest of the geosyncline, which was deformed only once in the Hudsonian orogeny.

The iron ores that are particularly characteristic of, but not confined to, Aphebian

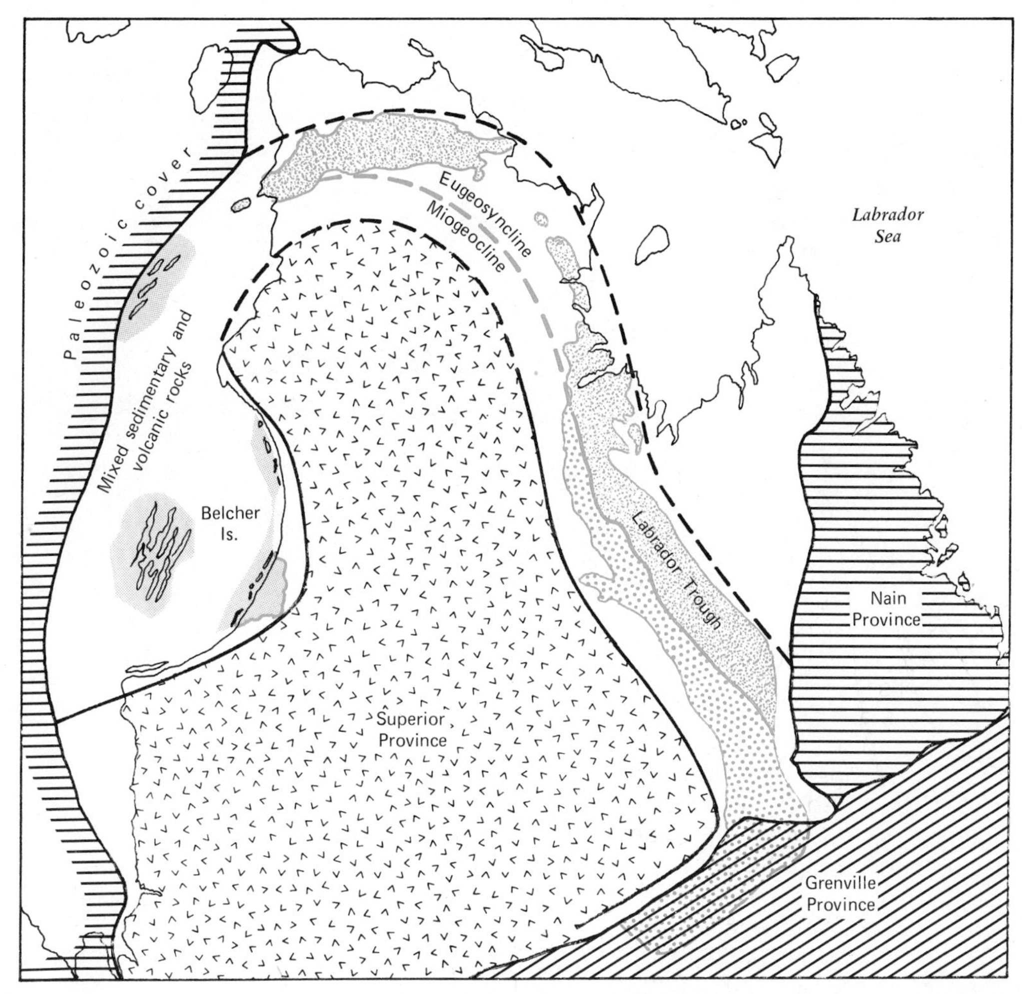

FIGURE 9–9 **Circum-Ungava geosyncline. (After E. Dimroth; courtesy of the Geological Survey of Canada.)**

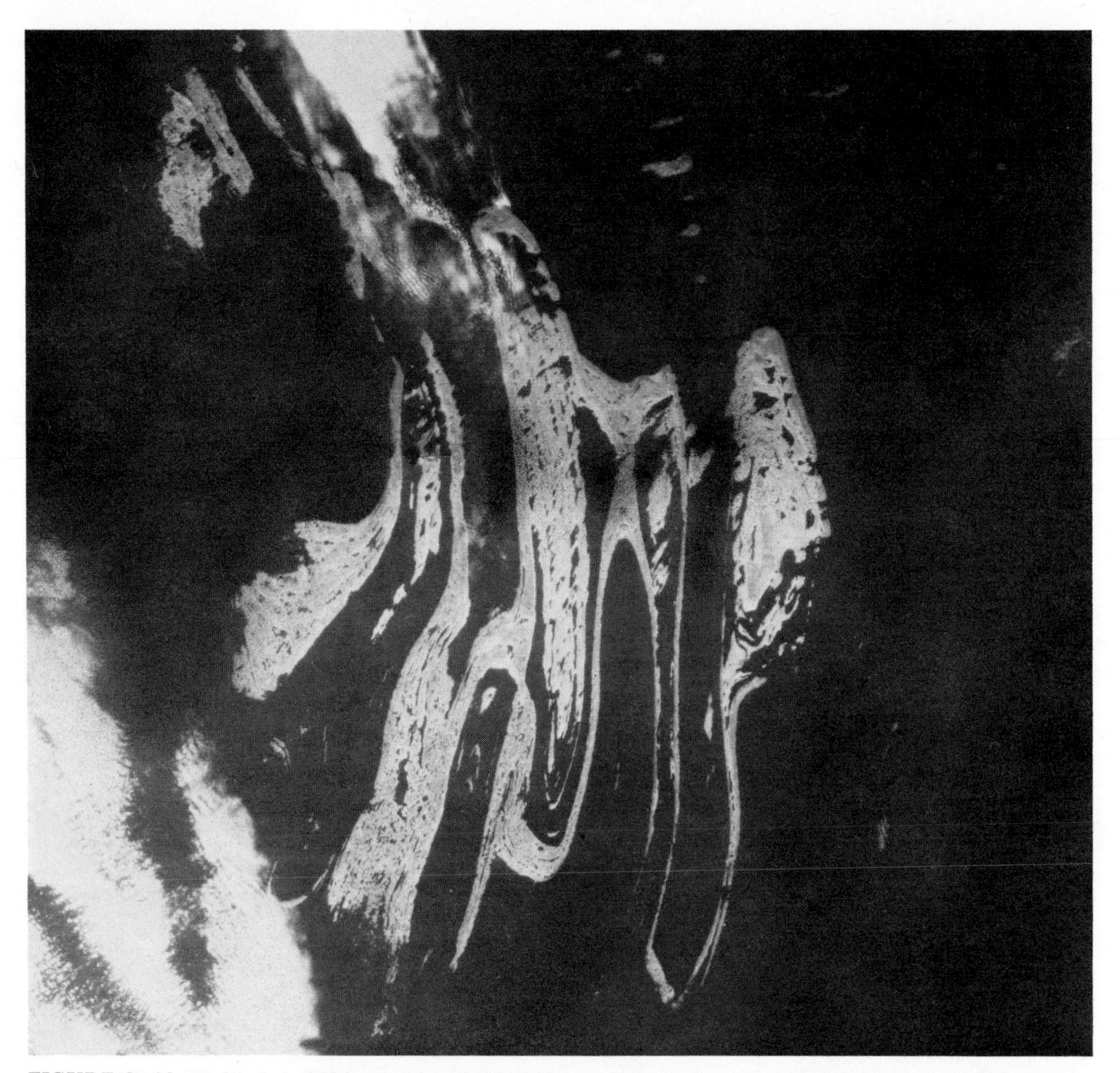

FIGURE 9–10 **Folded Aphebian rocks of the Belcher Islands in central Hudson Bay. Resistant beds in the succession form islands of intricate shapes; less resistant layers are covered by the waters of the bay. Light areas at lower left and top center are clouds. (Composite of two LANDSAT images; courtesy of the Canada Center for Remote Sensing.)**

rocks occur in sedimentary units called "iron formation." Iron formation consists of alternating thin beds of red chert (jasper) and an iron oxide, either magnetite or hematite (Fig. 9–12). The significance of this sediment as an atmospheric indicator has already been discussed (Chapter 8).

The rocks of the area south and west of Lake Superior belong to the southern province and are of Aphebian age. The iron formations of this region have been North America's main source of iron ore for more than a century. Although the richest ores are nearing depletion, the mines continue to produce more than 105,000 tons yearly, about 70 per cent of the iron ore pro-

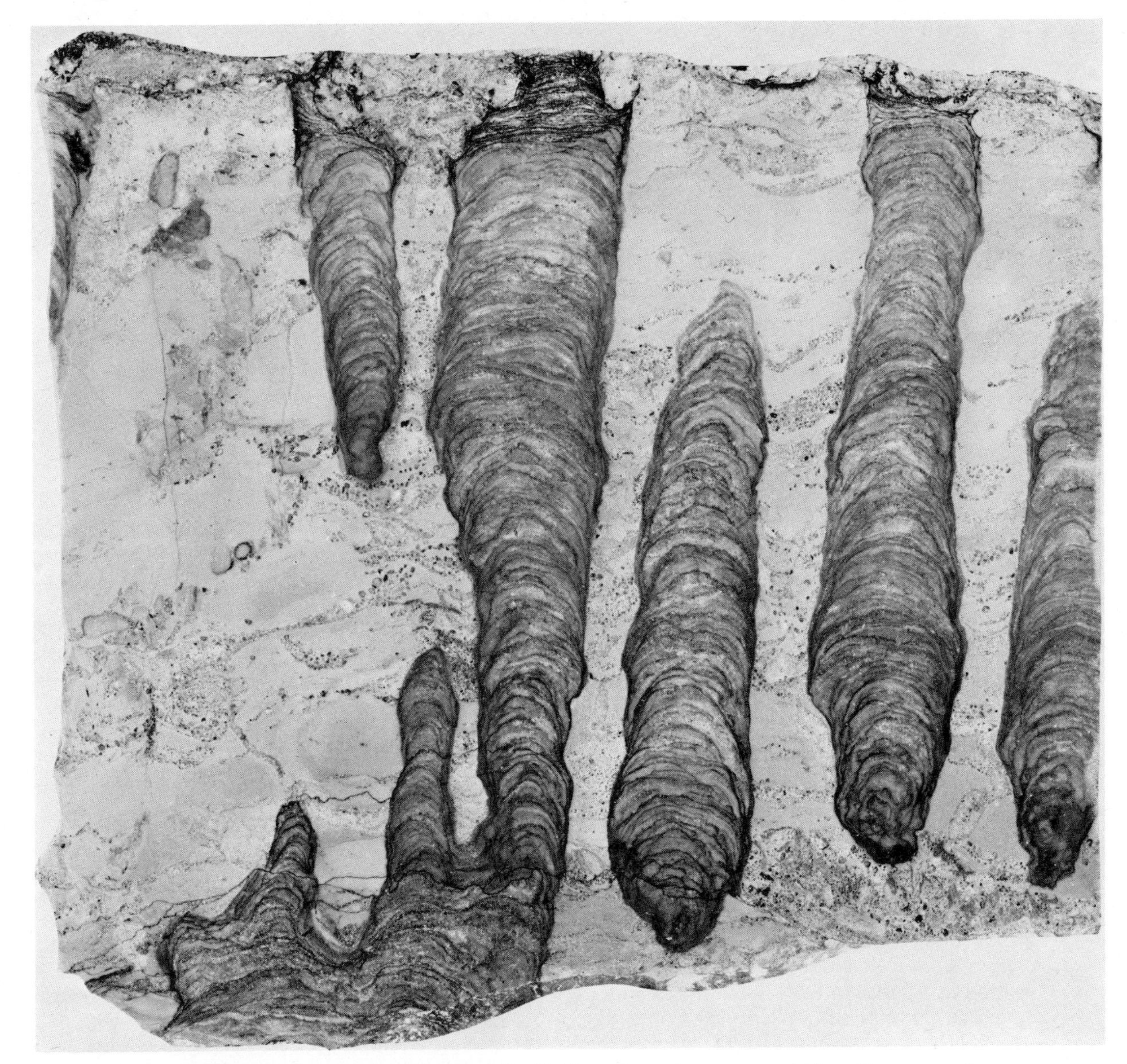

FIGURE 9–11 **Stromatolites from Manitounuk I., Hudson Bay, about natural size. (Courtesy of H. Hofmann and the Geological Survey of Canada.)**

duced in the United States. The most productive mines in this district are located in the Mesabi Range (Fig. 9–13), where the cherty iron oxides, iron silicates, and iron carbonates of the iron formations were weathered during early Proterozoic time under climatic conditions that favored the dissolution of silica by groundwater and the concentration of iron oxides. Until a few years ago only the iron formation that had been enriched by weathering was mined at Mesabi, but mining of the leaner, silica-rich formation (taconite) now predominates. Similar sedimentary rocks were deposited at the other iron ranges shown in Fig. 9–13. The rocks of this area, like the Huronian rocks

FIGURE 9–12 **Highly contorted banded iron formation, Beresford Lake, Manitoba. (Courtesy C. H. Stockwell and the Geological Survey of Canada.)**

north of Lake Huron, were folded about southwest–northeast axes and intruded by granites in the Hudsonian orogeny at the close of Aphebian time.

The Coronation geosyncline can be traced across the Slave and Bear provinces in the northwestern part of the shield. The limestones in the miogeoclinal suite of this geosyncline show some of the best developed stromatolitic structures in Aphebian rocks.

Attempts have been made to interpret the sedimentary rocks of the Aphebian Era in terms of the change in the oxygen content of the atmosphere. For instance, the lower part of the Huronian Supergroup shows little sign of oxidized iron minerals, but the upper part is rich in iron oxides. Unfortunately, such evidence is equivocal, for even in today's oxygen-rich atmosphere reducing conditions are produced over large areas where decaying organic matter consumes the available oxygen. However, some explanation is needed for the fact that, not only in the Canadian Shield but throughout the world, the major deposits of iron ores occur in Aphebian rocks.

HELIKIAN ROCKS

The highly irregular trends of structures in the Grenville province of the shield suggest that it has been subject to many episodes of deformation by forces acting successively in different directions. Isotopic and structural evidence of older orogenies and criteria for the identification of Archean and Aphebian rocks have been obscured or obliterated by the "overprint" of the youngest Proterozoic orogeny, the Grenvillian. The rocks of this province have been more extensively changed by metamorphism than those of adjacent provinces and, therefore, less is known of their original nature. Much of the area of the province is underlain by granitic gneisses and only in a few places can sedimentary rocks be positively identified. The Grenville Group consisting of perhaps 6000 m of marble, quartzite, and gneiss is one of these sedimentary sequences. It is generally considered to be of Helikian age, but it could be older.

The Keweenawan Group is the youngest sedimentary unit of the Lake Superior area. It occupies a belt stretching southwest from the peninsula in Lake Superior, from which it takes its name, to where it is covered by Paleozoic rocks at the shield edge. By means of gravity measurements, the Keweenawan Group can be traced 1600 km beneath the Paleozoic and Mesozoic platform sediments, from Michigan to Kansas. The group consists of 15,000 m of lavas, feldspathic sandstones, quartz sandstones, and shales. The lavas were largely gas-charged basalts that flowed repeatedly from fissures in the earth, building up thousands of meters of vesicular and amygdaloidal flows. The red color, cross-bedding, and

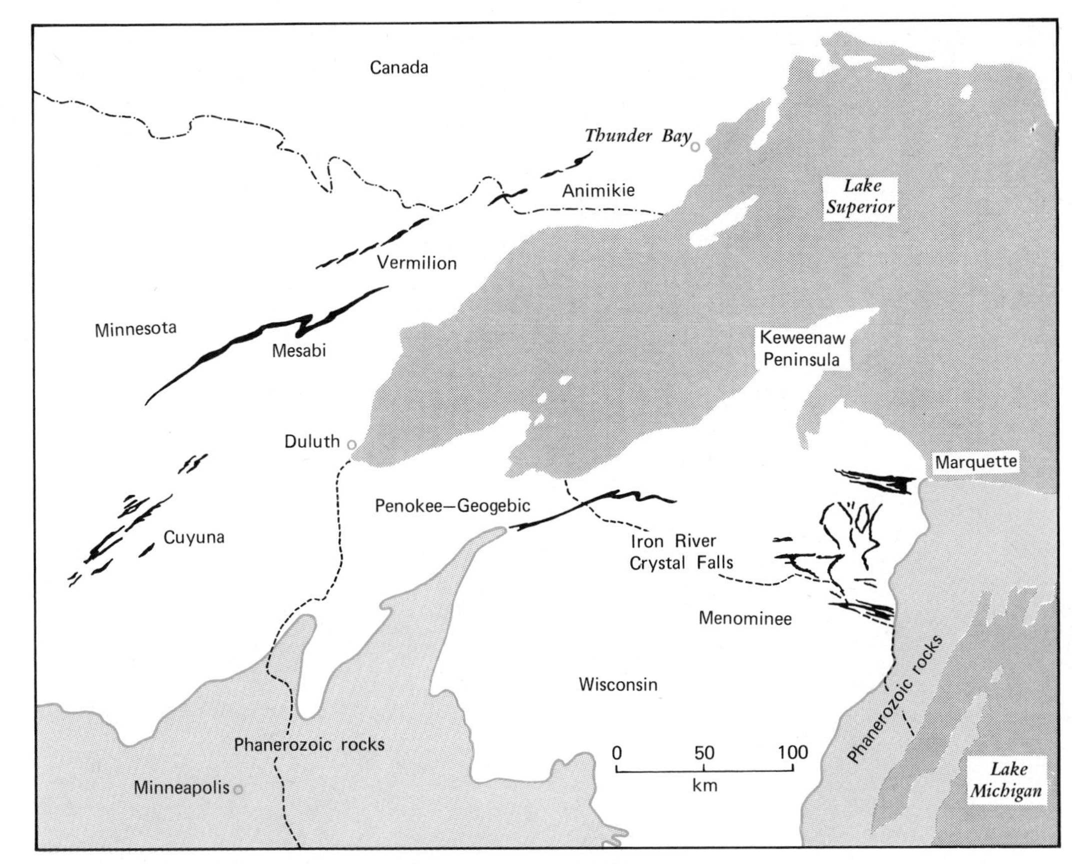

FIGURE 9–13 **Iron ranges in the Lake Superior area.**

mud cracks in the sediments interbedded with the basalt flows indicate that the group was deposited under continental conditions. It occupies a trough formed by downfaulting of the shield after the Hudsonian orogeny. Deposits of metallic copper within the flows and conglomerates of the Keweenawan Group were worked by the Indians long before white people came to the area. Copper production from the mines of the Keweenawan Peninsula reached a peak of 115,000 tons in 1905, but has since declined to insignificant amounts. The Keweenawan Group is unconformably overlain by cleaner red sandstones (Jacobsville and equivalents) that are locally as thick as 2300 m and may be Hadrynian in age. They are difficult to distinguish lithologically from Upper Cambrian sandstones that overlie them, although the disconformity between them must represent at least Early and Middle Cambrian time.

Other Helikian units of the shield are thick, undeformed successions of quartz sandstones in northern Saskatchewan and the Northwest Territories.

Hadrynian rocks are almost completely missing on the Canadian Shield, but accumulated to great thicknesses in the marginal geosynclines. They are now exposed in the Cordilleran and Appalachian mountains and are discussed in Chapters 14 and 17.

PRE-PALEOZOIC ROCKS BENEATH THE PLATFORM

The rocks of the shield extend toward the margins of North America beneath the cover of Paleozoic rocks on the platforms. The structural provinces can be identified by isotopic dating of small fragments brought up from oil wells that penetrate through the complete platform cover. Some of the boundaries between the provinces can be traced beneath the cover by geophysical measurements. The buried boundary between the Churchill and Central provinces can be traced by anomalies in gravity. The subsurface trace of the Grenville-Superior boundary has been followed southward through Michigan and Indiana on the basis of earthquake epicenters. Archean and Proterozoic rocks are also exposed by faulting and folding in the marginal mountain chains of the continent and are discussed further in chapters on these mountain systems.

Figure 9–14 is a map of the present state of our knowledge of the provinces beneath the platform. The Grenville province

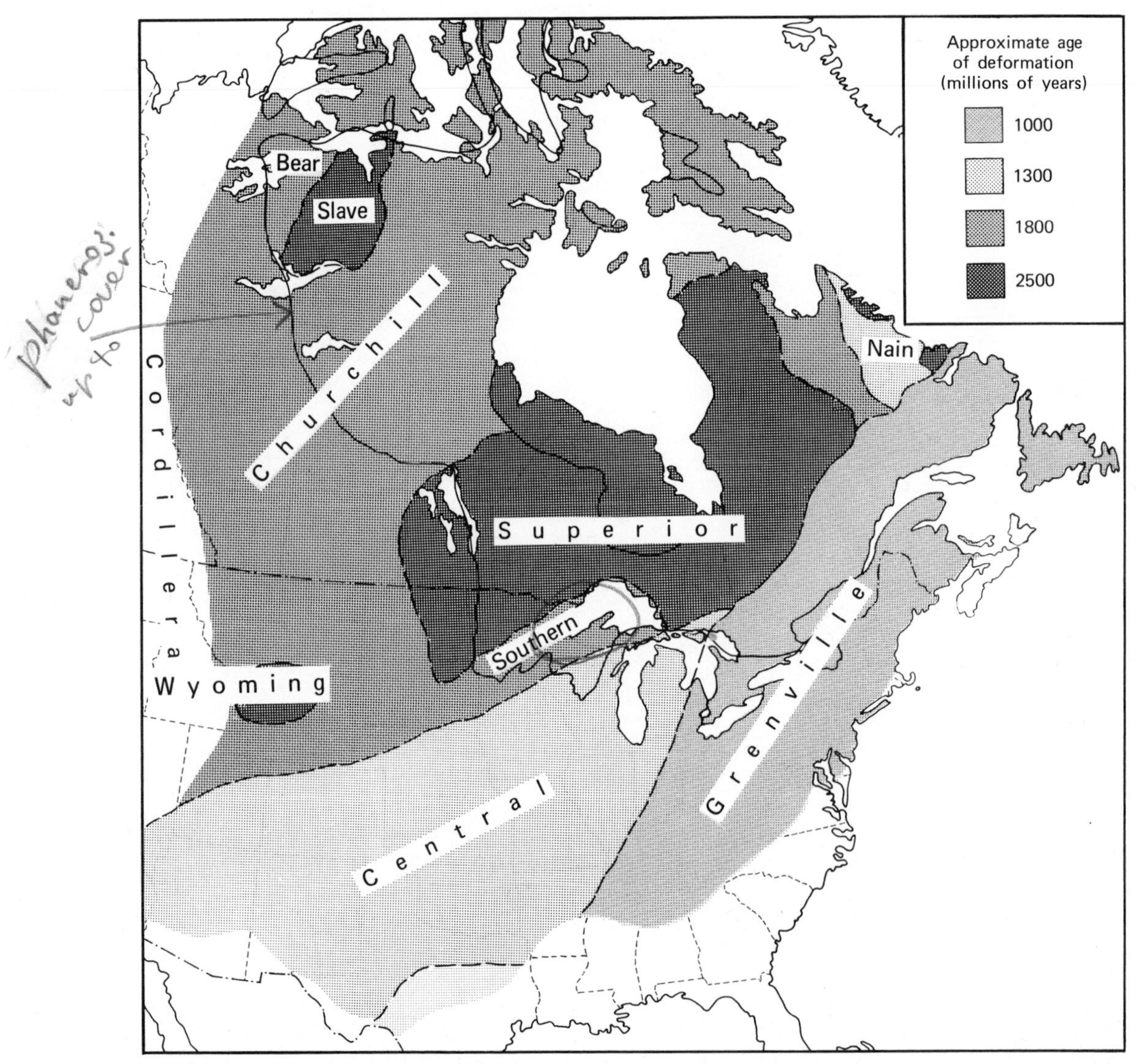

FIGURE 9–14 **Structural provinces of the Canadian Shield traced below the platform sediments by boreholes, geophysical measurements, and observations in marginal uplifts of the shield rocks.**

continues along the eastern and southern sides of the United States, parallel to the Appalachian Mountain system. Much of the central part of the United States is underlain by the Central province, which has a limited outcrop on the shield itself. Rocks of the Central province were deformed between 1.2 and 1.5 billion years ago in what is called the Elsonian orogeny in Canada, or the Mazatzal orogeny in the southwestern states. The Superior province with its Archean rocks does not extend much farther than the southwest boundary of the shield, but a small area at the margin of the Cordillera, called the Wyoming province, is also composed of Archean rocks metamorphosed in the Kenoran orogeny.

METALLOGENETIC PROVINCES

The distribution of mineral deposits in the Canadian Shield is not random. Each province is characterized by the presence of certain suites of elements or minerals. Knowledge of these distributions allows exploration geologists to focus their attention on an area in which the chances of finding the minerals they seek are good.

The mineral typical of the Bear province is pitchblende. The uranium deposits of Great Bear Lake have been known for many years, but thousands of small deposits of similar mineralogy dot this province. Although several base metals have been found, gold is the mineral typical of the Slave province to the south, where uranium is practically unknown. The Superior region is Canada's richest structural province, and contains many mines from which a wide variety of precious and base metals has been extracted. The Grenville province is noted for its deposits of magnetite, particularly a variety rich in titanium. Since the provinces differ in history and structure, it is not surprising that they are characterized by different suites of ore minerals (Fig. 9–15).

CONTINENTAL ACCRETION

James Dana proposed, as a corollary to his geosynclinal theory, that continents grow by the acquisition of geosynclines welded on to their peripheries during mountain building. Since the late 19th century various modifications of this idea have been proposed. According to a recent model the continents were much smaller in Archean time, each made up of a series of granitic nuclei a few hundreds of kilometers across. A suggestion as to the location of these nuclei within the Superior province is shown in Fig. 9–16. The nuclei were welded together in Archean time by the deformation of greenstone belts between them. Later, they grew on the periphery by the incorporation of marginal geosynclines developing between the fused nuclei and the oceanic crust. The transformation of the roots of the geosynclines, when they are thrust deep into the crust and granitized, is believed to add to the sialic continental nuclei. Ultimately the source of the new granitic crust must be emanations from the mantle in the form of granitizing fluids, granitic magmas, or lavas, for the sedimentary erosion products from the granitic protocontinents in the geosynclines are not adding new material to the nuclei from which they were derived but are being recycled. The results of these processes should be that each continent has a nucleus or nuclei of great age away from which the successive, consolidated geosynclinal belts are younger.

Examination of a map of the Archean and Proterozoic provinces (Fig. 9–14) shows that the provinces decrease in age of deformation away from the nucleus of the Superior province, but the pattern is far from regular. South of the Superior province successive provinces underwent major deformation 1800, 1500, and 1000 million years ago. However, in other directions, the sequence is irregular, and isolated areas whose major deformation was Kenoran exist on the periphery of the craton.

Continental nuclei may also have grown originally by addition of granitic material derived from the mantle, not to their margins but to their undersides. This underplating process is poorly understood at present and difficult to investigate. It must have been confined largely to Archean time for crustal thickness and processes of deformation

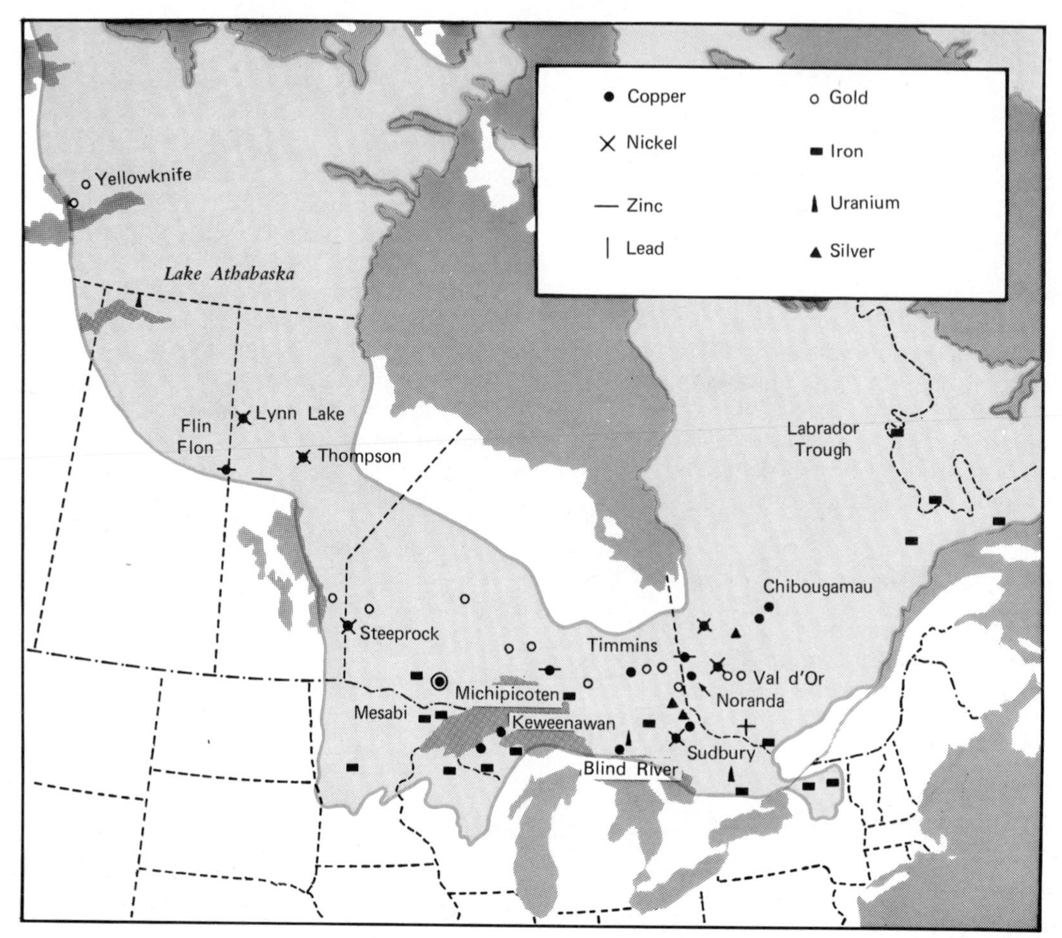

FIGURE 9–15 **Map of some major mineral deposits of the Canadian Shield.**

have been essentially constant from early Proterozoic time to the present.

Although the Archean and Proterozoic record of the origin and growth of the continents is obscure and still open to different interpretations, the general trend of shield history can be interpreted as one of increasing stability of the crust. Archean rocks reflect a thin crust over a hot, unstable mantle and relatively small continental nuclei. Aphebian rocks preserve the record of the first true geosynclines established marginal to the enlarged protocontinents. Helikian rocks, except in the Grenville province, are relatively undeformed sediments and lavas deposited on a stabilized craton cut by high angle faults. With the Grenville orogeny, the whole shield achieved a stability that has lasted to the present day. Since then it has acted as a cratonic area warped upward or depressed, but undeformed and without geosynclines.

ARCHEAN AND PROTEROZOIC FOSSILS

The Archean and Proterozoic eons have been characterized as the "age of microscopic life" for, apart from the stromatolites, all fossils so far found in rocks of these ages have been visible only under the microscope. Most of these fossil microorganisms have been found associated with stromatolites, and many of them were entombed in black cherts that preserve their

The oldest stromatolites are about 3000 million years old and also occur in the South African Archean succession in the Pongola System. Other Archean stromatolites have been found in the Steeprock Group in western Ontario. Further evidence of life in the Archean consists of trace quantities of complex organic compounds that have persisted through billions of years, and diffuse carbonaceous matter whose isotopic ratio of carbon 13 to carbon 12 is comparable to carbon of known organic origin.

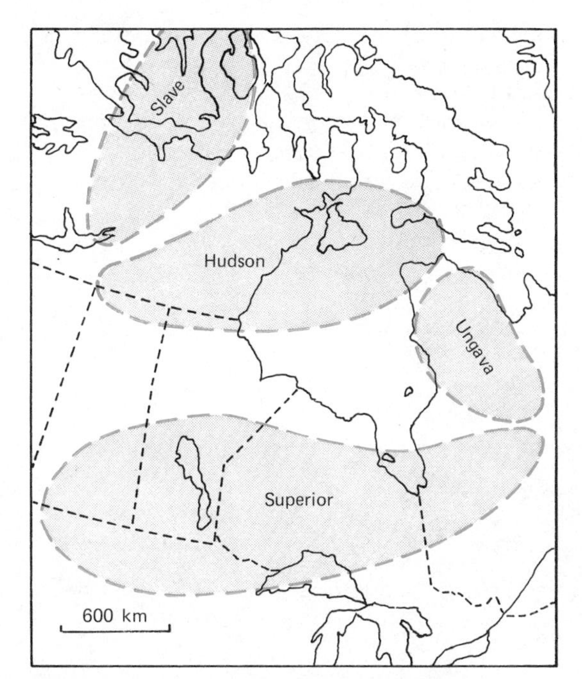

FIGURE 9–16 **Four postulated Archean protocontinents. The outlined areas represent the approximate relative sizes and position at some unspecified stage in the growth of the Archean crust. (After A. M. Goodwin; courtesy of the Geological Association of Canada.)**

delicate form with startling clarity. Micro-organisms have, so far, been recovered from about 30 localities in pre-Paleozoic rocks.

Archean Fossils

In the last chapter the emergence of the first prokaryotic cell from the primordial "soup" was discussed. Tiny spherical and ellipsoidal bodies about 20 micrometers across (Fig. 8–1A) from the Swaziland Group of the Transvaal in South Africa have been claimed to be the oldest fossils of such organisms. Although they bear considerable resemblance to bodies of non-biological origin and to inorganic bodies found in meteorites, these spheroids have been widely accepted as the oldest evidence of life on earth. The Swaziland rocks have been dated by different investigators as between 3300 and 3500 million years old.

Proterozoic Fossils

Although stromatolites are found in Archean rocks they do not become abundant until the beginning of Proterozoic time. So abundant and diversely built are the Proterozoic stromatolites that Russian geologists, in particular, have tried to show that they are valuable as guide fossils. However, the dependence of the shapes and sizes of stromatolites on the conditions under which they grew casts some doubt on their use for paleontological correlation.

The first unequivocal fossils of the micro-organisms themselves, rather than the stromatolitic structures that they built, are found in rocks of Aphebian age. The black cherts of the Gunflint Formation exposed along the north shore of Lake Superior at Schreiber have yielded a large group of prokaryotic organisms, some of which are illustrated in Fig. 8–1. Included in the Gunflint biota are cellular filaments that are the remains of blue-green algae, bacteria, and some micro-organisms of complex structure but doubtful affinity (Fig. 8–1F, G). The Gunflint micro-fossils are about 1900 million years old. Another important locality of Aphebian age in the shield is on the Belcher Islands in Hudson Bay (Figs. 8–5 and 9–11).

Considerable controversy surrounds the identification of the earliest eukaryotic fossils. When cells of all types are dried, the cell contents form a dark concentration of organic matter within the cell membrane (Fig. 9–17) that may look like the remnants of the nucleus, which is characteristic of the more advanced types of cells called eukaryotes (see Chapter 8). Some paleontologists claim that the first eukaryotic cells can be identified in the Gunflint fossils; others claim that they occur in the Beck Springs Dolomite of California (about 1300 million years old); and still others claim that they occur in the Bitter Springs Chert of Australia (about 900 million years old). The last of these formations contains beautifully preserved cells that appear to have nuclei, and some seem to have been caught in the act of cell division (Fig. 9–17). The origin of eukaryotic organisms cannot be placed accurately in time as yet, but the event is most likely to have taken place near the beginning of the Helikian Era.

Organisms with different kinds of cells

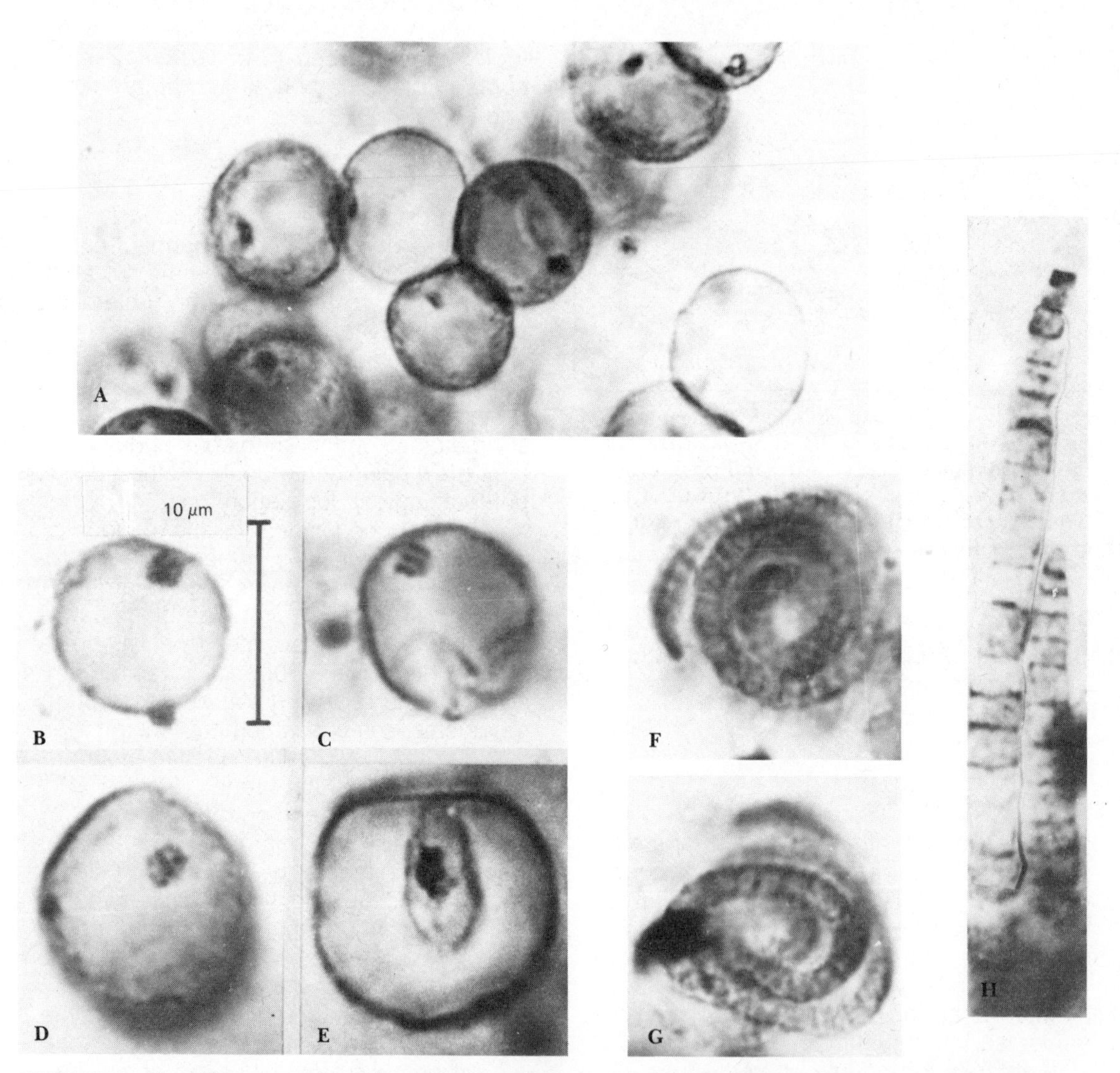

FIGURE 9–17 **Micro-organisms from the Late Proterozoic Bitter Springs Chert, Australia, all magnified about 2500 times. *a*. Cells of the green algae *Glenobotrydion aenigmatis* containing spots suggestive of intracellular structures. *b–e*. Single cells with internal granular areas suggestive of organelles and membranes. *f*, *g*. Coiled blue-green alga, *Contortothrix vermiformis*. *h*. Filamentous blue-green alga, *Cephalophytarion grande*. (Courtesy of J. William Schopf, *Journal of Paleontology*, and D. Reidel Publishing Co.)**

organized for different functions are called **metazoans;** those in which life functions can be carried out within a single cell are the **Protista.** Throughout the Archean and Proterozoic eons we have found no evidence of metazoans. For almost 2.5 million years the Protista were the only living things. About 700 million years ago in the Ediacaran Period at the start of the Paleozoic, the first metazoans appear and with them the first fossils larger than microscopic size.

SUMMARY

The Canadian Shield forms a continental nucleus of pre-Paleozoic rocks surrounded by flat-lying platform rocks and the deformed rocks of the marginal mountain chains.

Most of the area of the shield is underlain by metamorphic and igneous rocks. The shield includes the oldest rocks in the world (3800 million years) and preserves a record of five-sixths of geological time.

The adjustment of the topography to bedrock resistance and the similarity of the surface emerging from the covering platform sediments to that of the central part of the shield indicates that the characteristic low relief developed in pre-Paleozoic time.

Pre-Paleozoic time is divided into Archean and Proterozoic eons. The Proterozoic Eon is divided into three eras: Aphebian, Helikian, and Hadrynian, on the basis of orogenies at 1800 (Hudsonian) and 1000 (Grenville) million years ago.

The Archean rocks of the shield consist of complex synclines containing thick successions of graywackes and lavas downfolded into gneisses and folded during the Kenoran orogeny of 2500 million years ago.

The Archean crust was probably thinner and more mobile than the modern crust, and the continental segments were less extensive. Not until after the Kenoran orogeny can geosynclines and areas of cratonic deposition be distinguished in the rock record. Aphebian rocks of the shield, particularly those of the Great Lakes region, are rich in iron-bearing sediments.

The Canadian Shield can be divided into nine structural provinces on the basis of the pattern, degree, and age of deformation. The structural provinces can be traced beneath the Paleozoic and younger platform rocks. Much of the western platform of the United States is underlain by the Central Province, which was deformed between 1200 and 1400 million years ago. Each province of the shield is characterized by a particular set of ore deposits.

Since the time of the Grenville orogeny, the crust of the shield has been stable except for broad vertical movements.

The oldest traces of life are bacteria from rocks of South Africa about 3400 million years old. Stromatolites formed by early photosynthetic plants are found in rocks as old as 3100 million years. The recognition of the first eukaryotic cells has proved to be difficult due to the tendency of procaryotic cells to develop dark bodies like nuclei of eucaryotic cells during drying and preservation.

QUESTIONS

1. What features of Archean rocks separate them from Proterozoic and younger rocks and suggest that the tectonic behavior of the crust was different before the Kenoran orogeny?
2. How would you identify the original crust of the earth?
3. What important metallic resources are found in the Canadian Shield and where are the major areas of production of each?
4. By consulting reference works such as encyclopedias and other texts on regional geology, sketch on an outline map of the world the areas occupied by the pre-Paleozoic shields. Which

of the continents has the highest proportion of pre-Paleozoic rocks exposed? Which has the least?

SUGGESTIONS FOR FURTHER READING

BAER, A. J., ed. "Symposium on Basins and Geosynclines of the Canadian Shield." *Geological Survey of Canada,* Paper 70–40, 265 p., 1970.

CLOUD, P. "Beginnings of biospheric evolution and their biochemical consequences." *Paleobiology,* 2: 351–387, 1976.

CONDIE, K. C. "Archean Magmatism and crustal Thickening." *Bulletin Geological Society of America,* 84: 2981–92, 1973.

GLOVER, J. E., ed. "Symposium on Archean Rocks." *Geological Society of Australia,* Special Publication 3, 469 p., 1971.

KING, P. B., "Precambrian Geology of the United States; and explanatory Text to Accompany the Geologic Map of the United States." *U.S. Geological Survey,* Professional Paper 902, 1–85, 1976.

MOORBATH, S. "The Oldest Rocks and the Growth of the Continents." *Scientific American,* 236: 92–105, 1977

RANKAMA, K. ed. *The Precambrian.* New York: Interscience Publishers, 1963–1968.

SCHOPF, J. W. "Precambrian Paleobiology: Problems and Perspectives." *Ann. Review Earth and Planetary Sciences,* 3: 213–249, 1975.

STOCKWELL, C. H., and others. "The Geology of the Canadian Shield," in "Geology and Economic Minerals of Canada." *Economic Geology Report* 1, 45–150, 1968.

WALTER, M. R. "Interpreting Stromatolites." *American Scientist,* 65: 563–571, 1977.

CHAPTER TEN

Over large areas of the North American craton deposition of a lasting sedimentary record stopped at the end of the Aphebian Era and did not resume until Late Cambrian time, over a billion years later. Flat-lying Helikian sediments cover some areas of the Northwest Territories, Saskatchewan, and the Lake Superior Basin and contorted and metamorphosed rocks of probable Helikian age underlie the Grenville province, but over four-fifths of the Canadian Shield the youngest rocks are of early Proterozoic age. We cannot be sure that sediments did not accumulate during this billion years on the broad expanse of Archean and Proterozoic rocks stretching from Greenland to Arizona, but none remain to record conditions of that interval. Sedimentary rocks of late Helikian, Hadrynian, Ediacaran, and early Cambrian ages were deposited only along the edges of the cratons in the miogeoclines, and not until the beginning of late Cambrian time did the sea entering the cratonic interior of the continent again lay down a permanent stratigraphic record. During this immense period of erosion much of the shield achieved its present subdued topography, and the mountains formed by Archean and Proterozoic orogenies were reduced to plains of low relief mantled with thick residual regoliths (soils) (Fig. 10–1). Probably only in the Grenville province, where the youngest of the Proterozoic mountains had been raised, was there local relief of more than a few hundred meters. The local relief on the surface at the base of the Paleozoic systems can be appraised from maps showing the depth of closely spaced wells penetrating to that horizon, but the regional form of the surface has been changed by broad movements of the craton since it was made.

After the Grenville orogeny one billion years ago, the shield was not again subject to mountain-building forces. It responded to stresses in the crust by up-and-down movements of regional scale, accompanied locally by high-angle faulting. When the craton subsided below sea level the water from the oceans formed shallow seas over the interior of the continent. Such inward movements of the sea are called **transgressions**, the outward ones are called **regressions**. The shallow seas that resulted from transgressions are called **epeiric** (Greek: epeiros, a continent or mainland) or **epicontinental** (Greek prefix: epi, upon). Deposits from the successive epeiric seas accumulated on the craton to form the present crescent-shaped platform underlain by flat-lying sedimentary rocks that thicken away from the shield. At times the whole craton was covered by thin sedimentary deposits of such seas, but when they withdrew erosion soon stripped away most of the veneer,

THE SAUK SEQUENCE

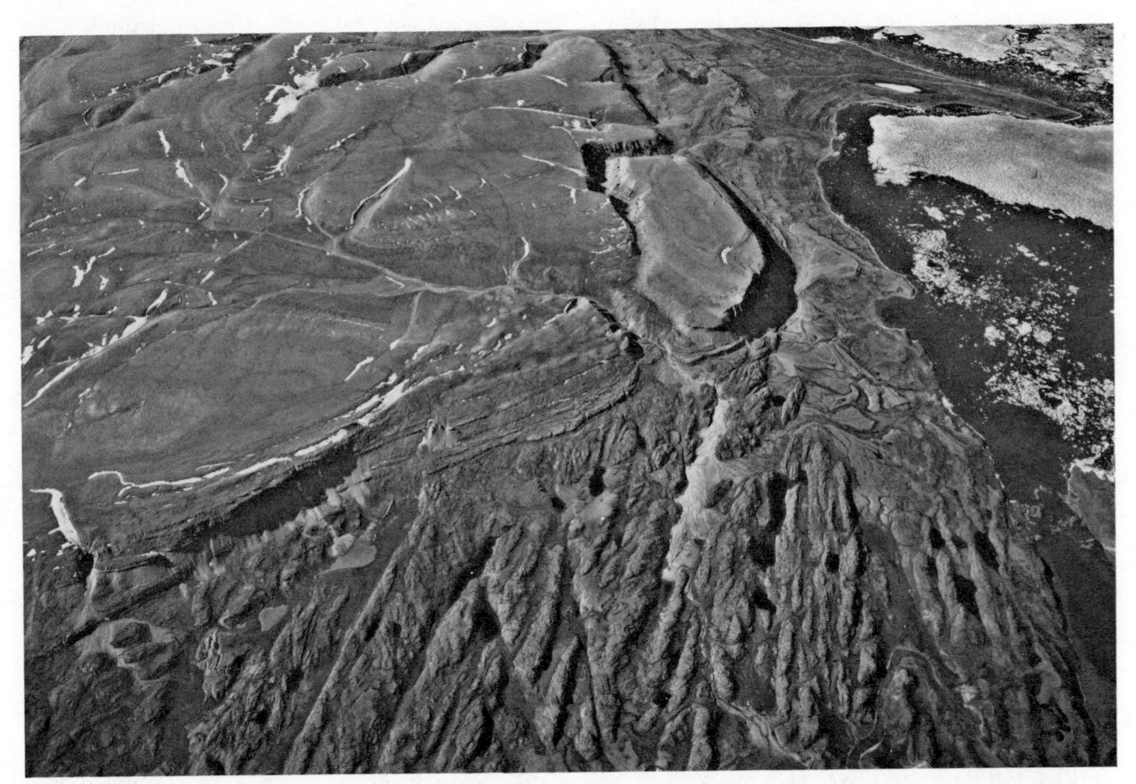

FIGURE 10–1 **Angular unconformity between flat-lying beds of Cambrian age and the folded metamorphic rocks of pre-Paleozoic age on the north coast of Devon Island, Arctic Canada. The unconformity marks the boundary between the Canadian Shield and the platform. (Original photo supplied by the Surveys and Mapping Branch, Dept. of Energy, Mines and Resources, Canada.)**

leaving only remnants scattered on the surface. The largest of these masses occupied the Hudson Bay lowlands. Among the remnants on the shield, limestones of Middle Silurian and Late Ordovician ages are most common showing that these were times of maximum transgression.

The tectonic behavior of the craton contrasted with that of the bordering geosynclines. Whereas subsidence and sedimentation were rapid and nearly continuous in the geosynclines, they were slow and frequently interrupted on the craton. This is reflected in the thinness of the sediments that cover the craton; their composition of clean sandstones, limestones, and evaporites; and the prevalence of unconformities interrupting the sedimentary record. The products of erosion supplied as sediments to the epeiric seas by cratonic sources were the result of deep weathering of continental rocks fundamentally similar to granite. The sediments supplied to bordering geosynclines were derived by the rapid mechanical erosion of complex source areas, some of oceanic type rocks.

ARCHES AND BASINS

Beneath the platform the strata are not strictly horizontal but undulate into broad synclines, anticlines, basins, arches, and domes. Dips on the limbs of these folds are very low, measurable in meters per kilometer rather than degrees. The Michigan Basin is a good example of a basin in Paleozoic rocks near the edge of the shield (Fig. 10–2). The geological map shows that the strata dip toward the center of the basin from all sides and resemble a set of stacked measuring spoons. The Cincinnati Arch serves as an example of a cratonic arch. Its axis can be followed southward from Ohio to Tennessee, where it is generally known as the Nashville Arch or Dome. Ordovician rocks are exposed in the breached crest of the anticline and Pennsylvanian rocks occupy the basinal areas on either side.

The present structure of the Michigan Basin and Cincinnati Arch is clear, but the history of these, and other cratonic structures, has been a subject of controversy. The formation of an arch may continuously accompany deposition, may be episodic, or may follow entirely the deposition of the rocks that form it. When arching accompanies deposition, the arch axis must subside more slowly than the surrounding shelf or basins. Because the thickness of the sediments that accumulate is proportional to the subsidence of an area (see Chapter 5), each layer of sediment will be thinner over the arch (Fig. 10–3A). If an arch is formed after deposition, the arched sedimentary strata will have been deposited in a shelf environment of uniform subsidence and will be of uniform thickness. The structure is later gently folded into an anticline and eroded to a flat surface (Fig. 10–3B). The arch may also be formed by intermittent uplift and truncation of beds, producing a series of unconformities (Fig. 10–3C). The problem of whether these cratonic structures were active during deposition requires careful study.

One of the tools at a geologist's disposal for determining whether a cratonic structure was active during its deposition is the isopach map (Chapter 4). If the upper and lower surfaces of the interval mapped on the isopach map are conformable, that is, if no erosion took place either before or after deposition of the beds involved, and the supply of sediment was adequate to make wavebase the control of sedimentation, then the map would show the relative subsidence of the area of deposition. If the surface below the sedimentary strata of the map is an erosional one, with sink holes or river valleys that are filled by the sedimentary layer, the isopach map will show thickening unrelated to subsidence in these depressions. If erosion has formed valleys in the upper surface of the unit after its deposition, the isopach map will show thick and thin areas independent of the rate of subsidence during deposition. Because erosion surfaces between strata are difficult to detect through the study of samples brought up in the drilling of an oil well, the history of cratonic arches worked out from such subsur-

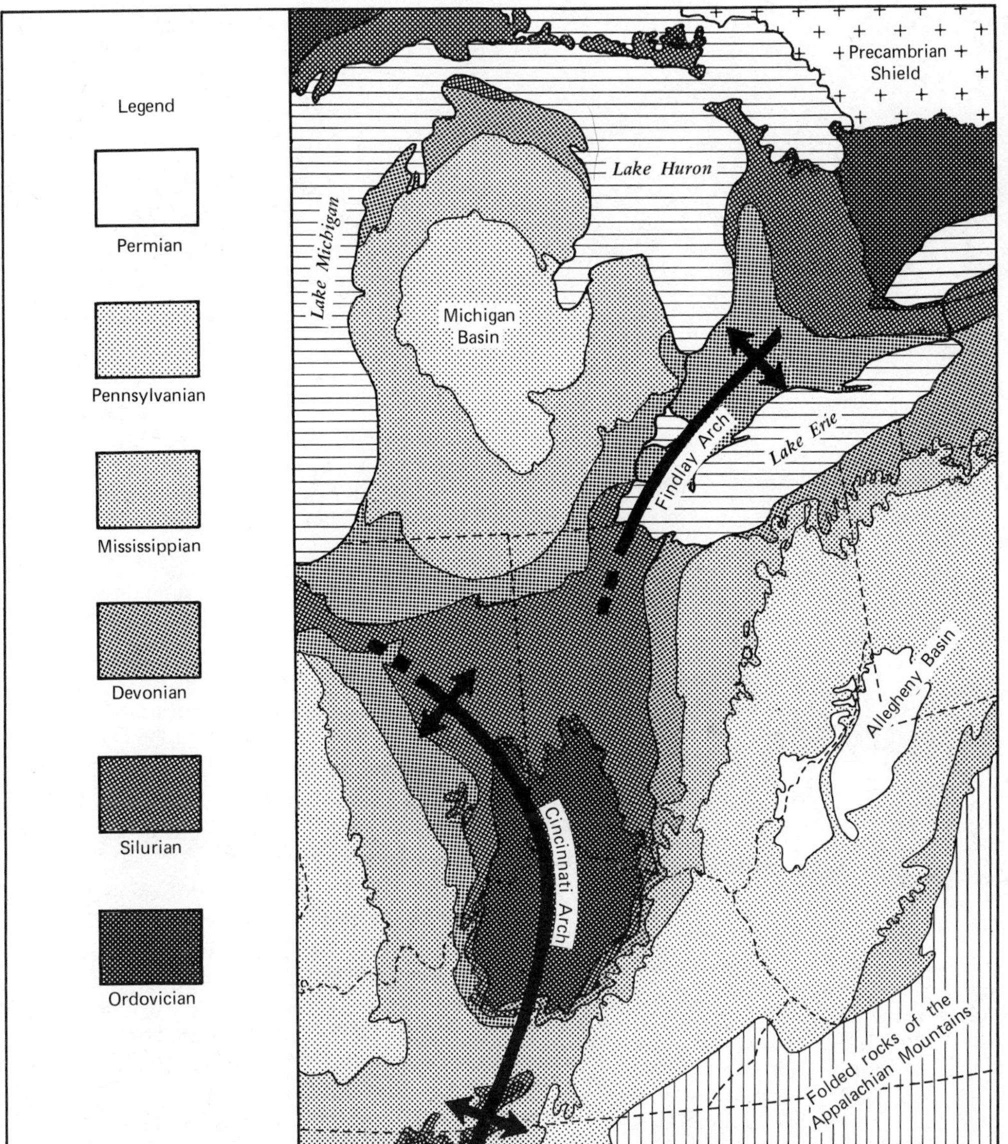

FIGURE 10–2 **Geological sketch map of the region of the Cincinnati arch showing the pattern of outcropping sedimentary strata around cratonic arches and basins. (Simplified from the Geological Map of North America.)**

face studies tend to be in dispute. Figure 10–4 is an isopach map of the Nashville Dome on the southern border of Fig. 10–2, showing that the Hermitage Formation thins over the axis. As both contacts of the Hermitage Formation in the Nashville area are erosion surfaces, the isopach map may, in this case, reflect the uplift and erosion of the arch before and after deposition of the formation and may indicate nothing about its rate of subsidence during Hermitage deposition. Thinning of the Cataract Formation over the Findlay Arch, that appears to be due to subsidence control, is shown in Fig. 4–8.

The distribution of formations on an ancient erosion surface has also been used as evidence for arching prior to distribution of beds over the unconformity. Maps that show the distribution of formations on buried un-

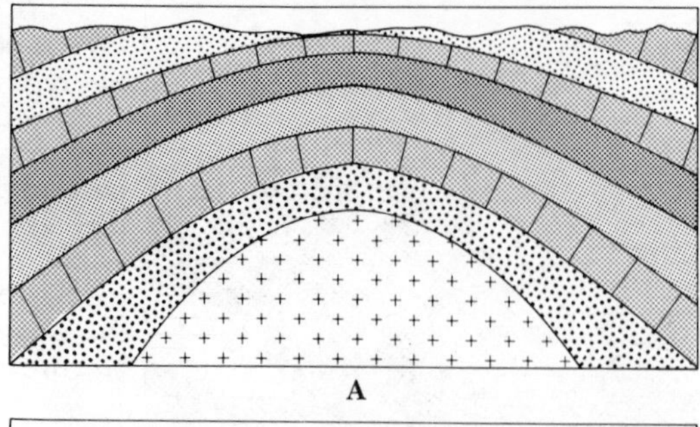

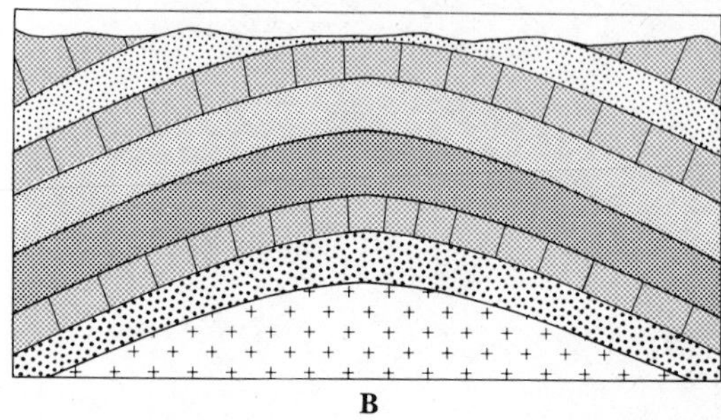

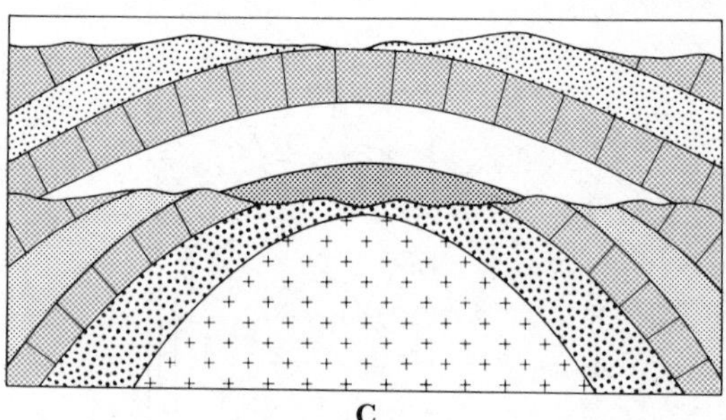

FIGURE 10–3 **Cross-sections of three types of cratonic arches. A. The strata thicken away from the crest of the arch, and the dip of the limbs therefore increases downward, indicating that the area of the arch subsided less than the surrounding basins during deposition of the sediments. B. The thickness of the strata does not change over the arch, indicating that the structure was formed after sedimentation stopped. C. A complex arch that was active during sedimentation, arched, eroded, and arched again after the deposition of a sequence of sediments.**

conformity surfaces are a type of paleogeologic map. If such a map shows an area of older strata between successively younger bands of strata, the geologist may conclude that this pattern reveals an arch whose uplifted axis has been eroded down to the older formations like the Cincinnati Arch.

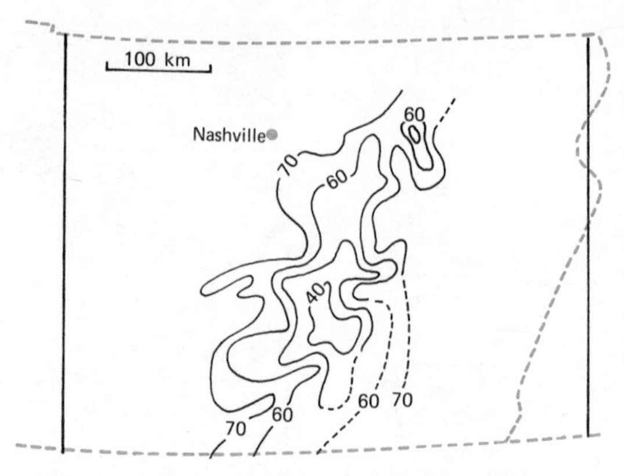

FIGURE 10–4 **Isopach map of the Hermitage Formation of central Tennessee in the region of the Nashville Arch. Contours in feet. (After C. W. Wilson; courtesy of the Geological Society of America.)**

However, a similar pattern of formations may be formed on an erosion surface by a valley of regional size eroded into flat-lying formations below an unconformity (Fig. 10–5).

The problem of the past activity of arches and basins may be approached from another direction. Subsidence and uplift affect not only the thickness of the sediments accumulating, but also the nature of the sediments. Commonly, areas of slow subsidence were characterized by shallow-water sediments showing signs of occasional emergence, such as mud cracks and flat pebble conglomerates, and were shoals where corals and other organisms that secrete calcium carbonate were concentrated. In contrast, basins of more rapid subsidence contain deposits of black shales and evaporite minerals, such as anhydrite and halite, that result from the restriction of circulation of sea water. Later we shall see that the Tathlina Arch localized a barrier reef in Devonian time and the Findlay Arch influenced the deposition of evaporites in late Silurian time.

In summary, isopach maps, cross-sections, and paleogeologic maps may help in the interpretation of the history of cratonic arches, but must be used with caution. The

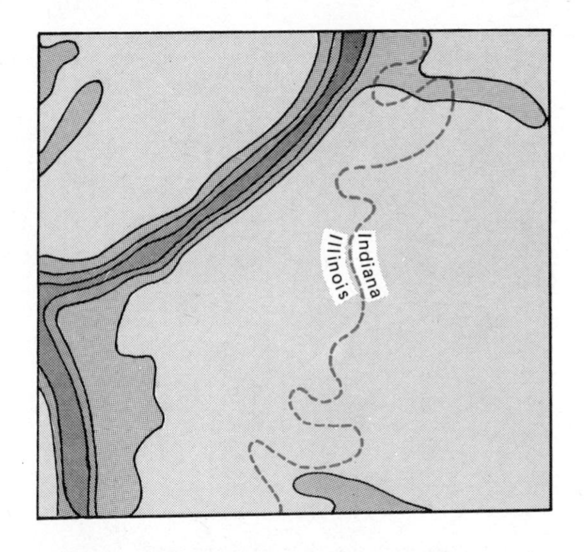

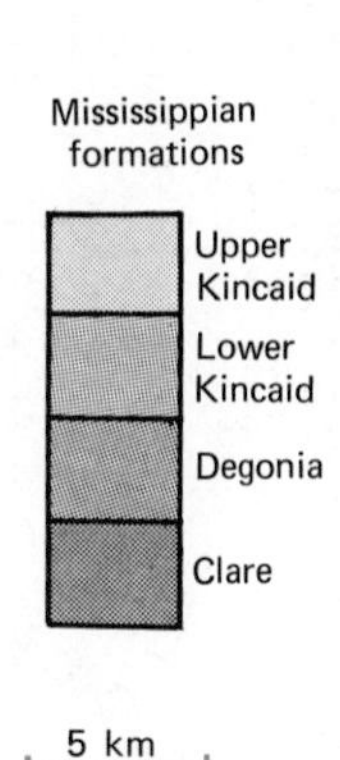

FIGURE 10–5 **Paleogeologic map of the sub-Pennsylvanian surface in southern Illinois showing the pattern of formations produced by channels cut into the Mississippian formations below the unconformity. The pattern of older formations exposed between successively younger ones is much like that produced by the arching and truncation of strata. (After R. Siever, Illinois Geological Survey.)**

most useful indication of the tectonic behavior of an arch or basin is its influence on the sediments accumulating around it.

THE EARLY PALEOZOIC PLATFORM

Early in the Paleozoic Era the arches and basins that became prominent later were not sharply differentiated from the surrounding shelf areas (Fig. 10–6). Peripheral to the shield was a group of basins separated by radial arches. South of the shield the Michigan and Illinois basins formed a common area of subsidence in Cambrian time but were separated later by the Kankakee Arch. The large basin of the early Paleozoic beneath North Dakota, Montana, and the Canadian Prairie provinces is called the Western Canada Basin. Separating these areas of subsidence was a broad, persistent structure called the Transcontinental Arch that can be traced from Wisconsin to Arizona. The broad arch extends from the salient in the shield border in central Minnesota and projects as a hill of Proterozoic rocks through the Mesozoic and Cenozoic platform sediments in South Dakota to form the Sioux Uplift. The arch is a complex structure and branches northwest and southeast to form the Cambridge Arch of western Nebraska and the Kansas Arch of eastern Kansas. The exposure of early Proterozoic rocks in the mountains of Colorado and southern Arizona may be related to their high position along this uplift, but along most of its length the structure lies deeply buried beneath Mesozoic and upper Paleozoic rocks. Although no lower Paleozoic rocks extend across the crest of the northern sector of the arch (see the facies maps Figs. 10–11, 11–1, 11–14, 11–17), the continuity of facies patterns from one side of the arch to the other, and the absence of thinning (other than by erosion) in formations as they approach its crest, is evidence that early Paleozoic seas crossed the arch and that their deposits were later removed by erosion. In Colorado some early Paleozoic strata extend across the arch at a structurally low section called the Colorado Sag.

The Western Canada Basin is bordered on the north by the Peace River and Tathlina arches. During much of early Paleozoic time the Meadow Lake Escarpment crossed northern Saskatchewan and appears to have affected the spread of early Paleozoic seas in this sector. A prominent line of normal faults crosses the cratonic shelf in northern Kentucky and constitutes the Rough Creek Fault zone. Thickening of the Cambrian succession by 700 m across this fault indicates that it was active in Cambrian time.

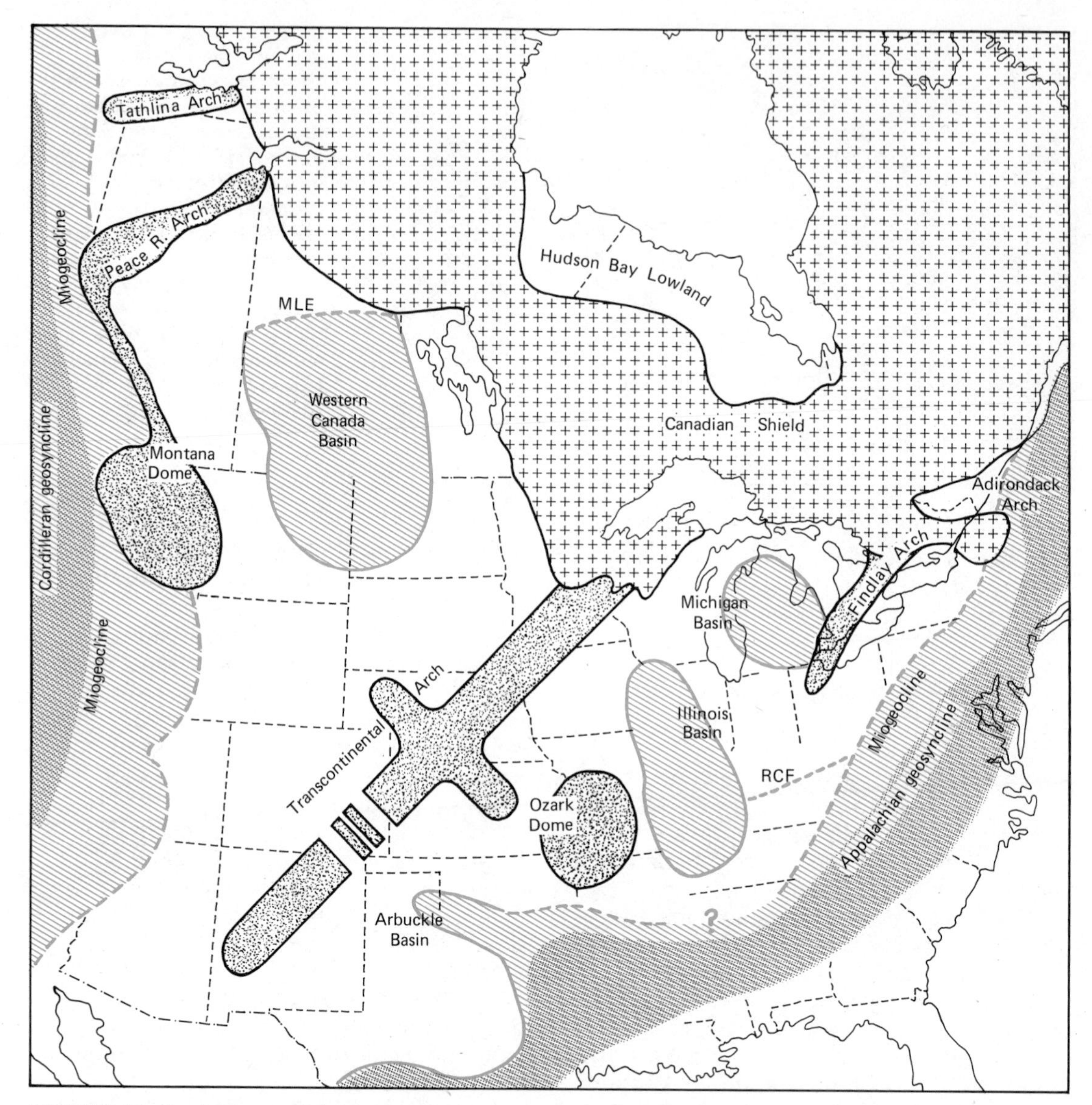

FIGURE 10–6 **Arches and basins active in the early Paleozoic Era in the North American craton. MLE = Meadow Lake Escarpment. RCF = Rough Creek Fault Zones.**

STRATIGRAPHIC DIVISIONS OF CRATONIC ROCKS

The cratonic rocks of North America have for years been divided into the systems and series of the standard relative time scale. This division, as discussed in Chapter 4, reflects the main stratigraphic divisions of northern Europe. Geologists in North America soon realized that the boundaries of the systems determined by paleontological correlation with Europe have little stratigraphic significance, and that most of the major breaks in deposition in North America do not correspond to the period boundaries. This does not mean that the standard time scale cannot be applied to North American rocks or that it is difficult to apply here, but merely that it is a scale independent of the major physical events of cratonic sedimentation. A stratigraphic division of the cratonic rocks can be made, however, on the basis of transgressions and regressions, and can be thought of as the relative time scale that might have been adopted if stratigraphic geology had originated in North America.

The stratigraphic record of the central part of the continent is full of unconformities. They are marked by missing fossil zones, surfaces of relief, basal conglomerates, ancient soils, zones of leaching, etc. When the cratonic basins were explored by drilling, geologists could trace these unconformities laterally to determine which were regionally persistent and which died out into regions of continuous deposition. They found that a few unconformities, ones not easily recognizable at a given locality, extended across basins and arches, and that about six extended across the whole continent. Figure 10–7 is a cross-section of the craton drawn to a highly exaggerated vertical scale to show the division of part of the sedimentary record by such regional unconformities. The stratigraphic units that are bounded by these regional unconformities are called **sequences.** Lawrence Sloss recognizes three unconformities as useful for defining four sequences in the Paleozoic Erathem.

Each Paleozoic sequence was deposited as the sea transgressed an erosion surface, reached full flood level, and then retreated. Neither the upper nor the lower boundaries of the sequences are of the same age throughout the craton. The transgressing sea is present for a longer time at the margins than in the center of the craton because it starts at the edges, spreads inward to the center only at its greatest flood, and retreats again to the margins. For this reason the time span represented by a sequence will differ from place to place. For example, the Sauk sequence, the lowest of the Paleozoic sequences, is represented in Wisconsin by Late Cambrian sedimentary rocks only, but in eastern Tennessee it includes rocks ranging in age from earliest Paleozoic to Early Ordovician.

According to Sloss, the unconformities that divide the Paleozoic cratonic rocks into sequences record three periods of withdrawal of the sea from the craton: at the end of Early Ordovician time, near the end of Early Devonian time, and at the end of Mississippian time. These unconformities separate the Sauk, Tippecanoe, Kaskaskia, and Absaroka sequences (Fig. 10–8). Younger rocks are divided into the Zuni and Tejas sequences by unconformities that record regressions in Early Jurassic time and at the end of the Cretaceous Period.

The relative thickness of transgressive

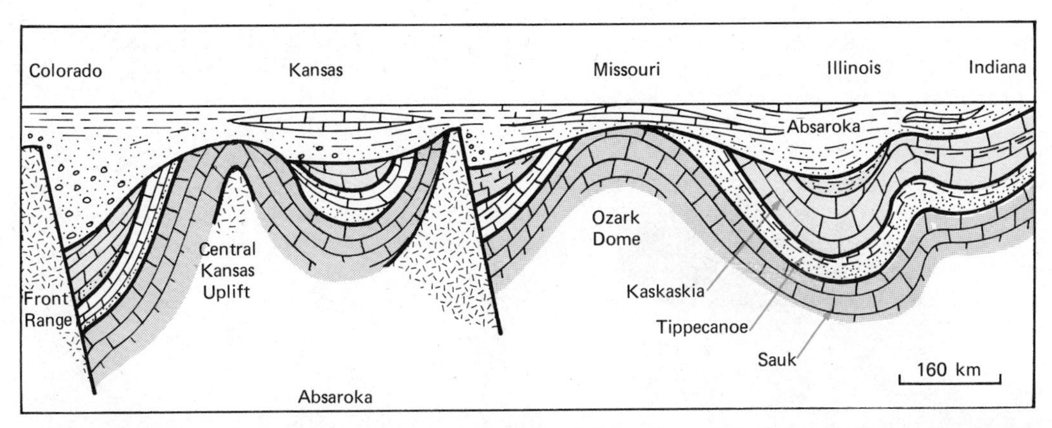

FIGURE 10–7 **Diagrammatic cross-section of the Paleozoic rocks of the platform showing the nature of the unconformities bounding the sequences. Vertical exaggeration extreme. (After L. L. Sloss; courtesy of the Geological Society of America.)**

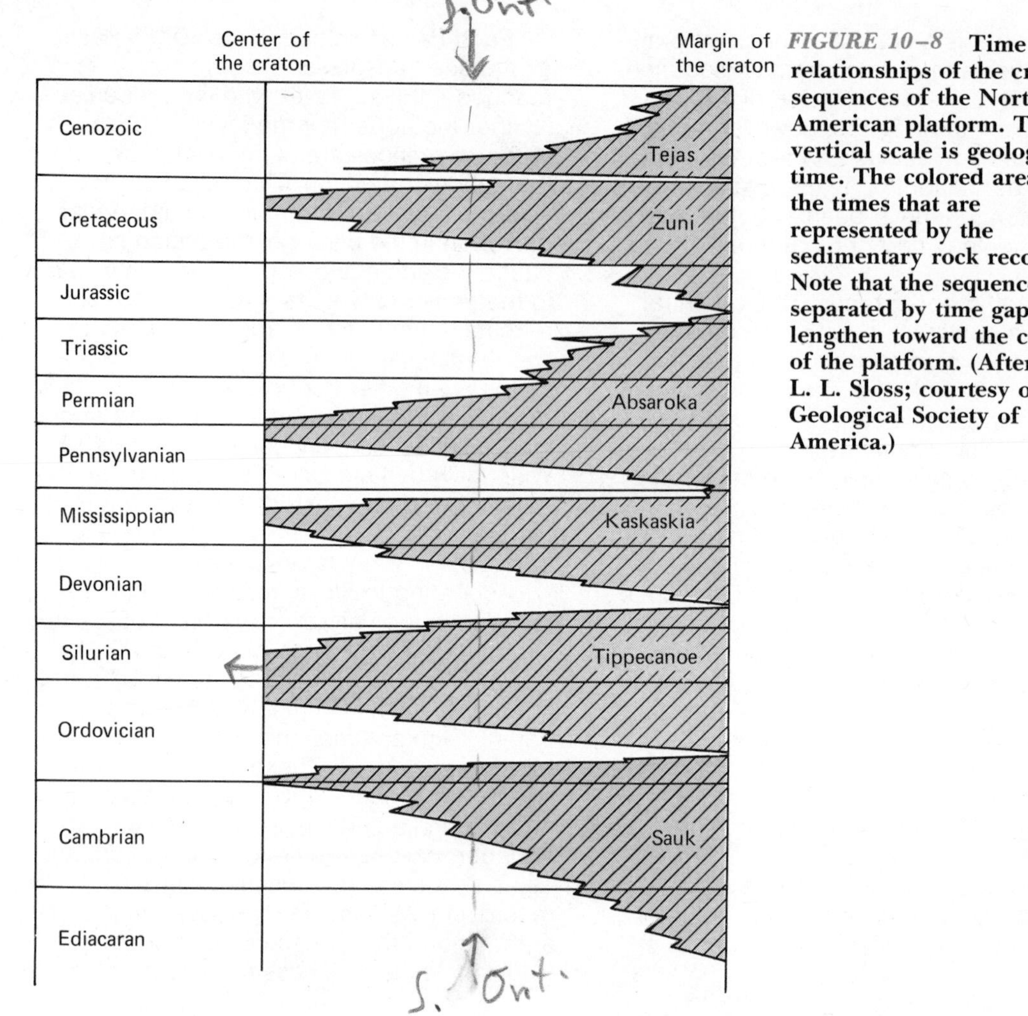

FIGURE 10–8 **Time relationships of the cratonic sequences of the North American platform. The vertical scale is geological time. The colored areas are the times that are represented by the sedimentary rock record. Note that the sequences are separated by time gaps that lengthen toward the center of the platform. (After L. L. Sloss; courtesy of the Geological Society of America.)**

and regressive deposits differs from sequence to sequence. The Sauk and Tippecanoe sequences have thick deposits laid down as the sea transgressed, and insignificant regressive deposits. The Absaroka, Zuni, and Tejas sequences have extensive and thick regressive deposits. When regression is a response to the uplift of the continent or a drop in sea level, deposits of the regressing sea are brought into the zone of erosion as the sea recedes. When regression is caused by building out of the land, as in a delta where rivers are delivering more sediment than the basin can accommodate below wave base, regressive deposits may be retained in the stratigraphic record.

LATE PROTEROZOIC ROCKS OF THE CONTINENTAL MARGINS

Proterozoic rocks along the western edge of the continent consist of the Belt Group of late Helikian age and the overlying Windermere Group of Hadrynian age and their correlatives. The Belt Group is exposed in the area of the Montana Dome of western Montana, northern Idaho, and adjacent British Columbia and forms the towering cliffs of Glacier and Waterton Lakes national parks (Fig. 10–9). These rocks and their correlatives in the Uinta Mountains of Wyoming and the Grand Canyon are further considered in Chapter 14. About 1 billion years ago they were gently folded, tilted, and

FIGURE 10–9 **Rocks of the Belt Group exposed in the mountains of Glacier National Park. Looking toward the plains across Edwards Mountain. (Official U. S. Air Force Photograph.)**

eroded to a low relief at the close of Helikian time.

In the closing era of the Proterozoic, the Hadrynian, the sea transgressed eastward from deeper water west of the Montana Dome and deposited the Windermere Group of eastern British Columbia and northern Washington. The group consists of about 200 m of shaly and sandy sediments. The Toby Conglomerate at the base of the group is composed of a great variety of rock fragments, and includes bedded argillites containing isolated pebbles that appear to have been dropped on to the site of deposition from above. This type of pebbly mudstone is now found in polar seas where floating icebergs release coarse glacial debris on fine sediments accumulating on the ocean floor. These argillites with "dropstones" are interpreted as indicative of glacial conditions in Hadrynian time on the landmass of the Montana Dome. The ice sheet is believed to have supplied icebergs that drifted westward in the initial Windermere seas.

Indications of a glacial climate in Hadrynian time are not confined to Washington and British Columbia (see also Chapter 7), but have been reported in North America from California, Utah, Wyoming, Alaska, Newfoundland, Virginia, and also from eastern Greenland. Tillites, as the consolidated deposits of ancient glaciers are called, and sediments associated with ancient glaciations, such as varved clays and pebbly mudstones, have been found in Hadrynian strata on practically all continents. Their wide distribution and apparent lack of relationship to either present or paleomagnetically determined past latitudes are puzzling. Possibly, this time was one of worldwide low temperatures; possibly, some of the deposits near paleoequators are incorrectly interpreted as tillites, for deposits of mudflows and turbidites can closely resemble tillites.

At the other side of the continent the rocks of the Great Smoky Mountains and the Blue Ridge record Hadrynian events. The lowest rocks in the succession in eastern Tennessee are 10,000 m of sandy sediments with abundant feldspar, called the Ocoee Group. The lithology of these sediments indicates derivation by rapid erosion from a granitic terrane. In Virginia the Mount Rogers Formation, probably a correlative of the Ocoee Group, contains conglomerates that have been interpreted as tillites of the Hadrynian glaciation.

BASAL BEDS OF THE SAUK SEQUENCE

The sediments that represent the beginnings of the first major Paleozoic sedimentary cycle—the Sauk Sequence—are the Chilhowee Group in the Appalachians and the quartz sandstones that overlie the Windermere Group in the Cordillera. The Chilhowee Group consists of quartzites and shales 2000 m thick, the top of which is of Cambrian age and the base of Ediacaran age.

The Ediacaran sandstones of the Cordillera reach great thicknesses (Fig. 10–10) and are also conformable above with Cambrian beds. The thickening of the sandstones away from the shield and the direction of cross-bedding in the sandstone indicate that these sediments were derived from the erosion of the shield. To determine the direction of supply of a sandstone, the geologist measures the direction of dip of as many cross-beds as possible in as many outcrops as possible. These measurements are treated statistically to produce an average direction for each outcrop or group of outcrops, and these averages are placed on a map. In a river system or under shallow water the transporting currents rarely run directly down the regional slope, but meander or are deflected by irregularities of topography and shoreline. Yet studies of recent cross-bedded sediments indicate that the average direction of dip of many cross-beds closely approximates the direction of transport of the sediment.

At the beginning of the Paleozoic Era all the rocks of the shield were exposed over the whole craton and remained so for several hundred million years, until the Sauk Sea spread out of the miogeoclines at the beginning of Late Cambrian time. Extensive weathering of this region of low relief must have formed a thick, sandy soil covering the gneisses, schists, granites, and sedimentary rocks. The sandy sediments at the base of the Paleozoic in the Cordillera and in the Ocoee and Chilhowee Groups appear to have been derived from this blanket of weathering products.

The Sauk Sea spread from the miogeoclines across the western shelf in Middle Cambrian time, and by the beginning of Late Cambrian time it covered the southern half of the craton, stretching from Montana to New York (Fig. 10–11). As the sea transgressed it first laid down a group of cross-bedded sandstones over the unconformity developed on Proterozoic and Archean rocks. Rivers from the shield swept quantities of sand to the advancing sea. The sediment was sifted and reworked by the waves to form a clean, sandy sea floor for miles behind the advancing coastline. The surface of the shield now exposes 80 per cent granite (including granite gneiss), and the same proportion probably was exposed in Cambrian time. When this bedrock was weathered, clay was formed from the decomposition of the dark minerals (hornblende, pyroxene, biotite, etc.) and the soda-lime feldspar of the granite, but the chemically stable quartz grains and, in part, the potassium feldspar, were left unchanged. The average igneous rock after weathering pro-

FIGURE 10–10 **Ediacaran and early Cambrian quartzites at Postern Peak in the Canadian Rockies. A thickness of almost 1200 m of these sedimentary rocks at the base of the Sauk sequence is exposed in this cliff. (Courtesy of the Canadian National Railways.)**

duces 79 per cent shale, 13 per cent sandstone, and 8 per cent soluble materials, eventually deposited as limestone and evaporites. Weathering of shield rocks sufficient to produce the great quantities of sand in the late Proterozoic and early Paleo-

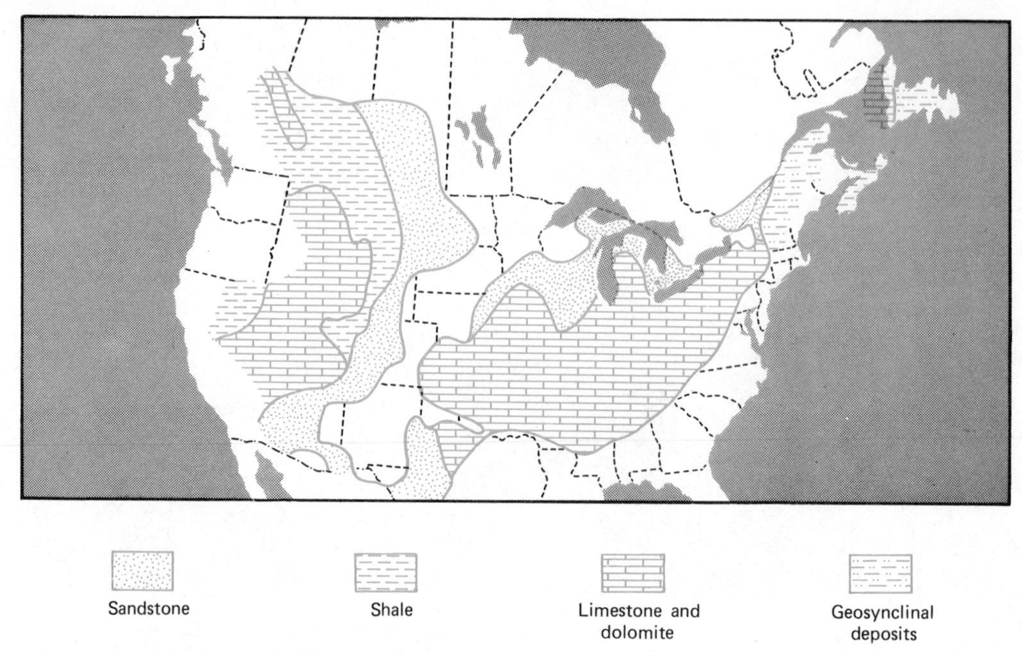

FIGURE 10–11 **Facies of the Upper Cambrian Series in central North America.**

zoic successions must have produced a great deal more clay, yet shales are uncommon in this interval. The clay may have been sifted from the sand by waves in the epeiric sea and been transported to the deeper water of the ocean basins or geosynclines, or may have been winnowed by the wind from the sand in coastal dunes and blown into the deep oceans.

Base of the Cambrian System

By convention the base of the Cambrian System is drawn at the first appearance of fossils of animals of complex organization, commonly brachiopods, trilobites, or archaeocyathids. This horizon occurs in both the eastern and western edges of the craton, within a thick succession of sandstones that shows little physical stratigraphic change at the boundary (Fig. 10–10). Fossils of metazoans do occur in the Ediacaran System, as is discussed below, but they are rare, and in most sedimentary successions of this interval the sudden appearance of the fossils of complex life records a dramatic event.

The boundaries of any time stratigraphic unit, such as a system, are time planes. The problems encountered by stratigraphers in defining the boundaries of time stratigraphic units on the basis of fossils are illustrated by the controversy over the placement of the base of the Cambrian System. Suppose that complex organisms bearing preservable hard parts evolved in a particular area of the ocean and spread rapidly around the world. In the basal Sauk sandstones of the northern Rocky Mountains fossils have been found only in thin shale lenses. Either the organisms that lived during the time of deposition of the sandstones were destroyed before preservation or the sandy sea bottom was not a suitable place for the animals to live. In areas where shale was accumulating, the evolution of complex organisms would be recorded immediately as fossils, but in areas where sand was accumulating, their appearance in the paleontological record would be delayed to a younger shale stratum. The placement of the horizon of oldest complex fossils in any given section will also depend on the care and persistence of geologists in finding the lowest occurrence of such fossils; the longer they look for fossils in a poorly fossiliferous succession, the lower they may find them. Influenced by factors of rate of dispersal, facies control, and collector persistence, the lowest record of complex fossils in widely separated successions of different facies is not likely to mark a time plane. Faced with the difficulties of defining a time plane for the base of the Cambrian, some geologists have used the first break in sedimentation or unconformity below the appearance of complex fossils as a convenient boundary with no pretensions of time significance. The problems of defining the base of the Cambrian is similar to those that surround the definition of any system boundary by fossils, but as this horizon marks an important event in the evolution of life as well as one commonly regarded as the beginning of the Paleozoic Era, it has received more attention than others. The discovery of a method of dating all kinds of sedimentary rocks in years would solve this and many other stratigraphic problems.

THE ORIGIN OF THE METAZOA AND THE CAMBRIAN FAUNA

The record of life preserved in Archean and Proterozoic rocks is entirely one of the PROTISTA, animals that maintain all their life functions within a single cell. Most protistan fossils are microscopic, and only where blue-green algal layers trapped sediment to produce stromatolites did protistans produce large structures. At the end of the Proterozoic Eon eukaryotic organisms were abundant, but animals composed of different cells specialized for various functions—the Metazoa—are not known to have existed. Tubes that appear to have been made by metazoans have been described from Proterozoic rocks (Fig. 10–12), but these objects, on further examination, have been

FIGURE 10–12 **Markings on the upper surface of a ripple-marked bed of Proterozoic quartzite (Huronian of Ontario). The regularity of the spindles and their apparent segmentation at first suggested they were formed by metazoans, but they are now believed to be of inorganic origin. (Courtesy of H. J. Hofmann and the Geological Survey of Canada.)**

attributed to inorganic processes. The first records of metazoan life are simple burrows of worm-like animals in clastic sediments of early Ediacaran age. In slightly younger rocks the burrows are more complex, and still higher in the succession are joined by traces of metazoans that moved on the sediment surface but left no other record of their form. The first impressions of the bodies of early metazoans occur in Ediacaran sediments just below the Cambrian boundary in several places in the world.

The Ediacaran fauna is best known from the Pound Quartzite in the Ediacara Hills of South Australia. The fossils occur as impressions on fine siltstone about 500 m below the base of the Cambrian System. This Ediacara assemblage is unlike anything in Cambrian or older Proterozoic beds. It contains several genera of jellyfish, two genera of sea fans (coelenterates; soft corals), two of segmented worms (annelids), and two that cannot be assigned to any known phylum but may be allied to the echinoderms (Fig. 10–13). Elements of this fauna are found also in rocks of Ediacaran age in Leicestershire, England, in southwest Africa, in the Soviet Union, and in Canada. The best preserved North American fossils of this age have been described from the Conception Group of eastern Newfoundland and consist of impressions of jellyfish and spindle-shaped organisms that are probably related to sea fans (Fig. 10–14). The beds of Ediacaran age in east-central British Columbia contain several complex trace fossils that must have been formed by metazoans and are thought by some to be trilobite tracks (Fig. 10–15).

The scattered occurrence of these rare fossils of soft-bodied forms hundreds of meters below the lowest Cambrian fossils is in contrast to the abundance of the fossil fauna that appears in Early Cambrian rocks. The arthropods, represented mostly by trilobites, make up 32 per cent of the Early Cambrian fauna. The archaeocyathids, an extinct group of sponge-like animals of coelenterate affinity, compose 30 per cent; the brachiopods, 20 per cent; the mollusks, mostly gastropods, 10 per cent; the annelid worms, 3 per cent; and the protozoans, sponges, coelenterates, echinoderms, and doubtful forms, about 1 per cent each. Of the 10 invertebrate phyla commonly preserved as fossils, 7 are present in Early Cambrian time and all have representatives in the fossil fauna before the end of the Cambrian Period. The advance from single-celled organisms of the late Proterozoic to the complex metazoans with preservable hard parts of the Cambrian is a unique event in life history and demands a unique explanation.

The appearance of metazoans in the paleontological record is followed closely by the appearance of animals with hard parts that are readily fossilizable. The fauna of the Ediacaran Period is composed of soft-bodied organisms that left their record in trails, burrows, and impressions of their bodies in fine sediment. The fact that this fauna has been found in several places around the world in essentially correlative beds and that similar, or more primitive metazoans have not been found in older beds suggests that metazoans originated as burrowing, worm-like animals at the close of Hadrynian time. The attainment of the metazoan condition apparently lead to a rapid diversification in Ediacaran time during which the major patterns of organization of the invertebrate phyla were established. This diversification was soon followed by the development of the ability to secrete a skeleton that could be fossilized. The causes of these events have been the subject of intense research and speculation since the beginning of paleontology as a science.

L. Berkner and W. Marshall reasoned that complex life could not develop until oxygen in the atmosphere reached about 1 per cent of the present atmospheric level (the Pasteur Point), at which time respiration could be supported (see also Chapter 8). They believe that such a level was reached near the beginning of the Paleozoic Era and that its attainment triggered the rapid evolution of the metazoans from eukaryotic protistans. According to their theory the level of oxygen in the atmosphere did not reach present-day values until the middle of the Paleozoic Era.

FIGURE 10–13 **Fossils from the Ediacara Hills, South Australia. A. *Tribrachidium,* an enigmatic organism possibly related to the echinoderms. B. *Glaessnerina,* a sea fan. C. *Cyclomedusa,* a jellyfish. D. *Spriggina,* a segmented worm. E. *Parvancoria,* of unknown affinity. Various magnifications. (Courtesy M. F. Glaessner.)**

FIGURE 10–14 **Enigmatic fossils of Ediacaran age of the Conception Group, Avalon Peninsula, Newfoundland. The spindle-shaped forms are about 5 cm across. (Courtesy of S. B. Misra and M. M. Anderson and the Geological Society of America.)**

Kenneth Towe points out that the synthesis of skeletons has a high oxygen requirement and suggests that the sudden appearance of animals with hard parts could also be explained by changes in the atmospheric oxygen level. Donald Rhoads and John Morse investigated the validity of this hypothesis in the fauna of modern marine basins deficient in oxygen at depth. They found that in the depths the only organisms present were burrowing worms without hard parts, and that animals with skeletons lived only in the upper waters where the oxygen content in the water was more than that in equilibrium with 10 per cent of the present atmospheric level. They suggest that this distribution indicates that at the time when skeleton-secreting organisms appear, the oxygen content of the atmosphere was ten times that postulated by Berkner and Marshall. In Chapter 8 we saw that other paleontologists prefer an atmospheric model in which the present atmospheric level of oxygen is reached near the time of the origin of the eukaryotes in mid-Proterozoic time. Paleontologists need a geochemical method of determining oxygen levels of the past before this problem will be resolved.

Many other hypotheses have been proposed to explain the sudden appearance of animals with skeletons.

1. Early metazoans were small, soft-bodied, floating forms that had no need of skeletons. At the beginning of the Cambrian they discovered that the ocean floor in shallow water was an

FIGURE 10–15 **The trace fossil, *Diplichnites*, from beds of the Gog Group, Alberta, believed to be formed by the crawling of an arthropod. (Courtesy of F. G. Young and the Geological Survey of Canada.)**

advantageous place to live and, in need of support against wave turbulence, secreted skeletons.

2. Not until Cambrian time did active predators appears among metazoans. In response, those preyed upon developed hard, protective coverings that became fossils.
3. The chemistry of the seas before Cambrian time was not conducive to the secretion of shells. As we have seen, apart from oxygen content, its composition is unlikely to have been much different then.

In the absence of a fossil record of the splitting of the metazoans into the invertebrate phyla that appear in the Ediacaran and Cambrian periods, we can only rely on the comparative anatomy of these forms to suggest their relationship to one another and the order in which they might have appeared. Figure 10–16 is a schematic representation of the phylogeny of the invertebrate phyla on the basis of such analysis. At present no explanation of the radiation of invertebrate types in Ediacaran time or of their acquisition of skeletons in early Cambrian time is wholly satisfactory.

THE SAUK TRANSGRESSION

In the epeiric seas that covered an increasing proportion of the continent as Cambrian time passed, sediments accumulated in three concentric facies belts. The inner, near-shore belt consisted of light-colored sands, silts, and muds with thin limestone beds. The intermediate belt, which lay seaward of the first, was the site of accumulation of pure limestones and dolomites of many types. The outer belt, near the marginal geosynclines, was characterized by dark-colored silts, muds and sands, impure limestones, and chert deposited in deeper water. During Cambrian time these facies belts shifted back and forth across the craton so that the various lithologies now interfinger. The pattern of sedimentary facies in a transgressing sea is beautifully illustrated along the walls of the Grand Canyon (Fig. 10–17). While the Lower Cambrian Tapeats Sandstone was being deposited in the shore zone, the Bright Angel Shale was accumulating in deeper water offshore. In the walls of the canyon, the beds of the Tapeats Sandstone can be traced laterally into the Bright Angel Shale, which in turn grades into the Muav Limestone. The Bright Angel Shale is a good example of a formation that represents different intervals of deposition at different places. As the shoreline moved slowly eastward, the site of shale deposition moved eastward with it so that the formation, which is partly of Early Cambrian age in the western part of the Canyon, is wholly of Middle Cambrian age in the eastern part. Formations with different ages in different places are products of every transgressing sea and are, therefore, commonplace in stratigraphy, but their relationships are more obvious in the deposits of the craton where transgressions are slow. The regional picture of the facies changes from sandstone to

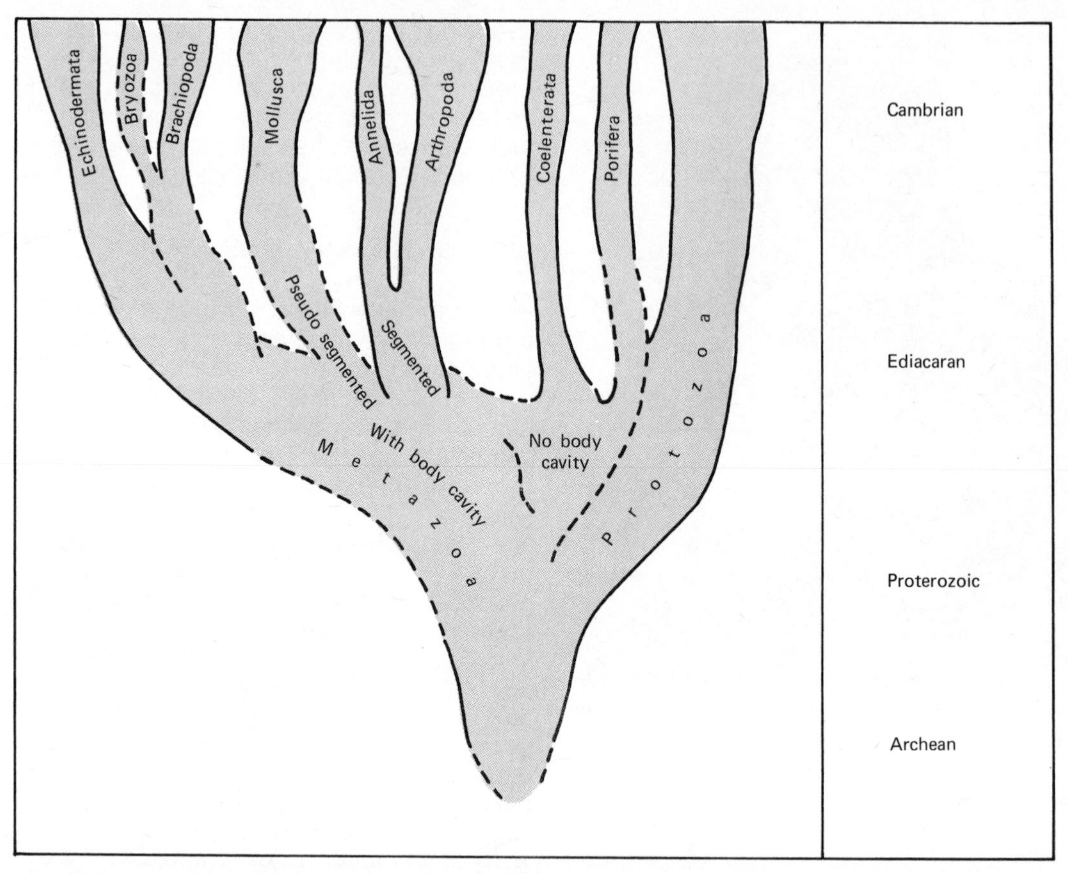

FIGURE 10–16 **Phylogeny of the major invertebrate phyla. The time scale on the right is intentionally vague. Dotted lines indicate an absence of record. (Suggested by the work of J. W. Valentine.)**

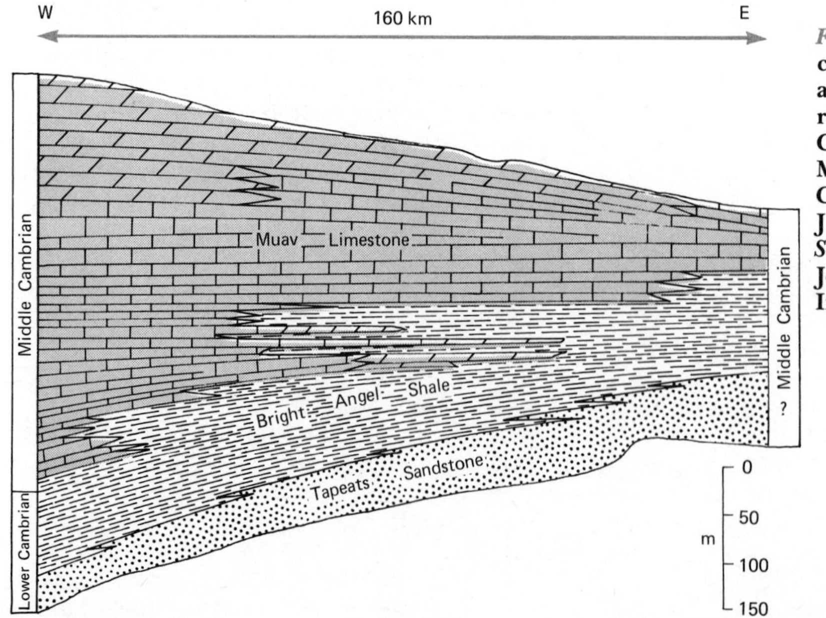

FIGURE 10–17 **Facies changes in the Middle and Upper Cambrian rocks of the Grand Canyon. (After E. D. McKee, redrawn from C. O. Dunbar and J. Rodgers, *Principles of Stratigraphy*. New York: John Wiley and Sons, Inc., 1957.)**

shale to carbonate in the Sauk transgression from the west is shown in Fig. 10–18.

The sediments deposited in the Sauk Sea at its greatest extent in Late Cambrian time are shown in Fig. 10–11. The sandstone facies extended as a collar about the edge of the Canadian Shield from Newfoundland to Alberta. Beneath the Upper Mississippi drainage basin, the sandstone that is exposed along the border of the shield in Wisconsin and Minnesota grades into limestone and dolomite that is exposed in the Ozark, Llano, and southern Oklahoma uplifts. Well records show that only under the western plains does a shale facies intervene between the sandstone and the limestone. This lateral facies change from sandstone directly into limestones, with shale omitted, is common in sediments of the shelf environment. In the limestone seas of the southern part of the continent, the water was warm and clear and a precipitate of calcium carbonate accumulated undiluted by detrital sediments from the distant land. (The term "detrital" refers to sedimentary particles that have been eroded from other rocks). Animals that extracted calcium carbonate from the sea water for their shells had not evolved in sufficient numbers at that time to form organic limestones. For this reason the limestones are believed to be either inorganic precipitates from saturated sea water or precipitates formed by marine algae. Common textures of these Cambrian limestones are: **oölitic** (composed of spheres of concentric structure the size of fish eggs), **pisolitic** (spheres the size of peas), and **pelletoidal** (formed of ellipsoidal, structureless pellets made by animals that pass sediment through their bodies for nourishment). Fossils of marine algae are common in the limestone. Figure 10–19 shows a large mass of algal limestone embedded in Late Cambrian sedimentary rocks in central Texas. Stromatolites like those in Proterozoic rocks are also a con-

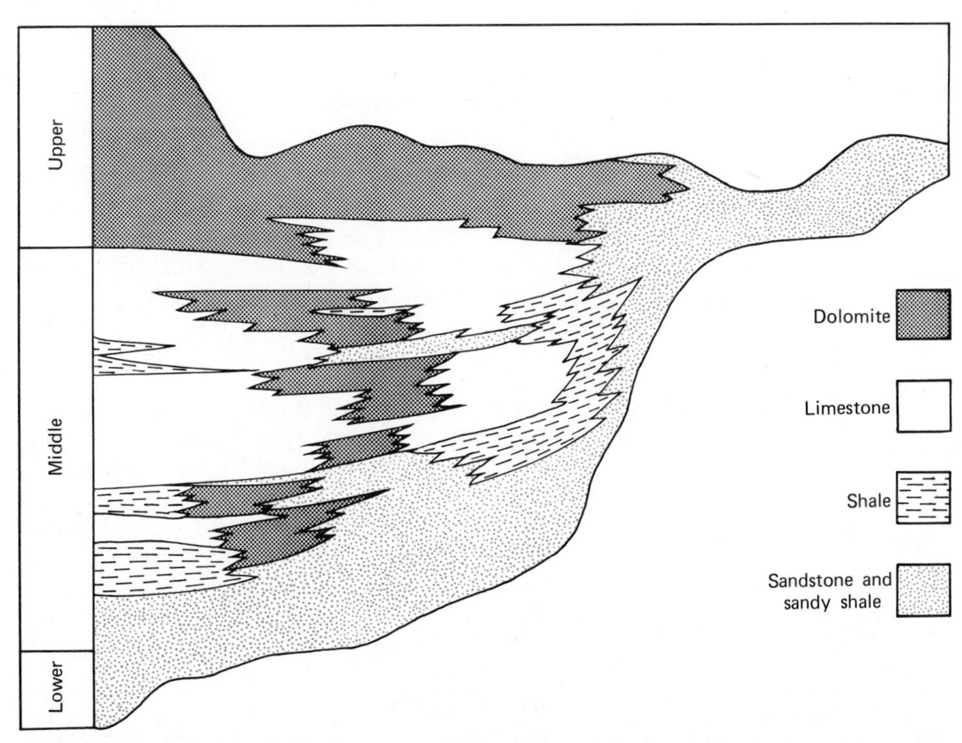

FIGURE 10–18 **Facies diagram of the Cambrian System from southern Utah to southwest Colorado. Maximum thickness 700 m. (Modified from *Geologic Atlas of the Rocky Mountain Region*.)**

FIGURE 10–19 **Reef of algal limestone embedded in Upper Cambrian shale, Llano River, Texas. (Courtesy of P. E. Cloud, Jr., from P. E. Cloud, Jr., and V. E. Barnes, *University of Texas Publication 4621.*)**

spicuous feature of late Sauk carbonates (see Fig. 10–23).

The pattern of Late Cambrian sedimentation persisted into Early Ordovician time with little change. Sand was laid down on the edges of the shield, and limestone continued to accumulate over most of the southern and marginal parts of the continent in Early Ordovician time. No physical break marks the boundary of the periods in the sandstone or limestone units, but Early Ordovician strata can be distinguished from Late Cambrian strata on the basis of fossils.

CAMBRIAN LIFE

Although during the Cambrian Period all the major invertebrate phyla appear in the fossil record, most of them are represented by rare specimens of a few genera. Nearly all Cambrian fossils belong to either the trilobites, brachiopods, or archaeocyathids.

Archaeocyathids

These animals form an extinct group, possibly of phyletic rank, whose characters place them between the sponges and the corals. Individual archaeocyathids are hollow, cone-shaped, and consist of inner and outer porous, calcareous walls separated by a series of radial septa (Fig. 10–20). The animals were abundant in certain parts of the Early Cambrian sea floor, and in places built small mounds that have been called reefs. They became rare in the Middle Cambrian Epoch, and by the end of the epoch they were extinct. The archaeocyathids are commonly the first fossils to appear above the Ediacaran–Cambrian contact where the transition takes place in the limestone facies, and were the first animals to secrete a strong calcareous skeleton.

Brachiopods

Most of the brachiopods of the Cambrian Period were inarticulates. These primitive

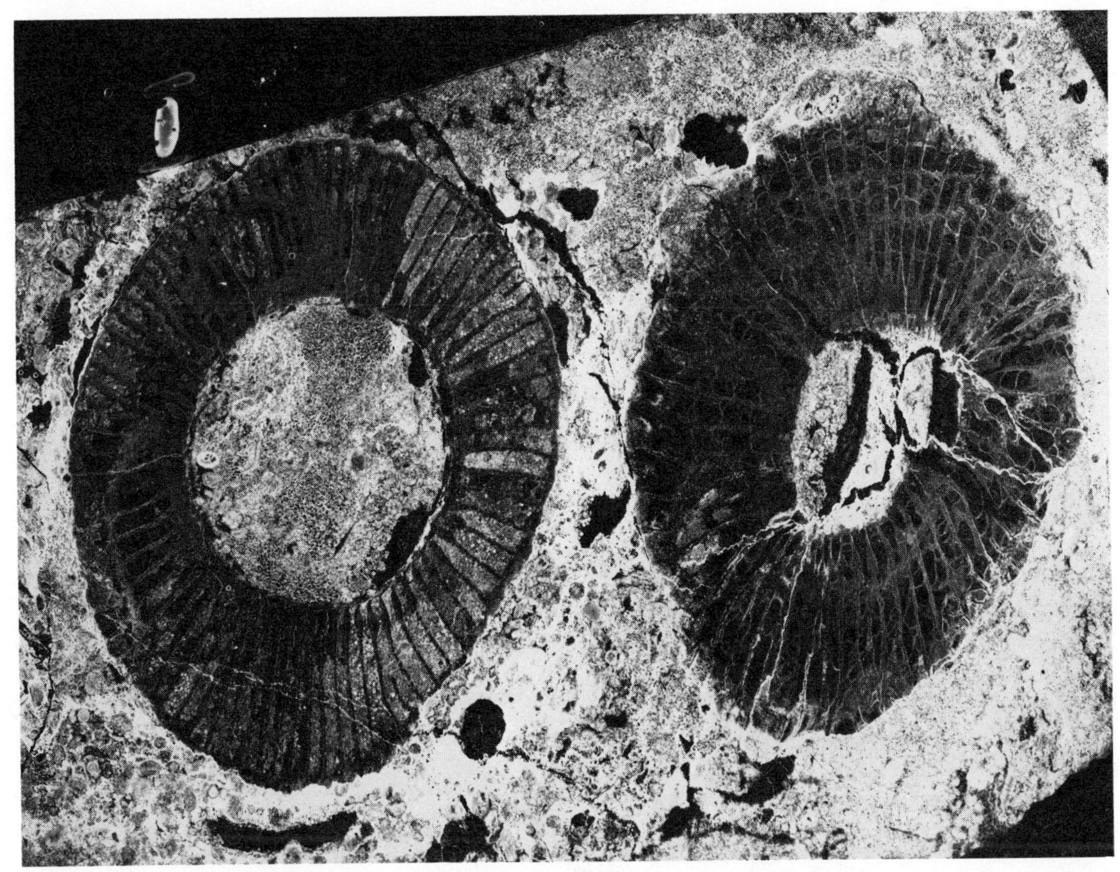

FIGURE 10–20 **Photomicrograph of a transverse thin section of two archaeocyathids, *Coscinocyathus* (*left*) and *Pycnoidocyathus* (*right*). Note the inner and outer walls and the central cavity. (Courtesy of V. J. Okulitch.)**

types secrete a pair of valves made up of layers of chitin and calcium phosphate held together by muscles attached to the interior of each valve, but lack a mechanism to lock the valves together. The Cambrian inarticulates were mostly small, conical shells now turned brown or black by the carbonization of the organic constituents.

The articulate brachiopods appear in Lower Cambrian strata along with the inarticulates, but did not become abundant until the Ordovician Period. These brachiopods secrete two valves of calcium carbonate with an interlocking hinge mechanism involving teeth and sockets to control their opening and closing.

Trilobites

The trilobites form an extinct class of the phylum Arthropoda. They are characterized by a longitudinally three-lobed, transversely segmented, chitinous carapace (or dorsal shield). Although most of them ranged between 2 and 10 cm in length, they included the largest animals of the Cambrian Period, and specimens of the genus *Paradoxides* attained a length of 50 cm. One-third of the Cambrian fossils described are trilobites. The first trilobites that appear at the base of the Cambrian System differ from most later ones in that the facial suture, along which the head shield splits to allow easy molting, followed the frontal margin and was therefore not visible from above (Fig. 10–21). The central raised part of the head shield, known as the **glabella**, was more extensively segmented than in later trilobites. There were as many as 45 body segments in some early genera, whereas most later trilobites had 10 to 12. These early trilobites have been placed in the suborder OLENELLINA after the most familiar member of the group, the genus *Olenellus.* Because the olenelline fauna is distinctive, restricted to the rocks of the Early Cambrian Epoch, and nearly worldwide in distribution, its presence makes the identification of Lower Cambrian beds relatively easy. The Early Cambrian fauna is found in so many parts of the world that it is referred to as a cosmopolitan fauna, although some of the member genera apparently preferred to live on the floor of geosynclines and others in the shelf seas. Middle and Upper Cambrian trilobites formed assemblages, each of which is largely confined to a geographic area or faunal realm. Marine faunal realms can also be identified in the oceans of today.

The Middle and Late Cambrian trilobite faunas of the world can be grouped into three faunal realms of which two are represented in North America. In each realm the fossil population has a certain unity and an overall dissimilarity to those of neighboring realms. The Cambrian trilobites of North America and Europe occupy realms that correspond to the main tectonic divisions of the continents into geosynclines and cratons. The geosynclinal realm includes mud-loving trilobites that flourished in the deeper waters of the geosynclines and now appear in the Appalachian rocks of eastern New England and the Acadian region of the Maritime provinces, and also on the west

FIGURE 10–21 ***Paedumias nevadensis*, an olenelline trilobite, closely related in form to *Olenellus*. (Courtesy of V. J. Okulitch.)**

coast. In the Middle Cambrian Series *Paradoxides* is the common trilobite in this realm: *Olenus* characterizes the Upper Cambrian Series (Fig. 10–22). The trilobites of the cratonic realm lived in shallow, clear, sandy epeiric seas. During the Middle Cambrian Epoch this fauna was characterized by *Albertella* and *Bathyuriscus* in the western interior and by many large-tailed genera. They were replaced in Late Cambrian time by *Dikelocephalus* and *Crepicephalus* (Fig. 10–23) and by a host of other trilobites.

The trilobites reached their peak in Late Cambrian time, and thereafter declined to extinction at the close of the Paleozoic Era. The earliest trilobites apparently crawled on the sea floor or swam for short distances in the shallow water. From this beginning the group diversified to become mud burrowers, active swimmers, passive floaters, and crawlers.

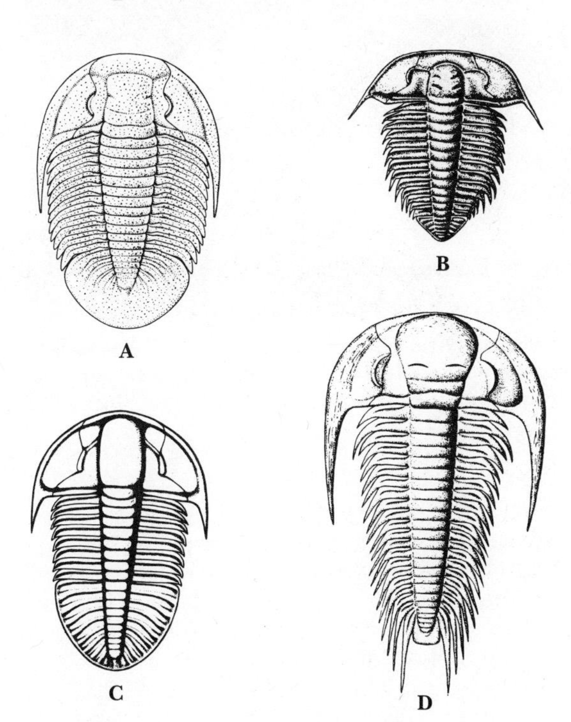

FIGURE 10–22 **Cambrian trilobites. A. *Tellerina*. B. *Olenus*. C. *Ogygopsis*. D. *Paradoxides*. A and B are Late Cambrian genera; C and D are Middle Cambrian. (From *Treatise on Invertebrate Paleontology*, courtesy of the Geological Society of America and the University of Kansas Press.)**

The Burgess Shale Biota

As the shoreline of the transgressing Sauk seas moved eastward from the Cordilleran miogeocline, sandstone accumulating there in early Cambrian time was succeeded by interbedded shale and carbonate in Middle Cambrian time and by carbonates alone in Late Cambrian and Early Ordovician time (Fig. 10–24). Middle Cambrian rocks are referred to as the Albertan Series, because they are grandly exposed in Alberta along the continental divide, where they form a rugged terrane with towering cliffs of over 1500 m. The Cambrian geology of this area was first described systematically by the American paleontologist Charles D. Walcott. In 1910 he made one of the most significant paleontological discoveries of all time on the trail through the Burgess Pass. This trail starts from the town of Field and zigzags up the steep slope of the Kickinghorse Valley to a height of 2000 m. From Burgess Pass it swings eastward toward Emerald Lake along the flank of a knife-edge ridge, joining the twin peaks of Mt. Wapta and Mt. Field. On the trail below this ridge one of the pack horses in Walcott's train overturned a slab of argillite, revealing fossils pressed like lustrous film on the bedding planes. This slab from the scree was easily traced back up the slope on "Fossil Ridge" to the bedrock from which it had fallen. In subsequent quarrying Walcott removed a large volume of the shale lens from which the slab had fallen and shipped it to Washington where he could study its fossils at leisure.

The shale, which was later called the Burgess Shale by Walcott, contains beautifully preserved fossils of soft-bodied animals that would usually decay on the ocean floor before being buried. The delicate hairs on the bodies of worms, the antennae of crustaceans, the legs and gills of trilobites, even the internal organs of some animals, are all preserved in this shale (Fig. 10–25). Structures and organisms that are nowhere preserved in younger deposits are revealed

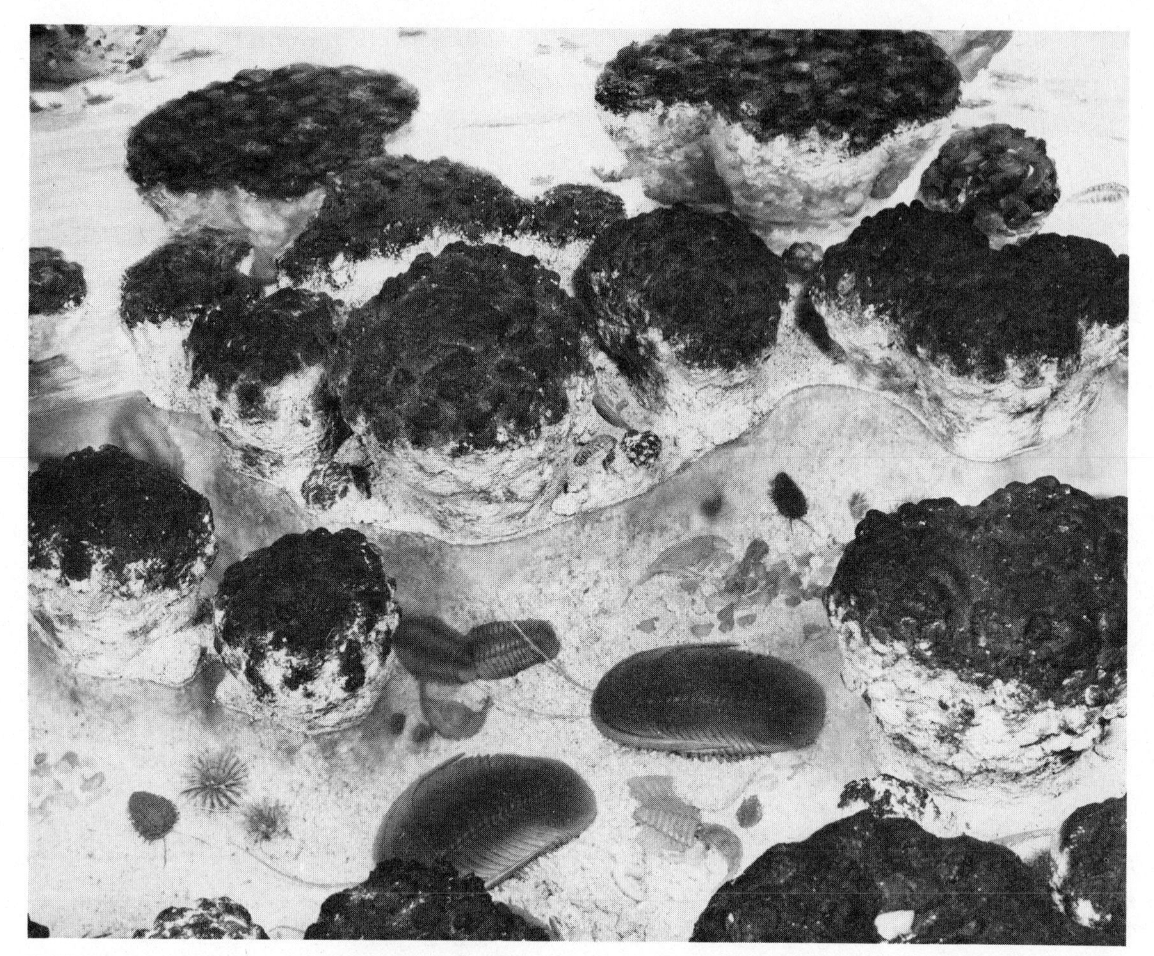

FIGURE 10–23 **Reconstruction of a Late Cambrian stromatolitic environment in central New York. The stromatolites, named *Cryptozoon*, are partially exposed at low tide. In the channels between them are the trilobites *Dikelocephalus* (the largest), *Prosaukia* (intermediate), and *Saratogia* (the smallest). (Courtesy of the New York State Museum.)**

to perfection in the Burgess Shale, More than 70 genera of plants and animals have been identified from this biota, and of these not more than half a dozen are known from elsewhere. New families and even orders, had to be designated to accommodate these genera. Among these fossils are more than a score of species of crustaceans whose general aspects are those of trilobites but which, for various technical reasons, cannot be included in that class.

Because the Burgess Shale is unique, much effort has been spent on understanding the conditions of Middle Cambrian time that may have produced it. The shale is a small lens in the Stephen Formation, a couple of meters thick and a few meters across. To the east the argillaceous Stephen Formation is replaced abruptly by the massive dolomite of the Cathedral Formation along what appears to have been a submarine cliff about 120 m high. The dolomite was apparently deposited as an algal limestone on a platform facing a basin of deeper water where the Stephen Formation accumulated. Unlike organisms that settle out of the water on to the sediment surface below, the Burgess Shale fossils do not lie flat on the

FIGURE 10–24 **The rocks of such peaks as Mount Edith Cavell in the Canadian Rockies are composed almost entirely of Cambrian strata. (Courtesy of the Canadian National Railways.)**

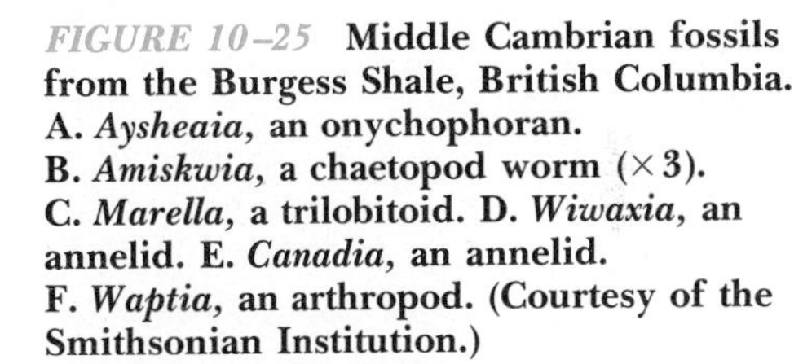

FIGURE 10–25 **Middle Cambrian fossils from the Burgess Shale, British Columbia. A. *Aysheaia*, an onychophoran. B. *Amiskwia*, a chaetopod worm (×3). C. *Marella*, a trilobitoid. D. *Wiwaxia*, an annelid. E. *Canadia*, an annelid. F. *Waptia*, an arthropod. (Courtesy of the Smithsonian Institution.)**

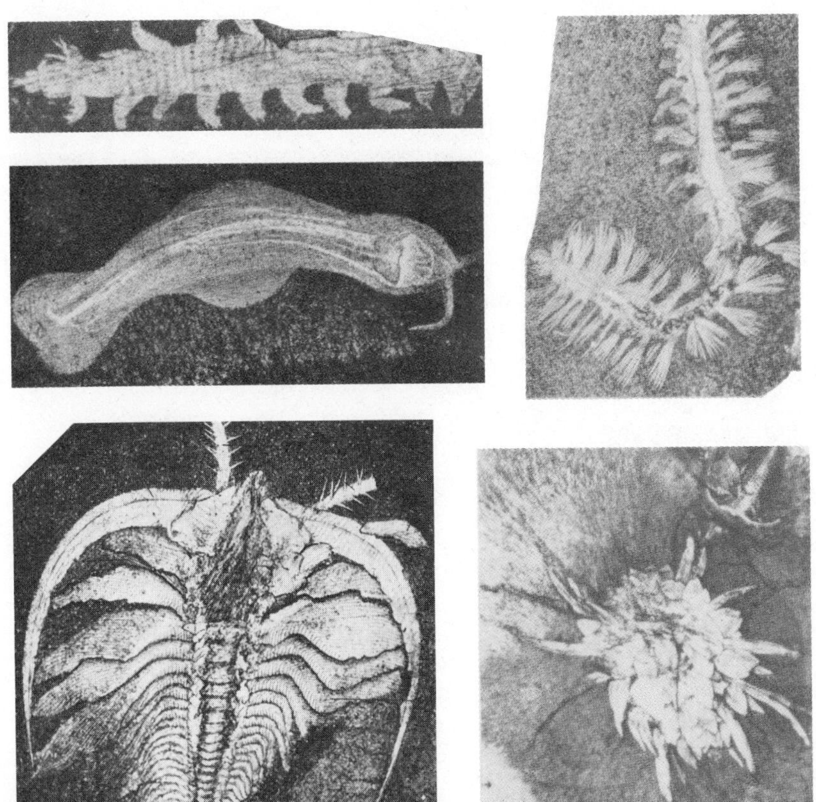

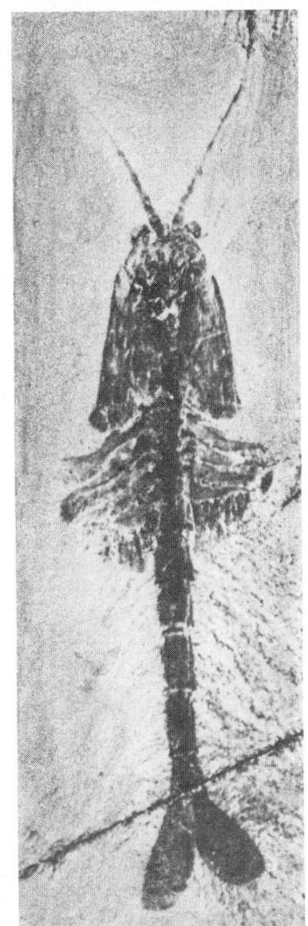

bedding planes but appear to cross them. This has suggested that the lens represents a mass of water-charged muddy sediment that has slumped off the slope to the east and, with its enclosed organisms, became buried in the deeper water of the basin. Under such conditions of rapid burial in an oxygen deficient basin, scavengers would not be able to destroy the soft parts of the organisms, and bacterial decay would be inhibited. Through these unique conditions paleontologists have been given a glimpse of the hosts of soft-bodied organisms that have likely been present in all past ages but, through lack of preservable hard parts, have not otherwise left a trace of their existence.

SUMMARY

The first epicontinental or epeiric seas of the Paleozoic Era transgressed the craton over a surface that had lain bare to weathering and erosion for hundreds of millions of years.

The platform surrounding the shield is subdivided into arches and basins whose vertical movements, uplift or subsidence, were continuous or intermittent.

The relationship between the basins and uplifts of the platform may be studied by means of isopach maps, stratigraphic cross sections, distribution of erosion surfaces, paleogeologic maps, and facies maps.

Platform rocks of North America are interrupted by five regional unconformities that divide the beds into six major sedimentary units called sequences.

Hadrynian sedimentary rocks are largely confined to the Appalachian and Cordilleran areas. Evidence of Hadrynian glaciation near the end of the Proterozoic Eon is widespread in these marginal mountain chains and in many other places in the world.

The basal beds of the Sauk Sequence are of Ediacaran age and consist of sandstones derived from the deep erosion of the Canadian Shield.

Evidence of the Ediacaran fauna of primitive metazoans is widely scattered around the world as burrows and feeding traces, but impressions of the soft-bodied organisms themselves are rare fossils.

Both the origin of metazoans and the secretion of skeletons in invertebrates may be related to changes in the oxygen content of the atmosphere near the beginning of the Paleozoic Era.

The transgression of the Sauk Sea from Ediacaran to Late Cambrian time caused a slow inward shifting of sand, shale, and carbonate facies belts.

Algae contributed extensively to limestones of the Sauk Sequence, but lime-secreting organisms were not present in sufficient abundance to make the extensive contribution they made to younger limestones.

Cambrian faunas are dominated by the archaeocyathids (confined almost entirely to the Lower Cambrian), brachiopods, and trilobites, although nearly all the major invertebrate groups with preservable hard parts are represented by the beginning of Ordovician time.

The base of the Cambrian System is drawn at the lowest appearance of trilobites or archaeocyathids, although this appearance is unlikely to represent a time plane.

The Middle Cambrian Burgess Shale preserves about 70 species of soft-bodied organisms unknown elsewhere that give us an insight into the diversity of early Paleozoic life.

QUESTIONS

1. By reference to the facies maps of the Paleozoic systems represented by Figs. 10–11, 11–1, 11–14, 11–17, 12–21, and 18–18, determine approxi-

mately what proportion of the craton was covered by deposits of the epeiric seas of these episodes. Assign each map to its appropriate sequence and approximate position within the sequence (e.g., Late Kaskaskia). What types of sediment were laid down in the central area of the craton at each of the times?
2. Summarize the criteria by which you could distinguish between a cratonic arch that had been active throughout the deposition of the sediments that overlie it, and one that had been formed simply by arching after the deposition of the platform sediments.
3. What are the sedimentary structures that a stratigrapher uses to recognize unconformities in a succession? How is the time value of the record missing at an unconformity estimated by geologists?
4. Compare and contrast the terms sequence, group, system, and period.

SUGGESTIONS FOR FURTHER READING

EARDLEY, A. J. "Central Stable Region of the United States," Chapter 5, in *Structural Geology of North America,* 37–62, New York: Harper and Row, 1962.

FRITZ, W. H. "Geological Setting of the Burgess Shale." *Proceedings North American Paleontological Convention,* Pt. I, 1155–1170, 1971.

LOCHMAN-BALK, C. "Cambrian of the Craton of the United States," in *Cambrian of the New World,* C. H. Holland ed. New York: Wiley–Interscience, 79–167, 1971.

SLOSS, L. L. "Sequences in the Cratonic Interior of North America." *Geological Society of America Bulletin,* 74: 93–111, 1963.

STANLEY, S. M. "Fossil Data and the Precambrian Cambrian Evolutionary Transition," *American Journal of Science,* 276: 56–76, 1976.

VALENTINE, J. W. "An Approach to an Ecological History of the Marine Biosphere," Chapter 10 in *Evolutionary Paleoecology of the Marine Biosphere,* Englewood Cliffs, N. J.; Prentice Hall, 409–451, 1973.

CHAPTER ELEVEN

The sea withdrew from the craton at the close of the Early Ordovician Epoch for a considerable time. We have no way of knowing the length of the interval in years, but the great differences in the faunas of Sauk and Tippecanoe rocks suggests that millions of years must have elapsed during which the world fauna evolved a new aspect. The shallow water invertebrate community that first appeared with the Tippecanoe transgression in the Middle Ordovician Epoch dominated epeiric seas until late Devonian time, when it was overtaken by catastrophe.

MULTICYCLE SANDSTONES

At the margin of the Canadian Shield in the upper Mississippi Valley, the transgressing Tippecanoe sea resorted and reground the grains of the Late Cambrian and Early Ordovician sandstones that had been laid down a few million years before. Rivers from the north contributed additional sand eroded from the crystalline rocks of the shield. As the grains became smaller and more nearly round, they were washed outward from the shore on the shallow sea floor and spread over a large area that is now the west side of the Mississippi Valley (Fig. 11–1). We deduce these conditions from the detailed study of the St. Peter Sandstone. The formation is composed of extremely well-rounded and well-sorted grains of quartz which, though possibly the result of wind action, could be equally the result of long-continued marine abrasion. The quartz content of the sand is about 99 per cent, so high that it is extensively used in glass making. Such purity cannot be attained in a single cycle of weathering, transportation, and deposition. Sandstones which are the product of the weathering of an older sandstone are said to be second cycle; those which are derived from second cycle sandstones are said to be multicycle.

Sandstones can be classified according to the "maturity" impressed on them by the agents of weathering, erosion, and deposition. Those that have a high proportion of poorly sorted, angular grains of mechanically and chemically unstable minerals are said to be immature. Those sandstones which are aggregates of rounded and sorted grains of chemically and mechanically stable minerals are said to be mature, for they have been subject longer to the agents that decompose, break, sort, and round the sedimentary particles. In this classification the St. Peter Sandstone might be called supermature, for nearly all the minerals except well-rounded and well-

THE TIPPECANOE SEQUENCE

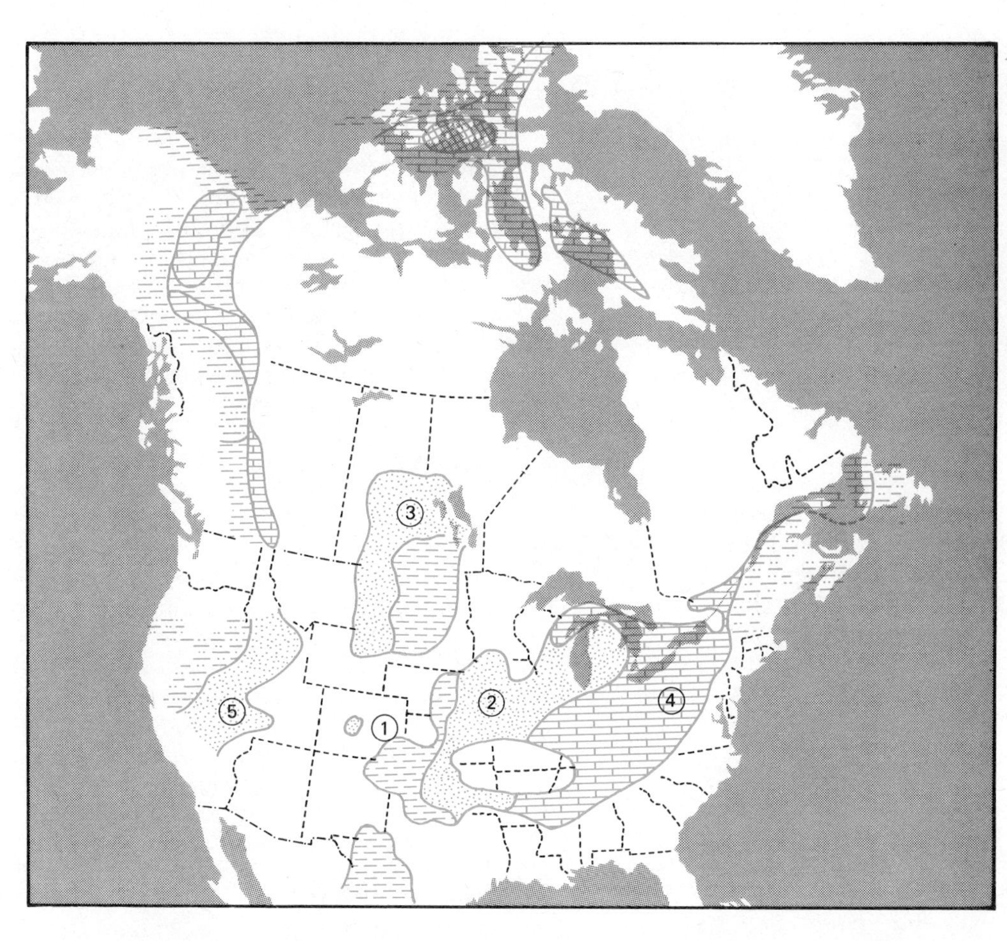

FIGURE 11–1 **Facies of the lower part of the Middle Ordovician Series (Chazy and Black River groups and their correlatives) in North America. 1. Harding Sandstone. 2. St. Peter Sandstone. 3. Winnipeg Sandstone. 4. Chazy and Black River limestones. 5. Swan Peak and Eureka sandstones.**

Sandstone

Shale

Limestone

Geosynclinal deposits

Evaporites

sorted quartz grains have been eliminated.

Sedimentary deposits like the St. Peter Sandstone are called blanket deposits because they cover wide areas but are limited in thickness. The St. Peter Sandstone is generally about 30 m thick (although it reaches 130 m in certain areas), and covers an area of approximately 8000 square kilometers.

Basal Tippecanoe sandstones west of the Transcontinental Arch are slightly younger than the St. Peter because the transgression there took place later. Among these sandstones are the Winnipeg and Deadwood sandstones that form the basal sediments of the Western Canada Basin, and the Harding Sandstone of central Colorado, which is important for its fossils of early vertebrates.

ORIGIN OF VERTEBRATES

The Harding Sandstone is a Middle Ordovician formation in the Rocky Mountains of Colorado. It was deposited along a shoreline and contains fragments of marine shells, evidence of the burrowing of littoral organisms, and fragments of the most primitive fish known.

The Agnatha

Although the fossils from the Harding Sandstone are commonly not well preserved (Fig. 11–2), the scattered remains are similar to later forms whose anatomy is well known. These early fish were bottom-living animals, completely covered with thick, bony plates and scales. The mouth, on the lower ventral surface, lacked true jaws, and they apparently fed by sucking in loose detritus from the bottom sediments. For this reason the class to which they belong is called the AGNATHA (Greek: no jaw). Fine food particles sucked into the mouth were filtered from the water as it was expelled through numerous pairs of gill slits. The major organs of sight and smell were located on the top of the skull. The most primitive of the agnathans lacked paired fins, but some of the later members of the group developed pectoral fins of a primitive sort (Fig. 11–3).

In some Paleozoic agnathous fish the skull was solidly ossified and shows cavities for the brain and major cranial nerves similar in general pattern to those of all higher vertebrates. These fish had a notochord, a longitudinal, strengthening rod running from the head to the tip of the tail, an axis on which the muscle fibers in the trunk and tail could act to move the body in a sinuous motion for swimming. They also must have had a spinal chord (a dorsal, hollow, nerve chord) passing above the notochord from brain to tail. The heavily armored agnathans are considered to be the ancestors of the modern jawless fish, the lamprey and the hagfish, that have lost all trace of bone and have become specialized as scavengers and parasites on more advanced fish.

Ancestors of the Vertebrates

The phylum CHORDATA, which includes the fish, is the last of the major phyla to appear in the fossil record. The search for fossil predecessors of the fish among the invertebrate phyla has led to a variety of hypotheses on the origin of this group, but to no definite ancestors. Three important characteristics of the most primitive vertebrates—paired gill slits, a notochord, and a dorsal hollow nerve chord—are also found in two invertebrate groups, the cephalochordates and the urochordates. The fish-like *Amphioxus* is a living representative of the former. However, the cephalochordates lack bone, and the group is without fossil representatives. The urochordates, sometimes called the sea squirts, are also boneless and have no fossil record, but show in their embryology a close link with the echinoderms. The facts that the echinoderms secrete hard parts in the form of plates that have an open, porous structure like that of bone, and that some of their primitive members have a superficial resemblance to agnathans, have suggested to some that the ancestors of the vertebrates are to be found in the phylum ECHINODERMATA. In particular, stalked echinoderms called carpoids

FIGURE 11–2 **Articulated plates of part of the dorsal shield of *Astraspis desiderata*, one of the oldest known vertebrates, a jawless ostracoderm from the Harding Sandstone, approximately × 2. (Courtesy of Tor Ørvig.)**

Fragmentary remains of the phosphatic dermal armor of ostracoderms have recently been described from strata of Late Cambrian age from northeastern Wyoming. Other fragmentary remains are reported from widely scattered sites of Early Ordovician age in North America, Spitzbergen, and eastern Greenland. Impressions of entire fish are known from the earliest Middle Ordovician in Australia. All of these ostracoderms have been found with typically marine invertebrate fossils.

(Fig. 11–4), have been proposed as close relatives of the agnathans, but serious difficulties have been raised to the acceptance of this hypothesis. First, the echinoderms secrete carbonate plates and the agnathans, phosphate plates. Second, the structure that forms the tail, or stalk, in carpoids is comparable to the structures at the anterior, or head, end of a chordate. The degree of specialization of the early agnathans indicates that they diverged from some group of invertebrates in the initial diversification of the metazoans in earliest Paleozoic time, and further fossil evidence is needed before the invertebrate groups from which they originated can be identified.

The earliest vertebrates are specialized above the level of any known relatives by the development of a brain, paired sense organs, and bone. The development of a distinct brain and the related organs of sight, smell, and balance may have been associated with the migration of the ancestors of the vertebrates from a relatively calm, offshore environment to a high-energy littoral environment, rich in oxygen and food

A

B

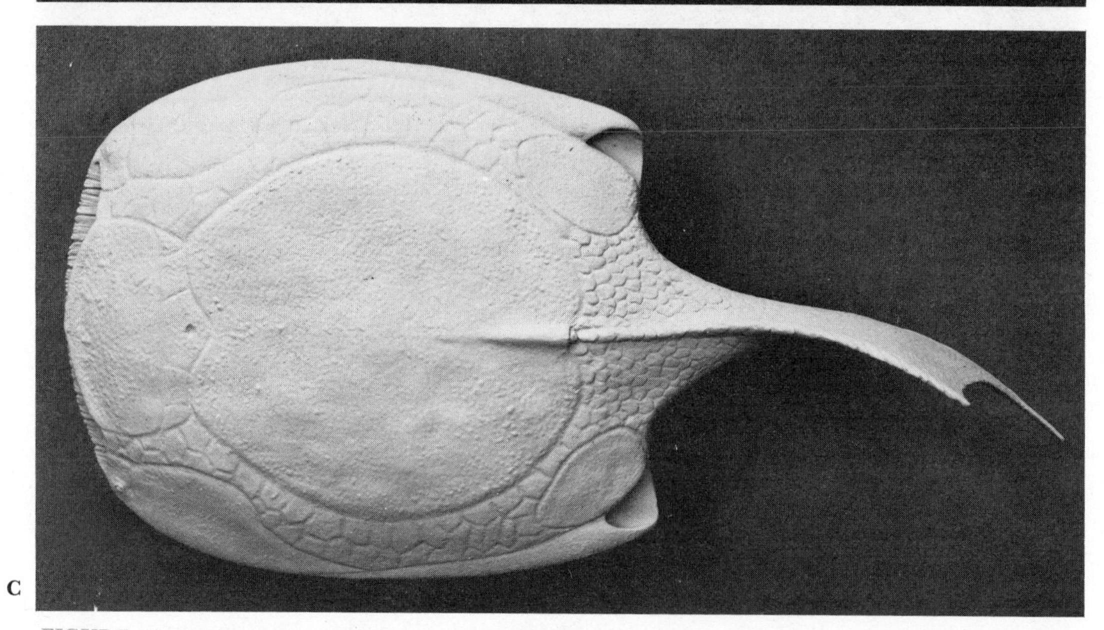

C

FIGURE 11–3 **Agnathous fish from Middle Paleozoic rocks. A. *Hemicyclaspis* (Lower Devonian), a bottom dwelling form with rigid anterior region and flexible, scaled trunk and tail. This dorsal view shows the eyes and pressure-sensitive areas of the head. B. *Pterygolepis* (Lower Devonian), a finely scaled genus, probably capable of more active swimming. The external gills open through a series of holes behind the head. C. *Drepanaspis* (Lower Devonian), a marine form with a much flattened head and thorax and a small flexible tail. This genus has a single pair of openings for the gills. (Courtesy of the American Museum of Natural History.)**

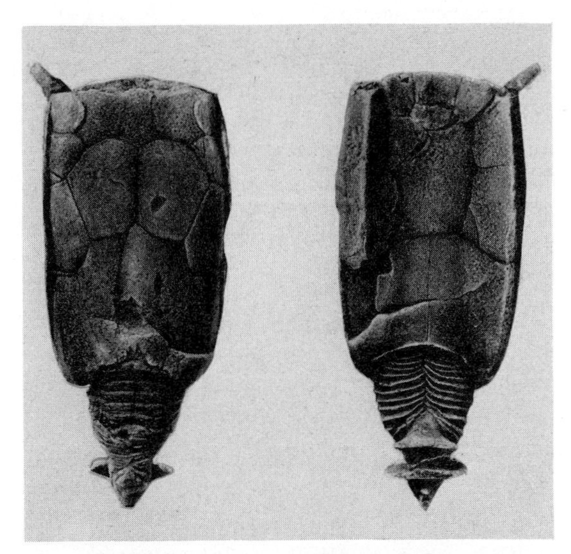

FIGURE 11–4 **The carpoid *Enopleura* showing upper and lower surfaces. One arm is missing, the other broken. Note the flexible "tail" ending in an "anchor." (Courtesy of K. E. Caster.)**

particles, but requiring greater coordination and control for swimming in turbulent water. Bone, a substance unique in structure and chemistry to the vertebrates, may have evolved at this stage. Bone developed initially at the surface of the body, as shown by the plated agnathans, and later came to replace the initially cartilaginous internal skeleton. Bone, which is approximately calcium hydroxyphosphate or hydroxyapatite ($Ca(OH)PO_4$), may have been secreted in early vertebrates as a reservoir of phosphate, and later developed as an alternative to the calcium carbonate skeletons common in the invertebrates. The armored agnathans that occupied littoral and brackish environments in Ordovician and Silurian time invaded typical marine and freshwater deposits in the Devonian, but became extinct at the end of that period. Of their boneless descendants, the lampreys, the oldest known fossil is from a Late Pennsylvanian marine deposit.

CHAZY LIMESTONE AND THE FIRST CORAL REEFS

While the basal Tippecanoe sandstones were accumulating in the central and western part of the craton, limestones and interbedded shales that constitute the Chazy Group were being deposited along the eastern margin from Alabama to Quebec (Fig. 11–1). The fauna and lithology of the Chazy limestones are in marked contrast with the fauna and lithology of the limestones of the Sauk sequence. Much of the Chazy limestone can be variously described as **calcarenite**, fossil-fragment limestone, skeletal limestone, or **bioclastic** limestone. These terms imply that the rock is composed of wave-broken fragments of the hard parts of lime-secreting organisms. Plates of cystoids and crinoids, and the shells of brachiopods, notably genera called *Rostricellula* and *Mimella,* are important constituents of the Chazy Limestone. The particles were moved by waves in the shallow water of the epeiric sea until they were broken or ground to sand size. Pre-Tippecanoe limestones are only rarely of this nature, but post-Sauk limestones are commonly composed of fossil fragments (Fig. 11–5).

The Chazy rocks of the Lake Champlain and southern Quebec regions form low domes in the flat countryside. In these domes the limestone is poorly bedded compared with the equivalent beds between the domes, and is crowded with the fossils of stromatoporoids, corals, bryozoans, and algae (Fig. 11–6). The fossil fauna of these limestone domes is comparable to that of modern reefs and suggest that the domes are ancient reefs. In the nautical definition a reef is a projection of the sea floor so close to sea level that it is an obstruction to navigation. To a geologist a **reef** is a mound, constructed on the sea floor by organisms, which rises to the zone agitated by the waves. The importance of this rise close to sea level is that mounds on the sea floor which do not penetrate wavebase have little effect on the sediments accumulating in the vicinity. When the reef penetrates the zone of the waves, it reaches an environment of more abundant food, light, and nutrients, but it is also subject to the erosion of the waves. A true organic reef should have a fan around its base of particles eroded from the mound. The reef mound will also incorporate a large quantity of this debris between the skeletons of the reef-building ani-

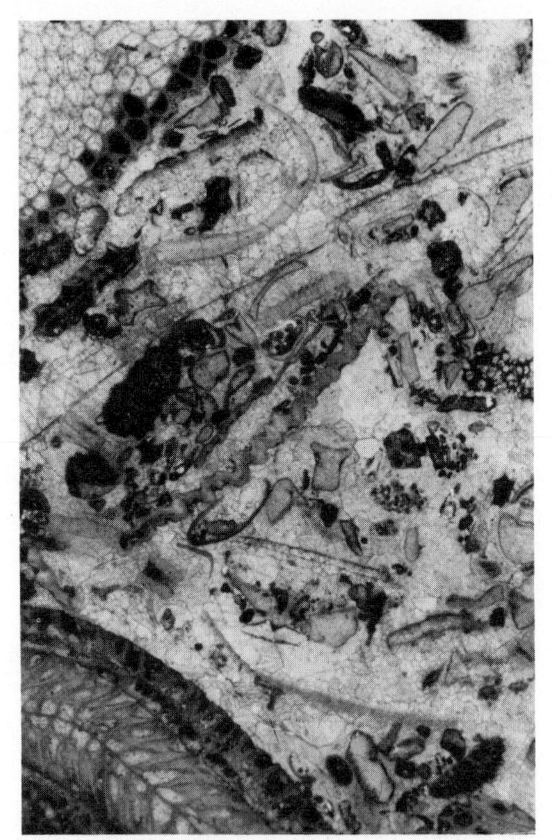

FIGURE 11–5 **Middle Ordovician limestone composed of fossil fragments—largely echinoderm plates, brachiopods, and bryozoans—in thin-section magnification about ×20. Basal Trenton Group, Quebec.**

mals. The reefs of the Chazy Group are small, rarely more than 100 m across and a few meters high, but the amount of carbonate fragments associated with them suggests that they grew in the zone of wave action. They were the first reefs in the stratigraphic record to be built of corals and stromatoporoids, a combination of organisms that was to dominate the reef community until the middle of Paleozoic time.

THE MID-PALEOZOIC REEF COMMUNITY

Stromatoporoids and Corals

Stromatoporoids are an extinct group of sponges that secreted a laminate, hemispherical, cabbage-shaped, or dendroid skeleton composed basically of laminae concentric with the growth surfaces and pillars secreted at right angles to the laminae (Fig. 11–6). Most stromatoporoids were about the size of a fist but some grew to 5 m in diameter. The surface of their skeletons is commonly marked by stellate grooves or canals that are traces of the water canal system of the sponge. Laminar and hemispherical stromatoporoids contributed their skeletons to the building of the reef edifice and bound loose constituents of the reef community together in turbulent water. Stick-like and branching stromatoporoids grew in more sheltered, lagoonal environments.

Rare fossils with some of the characteristics of corals have been found in rocks as old as the Middle Cambrian, and the fossil sea fans of the Ediacara fauna indicate that the COELENTERATA, the phylum to which the corals belong, originated near the beginning of Paleozoic time. As the Tippecanoe sea transgressed the craton, the coelenterates underwent an explosive diversification that gave rise to two major groups, the tabulate and rugosan corals (classes TABULATA and RUGOSA). The tabulate corals are all colonial and consist of closely spaced tubes, each of which once housed an individual polyp. The tubes, or corallites, have developed horizontal partitions (tabulae) and poorly developed radial partitions (septa). Most tabulate colonies are hemispherical, although some are branching and some laminar. The class reached its peak in the middle of the Silurian Period when such genera as *Favosites,* the honeycomb coral, and *Halysites,* the chain coral, were important constituents of the reef biota (Fig. 11–7).

Unlike the tabulate corals, the rugosans had their septa arranged in four-fold symmetry. For this reason they are also referred to as the tetracorals. The rugosans formed solitary, horn-shaped skeletons as well as colonies. They appear in the stratigraphic record soon after the tabulates, but do not form as important an element of the early Paleozoic reef community. The most common Ordovician solitary rugose coral is the

FIGURE 11–6 **Stromatoporoid in the Chazy Limestone, Isle La Motte, Vermont. From a laminar base the stromatoporoid has grown upward into a dome-like structure. (Courtesy of E. W. Mountjoy.)**

genus *Streptelasma,* in which the septa tangle in the center to form a vesicular mass. *Foerstephyllum* is a typical genus of colonial rugose corals from Ordovician rocks.

Brachiopods and Bryozoans

The tabulates and stromatoporoids constructed an environment in shallow water where many other invertebrates could thrive. The most important of these were the articulate brachiopods. The inarticulate brachiopods dominated Cambrian faunas, but by the end of Ordovician time the articulate brachiopods, an unimportant group in Cambrian rocks, had diversified into 54 families. Not only were the brachiopods abundant enough in mid-Paleozoic seas to be important contributors to the formation of limestones, but they evolved rapidly enough to be important aids in correlation and diversified into so many environments that they are useful in paleoecological reconstructions (Fig. 11–8).

The articulate brachiopods can be divided into six subgroups (orders or suborders, depending on the taxonomist). The earliest is the ORTHIDA which had shells of long straight hingelines and conspicuous radial ornamentation. *Platystrophia* and *Res-*

FIGURE 11–7 **Tabulate corals.** ***Left: Favosites*** **(Silurian).** ***Right: Halysites*** **(Silurian). Both are common constituents of Silurian reefs.**

FIGURE 11–8 **Middle Ordovician brachiopod** ***Hebertella*** **crowded on a slab of Chazy Limestone. (Courtesy of the Redpath Museum.)**

serella (Fig. 11–9B) are two common genera in Middle Ordovician rocks. The orthids are found in Cambrian rocks but reached their greatest diversity and numbers in the Ordovician Period. A second group, the STROPHOMENIDA had shells that were generally concavo-convex with a straight hingeline and relatively weak radial ornamentation. *Rafinesquina* is one of the most important Ordovician genera (Fig. 11–9E). A third group, the PENTAMERIDA, had a short hingeline and relatively smooth biconvex shells with a peculiar internal plate structure to accommodate the attachment of muscles. These brachiopods began in the Middle Cambrian but were not common until the Silurian Period. Some Middle Silurian limestones are made up entirely of the large shells of the best-known genus, *Pentamerus* (Fig. 11–9F). A fourth group, known as the SPIRIFERIDA, had a wide variety of shell forms, but all the members have an internal, calcareous, spiral structure for the support of the breathing apparatus (Fig. 11–9H). Spiriferids appeared in the Ordovician Period, became very abundant in the Devonian, and lasted until the Jurassic Period.

The RHYNCHONELLIDA were short-hinged, radially plicate forms, biconvex in most

FIGURE 11–9 **The major groups of Paleozoic brachiopods. A. *Lingula*, an inarticulate. B. *Platystrophia*, an orthid. C. *Dinorthis*, an orthid. D. *Leptaena*, a strophomenid. E. *Rafinesquina*, a strophomenid. F. *Pentamerus*, a pentamerid. G. *Eospirifer*, a spiriferid. H. *Mucrospirifer*, a spiriferid. I. *Cassidirostrum*, a rhynchonellid. J. *Cranaena*, a terebratulid. (Courtesy of the Redpath Museum and the Smithsonian Institution.)**

genera (Fig. 11–9I). *Lepidocyclus* is a characteristic Ordovician genus. They were abundant in the middle of the Paleozoic and again in the middle of the Mesozoic Era, after which they declined to the few genera that are alive today. The TEREBRATULIDA is a group of almost wholly smooth, short-hinged shells with a breathing and feeding apparatus carried on a calcareous loop (Fig. 11–9J). They began in Silurian time but were never abundant in Paleozoic beds.

Some of the Paleozoic representatives of the phylum BRYOZOA secreted massive, calcareous skeletons and were important constituents of the reef biota. Today the bryozoans are found on reefs largely beneath coral fronds or in crevices, but the Paleozoic representatives appear to have been capable of competing with the corals and stromatoporoids for living space or building small reef-like mounds on their own.

Community Analysis

To understand the environment of deposition of any fossil assemblage, such as the Middle Ordovician reef community, the paleontologist may try to reconstruct the roles played by each of the organisms that have been fossilized together. In making this community analysis, the paleontologist must determine how each organism lived (burrowing in the sediment, moving about on the surface, attached to the surface, etc.) and what sort of food was used by comparing the fossil organisms with their living counterparts. Invertebrates can be classified into groups on the basis of their method of obtaining food. Many, like the sponges and stromatoporoids, are filter feeders, passing large quantities of sea water over an apparatus to extract the micro-organisms that swim or float in it. Corals and bryozoans trap organisms suspended in the water by means of tentacles. Deposit feeders eat sediment to extract dead organic detritus in its interstices. Other invertebrates are scavengers (trilobites and modern lobsters) eating dead organisms, or predators taking live prey. In all communities the bacteria act as decomposers, breaking down dead organic matter to provide nutrients to keep the cycle of life going.

Figure 11–10 is a reconstruction of the trophic structure of the Ordovician reef community present in the Chazy Limestone. Most of the fossil organisms are believed to have fed on microscopic plants (phytoplankton) or animals (zooplankton) living in the sea water. When these organisms die they are eaten by scavengers and their remains decayed by bacteria. Two elements of the Chazy reef fauna as yet unmentioned belong to the phylum MOLLUSCA, the nautiloids and the gastropods.

Molluscs

The molluscs are poorly preserved in most Paleozoic sedimentary rocks because they used the form of calcium carbonate called aragonite extensively in their shells. Aragonite is calcium carbonate, like calcite, but is less stable and more soluble in circulating ground water. As a result shells composed of aragonite are commonly represented in the fossil record only as molds and casts. The nautiloids were octopus-like animals, living in relatively large, chambered shells. They first appeared in Late Cambrian time and by Middle Ordovician time were common in epeiric seas (Fig. 11–11). Some were crawlers, others probably swimmers. Most appear to have been predators eating other invertebrates. A buoyant gas that filled the chambers of the shell facilitated movement of the large cones. Many early Paleozoic nautiloids had straight, conical shells some of which reached lengths of 9 m, but some were curved and some coiled tightly in one plane like a watch spring. This latter type was rarely more than 30 cm in diameter.

Marine gastropods (snails) are basically adapted to crawling on sediment and rock surfaces, grazing on algae, or eating detritus. Their appearance in Cambrian time and subsequent rapid growth to abundance, has been credited by some paleontologists with the collapse of the stromatolite community that dominated carbonate environments in Proterozoic time. Only where environmental conditions exclude gastropods are stromatolites forming at the present time. By eating the blue-green algae that form the sediment-trapping layers of stromatolites, the gastropods could effectively prevent the structures from growing. The decline of the stromatolites may also have been caused by competition with the rapidly rising corals and stromatoporoids.

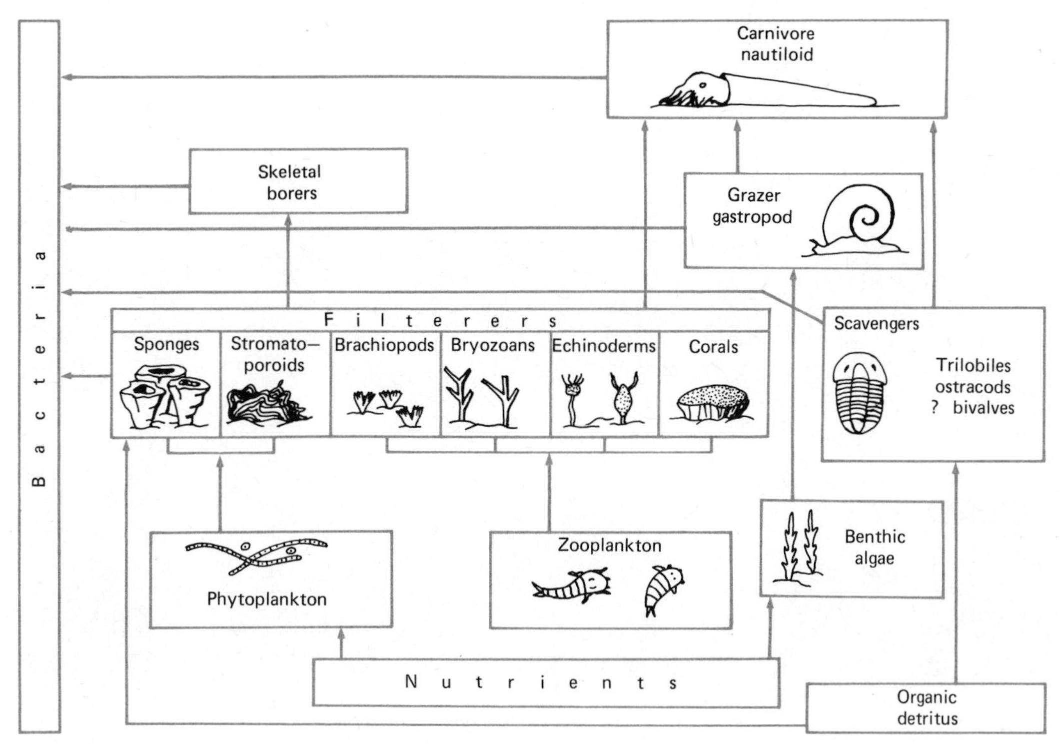

FIGURE 11–10 **Trophic analysis of a Chazyan reef community at Isle La Motte, Vermont, showing the roles played by the elements of the fauna that are preserved, their relationship to each other and to organisms that must have existed in the community but are not preserved (such as bacteria). (Modified from D. F. Toomey, R. M. Finks, and U. Kapp.)**

The BIVALVIA (also referred to as PELECYPODA) appear in Ordovician rocks and are locally abundant in fine, clastic sediments deposited near shorelines. They are neither well-preserved, abundant, nor useful as index fossils in most mid-Paleozoic limestones.

Trilobites

By Middle Ordovician time the trilobites, which has reached their peak in the Late Cambrian, were declining. However, the great variety of these arthropods found in Ordovician rocks shows that they were adapted to many different marine environments. Mud-burrowers, such as *Cryptolithus* (Fig. 11–12C), lost their eyes and developed a broad, frontal brim; floaters became spiny and small, increasing in surface area but not in weight; and crawlers remained smooth and generally large (*Isotelus*). Whole trilobite fossils are relatively rare, for the exoskeleton was held together, like that of other arthropods, by soft tissue that rotted on death. Periodic molting as the individual grew provided about a dozen exoskeletons, most of which almost immediately became dissociated into separate body segments and fragments.

Echinoderms

Most mid-Paleozoic limestones are crowded with small discs with radial marking and a central hole that commonly has five-fold symmetry. These are scattered plates that, when strung together by organic tissue, made up the stalks and feeding arms of echinoderms belonging to the classes CYSTOIDEA and CRINOIDEA. The cystoids appear in Cambrian rocks, flourished during the Ordovician, and declined in middle and late Paleozoic time. The crinoids became important in the Ordovician Period and remained so for the rest of the Paleozoic Era.

FIGURE 11–11 **Restoration of the sea floor of Middle Ordovician time. In the foreground a large, straight nautiloid lies on the sea floor with tentacles extended and a coiled one crawls on the sea floor in the right corner. Crinoids wave in the open water and green algae bend with the wave action. The knoll on the left is made up of colonial corals. A colony of graptolites is suspended in the water from a float above it. (Diorama by Paul Marchand; courtesy of the Redpath Museum.)**

These attached echinoderms, the sea lilies of modern oceans, lived in great undersea gardens and in death contributed their plates to form crinoidal sands that filled into the spaces and banked up around the periphery of many middle Paleozoic reef mounds. Crinoids and cystoids were suspension feeders, catching micro-organisms from the water in arms that channelled food to a mouth at their base.

TIPPECANOE EPEIRIC SEAS

Sedimentation on the east side of the craton during the Middle and Late Ordovician epochs was greatly influenced by the uplift of the Appalachian geosyncline in the Taconic orogeny, which will be discussed in Chapter 17. Fine grained sediment eroded from the rising Taconic Mountains made the upper part of the Ordovician System east of the Transcontinental Arch more argillaceous than correlatives west of the arch. The argillaceous Middle Ordovician limestones from Virginia to Quebec, and the Upper Ordovician limestones of the Cincinnati Arch region contain many beautifully preserved fossils which often weather free of their matrix (Fig. 11–13).

The limestone that overlies the basal Tippecanoe sandstones north and west of the Transcontinental Arch forms a thin, extensive blanket and contains a distinctive fauna. Segments of this limestone body are given different names, such as Bighorn (Wyoming), Montoya (New Mexico), Beaverfoot (Alberta), and Red River (Manitoba), but over much of the west side of the continent (Fig. 11–14) the limestone is lithologically

FIGURE 11–12 **Early Paleozoic trilobites. A. *Isotelus*, a crawler on the sea floor. B. *Flexicalymene*, also probably a crawler. C. *Cryptolithus* (×2), probably a mud-burrower as it lacks eyes. D. *Triarthrus*, possibly a swimmer as it is commonly found in dark shales lacking benthonic fossils. (Courtesy of the Smithsonian Institution and the Redpath Museum.)**

uniform and contains everywhere a fauna of large corals, nautiloids, and brachiopods. That these limestones formed a continuous sheet over much of the western and northern parts of the craton is indicated by many erosional remnants, or outliers, scattered throughout the Canadian Shield. These limestones were laid down in the most extensive of North American epeiric seas at the height of the Tippecanoe transgression.

Paleoclimatology

The lithology and fauna of the beds west of the Transcontinental Arch, from Death Valley, California, to the shores of Greenland, are essentially uniform. This poses several problems for the paleogeographer and paleoclimatologist. Similar elements of the fossil fauna are found in regions whose mean July temperature now ranges from 90°F (33°C) to 40°F (5°C). Regions so greatly different in climate have no faunal similarity today and probably did not have in the past. Such faunal distributions have led many geologists to consider polar wandering as an explanation (Chapter 7), but the following alternatives should also be kept in mind:

1. Climatic zones which roughly follow lines of latitude may not have been as well differentiated in Ordovician time

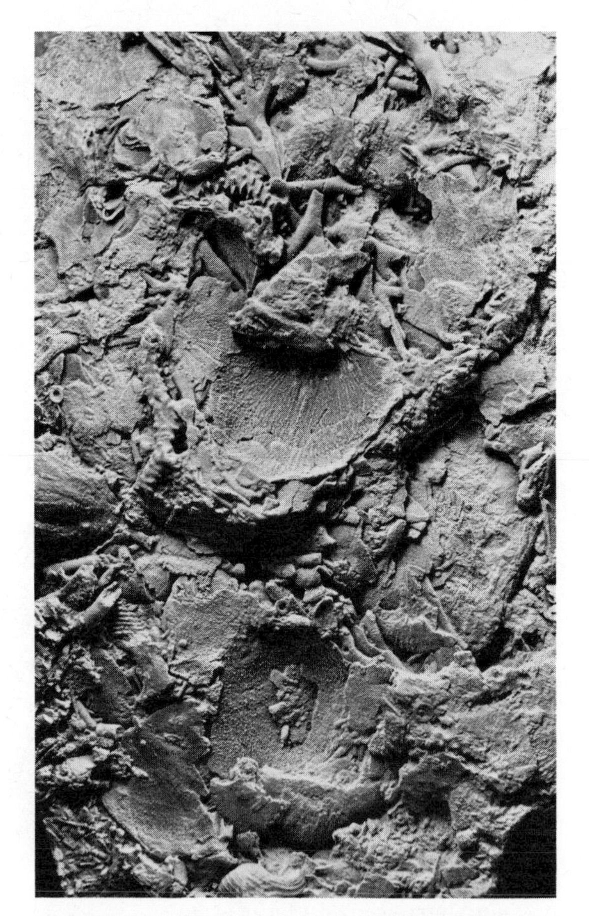

FIGURE 11-13 **Weathered bedding surface of Late Ordovician limestone from Cincinnati, Ohio, showing the abundance of fossils.**

as they are at present, so that the temperature of the epeiric sea could have been approximately the same at latitude 35° north as at 70° north. This is difficult to reconcile with what we know of the present condition of the ocean.

2. A strong, warm current, such as the present Gulf Stream, might have kept the marine environment of the epeiric sea uniformly warm even in the north. This hypothesis cannot be tested, however, for we know of no comparable modern epeiric sea in which such currents might be found.

The distribution of the limestone and its fauna is easily explained if the climatic zones of Ordovician time did not cross the North American continent as they do at present, but were so arranged that the area of deposition fell within the tropical zone of Ordovician time. The abundant fauna of large invertebrates suggests a warm water environment near the equator, and this is remarkably confirmed by the paleomagnetic evidence (Fig. 7-4) which locates the Ordovician equator across North America from Hudson Bay to the Gulf of California.

Conditions in Epeiric Sea

The rocks, fossils, and sedimentary features of the Late Ordovician blanket limestone all indicate that the epeiric sea of this time was shallow and that its floor was continually in the range of wave and current motion. Minor warping of the crust or change in sea level would move the shoreline hundreds of kilometers and either lay bare the sea floor or inundate the land. The position of the shoreline during any recognized time unit cannot be mapped accurately for it was continuously changing. When animals died their skeletons were either ground up or broken up by scavengers, borers, and wave motion to form lime-sand and lime-mud, or more rarely they were buried intact in the calcareous debris of other shells. Because subsidence over the craton was so slow, calcareous debris was reworked by the waves so that most of the myriads of shells were reduced to fragments before final burial took place.

Alan Shaw believes that the slope of the epeiric sea floor from the central craton to the edge of the platform was less than 2 m per km. M. L. Irwin suggested that it may be as low as a few centimeters per kilometer. Irwin described sedimentation in such seas in terms of three zones (Fig. 11-15). In the middle zone, tens of kilometers wide, the sea bottom is agitated by waves and tides and the characteristic sediments are limestone composed of fossil fragments and oöliths. In the nearshore zone the water is extremely shallow for hundreds of kilometers and is disturbed only during extraordinary storms, for tide and wave motion are dissipated by friction in the middle zone. Much of this zone is periodically exposed by the

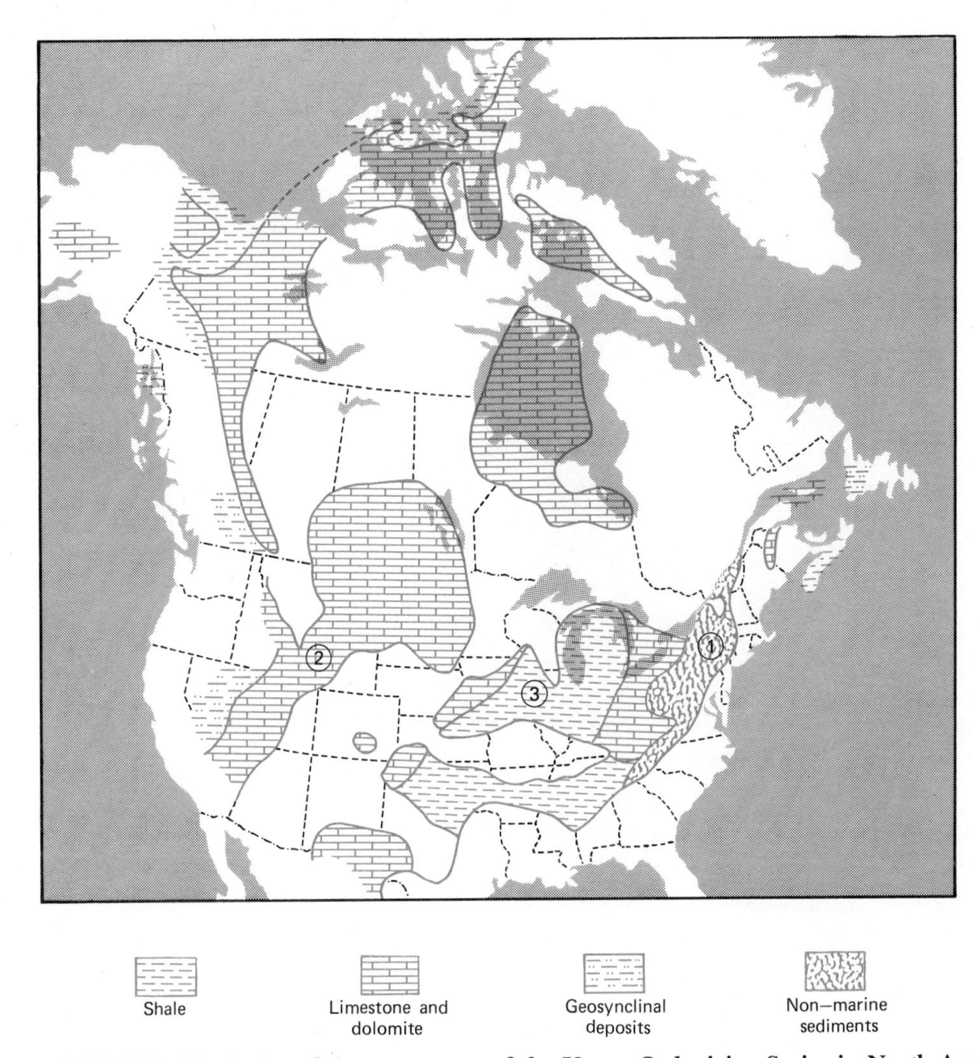

FIGURE 11–14 **Facies of the upper part of the Upper Ordovician Series in North America. Rocks of the Richmond Stage have been plotted where precise correlations are available; in other places all Upper Ordovician rocks have been plotted. 1. Queenston deltaic beds. 2. Bighorn Limestone. 3. Maquoketa Shale.**

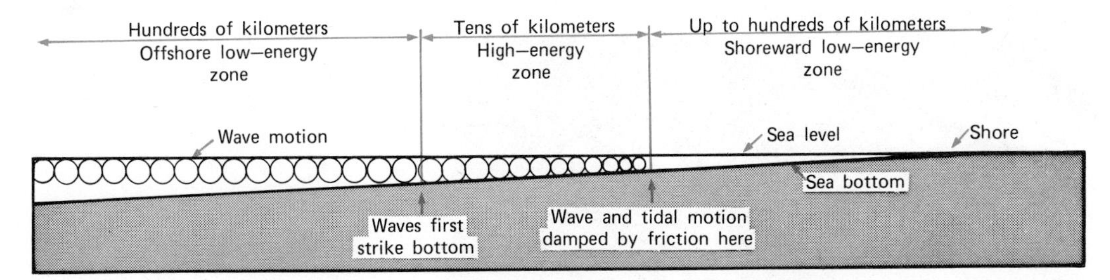

FIGURE 11–15 **Three zones of sedimentation in epeiric seas. (After M. L. Irwin; courtesy of the American Association of Petroleum Geologists.)**

recession of the sea and dried out to form a wide, low, coastal plain. Such coastal plains just above sea level along the Persian Gulf are today called **sabkhas** (Fig. 11–16). Evaporation draws seawater and groundwater from lower layers up to the surface layers and concentrates its salts there. Evaporite minerals, such as anhydrite, grow to form layers just below the surface, and calcitic muds are replaced by dolomite as the sea water bearing magnesium ions moves through them. Surfaces of modern sabkhas near the sea are commonly covered with a slimy algal growth.

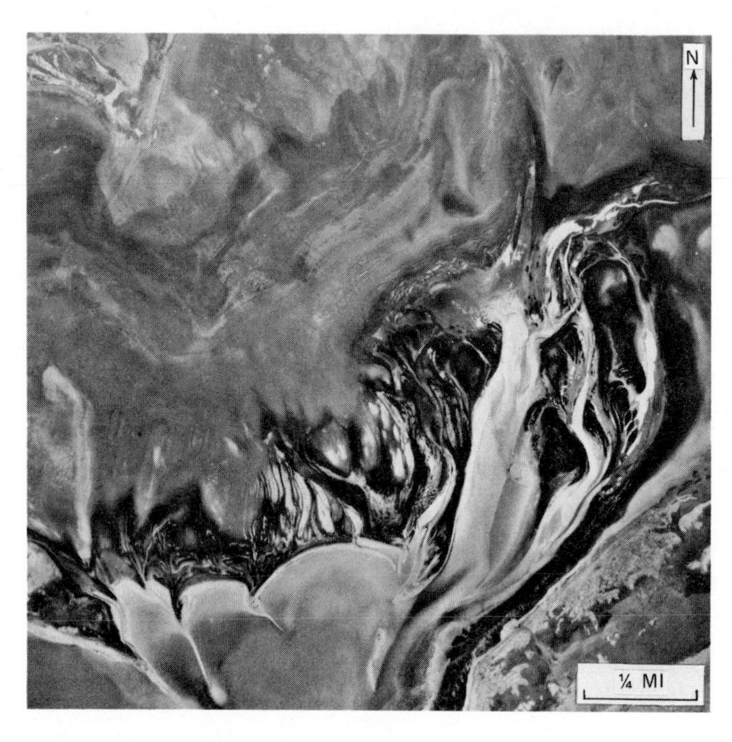

FIGURE 11–16 **Aerial photographs (above, vertical; below, oblique) of the sabkha environment along the shore of the Persian Gulf at Sabkha Faishakh. Note the low relief and the tidal channels. (Courtesy of L. V. Illing and the Society of Economic Paleontologists and Mineralogists.)**

In Middle Silurian time rocks that appear to have been formed in such an environment were laid down in the epeiric seas that flooded the craton. Like the Ordovician blanket limestones below them, the Silurian carbonates are also found in erosional outliers of the Canadian Shield (Fig. 11–17). Unlike the Ordovician rocks much of the Silurian carbonate is dolomite and contains structures indicative of a shallow, drying environment, such as molds of hopper-shaped salt crystals, stromatolitic algal structures, interbeds of anhydrite, and collapse breccias formed where evaporite minerals have been dissolved out by groundwater after deposition. Some of the Silurian dolomites appear to have been deposited in an environment like that of the modern Persian Gulf sabkhas, but in places the water was deeper and small coral–stromatoporoid reefs grew amid fossil fragment sands.

Paraconformities

The section of Silurian rocks along the Niagara Escarpment was used in Chapters 3 and 4 as an illustration of the principles of

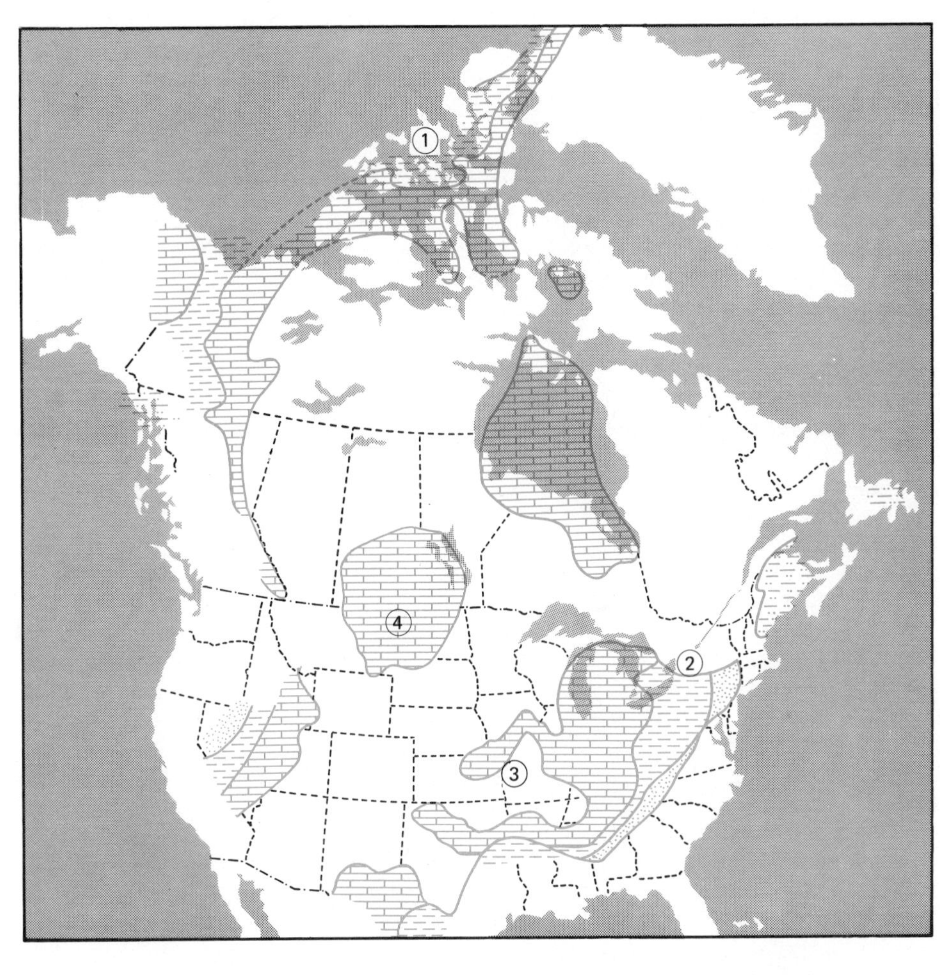

FIGURE 11–17 **Facies of the Middle Silurian Clinton Stage in North America. Note the clastics derived from the Appalachian highlands produced in the Taconic orogeny. 1. Cape Phillips Shale. 2. Clinton Group. 3. Lower Hunton Group. 4. Interlake Group.**

correlation. The extensive Silurian carbonates that occur north and west of the Transcontinental Arch appear to be equivalent to the Clinton and Albemarle groups of the Silurian section at Niagara. Between the youngest Ordovician rocks and the overlying Silurian rocks beds equivalent to the Cataract Group of the eastern part of the the continent are missing west of the Transcontinental Arch, yet there is little evidence at the Ordovician–Silurian contact of erosion that must have lasted for millions of years. Such a break in the expected stratigraphic succession, hidden by the parallelism of adjacent strata and lack of evident erosion, is called a **paraconformity.** However, some geologists believe that the paleontological dating of the Silurian beds is not adequate to demonstrate a break in sedimentation. They suggest that lack of an irregular erosion surface, the absence of truncation of the beds below the surface, and the remarkable parallelism of key silt beds above and below the contact, make it unlikely that the basin was subject to erosion during millions of years of Early Silurian time while the Cataract Group was being deposited to the east; but that, more likely, deposition was continuous and the fossils equivalent to the Cataract Group have not been recognized or preserved in the thin sequence that was deposited.

Other paraconformities have also been subject to disagreement. In southern Oklahoma the Late Silurian Henryhouse argillaceous limestone seems to be overlain conformably by the Early Devonian Harragan Formation of almost identical lithology. Although no sign of an erosion surface exists between these two formations, paleontologists have determined, by painstaking work, that the upper stages of the Silurian System are missing and that the two formations must be separated by a paraconformity. Geologists working with evidence from boreholes in the basin of central Oklahoma have found in the regional relationships no evidence of such a period of erosion between the Harragan and Henryhouse formations.

Although stratigraphers and paleontologists disagree about these two paraconformities, no one doubts that paraconformities do exist and that they are very difficult to detect except by fossils. The best known paraconformity in North America is exposed in the region of Louisville, Kentucky, where Middle Silurian limestone is overlain without a break by Middle Devonian limestone. Fossils that are common on both sides of the paraconformity, accurately date the limestones and leave no doubt that Late Silurian and Early Devonian rocks are missing from these sections; yet no sign of erosion or irregularity of bedding betrays the hidden boundary. The existence and significance of paraconformities will be demonstrated only when all types of sedimentary rocks can be dated accurately in years.

SILURIAN ORGANIC REEFS

In Silurian time the belt of the tropics traversed the Great Lakes area of North America. The sea water was warm and the coral-stromatoporoid community thrived and built reefs around the Michigan Basin. Although this community appeared first in Middle Ordovician time, not until Middle Silurian time were conditions of fauna and climate favorable for the extensive growth of its members (Fig. 11–18).

Reefs in the stratigraphic record are composed of carbonate rocks (limestone and dolomite) that can be distinguished from the sediments that surround them by their content of fossils in position of growth, lack of bedding, high porosity in the form of holes or vugs, and their association with limestone breccias and conglomerates. Reefs are composed of two elements: framebuilders that form a rigid, interlocking, wave-resistant framework of organisms in position of growth and which may volumetrically be only a minor part of the reef, although structurally the most important part; and organically-derived debris that occupies the space between the framebuilders and is bound by them into a solid mass. In the middle Paleozoic the framebuilders in-

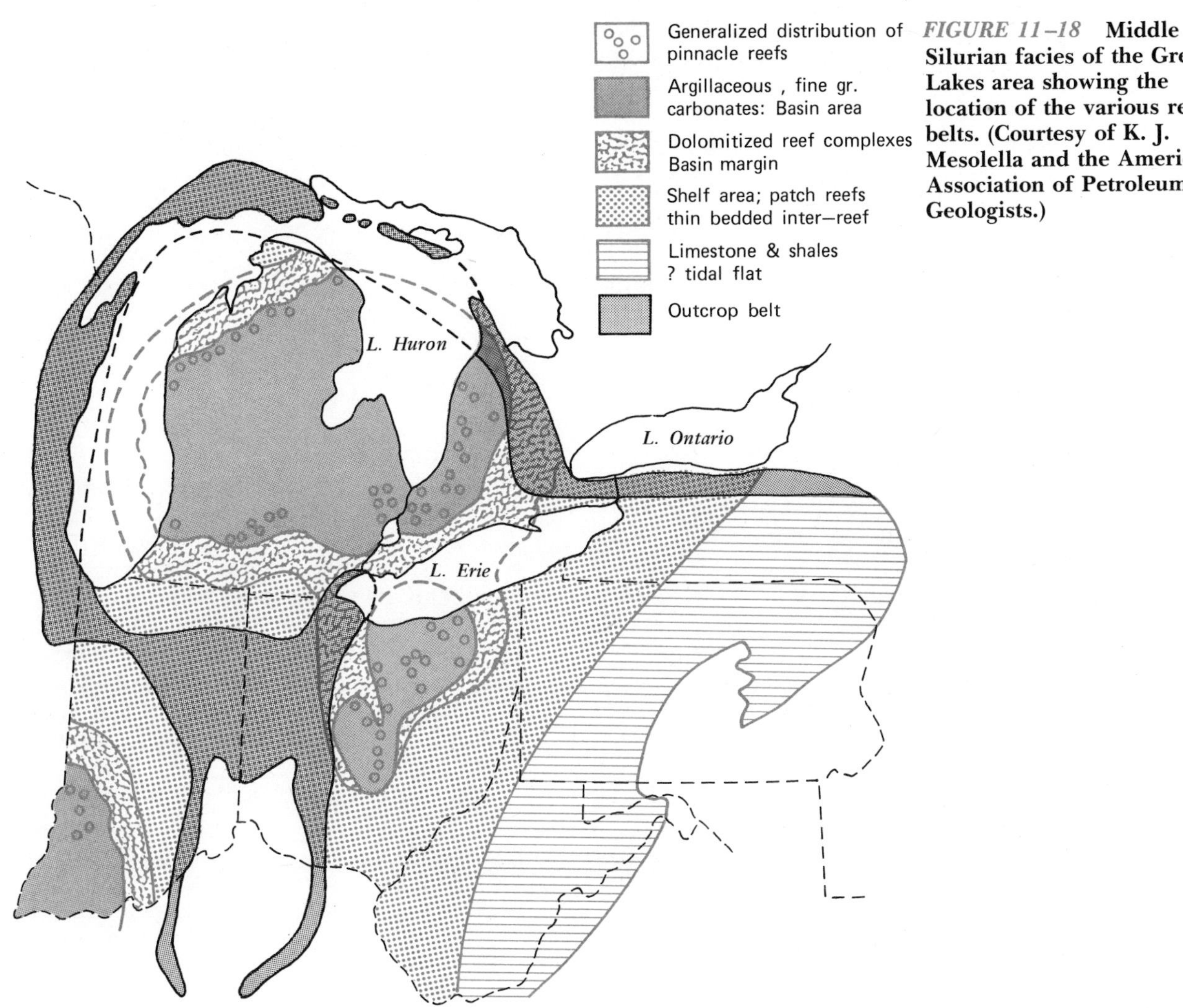

FIGURE 11–18 **Middle Silurian facies of the Great Lakes area showing the location of the various reef belts. (Courtesy of K. J. Mesolella and the American Association of Petroleum Geologists.)**

cluded the stromatoporoids, tabulate corals, calcareous algae, sponges, and a few other groups of only local importance. The spaces between the framebuilders are partially filled with fragments—torn loose and ground up by the surf—of the shells of animals that live in the reef but are not builders, such as brachiopods, gastropods, bivalves, and crinoids. In most reefs this granular debris does not fill the framework and the reef core is left highly porous. Boring organisms weaken, undermine, and break up the reef. The breaking waves tear off blocks which accumulate around the mound as an underwater talus deposit. This reef talus may come to be larger in volume than the reef core itself.

Although the surrounding and covering muds lose their moisture and are compacted to less than half their original thickness under the weight of the accumulating sedimentary layers, the reef core does not change in thickness and the beds above are flexed or draped over the reef. The steep dips of the beds that flank organic reefs are due not only to the original slope around the reef but also to this compaction.

The shape of reefs is controlled largely

by subsidence and the effort of the reef organisms to keep themselves at, or as close as possible to, sea level. If the floor on which the reef is being built sinks relative to sea level, the organisms tend to grow upward, and consequently build the reef vertically to keep its top near sea level, the zone of optimum light, food, and oxygen. In continually compensating for subsidence they may build a column of reef rock hundreds of meters high. This column later appears to penetrate the surrounding limestone beds, although the beds actually accumulated progressively alongside the reef. The Silurian reefs of the Michigan Basin grew in a zone around the basin where subsidence conditions were optimum. The reefs on the inner side of the reef tract were built into thin, spindly columns, for the energy of the growing animals was concentrated in keeping the growing face of the reef near sea level. These reefs have been called "pinnacle reefs." However, the reefs on the outer side of the tract near the shelf are stubby and spreading, for in the zone of slower subsidence the energy of the organisms could be expended in lateral growth rather than vertical growth. On the margins of the basin the reefs combined into a more or less continuous barrier reef (Fig. 11–19).

Like modern reefs, many ancient reefs show a distinct zonation of animals that lived on them. Different types of animals had different ecological requirements: some liked the surf zone on the windward side of the reef; others preferred the quiet waters of the lagoon, sheltered from the waves by the reef; and others preferred deeper water. Just as a map can be made of a modern reef showing the domains characterized by particular faunas, so maps can be made of the animals that grew in ancient reefs. From these maps such features of the former environment as the direction of the winds and currents can be deduced. The fauna and lithology of the Thornton Reef complex exposed in a quarry in northeastern Illinois have been mapped in detail. The core of framebuilding animals is small compared to the large area of detritus boken from this core and carried leeward to the lagoon and backreef bank (Fig. 11–20). These asymmetrical reefs preserve a record of the winds and currents of the past; if sufficient analyses of such reefs around the Michigan Basin were made, the geologist might be able to relate the ancient equator that was nearby to the ancient trade winds systems which flanked it.

Faunal zonation of the Niagaran reefs can be detected in vertical section as well as in plan. The fauna of the reef changes as the reef grows upward from the zone of quieter, deep water to the zone of wave tur-

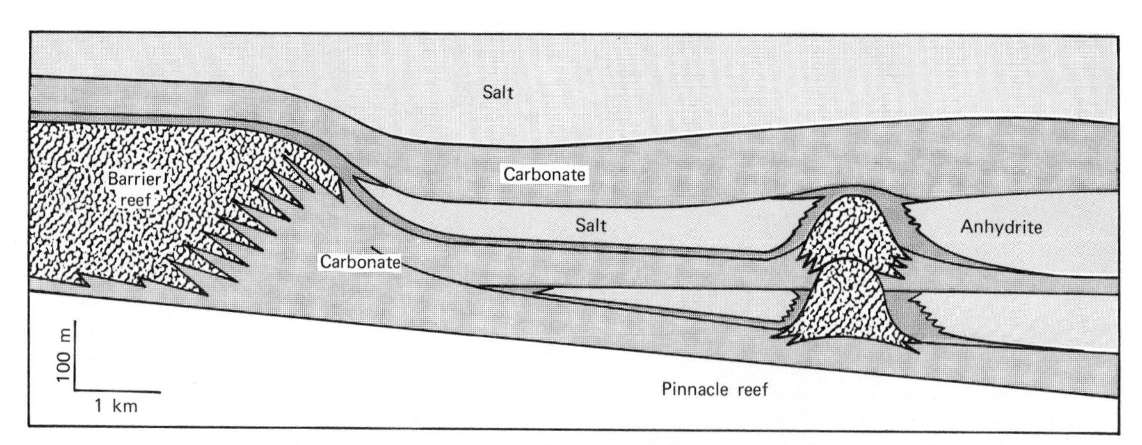

FIGURE 11–19 **Generalized cross-section of Niagaran and lower Salina facies in northern Michigan showing the relationship of the barrier and pinnacle reefs to the edge of the basin and to the evaporites. (Courtesy of K. J. Mesolella and others and the American Association of Petroleum Geologists.)**

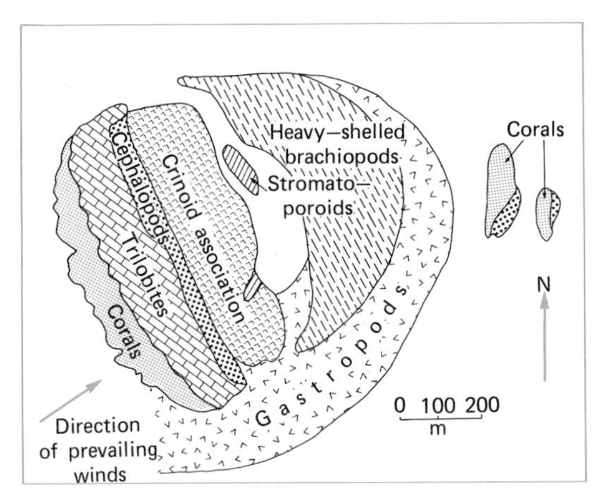

FIGURE 11–20 **Thornton Reef Complex (Middle Silurian, Illinois). The geographic distribution of fossils in the vicinity of the reef is a reflection of a southwesterly wind direction. (After J. C. Ingels; courtesy the American Association of Petroleum Geologists.)**

bulence, and animals capable of withstanding the surf attack become more abundant (Fig. 11–21). Massive stromatoporoids and tabulate corals increase in numbers, strong box-crinoids replace their more delicate relatives, and trilobites and molluscs increase in numbers. The type of ecological succession shown by the Silurian reefs of the Great Lakes area is common to many reefs in other parts of the stratigraphic record (Fig. 11–22). A phase of stabilization, during which attached echinoderms form a solid base for the development of the framebuilders, is followed by a stage of colonization. The maximum diversity is reached in the next stage, after which diversity declines as the fauna becomes dominated by the few organisms that are capable of withstanding the extreme environment in the surf as the reef grows in shallow water.

Because coarse organic fragments and framebuilders accumulate in a helter-skelter manner, reefs are commonly very porous and are therefore good reservoirs for the entrapment of oil. The Middle Silurian reefs of Ontario, Michigan, and Illinois have produced small quantities of oil and gas. In Illinois the reefs are surrounded, and their oil sealed in, by flanking argillaceous limestone; in southern Ontario, however, they protrude several hundred feet from the top of the Middle Silurian Guelph Dolomite and are surrounded by salt and gypsum deposits of the succeeding Salina Group.

FACIES IN AN EVAPORITE BASIN

The association of organic reefs with evaporites is surprising, yet commonplace in the stratigraphic record. It is surprising because sea water must be highly saline before salts can be precipitated, and organnisms that build reefs cannot live in such saline water. Where evaporite deposits of

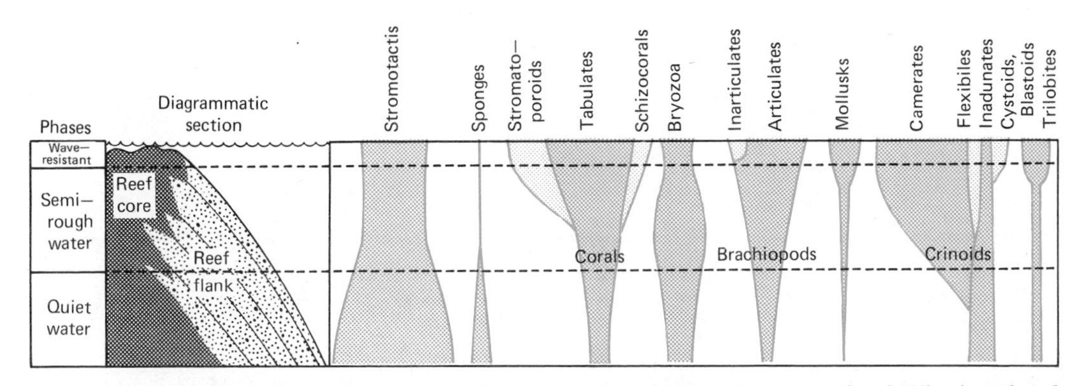

FIGURE 11–21 **Vertical changes in the faunas of the Middle Silurian reefs of Illinois. The changes in abundance as the reefs grew from the quiet, to the semi-rough, to the wave-resistant stage are shown by the widths of the areas on the right. (After H. A. Lowenstam; courtesy of the Geological Society of America.)**

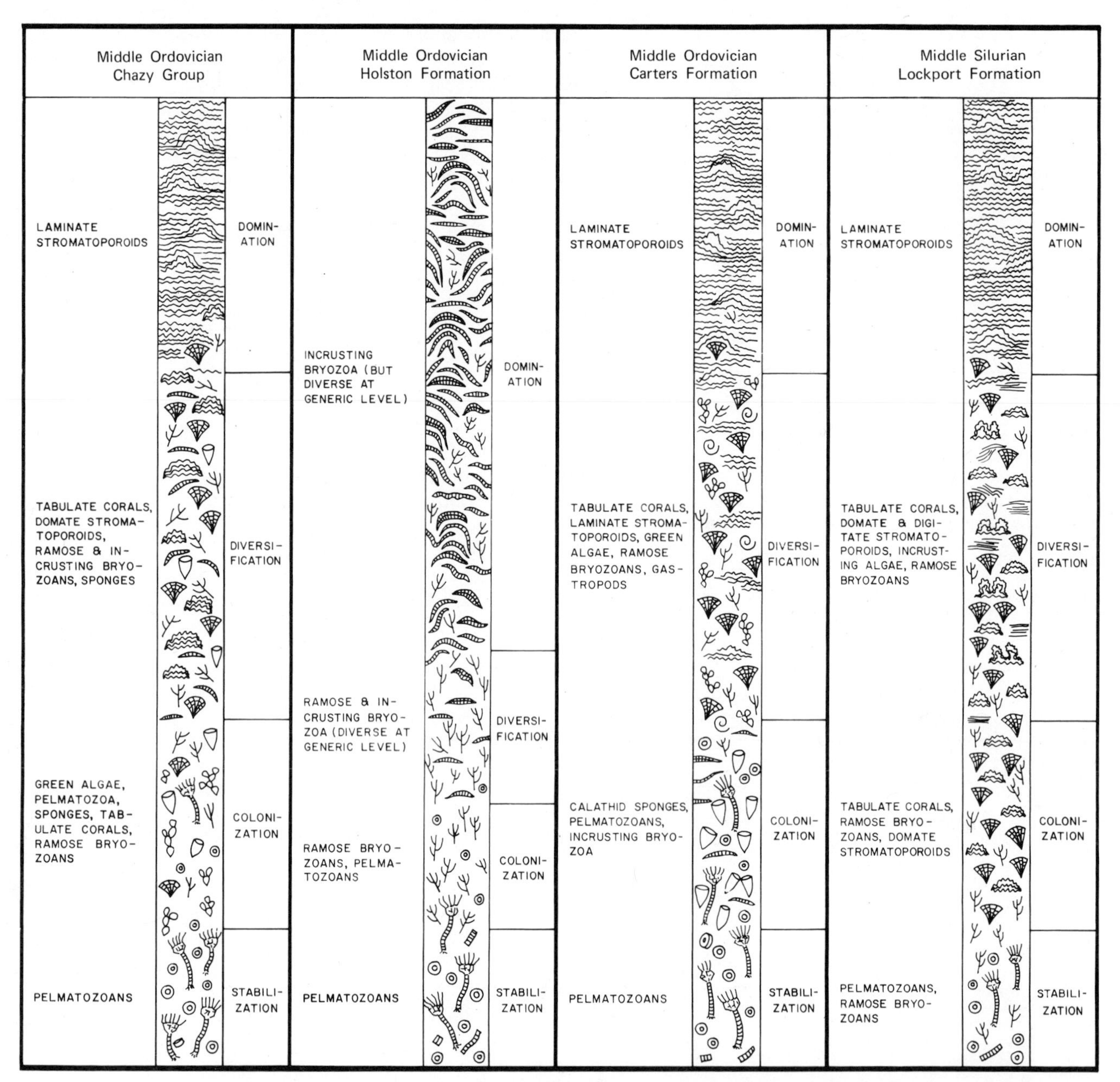

FIGURE 11–22 **Comparison of the developmental stages in four ancient reef masses showing the vertical succession of the reef biota. (Courtesy K. R. Walker and L. P. Alberstadt from *Paleobiology*, The Paleontological Society.)**

anhydrite, gypsum, and halite surround reefs built by organisms and are found infilling spaces in the reef itself, some explanation of this association is required.

As the Tippecanoe Sea ebbed from the craton in Late Silurian time deposition of gypsum, anhydrite, and halite took place in Michigan, Ohio, and New York. Mixing by tidal and wind currents normally keeps sea water within a few parts per thousand of normal salinity. If sea water is to be concentrated by evaporation to such a state that it can no longer hold its dissolved salts, this mixing must be stopped. Restriction of circulation may be caused by a shallow sill, or lip, across the mouth of a cratonic basin, a line of patch reefs, or a barrier reef (Fig. 11–23). In the Michigan and Ohio salt basins the restriction to sea water circulation was probably a submerged barrier on which reefs grew; in the New York Basin the influx of fine clastic sediments from highlands to the east probably also affected the pattern of sedimentation.

The Late Silurian climate must have been hot and dry, for sea water in the basins evaporated and the concentration of the salts in them increased. As water was lost to the basin by evaporation, more was drawn in from the ocean over the lip, but little of the heavy concentrated brine escaped. In this way the concentration of the brine in the basin increased until the salts could no longer be held in solution and crystallized out as small particles held up by the surface tension of the water. When the crystals grew too large to be held up, they fell through the brine and collected on the basin floor, building up a layer of evaporites. The order and nature of the salts crystallizing from sea water depends on their solubility, their original concentration in sea water, and local conditions in the basin. Generally, the first precipitate as the concentration increases is calcium carbonate, for normal sea water is saturated with it. With further evaporation and concentration anhydrite appears, and at still higher concentrations, halite. As the amount of water entering the basin changed through movements of the barrier and the rate of evaporation and freshwater runoff from rainfall changed, the type of evaporites being deposited at a particular section of the basin shifted laterally and limestone, anhydrite, and salt layers interfingered. Commonly, a cyclical repetition of evaporites is found in such basins as the Michigan and reflects cyclical changes in climate or cyclical warping of the margins. Figure 11–24 shows an increase in the proportion of halite toward the center of the Silurian basins and a zone of anhydrite separating the halite from the carbonate facies accumulating around the basin margins. The areas where tongues of the carbonate have intruded the evaporites of the basin have been interpreted as channels where normal sea water entered, diluting the more saline brines.

The mechanism postulated above to account for the Late Silurian evaporite deposits of the Salina Group contrasts with those suggested for the deposition of evaporites in the Middle Silurian dolomites in the Western Canada Basin. The first is referred to as the deep basin model, the second, as the sabkha model. Following descriptions of conditions in Persian Gulf sabkhas, some stratigraphers explained all ancient evaporites as sabkha deposits, including those

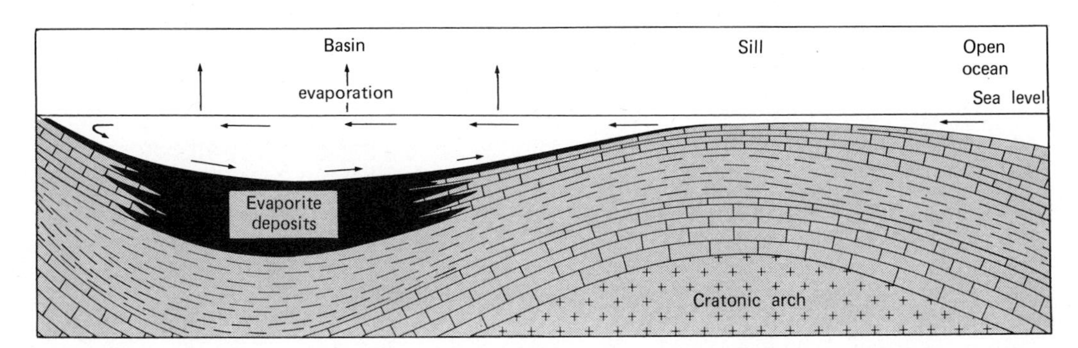

FIGURE 11–23 Cross-section of a cratonic basin and arch illustrating a model for the deposition of evaporites. The shallow-water sill restricts the inflow of normal sea water to the basin allowing its water to increase in concentration with evaporation. Note the thickening of the formations in the center of the basin.

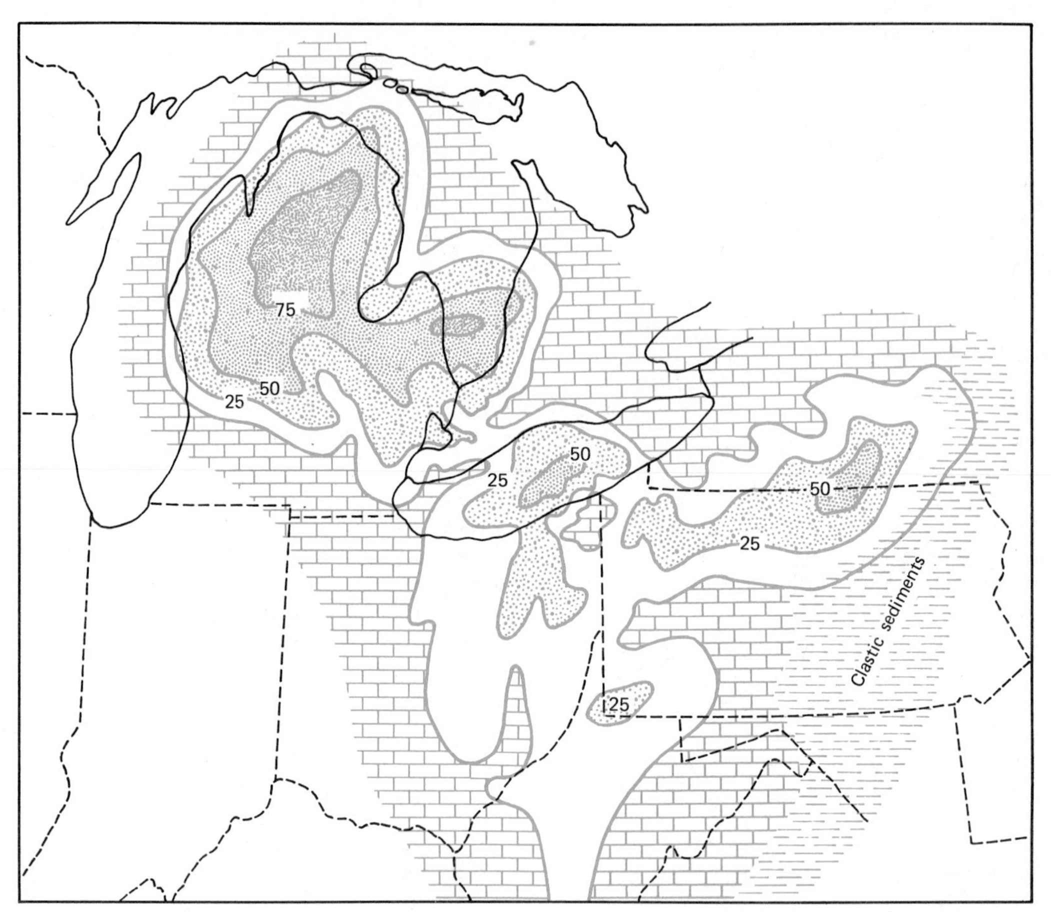

FIGURE 11–24 **Evaporite facies of the Upper Silurian Series in the Great Lakes area. The contours show the percentage of halite in the evaporite sequence. The basins are surrounded by areas of carbonate deposition. (After L. A. Briggs and H. L. Alling; courtesy of the American Association of Petroleum Geologists.)**

of the Silurian System in the Michigan Basin. Some ancient evaporites have been deposited in sabkha-like environments but others can only be explained as originating in deeper, silled basins.

Reefs and Evaporites

Three explanations for the relationship of the reefs of the Late Silurian basins and their enclosing evaporites have been suggested. First, after reef growth ceased the evaporites filled in around them and the two facies were never contemporaneous (Fig. 11–25A). Second, the reefs somehow continued to grow in surface waters while evaporite salts were being deposited in the heavier (or denser) water of the lower parts of the basin (Fig. 11–25B). Third, the periodic growth of the reefs alternated with episodes of evaporite deposition around them. As most of the reefs have been studied only through the examination of well cores and cuttings, the exact nature of the contacts and extent of interfingering of the reef rock with adjacent sediments is not clear. Possibly all three of these relationships occur in some reef tracts.

Early Devonian limestones and dolomites form the youngest deposits of the Tippecanoe sequence. These deposits are present only on the edges of the craton as the sea by this time had regressed to a mar-

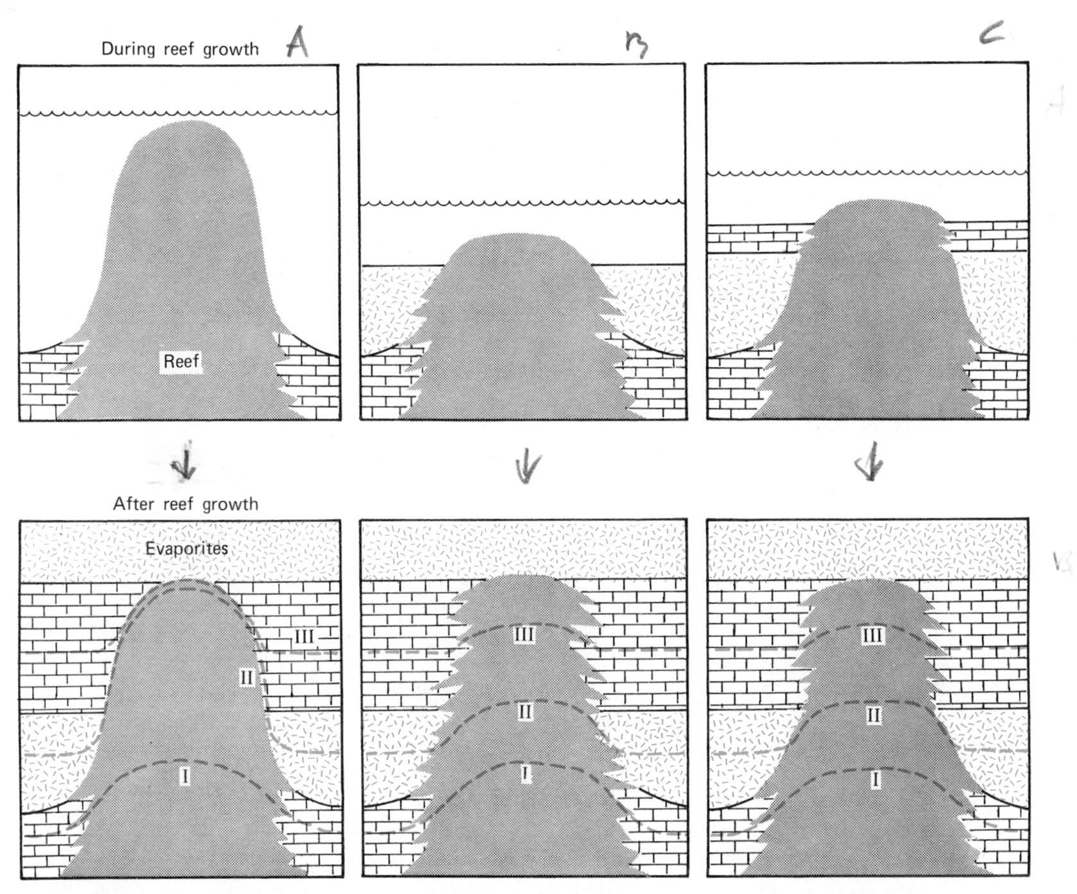

FIGURE 11–25 **Three models of the relationship between reefs and evaporites in the Upper Silurian basins of the Great Lakes area. In A the growth of the whole reef was followed by evaporite and then limestone deposition. In B both evaporite and the following limestone deposition were contemporaneous with the growth of the reef. In C evaporite deposition took place during a pause in reef growth following the first stage, and the succeeding limestone was contemporaneous with the second stage of reef growth. In the lower diagrams I, II, and III designate successive time lines through the stratigraphic sections. (Modified from K. J. Mesolella, J. D. Robinson, L. M. McCormick, and A. R. Ormiston.)**

ginal position at the end of the Tippecanoe cycle of sedimentation.

SUMMARY

Sandstones basal to the Tippecanoe sequence, such as the St. Peter, are composed almost entirely of well-rounded and well-sorted quartz grains that have passed through several cycles of weathering and erosion and deposition.

The oldest well-documented vertebrate fossils are found in one of these basal sandstones in Colorado and belong to the class Agnatha. These heavily armored, fish-like animals were bottom-dwellers sucking in loose detritus through a mouth lacking in jaws.

The ancestry of these initial vertebrates within the invertebrates is obscure because when they first appear they are highly specialized to a life in a high-energy, littoral environment.

Organic reefs are mounds on the sea floor constructed by organisms that grew into the zone agitated by the waves. The mid-Paleozoic reef community that first appeared in the Chazy Group at the base of the Tippecanoe sequence was dominated by the stromatoporoids and the corals.

Brachiopods are among the commonest invertebrate fossils in Paleozoic platform sediments. The more abundant articulates are divided into 6 orders.

Community analysis of fossil collections involves the assignment of each fossil organism to its position in the food chain of the environment in which the community lived.

Uniformity of lithology and fauna in a sedimentary deposit occupying a zone markedly transverse to modern climatic belts is evidence of a shift in polar positions.

Because waves lose energy by friction along gently shelving sea bottoms, shallow epeiric seas may have a nearshore zone of quiet water in which evaporation is high and low coastal plains called sabkhas periodically flooded by the sea on which dolomite and evaporites were deposited.

Paraconformities are gaps in the stratigraphic record that are difficult to detect lithologically, but they can be detected by a discontinuity in the succession of fossils.

A relatively small part of organic reefs is composed of framebuilders and a large part of fragments that have been broken from the reef by the surf and deposited within and around the growing core.

The shape of reefs is related to subsidence, growth potential, marine currents, and wind and wave directions.

Mapping of the geographic distribution of organisms in ancient reefs may give information about the wind and wave directions of the past. The vertical distribution of organisms may reflect the upward growth of the reef from colonizing organisms to wave resistant ones.

Late Silurian basins in New York and Michigan were restricted in their access to the open sea and evaporite minerals accumulated in relatively deep water.

Detailed examination of the contacts between evaporites and adjacent reef rocks are necessary to determine whether reef growth preceded evaporite deposition or was contemporaneous with it.

QUESTIONS

1. How do we know that Paleozoic reefs composed of animals now extinct did not grow in deep water?
2. What predators were present in Middle Ordovician seas that might have fed on the early Agnatha?
3. Make a table to show the food-gathering roles characteristic of the major groups of the early Paleozoic invertebrates. Such a table is a trophic classification.
4. Compare and contrast the evaporite depositional model of a sabkha with that of a deep-water basin, particularly with reference to the Silurian evaporites of North America.
5. Draw idealized facies maps showing the pattern of facies you would expect to find where evaporites are being deposited in a deep basin and in a coastal sabkha.

SUGGESTIONS FOR FURTHER READING

FISHER, J. H. ed. *Reefs and Evaporites—Concepts and Depositioned Models.* American Association of Petroleum Geologists Studies No. 5, 1977.

MESOLELLA, K. J., ROBINSON, J. D., McCORMICK, L. M., and ORMISTON, A. R. "Cyclic deposition of Silurian Carbonates and Evaporites in Michigan Basin." *American Association of Petroleum Geologists Bulletin,* 58: 34–62.

Discussions, 59: 535–541, 1974.

SCHMALZ, R. F. "Deep water evaporite deposition: a genetic model." *American Association of Petroleum Geologists Bulletin,* 53: 798–823, 1969.

CHAPTER TWELVE

During the interval between the withdrawal of the Tippecanoe Sea and the transgression of the Kaskaskia Sea the rising arches of the craton were eroded and the tilted edges of the sedimentary layers were truncated on the margins of the basins. When the returning seas converged on the center of the continent they laid down an initial stratum which, over wide areas, cuts across the bedding in the formations below. The dip of these lower beds is so low that the truncation cannot be seen in any one outcrop or detected in any single well record, but is revealed in regional relationships. In the mid-continent region of Oklahoma and Missouri the basal beds of the Kaskaskia sequence overlie beds ranging in age from Early Ordovician on the crest of the arches to Early Devonian in the centers of basins. The overlap is illustrated by a **paleogeologic map,** which shows the geology of an area as if the beds above a particular unconformity were stripped away (Fig. 12–1).

WORM'S EYE MAPS AND REWORKED SEDIMENTS

As the Kaskaskia Sea transgressed the North American platform, the first beds deposited were of late Early Devonian age. Over large areas these basal beds are clean quartz sandstones, similar to the basal Sauk and Tippecanoe sandstones, but in other areas the basal beds are limestones. Maps such as Fig. 12–2 can be plotted to show the lithology of the beds immediately above an unconformity, and have been called "worm's eye maps" because they depict the lower surface of a stratigraphic unit as if viewed from below. The largest bodies of sandstone at the base of the Kaskaskia sequence constitute the Oriskany Sandstone of New York and Pennsylvania and the Ridgely Sandstone farther south. Like the St. Peter, the Oriskany and its correlatives are blanket sands and are composed almost entirely of rounded, frosted quartz grains. The Oriskany is an important source of glass sand and a reservoir of many gas fields. Even where the basal Kaskaskia sediments are limestone, they contain thin beds of quartz sand and abundant rounded quartz grains "floating" in the carbonate, that is, isolated so that the grains do not support themselves. The occurrence of this sand suggests transportation by the wind and indicates the extent of erosion along the pre-Kaskaskia surface.

The source areas for the basal Kaskaskia sands were highlands in the Appalachian geosyncline and Cambrian and Ordovician sandstones along the margins of the Ozark Dome and the edge of the Canadian Shield in Wisconsin. The total absence of similar

THE KASKASKIA SEQUENCE

sand beds and "floating" grains in Silurian carbonates below the unconformity shows that these source areas were not being eroded during Silurian time and must, therefore, have been covered by Middle Silurian seas and by the sediment deposited in

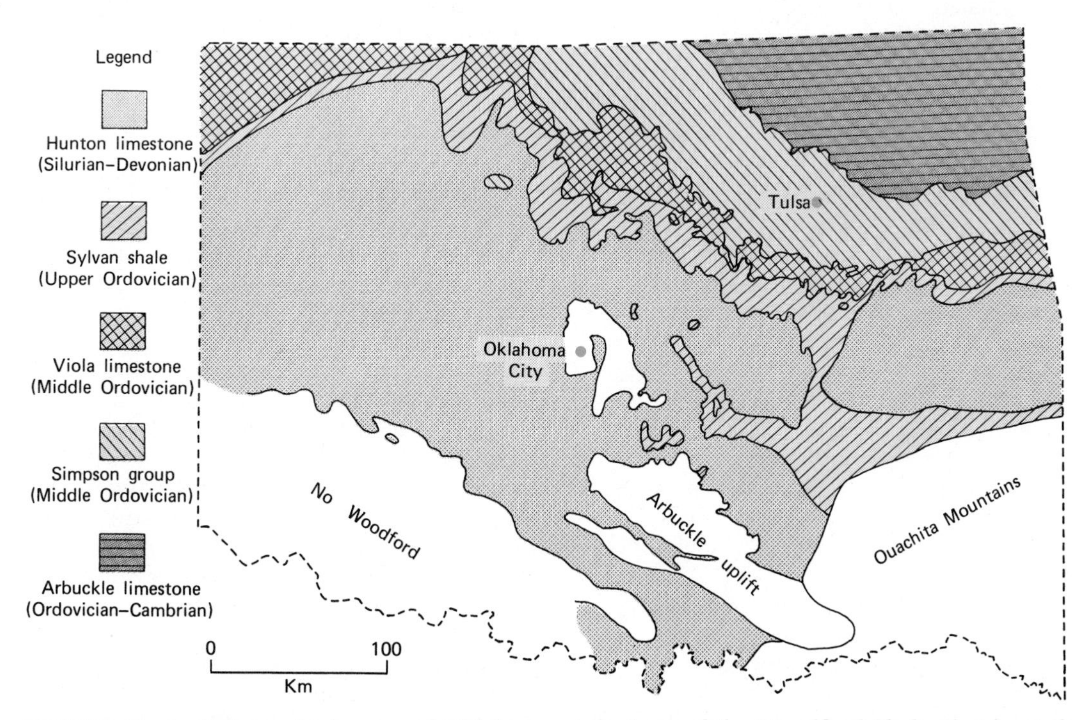

FIGURE 12–1 **Paleogeologic map of Oklahoma at the base of the Woodford Shale, that is, at the sub-Kaskaskia unconformity. The map represents the geological formations that would appear at the surface if the rocks younger than Devonian were removed. The sub-Kaskaskia surface crosses beds ranging in age from Ordovician to Devonian (After R. S. Tarr; courtesy of the American Association of Petroleum Geologists.)**

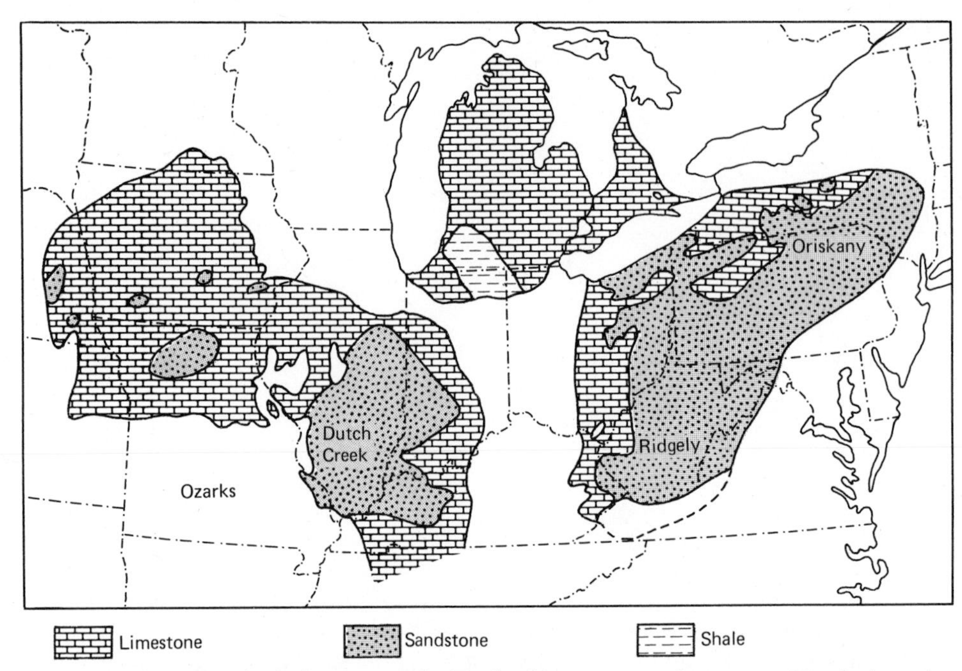

FIGURE 12–2 **Basal sandstones of the Kaskaskia sequence in eastern North America. On this "worm's eye" map the lithology of the beds immediately above the sub-Kaskaskia unconformity is plotted. (After C. F. Burk, in *Lithofacies Maps,* New York: John Wiley and Sons, Inc., 1960.)**

those seas. Because the sand does not appear below the unconformity, we deduce that the Silurian sedimentary cover was intact until the end of Tippecanoe deposition when the source areas were uplifted, and it was stripped away by erosion before the Kaskaskia transgression. Not only can the occurrence of certain sediments in the stratigraphic record lead to important conclusions about the history of their source areas, but also their absence may be significant to the historical geologist.

Presentations of the rock cycle in elementary textbooks may give the impression that sedimentary rocks are largely derived from the erosion of igneous rocks. However, two-thirds of the continental areas of today, which are the sources for modern sediments, are areas of sedimentary rocks and this proportion has changed little since the beginning of the Paleozoic Era. Not only is the main source of particles for the formation of clastic sedimentary rocks other sedimentary rocks, but most sediments are the result of two or more cycles of reworking. It has been estimated that 80 per cent of the grains in sandstones have been derived from pre-existing sediments. In Chapter 11 we saw that continued recycling results in supermature sandstones characteristic of platform areas.

The source area of a sediment may be determined from its content of diagnostic mineral grains. Quartz is not much help in such provenance studies for, although various kinds of quartz can be identified, most quartz grains are undistinguished and could come from the erosion of sedimentary, igneous, or metamorphic rocks. The rarer and more interesting grains are usually heavier and can be separated from the quartz by use of bromoform or other heavy liquids. In such liquids the quartz will float and the heavier minerals sink. Identification of these heavier materials may then give a clue to the primary source area of the sandstone, for some, such as sillimanite and kyanite, are peculiar to metamorphic sources, whereas others, such as rutile and zircon, point toward igneous sources.

SOLUTION OF EVAPORITES

In Middle Devonian time much of the craton east of the continental arch was under shallow, warm, clear seas in which limestone was deposited. In these waters corals thrived and built small scattered reefs. Conditions of subsidence that had led to deposition of salt and anhydrite in the Michigan Basin in Silurian time returned in Middle Devonian time, and more salt beds were laid down. The restriction of water circulation was apparently due to a sill formed by differential subsidence, for no barrier of reefs surrounded the basin. Silurian salt has been extensively leached from the edges of the basin, causing the Devonian strata laid down above it to collapse into caverns and form a spectacular breccia (Fig. 12–3). Some blocks in the Mackinac Breccia, at the Straits of Mackinac between lakes Michigan and Huron, are over a hundred meters across.

Not until the oil drill had probed the centers of cratonic basins did stratigraphers realize the abundance of evaporite beds, particularly salt, in the stratigraphic record, for salt beds never come to the surface in a humid climate—they are dissolved by rain and groundwater faster than erosion wears down the adjacent beds. The overlying strata settle or collapse into the space leached out and the gap in the section is hidden. Middle Devonian evaporites in Manitoba are obscured like the Silurian and Devonian salt in Michigan. The zone where the salt beds might have come to the surface is marked by saline springs and collapse structures much like the Mackinac Breccia, but salt, gypsum, and anhydrite cannot be found at this horizon except by drilling in the basin away from the exposed edge. Even in the depths of sedimentary basins, dissolution of evaporites causes collapse along vertical faults in the overlying strata.

DEVONIAN REEFS OF WESTERN CANADA

Under the northern Great Plains Devonian rocks record the complex interplay between basin subsidence, the growth of reefs, and the accumulation of evaporites. Because these Devonian reefs localize Canada's major oilfields and contain much of her petroleum reserves, they have been extensively drilled from Montana to the Arctic Circle. Most of the information on the history of the area comes from drilling records, but exposures of rocks of this period occur in the front ranges of the Rocky Mountains (Fig. 12–4) and can also be examined in exposures along the margins of the Canadian Shield.

After an initial phase of restricted seas and evaporite deposition at the boundary between Early and Middle Devonian time (Fig. 12–5A), the transgressing Kaskaskia Sea moved southward and occupied the Western Canada Basin (Fig. 12–5B). Evaporite deposition then stopped as open circulation of oceanic water allowed many patch reefs to form in Saskatchewan and Manitoba. A patch reef is a roughly circular reef usually less than a kilometer across that is isolated from the land and commonly grows in shallow, low-energy environments. Toward the end of the Middle Devonian Epoch patch reefs growing in the region of the Tathlina Arch formed the Presqu'ile Bar-

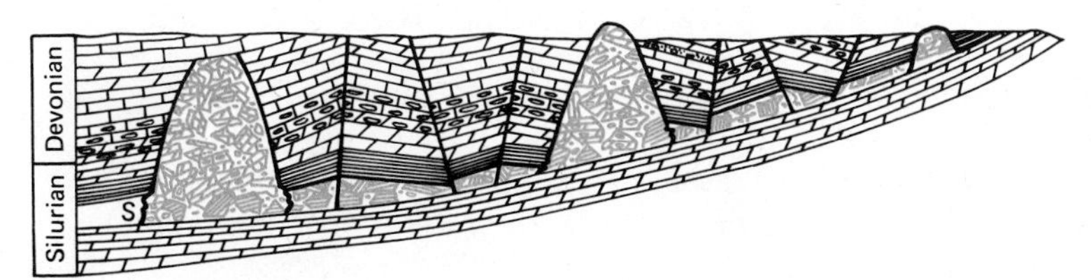

FIGURE 12–3 **Diagrammatic cross-section of the Mackinac Breccia showing how solution of Silurian salt (S) has resulted in the collapse of the higher Silurian and Devonian rocks. (From K. K. Landes; courtesy of the Michigan Department of Conservation.)**

FIGURE 12–4 **Late Devonian reefal carbonates embedded in dark shales along the front of the Flathead Range, southern Canadian Rocky Mountains. (Photo No. 154552. Courtesy of R. A. Price, Geological Survey of Canada.)**

rier Reef, a nearly continuous wall of growing organisms (Fig. 12–5B). The barrier restricted the supply of oceanic water to the basin and converted it into a gigantic evaporating pan within which the stromatoporoids and tabulate corals that had been building patch reefs died and the water became increasingly saline. From this great saline basin the Prairie Evaporite, which underlies most of southern Saskatchewan and northeastern Alberta to thicknesses up to 300 m, was laid down. At first enough sea water was passed southward over the restricting barrier reef to replenish the brine from which halite and anhydrite precipitated. Eventually, as inundations of sea water over the barrier became infrequent, the concentration of the brines on the drying salt flats reached the point where potassium salts (the most soluble in the ocean) precipitated out over a large area. The upper 30 to 70 m of the Prairie Evaporite contain about half of the world's potash reserves or about 50 billion tons of potassium salts. Many mines in southern Saskatchewan extract the salts primarily for their value in fertilizers.

At the beginning of Late Devonian time the sea spread farther into the basin and covered the West Alberta Arch which, in Middle Devonian time, separated the basin from the Cordilleran geosyncline. Evaporites continued to be deposited intermittently in the basin but the environment of deposition changed. The Prairie Evaporite had been deposited in a relatively deep basin, like the Late Silurian evaporites of the Michigan Basin, but Late Devonian evaporites and dolomites of the Western Canada Basin appear to have been deposited in a sabkha hundreds of kilometers wide. The reefs along the Tathlina Arch continued to grow, but now separated shale accumulating to the north from limestone accumulating to the southeast. South of the Peace River Dome, in the area known as Swan Hills, another group of patch reefs is now surrounded by argillaceous limestone.

With further Late Devonian transgression the shale facies that had been confined to the northeast corner of the basin moved into central Alberta. The line of abrupt transition from the shale to the carbonate facies (sometimes called the "carbonate front") then swung in a broad arc across southern Alberta (Fig. 12–5D). Reef-building orga-

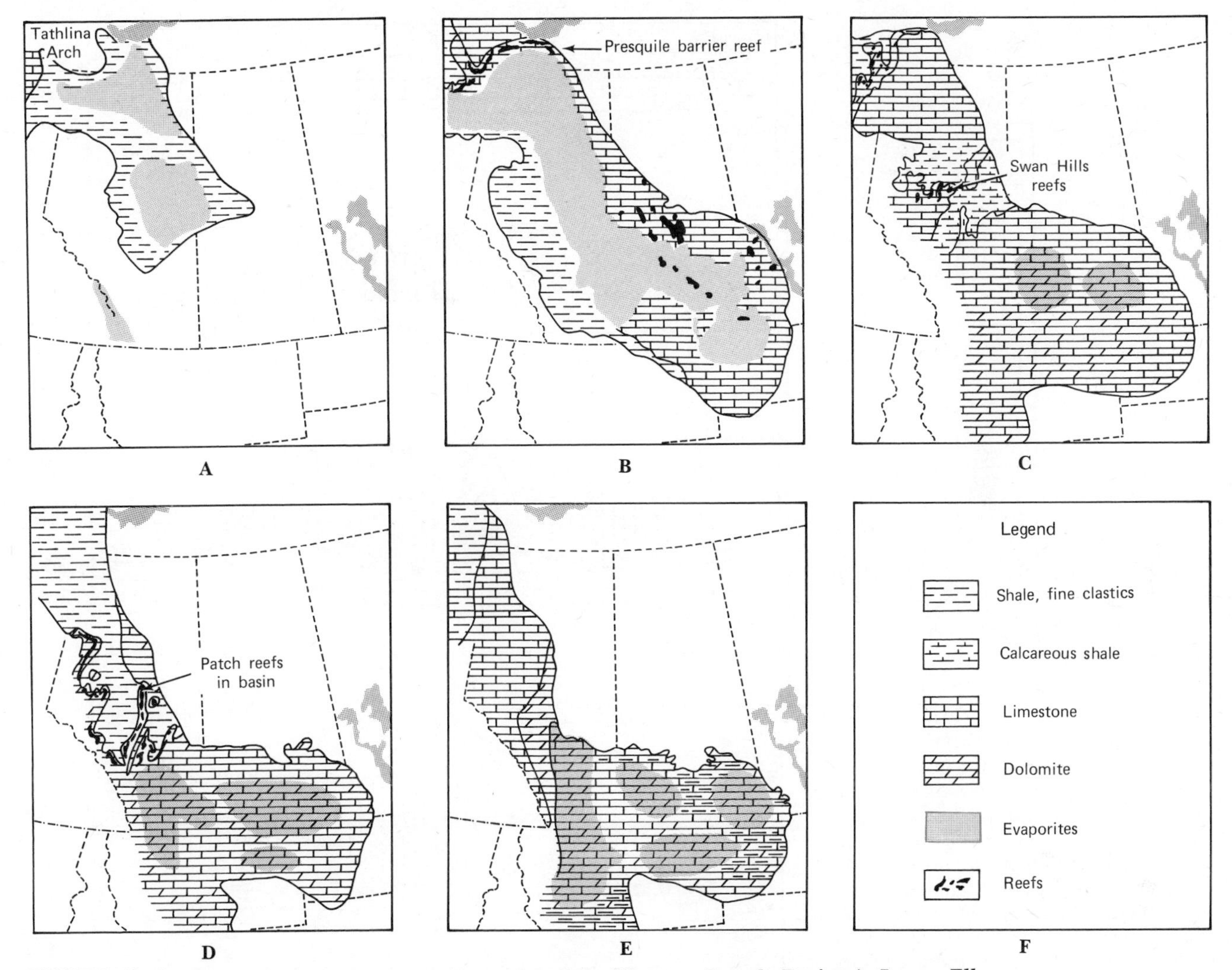

FIGURE 12–5 **Facies maps of the Devonian rocks of the Western Canada Basin. A. Lower Elk Point Group. B. Upper Elk Point Group. C. Beaverhill Lake Formation. D. Woodbend Group. E. Palliser Formation and equivalents. (From information in *Geological History of Western Canada*; courtesy of the Alberta Society of Petroleum Geologists.)**

nisms thrived along the edge of this front and along low ridges extending northeastward from it into the basin. Corals, stromatoporoids, and algae at the front built a discontinuous barrier reef, but those growing along the ridges in the shale basin built patch reefs (Fig. 12–6). Behind the barrier reefs a broad, shallow lagoon stretched south and eastward toward Wyoming and Manitoba, grading into coastal sabkhas.

Reef growth stopped in Alberta in the middle of the Late Devonian Epoch for reasons that are discussed below. The youngest Devonian strata on the west side of the basin are shallow-water limestones and dolomites deposited on a stable shelf with environments much like those of the Great Bahama Bank today. The interior of the basin continued to receive shallow-water evaporites and dolomite in a sabkha-like environment until the end of the period.

Petroleum Resources and Ore Deposits

Most of the reef rock is unbedded, coarsely crystalline dolomite. This dolomite was not secreted by the reef-building organisms but

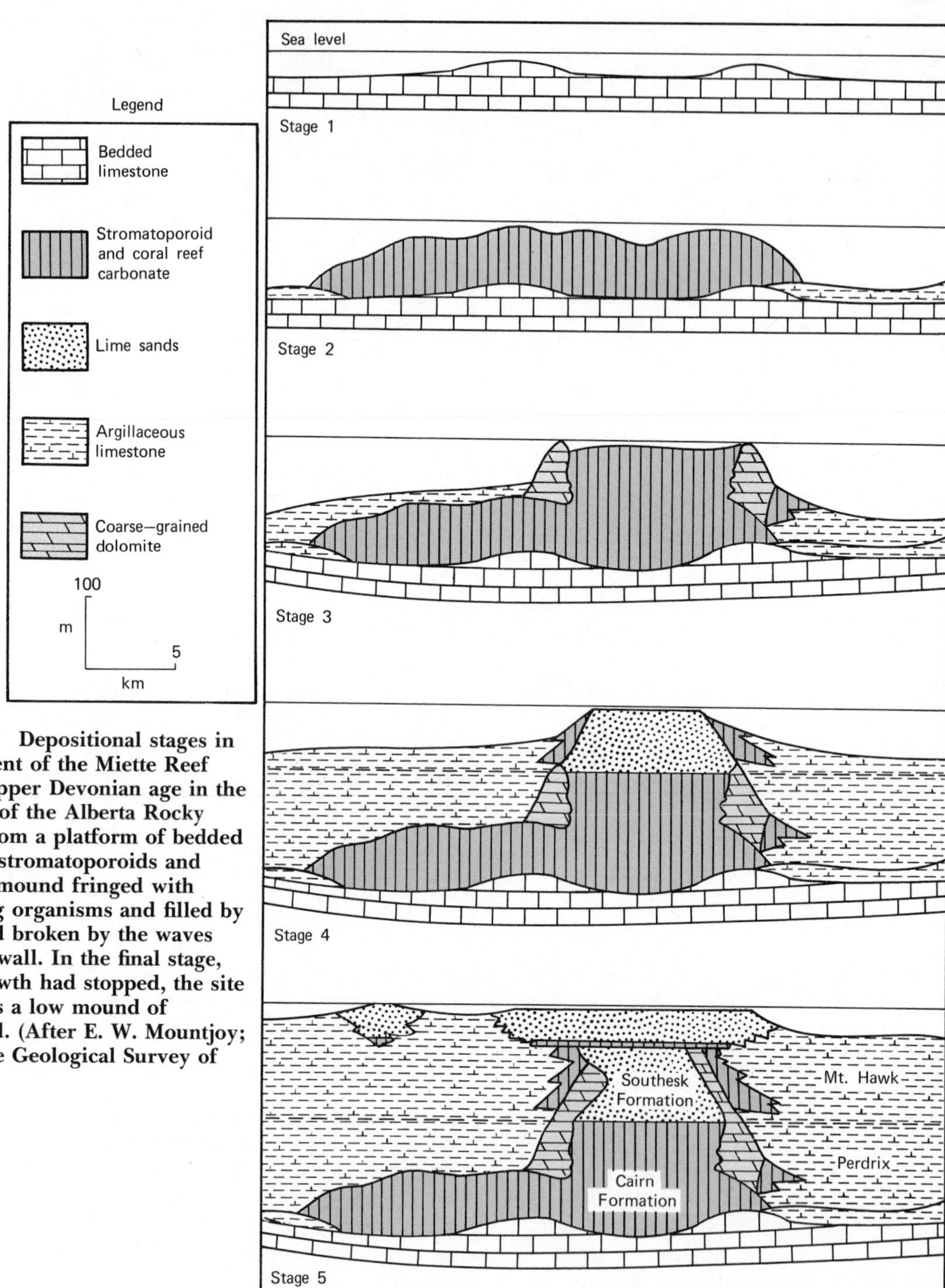

FIGURE 12–6 **Depositional stages in the development of the Miette Reef complex of Upper Devonian age in the Front Ranges of the Alberta Rocky Mountains. From a platform of bedded limestone the stromatoporoids and corals built a mound fringed with frame-building organisms and filled by carbonate sand broken by the waves from the reef wall. In the final stage, when reef growth had stopped, the site of the reef was a low mound of carbonate sand. (After E. W. Mountjoy; courtesy of the Geological Survey of Canada.)**

has replaced the original limestone structures produced by stromatoporoids and corals. In some of the reefs original textures and fossils have escaped pervasive dolomitization (Fig. 12–7). The porosity of the reef rock that makes it a good reservoir for oil and gas is formed partly by the haphazard growth of the framebuilding organisms and partly by the enlargement of voids during dolomitization. The dolomitization probably took place as sea water, incorporated in adjacent shales, was forced through the porous limestone, replacing some of the calcium ions with magnesium. When clays are deposited, the pore space between the grains is occupied by sea water and may make up 50 per cent of the volume of the sediment. When the clay is compacted under the weight of sediment accumulating above and becomes shale, its porosity is reduced to about 10 per cent and great quantities of water are squeezed out. The reefs within the shales act as rigid porous bodies into which the water is squeezed and through which it moves upward to escape eventually through overlying beds to the surface. The flow of water from the shales carried with it not only magnesium and other salts leached from the shales, but also droplets of hydrocarbons derived from the decay of micro-organisms that were entombed in the shales as they accumulated. The reefs acted as traps for these hydrocarbons and became the reservoirs for many of the oil fields that now underlie Alberta. Nearly all the patch reef and barrier reef tracts mentioned in the preceding paragraphs have produced oil and gas except those in the Middle Devonian of Saskatchewan, where interreef areas are occupied by evaporites and not the organic-rich, dark shales that are the source beds of petroleum.

As basinal shales compact, the "flushing" of waters through the "plumbing" provided by reef bodies may also produce concentrations of metals in the reefs. One of the largest lead–zinc deposits in the world is located in the Presqu'ile Barrier on the south side of Great Slave Lake at Pine Point. Metals leached as chloride complexes from the shales in the basin are believed to have been carried northward up the dip of the basin flank as the shales compacted, and precipitated as sulfides in the reef bodies where sulfide ions were available through the action of sulfate-reducing bacteria on adjacent anhydrite de-

FIGURE 12–7 **Core of reservoir rock taken from the Fenn Oil Field, Alberta, to show the nature of the Devonian reef reservoirs.**

posits. This association of lead (galena) and zinc (sphalerite) sulfides with reefs is not confined to the Middle Devonian of western Canada but is common throughout the stratigraphic record.

The Demise of the Middle Paleozoic Reef Community

The Middle Devonian Epoch and the early part of the Late Devonian Epoch were times of luxuriant reef growth in many parts of the world. In southern England, Belgium, central Europe, Australia, and Russia the stromatoporoid–tabulate coral community flourished and reached its greatest diversity. Yet, before the end of Devonian time hardly a stromatoporoid can be found in the stratigraphic record and the typical patch and barrier reefs of the Paleozoic Era were extinct. In Mississippian platform sediments algae and cryptostome bryozoans continued to produce isolated carbonate mud mounds, and in the Pennsylvanian Period leaf-like algae and chaetetid sponges produced small reefal bodies, but reefal structures comparable in size to those of the Devonian Period were not built again until Permian time. Large groups of brachiopods, such as the orthids and pentamerids, became extinct in Late Devonian time. The stromatoporoids themselves appear to have struggled through the late Paleozoic episode of adversity and made a weak comeback in the Mesozoic, but never again attained their former abundance and diversity.

Paleontology records similar extinctions, some of which are discussed later in this book, but none of which is easily explained. In North America a stratigraphic break and a brief interval of clastic sedimentation marks the end of reef growth; in Europe the reef facies is replaced by black, deep-water shales. Such changes in sedimentation are sufficient to explain sudden local changes in fauna as animals migrate to a more favorable environment, but are insufficient to explain why a flourishing community should disappear and not return when conditions again became hospitable. This crisis in invertebrate life is not reflected in the history of the vertebrates that inhabited these Devonian seas.

THE AGE OF FISH

Placoderms

A few fossils of fish with jaws, representing the ancestors of all the more advanced vertebrates, are known from Silurian rocks of North America and Europe. The most primitive of the jawed vertebrates, placoderms, resemble primitive agnathans in their heavy armor and bottom-dwelling habitat. The earliest forms, like the agnathans, may even have lacked paired fins. The presence of jaws, however, gave them the potential for much greater development as predators and by the end of the Devonian Period they had grown to body lengths of up to 2 m. The placoderms were considerably varied in their way of life. Some groups had flattened bodies and resembled skates and rays. Others, seen in Fig. 12–8, had external armor covering the fore-limbs, giving them a resemblance to arthropods. This group lived on the bottom and filtered sediments like their jawless ancestors. Most placoderms were actively swimming, deep-water predators, with wide-gaping jaws armed with sharp bony plates (Fig. 12–9).

Despite the general similarity of primitive placoderms to agnathans, no known jawless vertebrates are likely ancestors for jawed fish. The specialized embryological development of jaws, gill supports, and elements of the brain case in jawed vertebrates suggests a period of evolution prior to the appearance of even the oldest known agnathans. The early evolution of jawed fish may have occurred in fresh water, but few definitely freshwater deposits are known for this critical period of vertebrate evolution.

Sharks

Placoderms became extinct at the end of the Devonian Period. Early representatives possibly gave rise to the sharks in which the bony skeleton, except for spines and teeth, is completely absent. The cartilaginous skeleton gives them a specific gravity only slightly greater than that of water. The sharks are distinguished from higher bony fish by the absence of a swim bladder; this necessitates continuous swimming by all pelagic members of the group to maintain

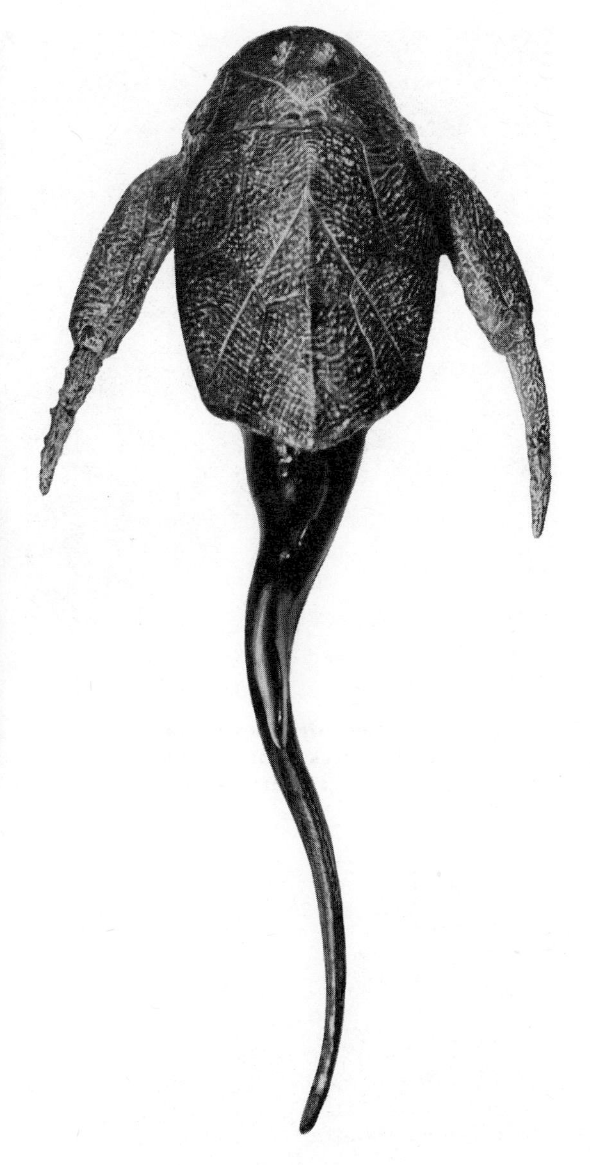

FIGURE 12–8 **The Upper Devonian placoderm *Bothriolepis*. *Left*, a fossil; *right:* a restoration, with a long fleshy tail. Fossils of *Bothriolepis* are found by the thousands on the north shore of the Bay of Chaleurs in eastern Quebec. These forms lived on the bottom, collecting food from the sediments.**

their position in the water. The absence of a swim bladder may have been a major factor in limiting the radiation of sharks, and in the development—as a result of limited buoyancy control—of numerous bottom-dwelling groups, such as skates and rays (Fig. 12–10). Because they have no bony skeleton, whole sharks are rare as fossils, but dissociated teeth are not uncommon. However, where calcification of cartilage has taken place fossils of sharks may provide a more complete view of their skeletons.

Among the best preserved sharks are specimens of the genus *Cladoselache* from the Upper Devonian Cleveland Shale (Fig. 12–11). Pyritization has led to the preservation of microscopic details of muscle and kidney tissue as well as the general body and fin outlines. This Devonian fish is clearly recognizable as a predacious shark, despite primitive features of the jaw support and fin structure. Throughout the history of the group sharks have lived primarily in marine waters, although a major freshwater group, the xenacanths, were common in the Carboniferous and Permian periods.

FIGURE 12–9 *Coccosteus*, a Devonian arthrodire, seizes one of a school of the acanthodian *Diplacanthus*. The neck joint that allowed placoderms to raise the skull is well shown. (Photograph by C. W. Stearn, by permission of the Trustees of the British Museum (Natural History).)

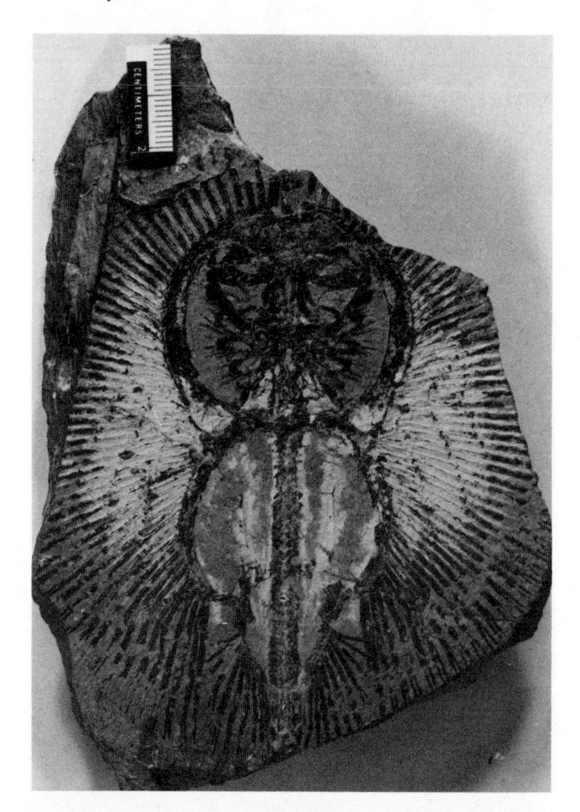

FIGURE 12–10 Fossil of the Cretaceous skate, *Cyclobatis*. The body is much flattened in this group of bottom-dwelling fish related to the sharks. (Courtesy of the Redpath Museum.)

FIGURE 12–11 **Upper Devonian seascape. The primitive shark *Cladoselache* eyes a few primitive ray-finned fish. (Courtesy of the Buffalo Museum of Science.)**

Bony Fish

All modern fish other than sharks and agnathans, are included in the Class OSTEICHTHYES, the bony fish. The most primitive bony fish, a group called acanthodians, are only preserved as scraps and spines in Late Silurian sediments. Better preserved Devonian acanthodians (Fig. 12–9) are small, fusiform fish with trunk and tail covered with small scales and head covered with a mosaic of small bony plates. The fins were reinforced on their leading edges by long, bony spines. Primitive forms may have several pairs of fins in addition to the normal pelvic and pectoral ones. The general anatomy of the acanthodians makes them plausible ancestors for modern bony fish, although close relationship has not been established. The group died out in the Permian Period.

The modern groups of bony fish appeared in Early Devonian time. In contrast with the cartilaginous sharks, the bony fish (OSTEICHTHYES) are characterized by possession of a swim bladder. The bladder developed very early in the evolution of the group from a pair of open pouches at the front end of the digestive tract. Although lost in some forms and developed as a respiratory structure in some primitive lineages, the swim bladder has served in most forms as a method of attaining positive buoyance. In primitive groups the swim bladder is open to the gut and air could be taken in and released through the mouth. In later forms the bladder is closed and oxygen pressure is controlled by release or absorption in the blood stream. The presence of the swim bladder and accompanying sophistication of fin structure are important factors in the numerical success and wide diversification of bony fish. By the end of Pennsylvanian time they had developed a wide range of body forms, indicating adaptation to a much wider range of habitat and feeding strategies than the sharks. Originally they may have been predominantly freshwater forms, but later invaded a wide range of marine environments. Bony fish continued to diversify throughout the Mesozoic and Cenozoic eras and today, with about 20,000 species, may be considered the most successful of all vertebrates.

ORIGIN OF TETRAPODS

Freshwater deposits are absent or have not been recognized in the early Paleozoic record. Some of the early fish groups probably evolved in this environment, but of this we have no record. Lakes and streams are of short duration, geologically speaking; are of limited extent; usually not the sites of extensive deposition; and are less likely than marine beds to form a part of the permanent stratigraphic record. On the other hand conditions during some periods of the earth's history may have been more conducive to the formation of extensive freshwater deposits than others. Such conditions appear to have existed during the Devonian Period and they resulted in a rich record of fossil fish.

Sarcopterygians

The bony fish, or Osteichthyes, are composed of two major groups. The actinopterygians, or ray-finned fish, make up the vast majority of modern fish. Four living genera are all that remain of the once very important second group, the sarcopterygians, or lobe-finned fish. In the ray-finned fish (Fig. 12–12) all the fins—paired and median—are supported by a series of fine, parallel rays, and the fin movement is controlled by muscles situated primarily in the body wall. In the lobe-finned fish (Fig. 12–13) each fin has a short bony axis at its base, with small rays only at the periphery. Much musculature is within the fin itself, attached to the internal skeleton. The fin structure is related to the habitat of the fish. Ray-finned fish are generally adapted to life in open water and propulsion is typically by the musculature of the trunk and tail, with paired fins control-

FIGURE 12–12 **The ray-finned fish *Priscacara* from the Eocene of Wyoming. (Courtesy of the Redpath Museum.)**

FIGURE 12–13 **On the margin of a Devonian freshwater pond, the lobe-finned crossopterygian *Eusthenopteron* clambers up on the bank with the help of its strong, bony fins. (Courtesy of the American Museum of Natural History.)**

ling steering and depth. The lobe-finned fish were adapted typically to life near, but not on, the bottom, where the ventrally directed muscular fins could be used to push the fish against the bottom. Some early members of the group are marine, but the majority have been found in freshwater deposits. Like other bony fish the sarcopterygians almost certainly had an open swim bladder, but in this group it played an important role in respiration as well as in controlling buoyancy. Bodies of fresh water are subject to periodic oxygen depletion, particularly if they are rich in plant and animal debris, and are subject to frequent and rapid changes in water level. Because these conditions favored fish that could use the swim bladder as a lung, Devonian freshwater deposits are dominated by sarcopterygians rather than actinopterygians. The most important of several groups of lobe-finned fish are the DIPNOI, or lungfish, and the crossopterygians. Both appear first at the beginning of the Devonian Period, already clearly distinct from one another as well as from the ray-finned fish. The lungfish are highly specialized from their first appearance in the fusion of the upper jaw to the skull and the presence of crushing, or slashing, tooth plates on the palate and the inner surface of the lower jaws. Despite their peculiar dentition, living lungfish have a broad dietary spectrum. Today there remain three lungfish genera; in the Paleozoic and Mesozoic this group was relatively common throughout the world. The African and South American forms can burrow into the mud and remain safe from dessication for over a year. This habit was certainly developed by Early Permian time, for numerous fossil burrows with lungfish still inside have been discovered in rocks of this age and possible burrows have been discovered in rocks as old as Devonian.

Of the crossopterygians, one genus, *Latimeria* (Fig. 12–14), survives today, surprisingly in fairly deep marine water off the east coast of Africa.

The most important group of crossopterygians were the rhipidistians (Devonian to Permian). These forms were the dominant freshwater predators of the late Paleozoic; the largest reached a length of over 2 m. The rhipidistians had the general sarcopterygian adaptations to life near the bottom of shallow pools, such as heavy muscular fins and lungs, but also possessed internal nostrils. The opening of the nasal sac within the mouth permitted them to breath without opening the mouth and to use the mouth and throat as a pump to force air into the lungs. One of the best known members of the group was *Eusthenopteron* (Fig. 12–13) from the Upper Devonian of eastern Canada.

Labyrinthodonts

In a bed at the top of the Devonian System in East Greenland the remains of rhipidistians are accompanied by those of a primitive, but unquestionably terrestrial, amphibian—*Ichthyostega* (Fig. 12–15). All features of the skeletal anatomy of these primitive amphibians indicate that they evolved from rhipidistian fish. Nearly every bone of the skull has its counterpart in the rhipidistian skull and the basic limb and girdle structure common to all terrestrial vertebrates was clearly established in these amphibians.

The evolution of the amphibians from their aquatic ancestors has been attributed to various environmental factors. The extensive red sediments in which the early amphibians and rhipidistians are found have been thought to indicate seasonal aridity. Periodic drying of ponds would favor the rhipidistians that had fins (limbs) that were more efficient on land in order to move from pond to pond. However, similar red sediments are forming today under conditions of relatively constant high humidity, and the ancestral amphibians could only have sur-

FIGURE 12–14 ***Latimeria*, a modern coelacanth. This group was once believed to have become extinct in the Mesozoic Era, but recently has been found in the Indian Ocean. Notice the stout fin bases, typical of the lobe-finned fish. (Courtesy of J. L. B. Smith.)**

FIGURE 12–15 **Skull of the oldest known amphibian, *Ichthyostega*, from the uppermost Devonian of East Greenland. (Photograph of a cast; courtesy of the Redpath Museum.)**

vived if similar damp areas on land were continuously available. The food of the early amphibians is also a problem. Land plants (p. 259) had evolved by this time, but the dentition of the early amphibians indicates that they were strictly carnivorous. Invertebrates had emerged on the land before Late Devonian time but were apparently not common. The early amphibians were certainly tied to the water for reproduction, probably had to remain damp to avoid dessication, and may have eaten aquatic invertebrates and fish that were swimming or stranded at the margins of ponds. Despite its well-developed limbs, *Ichthyostega* retained a fish-like tail for aquatic locomotion.

The earliest amphibians are termed "labyrinthodonts" because the dentine of their teeth, as seen in cross-section, is infolded in labyrinthine patterns. The adaptive significance of this feature is unknown, but this pattern occurs also in the rhipidistians and the most primitive reptiles. Labyrinthodont amphibians dominated the terrestrial landscape from the Late Devonian into the Early Permian Epoch. Like giant analogues of today's frogs and salamanders, they were sluggish, semi-aquatic, and carnivorous. Most, if not all, had aquatic larval stages with external gills, but grew gradually to maturity without marked metamorphosis.

With the evolution of reptiles in the Late Carboniferous and Permian, the more terrestrial labyrinthodont lineages became extinct. During Late Permian and Triassic time the labyrinthodonts underwent a major radiation as strictly aquatic animals, some even retaining external gills into the adult stage and reducing the limbs to the extent that they would have been useless in moving the animal on land. Many reached large size with skulls nearly a meter in length (Fig. 12–16). One group with very long skulls, apparently adapted to fish eating, became adapted to living in Early Triassic marine waters. Aquatic labyrinthodonts reached the height of their diversity in Late Triassic time. Their record in younger sediments is confined to recent discoveries in Lower Jurassic rocks of Australia. The apparent sudden decline may reflect the scarcity of freshwater sediments of Early Jurassic age.

Although large and numerous, labyrinthodonts are not the only known Paleozoic amphibians. Contemporary with them are several groups of smaller, apparently relatively rare forms collectively termed "lepospondyls." These forms exhibit a great range of skeletal and adaptive diversity. Some were greatly elongate, snake-like forms, with no trace of limbs or girdles, and others resembled salamanders or lizards (Fig. 12–17), yet others are totally without living analogues. All have relatively simple, spool-shaped vertebrae, in contrast to the more primitive pattern exhibited by labyrinthodonts in which each vertebra is made up of several discrete ossifications. No lepospondyl is known beyond the Lower Permian. This is somewhat surprising, since

FIGURE 12–16 **Restoration of the giant Triassic amphibian, *Mastodonsaurus*. The skulls of such animals reach a meter in length. (From J. Augusta and Z. Burian, *Prehistoric Animals*. London: Spring Books, 1956.)**

FIGURE 12–17 **Lepospondyl amphibians from the Carboniferous coal swamp deposits of Czechoslovakia showing several distinct body patterns: a limbless aistopod, a newt-like nectridean, and at the base a long-bodied microsaur. The animal with a large, flat head is the larval stage of a labyrinthodont. (From J. Augusta and Z. Burian, *Prehistoric Animals*. London: Spring Books, 1956.)**

they occupied a vast range of habitats, both terrestrial and aquatic, and most of the adaptive zones that they occupied were not repopulated by either reptiles or other amphibian genera until late in the Mesozoic. Their extinction introduced a long hiatus in the record of small amphibians.

MISSISSIPPIAN SYSTEM

Chattanooga Shale Problem

At the close of the Devonian Period the seas withdrew from the craton for a short time. Over the whole interior of the continent a marked change in sedimentation took place with the appearance at the Devonian–Mississippian boundary of a remarkably persistent shale called the Chattanooga in the eastern states, but known by a variety of local names. Although it is less than 10 m thick, it can be followed from Alabama to Michigan, from New York to the Midwest to Oklahoma, and across the Transcontinental Arch into Montana and Alberta (Fig. 12–18).

The Chattanooga Shale and its correlatives are bituminous, non-calcareous shales, with 15 to 20 per cent organic matter, a relatively high content of uranium, and a low content of fossils. The fossils include enigmatic, chitinous, jaw-like objects called **conodonts**, inarticulate brachiopods, and a few plants. The meager evidence indicates that the lower part of the formation is Devonian and the upper part is Mississippian in age. The shale has been dated by the lead

FIGURE 12–18 **The extent of the Chattanooga Shale and its equivalents in the United States. (Compiled from several sources.)**

method as 350 million years old, a date that is a valuable correlation point between the relative and absolute time scale.

Reconstruction of the conditions under which the shale was deposited is difficult. The brachiopods, scour channels, silt layers, and continuity upward with typical shallow-water sediments all indicate that it was deposited in shallow water. The high content of organic matter and restricted fauna suggest that it accumulated in stagnant water in which rotting organic matter produced reducing conditions. The site of deposition may have been broad, coastal lagoons receiving plant detritus from adjacent lands and separated from the sea by barrier bars. The transgression of the sea across the platform might have carried such a coastal environment over much of the continental interior, but the transgression must have been rapid, for the Chattanooga facies appears to be about the same age wherever it is found, unlike typical transgressive deposits such as the basal Sauk sandstones.

Fossil Fragment and Oölitic Limestones

Mississippian rocks of North America are divided into four series based on subdivisions of the type section of the system along the upper Mississippi Valley of Illinois and Missouri. The most persistent formation of the basal series, called the Kinderhook (Fig. 12–19), is the Chattanooga Shale; the other formations are of local extent. The overlying Osage Series is characterized by cherty limestones, and the Meramec Series by limestones made of fragments of shells, microfossils, and oölites. **Oölitic limestone** has the texture of fish roe and is composed of myriads of tiny spheres of calcium carbonate. In salt-water localities where oölitic limestone is being formed today, such as the Bahama Bank, the northern Red Sea, and the Great Salt Lake, the water is clear, shallow, and moved by currents, and more saline than normal. Calcium carbonate is precipitated chemically in concentric layers around nuclei to form minute spheres. Some of the limestones of the Meramec Series are made of these oölites, but many are com-

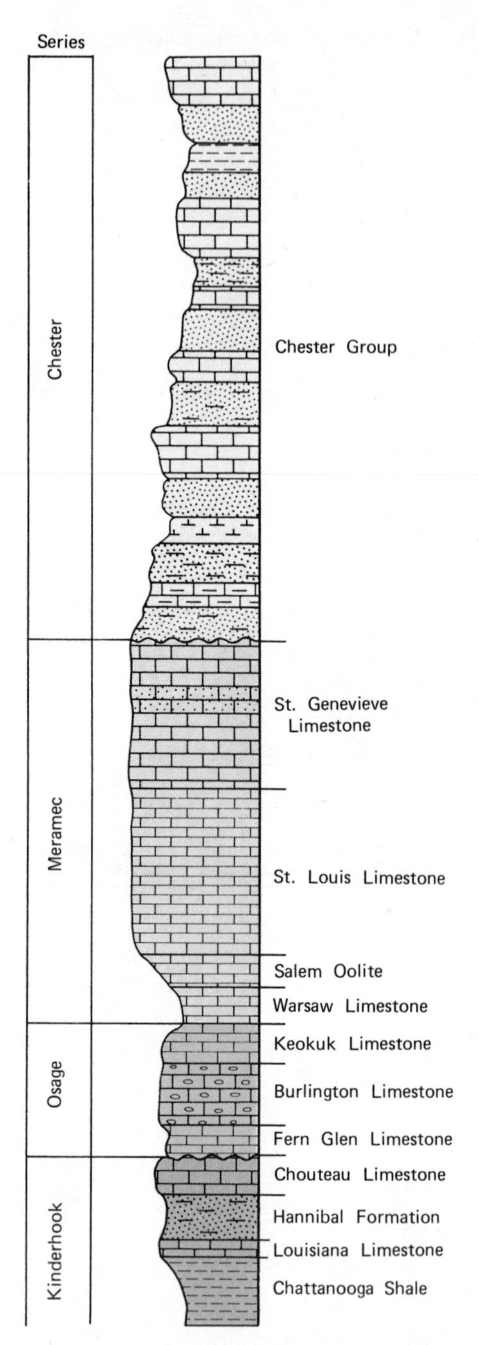

FIGURE 12–19 **Stratigraphic section of the Mississippian System in the type region. Total thickness is about 750 m. (Modified from J. M. Weller and others.)**

posed of sand-sized grains of shells that have been broken, sorted, and rounded by the waves in shallow water (Fig. 12–20). In some of the limestones the shells of the small, rounded foraminifer *Endothyra* are abundant. These "sandy" carbonate rocks make excellent building stone and are extensively quarried in Indiana, particularly around Bedford. The many plants operating in Indiana produce about 9 million dollars worth of shaped limestone block yearly, and make this state by far the largest producer of dimension stone in the United States. "Indiana Stone" is popular for exterior trim on buildings because of its pleasant, light-cream color, the ease with which it can be cut into complex patterns to form bas-reliefs, and its regular, compact texture.

Among the main constituents of limestones made of fossil fragments are the columnals and plates of crinoids. **Crinoids** reached their greatest abundance in the warm, shallow, Mississippian seas of the Midwest and western Canada. We can imagine they grew in extensive submarine meadows, rippling in the waves on their long, thin, graceful columns like garden flowers in the wind. On rare occasions they were preserved whole, flattened to the bottom by a storm and quickly covered by lime muds. Where this type of preservation has taken place, for instance at Crawfordsville, Indiana, and Legrand, Iowa, collectors have quarried huge slabs containing hundreds of perfect crinoids. Unfortunately elsewhere, the organic matter connecting the plates and columnals of the stem nearly always rotted in the water and the plates were scattered and sorted by the waves to form crinoidal limestones. Such limestones appear to be coarsely crystalline, for each plate and columnal is secreted as a single crystal of calcite and breaks along its rhombohedral cleavage, producing flat, glistening surfaces. Where the fragments were carried by currents, these limestones may be either cross-bedded or ripple-marked. In the Western Canada Basin and Rocky Mountains of Montana the Mission Canyon Limestone of Osage age contains a high proportion of crinoidal beds (Fig. 12–21). Correlative and lithologically similar formations form prominent cliffs in the eastern Cordillera, from the Grand Canyon of Arizona to the Northwest Territories, and are hundreds of meters thick. The number of crinoids required to supply such a deposit staggers the imagination.

Granular limestones, such as those of the Osage and Meramec Series, may have modest initial porosity caused by poor packing of irregular grains. When such limestones are dolomitized, weathered, and leached, the pores are enlarged and the rock becomes a satisfactory petroleum reservoir. Such secondary changes in Middle Mississippian fossil-fragment limestones have produced reservoirs in which oil and gas are trapped in the Canadian Rocky Mountains, the western states, and the Prairie provinces.

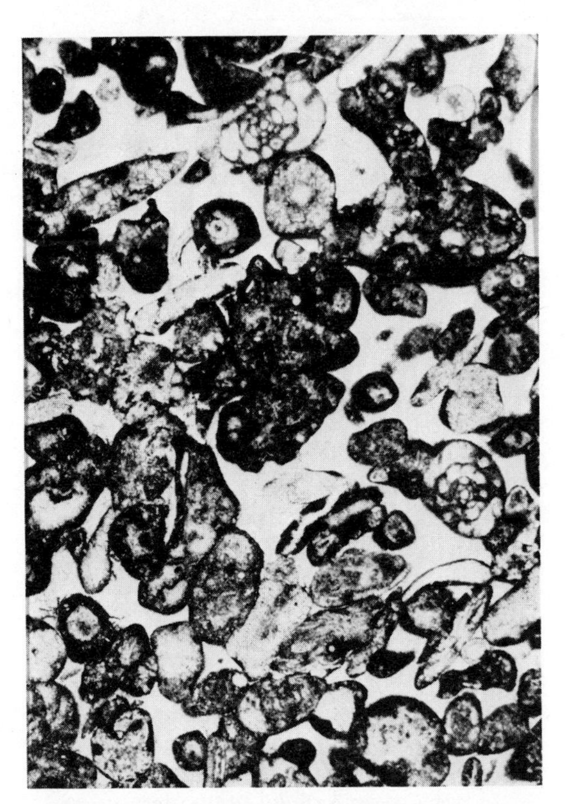

FIGURE 12–20 **Photomicrograph of the Salem Limestone showing its granular texture and microfossils (magnification $\times 15$). (Courtesy of the Illinois Geological Survey.)**

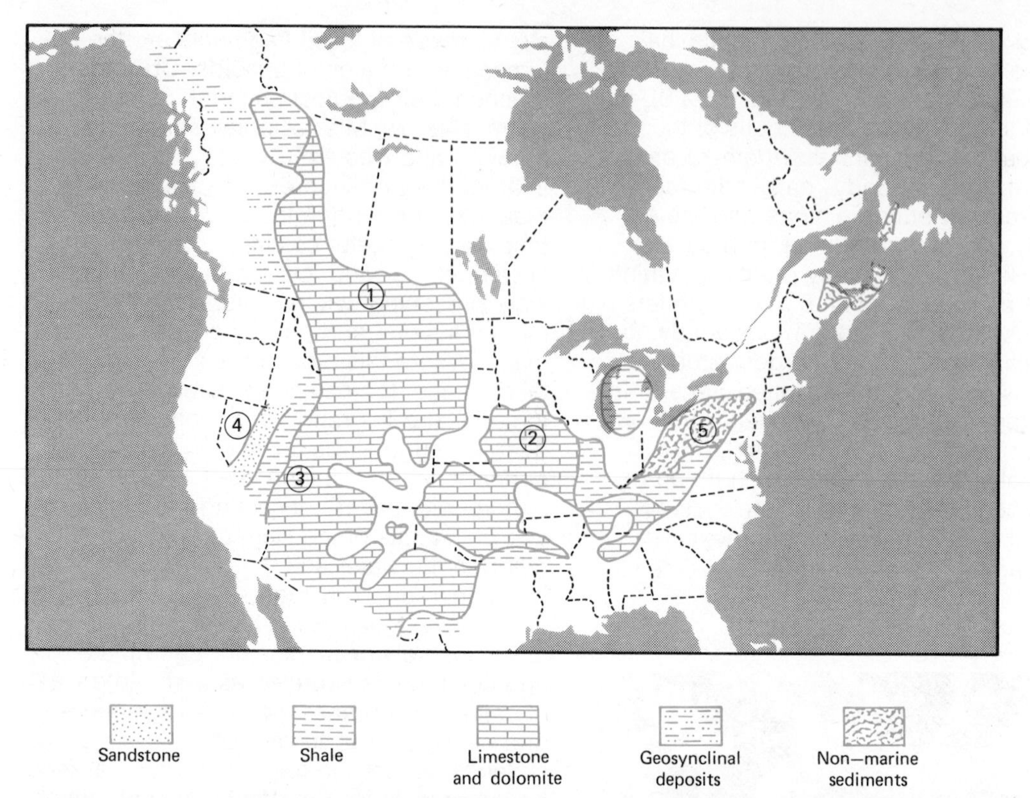

FIGURE 12–21 **Facies of the Middle Mississippian Osage Series in North America. 1. Mission Canyon Limestone. 2. Osage Series. 3. Redwall Limestone. 4. Chainman Shale and Diamond Peak Sandstone. 5. Pocono Sandstone.**

Retreat of the Kaskaskia Sea

The Chester Series is the youngest of the Mississippian System. The restriction of the Chester Series to the middle of cratonic basins and the margins of the craton is partly a reflection of the withdrawal of the Kaskaskia Sea at the end of Mississippian time and partly due to the period of erosion that followed deposition of the series and removed part of the last layers deposited by the regressing sea. The regressive deposits of neither the Sauk nor the Tippecanoe sequences is preserved, but the slower regression of the Kaskaskia Sea, the continuing subsidence of the basins, the uplift of the Canadian Shield, and perhaps the shorter period of erosion before the next transgression, could have been the reasons for the preservation of the record of the Late Mississippian regression.

The Chester Series of the Mississippi Valley consists of a persistent basal sandstone overlain by cyclically alternating sandstone and limestone formations which are laterally not persistent. Since these sands are reservoirs for petroleum in the center of the Illinois Basin, they have been extensively drilled. Detailed maps of the thickness of these sandstones show that they are thickest along sinuous branching trends, the shapes of which suggest the fillings of stream channels (Fig. 12–22). The direction of the dip of cross-bedding in these Chester sandstones is recorded in Fig. 12–23. In southern Illinois and adjacent states, the currents which deposited the cross-bedded sandstone throughout most of Chester time flowed from the northeast to the southwest. In individual sandstone formations this direction is closely correlated with the geometry of the sand distribution.

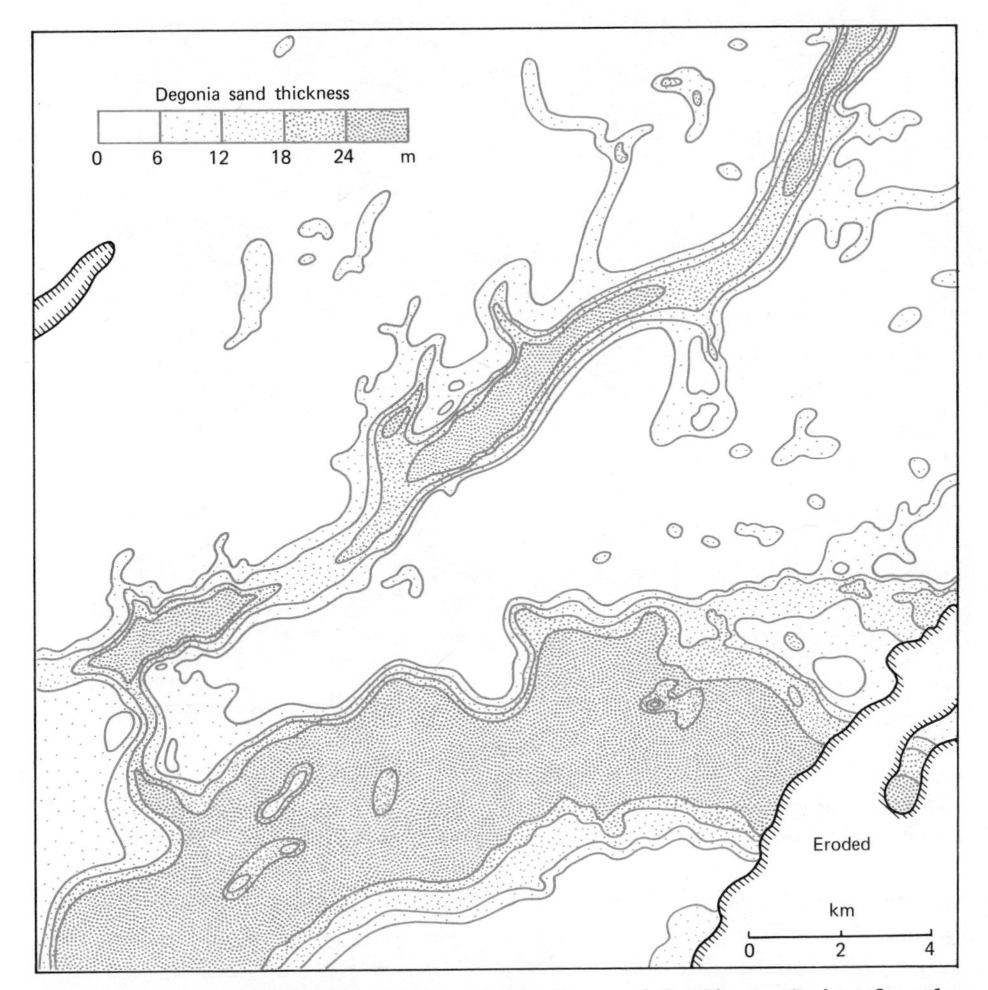

FIGURE 12–22 **Thickness of the Degonia Sandstone of the Chester Series of southern Illinois. This detailed isopach map shows that the sandstone is distributed like a channel filling in a dendritic pattern. The direction of the "tributaries" suggests that the sand was transported from the northeast and east. (From D. H. Swann; courtesy of the American Association of Petroleum Geologists.)**

Similar studies of cross-bedding and channel shape of younger systems in the same region confirm that the river system arising between the Canadian Shield in Quebec and the highlands of the northern Appalachians persisted through the Paleozoic and into the Cenozoic Era.

SUMMARY

Regional relationships at unconformities can be illustrated by paleogeologic maps that show the extent of warping and erosion between the deposition of the lower and upper sequences and worm's eye maps that show the initial sediments deposited above the unconformity.

Evaporites, especially halite, once deposited, are subject to dissolution by groundwater, particularly on the flanks of sedimentary basins but also in the center, resulting in collapse and brecciation of overlying beds.

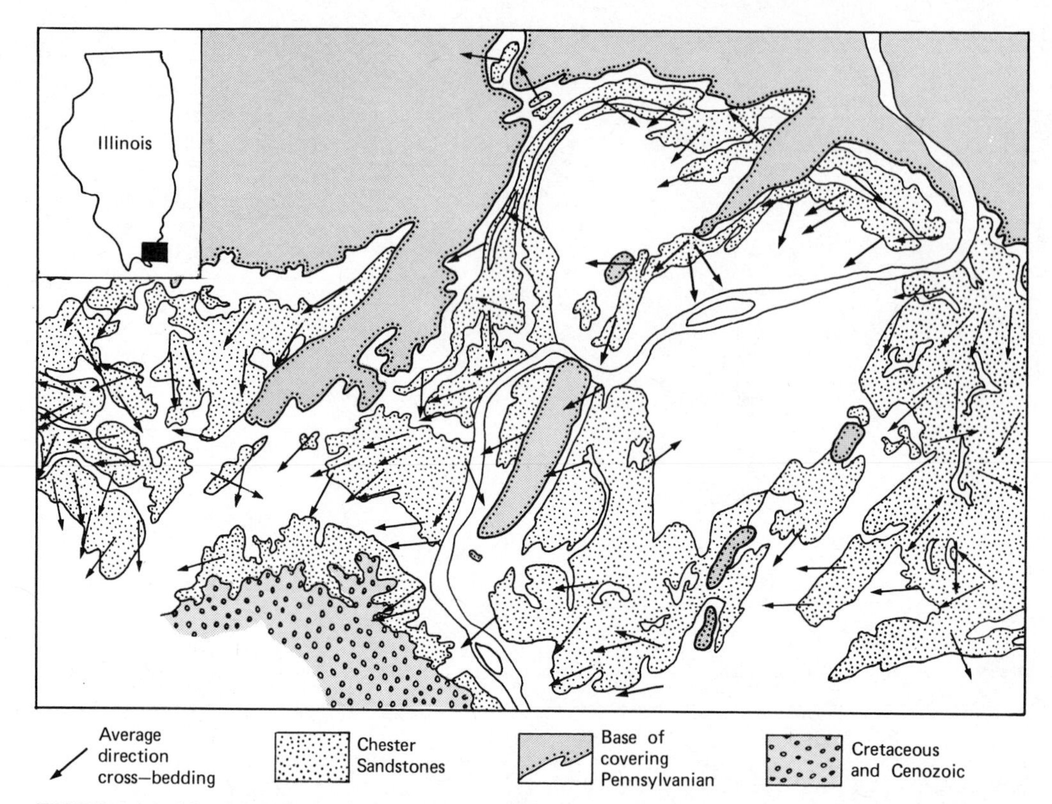

FIGURE 12–23 **Direction of cross-bedding in Chester sandstones in southern Illinois and Kentucky. (After P. E. Potter and others; courtesy of the American Association of Petroleum Geologists.)**

Middle Devonian reefs, limestones, and evaporites in the Western Canada Basin were controlled by a barrier reef at the Tathlina Arch. The Late Devonian Sea transgressed farther into the basin, bringing the shale facies into Central Alberta. Reefs grew along the edge of the southern Alberta shelf and along a ridge extending out into the basin.

Squeezing of water from compacting shales deposited in basinal areas between reefs is believed to transfer dolomitizing solutions, petroleum, and metallic ions from the basin sediments into the porous reef limestones.

Many members of the mid-Paleozoic reef community became extinct in a period of crisis of Late Devonian time.

The first jawed fish, members of a group called placoderms, appear in late Silurian rocks. Early placoderms might have given rise to sharks, which are characterized by cartilaginous skeletons and lack of a swim bladder. Primitive bony fish also appear in Late Silurian sediments. They diversified rapidly in Devonian time.

The ancestors of land animals were a group of bony fish called sarcopterygians, which had fins attached to a stout basal lobe and well-developed swim bladders that could act as lungs. The first land vertebrates, the labyrinthodont amphibians, appeared at the end of the Devonian Period and dominated Mississippian land faunas.

The Chattanooga Formation and its western correlatives form a unique platform deposit of highly bituminous, uranium-bearing shale probably deposited in widespread coastal lagoons.

Limestones composed of fossil fragments are characteristic cratonic blanket deposits, particularly of the Mississippian Period, when crinoids reached their acme. When leached or dolomitized, they form good reservoirs for oil or gas.

Chester sandstones form the regressive deposits of the Kaskaskia Sea in the Illinois Basin. The direction of their transport can be determined from studies of cross bedding and the shape of channel sands.

QUESTIONS

1. Describe each of the methods mentioned in Chapters 11 and 12 by which a stratigrapher can determine the source area of a clastic sedimentary deposit.
2. What function was served by the limbs of sarcopterygian fish and the swim bladders of bony fish and how are these functions modified in the first tetrapods?
3. The distribution of facies in the Redwater reef, one of the late Devonian reefs of Alberta, suggests that the wind approached the reef from the northeast. What winds were these and what was their relationship to the Devonian paleoequator?
4. Make a chart to show the evolutionary relationships between the following vertebrates discussed in this chapter: agnathans, placoderms, sharks, acanthodians, osteichthyes, sarcopterygians, dipnoi, crossopterygians, labyrinthodonts, and lepospondyls.

SUGGESTIONS FOR FURTHER READING

BASSETT, H. G. and STOUT, J. G. "Devonian of western Canada, *in* International Symposium on the Devonian System," Oswald, D., ed., *Alberta Society of Petroleum Geologists,* 2: 717–752, 1967.

FULLER, J. G. C. M. and PORTER, J. W. "Evaporites and carbonates: two Devonian basins of western Canada." *Canadian Petroleum Geologists Bulletin,* 17: 182–193, 1969.

JACKSON, S. A. and BEALES, F. W. "An aspect of sedimentary basin evolution: the concentration of Mississippi type ores during late stages of diagenesis." *Canadian Petroleum Geologists Bulletin,* 15: 383–433, 1967.

SUMMERSON, C. H. and SWANN, D. H. Patterns of Devonian sand on the North American craton. *Geological Society of America Bulletin,* 81: 469–490, 1970.

THOMSON, K. S. "The biology of the lobe-finned fishes." *Biological Reviews* 44: 91–154, 1969.

CHAPTER THIRTEEN

The Carboniferous System in the North American Platform is split by a regional unconformity that is the basis for its division into the Mississippian and Pennsylvanian systems, or subsystems, and marks the beginning of the Absaroka sequence. The surface marks not only an interval of withdrawal of epeiric seas from the platform, but also the beginning of a sedimentary regime much different from the platform regimes of the early and middle Paleozoic, which were dominated by carbonates and evaporites. Uplift and erosion of highlands peripheral to, and within, the platform resulted in the spread of clastic sediments over the cratonic basins from all sides.

The six stages generally recognized as subdivisions of the Pennsylvanian System in the United States are as follows:

Virgil
Missouri
Des Moines
Atoka
Morrow
Springer

Sediments of the first two stages are confined largely to the margins of the platform. The major Absaroka transgression took place in Atoka–Des Moines times when the Cherokee Shale was widely spread out over the southern platform. The Pennsylvanian beds cross the eroded edges of early Paleozoic sediments at the flanks of the cratonic uplifts to produce a striking unconformity, the regional aspect of which is well shown in Fig. 10–7. In southern Illinois the Chester Series was deeply eroded during the pre-Absaroka interval. Channels up to 100 m deep were cut into the flat-lying beds and subsequently filled with basal Pennsylvanian rocks. By plotting detailed paleogeologic maps from well records, the incision of these channels into the lower formations can be detected and the direction of the streams and their tributaries can be accurately plotted (Fig. 10–5). The direction of drainage is the same as that indicated by the cross-bedding in the Chester sandstones, that is, toward the southwest.

CYCLOTHEMS

Although now missing from cratonic arches, Pennsylvanian sedimentary rocks once formed a continuous sheet stretching from the Appalachian Mountains to Colorado and from North Dakota to Texas. Deposition over this vast area of the platform took place in cyclical repetitions of marine and non-marine strata. Such cyclical repetitions, in which different sedimentary rocks occur repeatedly in a larger stratigraphic unit, are called **cyclothems.** Since cyclothems were

THE ABSAROKA SEQUENCE

first described geologists have been trying to reconstruct the peculiar conditions under which they were deposited.

The Typical Illinois Cyclothem

Figure 13–1 represents the idealized succession of lithologies of a cyclothem in the

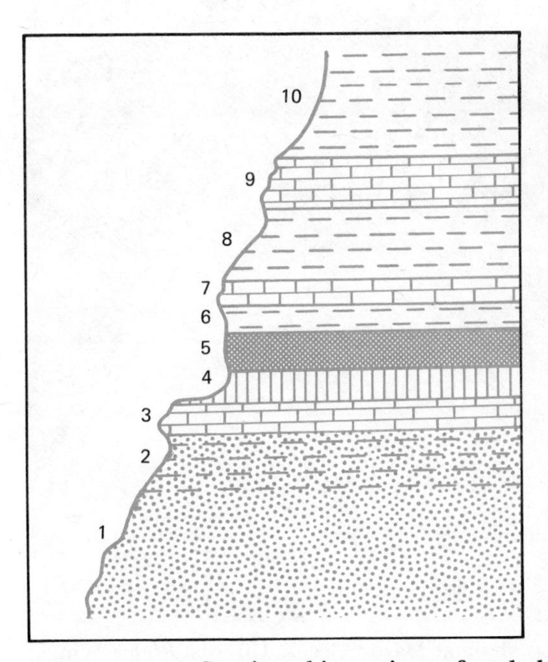

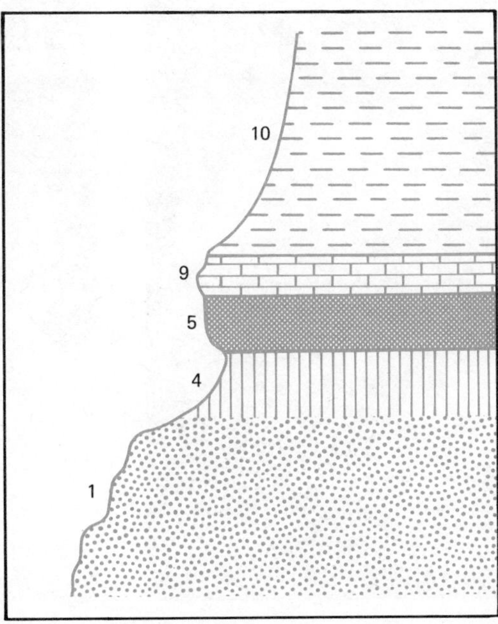

FIGURE 13–1 **Stratigraphic sections of cyclothems. *Left:* Typical Illinois cyclothem with all ten members well developed. *Right:* A cyclothem in which several members have not been deposited. (Modified from J. M. Weller; courtesy of the Geological Society of America.)**

Illinois Basin. The cycle starts at the unconformity and contains ten members. The first member (1) consists of a massive, cross-bedded sandstone showing signs of having been deposited rapidly near the shore under deltaic conditions. It is commonly overlain by a sandy shale (2) and limestone (3), containing plants and freshwater invertebrates, respectively. The fourth member (4) is a clay, called the underclay, that has been extensively leached so that only silica and clay minerals remain. It commonly contains tree roots and is overlain by the coal member (5). Coal is economically the most important member of the cyclothem, but many of the coal seams are less than a meter thick and are, therefore, uneconomical to mine. In Illinois fifty coal seams have been found in the Pennsylvanian System, and in Pennsylvania, more than one hundred. Some of the finest Pennsylvanian fossils have been found in the shale that overlies the coal (6). On Mazon Creek, near Morris, Illinois, calcareous concretions weather from shales in this stratigraphic interval and accumulate like pebbles in the

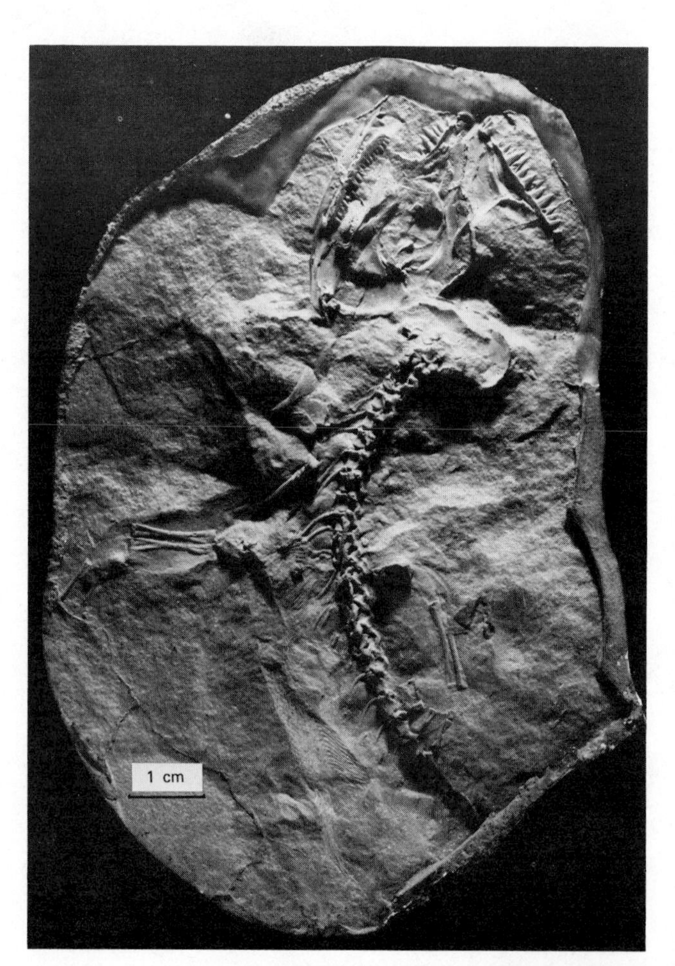

FIGURE 13–2 **Fossils from nodules in Pennsylvanian shale at Mazon Creek, Illinois. *Right:* Wings of a dragonfly-like insect showing the preservation of the color pattern. *Left:* Mold of the skeleton of the fossil reptile *Cephalerpeton.* The impression of the forelimbs can be seen on the left beyond the limits of the bone. (Courtesy F. M. Carpenter and the Museum of Comparative Zoology.)**

creek bed. Many species of fish, amphibians, plants, insects and other arthropods, and molluscs have been preserved in these nodules, some so faithfully that impressions of the soft parts can be seen (Fig. 13–2).

Members 6 to 10 are marine members and consist of alternating units of limestones and shales with a rich marine fauna of invertebrates. The cycle ends with an erosion surface, the channels of which may cut deeply into the sediments that were just deposited. The deposition of another deltaic sandstone above the erosion surface begins another cyclothem. Few cyclothems contain all ten members; usually several are missing in any given cycle.

The interpretation of the sequence of events that led to the succession of beds in a midwestern cyclothem is relatively simple. Members 1–5 were deposited in non-marine environments and represent a period in which a deltaic swamp was built out at the edge of a regressing sea. Members 1 and 3 are the deposits of lakes on the growing deltaic plain. The underclay of Member 4 has been thought to be the soil on which the swamp developed, but not all coal geologists agree with this theory. (The conditions in these coal swamps are considered in detail below.) With the transgression of the sea during the deposition of Member 6, the plants of the swamp were drowned and the sea began the deposition of the marine members of the cycle. At the end of the cycle, a rapid retreat of the sea was followed by subaerial erosion, or the cutting of channels below shallow water, and the formation of the unconformity. Each cyclothem represents a retreat and an advance of the sea; this much is clear, but whether these advances and retreats were caused by structural behavior of the basin, eustatic movement of sea level, changes in climate, or compaction of the coal-bearing succession, or a combination of these factors remains a problem.

Regional Variations

If observations of cyclothems are extended westward and eastward from Illinois, we find that the individual members of the cycles are remarkably persistent and some of the coal beds, although less than a meter thick, can be traced for hundreds of kilometers. The Colchester (No. 2) Coal of Illinois and its correlatives have been mapped from Kansas to Ohio, a distance of about 1100 km and, if the lower Kitanning Coal is the same layer in the Allegheny Basin, this coal bed can be traced a further 900 km. No modern counterpart of such an extensive swamp exists today. The individual members seem to be of the same age over the whole area of deposition, but evidence that they do, or do not, cross time markers is difficult to find because closely spaced, time-parallel horizons have not yet been defined in this succession. Regional variations take place in the lithology of the cyclothems as shown in Figure 13–3. In the lower part of the Brereton Cyclothem in Illinois the beds are channel and deltaic sands deposited at the sea edge, but in Kansas and Oklahoma they are marine limestones. Although the cyclicity of the sedimentary succession remains, in Kansas and Oklahoma the marine members are much thicker than the non-marine members, and in Pennsylvania the opposite is true—the non-marine members predominate to the virtual exclusion of the thin, marine members. In the mid-continent region the coal beds are very thin or absent in some cycles; in Pennsylvania and West Virginia they are thickest and most abundant. The regional variations in cyclothems are reflections of the presence of the sea during most of the Pennsylvanian Period in Kansas, its general absence in the Appalachian area, and its presence for about half the time in the area between the two.

In addition to those features already mentioned, any theory of the origin of cyclothems must account for the proximity of all the environments of deposition to sea level, the intercalation of cyclothems at the margins of the craton, and the occurrence of similar cycles in other continents. The faunas of the marine limestones of the cyclothems show that even these members were deposited in shallow water, and paleontologists have estimated that the sea on the craton never reached a depth greater than 70 m during Pennsylvanian time. In the

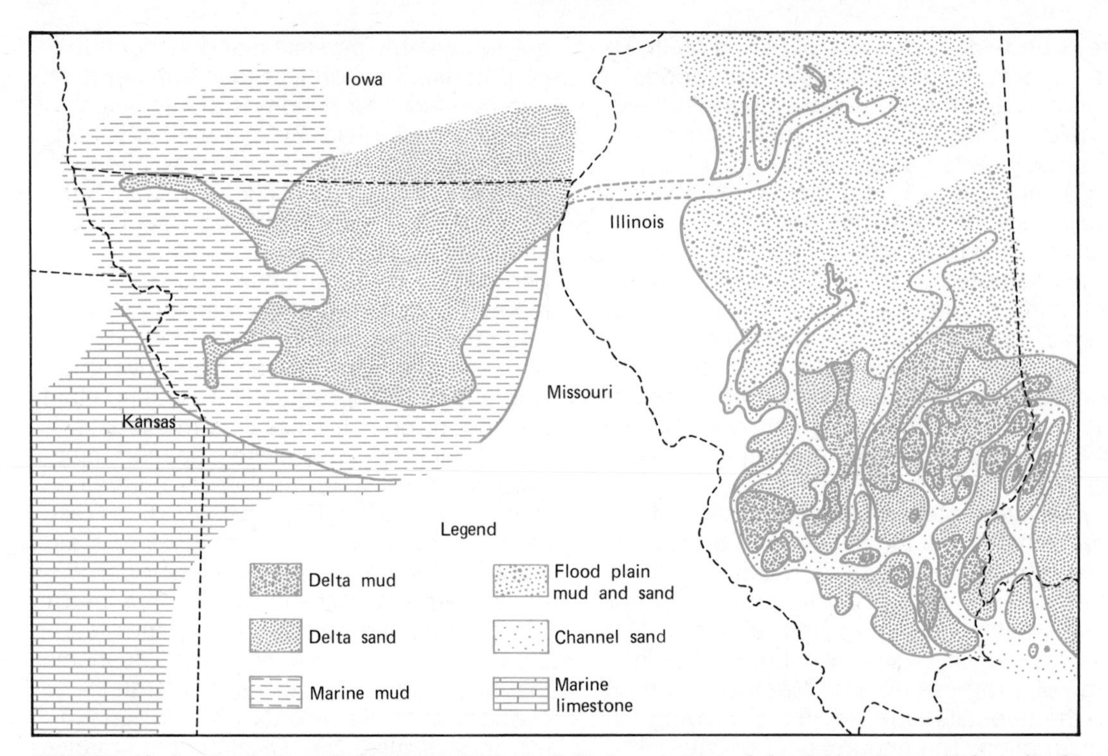

FIGURE 13–3 **Environmental reconstruction of the lower part of the Brereton cyclothem in the mid-continent region. The unpatterned areas were largely low-lying lands. A channel sand has been extended across the Mississippi Valley to supply the delta in northern Missouri, although Pennsylvanian deposits have been eroded now from the valley. (Modified from H. R. Wanless and others; courtesy of the Geological Society of America.)**

middle of the basins and at the edge of the craton where the Pennsylvanian System thickens, coal seams and other members of the cyclothems may split and additional cyclothems may be introduced. In Pennsylvania, where the system is thick, many more cyclothems can be counted than in Illinois or Kansas. Finally the cyclothemic repetition of beds found in North America is common in sedimentary sequences of late Paleozoic age on many continents of the world; therefore, a mechanism proposed to explain North American cyclothems must be applicable in Europe and elsewhere.

Theories of Origin

Many theories have been proposed to explain cyclothems, but only three will be considered here. One group of stratigraphers holds that the cyclic transgression and regression of the sea over the continental interior was caused by alternating episodes of depression and uplift of both the source area of the clastic sediments and the basin itself. During the numerous depressions, the alluvial plain was inundated by the sea; during each uplift the emergence of the plain and its aquisition of newly eroded sands resulted in the movement of the shoreline hundreds of kilometers seaward. As the basin subsided and the highlands of the source were reduced by erosion, the sea slowly encroached on the delta, edging the shoreline landward until another uplift of basin and source forced it back. The pulsating behavior of the platform during the formation of the cyclothems is related by this theory to the general instability of the continent in late Paleozoic time and to orogenic episodes in the bordering geosynclines.

The cyclical transgressions and regressions have also been ascribed to variations in sea level and a steadily subsiding basin. Proponents of this hypothesis believe that sea level rose and fell throughout the world —perhaps more than 100 times—in the rhythm of the cyclothems.

A third theory accounts for the cyclical variations in sedimentation by cyclical variations in climate. Climatic variations would affect the precipitation, vegetation, rate of weathering, and, most important, the discharge of the streams. Increase in the discharge would bring more sediment to the basin, building out the delta on which the coal swamps became established. Channel cutting, even below sea level, could accompany an increased discharge and could produce local unconformities. As subsidence and compaction carried the surface of deposition downward, and as discharge decreased, the sea would transgress the delta and begin the marine part of the cycle.

The source areas for the sediments of the cyclothems can be located by detailed studies of facies and sedimentary structures, such as cross-bedding in the sandstones. Detailed mapping of a single stratigraphic horizon over large area by Harold Wanless and his associates has produced maps like Fig. 13–4. The channels of southern Illinois and Ohio are traces of a delta of the major river flowing southwesterly across this part of the continent, that was active in the formation of the Chester sand channels. Other channels brought sediment from the uplifted basement rocks of the Nemaha Ridge. Between the alluvial plains built in

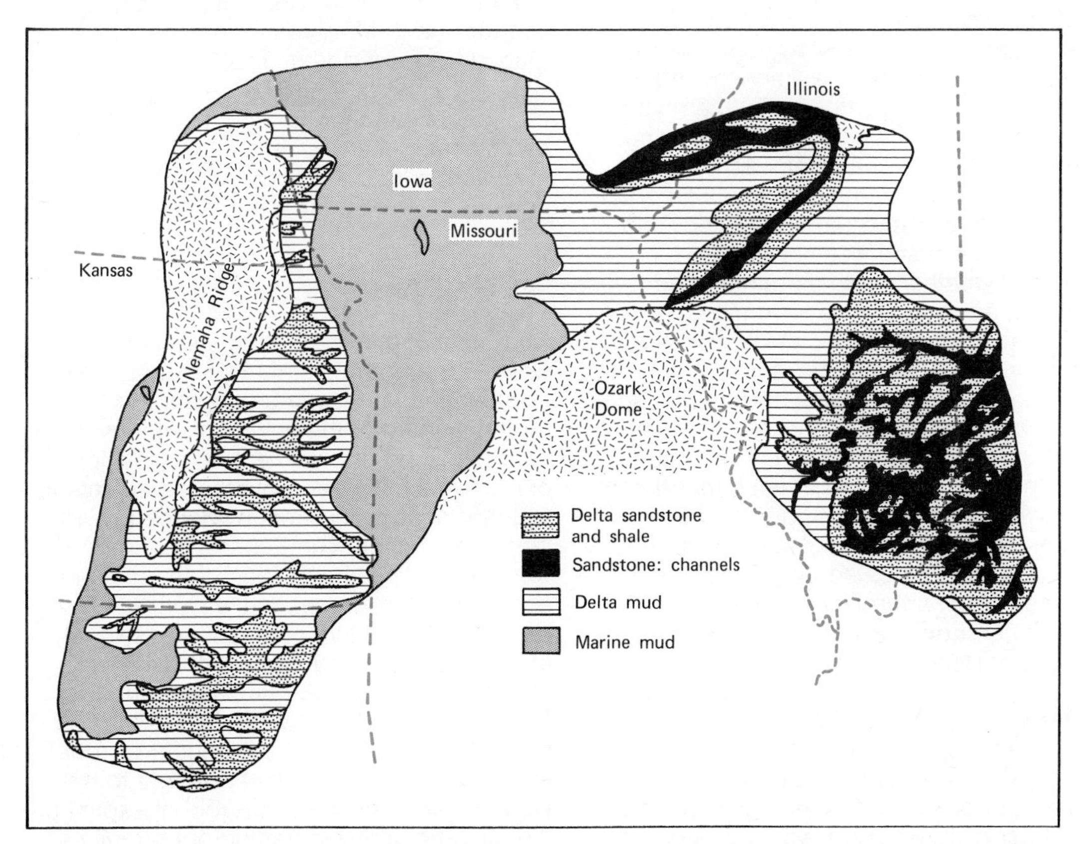

FIGURE 13–4 **Environmental reconstruction of the time of deposition of the bed below the Colchester (No. 2) Coal and its correlatives in the Illinois and mid-continent areas. (After H. R. Wanless and others; courtesy of the Geological Society of America.)**

front of these sources, marine muds accumulated in a shallow sea. The sands of the Allegheny Basin of Pennsylvania and West Virginia were supplied from the east, from highlands that were forming in the Appalachian geosyncline. From the south, highlands arising at this time in the Ouachita geosyncline supplied clastics to the edge of the platform. The continental interior seems to have been largely surrounded by source areas in late Paleozoic time. The portals by which the sea entered this lowland are not obvious, but entry from the southwest through gaps in the cratonal uplifts seems most likely.

GLACIATION IN THE SOUTHERN HEMISPHERE

Those favoring sea-level control of cyclothemic sedimentation have invoked the expansion and contraction of ice caps in the Southern Hemisphere as a mechanism for raising and lowering sea level in the rhythm of the cyclothems. During glaciation in the Pleistocene Epoch, sea level was lowered by over 100 m. When the climate was cold enough some of the water evaporated from the ocean by the sun was not returned in the normal weather cycle as rainfall and runoff, but accumulated to form ice caps on the continents. When the ice sheet melted in warmer, interglacial interludes, the water flowed back into the oceans and normal sea level was re-established. This fall and rise of sea level was repeated as the ice caps expanded and contracted at least four times during the Pleistocene ice age; in order to account for late Paleozoic cyclothems, it must have been repeated between 50 and 100 times. Such abundant glacial cycles have not been discovered in the late Paleozoic glacial deposits of the Southern Hemisphere.

The evidence of late Paleozoic glaciation in the Southern Hemisphere includes tillites, striated cobbles, polished and striated bedrock, roche moutonées, fluvial–glacial deposits, etc., spread over large areas of South America, Africa, India, Australia, and Antarctica. The distribution of these glacial features is indicated in Fig. 1–17 and was used by early proponents as a strong argument for continental drift, for only if continents are brought together does the glaciated area form a reasonable "ice cap." Unfortunately, the glacial deposits rarely contain diagnostic fossils and are, therefore, difficult to date and correlate between outcrops. In South America at least 10 glacial episodes, and perhaps as many as 17, have been recognized. The slow motion of the southern continents over the South Pole (Fig. 13–5) caused the interval of glaciation to occur later in Australia than in Central Africa. In South Africa glaciation may have started as early as Late Devonian time, but in general it was of Carboniferous age. In South America glaciation appears to have begun in Mississippian time and was over by the middle of the Permian. In Australia the maximum extent of glaciation occurred in the Early Permian and some glacial ice persisted until Late Permian time.

The temporal equivalence of Carboniferous glaciation near the South Pole and the cyclothemic coal swamps near the equator is remarkable, but a cause-and-effect relationship has not been established.

COAL

Coal, one of the great energy sources of the world, is "stored sunlight." The search for additional coal resources depends on an understanding of the environment in which coal forms. Its environment of deposition can be reconstructed from the sediments in which it occurs and from evidence provided by the coal itself. The Pennsylvanian Period was a time of accumulation of the most important coal resources of the Northern Hemisphere. Unlike petroleum resources which are distributed through most of the Phanerozoic systems, commercially valuable coal is found almost entirely within late Paleozoic and late Mesozoic rocks. Not until Carboniferous time had land plants evolved to the stage where they were capable of exploiting the climatic and geographic opportunities for rapid growth and diversification offered by the northern continents.

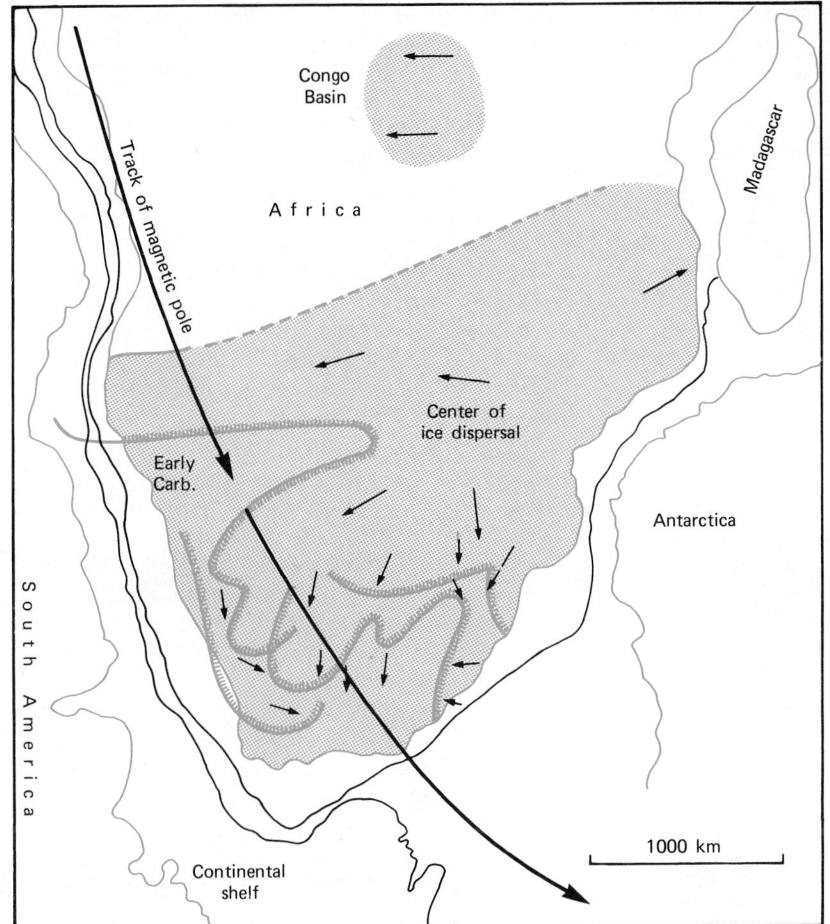

FIGURE 13–5 **Extent of late Paleozoic glaciation in Africa and its relationship to the position of the south magnetic pole. Directions of ice movement are schematically shown by colored arrows and the position of successive ice margins by hachured lines. The other continents composing the Gondwanaland part of Pangaea have been assembled around South Africa at the edges of their continental shelves. (Modified from diagrams by J. C. Crowell and L. Frakes.)**

Rise of Vascular Plants

Fossils of primitive photosynthetic organisms are known from rocks 3.1 billion years old and may have existed earlier. The early Archean fossils are prokaryotic blue-green algae, still represented in the seas today by individual cells or gigantic colonies. In stromatolites they formed conspicuous structures in Helikian and Hadrynian rocks, and survive in diminished numbers to the present day. The algal cells themselves are fossilized only under exceptional conditions. More advanced, eukaryotic algae have existed for more than 1 billion, possibly nearly 2 billion years (Chapter 8).

Eukaryotic algae are divided on the basis of their cellular structure and the dominant pigment into four groups: red algae (RHODOPHYTA), brown algae (CHROMOPHYTA), and green algae (CHLOROPHYTA). The red algae secrete massive encrusting and branching calcite skeletons that are important contributors to the reef biota from early Paleozoic time to the present. Some green algae make important contributions to freshwater limestones formed in lakes, and the marine group known as the dasycladaceans were locally important contributors to lagoonal limestones. The microscopic diatoms and coccoliths secreted siliceous and calcareous tests, and their contributions to deep water sedimentation is discussed in Chapter 19.

Until late Silurian time the land was bare of vegetation, with the possible exception of the rock-clinging lichens and a few questionably terrestrial plants. Cambrian, Ordovician, and probably Silurian landscapes must have been as bleak as those of desert regions today.

Plants faced a number of serious problems in adapting to life on land:

1. Support of the stem against the forces of gravity. (In water support was supplied by the buoyancy of the medium.)
2. Water conservation in the dry air.
3. Controlled passage of respiratory and photosynthetic gases, O_2 and CO_2, without constant water loss.
4. Distribution of reproductive cells in the air.
5. Transport of water and nutrients, available to non-vascular plants in the water surrounding the entire plant.

The cellular and biochemical nature of terrestrial plants indicate that, of the algal groups, they are closest to the CHLOROPHYTA. There are, however, no fossils of green algae that could be said to represent the group that gave rise to land plants.

Primitive Vascular Plants

The oldest specimens of terrestrial plants can be unequivocably separated from aquatic forms by the presence of **vascular tissue**—special cells evolved to conduct water and nutrients from the substrate through the stem. The vascular tissue forms a semi-rigid bundle of cells running the length of the axis of the plant, producing an internal "skeleton" to support it on land. The external surface of the stem was made more-or-less watertight by a waxy cuticle layer secreted by the superficial cells. Specialized cells at the surface of the plant could be expanded or contracted to permit the transfer of respiratory gases through openings called **stomata** without continuous water loss. Reproductive cells, the **spores,** were located in inclosed structures called **sporangia,** located at the ends of the stems where they could be dispersed into the air.

In 1859 J. William Dawson recognized in Devonian sediments of the Gaspé Peninsula of Quebec a branching plant which he called *Psilophyton* and which he believed was the first land plant. Since Dawson's discovery similar fossil plants have been described from beds of latest Silurian to Late Devonian age. They were long referred to as the Psilophytales, but are now called the Rhyniales (Fig. 13–6).

The simplest and apparently the most primitive of the Rhyniales is *Cooksonia* from the Upper Silurian of Czechoslovakia. Another genus, *Rhynia* (from which the group received its name), from the Middle Devonian of Scotland, was found preserved in a silicified Devonian peat bog in complete anatomical detail, together with a few ticks and primitive spiders. The underground root-stock, or **rhizome,** was covered with many hair-like structures. The stems were at least 20 cm long and up to 6 mm thick, and were completely bare without traces of leaves or spines. The sporangia, or spore cases, were carried on the tips of the branches, and some have yielded well-preserved spores.

The very simple type of vascular plants represented by the Rhyniales apparently did not survive the Devonian (unless the modern genus *Psilotum,* of doubtful affinity, is included in this group).

In Early Devonian time diversification of plant types, perhaps all derived from the Rhyniales pattern, gave rise to a vast assemblage that continued through the Carboniferous and Permian. Three prominent groups of the late Paleozoic flora are the LYCOPSIDA, SPHENOPSIDA, and PTEROPSIDA. These persisted as abundant constituents of the world flora until the close of the Paleozoic Era when the lycopsids and sphenopsids approached extinction. Only a handful of genera persist today. The pteropsids, on the other hand, have remained important members of the world flora until the present, and may be related to the seed-bearing, and hence the flowering, plants of modern times.

Lycopsida

The lycopsids, sometimes called the scale trees, are a group represented today only

FIGURE 13–6 **Habitat group showing land plants along an Early Devonian shore. In the foreground is *Psilophyton* with its curled tips, accompanied by primitive low ferns. Large tree-sized ferns (*Aneurophyton*) are shown with bulbous trunk bases. The apparently leafless trees on the left are lycopsids. (Mural by G. A. Reid; courtesy of the Royal Ontario Museum.)**

by the common creeping ground pine, *Lycopodium* (Fig. A-3), and two or three other genera. The earliest lycopsid, *Baragwanathia,* from the Devonian of Australia, is somewhat larger than *Lycopodium* but otherwise similar. *Asteroxylon,* a well-known member of the Rhynie flora, had both twofold and irregular branching. The lower parts of the stem, at least, are clothed in an abundance of spirally arranged, leaf-like structures, possibly the forerunners of the spiral leaves of later lycopsids. Early Devonian lycopsids resemble *Lycopodium* in size, but in the Carboniferous coal swamps they developed into trees more than 30 m high (Fig. 13–7). Most of them had straight, high trunks that were branched only near the top. The younger branches were thickly set with long, narrow leaves that reached lengths of about 1 m. The roots of the lycopsid trees were small compared with the bulk of the tree and spread laterally rather than downward. The tree reproduced by means of spores that were usually housed in cases at the ends of the branches. *Lepidodendron* grew to heights of about 35 m and was extensively branched at the top (Fig. 13–7). Leaves left scars on the bark where they were shed in a pattern similar to that of a snake's skin (Fig. 13–8). Another important lycopsid of this time was *Sigillaria* (Fig. 13–8 and 13–14). Its sword-like leaves up to 1 m long grew in vertical rows from the trunk and left scars of octagonal outline. The tree grew to heights comparable to that of *Lepidodendron,* but was either unbranched or branched once near the top where the leaves of the mature plant appeared.

FIGURE 13–7 **Reconstruction of a Pennsylvanian coal forest. Tall branched *Lepidodendron* trees dominate the scene. Unbranched *Sigillaria* trees with tulip-shaped crowns of leaves and *Calamites* with whorls of leaves surrounding the trunks and branches are prominent in the swamp. A dragonfly skims across the water, and an amphibian basks in the sun. (Mural by P. R. Haldorsen; courtesy of the National Museum of Canada.)**

Sphenopsida

The sphenopsids (or arthrophytes) are spore-bearing plants characterized by jointed stems. They are represented today by a single genus, *Equisetum,* the horsetail, or scouring rush of railway embankments and swampy places (Fig. A–4). The main stem of the sphenopsid is usually longitudinally ribbed and is divided into segments by joints, from which circlets of branches or leaves are given off. Spore cases form at the tips of the main branches. Although modern horsetails are less than a meter high, their ancient representatives grew to heights of nearly 30 m (Fig. 13–7) and were minor contributors to the formation of coal. The first sphenopsids occur as small fossils in the Rhynie Chert, but in the Pennsylvanian coal swamps the most important genus *Calamites* (Fig. 13–7) grew to the size of a tree. Most of the stem fossils are internal molds to which coalified bark material adheres in some specimens (Fig. 13–8 and 13–9). Like their modern representatives, *Calamites* was either hollow or filled with pith. The circlet of leaves was united at their bases into a continuous sheath surrounding the parent branch, from which they easily became separated. When only the leaves of *Calamites* are preserved, they are given names such as *Annularia* or *Asterophyllites* (Fig. 13–10).

Pteropsida

A large part of the Carboniferous foliage had a pattern resembling the leaves of modern ferns. This kind of foliage grew on woody structures and was associated with reproductive organs indicative of a variety of groups, some of which were much more specialized than living ferns. The oldest known true ferns, the PTEROPSIDA, are found

FIGURE 13–8 **Bark of lycopsid trees with characteristic leaf-scar patterns.** ***Left: Sigillaria. Right: Lepidodendron.*** **(Courtesy of the Redpath Museum.)**

in beds of Late Devonian age. They reproduce by spores that are shed from spore cases on the undersides of the leaves.

Plants of the Middle and Upper Devonian Series were more advanced than the Lower Devonian Rhyniales in several ways. They evolved true leaves, either according to a fern-like pattern, or as long, narrow blades. They developed a woody stem by the accumulation of additional vascular tissue, which allowed the evolution of a tree-like or arborescent habit. Seeds and pollen developed to replace the spores as a means of reproduction. A spore is a single cell that is cast off by the plant and grows as it can, but a seed is fertilized and grows in the ovary for some time until it is released to form a new plant. These advanced features formed the basis for the differentiation of the land plants in the Carboniferous. Only a single innovation, that of the enclosed seed and embryo of the flowering plants, was yet to be perfected.

FIGURE 13–9 **Part of a *Calamites* stem showing the vertical fluting and annular constrictions. (Courtesy of the Field Museum of Natural History.)**

Progymnosperms

From Middle and Late Devonian beds numerous fossils are known which reproduced by spores, like ferns, but had a woody structure resembling that of gymnosperms (the conifers and cordaites). These are called the Progymnosperms and are thought to be ancestral to the seed plants of the Carboniferous. The genus *Callixylon,* from Upper Devonian rocks, is known to have had a trunk up to $1\frac{1}{2}$ m in diameter. It probably stood several meters tall, although the longest trunk known measures only 3 m. The leaves associated with *Callixylon* are called *Archaeopteris* and resemble those of a fern (Fig. 13–11). An earlier genus, *Aneurophyton* (late Early Devonian) shows a more primitive stage in the development of fern-like foliage. In this genus the frond gives the appearance of a fern in which all the flattened leaf-like tissue is missing and only the stems are present (Fig. 13–6).

FIGURE 13–10 ***Annularia,* whorls of leaves that have fallen away from the constrictions on the branches of *Calamites* or other arthrophytes. (Courtesy of the Field Museum of Natural History.)**

Pteridospermales

Other important constituents of the coal swamp flora were three groups of seed plants, the Pteridospermales (seed ferns), the Cordaitales (cordaites), and the Coniferales (conifers).

The pteridosperms (seed ferns) were entirely fern-like in their leaves and stems but,

FIGURE 13–11 ***Archaeopteris*, the fern-like foliage of the arborescent progymnosperm *Callixylon*. (Courtesy of the Redpath Museum.)**

unlike any plant so far described, they reproduced by means of seeds and pollen. Although the seed ferns are now extinct, the more complex cycads and the angiosperms, that later came to dominate the world flora, are believed to have evolved from them. Unless the seed-bearing organs are preserved, some fern-like fossils, such as *Neuropteris* and *Pecopteris* (Fig. 13–12), cannot be identified as either pteropsids or pteridosperms. *Glossopteris* (Fig. 13–13) is a well-known seed fern that occurs only in the Southern Hemisphere. The distribution of the *Glossopteris* flora is one of the evidences favoring the union of South America, Africa, India, and Australia into a vast southern continent. Both the true ferns and the seed ferns grew in late Paleozoic time into trees 10 to 15 m high, like the modern tree ferns of the tropics. The pteridosperms persisted into the Jurassic Period, but they were never a major constituent of the land flora after Pennsylvanian time.

Cordaitales

These were a primitive order of trees related to the modern conifers. They had long, blade-like leaves and bore clusters of naked seeds. Most cordaites were trees with wood like the conifers, but with a central core of pith. They appeared in the Pennsylvanian Period and rapidly became one of the most abundant seed-bearing plants. *Cordaites* grew to a height of 30 m in the Pennsylvanian coal swamps. The upper part of the trunk was divided into several branches that bore leaves up to 10 cm wide and 100 cm long. Of all the plants so far mentioned only the ferns and the cordaites persisted in force through the Permian Period (Fig. 13–14), but the latter died out in the early part of the Mesozoic Era. They are thought to have given rise to the conifers during the Pennsylvanian Period.

Coniferales

The chief characteristics of the conifers are cones that bear naked seeds and a trunk composed solidly of wood (Fig. A–6). In these two features they differ from the cordaites. In the Pennsylvanian they became the dominant kind of tree, yielding only in the Cretaceous to the spread of the angiosperms. *Walchia* (Fig. 13–14) is a genus of Permian time.

ENVIRONMENT OF COAL DEPOSITION

Traces of twigs and leaves found in coal are evidences that it is a product of plant growth. If a slice of soft coal is ground so thin that light will pass through it, the microscope reveals that it is composed of amorphous, brown, organic matter in which are embedded plant spores and shreds of plant matter still bearing the cellular structure of wood or leaves (Fig. 13–15). Coal is, therefore, considered to be highly condensed, partly decayed plant matter, compressed to a compact, dense bed from a porous tangle of branches, roots, trunks, leaves, and their debris. The degree of compaction and metamorphism defines various "ranks" of coal,

FIGURE 13–12 Reconstruction of the seed fern *Neuropteris*. Notice the fern-like foliage and the seeds hanging from the tips of the branches. (Courtesy of the Field Museum of Natural History.)

FIGURE 13–13 Leaves of *Glossopteris*, a member of the Pteridospermales, common in the southern continents that formed Gondwanaland. (Courtesy of R. D. Gibbs.)

FIGURE 13–14 **Diorama of a water hole in the arid part of north-central Texas during Permian time. The large tree in the center foreground is *Cordaites*. Two tall *Sigillaria* trees grow nearby. At the extreme right is a tall *Lepidodendron* and a *Sigillaria* with a divided top. Of intermediate height are two tree ferns; small bush-like ferns form a thicket nearby. Primitive mammal-like reptiles with sails (*Dimetrodon* at left, *Edaphosaurus* at right) and smaller amphibians are the only animals present. In the distance are primitive conifers, probably *Walchia*. (Courtesy of the New York State Museum.)**

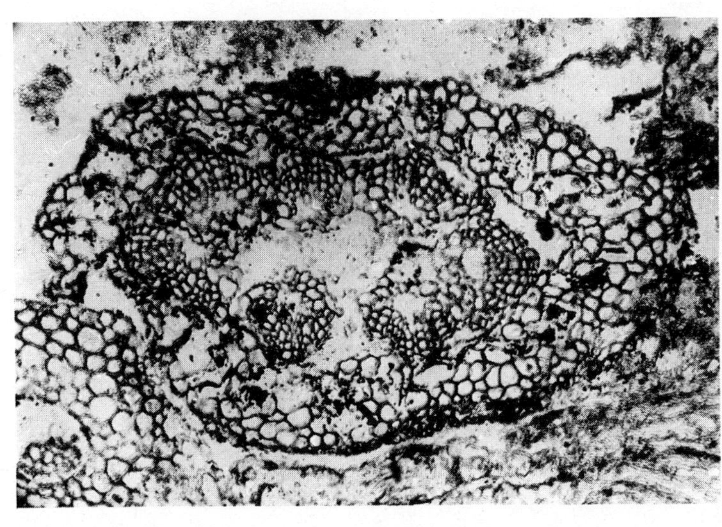

FIGURE 13–15 **Cellular structure in branches of the coal forest plant *Calamites* preserved in a calcareous concretion, or coal ball, from eastern Kentucky, ×100. (Courtesy of J. M. Schopf and the U. S. Geological Survey.)**

which in order of progressive change from accumulated vegetation, such as peat, are: **lignite, bituminous coal,** and **anthracite.** The microscope reveals that the plants of the coal forest grew under conditions like those of subtropical coastal swamps of the present day. The absence of rings in the wood of the trunks indicates very little variation in seasonal temperature. The large leaves and big breathing pores (stomata) indicate a moist, equable climate. Unlike those of our upland types, the roots of the coal swamp trees did not reach downward into the soil in search of water, but spread out laterally just beneath the surface.

Detailed studies of the relationship of coal to the enclosing sediments show that not all coals were deposited in the same environment. Some coals were formed in flood-plain swamps on deltas laced with distributary channels now represented by sandstone lenses; others formed as fillings of alluvial channels by plant debris; still others formed in estuaries or in coastal lagoons behind barrier islands. Environments that duplicate the Pennsylvanian coal swamps are found in the delta of the Mississippi River and the Florida Everglades. In the former, vegetation accumulates as peat in coastal basins, between the distributary channels of the birdfoot delta and in the broad flood plains of the inland delta. In the Everglades of southern Florida thick layers of peat are accumulating in brackish mangrove swamps along the coastline and further inland in freshwater swamps. Sediments with marine fossils immediately above the coal beds suggest that the coal formed near sea level and that some of the trees may have been able to tolerate salt water, as do modern mangroves.

The deltaic plain on which the forest of the Pennsylvanian coal swamps grew was bordered on the south and east by highlands in the Appalachian and Ouachita geosynclines, on the north by the low swell of the Canadian Shield, and on the west and southwest by cratonic uplifts. At times the whole plain was green with living plants, ranging in size from scale trees a hundred meters high to ground-trailing ferns. The trees grew on a mush of organic material resulting from the semi-decay of earlier plants, and on death made their own contribution to the accumulating peat layer. Bacterial decay was limited by lack of oxygen in the stagnant water, and branches and leaves were only partly decayed before being buried by other material. Rainfall must have been high but storms rare, since the trees were not strongly enough rooted to stand against strong winds. As the peaty deposit sank deeper under the increasing weight of the accumulating strata, water was pressed from it until every 10 m of wet, porous, peaty material was converted into about 1 m of solid coal.

The Coal Fields

Erosion has attacked the sheet of Pennsylvanian sediments that once stretched continuously from the Appalachians to Kansas, breaking it into segments that are now the Allegheny, Michigan, Illinois, and Mid-Continent coal fields (Fig. 13–16).

The plateau country of Pennsylvania, West Virginia, Ohio, Kentucky, Tennessee, and Alabama is underlain by one of the largest bituminous coal fields in the world. The most valuable seam in the Allegheny Basin is the Pittsburgh Coal, which is about 5 m thick near Pittsburgh and potentially productive over an area of approximately 15,000 square kilometers. Mines in the basin produce about 440 million metric tons annually, valued at 3.3 billion dollars; yet, the remaining reserves of the area are estimated at 330 billion metric tons.

The Cincinnati Arch separates the Allegheny Basin from the Illinois Basin coal fields. In Indiana, Illinois, and western Kentucky the coal beds are thinner, and fewer of the seams are worth mining. This trend to thinner and fewer coals in the cyclothems is continued into the Mid-Continent field. In areas where the cover of sediment or drift is thin, it is removed and the seam is taken out with power shovels. Unfortunately, extensive disruption of the countryside by this strip mining of the coal has spoiled the natural beauty and ruined potential farming of many areas of the Midwest and the hill country of the Appalachian states.

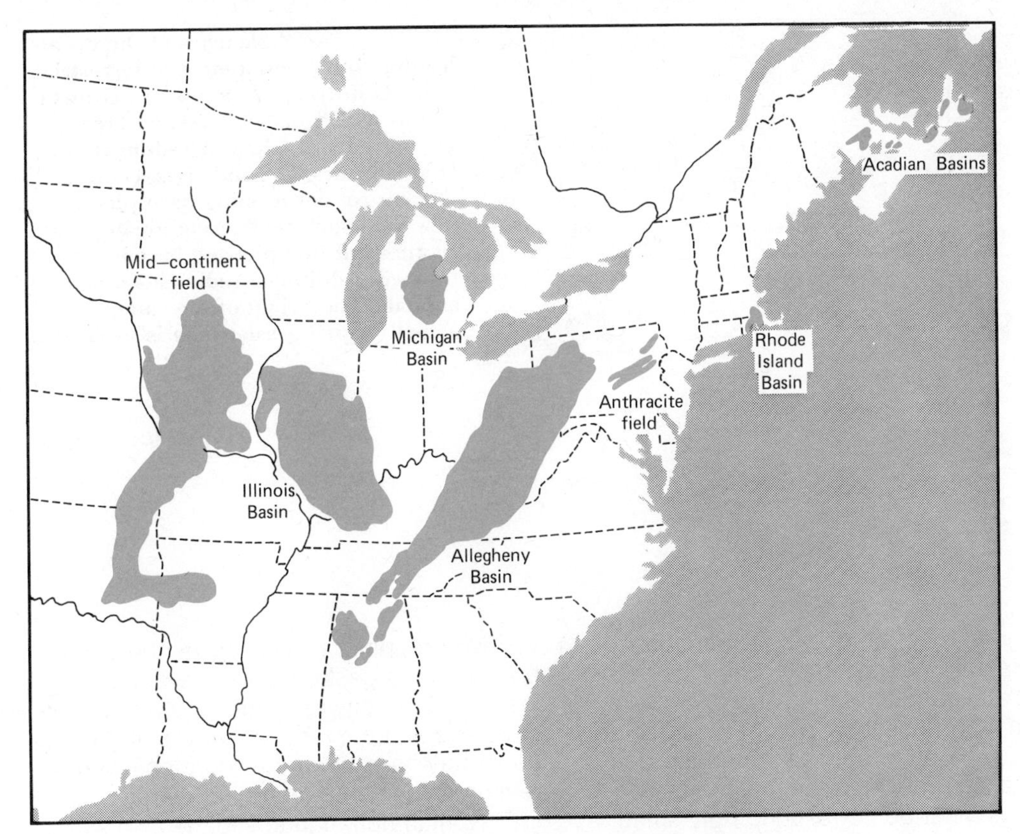

FIGURE 13–16 **North American coal fields of Pennsylvanian age.**

ORIGIN OF REPTILES

The forest in which the coal deposits at Joggins, Nova Scotia, were laid down was dominated by giant lycopsids, such as *Lepidodendron* and *Sigillaria.* The rigid tissue of these trees was limited to the bark; the central part was pulpy tissue that decayed rapidly. Occasionally, the swamps in which they grew were overwhelmed by sediments burying the bases of the growing trees to depths of several meters. The trees died and the tops fell over (Fig. 13–17), leaving hollow stumps up to 10 m long buried upright in the sediment. Animals and plants re-invaded a terrane marked by deep pits formed by the lycopsid stumps. Unwary or curious animals fell into these pits and once at the bottom could not get out. Their bodies were covered with sediment and vegetation washed in from the surface. The important feature of this strange type of preservation is that it led to the fossilization of fully terrestrial animals. Typically, the vertebrate record is biased toward aquatic forms because these lived where sediment was accumulating. The combination of geological and botanical factors at this time is fortunate, for it coincides with the emergence of the first truly terrestrial vertebrates —the reptiles.

Reptiles, like their descendants the birds and mammals, are capable of reproducing without the necessity of standing water for the deposit of eggs and the maturation of aquatic larvae. Among reptiles, birds, and mammals the egg and developing embryo are surrounded by extra-embryonic membranes that serve to support and protect the embryo and provide for water retention, the

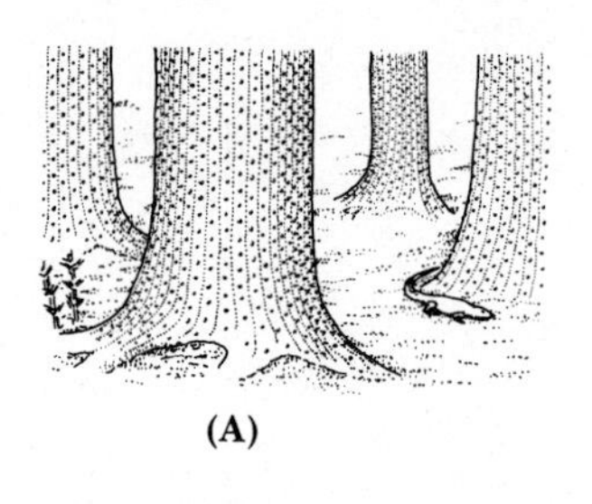

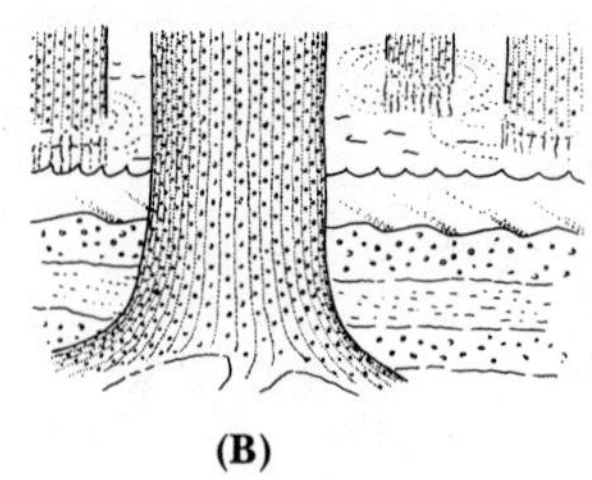

(C)

FIGURE 13–17 **Sketches of the events leading to the entrapment of terrestrial animals in lycopod stumps. A. Growth of forest of *Sigillaria* trees. B. Trees are buried and killed by sediments as land subsides. C. Land rises, centers of trees rot out forming deep pits in new land surface. D. Amphibians and reptiles fall into pits and become trapped. Additional sediments cover animals. The height of the stumps is actually much greater than is indicated.**

exchange of respiratory gases, nutrients, and metabolic wastes. The eggs of amphibians lack these membranes and must be small enough (less than 10 mm in diameter) to allow diffusion of gases through the general surface. The young that hatch from such eggs must be correspondingly small and require an aquatic environment for growth. Reptile and bird eggs are limited in size only by the physical limits of a hollow shell that is thin enough to be broken through by the emerging young. In the case of reptiles, the young are miniature replicas of adults when they hatch.

Reptiles as a group are typically terrestrial as adults. The development of a thick, cornified skin allows them to restrict water loss so that they escape the severe problem of dessication faced by most amphibians. The earliest reptiles, the captorhinomorph cotylosaurs (Fig. 13–18), resembled modern lizards in size and shape and probably had a similar way of life. The similarity of the structure of the jaws and teeth suggests a diet like that of the vast majority of living lizards who feed predominantly on insects. Several factors appear to have contributed to the origin and early success of the reptiles: their attainment of a light, agile skeleton and a terrestrial pattern of reproduction, and the coincident spread of the insects.

Two groups of invertebrates, the arthropods and the gastropods, were able to leave aquatic environments and live on land in Paleozoic time. The first terrestrial arthropods may have been the scorpions whose remains have been found in Silurian rocks. In Devonian non-marine sediments fossil ticks and questionable spiders have been discovered. The coal forest supported a large population of insects, including cricket-like forms, cockroaches, and an extinct order with non-folding wings that resembled a modern dragonfly (Fig. 13–2, 13–7). These were the first animals to solve the problems of flight, and in doing so they opened for exploitation an entirely new environment.

The reptiles must have evolved from the labyrinthodont amphibians, but little is known of their immediate ancestors because terrestrial deposits that might contain transitional forms are unknown at the Mississippian–Pennsylvanian boundary.

The captorhinomorph cotylosaurs remained a restricted group of small reptiles throughout the Pennsylvanian Period. Within that period, however, they gave rise to more specialized groups which emerged as the dominant reptiles of the late Paleozoic, Mesozoic, and Cenozoic time. The early divergence of the major reptile groups is re-

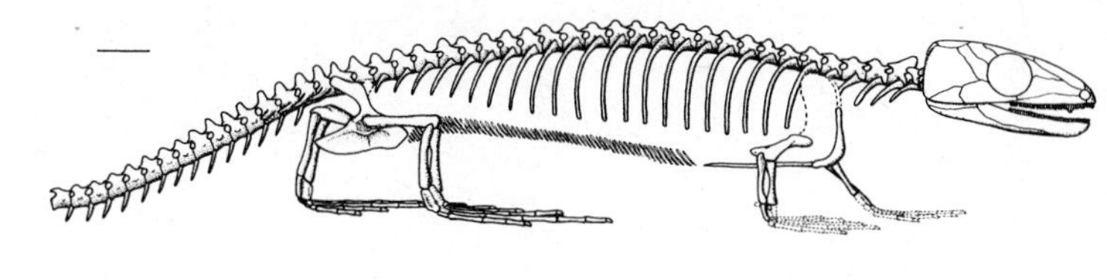

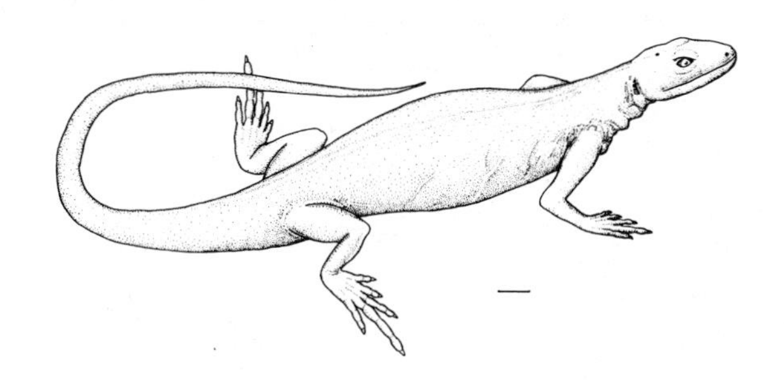

FIGURE 13–18 **Reconstructions of the skeleton and body form of *Hylonomus lyelli*, the oldest known reptile. Scales are 1 cm in length. From the fossil tree stumps of Joggins, Nova Scotia.**

flected in the changes in the configuration of the jaw muscles, as shown by the development of openings in the back of the skull. In small primitive reptiles the skull roof behind the eyes forms a continuous covering over the jaw muscles (**anapsid** condition). Specialization involving different diets led in several lineages to an increase in body size which required either a disproportionate growth of the skull or a reorganization of the jaw muscles and the bones to which they were attached. In one group (**synapsids**) an opening developed low in the cheek region (Fig. 13–19). This condition made possible the development of jaws and teeth capable of breaking down food to very small particles. The turtles formed an opening at the back of the skull, permitting the jaw muscles to expand into the neck region. The vast majority of reptiles, fossil and Recent, evolved from a condition (**diapsid**) in which there are two pairs of openings, one at the top of the skull and the other in the cheek. This pattern is shown in a primitive manner by one living reptile, the tuatara *Sphenodon* from New Zealand, and in modified form in crocodiles, lizards, and snakes. All these basic patterns had been initiated by the end of the Pennsylvanian Period.

While the early stages of reptilian evolution are recorded largely in the lowlands bordering the Carboniferous coal swamps, the fossil record of the next stage is seen largely in the deltaic deposits of southwestern North America, which will be considered in a later section. These deltas developed at the mouths of rivers flowing from the cratonic highlands that arose in the late Paleozoic.

CRATONIC HIGHLANDS AND BASINS

Yoked Basins

In Pennsylvanian and Permian time, segments of the southwestern corner of the craton were strongly uplifted into arches and domes that were eroded and supplied coarse detritus to the basins nearby. This fragmentation of the platform area near the south end of the Transcontinental Arch may have been connected with growing tectonic unrest in the Ouachita and Cordilleran geosynclines which embrace this corner of the craton. The subsidence of some of the basins in response to the uplift of adjacent highlands suggests that transfer of material by erosion and sedimentation from the high segment of lithosphere to the low segment is compensated for in the asthenosphere by a transfer of plastic material from the low to

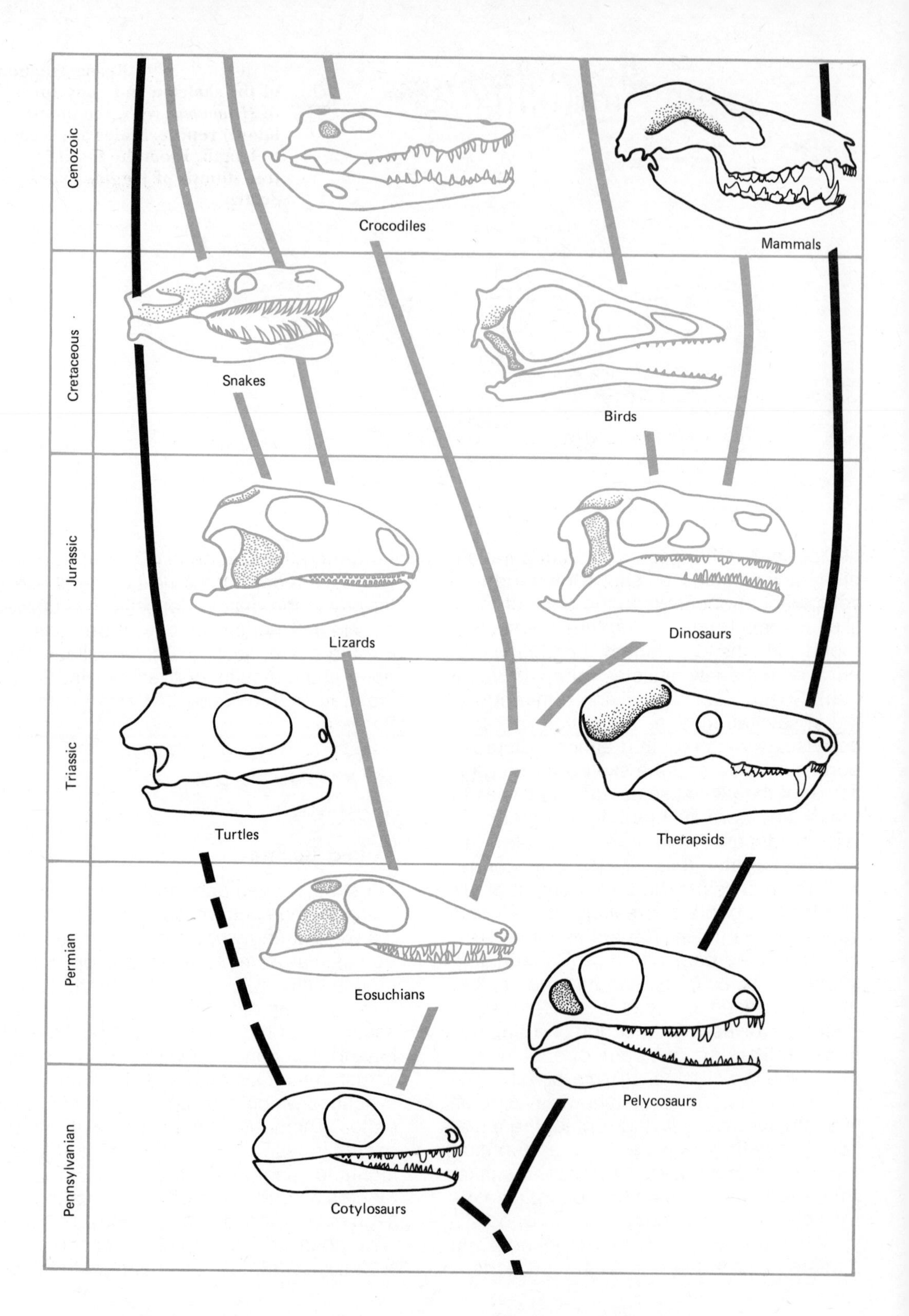
Cenozoic
Cretaceous
Jurassic
Triassic
Permian
Pennsylvanian
Crocodiles
Mammals
Snakes
Birds
Lizards
Dinosaurs
Turtles
Therapsids
Eosuchians
Pelycosaurs
Cotylosaurs

FIGURE 13–19 **Phylogeny of reptiles and their descendants showing the distribution of different patterns of temporal openings (shaded). The primitive condition (anapsid), without any temporal opening, is shown by the cotylosaurs. Turtles are usually considered to be anapsids, although they have openings for the temporal musculature developed at the back of the skull. Synapsids, or mammal-like reptiles—the pelycosaurs and therapsids—have a single opening located low on the cheek, a pattern retained by the mammals. Most reptiles, fossil and living, evolved from the diapsid pattern (colored) with two openings, one lateral and one dorsal. Lizards become specialized by the loss of the lower bar which resulted in a more moveable jaw. Snakes loosen the jaws to an even greater extent. Living crocodiles retain two distinct openings, as did most of their extinct relatives.**

the high segments. Such basins have been called **yoked basins** because their subsidence appears to be connected (or yoked) to the uplift of a neighboring block. The boundary between the basin and the uplift may be a simple flexure or a fault.

Figure 13–20 shows the principal basins and uplifts of Pennsylvanian time. The tectonic elements had diverse histories and not all were active at the same time. The group of uplifts along the Texas–Oklahoma border and in the Texas Panhandle have been called the Oklahoma Mountains; those in Colorado and New Mexico have been called the Ancestral Rockies, or the Colorado Mountains. Periodic uplift of the arches has produced complex facies changes in late Paleozoic sediments, unconformities, and structures, all of which are important in forming traps for oil and gas. The Absaroka rocks of Kansas, Oklahoma, and Texas yield a large part of the oil supply of the United States. The search for additional petroleum in these rocks is aided by an understanding of the history of the basins. Space is not available for consideration of the history of all the basins and uplifts; three areas are discussed: the Oklahoma Mountains and the Anadarko Basin, the Paradox Basin and its surrounding uplifts, and the Permian Basin of West Texas.

Oklahoma and Kansas

The Oklahoma Mountains consist of elongate domes of basement rock. A few of these domes form ridges and hills at the surface, but most are now deeply buried be-

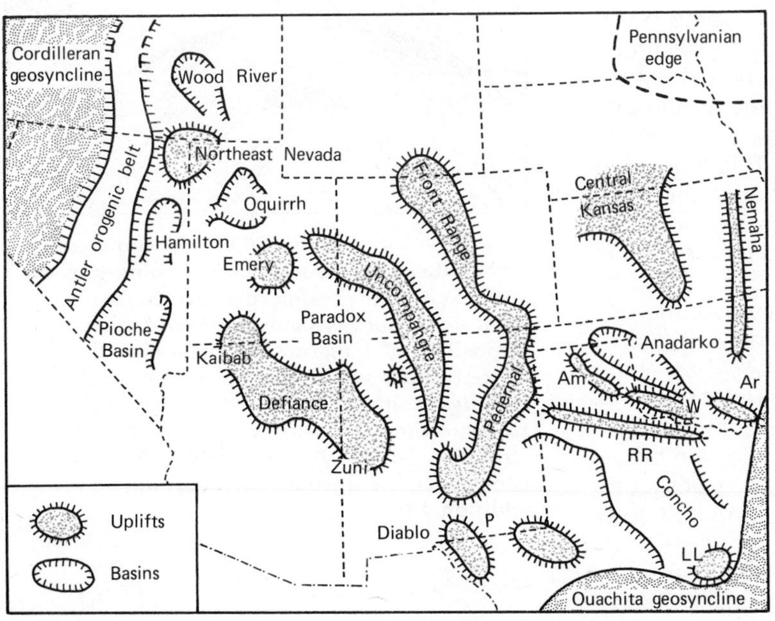

FIGURE 13–20 **Basins and uplifts of the southwestern states in the Pennsylvanian Period. Am. Amarillo Ridge, W. Wichita Uplift, Ar, Arbuckle Uplift, RR. Red River Uplift, LL. Llano uplift, P. Pecos Arch.**

neath Permian, Mesozoic, and Cenozoic sedimentary rocks and have been discovered by drilling in the search for oil. The elevation of the Oklahoma Mountains began during the Morrow Age and continued to a climax at the end of the Pennsylvanian Period. The first episode of uplift, referred to as the Wichita Orogeny, resulted in the stripping of up to 4500 m of pre-Pennsylvanian sedimentary rocks covering the granite basement of the Arbuckle Mountains. The eroded material was carried to the Ardmore and Anadarko basins, which accumulated 5000 m of sedimentary rocks in this episode. The progressive stripping by erosion of the crests of the rising anticline exposed older formations and is reflected in the type of sedimentary particles supplied to the basins. Although thin sandstone tongues spread from the mountains as far north as Kansas, most of the sediments were trapped in the rapidly subsiding basins. The sands from these uplifts, now folded by later movements, form reservoirs for many of the oilfields.

In Kansas the rise of the Central Kansas Arch and the Nemaha Granite Ridge took place in Early Pennsylvanian time. The Nemaha Ridge is an acute uplift of early Paleozoic rocks bounded on one side by a fault and stripped by erosion to its pre-Paleozoic core before it was covered by the basal shales of the transgressing Absaroka Sea. Many of the most productive oilfields of the mid-continent region derived their oil from what are known as bald-headed structures on the Nemaha Ridge. This type of trap results from the doming of strata containing a porous bed, their erosion to a more-or-less flat surface, and the sealing of the truncated edges of the reservoir bed by an impervious layer deposited over the unconformity. The sequence of events is illustrated graphically in Fig. 13–21. Two of the largest fields in cumulative production in the mid-continent area, the Eldorado and Oklahoma City fields, are located on bald-headed structures in which the major reservoir rock is the Upper Cambrian and Lower Ordovician Arbuckle Limestone.

The shale that seals the bald-headed structures in much of Kansas is the Chero-

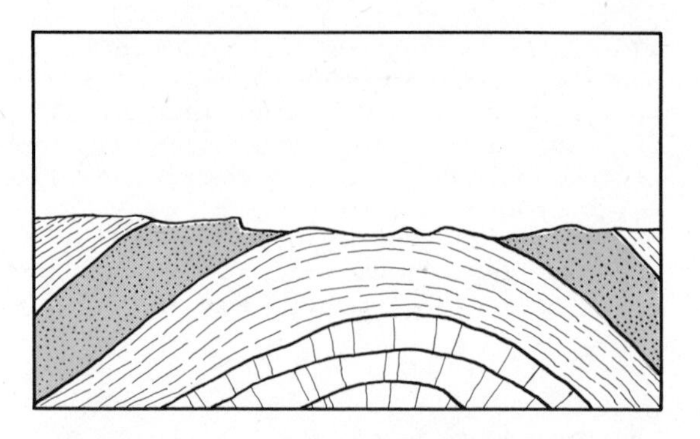

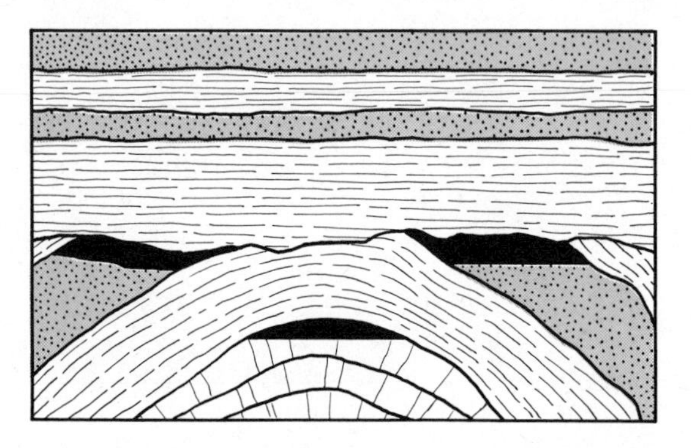

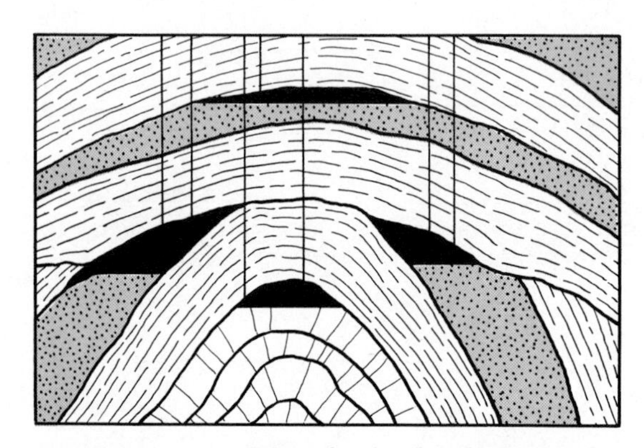

FIGURE 13–21 **Stages in the development of a "bald-headed" structure. *Top:* doming of the strata and their erosion to a surface that truncates a porous sandstone bed. *Middle:* Deposition of impermeable shale over the unconformity sealing the porous bed and making a trap for oil (black) migrating up the flanks of the fold. This is the "bald-headed" structure. *Bottom:* Renewed uplift deforms the beds above the unconformity forming additional traps.**

kee Formation (of Atoka–Des Moines age). This shale contains long, thin lenses of sand called "shoe-string sands." Most of the Cherokee sands are believed to have been bars analogous to those now forming along the Atlantic Coast of the United States, but others appear to have been channel fillings. They form the reservoirs for several productive oilfields in Kansas.

The Absaroka Sea reached its maximum size in Missouri time. The long regression that continued with many interruptions through the Permian and Triassic periods began in Virgil time. Although marine sediments continued to be deposited in Kansas and Oklahoma as late as Middle Permian time, the great epeiric sea that had stretched from the northern Appalachians to Oklahoma was slowly restricted; more clastic sediments entered from bordering uplifts, and before the end of the Paleozoic the sea was completely silted up.

Near the end of Pennsylvanian time (Virgil) and early in the Permian Period the Oklahoma Mountain system was strongly uplifted in what has been called the Arbuckle Orogeny. Up to this time the Oklahoma structures were simple and had been raised intermittently since Late Mississippian time. In this orogeny vertical movements still dominated, but the rocks broke along a series of faults into horsts and graben. The complexity of the structure along the north front of the Arbuckle and Wichita mountains has been revealed by deep drilling in southern Oklahoma. Anticlines in the basins that were formed in the Early Pennsylvanian Wichita Orogeny and then eroded were refolded and broken along high-angle faults. The complexity of stratigraphy and structure produced by this sequence of events is illustrated in the cross-section of the Eola Field (Fig. 13–22). Petroleum has been trapped in many of these anticlines in

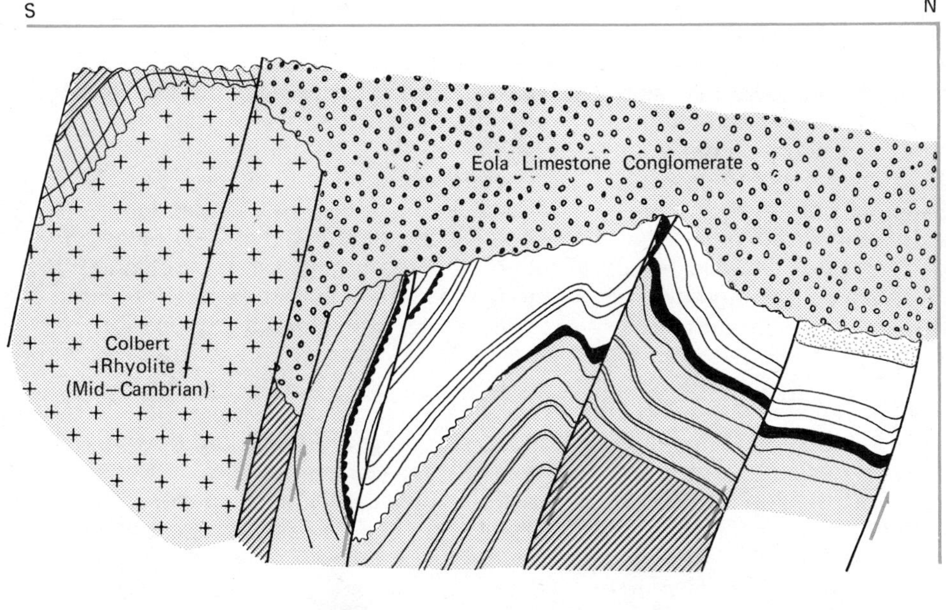

FIGURE 13–22 **Cross-section of the Eola Field, Oklahoma, showing the intensity of the deformation along almost vertical faults at the north side of the Arbuckle Mountains as revealed by drilling. Horizontal and vertical scales equal. (After B. H. Harlton; courtesy of the American Association of Petroleum Geologists.)**

beds that range in age from Ordovician to Permian, but mostly in Pennsylvanian sandstones. Some of the traps were formed in the crests of anticlines, others were formed where permeable beds had been brought against impermeable ones along faults, and still others were formed where sand tongues pinch out over the tops of buried hills.

In the western part of the Oklahoma Mountains, late Pennsylvanian deformation was reflected largely as uplift; little folding or faulting took place. The regolith that had been developed by the weathering of the granite basements rocks was washed off into the surrounding basins by the rejuvenated streams. This "granite wash" is one of the most important reservoirs in the Panhandle oil and gas field of northern Texas. Along the flanks of the Oklahoma Mountain uplifts Late Pennsylvanian and Early Permian sedimentary rocks are red, conglomeratic sandstone and shale derived from the erosion of the ranges. As the erosion products built up, the ranges were progressively buried by their own debris and had little relief by the end of the Permian Period.

The Four Corners Area

That part of the Colorado Plateau where the states of Arizona, New Mexico, Utah, and Colorado come together has been called the "Four Corners." Investigation of the late Paleozoic stratigraphy of this region was stimulated by the discovery of oil in the Paradox Basin. The late Paleozoic uplifts of Colorado, unlike those of Oklahoma, were rejuvenated in Cenozoic time and now form the highest peaks in the southern Rocky Mountains. The Front Range of Colorado is the successor to the largest of these uplifts, which extended from southern Wyoming to central New Mexico in late Paleozoic time. The Central Colorado Basin separated the Front Range uplift from the subparallel Uncompahgre Uplift to the west. In Northern Utah the Oquirrh Basin, the most rapidly subsiding of the late Paleozoic basins, received 7500 m of Pennsylvanian and Permian sandstone. Much of Wyoming was a shelf area where about 100 m of calcareous sandstone, variously known as the Quadrant, Tensleep, or Weber were laid down. These Pennsylvanian and Early Permian sandstones have large cross-beds that have been interpreted as dune structures. The downwind direction of dip of the cross-bedding has shown that for much of late Paleozoic time the wind blew from the north obliquely toward the paleoequator (see Fig. 7–4), which crossed the United States from New York to California. The winds recorded by the cross-bedding in the Carboniferous sandstones may have been the Trade Winds of 300 million years ago.

As soon as the cores of the Colorado uplifts were exposed by the erosion of the covering sediments, they began to shed vast quantities of arkosic debris. From Pennsylvanian to mid-Permian time a thick succession of these red, feldspathic sandstones accumulated in basins adjacent to the Front Range and Uncompahgre uplifts. East of the Colorado Front Range these late Paleozoic sandstones are tilted vertically to form the fantastically weathered rocks of the "Garden of the Gods."

The Paradox Basin

Transgression of the Absaroka Sea into the Four Corners area began in Early Pennsylvanian time (Morrow Age) over a surface of Middle Mississippian limestone that had developed a karst topography of sinkholes and caverns produced by solution. The clayey constituents of the limestone formed a red, lateritic, residual soil that was reworked by the transgressing sea and introduced into deep cavities in the limestone, forming a spectacular unconformity (Fig. 13–23).

In Des Moines time the western access of the sea to the Paradox Basin was restricted, and cyclical deposits of salt, gypsum, and anhydrite were laid down along its axis. Away from the center of the basin where the water was less saline, fossil fragment and oölitic limestones replaced the evaporites. Along the western margin of the basin, barrier and patch reefs grew, and talus and lagoonal deposits generally associated with such growth were laid down (Fig. 13–24). Within these carbonates several oil and gas accumulations have been

FIGURE 13–23 **Disconformity between Middle Mississippian Guernsey Limestone below (light in color), weathered into a highly irregular surface with sinkholes, and the red Pennsylvanian Hartville Formation above (dark), filling the depressions. Guernsey, Wyoming. (Courtesy of J. H. Rathbone and the Rocky Mountain Association of Geologists.)**

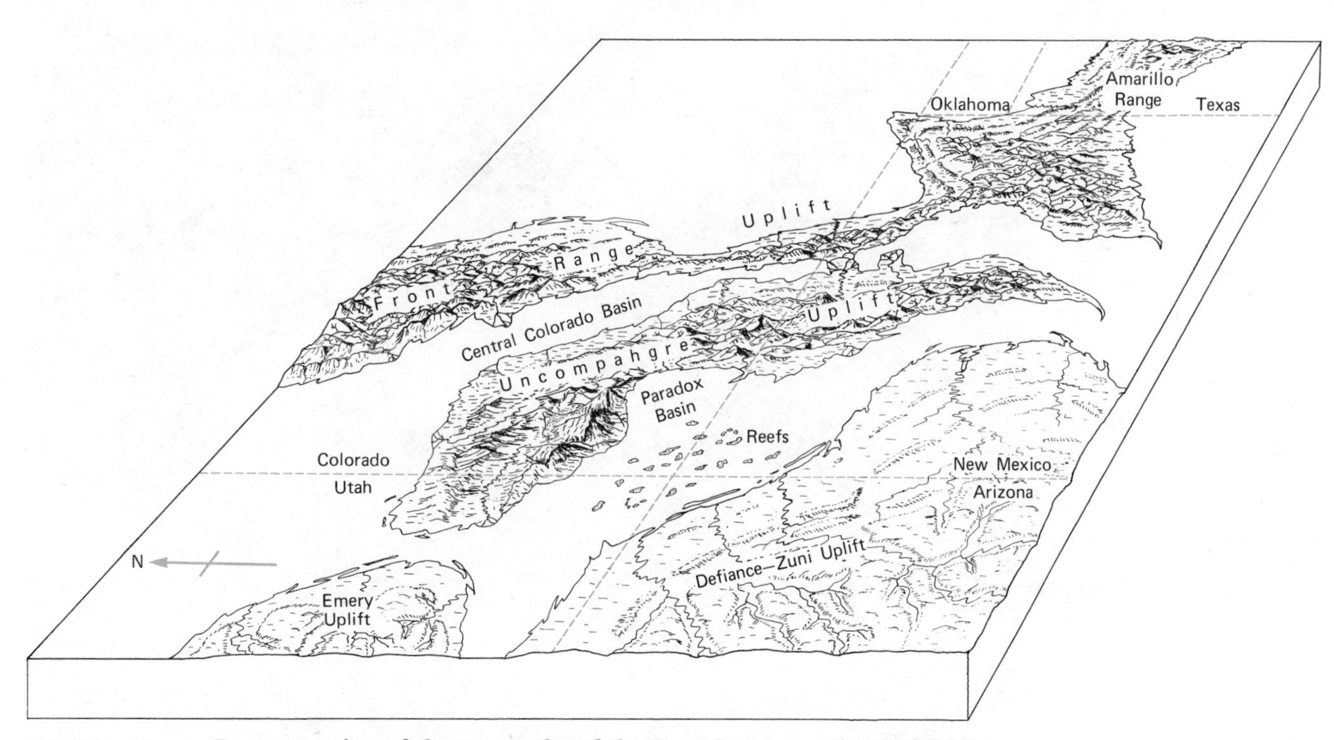

FIGURE 13–24 **Reconstruction of the geography of the Four Corners region in Middle Pennsylvanian time. The observer is looking from northern Arizona across Colorado. In the foreground evaporites accumulate in the Paradox Basin and reefs grow on the shelf on its southwest flank.**

discovered. The episode of evaporite deposition was brought to a close in Missouri time when a strong uplift of the Uncompahgre Arch along a fault shed a great wedge of arkosic sediments into the Paradox Basin, filling it above sea level.

At the northeast edge of the Paradox Basin near the Uncompahgre Uplift, Pennsylvanian beds are folded into anticlines (Fig. 13–25). Locally, on the crests of these structures, drills penetrated 4500 m of the salt sequence, but wells between the anticlines found the salt to be missing entirely. Geologists concluded that the salt, originally laid down in a bed or a series of beds, flowed into the cores of the anticlines just as glacial ice flows under pressure. The flow of masses of salt under pressure can be demonstrated in the laboratory and in many places in the world where salt has been injected like an igneous intrusion into sedimentary rocks. The movement of the

FIGURE 13–25 **The Uncompahgre Plateau and the Paradox Valley, western Colorado, imagery from the Landsat satellite taken from an altitude of 905 km. 1. Uncompahgre Plateau. 2. Colorado River. 3. Paradox Valley. 4. Black Canyon of the Gunnison River. (Courtesy of NASA.)**

salt into the cores of the anticlines is believed to have started soon after it was deposited (Fig. 13–26). The uplift of the Uncompahgre area late in the Pennsylvanian Period may have thrown the adjacent sediments of the Paradox Basin into gentle folds, triggering the movement of the salt, or the anticlines may have been localized by basement faults. The arkose eroded from the uplifts accumulated to a greater depth in the synclinal troughs than in the anticlinal crests, squeezing the salt laterally toward the region of lesser hydrostatic pressure in the anticlines. Some movement of the salt and growth of the anticlines seems to have persisted as late as Jurassic time.

Permian Basin of West Texas

Slowly the sea that had ebbed and flowed across the mid-continent region since Early Pennsylvanian time regressed southwestward, and by mid-Permian time Kansas and Oklahoma were the sites of continental deposition with coastal lagoons in which evaporites accumulated. The sea was then largely confined to West Texas and southern New Mexico, where it occupied a complex basin divided into three sub-basins by the Diablo Platform and the Central Basin Platform (Fig. 13–27). In Early Permian time the platform and the shelves surrounding the basins were the site of deposition of fossil fragment limestones, while the basins themselves received black, bituminous limestones and shales. Apparently, circulation of oceanic water to the deeper basins was restricted and as oxygen content was depleted, organic matter accumulated in a stagnant environment.

The boundary between the shallow water

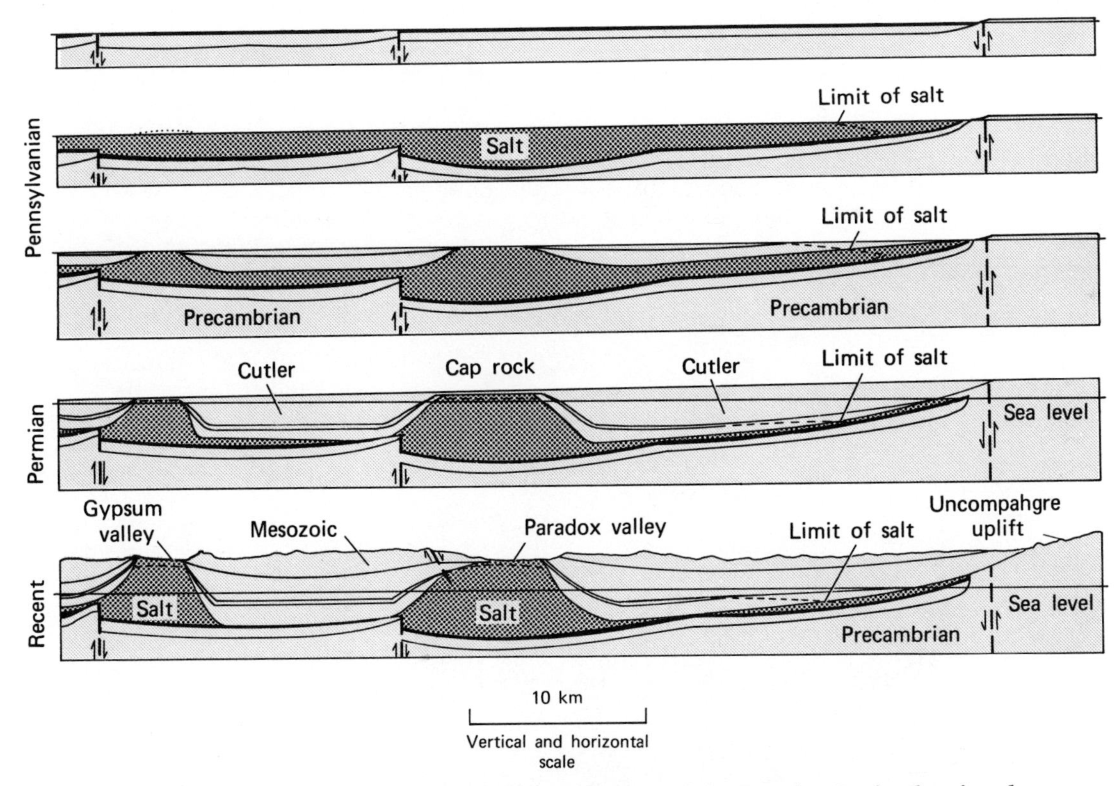

FIGURE 13–26 **Cross-sections of the salt-cored anticlines of the Paradox Basin showing the migration of the salt beginning in Pennsylvanian time soon after it was deposited and the suggestion that the structures are controlled by faults in basement rocks. (After D. P. Elston and others; courtesy of the American Association of Petroleum Geologists.)**

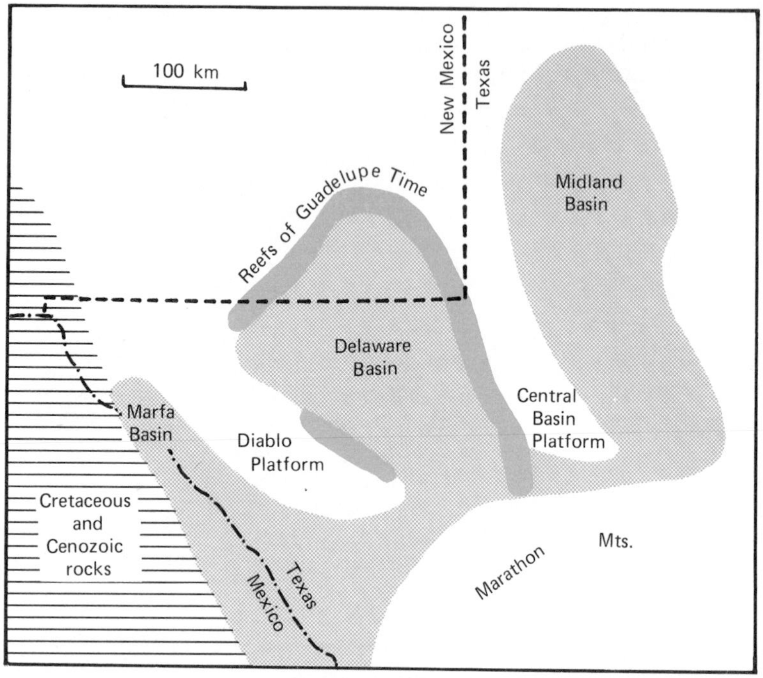

FIGURE 13–27 **Map showing the Permian basins of West Texas.**

of the shelves and the deeper water of the basin had been the site of rapid limestone deposition since Pennsylvanian time. As the relief between the basins and shelves increased in Middle Permian time, carbonate-secreting organisms were concentrated there and built a structure that is commonly referred to as a barrier reef. The Middle Permian Capitan Limestone, which formed along the margin of the Delaware Basin, now caps an escarpment along which the facies associated with the transition from basinal to lagoonal environments are grandly displayed (Fig. 13–28). Most of the

FIGURE 13–28 **Aerial photograph of El Capitan, Guadalupe Mountains, West Texas. The resistant limestone of the Capitan Formation forms the steep cliffs. The slopes below are underlain by brecciated limestones of the slope facies. Bituminous sandstones of the basin facies underlie the foreground. (Courtesy of Muldrow Aerial Surveys Corp.)**

Capitan Limestone is a very fine-grained, light-colored rock with a fauna of algae, sponges, and bryozoans. Much of it is without bedding, but toward the basin the unbedded limestone interfingers with a coarse limestone breccia that accumulated as a talus along the steep descent to the basin (Fig. 13–29). These talus beds contain fossils that have been replaced by chert, and many beautifully fragile specimens (Fig. 13–30B,C) have been prepared from it by dissolving the enclosing limestone in acid. When allowance is made for modification by compaction, the dip of the talus beds and the width of the talus facies indicates that the depth of the water in the Delaware Basin was about 700 m.

Lack of an obvious framework of reef-building organisms, the abundance of very fine grained carbonate, and the brecciated texture in the Capitan Limestone have cast some doubt that it was formed as a barrier reef. Some geologists classify it as an ancient bank, which implies an accumulation of skeletal carbonates formed by organisms that did not have the potential to erect a wave-resistant structure. It is difficult to determine whether the algae, sponges, and bryozoans (Fig. 13–31) were sufficiently common and strong to build a wave-resistant reef or only passively contributed their skeletons to the accumulation of a limestone bank along the basin margin. Certainly, limestones did accumulate along the basin margin to a greater thickness than elsewhere and formed a barrier restricting the circulation of sea water to the shelf behind. This barrier was not continuous, for sands washed southward across the lagoons past the barrier and accumulated in the stagnant, deeper waters of the Delaware Basin during Capitan deposition. Organic matter buried in these black, bituminous sandstones of the basins is believed to have been the source of the petroleum now found in the reservoirs of the surrounding shelves and platforms.

In the lagoonal environment behind the Capitan barrier the water was shallow. Near the platform edge carbonates were laid down, but near the shore the water was more saline and gypsum and anhydrite precipitated. In the lagoonal limestones of the Carlsbad Group near Carlsbad, New Mexico, solution by groundwater subsequently has produced one of the largest cavern systems in the world. Much of the Capitan Limestone that accumulated along the basin margins has been changed into coarsely crystalline dolomite. One theory ascribed the dolomitization to brines from the lagoon, enriched in magnesium through evaporation, percolating downward and basinward through the marginal limestones. Attempts

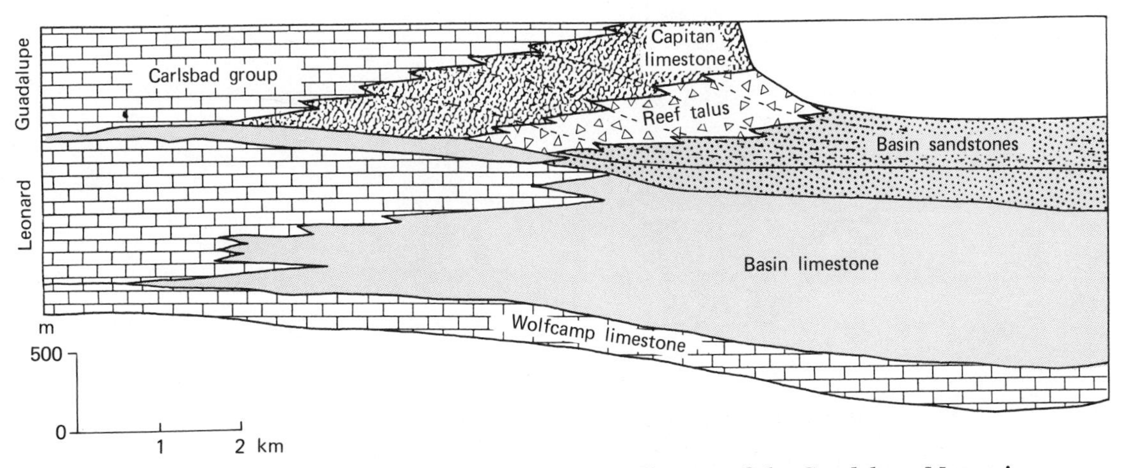

FIGURE 13–29 **Cross-section of the Middle Permian sediments of the Guadalupe Mountains showing the relationship of the basin-margin limestones to the other facies. (Modified from N. D. Newell and P. B. King.)**

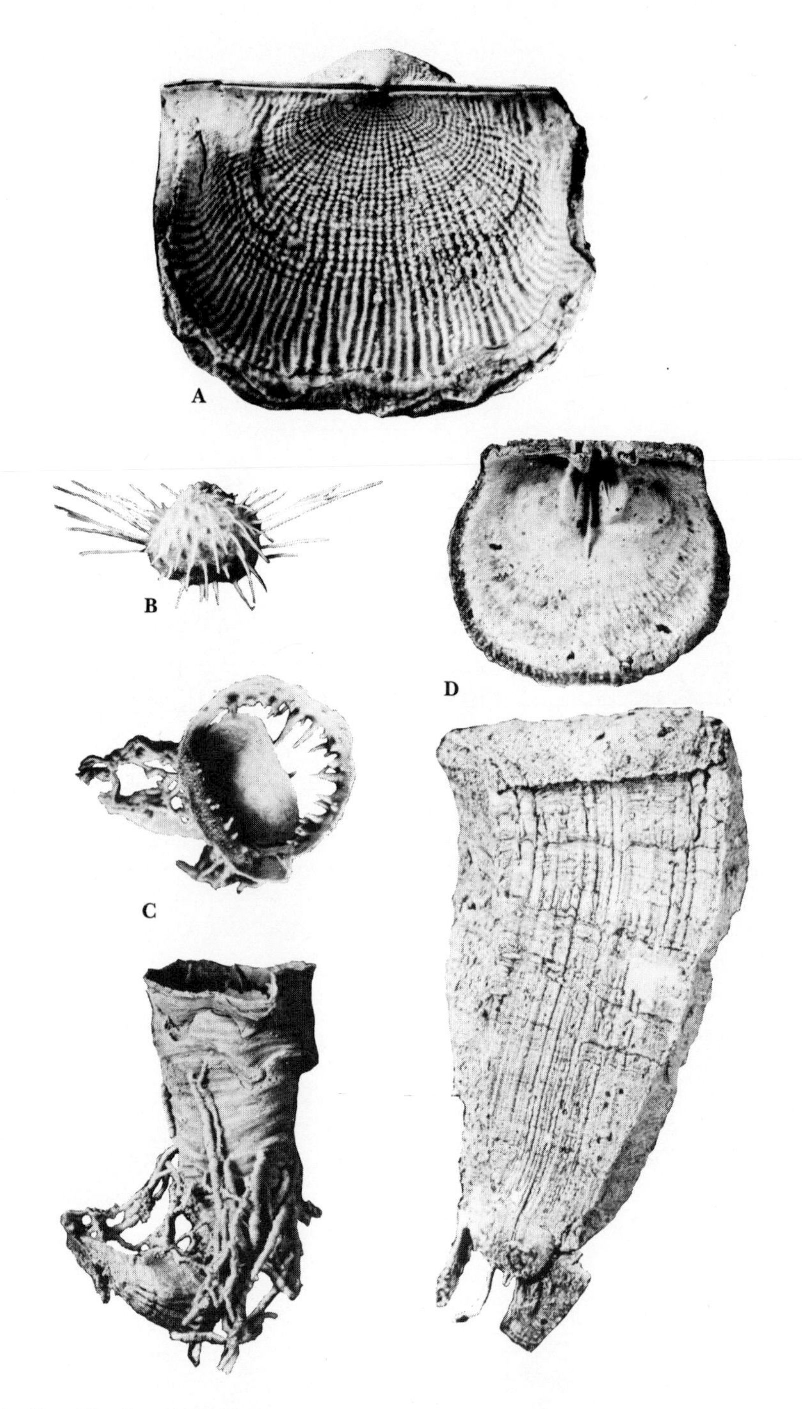

FIGURE 13–30 **Permian brachiopods. A. *Reticulata*, a productid. B. *Echinaurus*, another productid showing large lateral spines. C. *Prorichtofenia*, top and side views of the horn-like pedicle valve and the cap-like brachial valve. D. *Scacchinella*, with a cap-like brachial valve at the top that fits on to the horn-like pedicle valve. (Courtesy of the Smithsonian Institution.)**

FIGURE 13-31 **Reconstruction of the Middle Permian Capitan Limestone environment. The sea floor is covered with brachiopods, algae, sponges, fenestellid bryozoans, and corals. The large sponges are *Heliospongia* and the fish belong to the genus *Platysomus*. (From a diorama by T. L. Chase at the Permian Basin Petroleum Museum, Midland, Texas.)**

have been made to use this theory of "seepage refluxion" to explain all sedimentary dolomites, but it is certainly not strictly applicable to dolomites of wide areal extent or those associated with ancient sabkha environments. Dolomitization of limestone took place in the past in several different environments.

Until the end of Middle Permian time the evaporitic facies was largely confined to the lagoons behind the basin margin barriers. At this time the entrances, first to the Midland Basin and then to the Delaware Basin, were restricted by a sill or the growth of a reef across their mouths. The climate of Permian time must have been very dry for evaporation of sea water proceeded rapidly in the basins, the salinity rose, and the organisms forming the basinside barriers were killed. Eventually, through continued evaporation and deposition in Late Permian time, these two basins were filled to depths of about 600 m with anhydrite, halite, and potassium salts.

The Late Permian evaporites of the Permian basin are unique in the regularity and persistence of their fine laminations. The cyclical laminations begin as fine organic layers in the Middle Permian siltstones below the evaporite facies. They change upward to organically laminated claystone, organically laminated calcite, calcite-laminated anhydrite, and finally to anhydrite-laminated halite. Although the couplets average only about 2 mm in thickness, some can be traced by well cores over distances of over 100 km (Fig. 13–32). The thinness and lateral persistence of the laminae leave little doubt that these evaporites were laid down in relatively deep water. The couplets have long been considered to represent the deposits of a year, perhaps of a wet and dry season. If they are yearly, the 450 m of the laminated sequence must represent 260,000 years of deposition. Recently some doubt has been cast on the annual nature of the couplets by calculations that the degree of aridity required to produce them is unrealistically high.

The West Texas Permian Basin is a major petroleum province. Production comes from stratigraphic traps associated with the limestone facies changes at the basin margins, and also from deep structures formed in early Paleozoic rocks by Pennsylvanian folding and faulting along the Central Basin Platform.

THE END OF AN ERA

The change in the world marine community that took place at the close of the Permian Period was recognized by early geologists as the most important event in the history of invertebrates and formed the basis for the separation of the Paleozoic and Mesozoic eras. Many groups of invertebrates that had been successful in Paleozoic time became extinct in the Permian Period and were replaced by others at the beginning of Mesozoic time. The cause, or causes, of these extinctions is one of the major problems of paleontology. Major changes took place in the FORAMINIFERA, ANTHOZOA, BRACHIOPODA, BRYOZOA, AMMONOIDEA, ECHINODERMATA, and TRILOBITA.

Forams are rare in early Paleozoic rocks, and most foraminiferal tests are composed of sand grains. In the Devonian Period a group of these single-celled animals began to secrete tests of calcium carbonate, composed of many chambers coiled in one plane. Tests of this type, known as endothyroid, became abundant during the Mississippian Period and made important contributions to the skeletal limestones characteristic of the middle part of the system (Fig. 12–20). A family of forams, called the fusulines, evolved from the endothyroid stock at the beginning of the Pennsylvanian Period. The fusulines formed spindle-shaped tests composed of a wall coiled along the long axis of the spindle and enclosing a space divided from pole to pole by thin plates or septa (Fig. 13–33). Most fusuline tests are about the size and shape of a grain of wheat, but some grew to about 3 cm in length. Throughout the Pennsylvanian and Permian periods the fusulines evolved rapidly, multiplied astronomically, and spread globally, and as a result became the most satisfactory fossils for dating and correlating limestones and shales of late Paleozoic age. As the Permian Period grew to a close the fusulines became extinct.

Late Paleozoic coral faunas are dominated by solitary and colonial rugosans. The tabulates that were so important in the early Paleozoic had shrunk to an insignificant number. Although massive colonial rugosans are abundant in some beds, they do not appear to have had the reef-building potential of the early Paleozoic tabulates allied with the stromatoporoids. By the end of the Permian, all rugose corals were extinct.

Brachiopods flourished in the late Paleozoic seas as never before. The dominant group, the productids, arose from strophomenoid stock in the Devonian Period. Productids superficially resembled such early Paleozoic strophomenoids as *Rafinesquina,* but were adapted to living in soft, muddy ocean floors, and so evolved a spiny exterior to keep their shells from sinking in the mud and an extended anterior aperture to maintain contact with the water when the

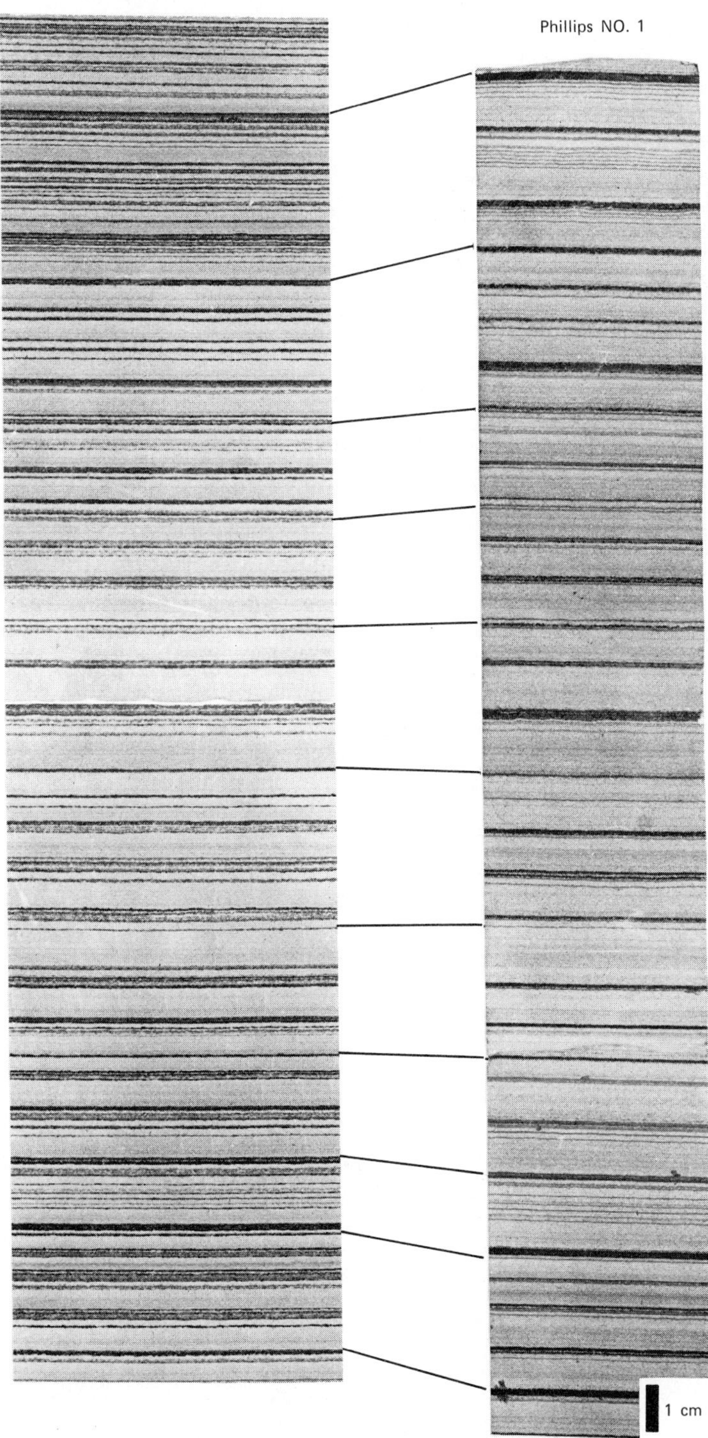

FIGURE 13–32 **Cores of varved evaporites of the Castile Formation from the Delaware Basin. The lines joining the two cores show correlation of couplets between two wells 91 km apart. (Courtesy of R. Y. Anderson and the Geological Society of America.)**

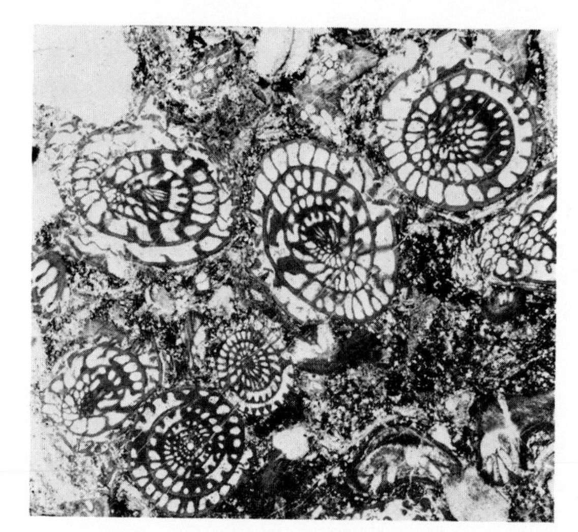

FIGURE 13–33 **Photomicrograph of Pennsylvanian limestone showing cross-sections of the fusuline, *Triticites*, × 10. (Courtesy of the Illinois Geological Survey.)**

rest of the shell was largely engulfed in soft sediment (Fig. 13–30). Some of them were cemented to rocks or other fossils by one valve which, as a result, grew larger and more irregular than the other. This trend in brachiopod evolution led in the Permian Period to such forms as *Richtofenia* and *Scacchinella,* so aberrant that they appear more like corals than brachiopods (Fig. 13–30). The productids did not survive the Permian Period.

Although the massive, stony bryozoans of the early Paleozoic persisted, the most conspicuous bryozoans of the late part of the era were the fenestellids, which secreted a colony in the form of lacey fronds (Fig. 13–34). Two of the four classes of Paleozoic bryozoans, including the fenestellid family, died out at the end of the Permian.

The AMMONOIDEA were cephalopods that appeared in the Devonian Period, diversified in the late Paleozoic, and became the most important molluscan group of the Mesozoic. The chambered shells of typical ammonoids are coiled in one plane like a watch spring (Fig. 13–35). They are divided into subgroups on the basis of the pattern of the suture, which is the line along which the chamber partitions meet the inside surface of the outer wall. The ammonoid suture evolved into increasingly complex forms: from the broadly curving goniatite type; to the ceratite type, with one side of the curves folded minutely; to the ammonite type, in which the entire suture is complicated by minute folds of several orders (Fig. 13–36). The goniatite group, which is the most abundant in Paleozoic rocks, became extinct in the late Permian invertebrate crisis.

The abundance of crinoids in the Mississippian sea has already been described (Fig. 13–37). The blastoids were the only other group to reach abundance in the late Paleozoic. Members of this group were rare until late Mississippian time when they gave rise to a rich fauna in rocks of the Chester Series, particularly characterized by the bud-like *Pentremites* (Fig. A–23). Blastoids are rare in Pennsylvanian rocks but are again abundant in Indonesia in Permian rocks. They, with two orders of crinoids, became extinct at the end of the Permian Period.

Since the Ordovician Period the trilobites had decreased in abundance and diversity. Their demise at the end of the Paleozoic Era, therefore, is not surprising.

North America is not a favorable area in which to study the invertebrate faunas of the era boundary, for the beds of latest Permian and early Triassic age are nearly everywhere non-marine. Throughout the world epeiric seas withdrew from the continental platforms at this time. Although along the trace of the ancient late Paleozoic and Mesozoic seaway (Tethys) in Iran, Pakistan, and China late Permian and early Triassic marine beds appear to be conformable and some beds contain mixtures of typically Permian and typically Triassic fossils, a stage appears to be missing in all sections so far investigated in detail. At present no stratigraphic section has been demonstrated to record continuous deposition across the era boundary. The history of vertebrates and plants shows no marked extinctions nor introduction of new forms at the end of the Paleozoic Era.

We may never be sure why certain large groups of organisms become extinct at particular times in geological history because

FIGURE 13–34 **Fenestellid bryozoans from Lower Mississippian rocks of Illinois. (Courtesy of the Smithsonian Institution.)**

FIGURE 13–35 **A Mississippian goniatite ammonoid, *Imitoceras*, showing the characteristic suture pattern. (Courtesy of the Smithsonian Institution.)**

FIGURE 13–36 **Permian ammonoids. *Left: Popanoceras,* with a ceratite suture. *Right: Timorites,* with an ammonite suture. (Courtesy of A. K. Miller and W. M. Furnish.)**

FIGURE 13–37 **A group of crinoids growing on the Mississippian sea floor. (Courtesy of the Smithsonian Institution.)**

we cannot reconstruct the total environments of the past in sufficient detail. Changes in the environment to which the animals and plants have become adapted may have been one major cause of the extinctions. If a species or family has become highly specialized to a particular environment, and that environment is wiped out by earth movements or changes in sea level, climate, ocean currents, etc., the group will die unless further adaptation is possible. Less specialized animals or plants may be able to evolve in another direction and become adapted to the new situation, but highly specialized organisms rarely retain this evolutionary flexibility.

At the end of the Paleozoic Era the continents were uplifted, and the epeiric seas in which the invertebrates flourished were restricted. Competition among the marine organisms became greater as they were crowded for room and food, and those that were highly specialized for life in a particular habitat and could not adapt to changing conditions lost out in the struggle for existence.

THE END OF A SEQUENCE

With the withdrawal of the Absaroka Sea in mid-Permian time from the western platform to the West Texas Basin and the miogeoclinal basins along the western platform edge, much of southwestern United States was covered by continental red siltstone and sandstone of Late Permian, Triassic, and Early Jurassic age (Fig. 13–38). Within these sparsely fossiliferous deposits the boundaries between the systems are hard to locate.

On the Colorado Plateau the Triassic System is represented by four formations: the Moenkopi at the base, followed by the Chinle, Wingate, and Kayenta. The lower formations were deposited by streams flowing from uplifts in southern Idaho and Arizona (Fig. 13–39) to a shoreline in Nevada. Water percolating along buried stream channels in the Chinle Formation has precipitated uranium salts that are mined extensively in the Colorado Plateau. Successful prospecting for this uranium depends in large part on the use of stratigraphic analysis and geophysics to locate promising channels.

Among the most interesting of the few fossils found in the Chinle sandstones are the trunks of petrified conifers (Fig. 13–40). The wood of the trees that once grew on the Chinle landscape has been faithfully re-

FIGURE 13–38 **Mitten Butte, Monument Valley, northeast Arizona. The cliff-forming beds are the Middle Permian De Chelly Sandstone; the less resistant beds below are the Organ Rock Shale. (Courtesy of J. H. Rathbone and the Rocky Mountain Association of Geologists.)**

FIGURE 13–39 **Facies map of the Upper Triassic Series in western North America.**

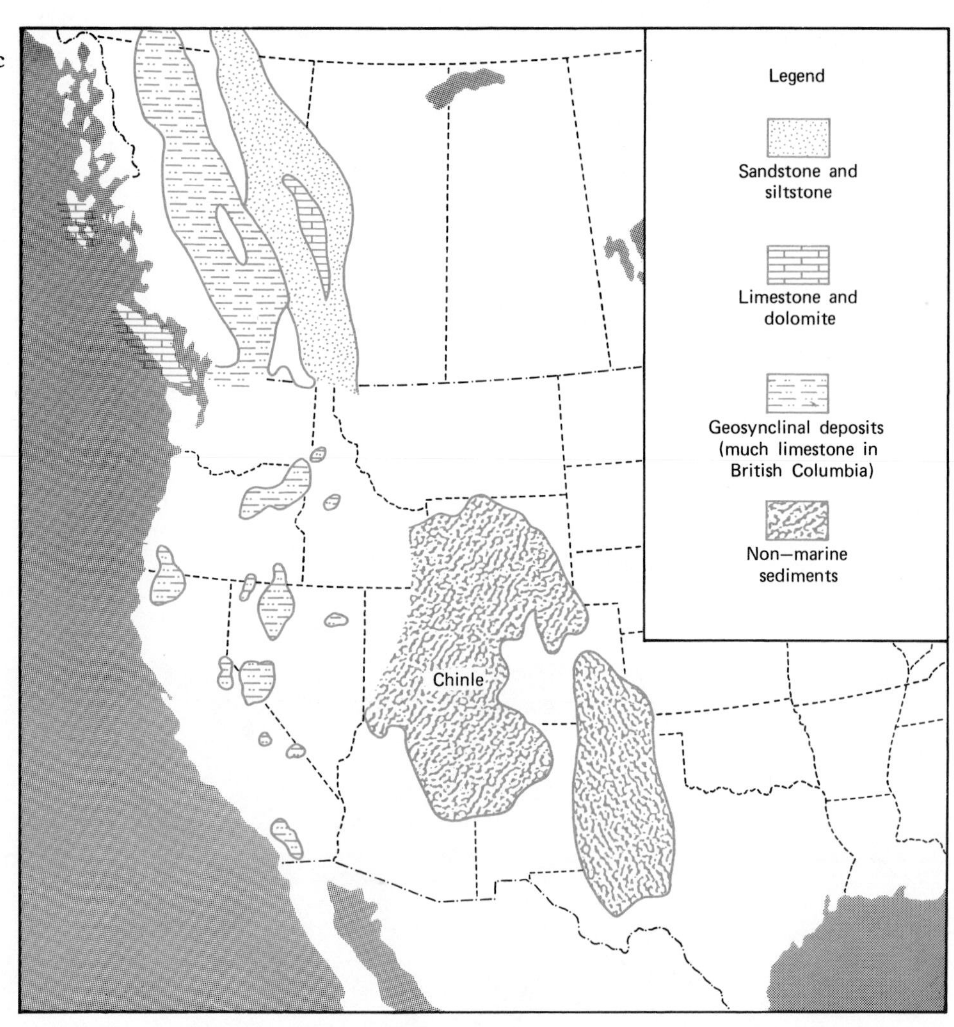

FIGURE 13–40 **Petrified logs of the Chinle Formation, Petrified Forest National Monument, Arizona. (Courtesy of the U. S. Geological Survey.)**

placed by silica-bearing solutions so that the microscopic structure is now preserved as brightly colored silica. The larger trunks are more than 30 m long and 3 m thick at the base. The Wingate Sandstone is characterized by cross-bedding on a large scale and is interpreted as a dune deposit. After a break in sedimentation marked by the silty and sandy red mudstone of the Kayenta Formation, this sandstone facies resumes through a thousand meters of section as the Navajo Formation, probably of Early Jurassic age (Fig. 13–41).

The desert sandstones of the Triassic–Jurassic boundary are the youngest deposits of the Absaroka sequence. The major cycle of sedimentation that began at the close of the Mississippian Period had taken more than three periods to complete. The redbeds that form the regressive deposits of the sequence are in contrast to the marine limestones and evaporites typical of earlier Paleozoic cycles.

FAUNA OF THE LATE PALEOZOIC–EARLY MESOZOIC REDBED FACIES

The Early Permian redbeds of north-central Texas and adjacent Oklahoma contain a rich fauna of amphibians and primitive reptiles. These sediments may represent a typical delta or flood-plain deposit, but must have accumulated in lowland areas close to water. The labyrinthodonts are represented by a wide range of aquatic and semi-aquatic forms as well as by animals that were among the most nearly terrestrial of all known amphibians. The end of deposition of this Lower Permian succession marks the end of the fossil record of most of the terrestrial groups of amphibians. The lepospondyls are represented by relatively rare fossils (except for the common aquatic form *Diplocaulis,* Fig. 13–42), but this, their last appearance in the fossil record, shows them

FIGURE 13–41 **Dune cross-bedding in the Navajo Sandstone of Early Jurassic age, Zion–Mt. Carmel Highway, Utah. (Courtesy of the U. S. Geological Survey.)**

to be still a remarkably diverse group. The primitive captorhinomorph cotylosaurs are rare, but the pelycosaurs (Fig. 13–42) derived from them are the dominant elements of the fauna. The pelycosaurs are the most primitive of the synapsids, or mammal-like reptiles. They originated and underwent diversification in Pennsylvanian time, but were few in number until the beginning of the Permian Period when the lineage gave rise to large and small terrestrial carnivores, semi-aquatic fish eaters, and large herbivores. Primitive pelycosaurs resembled, in a general way, large lizards with disproportionately large skulls. Specialized members of both the herbivorous line (*Edaphosaurus*, Fig. 13–42) and the large terrestrial carnivores (*Dimetrodon*) had greatly elongated vertebral spines that projected far above the back in the form of a comb and were apparently joined by a membrane to produce a "sail." Increasing the surface area of the body provided a rapid means of absorbing or dissipating heat. Pelycosaurs, like other reptiles, were not able to produce high body temperatures by metabolic means as do mammals and birds, but relied upon the heat of the sun or the earth to bring their bodies to an efficient temperature. This evidence of attention to temperature control is one of the few features of the pelycosaurs that reflect their affinities with mammals. In general their skeletal anatomy is that of primitive reptiles.

The sediments of the Texas–Oklahoma redbeds show a gradual tendency toward periodic, and finally general, dessication. The vertebrate fauna becomes sparse in the upper part of the Lower Permian series. In these beds are tantalizingly incomplete remains of the next level of the evolution of the mammal-like reptiles, the therapsids.

FIGURE 13–42 **A group of Lower Permian amphibians and reptiles. *Edaphosaurus* with nubbly dorsal spines in the left foreground. The other five reptiles are the carnivorous genus *Dimetrodon*. On the left, a group of small reptiles, and on the right the small aquatic amphibian *Diplocaulis*. (By Charles Knight; courtesy of the Field Museum of Natural History.)**

The Middle Permian gap in the paleontological record of the evolution of the synapsids in North America produces an unnaturally discrete boundary between the pelycosaurs and the therapsids, but an excellent record of the stages in the transition from the therapsids to the mammals is known from Russia and the southern continents. Although the skull of the early therapsids shows unquestionable affinity with that of the carnivorous pelycosaurs, the body is more advanced. In the evolution of therapsids the limbs were brought in under the body, permitting a much more efficient fore-and-aft movement of the legs than was possible with the sprawling gait of the pelycosaurs and other primitive reptiles. The more erect posture of the therapsids had a far-reaching effect on their physiology, for maintainence of an erect posture required a greater and more continuous output of energy than did a sprawling posture. This, in turn, led to a greater output of metabolic heat and, eventually, an internally-maintained body temperature. Modifications of the animal that would affect metabolic activity seem to be important in the evolution of many therapsid lines.

The therapsids, like the pelycosaurs, include herbivorous and carnivorous groups, but the therapsids show much greater diversification and several lineages have been recognized. The dominant herbivores in Late Permian time and through the Triassic are the dicynodonts (Fig. 13–43), a diverse assemblage in which most of the teeth are lost and replaced by a horny, turtle-like beak. The dentition is typically reduced to two large teeth in the upper jaw; their bases formed a sharp cutting surface against which the lower jaw moved. This adaptation was apparently very efficient for feeding on

FIGURE 13–43 **Mammal-like reptiles from the Early Triassic of Africa. Several carnivorous therapsids (*Cynognathus*) menace a plant-eating related form (*Kannemeyeria*). The bushes are primitive conifers. (By Charles Knight; courtesy of the Field Museum of Natural History.)**

a variety of plant materials, since there are thousands of dicynodont specimens representing a great variety of sizes and adaptive types. One wide spread form is the genus *Lystrosaurus,* known from China, India, southern Africa, and Antarctica.

Among the carnivorous therapsids there were several lineages that successively dominated the fauna, beginning in the Middle Permian and extending into the Early Jurassic. The most diverse and well known are the cynodonts (Fig. 13–43), which show progressive development toward mammalian anatomy.

The maintenance of the higher metabolic rate by these forms required large quantities of food broken down into small particles in the mouth to facilitate rapid digestion. Unlike most carnivorous reptiles, the cynodonts developed complex cheek teeth (in the position of mammalian molars and premolars) to cut and crush the food. Because the food was chewed for a long time, the therapsids had to separate the passage to the lungs from the mouth cavity, for which purpose a secondary palate was evolved. At the same time the lower jaw became modified so that most of the stresses were placed on a single bone, the dentary. In all tetrapods this bone supports the teeth, but in most reptiles the articulation with the skull and the place of muscle attachment are concentrated on other bones of the lower jaw. Throughout the evolution of the therapsids the dentary expands to provide more area of muscle attachment, until it becomes the principal bone in the lower jaw, and provides a new area of jaw articulation. These, and other related changes in the skull and body skeleton, result in the development of a pattern that is almost mammalian by the end of the Triassic Period. Surprisingly, the attainment of these mammalian features in advanced therapsids coincides with the almost complete extinction of the group.

The pelycosaurs were the dominant terrestrial vertebrates in the Early Permian Epoch, and the therapsids dominated from the Middle Permian to the Middle Triassic epochs. By Late Triassic time, however, the dinosaurs (to be discussed in a subsequent chapter) had attained a position of supremacy, and through competition and predation had eliminated all but a few minor groups of therapsids, none of which survived the Early Jurassic Epoch. By some time in the Middle or Late Triassic epochs one or more groups of small carnivorous therapsids had given rise to animals whose anatomy is classified as truly mammalian.

SUMMARY

In contrast to the carbonates typical of lower and middle Paleozoic sequences, late Paleozoic sedimentary rocks of the North American platform comprise cyclically repeated units of clastic sediments derived from the erosion of adjacent highlands.

A typical cycle begins with deposits, including coal, formed in nonmarine deltaic swamps and ends with deposits of a transgressing shallow sea. In these cycles of the Pennsylvanian System the marine members are thickest in the western parts of the platform and the non-marine members are thickest in the east and south near the sources of sediment.

The southern part of Gondwanaland was extensively glaciated during an interval from late Devonian to late Permian time.

The earliest vascular plants (those adapted to live on land) are found in Upper Silurian rocks. These forms, the Rhyniales, have creeping rootstocks and rising stems topped by spore cases.

The coal forests were dominated by the lycopods or scale trees. Other important plants in this environment were joint-stemmed plants like *Calamites,* seed and true ferns, and cordaites. The coal forest plants grew within, or on, the margins of a vast deltaic plain formed by sediment eroded from the Appalachian and Ouachita highlands. The compaction of the vegetable detritus condensed the organic mush into coal.

Reptiles evolved from labyrinthodont amphibians in mid-Carboniferous time. They were much better adapted to life on land than the amphibians and probably ate insects. Reptiles are classified according to the openings in their skulls.

Yoked cratonic basins have neighboring uplifts that supply them with sediments, and whose elevation seems to be connected with the subsidence of the basin through subcrustal transfer of matter.

The diverse structures, many unconformities, varied stratigraphic patterns, and petroleum traps that characterize yoked basins are illustrated by the sediments around the Oklahoma Mountains.

The yoked basins of the Colorado Mountains system were largely filled with arkosic sediment. The evaporites of the Paradox Basin illustrate the mobility of salt under the pressure of overlying sediments.

The Permian beds of West Texas show topographic control of carbonate deposition. Carbonate-secreting organisms were concentrated along bank margins, while evaporites accumulated in shoreward lagoons and dark shale was deposited in deep oxygen-deficient basins. Isolation of the basins from the open sea resulted in their conversion into evaporite basins that filled with varved limestones, anhydrite, and halite.

The end of the Paleozoic Era was a time of crisis and partial extinction for the forams, rugose corals, brachiopods, bryozoans, goniatites, blastoids, trilobites, and crinoids.

The final beds of the Absaroka sequence are non-marine desert sandstones of Late Permian, Triassic, and Early Jurassic age.

The mammal-like reptiles, or synapsids, dominated Permian and Early Triassic land faunas. Advanced synapsids, called therapsids, evolved toward mammals by bringing the legs beneath the body, developing complex cheek teeth, a secondary palate, a single lower jawbone, and possibly the control over internal body temperature that we call homeothermy.

QUESTIONS

1. For each of the 10 members of a typical cyclothem of the Pennsylvanian summarize its lithology, its environment of deposition, and the changes in a climatic cycle that could have started its deposition.
2. What evidence supports the causal relationship between late Paleozoic glaciation and the cyclothems and what evidence negates it?
3. Describe the stages by which coal is formed from decaying organic matter.
4. By consulting reference books (such as the U.S. Department of Mines Yearbooks) determine how much of the energy resources of your country, state, or province are supplied by the mining of coal. What proportion of this coal is mined within your area and what proportion is imported? For your country compute how long the present coal resources will last if used at the present rate.
5. What basic advances in adaptation to living on land do the reptiles have over the amphibians?
6. Consider the invertebrates that were affected by the late Permian crisis. What common adaptations or methods of obtaining food did they have that would suggest the nature of environmental change that caused the crisis? Are the suggested changes compatible with the continuity of plant and vertebrate life from Paleozoic to Mesozoic time?
7. Describe briefly the different types of traps and reservoirs for petroleum that are mentioned in Chapters 12 and 13.

SUGGESTIONS FOR FURTHER READING

ANDERSON, R. Y., DUNN, W. E., KIRKLAND, D. W., and SNIDER, H. I. "Permian Castile Varved Evaporite Sequence, West Texas and New Mexico." *Geological Society of America Bulletin,* 83: 59–86, 1972.

BANKS, H. P. *Evolution and Plants of the Past.* Belmont, California: Wadsworth Publishing Co. Inc., pp. 1–170, 1970.

CROWELL, J. C., and FRAKES, L. A. "Late Paleozoic Glaciation in the southern hemisphere (a series of papers on various continents)." *Geological Society of America Bulletin,* 80: 1007–1042, 82: 2515–2540, 83: 2887–2912, 1969–1972.

CYS, J. M., and ACHAUER, C. W. "Origin of Capitan Formation, Guadalupe Mountains, New Mexico and Texas." *American Association of Petroleum Geologists Bulletin,* 53: 2314–2323, 1971.

DAPPLES, E. C., and HOPKINS, M. E. ed. "Environments of coal deposition." *Geological Society of America Special Paper 114,* 204 p., 1969.

MCKEE, E. D., ORIEL, S. S., and others. "Paleotectonic Map of the Permian System." *U.S. Geological Survey Miscellaneous Geological Investigations Map I-450 and Professional Paper 515,* 1967.

MOTTS, W. S. "Geology and Paleoenvironments of northern segment, Capitan Shelf, New Mexico and Texas." *Geological Society of America Bulletin,* 83: 701–732, 1972.

ROMER, A. S. *Vertebrate Paleontology.* Chicago: University of Chicago Press, 1966.

SILVER, B. A., and TODD, R. G. "Permian cyclic strata, northern Midland and Delaware Basins, west Texas and southeastern New Mexico." *American Association of Petroleum Geologists,* 53: 2223–2257, 1971.

PART FOUR

the cordillera

CHAPTER FOURTEEN

REGIONAL SETTING

From a discussion of the shallow-water sedimentation and vertical movements of the North American craton, we now turn to an entirely different tectonic environment. From a consideration of the stratigraphic record preserved on top of a lithosphere plate, we turn to the examination of events and deposition along one of its borders. Long before lithosphere plates were recognized the continental margins were called mobile belts, because their deformed rocks preserve a record of repeated compressional movements that contrast to the basic stability of the continental platforms. We now believe the "mobility" of the marginal belts is an expression of the overall mobility of lithosphere plates.

The Cordilleran Mountain System spans half the world—from the Aleutian Islands to the tip of South America, a distance of 18,000 km. It is part of a continuous orogenic system that rings the Pacific Ocean, with segments in the Antarctic, New Zealand, the western Pacific Islands, and Siberia. The North American segment is about 10,000 km long, about 1600 km wide at its greatest width at latitude 40°N, and reaches its greatest height, 6,166 m, at Mount McKinley in Alaska (Fig. 14–1). This mountain system is the result of the interaction of the Pacific Ocean plates and the Americas plate over an interval of nearly a billion years. That this interaction continues to the present day is witnessed by earthquakes, volcanic eruptions, and the recent elevation of the Cordilleran Mountains.

Like marginal mountain chains described in Chapter 2, the Cordillera consists of a belt of crystalline, metamorphic, and igneous rocks along the oceanside, and a belt of deformed but little-metamorphosed sedimentary rocks between it and the platform. The igneous and metamorphic belt was formed by compression of, and intrusion into, sediments and lavas deposited on the Pacific oceanic crust in trenches and basins that are collectively called the Cordilleran geosyncline. The inner belt consists of rocks laid down in the miogeocline, the oceanward slope of the edge of the platform. The distribution of these miogeoclinal and geosynclinal rocks and the igneous and metamorphic belts of the Cordillera are shown in Fig. 14–2.

Geographically, the North American Cordillera is commonly divided longitudinally into three belts: the Western Ranges, the Interior Plateaus and Ranges, and the Eastern Ranges or Rocky Mountains (Figs. 14–3 and 14–4). The Western Ranges include the Coast Ranges of California and British Columbia, the Alaska Range, Cascades Range of Washington and Oregon, and the Sierra Nevada of California. The Western Ranges

THE PALEOZOIC ERA

FIGURE 14–1 **Mount McKinley (6,166 m), the highest peak in North America. (Courtesy of Bradford Washburn.)**

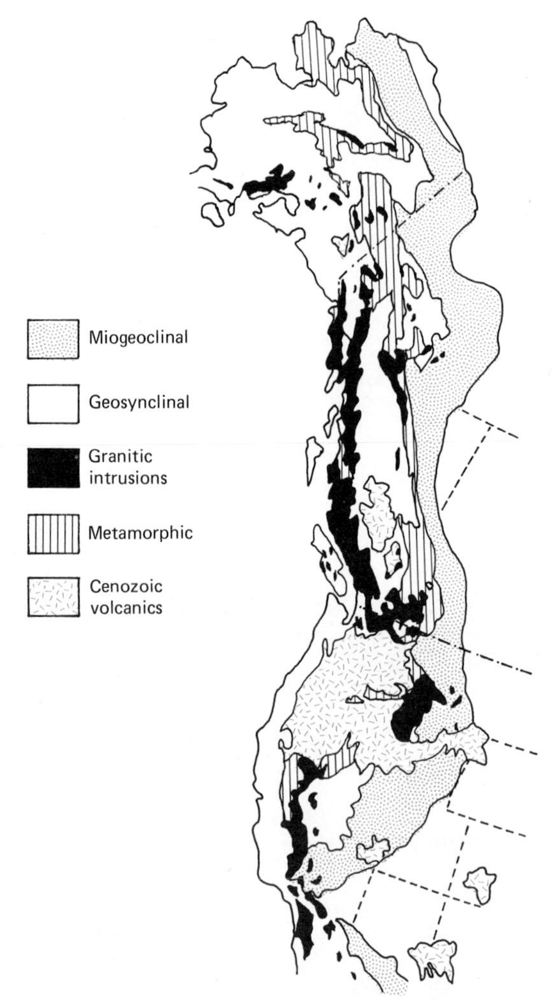

FIGURE 14–2 **Major rock types of the North American Cordillera. (Modified from J. W. H. Monger, J. G. Souther, H. Gabrielse, and P. B. King.)**

are characterized by batholithic intrusions, metamorphosed geosynclinal rocks, active volcanoes, and high relief. In the Interior Plateaus the elevation is great but the relief is less than in the Coast Ranges. Bedrock over most of the belt is like that of the Western Ranges, with the following exceptions. In Washington, Oregon, and southern Idaho repeated flows of mid-Cenozoic lavas have buried the Cordilleran structures to depths of hundreds of meters, forming the Columbia and Snake River plateaus. This cover and the Proterozoic rocks that occupy a large area along the Idaho–Montana border effectively bisect the Cordilleran geosyncline into northern and southern sectors. The Colorado Plateau is a large subcircular area underlain by flat-lying and little-deformed Paleozoic platform sediments surrounded by the deformed rocks of the Cordillera.

Within the Western Ranges and Interior Plateaus, basins are occupied by Cretaceous and Cenozoic sediments and lavas deposited after the main episode of deformation in the Cordilleran geosyncline in late Mesozoic time. These have been called **successor basins.** In addition, Cenozoic sedimentary rocks underlie a narrow continental shelf off the western coast of the continent.

The Eastern Ranges of the Cordillera, commonly referred to as the Rocky Mountains, are underlain by miogeoclinal and platform sediments. From Montana to Alaska these sediments have been broken by thrust faults and piled like shingles on the platform to the east. South of Montana thrust faulting is not prominent and the ranges of the Rockies have uplifted cores of pre-Paleozoic rocks surrounded by younger sediments.

A sedimentary record spanning a billion years gives evidence of the nature of the margin of the Americas plate in the Cordilleran Mountains. However, no certain knowledge of the nature of the Pacific Ocean Basin in the Paleozoic is possible because the oldest crust in the whole Pacific Basin is about 180 million years old. During development of the Cordilleran geosyncline sedimentary, volcanic, and batholithic material has been added to the western side of the Americas plate, presumably by the subduction of Pacific lithosphere. Some, or all, of the Paleozoic rocks of the western Cordillera may have been deposited in the Pacific Basin unknown distances south and west of their present positions. Whether or not sections of other continental plates, such as Asia, have been involved in this accretionary process is a matter of debate. It has been suggested that at one time the whole Pacific Coast of North America fitted against

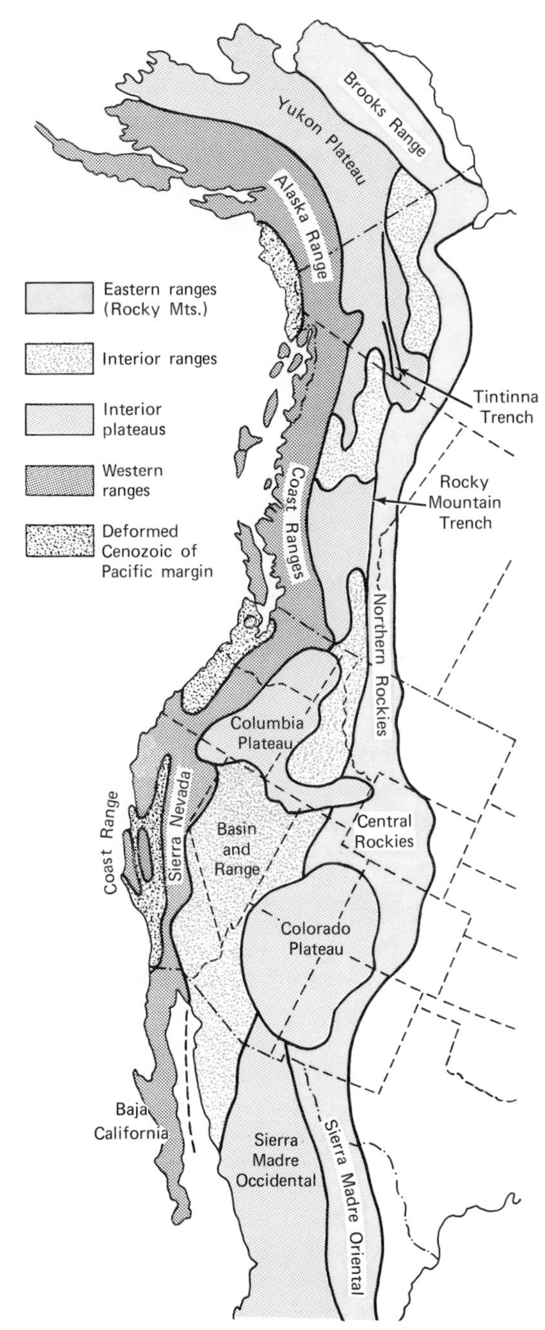

FIGURE 14–3 **Major physiographic divisions of the North American Cordillera.**

the eastern coast of Asia, from which it has drifted away. The similarity of late Paleozoic fossil faunas of the western coast of North America to the fossils of the Asiatic Tethyan region has suggested to some that Asia was at this time adjacent to the Americas plate, and when rifting took place early in the Mesozoic Era, a part of Asia was left behind and now forms the western coastal regions of North America. In late Paleozoic time the Atlantic Basin certainly was closed, but the history of the Pacific Basin in the Paleozoic still remains an enigma.

PROTEROZOIC ROCKS IN THE CORDILLERA

The basal beds of the Sauk sequence along the eastern margin of the Cordillera overlie a thick succession of clastic and carbonate sedimentary rocks of Helikian age that are known as the Belt Group in Montana, the Purcell Group in British Columbia, the Uinta Mountain Group and Big Cottonwood Formation of northern Utah, the Grand Canyon Group of northern Arizona, etc. (see Chapter 10). Belt Group sediments reach a thickness of about 10,000 m along the western edge of their extent and consist of siltstone, sandstone, and limestone. Some of the sediments show signs of shallow-water deposition, others appear to be turbidites deposited in deeper water. The limestones of the Belt Group commonly have abundant stromatolites that range in size between that of a cabbage and that of a room. Most of these forms have been given the generic name *Collenia.* The equivalent Purcell Group reaches thicknesses of 14,000 m and the Uinta Group 7000 m. These late Helikian sediments do not appear ever to have formed a continuous belt along the western side of the craton, but were deposited in a series of deep basins, the axes of which were oblique to the general trend of the Cordilleran structure. Their orientation appears to be controlled by the structure of the Canadian Shield. Before the deposition of the Hadrynian and overlying Paleozoic sediments, the Helikian sediments were moderately deformed between 1000 and 850 million years ago.

Hadrynian sediments were laid down along a continental margin which paralleled

FIGURE 14–4 **The Cordillera at the 49th parallel. Mosaic of imagery made by the Landsat satellite. The following features are identified by the numbers: 1. Vancouver. 2. the Coast Range. 3. Central Plateau of British Columbia. 4. Cascade Range of Washington. 5. Mount Baker, a volcano. 6. Lake Pend Oreille, Idaho. 7. Rocky Mountain Trench. 8. Calgary, Alberta. 9. Glacier National Park, Montana. (Portion of a photograph supplied by the Surveys and Mapping Branch, Department of Energy, Mines and Resources, Ottawa.)**

the modern Pacific coast but was about 700 km to the east of it (Fig. 14–5). The sediments of the Windermere Group, discussed briefly in Chapter 10, built out a continental shelf extending westward with sediment derived from the erosion of the shield. The prism of sediments reached a thickness of 7500 m along the 49th parallel. In the Northwest Territories, British Columbia, Washington, Utah, and Nevada basal Hadrynian rocks contain basaltic lavas, the extrusion of which has been attributed to the rifting of the lithosphere when continental blocks separate. The nature of the western block in this rifting is unclear, but the continental shelf that developed at the beginning of the Paleozoic (Fig. 14–6) appears to have been similar to the Atlantic continental shelf of the present day, that resulted from the rifting of the Atlantic in Mesozoic time.

If a geosyncline is the depositional basin at the margin of a lithosphere plate from which a folded mountain chain develops, then the Hadrynian sediments building out from the craton mark the beginning of the Cordilleran geosyncline and the start of a sedimentary–tectonic cycle. The result of that cycle is the modern Cordillera.

LOWER PALEOZOIC ROCKS OF THE CORDILLERAN GEOSYNCLINE

While the record of early Paleozoic sedimentation along the cratonic margin in the miogeoclines is good, rocks of this age are represented by only scattered outcrops and deformed rocks in the interior and western belts of the Cordillera. Areas in which early Paleozoic rocks are exposed include central Nevada, the Klamath Mountains of southwestern Oregon and adjacent California, western Idaho, the "Panhandle region" of Alaska, and central Alaska. Small patches of volcanic rocks of Devonian and possibly older age have been found in central Oregon and northwestern Washington.

The most intensely studied of these rocks occur in the Basin and Range province of central Nevada. The lowest rocks in this succession, of Cambrian age, consist of 1500 m of chert, argillite, and volcanic rocks. The Ordovician rocks are called the Valmy Formation in the Antler Peak area and consist of shaly sediments 2400 m thick, with some interbeds of sandstone, volcanics, and limestone. Thinner units of Silurian age, but of similar lithology, overlie the Ordovician in several of the ranges. Not all the sections of lower Paleozoic rocks in this belt are thick, however. In central Idaho the whole of the Ordovician and much of the Silurian System is represented by only 300 m of shales, and in central Alaska and adjacent Yukon Territory the Road River shales, less than 300 m thick, represent Ordovician, Silurian, and early Devonian time. Such relatively thin, uniform units representing long episodes of geological time

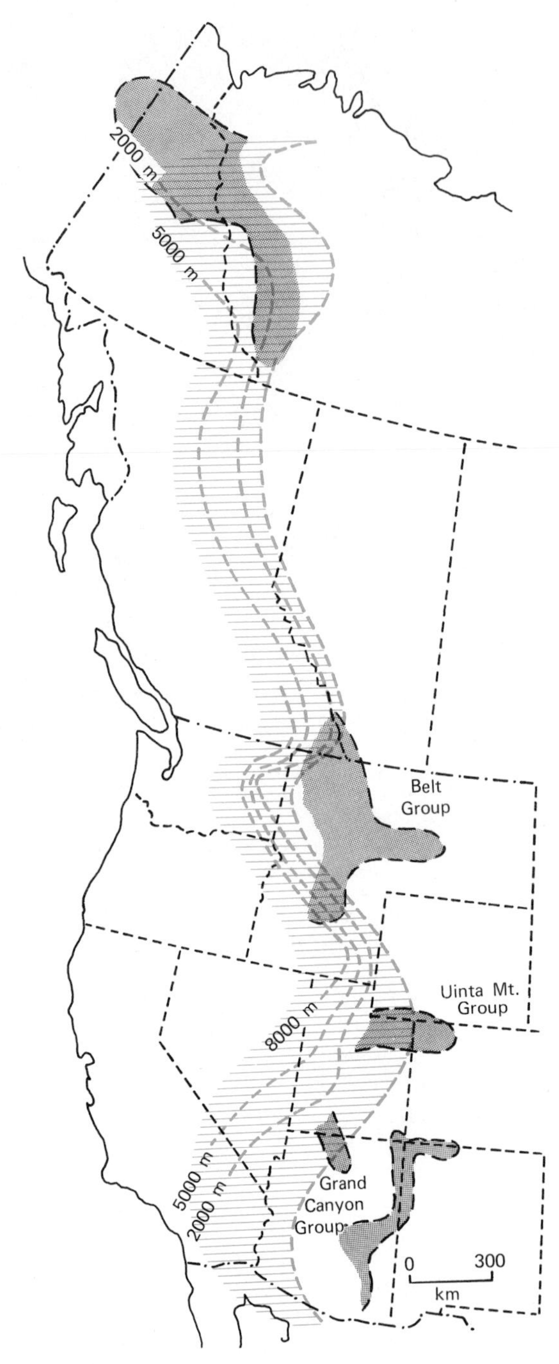

FIGURE 14–5 **Isopach map of Lower Cambrian, Ediacaran, and Hadrynian rocks with the areas of occurrence of Belt Group and equivalent rocks superposed. (After J. H. Stewart; courtesy of the Geological Society of America.)**

must be products of slow deposition under stable conditions, as would occur in relatively deep water. They are called "**condensed sequences.**"

One of the most complete sections of lower Paleozoic rocks in the western Cordillera is exposed in the "Panhandle region" of southern Alaska. Here graywackes, conglomerates, pillow lavas, and pyroclastic deposits of Ordovician, Silurian, and Devonian ages reach great thicknesses. The area is characterized by abrupt lateral changes in facies and lavas, ranging from rhyolites to basalts. Smaller bodies of intrusive igneous rocks have been isotopically dated as of Ordovician and Devonian age. These beds, and their correlatives in the Klamath Mountains, appear to have been deposited around a volcanic island arc system similar to the island arcs of the western Pacific Ocean today.

Lower Paleozoic rocks in the Cordillera west of the miogeocline can be divided into two facies belts. The western belt, of which the Alaska Panhandle is typical, is characterized by abundant volcanics and graywacke-suite sediments deposited in an oceanic trench and around a volcanic arc (Fig. 14–7). The eastern one, represented by the Nevada outcrops, is characterized by shale, bedded chert, and minor interbeds of sandstone, limestone, and volcanic rocks. The sediments of the eastern belt were probably deposited in a deep-water basin lying between the miogeoclinal continental margin and the volcanic arc. The chert beds in such a basin may be derived from the volcanic ash erupted from the island arc. Finely divided volcanic glass ejected explosively from volcanoes could devitrify on the sea floor, releasing silica into the sea water. If animals and plants which use silica, such as diatoms, radiolarians, or sponges, abound they will extract the silica to form opaline shells and leave them behind as a siliceous ooze on the ocean floor when they die. If no such organisms are present in the sea, the silica may form a gel on the ocean floor, which in time will consolidate into chert. Chert in geosynclinal assemblages occurs as bedded deposits or as nodules

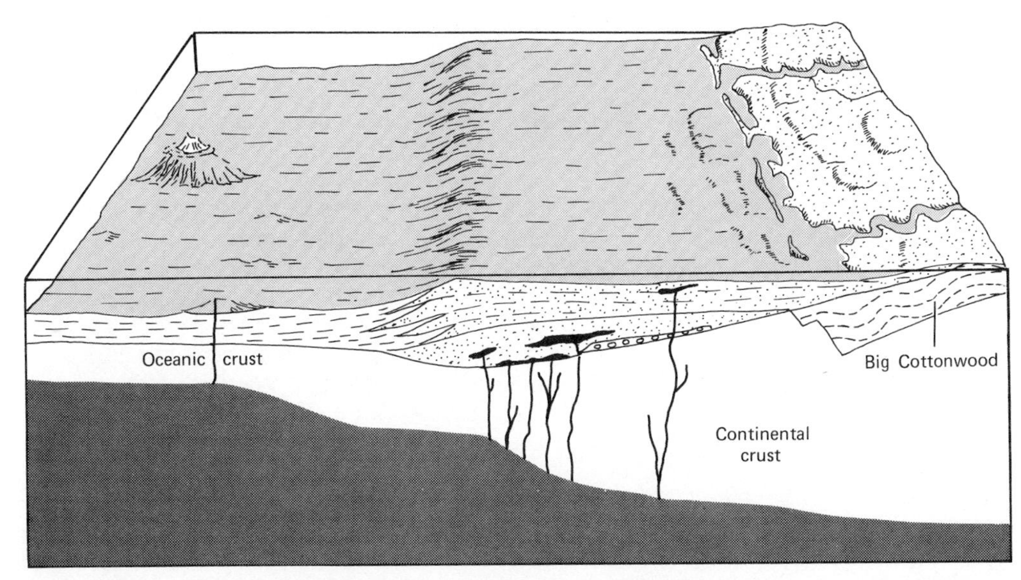

FIGURE 14–6 **Reconstruction of the continental margin of Early Cambrian time. Basaltic lava flows (black) originate from the base of the crust. A basal conglomerate, possibly of glacial origin, is shown. (After J. H. Stewart; courtesy of the Geological Society of America.)**

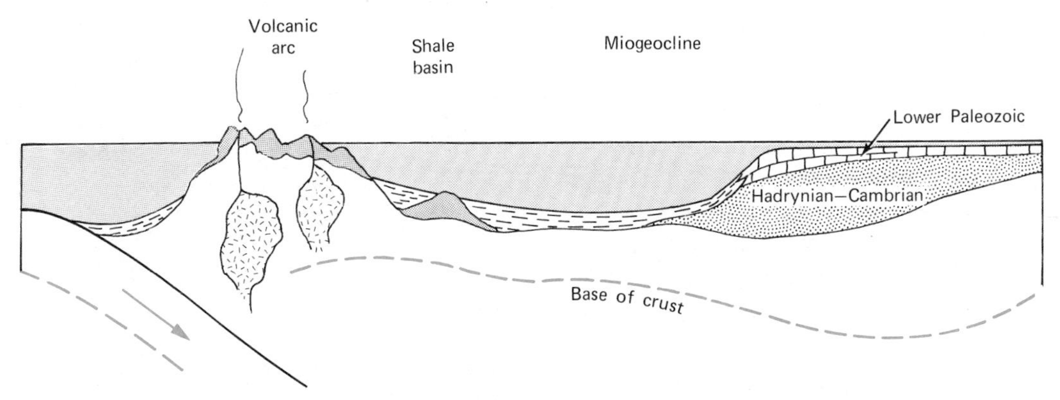

FIGURE 14–7 **Diagrammatic reconstruction of the island arc and basin of early Paleozoic time in the Cordillera. (Based on a diagram of M. Churkin.)**

within limestones or other sedimentary rocks. Outside geosynclines bedded cherts are rare, but nodular cherts are common. Unfossiliferous cherts have not necessarily formed inorganically, for the delicate opaline shells of micro-organisms secreting silica are easily destroyed by dissolution and reprecipitation phenomena before the siliceous deposit is consolidated into chert.

Palinspastic Restorations

Reconstruction of the environments of deposition in a geosynclinal area is hampered by metamorphism, deformation, and faulting of the rocks. In a complex mountain chain, such as the Cordillera, the older rocks of the geosyncline may have passed through three or four periods of deformation. To reconstruct the Paleozoic paleogeography of the Cordillera the stratigrapher must take into account that the sedimentary rocks have been moved laterally along low-angle thrust faults eastward from their place of deposition. Estimates of the distance through which the rocks have been thrust can be made and a map constructed to compensate for the movement on the faults. Such maps are called **palinspastic** (Greek: palin = backwards, spastikos = related to something drawn out). Figure 14–8 is a palinspastic map of the Basin and Range area distorted to restore lower Paleozoic rocks to their positions of deposition. Some of the faults shown are of Paleozoic age, others of Mesozoic age. The Roberts Mountain thrust is of Devonian age and, if a map were to be drawn as a base for plotting the original location of late Paleozoic sediments, its development would be ignored because these sediments were deposited after the lateral motion along that fault had taken place.

Graptolite Shale Facies

Fossils are scarce in the geosynclinal sediments of early Paleozoic age. In most of these shale and chert successions the only fossils useful for biostratigraphic zonation belong to a group of extinct organisms called graptolites, whose fossils look like pencil markings (Greek: graphein = to write, lithos = stone). Graptolites secreted a colonial skeleton consisting of a series of minute cups composed of chitin, which on preservation was carbonized to a thin, lustrous film. Primitive net-like or tree-like forms with an indeterminate number of branches, or stipes, appeared in the Middle Cambrian Epoch. From these forms the more advanced graptolite stock evolved in Early Ordovician time by a process of stipe reduction, passing through stages of 64, 32, 16, 8, 4, and 2 branches (Fig. 14–9). Colonies with four branches, either hanging from a central disc (*Tetragraptus*) or united back-to-back along an axis (*Phyllograptus*), are common and characteristic of Lower Ordovician rocks. By late Early Ordovician time the two-branched state was reached. At first the branches were spread apart like wings (*Didymograptus*), but later they became fused back-to-back to form a biserial colony (*Diplograptus*). In Silurian time most of the advanced graptolites consisted of a single branch (*Monograptus*), and by the Middle Devonian this group had become extinct.

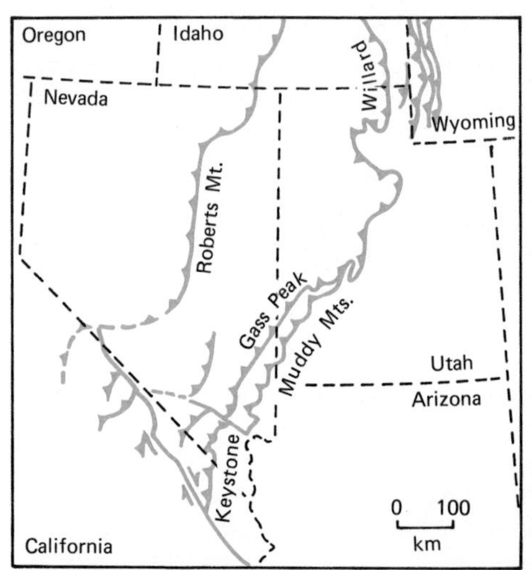

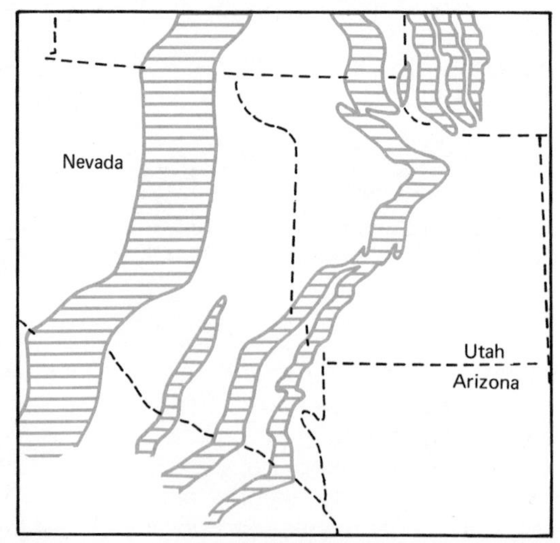

FIGURE 14–8 **Palinspastic map. *Left:* Major thrust faults of the southwestern states. *Right:* Palinspastic base prepared by restoring the fault blocks to their early Paleozoic positions. Note the distortion of the state boundaries used here as reference points. (After J. H. Stewart and F. G. Poole; courtesy of the Geological Society of America.)**

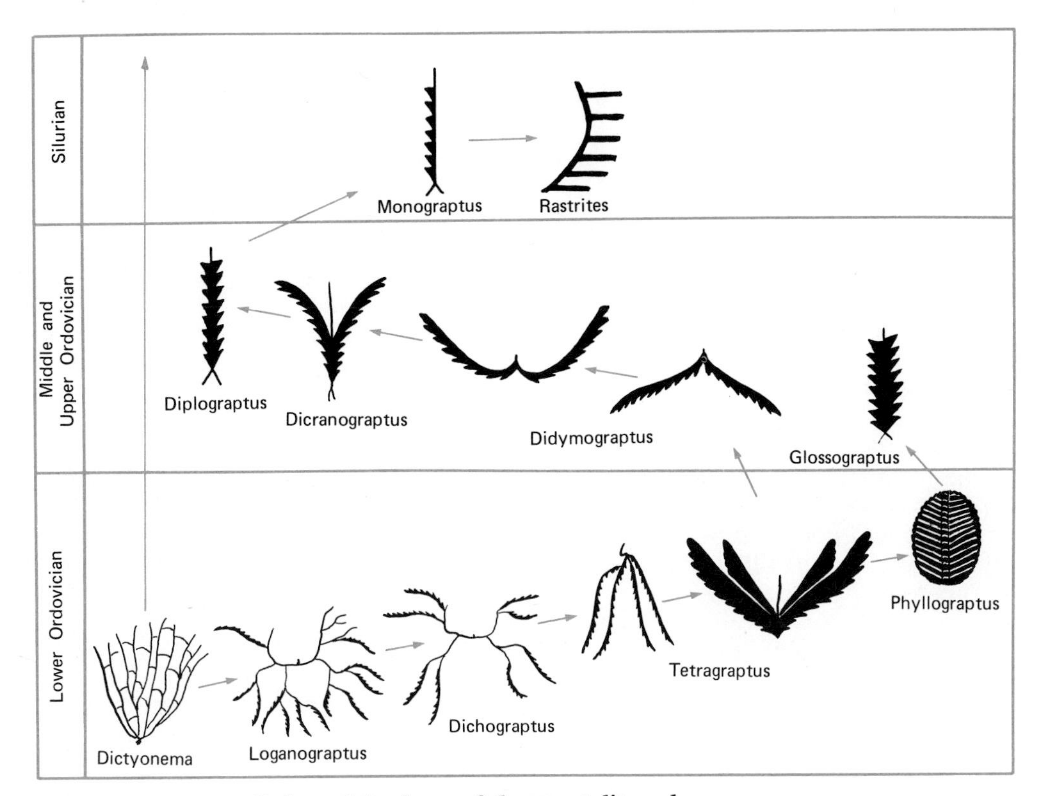

FIGURE 14–9 **Evolution of the form of the graptolite colony.**

The simple, net-like dendroid graptolites persisted into the Mississippian Period.

The graptolites probably floated free or were suspended from floats or seaweed near the surface of the ocean and were, therefore, distributed by winds and currents to all parts of the world. Because the group evolved rapidly and each new type spread quickly over a wide area, the graptolites make excellent guide fossils for Ordovician and Silurian rocks. Although they are found in some limestones and cherts, they are particularly characteristic of geosynclinal black shale of these systems. Such shales accumulated in deep, still water in oxygen-deficient basins where scavengers and bottom-living organisms in general were rare and bacterial decay was retarded, giving their delicate organic structures a chance of passing into the fossil record. Associated with them may be trilobites that swam in the sea and small inarticulate brachiopods that likely were attached to seaweed floating in the surface waters.

After decades of controversy concerning their affinity to living animals, the extinct graptolites are now generally regarded as members of the phylum PROTOCHORDATA. This group includes as living representatives the acorn worms and the small, branching marine organism called *Rhabdopleura* which secretes a chitinous tube on the surfaces of shells and rocks. The protochordates are thought by some experts to be relatives of the phylum CHORDATA that includes the vertebrates.

The principal fossils found in late Paleozoic black shales are goniatite cephalopods and the traces of pelagic fish.

ANTLER OROGENY

In Late Devonian and Early Mississippian time, the lower Paleozoic rocks along the length of the Cordilleran geosyncline were folded and thrust eastward against the continent in an orogenic episode named after the Antler Peak area of central Nevada. In Nevada the graptolitic shale facies of Silurian and Ordovician age was carried 150 km eastward over the miogeoclinal sediments on the Roberts Mountain thrust. In southeastern British Columbia late Paleozoic sediments unconformably overlie metamorphic rocks of probable pre-Paleozoic or early Paleozoic age of the Shuswap Group (Fig. 14–10). Potassium–argon dating shows that granitic gneiss in this area was intruded in Devonian time, and suggests that the extensive metamorphism of this area took place during the Antler orogeny. However, the possibility remains that some of the metamorphism is related to Mesozoic orogeny. The gneissic banding of the Shuswap Group dips outward from several centers, defining a series of domes that appear to be the result of forces directed upward in the orogeny.

Evidence of deformation and igneous intrusion in Late Devonian time is also found at the north end of the geosyncline in the Yukon Territory. In the northwest corner of North America two geosynclinal belts converge, the Cordilleran and the Innuitian. The Innuitian belt, spanning the Arctic Islands, was extensively deformed in Late Devonian time in the Ellesmeran orogeny. This period of mountain building and a contemporaneous one in the Appalachian geosyncline, the Acadian, will be discussed further in later chapters. The Antler orogeny was one of the late Devonian deformations that affected the three margins of North America.

In Alaska and Nevada the highlands raised in the Antler orogeny were eroded to produce large volumes of detritus that accumulated in adjacent troughs. In eastern Nevada clastic deposits comprising the Diamond Peak Sandstone and Chainman Shale (Fig. 14–11) of Mississippian age accumulated in a narrow, deep trough to a thickness of up to 4500 m. Most of these sediments were laid down by turbidity currents

FIGURE 14–10 **Veined gneiss of the Shuswap metamorphic complex, central Monashee Mountains, British Columbia. (Photograph No. 159330, courtesy of the Geological Survey of Canada, Ottawa.)**

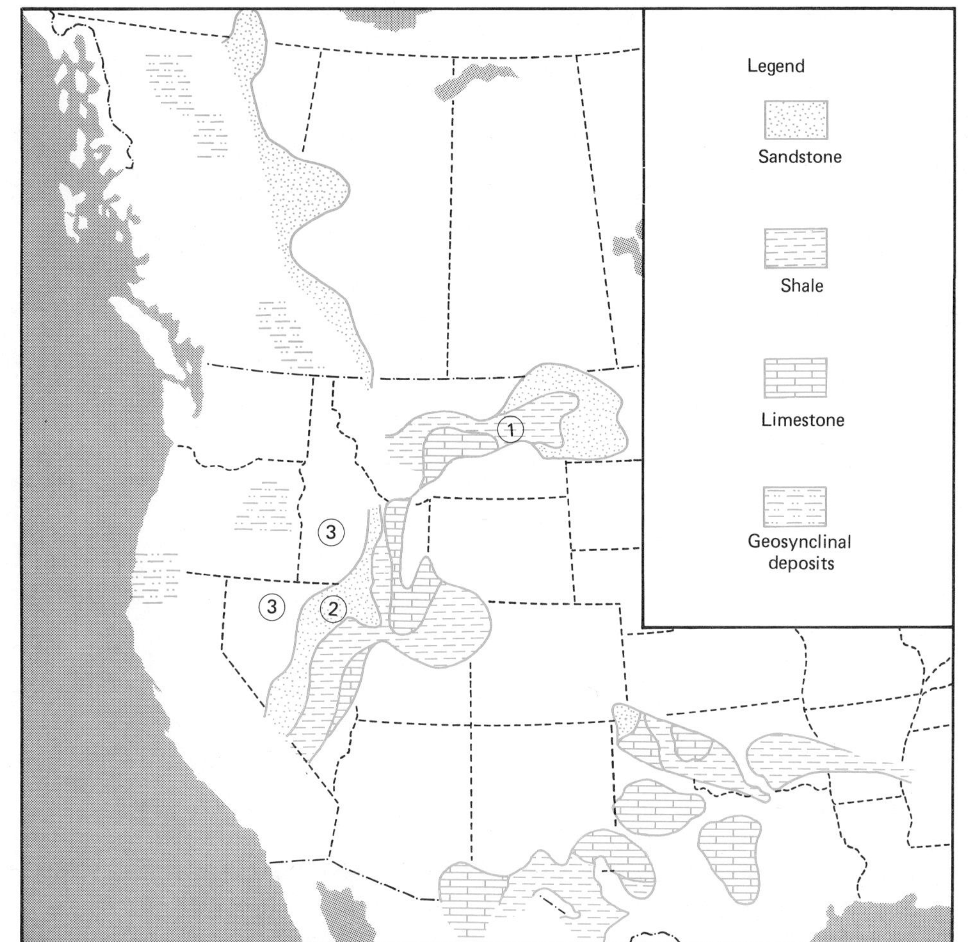

FIGURE 14–11 **Facies of the Upper Mississippian Chester Series in western North America. 1. Big Snowy Group. 2. Diamond Peak Sandstone. 3. Trend of the Antler orogenic belt.**

in deep water and can be classified with the flysch facies (see Chapter 2). By Pennsylvanian time the Antler source area must have been reduced in relief as limestone overlies the flysch succession over much of eastern Nevada.

In the northern sector of the geosyncline —Alaska, Yukon, and Northwest Territories —Late Devonian sediments consist of up to 2000 m of coarse clastics shed from uplifts in the geosynclinal zone. Strangely, in the central part of the Cordilleran belt in southern British Columbia and adjacent Alberta, little detritus was supplied from the Antler orogenic belt to the neighboring geosyncline.

The appearance of clastic sediments eroded from the Antler orogenic belt on the Cordilleran miogeocline marks a reversal in sediment supply. In early Paleozoic time clastic sediments supplied to the miogeocline came from the erosion of the Canadian Shield; from Late Devonian to Pennsylvanian time the major sources of supply at the north and south ends of the Cordillera were highlands raised in the geosyncline. The Antler deformation can be explained as a result of accelerated subduction along the island arc that bordered the west margin of the Americas plate through Paleozoic time. The oceanic crust exerted lateral pressure on the volcanic arc and the shale basin behind the arc, compressing and thrusting some of its sedimentary fill up and over the continental margin (Fig. 14–12). The Antler orogeny was the first of a series of major

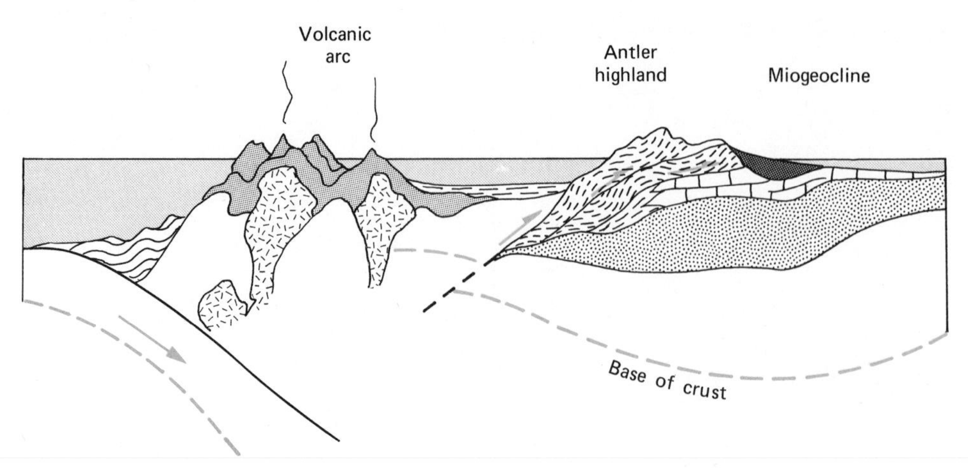

FIGURE 14–12 **Diagrammatic reconstruction of the Cordilleran geosyncline in Early Mississippian time showing the effects of the Antler orogeny. (After a diagram by M. Churkin and F. G. Poole.)**

events in which the present structure of the Cordillera evolved.

LATE PALEOZOIC SEDIMENTS AND OROGENY

Geosynclinal Sedimentation

The sedimentary record of late Paleozoic conditions in the Cordillera is much more complete than that of the early Paleozoic, for late Paleozoic rocks are exposed more widely, particularly in the northern Cordillera. South of the Columbia Plateau a deep-water basin in Nevada continued to receive shales and cherts of the Havallah and Schoonover formations. Late Paleozoic sediments and volcanics which reach great thicknesses in the Sierra Nevada and Klamath Mountains of California and Oregon and the Oregon–Idaho border region, can be interpreted as evidence of the continuing presence of an island arc on the oceanward side of the shale–chert basin.

In British Columbia late Paleozoic sediments are characterized by conglomerates and sandstones composed of chert grains. Throughout the whole width of the western Cordillera the large volume of basaltic lavas extruded is indicative of widespread volcanism. The distribution of volcanics and coarse sediments suggests that within the width of the present Cordillera there were several highlands, or volcanic arcs, rather than a simple offshore island arc. Thick successions of cherty, shallow-water limestone accumulated around the volcanic islands in the geosyncline throughout the Carboniferous and into Permian time.

At the Platform Edge

The bedded cherts in the Paleozoic geosynclinal successions of the Cordillera are believed to have been formed from silica derived through the devitrification of volcanic ash. Ash particles from explosive volcanoes in the Cordillera were also carried eastward by winds and influenced sedimentation on the miogeocline and adjacent platform. The chert content of Mississippian sediments increases systematically toward the Cordilleran geosyncline (Fig. 14–13) and also southward toward the Ouachita geosyncline, to be discussed in Chapter 18. Near the volcanoes the amount of silica available from submarine weathering of ash was sufficient to produce bedded cherts; farther from the source area where the supply was less, the silica took the form of nodular replacement of other sediments.

The Middle Permian Phosphoria Formation of western Wyoming and adjacent

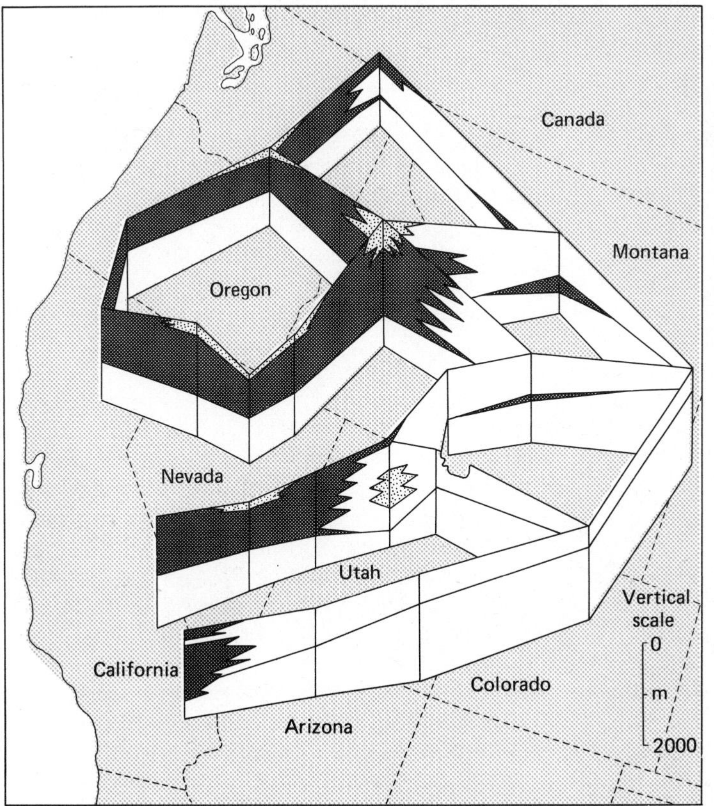

FIGURE 14–13 **Fence diagram of intersecting stratigraphic sections of the Mississippian rocks of the western United States showing the decrease in siliceous sediments (dark areas) in the system from the geosynclinal to the shelf facies. The stippled pattern represents sandstone; unpatterned areas are other sediments. (Courtesy of H. J. Bissell and the Society of Economic Paleontologists and Mineralogists.)**

Idaho may be an example of the localizing effect of the boundary between the shallow-water platform and the deepening water of the miogeocline. The formation consists of limestone, cherty limestone, bedded chert, phosphatic limestone, and phosphorite and reaches a maximum thickness of about 150 m. Phosphorite is an unusual, dark-gray rock with a concretionary pelletoidal texture and a composition of carbonate–fluorapatite ($Ca_{10}(PO_4,CO_3)F_{2-3}$). The conditions under which phosphorites were deposited are difficult to reconstruct and controversy exists over the extent to which deposits were formed by direct precipitation and by replacement of other carbonates on the sea floor. Similar deposits now forming off the coast of southern California appear to be localized where water saturated with phosphate wells up from the oceanic depths. Precipitation takes place in 100 to 500 m of water along the continental slope. The beds of the formation that are rich in calcium phosphate are mined by the fertilizer and chemical industries.

Ultramafic Rocks

Mafic igneous rocks are those that contain more dark-colored ferromagnesian minerals than light-colored quartz and feldspar. (The igneous rocks with a preponderance of light-colored minerals are said to be felsic.) Ultramafic rocks are composed almost entirely of dark-colored minerals, such as pyroxene and olivine. Such rocks are believed to form the upper part of the earth's mantle (see Chapter 1). Their occurrence in geosynclinal sequences has suggested to geologists that plate motion at times may open deep-reaching faults along which fluid rock from the mantle may be injected into the crust. The association of altered basaltic lavas (greenstones), chert, and ultramafic rocks with geosynclinal deposition was first pointed out by G. Steinmann (1926) and has come to be known as Steinmann's trinity.

Small bodies of ultramafic rock are found throughout the length of the Cordillera and appear, on the basis of isotope dating, to have been intruded over a long interval from Devonian to Jurassic time. Most of them, however, occur within the late Paleozoic succession and many were intruded during the Carboniferous or Permian periods. Ultramafics are commonly highly sheared, altered to serpentine, and displaced from their original site of intrusion along regional faults. The serpentinized ultramafic bodies are an important source of the fire-resistant mineral, asbestos, and supply the mines at Cassiar, British Columbia, and Jefferson Lake, California.

Somona–Tahltanian Orogeny

The Paleozoic cycle of orogeny and deposition in the Cordillera was brought to a close by an interval of orogeny that is poorly dated. In Nevada late Paleozoic rocks of the marginal basin moved eastward along the Golconda thrust, which overrides the Roberts Mountain thrust from the west. This major eastward movement is believed to be of early Triassic age. Over much of the Cordillera angular unconformities separate Permian rocks from Late Triassic rocks. In the Canadian Cordillera sediments of Early Triassic age are completely missing and those of Middle Triassic age are widely scattered, suggesting that the geosyncline was uplifted at this time. In British Columbia granitic intrusions occurred during the Triassic Period, and stratigraphic evidence suggests that deformational movement may have lasted from Early into Late Triassic time. In this area the term "Tahltanian orogeny" has been applied to this episode of deformation. In Nevada the name "Somona" is used for the movements close to the Permian–Triassic boundary.

Despite the fact that deformation was widespread in the Cordillera and sedimentary and volcanic successions in the geosyncline at the beginning of the Mesozoic Era were thick and coarse, little clastic material reached the miogeocline to the east, where limestone and siltstone are the main Permian and Triassic sediments (Fig. 14–14).

The climactic events in which the main structures of the Cordillera, as we know it today, were formed began in Middle Triassic time and extended to the early part of the Cenozoic Era. For this interval the interaction of the Pacific Ocean plate with the North American plate becomes clearer; the Paleozoic history of the plate margin relationships are still obscure and conjectural.

SUMMARY

The North American Cordillera is divided into the Western Ranges, characterized by batholithic intrusions, metamorphosed geosynclinal rocks, and active volcanoes; the Interior Plateaus, formed largely of similar rocks; and the Eastern Ranges, underlain by extensively thrust faulted miogeoclinal rocks.

Late Helikian Belt Group sediments were deposited to great thicknesses in troughs indenting the western edge of the continent. Hadrynian sediments of the Windermere Group were initial deposits of a continental shelf (miogeocline) that occupied the western margin of the platform during much of Paleozoic time.

Lower Paleozoic deposits of the western Cordillera were laid down around an offshore volcanic arc and in a deep basin between the arc and the miogeocline at the continental margin. The deep water shales of the basin contain many graptolites, an extinct group of small floating animals probably related to the protochordates. The spatial relationships between the facies have been altered by subsequent thrust faulting and folding, but can be restored in palinspastic maps.

In Middle and Late Devonian time the Cordillera was widely affected by the Antler orogeny in which the western facies belts were thrust over miogeoclinal sediments, some metamorphism and intrusion took

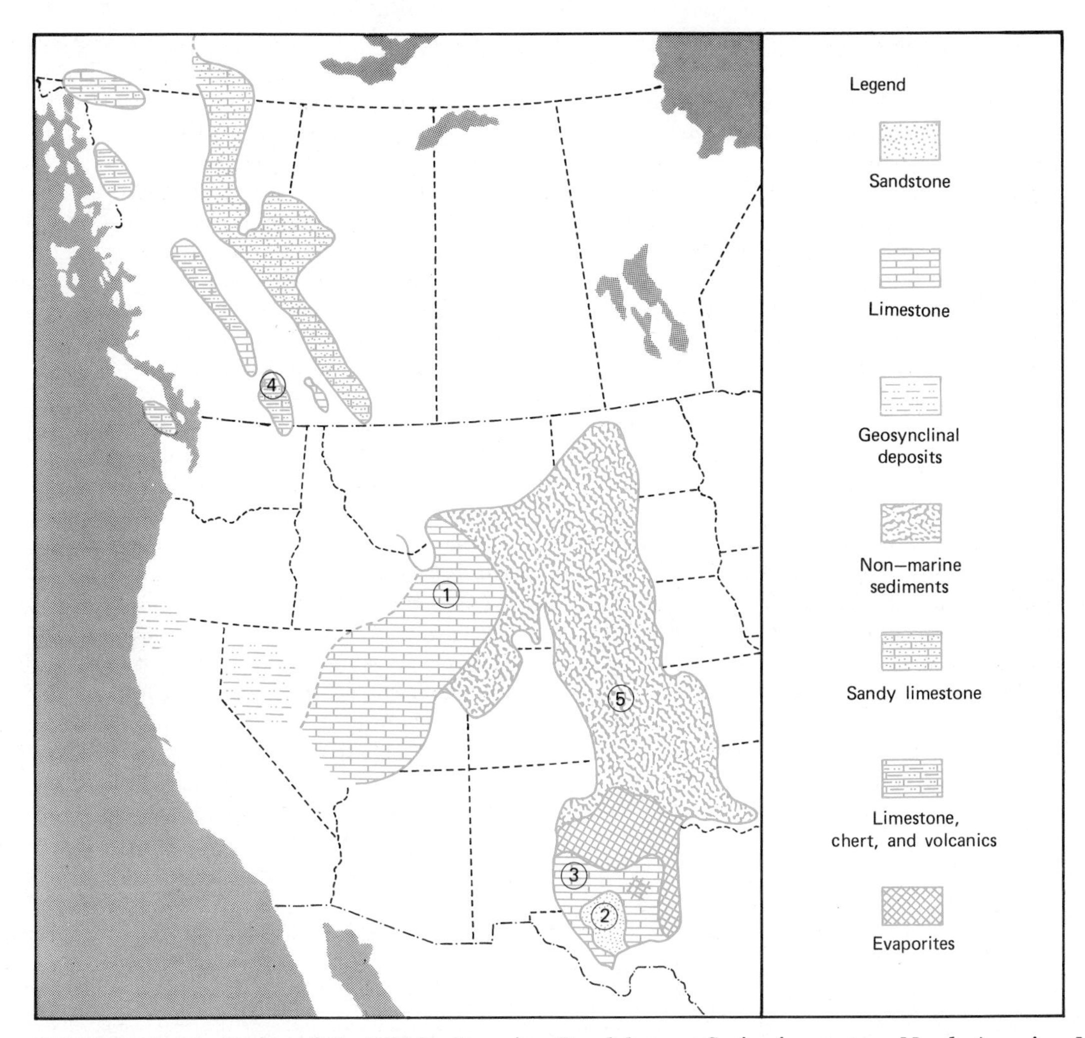

FIGURE 14–14 **Facies of the Middle Permian Guadalupean Series in western North America. Note the many areas of limestone in the Cordilleran geosyncline. 1. Phosphoria Formation. 2. Delaware Basin. 3. Carlsbad Limestone. 4. Cache Creek Group. 5. Lykins Formation.**

place, and highlands were formed that shed detritus into miogeoclinal areas.

Bedded and nodular cherts in Paleozoic sediments may be derived from the devitrification of volcanic ash and reflect periods of explosive volcanism in the Cordilleran island arc systems.

Small bodies of ultramafic rock are found throughout the late Paleozoic succession of the western Cordillera.

The end of the Paleozoic and the beginning of the Mesozoic eras was a time of widespread orogeny in the Cordillera.

QUESTIONS

1. Describe the origin of each of the following rock types common in geosynclinal areas: pillow lava, bedded

chert, greenstone, turbidite, and graywacke.

2. What is a palinspastic map and what information does a geologist need in order to construct one?
3. Why are graptolites common in deep-water shale deposits and why are brachiopods and corals rarely found?
4. What geological factors limit our knowledge of the Pacific margin of the North American plate in pre-Paleozoic and early Paleozoic time?

SUGGESTIONS FOR FURTHER READING

BURCHFIEL, B. C., and DAVIS, G. A. "Structural framework and evolution of the southern part of the Cordilleran orogen, western United States." *American Journal of Science,* 272: 97–118, 1972.

CHURKIN, M., JR. "Paleozoic marginal ocean basin: volcanic arc systems in the Cordilleran fold belt." *Society of Economic Paleontologists and Mineralogists, Special Publication 19,* 174–192, 1974.

DOUGLAS, R. J. W., and others. "Geology of Western Canada. in, Geology and Economic Minerals of Canada." *Geological Survey of Canada,* 367–488, 1970.

POOLE, F. G. "Flysch deposits of the Antler foreland basin, western United States." *Society of Economic Paleontologists and Mineralogists, Special Publication 22,* 58–82, 1974.

ROBERTS, R. J. "The Evolution of the Cordilleran fold belt." *Geological Society of America Bulletin,* 83: 1986–2004, 1972.

ROGERS, J. J. W., and others. "Paleozoic and Lower Mesozoic Volcanism and continental growth in the western United States." *Geological Society of America Bulletin,* 85: 1913–1924, 1974.

STEWART, J. H., and POOLE, F. G. "Lower Paleozoic and uppermost Precambrian Cordilleran Miogeocline, Great Basin, western United States. *Society of Economic Paleontologists and Mineralogists, Special Publication 22,* 28–57, 1974.

CHAPTER FIFTEEN

INTRODUCTION

Modern plate tectonic theory holds that folded mountain chains are the result of the interaction between lithosphere plates at a subduction zone, and that the structures produced by this interaction at any one time will depend on the rate of ocean-floor spreading, the nature of the oceanic plate subducted, the relative motion of the convergent plates, etc. Study of the history of mountain chains has led many geologists to conclude that a sequence of events, the geotectonic or geosynclinal cycle, is common to many of them. Stages in the cycle commonly recognized include the formation of a geosynclinal trough at the continental margin, preliminary compressional movements that raise volcanic highlands which supply sediment to the trough, and final orogeny in which the trough is collapsed against the continent. In their extreme form, hypotheses of the geotectonic cycle postulate that once the preliminary geosyncline has formed, orogeny proceeds through predictable stages to the formation of a mountain chain and, subsequently, a new trough forms on the oceanward side of the old one to start a new cycle. Cyclic patterns are not inherent in the plate tectonic explanation of mountain building, and the increasing popularity of plate tectonic concepts has led to a decreasing emphasis on orogenic models implying a geotectonic cycle.

In Mesozoic and early Cenozoic time the structures we now see in the Cordillera were formed in events that have collectively been called the Cordilleran orogeny (Fig. 15–1). As these orogenic events drew to a close in mid-Cenozoic time, the whole tectonic picture along the western side of the North American plate was changing as the East Pacific Rise spreading zone moved into the subduction zone that, in one form or another, had occupied the edge of the continent since the beginning of the Paleozoic Era. In one sense a tectonic cycle had come to a close; in another sense, the North American plate margin had only taken on a new configuration.

Within the Cordilleran orogeny many phases have been distinguished by local names, such as Columbian, Nevadan, Sevier, Inklinian, Laramide, etc. These names may be helpful locally in referring to an episode of deformation, but they should not obscure the fact that the raising of the Cordillera was a long process, the sum of many local events of relatively short duration recording the local responses of the crust to forces along the plate boundary.

Although in this chapter our attention is focused on the western side of the North American plate, we must not forget that events there were not independent of changes taking place on the eastern side of

CORDILLERAN OROGENY AND MESOZOIC LIFE

FIGURE 15–1 **Folded rocks of the Late Cambrian Arctomys and Waterfowl formations, Monashee Mountains, British Columbia. (Courtesy of R. A. Price and the Geological Survey of Canada, Ottawa, Photograph #154571.)**

the plate, where the Atlantic Ocean was beginning to open in Mesozoic time. In Late Triassic time North America began to separate from Africa and by Cretaceous time it was being moved westward against the Pacific crust as the North Atlantic spreading center added new crust to the east side of the Americas plate.

The climax of the building of the Cordilleran Mountain System took place during the last great transgression of the sea onto the North American platform. Into this inland sea erosion products of the rising Cordillera were poured eastward, eventually silting it up. The cycle of transgression and regression defined the fifth of Sloss' cratonic sequences, the Zuni.

The intervals of transgression of the epeiric seas across the craton appear to be correlated with the periods of orogeny at the margins of the plates. The correlation was first noted by Emile Haug (1900) and has recently been termed Haug's Law. The relationship has been attributed to the rise of sea level caused by continental rifting, the swelling of mid-ocean rift systems, and the displacement of sea water over the continents. Others suggest that the rise and fall of continental plates caused by the transfer of material beneath them in the asthenosphere is responsible for transgressions and regressions.

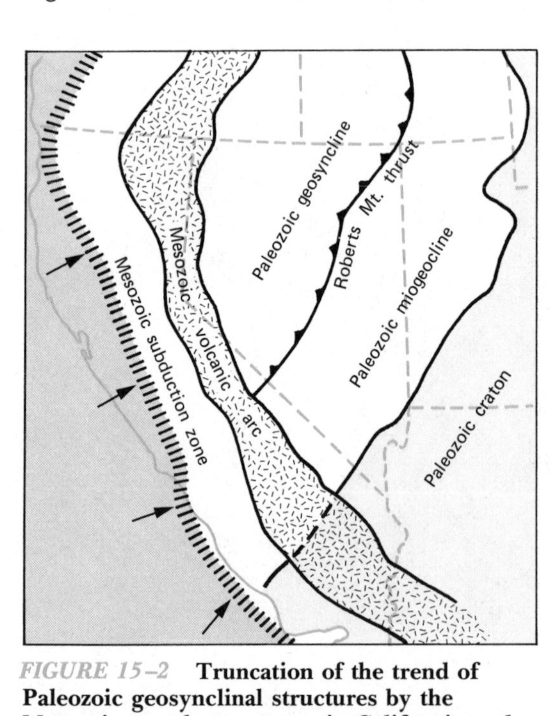

FIGURE 15–2 **Truncation of the trend of Paleozoic geosynclinal structures by the Mesozoic trench-arc system in California and Nevada. (Modified from B. C. Burchfiel and G. A. Davis; courtesy of the American Journal of Science.)**

JURASSIC–CRETACEOUS RECORD IN THE CORDILLERAN GEOSYNCLINE

Along most of the length of the Cordillera the pattern of sedimentation changed little from Permian to Jurassic time, but in southern California the trend of middle Mesozoic sedimentary facies crosses that of Paleozoic rocks at right angles. During the Paleozoic the depositional slope in Nevada and southern California was to the northwest and the facies belts trended at right angles to this, northeast–southwest (Fig. 15–2). However, the belt of Jurassic and Cretaceous sediments and volcanics was oriented in a northwest–southeast direction, parallel to the present coast but cutting across the trend of the older rocks.

Late Triassic and Early Jurassic Volcanics

The model of an offshore island arc separated from the miogeocline by a deep basin that is postulated to explain the distribution of Paleozoic rocks in the western Cordillera no longer accounts for the distribution of sediments and volcanics after the Somona orogeny at the beginning of Mesozoic time. Late Triassic and Early Jurassic rocks of the western Cordillera consisting of great thicknesses—up to 10,000 m—of volcanics and graywackes could be interpreted as deposited at their present positions and as evidence of a continuing island arc along the Pacific margin. However, another interpretation of these and also of the late Paleozoic volcanic successions is that they were deposited in various unknown parts of the Pacific Basin west of the present continental margin and have been carried eastward against the western margin of the Americas plate as the Pacific lithosphere was subducted. According to this hypothesis, the western Cordillera can be divided into large structural blocks that are now separated by major faults. The blocks have originated in various environments of the Pacific Basin and have been transported various distances to their present sites.

Up to 6000 m of oceanic pillow lavas of

Late Triassic age found in the Wrangell Mountains of southeast Alaska, on the Queen Charlotte Islands off northern British Columbia, and on Vancouver Island support this hypothesis. The magnetic poles in these tholeiitic basalts indicate that they accumulated 50° of latitude south of the rocks that form the North American craton. These determinations seem to indicate that the volcanics were extruded on the Pacific Ocean floor 5000 km (=50° lat.) south of their present position and have been carried north against the west margin of the Americas plate by the conveyor-belt action of the Pacific plate. This displaced terrane, which has been called Wrangellia after the Alaskan Mountains, was presumably formed by crustal spreading, but precisely where this rifting took place is not clear.

In British Columbia the Jurassic was a period of deformation, metamorphism, intrusion, and uplift as well as volcanism. In the northern sectors the sea drained from the basins between the growing mountains, and non-marine sediments were deposited by Late Jurassic time. In the southern basins the transition to non-marine deposition took place in Early Cretaceous time.

The Franciscan Mélange

Geosynclinal sediments of Late Jurassic and Cretaceous age are extensively exposed in the Coast Ranges and the western sectors of the Sierra Nevada of California. In the Western Sierra Nevada Jurassic sediments and volcanics are highly deformed and metamorphosed. On the east side of the Coast Ranges Late Jurassic and Cretaceous rocks, referred to as the "Great Valley" succession, consist of sandstone, siltstone, shale, and conglomerates which thicken to the west to 4000 m. These beds appear to have been deposited on a continental shelf and slope facing west. The sandstones deposited at the bottom of the slope are turbidites. The Franciscan complex of graywackes, siltstones, shales, and volcanic rocks and ultramafic intrusions, ranging in age from Late Jurassic to Eocene, occupies the California coast (Fig. 15–3) and is separated from the Great Valley succession by a major fault. The structure of the Franciscan complex is chaotic. Blocks of various lithologies, ranging from several meters to several kilometers across, are jumbled in a sheared, slaty matrix. The technical term for such a complex is "mélange" (may-lawn-je) (French: something mixed). The Franciscan complex, or mélange, contains the seaward correlatives of the Great Valley succession, deposited in deep water in a trench where the Pacific crust was being subducted throughout Late Jurassic and Cretaceous time. These sediments and lavas were deformed, more or less continuously, as they were smeared and plastered against the continental margin to the east (Fig. 15–4). The phrase "tectonic churning" has been used to describe this process and gives a dramatic word-picture of the surficial movements in a trench.

The processes in a subduction zone impress on the sediments and lavas accumulating there features by which the former location of the zone may be determined after it is gone. The suite of volcanic rocks associated with island arcs and trenches (usually referred to as the calc-alkaline suite) is distinctive chemically. The metamorphism imposed on the sediments is characterized by minerals which define the "blue schist" metamorphic facies. Because the minerals have been formed as the cool, oceanic lithosphere plate plunges deep into the subduction zone, they have properties characteristic of high pressure and low temperature conditions. The chaotic structure of the trench rocks has already been mentioned. The presence of ultramafic rocks may suggest faults along a subduction zone that tapped mantle material. All these features are present in the Franciscan rocks of the Coast Ranges of California and have been used for the identification of ancient subduction zones elsewhere in the Cordillera.

By Late Cretaceous time most of the area that is now the western and central Cordillera had been deformed, extensively intruded by igneous rocks, and elevated above sea level. Structurally and topographically the area was much like that of the Andes of western South America today.

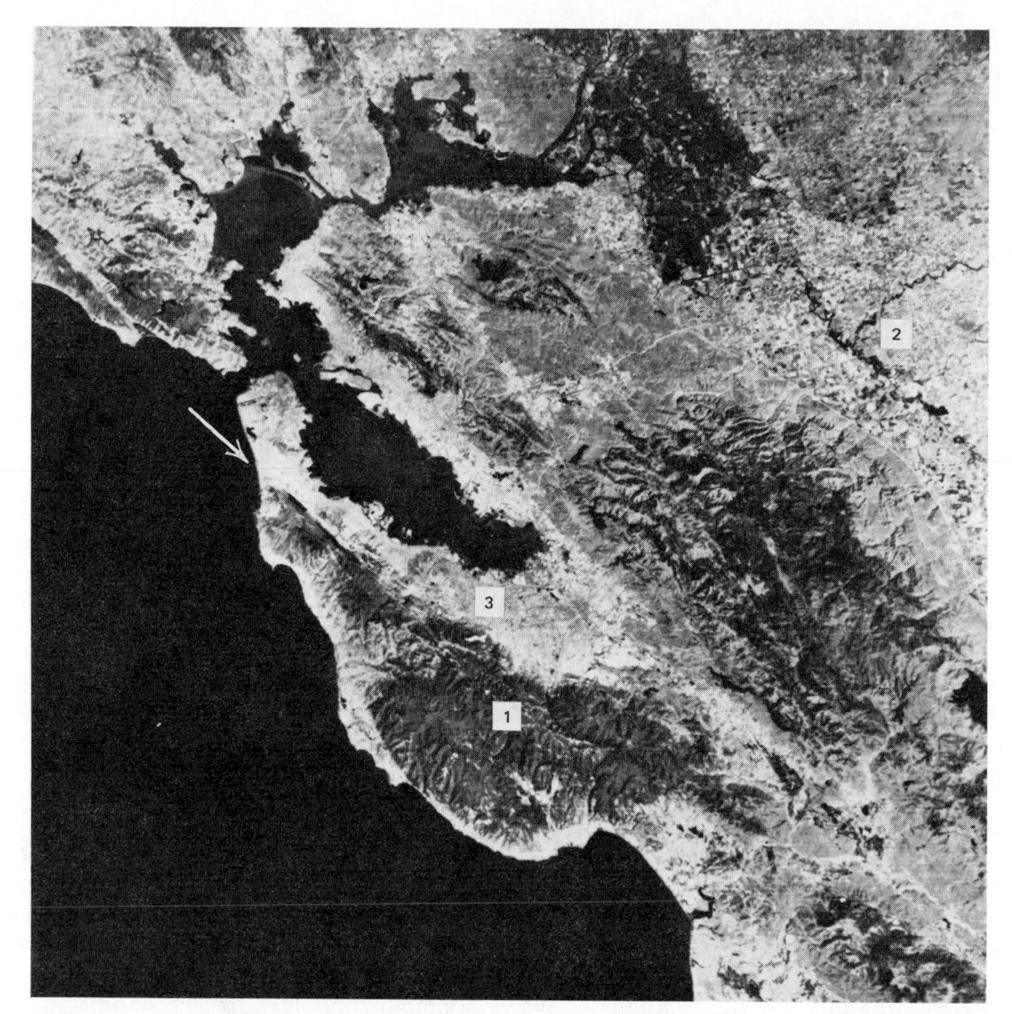

FIGURE 15–3 **California Coast Ranges (1) and Great Valley (2) at San Francisco (3). The arrow shows the intersection of the San Andreas fault with the coast line. Landsat imagery. (NASA Photographs.)**

High mountains capped by active volcanoes bordered a trench along the continental margin where subduction continued.

CORDILLERAN INTRUSIVE ROCKS

In the interval from Late Triassic to Late Cretaceous time the axis of the Cordilleran geosyncline was repeatedly intruded by granitic rocks. Once the large igneous masses were thought to have been intruded as bodies of magma hundreds or thousands of kilometers across in the Late Jurassic Epoch. Detailed petrographic studies of the batholiths and hundreds of isotopic age determinations have demonstrated that individual intrusions were relatively small and the emplacement of each took a relatively short time, but large masses of these smaller bodies were built up over an interval of hundreds of millions of years. The individual intrusions, or plutons, differ slightly in mineralogy and texture, but the average composition of the batholiths appears to be that of quartz diorite, a rock with about equal amounts of light and dark minerals and more than 15 per cent quartz.

Continuity

The present positions of the Mesozoic batholiths of the Cordillera are shown in Fig. 14–2. The Sierra Nevada batholith of eastern California is covered on the north by the Cenozoic lavas of the Modoc Plateau, an extension of the Columbia Plateau, but in windows through the lavas the zone of intrusion appears to swing eastward. The Idaho batholith, which occupies much of that state, is the most easterly of the large bodies. It is bordered by several smaller intrusives, such as the Boulder, Tobacco Root, and Phillipsburg batholiths. In British Columbia the belt of intrusion splits into the Coast Range and Omineca zones north of the international border. The former occupies a belt almost 1000 km long, from the 49th parallel to Alaska. The latter is defined by scattered intrusions along a zone of high-grade metamorphism in southeastern British Columbia and by the Omineca batholith of the northeastern part of the province.

Although the northern end of the Sierra Nevada batholith is separated at present by about 500 km of Columbia Plateau basalts from the southern end of the Idaho batholith, the two, and other Mesozoic batholiths of the Cordillera, may have originally formed a continuous belt of intrusion. Cenozoic events, to be discussed in the next chapter, have extended the crust through normal faulting in the Basin and Range region of Nevada and moved the two batholiths apart. Prior to this extension the two batholiths would have formed an intrusive belt that swung eastward around the state of Washington (Fig. 15–5).

Isotopic Dating

Potassium–argon age determinations on the large Cordilleran batholiths have established that the times of intrusion of individual plutons range from Late Triassic to early Cenozoic. In the Sierra Nevada five main episodes in the interval from Late Triassic to Late Cretaceous epochs have been distinguished. Unfortunately, the precision of many potassium-argon age determinations in insufficient to resolve closely spaced intrusive episodes. Some geochronologists claim that the present state of the data is sufficient to define for the Cordillera as a whole only two intervals of intrusion (Late Jurassic and Late Cretaceous), and to suggest a third at the end of Triassic time. Soon after their intrusion the igneous bodies were unroofed by erosion, as indicated by fragments of the intrusive rock that appear throughout Jurassic and Cretaceous sediments of adjacent basins. Warren Hamilton explains this early unroofing by postulating that batholiths do not crystallize deep in the crust, but the magma is emplaced near the surface, spreading out beneath an insulating blanket formed by ash and flows ejected from volcanoes fed from the magma. This batholithic model is not acceptable to all geologists, but the successive appearance of fragments of different granitic plutons throughout the sections of Mesozoic sedimentary rocks in Cordilleran basins is further evidence of the essential

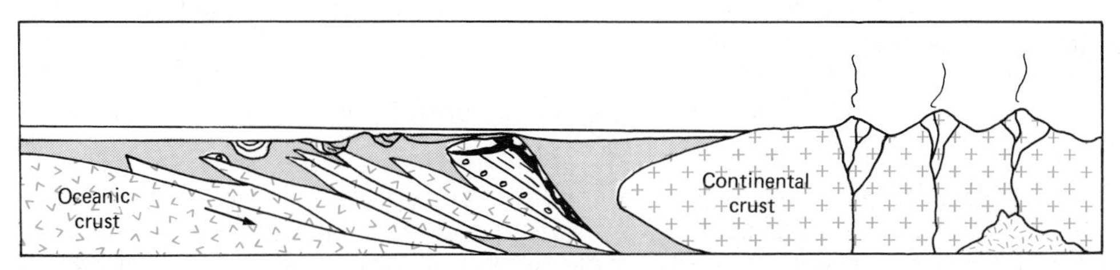

FIGURE 15–4 **A reconstruction in cross-section of the Cordilleran subduction zone in California during Late Cretaceous time showing the slicing off of basement blocks believed to be a mechanism in the formation of the Franciscan mélange. Blue schist metamorphism occurs along the contacts of the blocks. (Adapted from C. Maxwell; courtesy of the Geological Society of America.)**

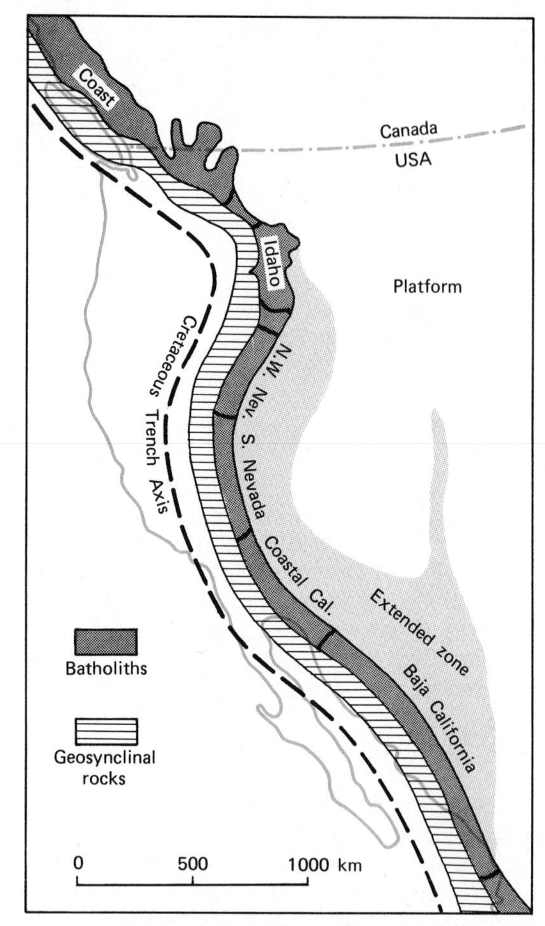

FIGURE 15–5 **Continuity of late Mesozoic batholiths. The map shows a reconstruction of the western margin of the continent in Late Cretaceous time, plotted by eliminating the extension produced in Cenozoic time by the Basin and Range faulting. The shapes of the batholiths are generalized. (After Warren Hamilton; courtesy of the Geological Society of America.)**

continuity of intrusion over much of middle and late Mesozoic time.

Ore Deposits

The great California Gold Rush of 1849 was set off by the discovery of placer gold concentrated in gravels from the erosion of gold veins emplaced in the Sierra Nevada during the Cordilleran orogeny. The gold of California is an example of the mineral wealth that was introduced in the Cordilleran belt by the intrusions of Mesozoic age. The ores of the Coeur d'Alene district of Idaho are associated with the Idaho batholith. The magmas of the Boulder batholith were accompanied by solutions that deposited the ores of the Butte district. The main pit at Butte is the fourth most productive copper mine in the United States, producing about 110,000 metric tons per year. The ore deposits of British Columbia, such as the Sullivan lead-zinc mine at Trail (production in 1972, 100,000 tons of each), the mercury of the Pinchi Lake area, the copper of the Britannia area, and many others owe their existence to the Cordilleran intrusive episode of the middle and late Mesozoic.

STRUCTURES PRODUCED BY CORDILLERAN DEFORMATION

Western and Central Cordillera

Along the site of the Cordilleran geosyncline sedimentary rocks and lavas have been intensely deformed, faulted, and metamorphosed so that their structure is difficult to resolve or to describe concisely. In the Sierra Nevada late Paleozoic and Mesozoic rocks are folded isoclinally, that is, the limbs of the folds are bent into subparallel positions so that the crests of anticlines and the troughs of synclines are obscured. All the beds dip with the same inclination toward the east. In the Klamath Mountains (Fig. 15–6), Sierra Nevada, and in southwestern British Columbia thrust planes dip eastward parallel to the dip of the fold limbs and to the east-dipping subduction zone along which, when they were formed, the Pacific Plate was underthrusting the continent.

Deformation of the Miogeoclinal Sediments

Although other episodes of deformation in Paleozoic and early Mesozoic time had affected the central and western Cordillera, not until Late Jurassic time did the deformation affect the rocks that had been deposited in the miogeocline and the platform. The thick sedimentary prism of the miogeo-

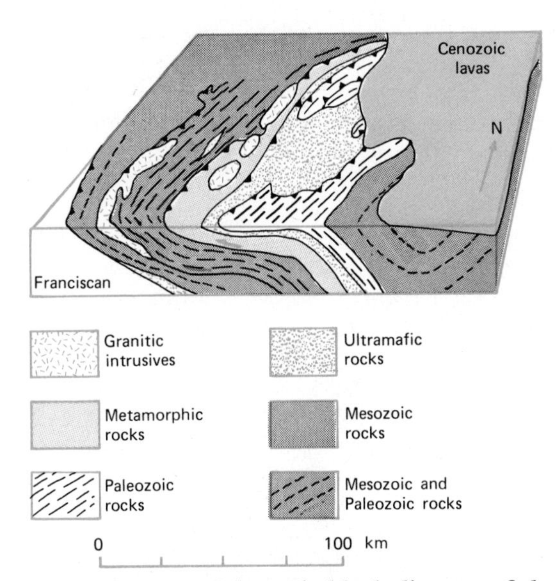

FIGURE 15–6 **Schematic block diagram of the main structures of the Klamath Mountains at the border of Oregon and California. (After G. A. Davis; courtesy of the Geological Society of America.)**

cline yielded to the forces directed eastward from the geosyncline by shearing from the Proterozoic and Archean basement and breaking along planes of weakness in the sedimentary succession parallel to the bedding. This style of deformation results in mildly deformed fault blocks separated by horizontal or low-angle thrust faults along which they have moved eastward for distances of tens of kilometers. The French word "décollement" (unsticking) has been used to describe this kind of deformation. Along décollement thrusts the upper fault plates slide on bedding planes lubricated by weak shales or evaporites. The mechanism of such faulting is difficult to explain, for calculations of the friction to be expected along such a fault plane and the strength of the upper plate of sedimentary strata demonstrate that laterally directed force applied to the back of a plate tens of kilometers wide and several kilometers thick would break it into many segments rather than transport it as a whole over the beds below. Geologists have, therefore, sought a force, such as gravity, that would act on every particle throughout the upper plate and not just on its western end. If the thrust plates were moved by gravity, then the western end must have been lifted by the orogenic forces to produce a slope down which the crustal slabs several kilometers in all dimensions slid for tens of kilometers. Given the proper lubrication, some geologists believe that slopes of as little as 2° would have been steep enough. Many structural geologists do not agree that this mechanism of "gravity tectonics" is adequate to explain the structure of the zone of low-angle thrusts that characterizes the cratonic margins of many fold belts; further research on this problem is necessary before the puzzle of such faults is resolved.

The zone of major thrust faulting that follows the Paleozoic miogeoclinal belt from southern California through southern Nevada and northwestern Utah to Idaho (Fig. 15–7) has been called the Sevier thrust belt (see also Fig. 14–8). Displacement along one of these thrusts has been estimated to be at least 32 km, and similar displacements have been postulated along others. The total crustal shortening across the belt has been estimated by Richard Armstrong to be between 65 and 100 km. The development of these thrusts took place in the interval between Late Jurassic and Late Cretaceous time, at the same time as the deposition of the Fransiscan complex and the intrusion of the Sierra Nevada igneous rocks.

Similar structures characterize the Canadian Rocky Mountains of Alberta and northwestern British Columbia. The ranges of the Canadian Rockies are composed of resistant limestones and quartzites of the Paleozoic miogeoclinal succession. Each range is a fault block bounded on the east by a steeply westward-dipping thrust fault along which it has moved eastward (Fig. 15–8). The steeply-dipping thrust faults flatten as they pass downward, approaching parallelism with the bedding, and passing eventually into a zone of décollement in which the whole sedimentary wedge sheared off the basement. Along the Lewis thrust the Proterozoic rocks of the Belt Group have been thrust as much as 65 km eastward over contorted Cretaceous shales to form

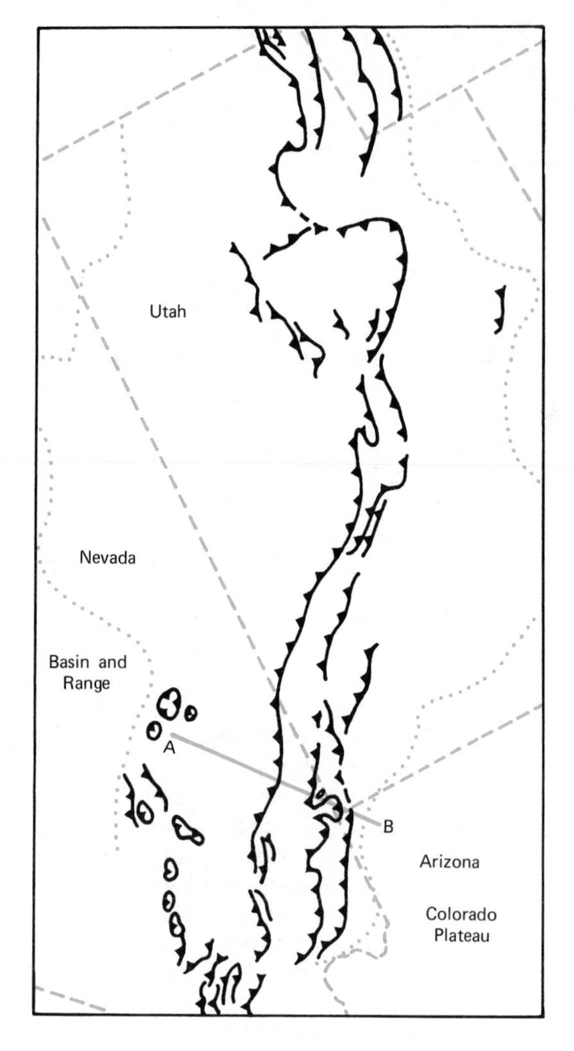

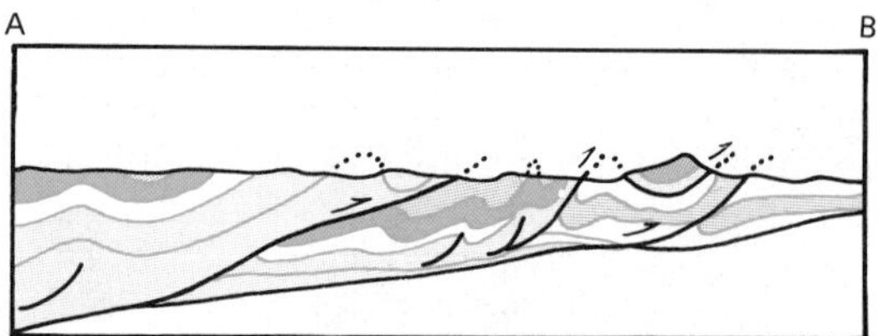

FIGURE 15–7 **Major thrusts of the Sevier thrust belt, Nevada and Utah. The cross-section has a vertical exaggeration factor of 1.5 and is reconstructed along the line A–B on the map. (After R. L. Armstrong; courtesy of the Geological Society of America.)**

the Lewis Range in Glacier National Park and Waterton Lakes National Park. The style of deformation can be followed northward up the eastern side of the Cordillera into the Brooks Range of Alaska.

The limestone ranges of the Canadian Rockies lie west of a narrow foothills belt in which Mesozoic and Lower Cenozoic sandstones and shales are sliced into many thin segments by thrust faults that branch from major décollement thrusts at depth (Fig. 15–8).

Although the style of deformation is similar to that of the Sevier orogenic belt, evidence for the age of deformation in the Canadian Rockies is less satisfactory. Raymond Price and Eric Mountjoy suggest that the stratigraphic and structural evidence best fit a model in which the first thrusts formed as early as late Jurassic time on the western margin of the Rockies, due to the spreading and uplift of the metamorphic–igneous core of the central Cordilleran zone, and during Cretaceous time the miogeoclinal sedimentary prism failed along progressively more easterly fault planes. The last movements along the foothills belt displaced strata of Paleocene age and probably took place in the Eocene Epoch. The deformation of the Canadian Rockies is pictured as taking place over an interval of 100 million years, the same interval in which intrusion and orogeny affected the western and central Cordillera.

Deformation of the Platform

In Canada the deformation associated with the Cordilleran orogeny dies out into the flat-lying platform rocks near the eastern margin of the Paleozoic miogeocline. South of Montana, where the Cordillera is widest, the zone of deformation extends across the edge of the craton to central Colorado and eastern Wyoming. In central Wyoming and southern Montana the style of deformation is much different from the décollement thrusting that characterizes the miogeoclinal zone (Fig. 15–9). The ranges have central cores of Proterozoic and Archean crystalline rocks surrounded by Paleozoic and Mesozoic sedimentary rocks that dip away on all sides, defining elliptical domes. The ranges are separated by broad basins floored by essentially undeformed early Cenozoic sediments. The margin of the uplifted pre-Paleozoic cores of the ranges may be an unconformity in places, in other places it may be a high-angle thrust fault, in still others a normal fault. The movement along the faults suggests that vertical uplift, rather than horizontal compression, was the dominant force in the formation of the ranges.

The Rockies of central Colorado and New Mexico are separated from the Sevier zone of Cretaceous thrusts by the Colorado Plateau. The deformation of this segment of the Rockies was greatly influenced by the late Paleozoic basement uplifts of the Colorado Mountains. During early Mesozoic time these uplifts, such as the Uncompahgre and Front Range (Chapter 13), were lowered in relief by erosion and progressively buried in their own debris. The Front Range Uplift was rejuvenated in Late Cretaceous and early Cenozoic time and eventually elevated to a height of more than 4,300 m to form the highest peaks in the Rocky Mountains. The Front Range structure, which had been a simple arch in late Paleozoic time, was uplifted into two parallel ranges with a valley comprising North, Middle, and South Parks between them (Fig. 15–10). Although high-angle thrust faults are present in the Front Range, these may be related to dominantly upward forces. Early Cenozoic vertical uplift in the Sangre de Cristo Range (Fig. 16–21) has been estimated at 3 to 5 km. On the east side of the Front Range Uplift, Paleozoic and Mesozoic strata are tilted into a vertical position and eroded to form the spectacular scenery of the "Garden of the Gods." On the western side of the Front Range and the Sangre de Cristo Range the structure is more complex and eastward-dipping thrust faults cut basement and Paleozoic rocks. Some geologists have viewed this thrusting as evidence of lateral compression, others as adjustments of surface strata to deforming forces that were fundamentally vertical.

The subcircular Colorado Plateau is underlain by gently warped and faulted Paleozoic and Mesozoic sedimentary rocks which appear to have acted as a rigid block during the Cordilleran orogeny.

The structure of the Cordillera changes near the Mexican border along a line that has been called the Texas lineament. Along this line, the origin of which is uncertain, the zone of thrusting characteristic of the miogeocline appears to have been shifted eastward several hundred kilometers to the Sierra Madre Oriental of northern Mexico. This range consists of Mesozoic limestones and shales broadly folded and cut by low-angle thrust faults. The western side of the Cordillera at this latitude is covered by thick Cenozoic lava flows that make up the Sierra Madre Occidental.

Laramide and Nevadan Orogenies

Within the last 20 years our understanding of the timing of Cordilleran orogeny has changed radically. Twenty years ago geologists generally agreed that deformation and intrusion of the western and central Cordillera was accomplished in Late Jurassic time in an orogeny known as the Nevadan. The deformation of the eastern Cordillera, the Rocky Mountains, was thought to have been confined to the Late Cretaceous and early Cenozoic time and these movements were grouped as the Laramide orogeny (after the Laramie Range of southern Wyoming). We now realize that deformation of the trench–volcanic arc deposits occurred over a long interval, including much of Jurassic and Cretaceous time, and that deformation of the miogeoclinal prism continued through the Cretaceous Period into the Cenozoic Era. The terms "Laramide" and "Nevadan" have, therefore, lost much of their meaning.

CAUSES OF CORDILLERAN DEFORMATION

According to our current understanding of orogenic processes the Cordilleran deformation of the western side of the North American plate was caused by the stress placed upon it by the underthrusting of the Pacific plate along an east-dipping subduction zone. The evidence for the presence of

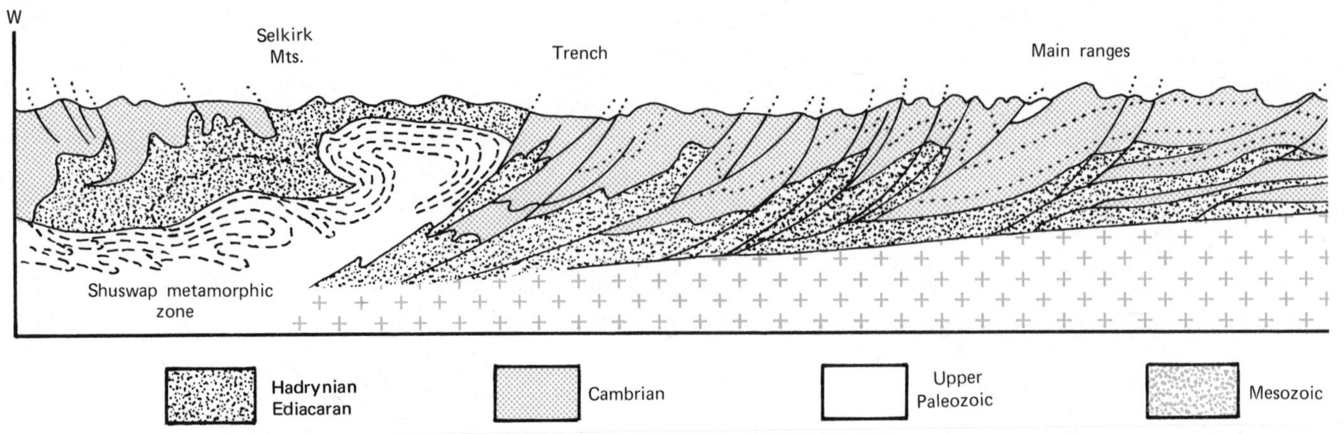

FIGURE 15–8 **Cross-section of the eastern Canadian Cordillera in southern British Columbia and Alberta. (After R. A. Price and E. W. Mountjoy; courtesy of the Geological Association of Canada.)**

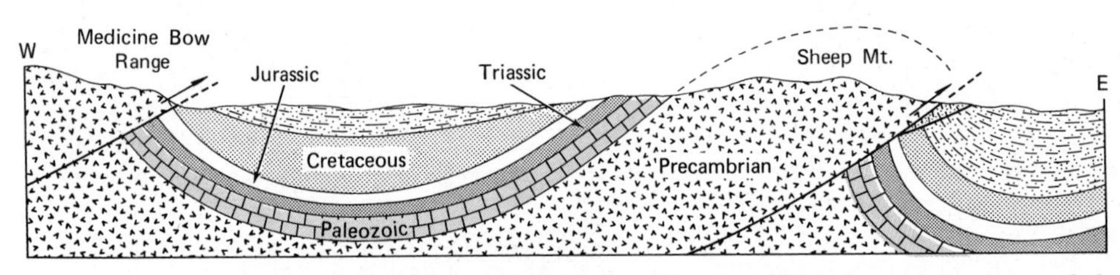

FIGURE 15–9 **Asymmetrical broken anticlines with crystalline basement cores of the Medicine Bow Range and Sheep Mountain in southeastern Wyoming. (After A. J. Eardley.)**

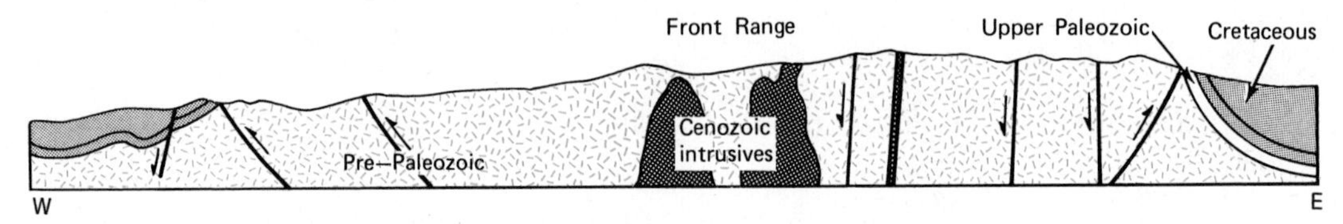

FIGURE 15–10 **Cross-section of the Front Range of Colorado. (After A. J. Eardley.)**

this zone during late Mesozoic time in approximately the position of the present coastline is provided by deformed trench sediments, blue schist facies metamorphism, and abundant calc-alkaline volcanism. In the last chapter the evidence was reviewed for the presence of such a subduction zone along the plate boundary during much of Paleozoic and early Mesozoic time when no orogeny of the magnitude of the Cordilleran occurred. Several factors could have caused the compressive stress generated by the convergent plate motions to be transmitted to the edge of the continent more strongly in late Mesozoic time. The rate of subduction may have quickened. Some "hard to swallow" feature of the oceanic crust, such as an inactive island arc or a line of sea mounts, may have clogged up the subduction zone. The westward movement of the North American plate as the Atlantic rift zone formed may have increased

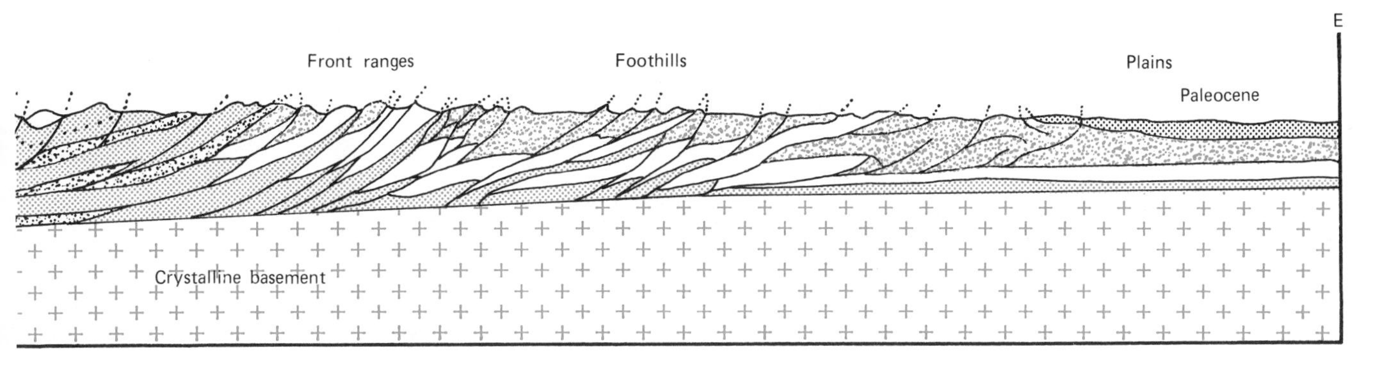

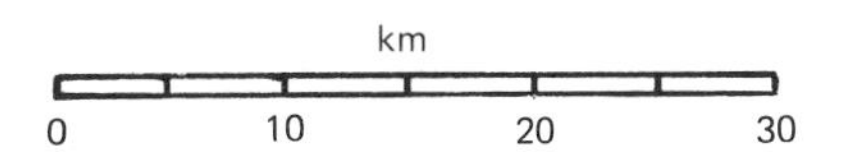

the relative rate of convergence of the two plates along the west coast subduction zone. Peter Coney has suggested that the major orogenic interval in the Cordillera between 180 and 80 million years ago (Early Jurassic to Late Cretaceous) was caused by the separation of North America from Africa (see Fig. 15–11). He believes that the change in the style of the deformation at the end of the Cretaceous is due to a change in the motion of the Americas plates as the North Atlantic Ocean between North America and Europe began to open.

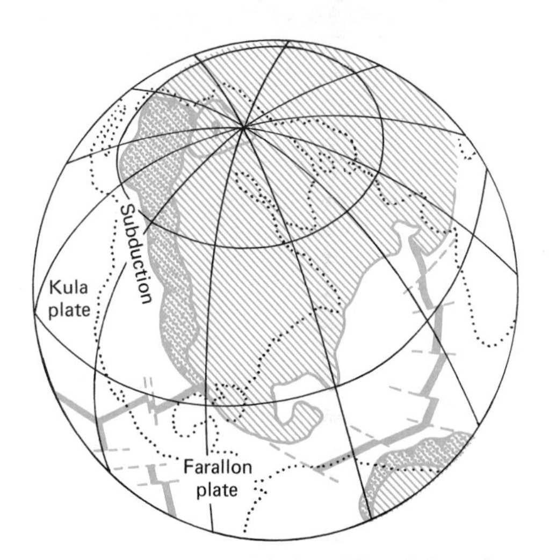

FIGURE 15–11 **Relationship of the plate boundaries of the North American plate in Late Cretaceous time, about 80 million years ago. (After P. J. Coney; courtesy of the American Journal of Science.)**

The pattern of magnetic anomalies on the ocean floor shows how much crust was added at a spreading zone between reversals in the earth's polarity. The anomalies have been numbered and dated in years by isotopic methods. By progressively subtracting the segments of the ocean floor identified by magnetic anomalies, and closing the spreading centers and restoring the crust swallowed in subduction zones, the geometric distribution of spreading centers, transform faults, and subduction zones of the past can be reconstructed. Such reconstructions for the eastern Pacific are hampered by the fact that nearly all of the lithosphere generated east of the East Pacific spreading center has been consumed in the Cordilleran subduction zone. This part of the lithosphere that has been subducted has been called the Farallon plate. Another plate, the Kula plate, has since disappeared down the Aleutian subduction zone (Fig. 15–11). We shall follow the development of the Pacific plate boundaries during the last 80 million years and their influence on the geology of the continent in the next chapter.

JURASSIC ROCKS OF THE WESTERN PLATFORM

The deformational history of a mountain belt is revealed in the structure of its ranges, the minerals of its metamorphic rocks, and the sequence of intrusion of its batholiths. The elevation of the mountains may be recorded in the sediments deposited during and after orogeny in intermontane basins, but this record is commonly interrupted by unconformities, obscured by faulting and folding, and quickly attacked by erosion. The sedimentary strata that accumulate on the flanks of a mountain chain may be thousands of meters thick near the mountains, but thin to a feather edge away from them and are, therefore, appropriately referred to as a **clastic wedge.** They preserve the best record of the elevation of the mountain chain.

During most of the Paleozoic and Triassic the western side of the platform received little clastic sediment. The shale and sandstone that reached the platform was eroded from the crystalline rocks of the Canadian Shield, from sedimentary rocks exposed along the edges of the shield, and from cratonic uplifts of basement rocks such as the Colorado Mountains. In Late Jurassic time a complete reversal of this pattern of sediment supply was effected by the Cordilleran orogeny and the building of the Cordilleran clastic wedge.

The Zuni sequence begins with the transgression of the sea from the north and through gaps in the islands of the Cordillera to the west across the desert sandstones of the Navajo and equivalent formations, which had been deposited at the close of the Absaroka sequence. Until Late Jurassic time the pattern of facies on the western platform was much like that of the Paleozoic; limestones accumulated along the miogeocline, fine clastics were introduced from the shield, and evaporites collected in areas like the Western Canada sedimentary basin.

Ammonoids and Belemnites

Many of the principles of correlating and dating sedimentary strata by means of fossils were developed through the study of marine Jurassic rocks of northern Europe. In the shaly rocks of Europe and similar rocks in the Jurassic System of the western platform, the most important guide fossils are the ammonoid cephalopods. In late Paleozoic time the ammonoids were largely goniatites, a group that died out at the end of the era. In Triassic time the ammonoids with sutures of the ceratite type underwent an explosive radiation, becoming abundant and diverse (Fig. 15–12). At the end of the Triassic Period some drastic change in the marine environment resulted in the near extinction of the whole ammonoid group; only one family survived to give rise to the ammonoids of the Jurassic and Cretaceous periods. These ammonoids had a suture line that was complexly patterned in both the forward- and backward-facing folds and is called the **ammonite suture.** The term "ammonite" should strictly be applied only to these middle and late Mesozoic ammonoids. They were swimming cephalopods, independent of sedimentary facies on the sea floor, and were carried by currents to all parts of the world (Fig. 15–13). The rapidity of their dispersal and evolution make them ideal fossils for correlating beds. Intensive collecting of ammonites from some European sedimentary successions has made

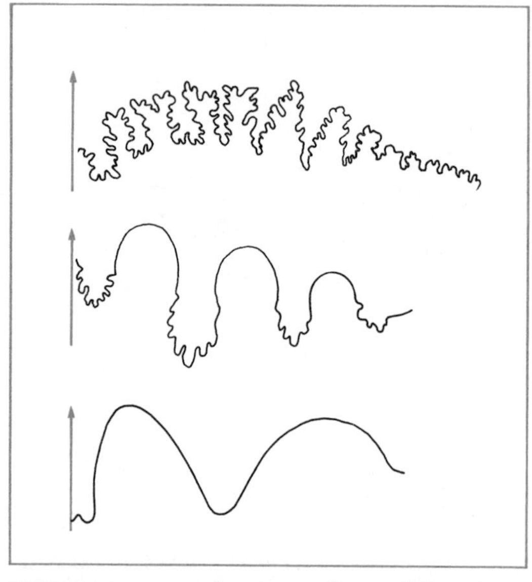

FIGURE 15–12 **The suture lines of three types of ammonoids, seen as if they were removed from the curved shell. The arrow points forward on the outside edge of the last whorl of the shell. *Top:* The ammonite *Pompeckjites* (Triassic). *Middle:* The ceratite *Bulogites* (Triassic). *Bottom:* The goniatite *Muensteroceras* (Mississippian).**

FIGURE 15–13 **Late Cretaceous marine habitat group. On the right, the large ammonite *Placenticeras;* in the foreground, the straight ammonite *Baculites.* Above *Placenticeras* are two squid-like belemnites. Although the ammonites are reconstructed as being on the bottom, they were normally swimmers. (Courtesy of the University of Michigan Museum.)**

possible the division of the strata into zones only a few centimeters thick. By means of ammonite ranges the Jurassic System is divided into 58 stages that can be recognized on a worldwide basis. Most ammonites were coiled in one plane like a watch spring. However, some grew into straight, hooked, screw-like, and irregular forms, particularly in the Cretaceous Period, presumably as adaptations to various swimming habits.

Among other fossils commonly found in Jurassic and Cretaceous shales and limestones are belemnites. The fossils of belemnites are sharp-pointed rods composed of solid calcite of the shape and size of cigars (Fig. A–18). From specimens in which the impressions of the soft parts of the animal have been preserved, paleontologists have learned that the rod formed part of the internal skeleton of an elongate, squid-like cephalopod (Fig. 15–13). Because they were fast swimmers and quickly dispersed throughout the world, the belemnites, like the ammonites, make excellent guide fossils for Mesozoic correlation.

Sediments from the Cordillera

In rocks of the Oxfordian Stage, the first stage of four constituting the Upper Jurassic Series, there is strong evidence that highlands in the Cordilleran geosyncline were acting as suppliers of sediment to the platform seas. The distribution of highlands shown in the facies map (Fig. 15–14) is

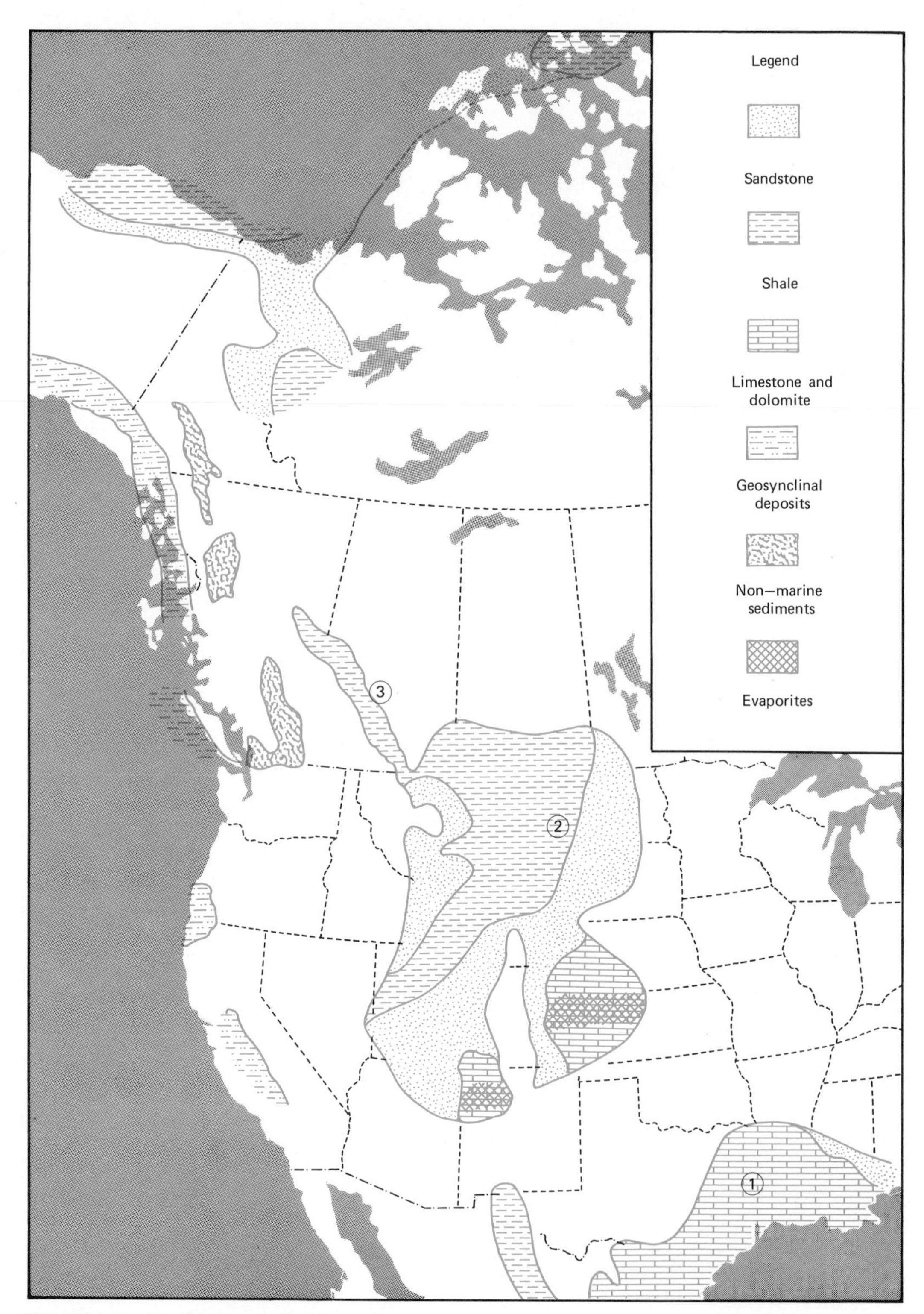

FIGURE 15–14 **Facies of the Upper Jurassic Oxfordian Stage in western North America. 1. Smackover Limestone. 2. Sundance Formation. 3. Fernie Shale. Much of the central Cordillera was uplifted at this time.**

partly conjectural, partly the result of analyzing the directions of sediment distribution in Jurassic sediments scattered throughout the Cordillera. At the beginning of Late Jurassic time sediments brought to the sea were deposited as marine sandstones and passed eastward into the shales accumulating on the platform. Gradually, the rate of sediment supply reached and surpassed the rate of subsidence of the miogeocline and the platform margin was silted up by streams, forming an extensive alluvial apron along the growing and advancing Cordilleran Mountain front. These sediments are entirely different from the pure quartz sandstones and shales derived from the Canadian Shield. The sandstones are rich in clay, mica, and unstable minerals, and may be either graywackes or arkoses depending on whether they were supplied by erosion of a metamorphic-volcanic terrane or by the erosion of a terrane where granitic batholiths had been unroofed. Sediments previously deposited in the geosyncline also contributed to the clastic wedge as they were deformed and uplifted in the orogeny.

Morrison Formation

By latest Jurassic time the whole western side of the platform in the United States had been silted up by sediment from the Cordilleran highlands. The brightly colored clays, silts, and fine sands laid down in the broad alluvial plain that developed east of the mountains today constitute the Morrison Formation. These beds contain a rich fossil biota that includes the largest dinosaurs (Fig. 15–15), primitive mammals, many plants, and freshwater invertebrates. Most of

FIGURE 15–15 ***Apatosaurus*, a Jurassic sauropod dinosaur from North America similar in many respects to *Brontosaurus*. [Courtesy of the British Museum (Natural History.)]**

the large specimens of dinosaurs displayed in American museums were collected from this formation by O. C. Marsh of Yale University and E. D. Cope of the Philadelphia Academy of Sciences. The bitter competition that grew between these two men in the collection and description of new specimens supplied geological historians with many anecdotes and enriched universities and museums with many specimens that would otherwise be still in the ground.

Sifting and screening of the fine sediments of the Morrison Formation has revealed in a few localities the tiny teeth and jaws of Mesozoic mammals. Teeth are the commonest of mammalian fossils because they are far more resistant to weathering than bone. Many fossil mammals are known only from their teeth. The molar teeth, in particular, are important to our understanding of the food eaten by animals and have been widely used in the classification of mammalian groups.

MESOZOIC MAMMALS

Pelycosaurs and therapsids occupied a broad spectrum of environments as the dominant reptilian carnivores and herbivores of Permian and Triassic time. The success of dinosaurs in the Mesozoic Era restricted the early mammals, their descendants, to a very minor role in the environment. All were small, none more than the size of a domestic cat, and most of the size of shrews and mice. They were probably noctural and of secretive habits. Some therapsid lineages may have developed this lifestyle by the Middle Triassic, in the course of adaptation to hunting insects.

Insulation

The small size of primitive mammals is related to the attainment of many other mammalian features. Therapsids, as judged by their posture, dentition, and possession of a secondary palate certainly has a high and reasonably constant body temperature. Retention of metabolic heat would have been facilitated by their relatively great body size. There is no direct evidence for the presence or absence of further mammalian features. The mammals of the Late Triassic and Early Jurassic (Fig. 15–16) are so small (30 to 40 g in weight) and have such a high surface-to-volume ratio that maintenance of a high body temperature without insulation would have been impossible. The earliest mammals, and their immediate predecessors among the therapsids, must have had a heavy coating of hair. The origin of this uniquely mammalian tissue is hard to explain. Isolated hairs could have evolved initially as touch-sensation structures in small forms that were nocturnal and spent much time in burrows or crevices where the sense of sight would be of little use. The whiskers of rodents and small carnivores, such as the domestic cat, play a similar sensory role in modern forms. Such structures over the whole body surface also would trap the surface air and insulate the body.

Teeth

A fundamentally mammalian tooth pattern had evolved by Late Triassic time. No true mammals have more than two generations of teeth (deciduous and permanent). In most groups the teeth are clearly differentiated into incisors, canines, premolars, and molars. The molars and premolars typically have a specific shape and occlusion between upper and lower teeth so that the food is efficiently cut or crushed prior to swallowing. This is in strong contrast to the typical reptilian pattern in which a large number of essentially similar peg-like teeth in both jaws are replaced many times in the life of an individual; up to half of the tooth positions may be empty at any one time. Specific relationships between upper and lower teeth could not evolve in the reptiles where such continuous replacement takes place. Food is swallowed in large chunks and takes a long time to digest (one meal may last a month for a snake). Advanced therapsids evolved quite complicated cheek teeth, but all apparently retained a basically reptilian habit of tooth replacement. They were generally of fairly large size and long life, and using their teeth in a reptilian fash-

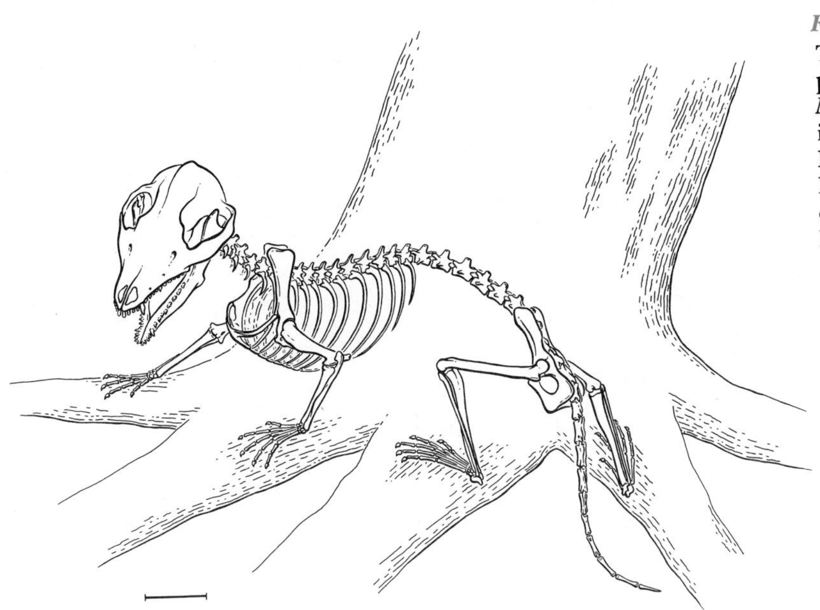

FIGURE 15–16 **Upper Triassic mammal modeled principally after *Megazostrodon*. Scale bar is 1 cm long. (From F. A. Jenkins and F. R. Parrington; courtesy of the Royal Society, London.)**

ion needed several sets to last throughout their life. The small size of the earliest mammals was probably correlated with a short life span. They would not have required, nor would have had time to produce, more than two sets of teeth, in which case a specific pattern of occlusion between upper and lower teeth could develop. Although Mesozoic mammals are very rarely found, despite an intensive search for their remains, several groups have been recognized on the basis of their distinctive molar patterns indicative of specialization to different food sources or methods of mastication. Two major lineages are currently recognized. The forms which show the occlusion of most of the tooth surface of the uppers on the lowers include a carnivorous group, the triconodonts (Fig. 15–17). They apparently also include a number of herbivorous lines with crushing dentition, such as the docodonts (a minor group), and possibly the multituberculates, an assemblage that had the general appearance of rodents and achieved the largest size of any Mesozoic mammals. The most primitive of living mammals, the egg-laying Australian monotremes, the echidna and platypus, may also have evolved from this line, but neither has a significant fossil record nor any teeth in the adult, which makes their phylogenetic position difficult to evaluate. They separated from other mammalian groups early in mammalian history, probably no later than Late Triassic time.

The second lineage in which the upper and lower molars tend to alternate with each other gave rise to all higher mammals. The molars in these forms are basically triangular in plan, with the apex of the triangle of the upper molar pointed inward toward the tongue and the apex of the lower molar facing outward toward the cheek. The front and back edges of the triangles form cutting or shearing surfaces as the teeth occlude. In the earliest Jurassic the angle at the apex is obtuse and the triangular nature is not at all evident, but in subsequent forms the angle becomes more acute and the tooth more obviously triangular. Mammals from the Jurassic with simple triangular teeth are called symmetrodonts. These were a small group of mouse- to cat-sized insectivores or carnivores. Late in the Jurassic the triangle was augmented in advanced groups by a posterior shelf, termed a "talonid" or "heel," on

FIGURE 15–17 **The skull of a tricondont mammal, still partly enclosed in matrix, from the Lower Cretaceous beds of Montana. (Courtesy of A. W. Crompton and F. A. Jenkins.)**

the lower tooth, into which fitted a newly developed cusp on the upper molar. These changes added a crushing surface to the shearing edges of the more primitive teeth. Mammals which show the early development of the talonid, known as pantotheres, are found in middle Late Jurassic to Late Cretaceous beds. This group remains relatively poorly known, but is of great importance since this primitive tooth pattern certainly gave rise to that of modern marsupial and placental mammals. These groups were clearly distinguished from one another by the Late Cretaceous, and less distinct ancestors are recognizable in Lower Cretaceous beds.

Reproduction

In addition to the development of the tooth pattern that was basic to all higher mammals, some among the Mesozoic mammals presumably initiated live birth and the suckling of the young. Although evidence from fossils is missing, primitive mammary glands may have evolved soon after the origin of hair. The development of hair as insulation requires simultaneous development of special compensatory methods of heat loss, such as sweating. Mammary glands are similar to sweat glands, and may have evolved initially to provide a warm moist area for the young at the surface of the mother's body. The nutritive functions apparently developed later.

Live birth, characteristic of the placentals, apparently developed late in mammalian evolution. The tiny Mesozoic mammals would have been so subject to heat loss that their young could not have maintained a high body temperature on their own but would have been dependent on the warmth of their parents at birth. This implies birth at a very immature, essentially embryonic stage, like the young of most marsupials. Only later in Cretaceous time, with the achievement of large size and suppression of ovulation during pregnancy, did the placental pattern develop.

Brain Size

Mammals are usually considered to have a much larger brain size in relation to body weight than have the reptiles. The brain of the therapsids is relatively little larger than that of other reptiles, and the Mesozoic mammals and living marsupials are only marginally superior to reptiles in the volume of the brain. Even among the placentals the brain remains small as late as early Cenozoic time. What did occur within the therapsids was the marked reorganization of the bony brain case. The bones that form the outside of the skull in primitive reptiles were modified for more efficient jaw musculature so that they grew down and over the narrow brain case, forming a strong external covering. The archaic internal brain case was then lost and much later, in the middle Cenozoic, the new external structure expanded as selection placed a greater premium on the development of the higher functions of the brain.

CRETACEOUS SYSTEM OF THE WESTERN PLATFORM

The greatest volume of sediment eroded from the rising Cordilleran Mountains is contained in the Cretaceous System (Fig. 15–18). The system in western North America can be divided into stages and zones by means of fossils, or into formations and facies by means of lithology. Rocks of this age in Europe are divided into zones on the basis of ammonites; these zones are grouped into stages, nine of which are commonly recognized as useful for worldwide correlation. Three of these stages—the Neocomian (oldest), Aptian, and Albian—are divisions of the Lower Cretaceous Series. The other six stages—the Cenomanian, Turonian, Coniacian, Santonian, Campanian, and Maestrichtian (youngest)—comprise the Upper Cretaceous Series. Because the system in the type section along the English Channel is divided into an upper chalk and a lower sandstone facies, the Cretaceous System, unlike most other systems, has no middle series. Comparison of ammonites found in the western states and provinces with those of Europe allows the paleontologist to make precise age determinations and correlations in Cretaceous rocks.

Cretaceous Facies

The Cretaceous sedimentary rocks of the Rockies and plains can also be described in terms of facies. A sedimentary succession as extensive and thick as this one contains a multitude of lithologies, but they can be grouped into four contemporaneously deposited facies: 1. marine limestones deposited in relatively deep water, 2. marine shale deposited in shallow water, 3. light-colored sandstone deposited at the shoreline, 4. gray shales and lenticular sandstones deposited on land in coastal swamps. As the shoreline moved back and forth across the edge of the platform, these facies belts followed, producing a complexly interfingering set of marine and nonmarine formations.

The limestone facies was deposited at times when little detritus reached the epeiric sea. Two limestone formations divide the shales of the Cretaceous System—the Niobrara Chalk above and the Greenhorn limestone below (Fig. 15–19). The Niobrara Chalk is known for the large, beautifully preserved vertebrate fossils that have been found in it. These include birds, swimming and flying reptiles, turtles, and fish. The unbroken and articulated state in which the specimens have been preserved indicates that the waters of the Niobrara Sea were untroubled by waves and currents. The chalk covers a large area of South Dakota, Nebraska, Kansas, western Oklahoma, and Texas. Northward into Canada the Niobrara and Greenhorn limestones lens out, but their positions are indicated in the thick shales by two zones of tiny, white calcareous spheres known in subsurface correlation as the "first and second white specks." The clean sandstone shoreline facies forms a narrow facies belt between the deltaic, terrestrial sandstones and the marine shales. The Dakota Sandstone, a transgressive deposit at the base of the Upper Cretaceous Series, is representative of this facies and forms prominent ridges or "hogbacks" where it is tilted up along the east side of the Front Range in Colorado (Fig. 15–20).

Oscillations of the shoreline were caused by the building out of the land in the form of deltas at times of rapid supply of sediments and the sudden transgression of the sea over these deltaic sands, perhaps due to sudden subsidence. The form typical of these regressive sandstone tongues, which project into the marine shale facies, is illustrated in Figs. 15–21 and 15–22. Bedding planes in the sandstone tongue are not parallel to its upper surface, but slope gently seaward like the foreset beds of a delta. On the alluvium, coastal swamps formed and the trees left a carbonaceous residue as thin beds of coal.

Each episode of delta-building closed

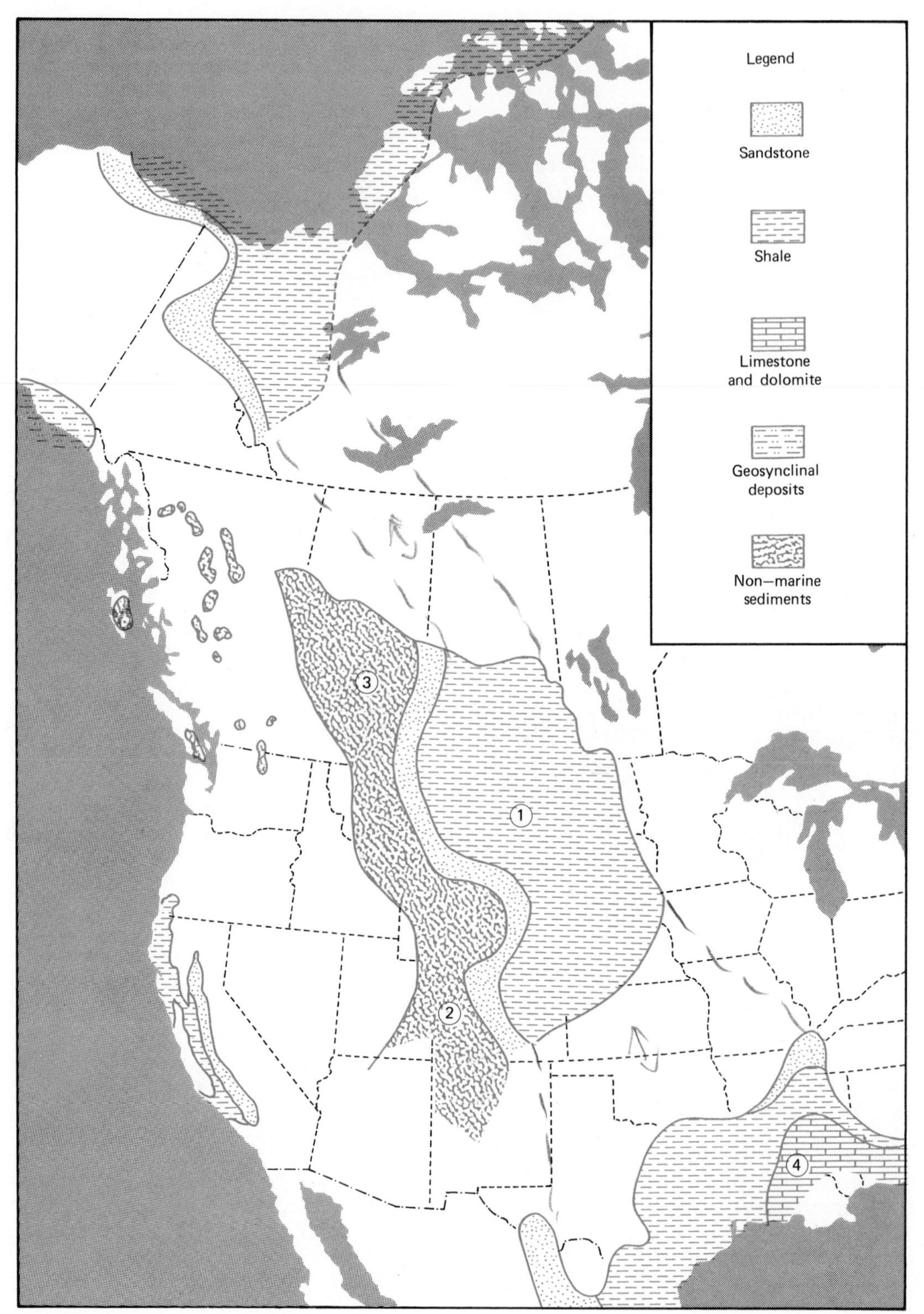

Legend
Sandstone
Shale
Limestone and dolomite
Geosynclinal deposits
Non–marine sediments
1
2
3
4

U. Cret.

FIGURE 15–18 **Facies of the Upper Cretaceous Montana Group and its correlatives in western North America. Most of the central and western Cordillera had been uplifted by this time and deposition was confined there to small intermontane basins. 1. Pierre Shale. 2. Mesaverde Sandstone. 3. Belly River Sandstone. 4. Taylor Limestone.**

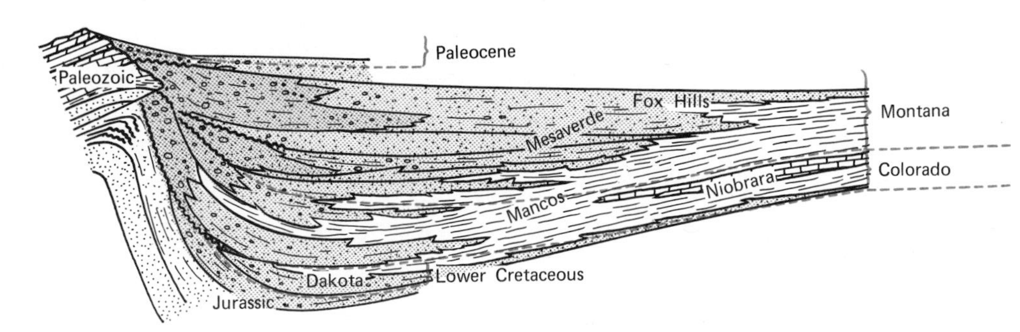

FIGURE 15–19 **Facies diagram of the late Mesozoic clastic wedge deposits of central Utah showing the sediments transported eastward from the rising Cordilleran highlands. (After R. L. Armstrong; courtesy of the Geological Society of America.)**

FIGURE 15–20 **"Hogbacks" along the Front Range of Colorado formed by the Lower Cretaceous Dakota Sandstone. (Courtesy J. H. Rathbone and the Rocky Mountain Association of Geologists.)**

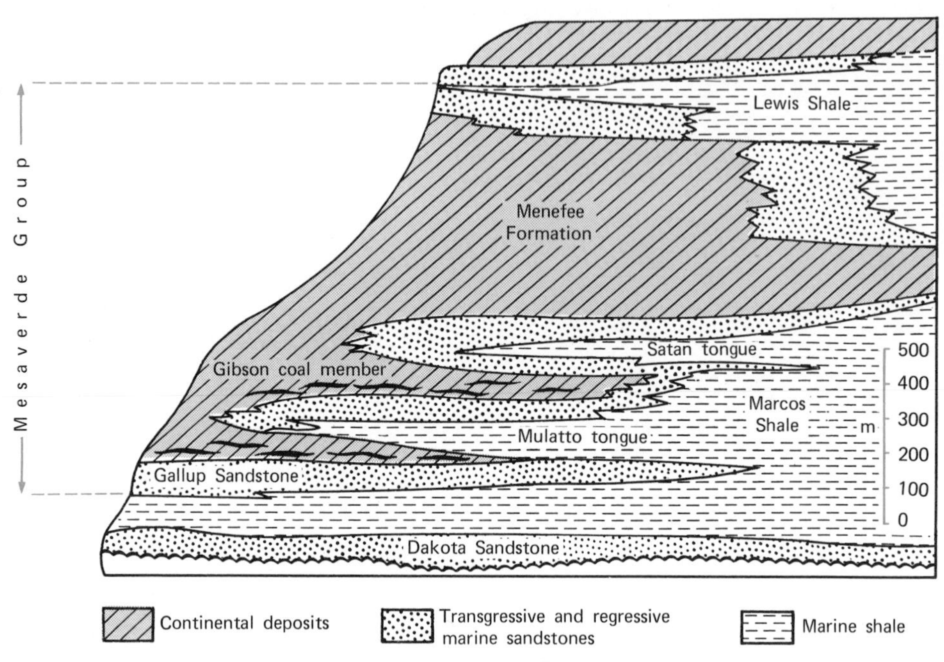

FIGURE 15–21 **Cross-section of the Upper Cretaceous facies of the San Juan Basin, Colorado. Black lenses represent coal. (After F. F. Sabins, Jr.; courtesy of the American Association of Petroleum Geologists.)**

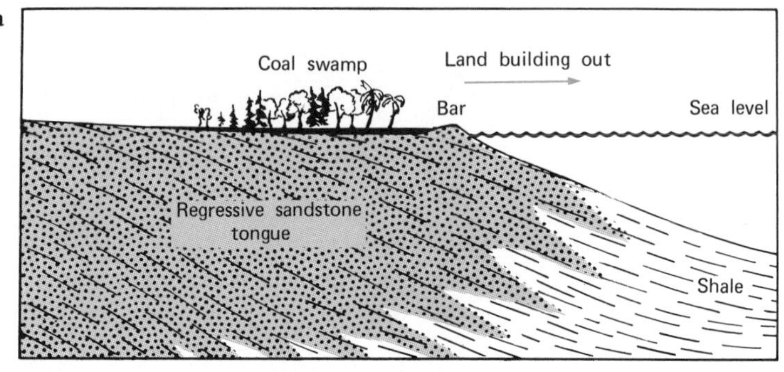

FIGURE 15–22 **Geometry of a regressive sandstone tongue showing the irregular, gradational lower border and the position of the coal deposition.**

with the sudden transgression of the sea over a swamp, killing the vegetation and setting the stage for the beginning of a new regressive cycle. These cyclical regressive deposits and their coal beds are comparable to the Pennsylvanian cyclothems.

Mesozoic Vegetation

The plants that contributed organic matter to the formation of coals in the regressive sandstones of the Cretaceous System were entirely different from those that formed the Pennsylvanian coal swamps. The coal swamp flora died out through Permian and Triassic time, perhaps due to increased worldwide aridity during this interval. The early Mesozoic flora was dominated by the true conifers, the cycads, and the cycadeoids. The cycads are still living and are known as sago palms, but the cycadeoids are extinct. Many of the cycadeoids grew from low, almost spherical trunks close to the ground and had palm-like leaves (Fig.

15–23). The cycads bear male and female cones, but the cycadeoids bore the first complex flowers.

The angiosperms make up the greater part of the visible modern flora and are familiar to us as the flowering plants. Fossil leaves of angiosperms appear suddenly in the Lower Cretaceous deposits of the western platform in great diversity and abundance. The first angiosperm fossils are remarkably modern in aspect and include such genera as the birch, elm, willow, etc. (Fig. 15–24). Because they appear all over the world at this time, paleontologists have suggested they were confined to the highlands during their early evolution, and only when conditions become appropriate for them to migrate to the lowlands were they preserved in the paleontological record in abundance. Paleobotanists are not agreed on the ancestors of the angiosperms, but the cycadeoids appear to be the most likely candidates.

Regressive Sandstones

In the plains region the Upper Cretaceous Series can be divided into four sets of beds representing transgression and four sets representing regression. The regressive stratigraphic units of sandstone and shale are hundreds to thousands of meters thick and project eastward into the marine shale that accumulated on the cratonic shelf during much of the period.

On a still larger scale, the whole of the

FIGURE 15–23 **Early Mesozoic landscape showing some aspects of the flora. In the marshy foreground are several kinds of *Calamites*, which had persisted from the Paleozoic Era. Ferns abound, and cycadeoids on bulbous and treelike stems were becoming abundant. In the distance are conifers. (From J. Augusta and Z. Burian, *Prehistoric Animals*. London: Spring Books, 1956.)**

FIGURE 15–24 **Leaf of *Liquidambar*, an angiosperm from Cretaceous rocks of western North America. (Courtesy of the Smithsonian Institution.)**

Cretaceous System is a regressive succession. The pattern of facies changes in cross-section is illustrated in Fig. 15–25. With many oscillations the shoreline was driven eastward by the increasing supply of clastics from the west until, by the end of the period, the whole epeiric sea had been converted into an alluvial plain 800 km at its greatest width. Early in the epoch the late Cretaceous epeiric sea had extended from Utah to the Great Lakes and from the Arctic Ocean to the Gulf of Mexico (Fig. 15–18). Its eastward extent is established by small outliers of Cretaceous strata resting on the Canadian Shield in the Mesabi district of Minnesota. By the end of the period, the sandstones of the Fox Hills, Hell Creek, and Lance formations and their equivalents had spread over the shale facies as far as the Dakotas. The Fox Hills Sandstone is the marine part of the regressive sandstone unit; the overlying Lance and Hell Creek formations are non-marine. The non-marine sandstones of the clastic wedge continued to accumulate into the Paleocene Epoch with little change, and the lack of a break in sedimentation at the era boundary caused considerable uncertainty in locating the beginning of the Cenozoic beds in the stratigraphic succession.

The load of sediments derived from the Cordilleran Mountains depressed the edge of the craton into a series of basins stretching from the Arctic Ocean to the Gulf of Mexico. Cretaceous parts of this wedge reach thicknesses of over 6000 m in western Wyoming, where sands were deposited close to the Cretaceous mountain front (Fig. 15–26). The wedge does not thin regularly eastward, but thickens locally into basins along the edge of the platform.

The regressive sandstone tongues of the Cretaceous clastic wedge are economically important, not only because they are commonly associated with coal deposits, but also because they serve as reservoirs for oil and gas. Large gas fields in the San Juan Basin of southern Colorado and northern New Mexico have reservoirs in the sandstone tongues of the Mesaverde Formation, which extend out into the Mancos Shale. Farther north the Dakota Sandstone has acted as a reservoir for petroleum in many fields in Colorado and Wyoming. Lower Cretaceous sandstones, such as the Viking, Bow Island, and "Glauconitic," yield both oil and gas in numerous fields in Montana and Alberta. The Cardium Sandstone lenses out eastward into Upper Cretaceous shales and forms a stratigraphic trap for oil at the Pembina Field, Canada's largest, with reserves of 750 million barrels.

Bentonites and Volcanism

The intrusion at depth of granitic batholiths in the Cordillera during Jurassic and Cretaceous time was preceded and accompanied by explosive volcanism at the surface. Lavas derived from granitic rocks are viscous and the style of eruption of volcanoes fed by such lavas is explosive, yielding quantities of ash. During Late Jurassic and Early Cretaceous time this volcanic activity sent great volumes of ash into the atmosphere, where it was carried eastward by the winds and dropped into the epeiric sea. The concentration of this ash in the Montana, Wyoming, and northern Colorado area is a result of the volcanoes associated with the intrusion of the Idaho batholith. Where a high proportion of ash was mixed with ter-

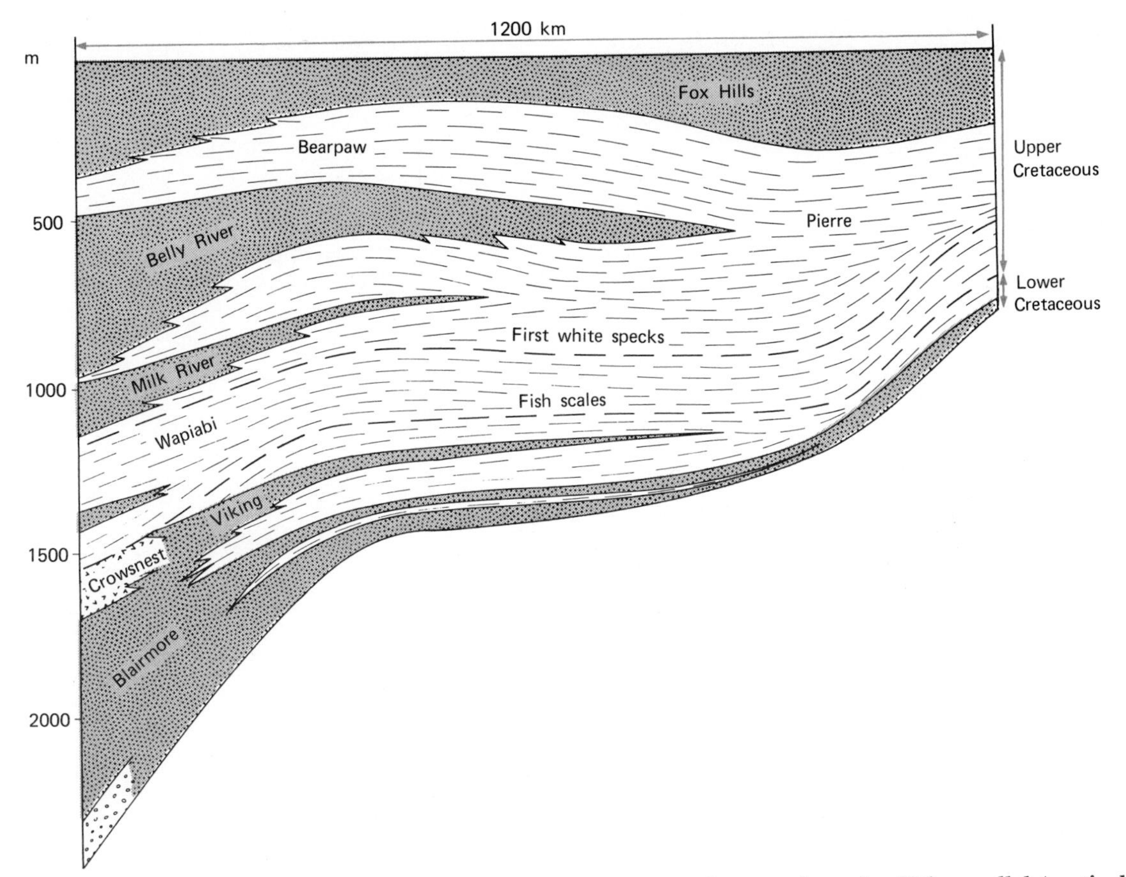

FIGURE 15–25 **East–west facies diagram of the Cretaceous System along the 49th parallel (vertical exaggeration extreme). Note the sandstone tongues spreading eastward at the beginning and end of the period. The white specks and fish scale zones are key beds used in subsurface correlation.**

restrial muds accumulating in the sea a porcellaneous, hard, siliceous shale resulted. The Mowry Shale, which reaches thicknesses of 600 m and covers large areas east of the Idaho batholith, is typical of such a deposit.

Major eruptions may eject so much ash that they produce discrete ash beds. Such eruptions are believed to be recorded in the Cretaceous System of the western platform in beds called **bentonites.** These are distinctively light-colored, unconsolidated sediments consisting largely of the clay mineral montmorillonite. They form beds that are commonly less than a meter thick but have been found up to 8 m thick. Because bentonite beds represent a single event of short duration and can be followed for hundreds of kilometers through the changing facies of the Cretaceous clastic wedge, they are invaluable for establishing correlation. The geometry of the bentonite beds is closely comparable to that of ash beds formed in modern times from the eruption of active volcanoes, and their position with respect to the Cordillera is evidence of Cretaceous westerly winds. By Cretaceous time the northwest movement of the continent with respect to the poles had brought Wyoming from the latitude of the Trade Winds, indicated by the cross-bedding of the Pennsylvanian sandstones (Chapter 13),

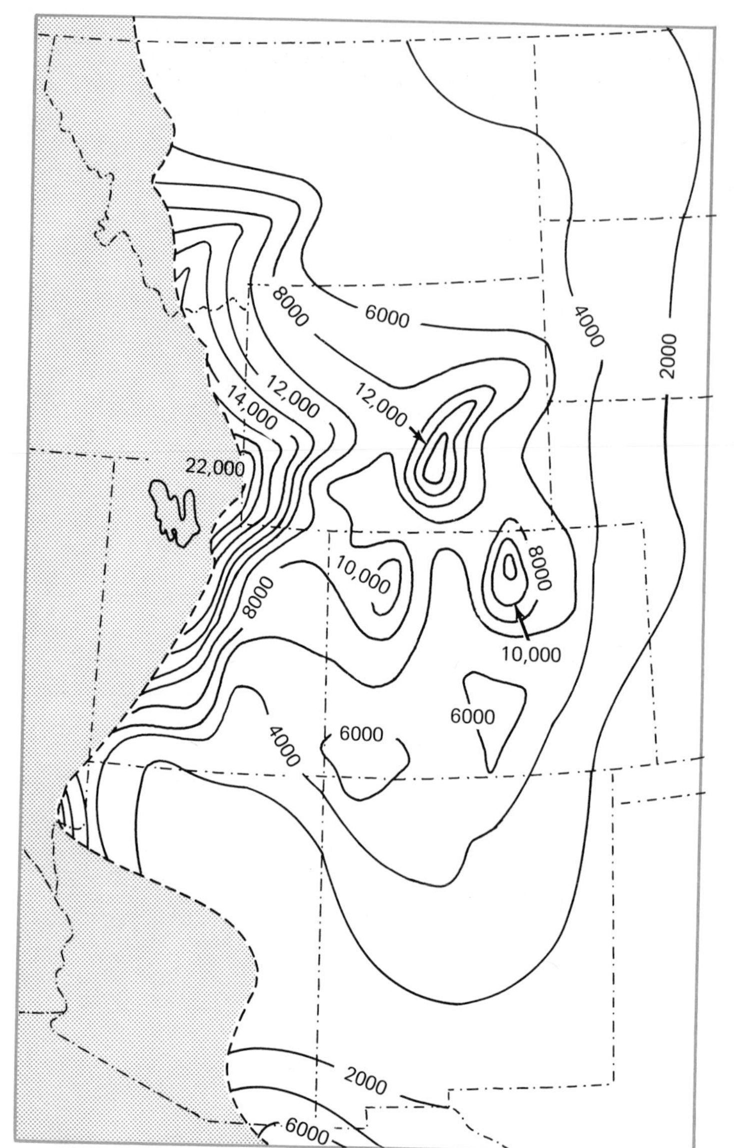

FIGURE 15–26 **Isopach map of the Upper Cretaceous Series in the western states. The contour are in feet and the contour interval is 2000 ft, except near the Great Salt Lake. (After W. C. Krumbein and F. G. Nagel; courtesy of the American Association of Petroleum Geologists.)**

to the temperate zone of the prevailing westerlies.

The montmorillonite of the bentonite seems to have been formed by the decomposition of glassy, ash particles on the ocean floor. When wet the mineral swells to a gooey mass several times its dry volume and, therefore, can be used as a filler or to give "body" to drilling muds in oil wells.

THE AGE OF REPTILES

The Mesozoic Era has been called the "Age of Reptiles," for during this 160 million years of earth history the largest and most abundant vertebrates of land, sea, and air belonged to this class. Their bones are found in continental and marine sediments of this age throughout the world and are particularly abundant in the Jurassic and Cretaceous sediments of the clastic wedges of western North America. The most important records of reptilian life of this continent are preserved in the Triassic redbeds of the southwestern states, the Morrison Formation of Late Jurassic age and the Late Cretaceous sandstones of the Great Plains and intermontane basins.

Turtles

Turtles are not known until late in the Triassic, and no forms are known that provide any information on the origin of the features that characterize this group. Other reptile groups developed a bony carapace, but never in the pattern that distinguishes turtles. The absence of lateral or dorsal temporal openings in the skull indicates that turtles diverged from the primitive reptilian stock, but nothing more specific can be said of their ancestry. Triassic turtles are primitive in not being able to retract their heads into their shells. Subsequent turtles developed one of two patterns. The head was retracted by bending it either sideways or vertically. Turtles with both patterns are found in Jurassic rocks. The former achieved their greatest success in the Mesozoic and early Cenozoic and are now limited to a few groups in the southern continents. The latter apparently differentiated at about the same time, but remain dominant today. Modern turtles show a long history of adaptation to aquatic and semi-aquatic environments, as well as to terrestrial habitats.

Early Diapsids

The Permian eosuchians (Fig. 13–19), a primitive reptilian order with two pairs of temporal openings in the skull, gave rise to the advanced reptiles that dominated Mesozoic faunas. The basic diapsid stock from which this adaptive radiation took place evolved in late Pennsylvanian time from cotylosaur ancestors with a lizard-like adaptation. This assemblage gave rise to the true lizards by the end of the Permian. Lizards are characterized by a highly mobile jaw which can gape widely, and small body size, both probably related to a diet of insects. Relatively few lizards are known from Triassic and Jurassic rocks, but most modern groups are recognizable by the end of Jurassic time.

Mosasaurs

The mosasaurs were members of the lizard clan that returned to the sea. In their long, streamlined body and tapering tail they resembled the sea serpents of mythology (Fig. 15–27). The head was elongated into a formidable jaw with many sharp teeth. The limbs were modified for balancing and steering, and the motion of the tail propelled the animal. The mosasaurs were confined to the Late Cretaceous seas and reached their greatest size in the 3 m long *Tylosaurus*.

Snakes

The group of lizards that gave rise to the mosasaurs also shows numerous features of the skull suggesting affinities with snakes. The earliest known true snake, already clearly a member of this group, is from the Lower Cretaceous of Argentina. Although many lizard groups have lost their legs, no forms are known that directly link lizards with snakes, and there is continuing dispute over the environment of the immediate ancestors of snakes. The burrowing habits of many primitive living snakes, as well as peculiarities of the eye and ear of snakes, suggest that the ancestors may have been subterranean.

Thecodonts—Prototypes of the Ruling Reptiles

Lizards, snakes, and the surviving diapsid *Sphenodon* are usually grouped in the subclass LEPIDOSAURIA. The other large diapsids are called archosaurs. They include the living crocodiles as well as a vast assemblage of Mesozoic animals, the dinosaurs and flying reptiles. The early archosaurs, called thecodonts, appear late in Permian time, having apparently evolved from the basic diapsid stock of the late Paleozoic.

FIGURE 15–27 **A Late Cretaceous seascape in western North America. The plesiosaur *Elasmosaurus* attempts to repel the attack of the fierce mosasaur *Tylosaurus*. Overhead *Pteranodon*, the largest of the pterosaurs, glides on a 8-m wing span. (From J. Augusta and Z. Burian, *Prehistoric Animals*. London: Spring Books, 1956.)**

Phytosaurs and Crocodiles

The immediate ancestry of the crocodiles remains obscure but the earliest members of the group, from Upper Triassic beds, appear to have been agile, terrestrial forms. The ecologic niche now occupied by crocodiles—floating or lying almost submerged in rivers and lakes ready to devour passersby—was taken in Triassic time by another group of archosaurs, the phytosaurs (Fig. 15–28). In exterior form the phytosaur was almost identical with the modern crocodile, but the external nostrils opened at the top of a raised area between the eyes rather than on the end of the snout.

Advanced Archosaurs

Some modern lizards, when pursued, abandon their normal four-footed gait and take to their hind feet, using their tail as a balance. Some thecodonts initiated a tendency to stand and walk on their hind feet rather than on all fours. These included the ancestors of the great reptiles of the Mesozoic Era, commonly called dinosaurs. With the change from the quadrupedal posture of their ancestors, the hind legs were lengthened and strengthened, and the forelegs were reduced. The hind limbs were moved directly under the body instead of being sprawled out as in the amphibians and most of the

FIGURE 15–28 **The Triassic phytosaur *Rutiodon*, a form resembling in habits the living crocodiles, but not closely related. (Courtesy of the American Museum of Natural History.)**

early reptiles, thus helping to support the greater weight.

DINOSAURS: MASTERS OF THE LAND

The suffix "-saur," common in scientific names of many reptiles, is derived from the Greek word for lizard, *sauros;* the prefix "din-" comes from the Greek word *deinos,* meaning terrible. The term "dinosaur" was coined by Sir Richard Owen in 1841 for the giant reptiles of the Mesozoic Era, which were then coming to the attention of the scientific world. When a fuller understanding of the nature of these remarkable fossils had been gained, paleontologists realized that these reptiles, which Owen had regarded as a single group, consisted of two distinct subdivisions, and the term "dinosaur," although still popularly used to include all huge Mesozoic land reptiles, has been discarded by some paleontologists as of little real taxonomic significance.

The separation of the dinosaurs is based on differences in the pelvic structure. The more primitive type of structure characterized the order known as the SAURISCHIA (reptile-hipped) and had a triradiate form like that of the thecodonts, from which the dinosaurs were derived early in the Triassic Period. The other order, the ORNITHISCHIA (bird-hipped), comprises reptiles with a hip structure like that of the birds. Figure 15–29 is a diagrammatic classification of the main subdivisions of the dinosaurs, showing their ranges and relationships.

Theropods

The first saurischians to appear were the carnivorous theropods of Upper Triassic time. A carnivorous way of life demands agility and speed, and they retained the thecodont's bipedal gait and many other features (Fig. 15–30). In fact, it is difficult to establish an exact separation between them and the thecodonts. Their forelimbs were small and probably served only for grasping food.

One lineage consisted of small, birdlike,

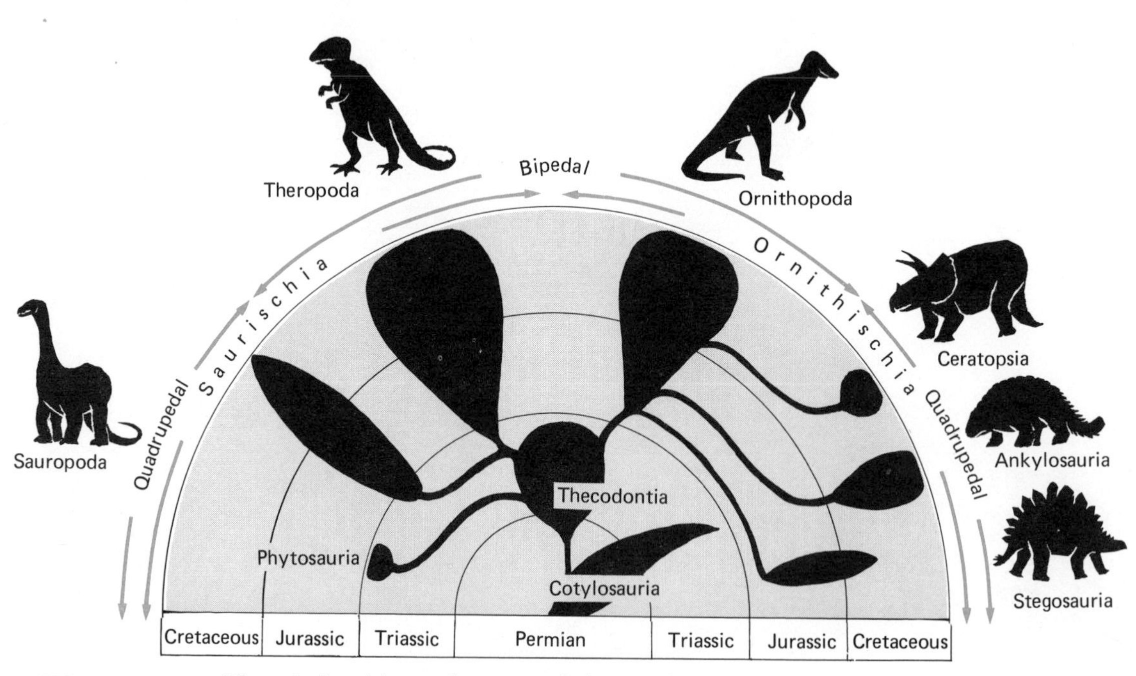

FIGURE 15–29 **The relationships and ranges of the subdivisions of the dinosaurs.**

FIGURE 15–30 ***Scleromochlus*, from the Triassic of Scotland. Long considered one of the thecodonts, which it resembles in many respects, it is now classified as a primitive saurischian dinosaur. The boundary between thecodonts and dinosaurs can be determined only with difficulty. (Courtesy of the British Museum (Natural History).)**

lightly built, hollow-boned, small-headed forms, with three digits on fore and hind feet. A typical but late genus is *Ornithomimus* (Fig. 15–31), which probably fed on insects and other small animals. A second lineage of theropods evolved into giants. Their hind limbs were massive, their head was enormously enlarged, and their jaws took over the functions of grasping and tearing food. Their useless forelimbs were reduce to non-functional vestiges. *Antrodemus* (Upper Jurassic) and *Gorgosaurus* (Cretaceous) were capable of attacking the largest of their contemporaries, even the 50-ton monsters described below. The largest theropod, a 15 m reptile called *Tyrannosaurus,* appeared in late Cretaceous time (Figs. 15–31 and 15–34). *Tyrannosaurus* held his massive head 7 m above the ground and weighed 6 to 8 tons. He was the greatest land carnivore that the world has ever seen.

Sauropods

Except for the theropods all dinosaurs were herbivores, or eaters of vegetation. The sauropods were the vegetarian branch of the SAURISCHIA and possibly originated from the heavier quadrupedal thecodonts. They evolved into the largest land vertebrates that ever lived, and have been exceeded in bulk among animals only by certain modern whales. Such late Jurassic giants as *Brontosaurus* and *Diplodocus* reached lengths of 25 m and weights of more than 50 tons. *Apatosaurus,* from Late Jurassic beds of western North America, was only 20 m long and weighed merely 40 tons (Fig. 15–15). Sauropods did not need

FIGURE 15–31 **A Late Cretaceous landscape with several kinds of dinosaurs. On the left, *Trachodon,* a duck-billed wader, is alarmed by the proximity of *Tyrannosaurus,* whose attention is momentarily riveted on the armored dinosaur *Ankylosaurus* below him. In the middle distance a small carnivore, *Ornithomimus,* seizes a bird. Flowering plants abound to the near exclusion of earlier types. (Diorama by Mrs. Moray Macnaughton; courtesy of the Redpath Museum.)**

speed or agility and developed solid, pillar-like legs to support their great bulk. The relative shortness of their forelimbs, a feature of all dinosaurs, was not marked. Their neck was long and their head comparatively small, not much larger than a horse's, with a small mouth equipped with peg-like and flat teeth for cropping and crushing vegetation.

Ornithopods

Among the ornithischian dinosaurs, those that retained the two-footed stance are grouped together as the ornithopods, some of which are referred to as the duck-billed dinosaurs because the front part of their mouth lacked teeth and was modified into a beak or bill (Fig. 15–31). The ornithopods appeared in the Late Triassic, by which time they had diverged markedly from the thecodont pattern, but reached their greatest diversity in the Late Cretaceous Epoch, when advanced members of the group grew to length of 10 and 12 m and weights of several tons. Their duck bills and webbed feet show that they lived near the water and probably used it as a refuge when attacked. Some Late Cretaceous ornithopods show peculiar extensions of the nasal passages into hollow crests at the top of the skull; these have been interpreted as air reservoirs that allowed the animal to stay submerged for relatively long periods, or as resonance organs used by the animal to intensify its call. They may also have assisted in species recognition. Three other ornithischian lineages are recognized—the stegosaurs, ankylosaurs, and ceratopsians—all of which reverted to a quadrupedal stance.

Stegosaurs and Ankylosaurs

The stegosaurs and ankylosaurs were the plated and armored dinosaurs, respectively. The stegosaurs, typified by *Stegosaurus,* were large, quadrupedal animals with a double row of upright, bony plates along the back and a short tail armed with several bony spikes (Fig. 15–32). The skull was small, and the brain was extremely small in relation to the size of the animal—far smaller, in fact, than the enlargement of the neural cord in the hip region that served to direct the operation of the hind limbs and tail. The stegosaurs were among the first ornithischians, appearing early in Jurassic time, but they were also the first major group of dinosaurs to become extinct.

The ankylosaurs were squat, slow-moving, primarily Cretaceous, armored reptiles. In *Ankylosaurus* the back was covered by thick, bony plates that gave protection as effective as that possessed by a turtle; in addition, along each flank was a row of bony spikes (Fig. 15–31). The tail ended in a great knob of bone that could be swung as a bludgeon. The offensive use of the tail by animals that are armored for defense was adopted by such distantly related animals as the porcupines, the glyptodonts (Pleistocene inhabitants of the Americas), the ankylosaurs, and the stegosaurs. Both the plated and armored dinosaurs had small, weak teeth and must have fed upon soft plants. *Polacanthus* (Fig. 15–33) was an early Cretaceous ankylosaur which de-

FIGURE 15–32 **The late Jurassic plated dinosaur *Stegosaurus*. The vegetation in the foreground and right middleground is made up of cycadeoids. (Courtesy of the Field Museum of Natural History.)**

FIGURE 15–33 ***Polacanthus,*** **a bizarre ankylosaur from Lower Cretaceous beds of Great Britain. [Courtesy of the British Museum (Natural History).]**

pended for defense upon enormous, solid spines along the back and smaller plates along the tail.

Ceratopsians

The ceratopsians, or horned dinosaurs, were a highly specialized, Late Cretaceous suborder of the ornithischians. In habitat and general appearance they were close to the rhinoceroses of the present day. The ceratopsians had a thickset, short, unarmored body supported on pillar-like legs. The short neck and massive head were covered by a great, bony shield. This acted as a place of attachment for large neck and jaw muscles, and also as a protection for the vital neck region. Most members of this group developed horns on the nose, above the eyes, and around the edge of the frill. The largest of the ceratopsians, *Triceratops,* appeared near the end of the Cretaceous (Fig. 15–34). It was 6 m in length and had a long horn above each eye and one on the nose.

In the Mongolian desert skeletons of the primitive ceratopsian *Protoceratops* were found associated with clutches of eggs and immature specimens. In some of the elongate eggs, laid to hatch in the warmth of the sun 100 million years ago, the remains of embryonic dinosaurs could be detected. Although dinosaur eggs are among the rarest of fossils, finds such as the Mongolian one leave little doubt that dinosaurs reproduced in this manner.

MARINE REPTILES

In mid-Paleozoic time, some vertebrate animals left the water and began the conquest of the land. Yet the advantages of life in the sea are so great that many groups of both reptiles and mammals later forsook their acquired land habitat and went back to an aquatic life. Six groups of reptiles—the mesosaurs, ichthyosaurs, plesiosaurs, mosasaurs, crocodiles, and turtles—made this transition, the last two incompletely. The limbs of all these animals were modified to fin-like paddles; their breathing was restricted to times when they were at the surface; and they adopted habits to assure that their young did not drown when they were born.

Ichthyosaurs

Of all the marine reptiles, none adapted better to life in the sea or became more fish-like in form than the ichthyosaurs. The typical ichthyosaur had a streamlined body like that of a tuna or salmon, with a long jaw armed with many sharp teeth for catching fish (Fig. 15–35). The tail was fish-like in form and function and propelled the ichthyosaur by its lateral motion. The backbone was bent down into the lower lobe of the forked tail (Fig. 15–36), and the upper lobe was supported by the skin. The limbs were modified to become balancing and steering fins, and the bones in them were reduced to small, close-fitting, polygonal plates. Well-preserved skeletons of ichthyosaurs with embryonic young within the body cavity suggest that these animals did not return to shore to lay eggs, as do marine turtles of the present day, but bore their young alive in the ocean, as do the whales. They were doubtless as completely aquatic as are the dolphins today; even temporary excursions on land must have been impossible. The largest ichthyosaurs grew to 10 m but many were only a meter long.

The ichthyosaurs appeared suddenly in mid-Triassic time, and no intermediate forms between them and any probable ancestors are known. They became extinct in the Late Cretaceous.

Plesiosaurs

The plesiosaurs evolved mechanisms for aquatic locomotion distinct from those of the ichthyosaurs. They had a short, thick body and a relatively short tail. The limbs were modified into four long paddles by which the animal may have rowed itself forward or backward on the surface of the ocean (Fig. 15–37). Alternatively, they may have used the paddles as wings to "fly" through the water, as do penguins.

One group of plesiosaurs had a long, flexible neck on which the small head could be maneuvered rapidly to catch fish or to fight. The largest of the plesiosaurs measured 15 m in length. The neck of *Elasmosaurus* (Fig. 15–27), one of the largest, was twice the length of the body and contained 60 vertebrae. Others had shorter necks and elongate heads. Ancestors of the plesiosaurs lived in the Triassic Period and suggest a transition from small terrestrial forms close to the main reptilian stock.

THE CONQUEST OF THE AIR

For 200 million years after the first animals climbed out on land from the lakes and streams, the air was populated only by the insects. The aerial environment was the last major one to be conquered by the vertebrates, because highly specialized animals had to evolve before the many problems of flight could be solved. A flying animal must have a strong skeleton to hold the enlarged muscles needed for flight, but its bones must be light so as not to be a burden. Both birds and reptiles solved this problem in part by developing hollow bones and a keeled breastbone for the attachment of the muscles that drove the wings. Furthermore, the limbs of a flying animal must be highly modified to support it in the air. The birds solved this problem by developing feathers to form wings; the reptiles and bats solved it by developing a membrane, stretched between the forelimb and the body, as a wing (Fig. 15–38). A flying animal must have a high rate of metabolism, so that it may maintain the strenuous actions of flying over a long period. The birds and the bats solved this problem by having a high, constant body temperature. There has been some question as to whether flying reptiles could have had a sufficiently high metabolism to sustain flight. The development of a bird-like skeleton and brain, however, indicates that they were otherwise well adapted to flight. This has led some paleontologists

FIGURE 15–35 **The ichthyosaur *Eurhinosaurus*. Air breathers, these animals had to come to the surface periodically; they probably resembled the modern porpoise in habits as well as in shape. (From J. Augusta and Z. Burian, *Prehistoric Animals*. London: Spring Books, 1956.)**

FIGURE 15–34 **The largest land carnivore of all time, *Tyrannosaurus*, faces the huge, horned dinosaur *Triceratops*. Both lived and became extinct toward the close of the Cretaceous Period. (Courtesy of the Field Museum of Natural History.)**

FIGURE 15–36 **Skeleton of an ichthyosaur showing the highly modified limbs with the finger bones reduced to small plates. Note the downward deflection of the tail. (Courtesy of the Redpath Museum.)**

FIGURE 15–37 **The plesiosaur *Cryptodeidus*. Two of these animals are seen from above, swimming at the water's surface, where they used their strong paddles to propel themselves. (From J. Augusta and Z. Burian, *Prehistoric Animals*. London: Spring Books, 1956.)**

to suspect that the flying reptiles may have attained a high, well-controlled body temperature, like the birds.

Pterosaurs

The flying reptiles of the Mesozoic Era are called pterosaurs. This highly specialized group arose from thecodont stock before the beginning of the Jurassic Period, though its relationship to other archosaurs is not clear. They persisted until shortly before the end of the Cretaceous Period; their extinction may have been caused by unsuccessful competition with birds. The pterosaur wing was a membrane stretched between the enormously enlarged fourth finger and the body and hind legs of the animal (Fig. 15–38). The fifth finger was lost, and the other fingers at the front of the wing were modified for grasping cliffs and trees, from which the animal probably hung. The hind limbs were feeble and probably could not support the pterosaur efficiently on land. Some pterosaurs, such as *Rhamphorhynchus,* had a long, trailing tail with a rudder-like enlargement; others, for example *Pterodactylus* (Fig. 15–39) and *Pteranodon* (Fig. 15–27), had none. *Pteranodon* was the largest of the flying reptiles and the largest animal, in span of wings, ever to fly. Although its body was only the size of a turkey, its

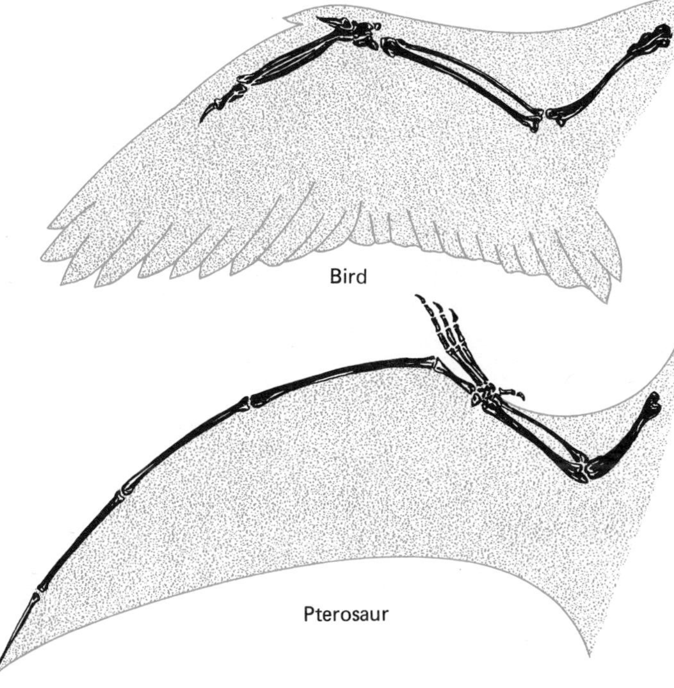

FIGURE 15–38 **The wing structures of a bat, a bird, and a pterosaur.**

wing span was more than 15 m. It was probably incapable of flapping such a long fragile wing efficiently, but flew largely by gliding on upcurrents of air, just as eagles and hawks do today. The head was elongated in front into a long beak for catching fish and projected behind into a blade-like crest. Many pterosaurs are believed to have glided and swooped over the sea, looking for surface-swimming fish, but an animal like *Pteranodon* would have had great difficulty rising into the air again after making a catch.

BIRDS

Birds are the last vertebrate class to appear in the fossil record. The oldest bird, *Archaeopteryx* (Fig. 15–40, 15–41), is known from the Upper Jurassic lithographic limestones of southern Germany. It was the size of a crow, but with a long tail and numerous sharp teeth. These exceedingly fine-grained rocks preserve not only the bones, but the impressions of the feathers as well. Were it not for these feathers, *Archaeopteryx* would probably not have been recognized as a

FIGURE 15–39 ***Pterodactylus,*** **a small pterosaur. Note the forward-pointing teeth, useful for impaling surface-swimming fish while the reptile was skimming over the sea. (Courtesy of the British Museum (Natural History).)**

FIGURE 15–40 ***Archaeopteryx,*** **the nearly perfect Berlin specimen in which the skeleton and feather impressions are preserved in fine-grained limestone. (Courtesy of J. H. Ostrom and the Humboldt Museum.)**

FIGURE 15–41 **Restoration of *Archaeopteryx,* an ancestral bird from the Upper Jurassic of southern Germany. The small reptile is *Compsognathus,* a coelurosaur dinosaur, thought to be close to the ancestry of birds. (Reproduced from J. Augusta and Z. Burian, *Prehistoric Animals.* London: Spring Books, 1956.)**

bird, for almost all features of the skeleton resemble those of small, bipedal dinosaurs. John Ostrom has recently argued that *Archaeopteryx,* and birds in general, evolved from primitive theropod dinosaurs, similar to the genus *Compsognathus.*

Birds have the highest metabolic rate of all vertebrates, and maintain a body temperature in excess of that of mammals. The dinosaurs, on the other hand, are commonly considered typical examples of lethargy and stupidity. The internal structure of the dinosaur bones and their upright posture suggest to some, however, that they had a high, constant body temperature. The active carnivorous dinosaurs, in particular, would have required a high, constant energy expenditure to maintain bipedal posture. Animals of large bulk, like the dinosaurs, can maintain high temperature with less expenditure of energy than smaller animals because their surface area to volume ratio, which determines the rate of heat loss, is low. Very few dinosaurs weighed less than 20 kg, and most weighed a great deal more. The dinosaurs ancestral to the birds must have been exceptional in their small body size, which may have been associated with specialization toward insect-eating. If these forms were warm blooded, as seems probable from their bipedal posture, their large surface-to-volume ratio would have made them subject to rapid heat loss, unless they evolved some form of insulation. Selection favored the elongation of the body scales, which would enclose dead air space, lessening heat loss.

That the ancestors of the birds were arboreal in habit has long been accepted. Longer scales would be selected because an increase in the surface area of the limbs and body would slow the animal's fall if it tumbled from the trees. Further increase in the extent of the scales would permit gliding, and eventually active flight.

Recent work by John Ostrom, however, indicates that the earliest known bird, *Archaeopteryx,* was not arboreal but rather a ground dweller. He argues that the typically bird-like foot of this animal developed for running, as did that of carnivorous dinosaurs. The highly efficient bipedal running of birds is quite unlike the quadrupedal gait of other groups that evolved as arboreal gliders, such as flying squirrels and gliding lemurs. Ostrom suggests that flight movements started in forms like *Archaeopteryx* for catching insects and other small terrestrial prey. Some living birds disturb the underbrush with their wings to dislodge insects. The same muscles are used as those responsible for the power strokes in flight. Ostrom argues that the flight feathers on the wings were elongated to serve as a net to trap insects. Active flight may have begun inadvertently, in short leaps after prey, when the arms (or wings) were brought down with enough force that the air passing over the leading edge of the arm generated lift. Further structural and behavioral modification of this system would be selected for, since longer jumps would assist in capturing prey.

Archaeopteryx was probably a poor flyer. The bones, although hollow, are not pneumatic and the large bone to which the large flight muscles of modern birds is attached, the sternum, is not present. Modern birds have a special pulley-like structure so that the major muscle for lifting the wing can be located ventrally, in the breast area. This arrangement had not evolved in *Archaeopteryx.* The heavy, toothed skull and the long, feathered tail suggest that *Archaeopteryx* had stable flight, very limited maneuverability, and hard landings.

Archaeopteryx is unlike all other birds in its primitive anatomy. It is the best, single example of a fossil linking two major groups. No other birds are known from Jurassic rocks. In Upper Cretaceous sediments, much more modern kinds of birds have been collected, although they are rare, and the few forms known are specialized aquatic genera.

Most of the modern orders of birds are represented by scattered remains throughout the Cenozoic. Fossils of birds are extremely rare throughout the history of the group. The record is biased toward the preservation of aquatic genera and large, flightless forms. Apparently only a single major radiation of bird took place, that which led to the groups living today. In the absence of placental carnivores in South America, giant flesh-eating birds evolved in the early Cenozoic. Genera such as *Andalgalornis* (Fig. 15–42) exceeded a meter in height.

THE END OF REPTILIAN DOMINANCE

At the end of the Mesozoic Era, sweeping changes in the environment of the reptilian faunas of the world resulted in the wholesale extinction of the saurischian and ornithischian dinosaurs, the pterosaurs, the plesiosaurs, the ichthyosaurs, and the mosasaurs after 100 million years of reptilian dominance of land, sea, and air. A change that would have affected the varied habitats occupied by the many reptiles listed above is difficult to imagine. Swamp dweller and upland browser, carnivore and herbivore, sea serpent and flying reptile, biped and quadruped, armored and unarmored forms were among those that failed to survive the end of the Cretaceous Period. With the exception of the stegosaurs, the entire assemblage seems to have become extinct within a few million years. Yet some reptiles, such as the crocodiles, turtles, and lizards, which seem to have occupied en-

FIGURE 15–42 ***Andalgalornis*****, one of the many large, mostly flightless, early Cenozoic birds of South America, about one-twelfth actual size. (Courtesy of the Field Museum of Natural History.)**

vironments similar to those of their extinct relatives, did survive to the present day. No single mechanism has been suggested that satisfactorily explains this phenomenon. Worldwide physical calamity is unlikely. In western North America, deposition was continuous from the Late Cretaceous into the Paleocene Epoch. Dinosaur remains become rarer and rarer in the Upper Cretaceous rocks, and near the end of the epoch vanish completely. Other surviving reptilian groups continue into the Cenozoic beds without interruption. Archosaur extinction can hardly have been a result of direct competition with mammals, for that group did not become dominant until after the archosaurs had disappeared.

Possible Causes of Extinction

The general elevation of the continents at the end of the Mesozoic Era may have modified the vegetation on which the herbivorous dinosaurs fed and may have chilled the seas until they were no longer tolerable to the marine reptiles. A drastic change in the vegetation, the emergence of the angiosperms, took place within the Cretaceous, but the herbivorous dinosaurs appear to have satisfactorily adapted to this change. The survival of the late Cretaceous crisis by the warm-blooded mammals and birds suggests that the crisis may have been a climatic one to which the reptiles would have been more sensitive. A similar evolutionary crisis of unknown origin marked the end of the Triassic Period, when six orders of reptiles, including the thecodonts and mammal-like reptiles, became extinct, as did numerous amphibian groups. Both of these crises for the reptiles were also crises for the ammonoids, but the connection between such remotely related animals is difficult to imagine.

If, as we suspect, the large herbivores—particularly the sauropods—were dependent on extensive areas of swamp and lake for support and food supplies, they would have been vulnerable to slight climatic or geological changes which might have drained these areas. Their extinction, which apparently did slightly precede that of the carnivorous forms, could have led to the death of the reptiles that preyed upon them. The extinction of such terrestrial herbivores as the ceratopsians, however, would seem to be an independent phenomenon, and the death of the totally marine plesiosaurs seems even less reasonably related.

Other suggestions as to the fate of the reptiles in late Mesozoic time have been made. There is no support for such concepts as racial senility, or loss of vigor, or a disease that attacked reptiles alone. The extinction of the dinosaurs and other groups of reptiles remains as much a problem as the dying of many invertebrate stocks at the end of the Paleozoic Era. Both probably resulted from the failure of specialized animals to evolve rapidly enough to adapt to changes in their environments.

SUMMARY

The Cordillera of today has been formed by the interaction of the Pacific and North American plates during Mesozoic and Cenozoic time along the subduction zone that had characterized this plate junction since early Paleozoic time.

The Cordilleran orogeny was a complex event, the sum of local events of short duration and local responses of the crust to forces generated along a plate boundary thousands of kilometers long.

Some of the early Mesozoic units in the western Cordillera may have been deposited many kilometers southwest of their present positions and subsequently plastered against the margin of North America by the movement of the Pacific plate into the Cordilleran subduction zone. Chaotic rock units, like the Franciscan Group of California, are characteristic of the trench environment and are associated with calc-alkaline volcanics, blue schist metamorphism, and ultramafic rocks.

The western Cordillera was intruded by granitic masses during the interval from Late Triassic to Late Cretaceous time. The large batholiths are made up of many individual igneous bodies emplaced over an interval of tens of millions of years. Metallic mineral deposits are associated with these intrusions.

In Cretaceous time the deformation affected the miogeoclinal belt breaking it into thrust slices and shearing them from the basement along décollement thrusts. These structures developed over an interval of tens of millions of years. The ranges of central Wyoming and Colorado east of the belt of miogeoclinal deformation have cores of pre-Paleozoic rocks that have been pushed upward through the platform sediments.

The sediments eroded from the rising Cordilleran mountains were deposited in a complex clastic wedge of Late Jurassic to Cenozoic age on the western side of the platform. Marine successions in the clastic wedge are correlated on the basis of ammonites and belemnites. The Morrison Formation of latest Jurassic age is famous for its fossils of dinosaurs and early mammals.

The first mammals arose in Late Triassic time from synapsid reptiles. They were very small and, to regulate their internal temperature, must have had hair for insulation and sweat glands for cooling. The latter may have given rise to mammary glands for feeding the young. Typical mammalian teeth are differentiated into several types, replaced only once, and have a specific pattern of occlusion.

Cretaceous rocks of the western platform are a clastic wedge composed of minor limestone, marine shale, light-colored shoreline sandstones, and siltstones and sandstones deposited in coastal swamps. With many oscillations these facies moved eastward as the supply of sediment from the Cordillera filled the inland sea. Coals in the regressive sandstone units were derived from conifers, cycads, and cycadeoids, which were joined by the angiosperms in Early Cretaceous time. Correlation within the Cretaceous clastic wedge is facilitated by key beds formed by volcanic ash called bentonite, which erupted from Cordilleran volcanoes.

The thecodonts, ancestors of the reptiles dominant in the Late Mesozoic, arose from diapsid reptiles in Permian time. They moved the hind limbs beneath the body and became bipedal. The thecodonts gave rise to the six groups of dinosaurs: theropods (carnivorous bipeds), sauropods (large quadrupeds), ornithopods (duck-billed bipeds), stegosaurs and ankylosaurs (armored quadripeds), and ceratopsians (horned quadripeds). Reptiles that invaded the seas in Mesozoic time include the ichthyosaurs, plesiosaurs, and mosasaurs. Reptiles also adapted to an aerial environment by evolving a membrane between the fourth digit and the body, thus forming a wing. Dinosaurs and pterosaurs may have had constant high body temperatures. The first bird, *Archaeopteryx,* may have been a ground-living animal that initially developed feathers for insulation and catching insects and subsequently used them for flying.

QUESTIONS

1. Trace the steps by which animals achieved control over their internal temperature from primitive reptiles to mammals.
2. What features of sedimentary rocks and volcanics can the geologist use to identify the location of a subduction zone that is now inactive?
3. Contrast the structure of the Cordillera in California (Western Ranges) with that of eastern Nevada (miogeoclinal) and that of central Wyoming (platform).
4. What advances were made by mammals over reptiles in food processing and in reproduction?
5. Compare the position and environment of deposition of the coals in the Cretaceous regressive sandstone tongues of the western states to those of the coals of the Pennsylvanian cyclothems of Illinois.

6. How were the theropods adapted to a carnivorous way of life and the ichthyosaurs for an aquatic existence?

SUGGESTIONS FOR FURTHER READING

ARMSTRONG, R. L. "Sevier Orogenic belt in Nevada and Utah." *Geological Society of America Bulletin,* 79: 429–458, 1968.

CONEY, P. J. "Cordilleran Tectonics and North American Plate motion." *American Journal of Science,* 272: 603–628, 1972.

CROMPTON, A. W., and PARKER, D. "Evolution of the Mammalian masticatory apparatus." *American Scientist,* 66: 192–201, 1978.

HAMILTON, W. "Mesozoic California and the underflow of Pacific mantle." *Geological Society of America Bulletin,* 80: 2409–2430, 1969.

LANPHERE, M. A., and REED, B. L. "Timing Mesozoic and Cenozoic plutonic events in circum-Pacific North America." *Geological Society of America Bulletin,* 84: 3773–3782, 1973.

OSTROM, J. H. "Some hypothetical anatomical stages in the evolution of Avian Flight." *Smithsonian Contributions to Paleobiology,* No. 27, 1–21, 1976.

ROMER, A. S. *Vertebrate Paleontology,* University of Chicago: Univ. Chicago Press, 3rd edition, 1966.

RUTLAND, R. W. R. "On the interpretation of Cordilleran orogenic belts." *American Journal of Science,* 273: 811–849, 1973.

WHEELER, J. O., ed. "Structure of the southern Canadian Cordillera." *Geological Association of Canada, Special Paper 6,* 1–166, 1970.

CHAPTER SIXTEEN

INTRODUCTION

In the eastern Cordillera the deformation of the crust by compressional movements was most widespread during Late Cretaceous time, and thereafter areas of such deformation were of only local extent. However, along the eastern front of the Cordilleran thrust belt in western Wyoming and Montana, Paleocene and Eocene sediments are commonly deformed and cut by thrust faults, showing that the compressional forces were still at work there until the middle of the Cenozoic Era. The volume and coarse texture of Paleocene sedimentary rocks within and east of the Rocky Mountains and of Eocene rocks within intermontane basins indicates a high relief of the mountains from which the sediments were eroded.

By the beginning of Cenozoic time nonmarine sedimentation in the Cordillera had almost totally replaced marine sedimentation. The Cenozoic stratigraphic record consists of sediments laid down in isolated basins surrounded by the growing mountains, and of clastic wedge deposits shed eastward as a vast alluvial apron across the present site of the Great Plains. The basins between the mountains have been referred to as "successor basins" because deposition in them followed the main episode of orogeny. In these successor basins lava flows and ash deposits may form as important a part of the filling as sedimentary rocks.

In tracing the history of the Cordillera in the Paleozoic and Mesozoic eras we have had to rely on the stratigraphic record to supply information about the landforms because the original rivers, mountains, basins, and highlands have been largely eliminated by subsequent erosion or covered by later deposition. The history of the Cenozoic Era involves the development of landforms that are still part of the landscape (Fig. 16–1). The shapes and geographic positions of these landforms are additional evidence for the interpretation of the Cenozoic history of the continent.

At present the Pacific margin of the Americas plate is formed by subduction zones and transform faults. Zones of ocean floor spreading occur along the East Pacific Rise and west of the coast from Oregon to Vancouver Island (Fig. 1–13). The pattern of magnetic anomalies on the ocean floors formed by spreading in the Cenozoic Era allows geophysicists to reconstruct the configuration of the western margin of the Americas plate and the adjacent Pacific Ocean floor during Cenozoic time. At the end of the Cretaceous Period the East Pacific spreading center was still hundreds of kilometers west of the Americas plate in the latitude of California, and was feeding litho-

THE CORDILLERA IN THE CENOZOIC

FIGURE 16–1 **The volcano Shishaldin, with Isanotski and Round Top peaks, Unimak, one of the Aleutian Islands. The Aleutian volcanism is a reflection of subduction along the trench to the south of the islands.(Navy Department Photograph 80-G-81322.)**

sphere into a subduction zone at the margin of the plate where the Franciscan mélanges were forming (Fig. 16–2). North of California the plate margin was either a zone of very oblique subduction (that is, one where the angle between the trend of the trench and the motion of the plate being consumed is small), or a transform fault merging northward with the Aleutian trench. During the Cenozoic Era the spreading center encountered the subduction zone in the latitude of California, and along this segment the Farallon plate was entirely subducted and the motion between the plates became a horizontal one (Fig. 16–3). The present relative motion between the Pacific plate and the Americas plate in California and west of British Columbia is described as being "right lateral." If the observer is standing on one plate, the other plate appears to have moved to the right. An isolated remnant of the East Pacific spreading center that has not yet encountered the Cordilleran subduction zone remains off the coast of Oregon, Washington, and Vancouver Island, and the subduction zone that once occupied the whole west side of the Americas plate is now restricted to a segment along that coastline and the Middle America trench off Mexico. The effects of the changing configurations of the west margin of the Americas plate on the geological evolution of the Cordillera are examined in this chapter.

In the space available we can only sample the Cenozoic history of the diverse regions of the North American Cordillera to emphasize the most important features of its development in the past 70 million years. In Canada and much of Alaska the Cenozoic history of the Cordillera has been obscured by the effects of Pleistocene glaciation. Successor basins that were filled with Cretaceous and early Cenozoic sediments and volcanics preserve some evidence of this history. In Wyoming and Colorado a fine record of the development of the Rocky Mountains has been preserved in the thick sediments of the successor basins, in the erosion surfaces of the ranges, and in the patterns of the drainage systems. In the Basin and Range province Cenozoic beds reveal the extent of explosive volcanism in the Cordillera, and the structures show the effect of tensional forces in the crust. In the Columbia Plateau the release of great quantities of basaltic lavas is illustrated. In California Cenozoic basin fillings record the active nature of the Pacific margin and the role of transform motion in the deformation of the crust. These areas will form the basis of the following discussion of the Cordillera in Cenozoic time.

WYOMING AND COLORADO

Early Cenozoic Events

Pebbles of pre-Paleozoic rocks in latest Cretaceous sediment of the Laramie Basin indicate that some of the ranges of the Rockies in Wyoming were elevated by the end of the Mesozoic Era to such an extent that streams had cut through the thick sedimentary cover to the basement rocks. The deformation in this segment of the Cordillera has resulted in vertical uplifts; local folds and reverse faults can be interpreted as secondary effects of the upward movement of the pre-Paleozoic cores of the ranges. The first elevation of the Wyoming–Colorado sector of the Rockies took place in the Paleocene Epoch, and by Eocene time the ranges may have had a relief comparable with that of the present day. The sediments from the erosion of these early Cenozoic mountains were deposited in adjacent basins as coarse conglomerates near the mountain fronts and as variegated sandstones and red banded shales farther out in the basins. The spectacular turretted scenery of the Bryce Canyon, Utah, is cut in the Eocene clastic sediments of the Wasatch Formation, eroded from the rising mountains (Fig. 16–4).

During Eocene time large lakes filled the Green River Basin of southwestern Wyoming and the Uinta Basin of eastern Utah and northwestern Colorado (Fig. 16–5). Although the deposits of these two basins are now separated, the two lakes probably joined around the eastern end of the Uinta Arch during some stages of their history. In the

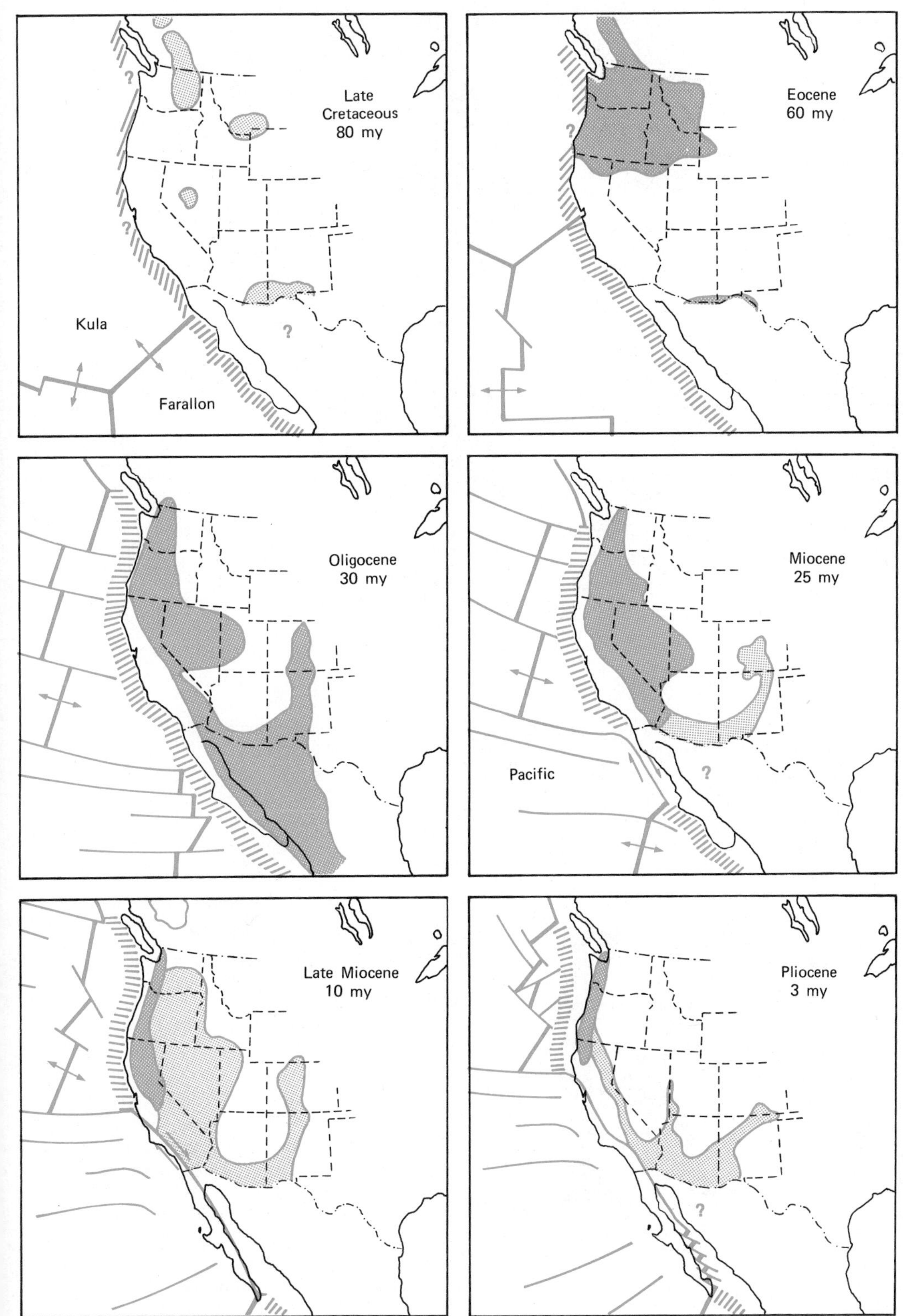

FIGURE 16–2 **Plate relationships and volcanism along the western side of the North American plate in late Mesozoic and Cenozoic time. The Cordilleran subduction zone is indicated by the hachured lines. Areas of volcanism are indicated by the colored pattern; the darker shade represents andesitic calc-alkaline volcanics, the lighter shade, basaltic volcanics. (Compiled from several sources, mainly T. Atwater, P. Lipman, H. J. Proslka, R. L. Christiansen.)**

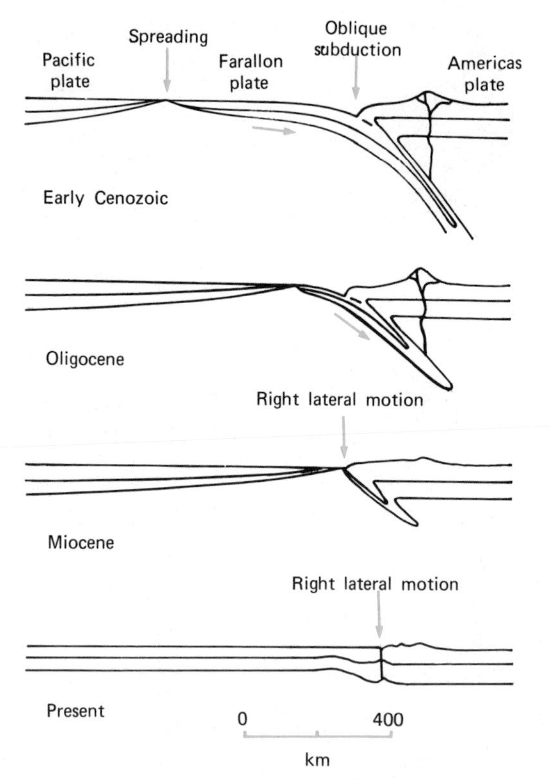

FIGURE 16–3 **Cross-sections of southern California and the adjacent Pacific Ocean during the encounter of the East Pacific Rise spreading-center and the Cordilleran subduction zone. The Pacific plate is considered to be fixed and the Farallon plate moves at 10 cm/year to the east. (After Tanya Atwater; courtesy of the Geological Society of America.)**

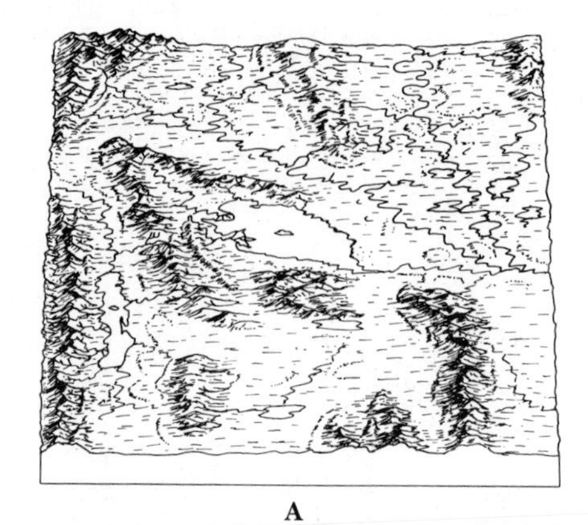

A

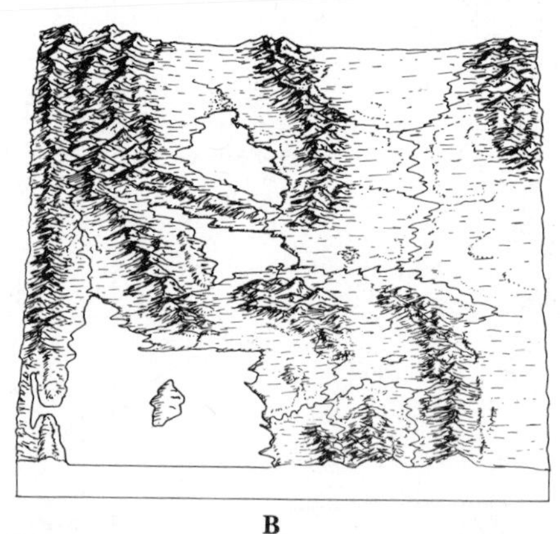

B

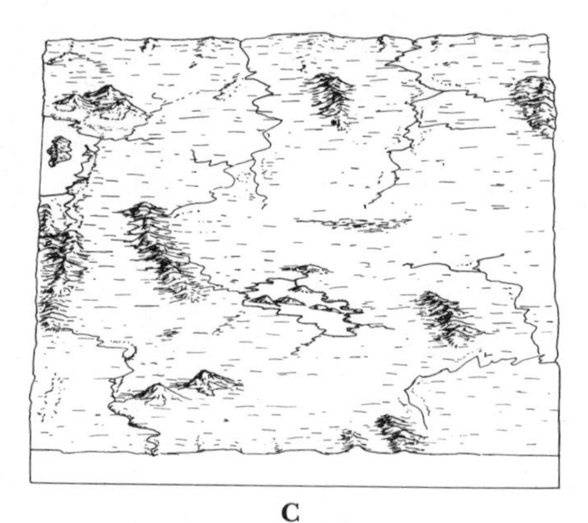

C

FIGURE 16–5 **Topography of Wyoming during the Cenozoic Era: A. Late Paleocene, when orogenic forces were still deforming the rocks and relief was high. B. Early Middle Eocene, when relief was greatest and extensive lakes were formed by the westward tilting of the area. C. Late Pliocene, when filling of the basins and erosion of the highlands brought the area to its lowest relief. (Modified from J. D. Love, D. O. McGrew, and H. D. Thomas.)**

FIGURE 16–4 **Beds of the Wasatch Formation eroded to form the "Mormon Temple," Bryce Canyon National Park, Utah. (Courtesy of the National Park Service.)**

lakes a great variety of sediments, including oil shale, algal limestone, phosphate rock, anhydrite, salt, and sodium carbonate (trona) were laid down in thin beds (Fig. 16–6). Oil shale is a light brown to black shale that contains hydrocarbons in the form of waxy spores and pollen grains. Under proper treatment these hydrocarbons can be distilled and used to prepare petroleum products. The petroleum that can be distilled readily from the minable beds of oil shale of the Green River Formation (that is, those yielding 30–35 gallons per ton, 125–150 liters per ton) have been estimated as 160 billion barrels.

The fine, regular laminations in these sediments could be annual deposits, or varves. Counts of the layers suggest that the 600 m of the Green River beds took 4 million years to be deposited. In order for evaporite minerals such as sodium carbonate, anhydrite, and dolomite to be associated with fossil-bearing, organic-rich shales, the lake in which the formation was deposited must have fluctuated greatly in level like the ephemeral lakes, called playa lakes, that occupy basins between the ranges in Nevada today. During arid times the lake water was concentrated to a brine by evaporation and confined to the center of the lake. Sodium carbonate was precipitated. During periods of greater rainfall the lake expanded greatly and algal blooms formed a large quantity of organic ooze on the bottom of the lake from which the waxy hydrocarbons were incorporated in the oil shale. From the fossil plants found in the Green River Formation paleobotanists have learned that the climate was temperate to sub-tropical, with an annual temperature in the range of 15° to 20°C (60° to 70°F) and a seasonal rainfall of 60 to 76 cm (24 to 30 in). Some of the most perfectly preserved fossil fish in the world are found pressed flat on the bedding planes of the Green River mudstones (Fig. 12–12).

Although Eocene sediments are thick and widespread in the basins between the mountains, they are missing from the wedge of sediments that spread eastward into the

FIGURE 16–6 **Thin, even beds of the Green River Formation, Bitter Creek Valley, Wyoming. (Courtesy of W. H. Bradley, U. S. Geological Survey.)**

plains from the Rockies. Possibly, regional warping of the Rocky Mountain area during early Cenozoic time resulted in the drainage from the mountains being deflected from an eastward course, or else the successor basins may have acted as such efficient sediment traps that little of the products of erosion escaped the mountain terrane.

Middle Cenozoic Events

Prominent erosion surfaces have been observed on many of the ranges of the central Rocky Mountains. These surfaces have different names in different ranges but all have been referred to collectively as the Subsummit surfaces, because the smooth summits of some of the ranges define other surfaces (Summit) formed by high altitude weathering about 600 m above the Subsummit surfaces (Fig. 16–7). At first the Subsummit surfaces were interpreted as remnants of a peneplain to which the Rockies had been reduced by mid-Cenozoic time. However, these surfaces can be interpreted more consistently as pediments, that is, as surfaces of erosion that slope outward from mountain ranges in semi-arid climates, like that of the Basin and Range (Fig. 16–8). In contrast to peneplains, pediments can be formed at relatively high altitudes and, if the Subsummit surfaces are remnants of pediments, their presence does not imply that the mountain belt was reduced to near sea level in mid-Cenozoic time.

The pediments that now surround the ranges of Nevada are continuous with the surfaces of deposition of the sediments that fill the basins. In mid-Cenozoic time the Subsummit surfaces were probably continuous with the basin fillings that accumulated between the ranges of the Rockies. During Oligocene and Miocene time agents of erosion steadily wore down the ranges and filled the intermontane basins, decreasing the relief and developing extensive ped-

FIGURE 16–7 **The Sherman erosion surface cut on the crystalline rocks of the Colorado Front Range, one of the Subsummit surfaces. Pikes Peak in the background rises above this surface. (Courtesy of the U. S. Geological Survey.)**

FIGURE 16–8 **Pediment surface east of Oro Grande, Mojave area, California. Mojave River is in foreground. (Airphotograph courtesy of John S. Shelton.)**

iments in the semi-arid climate (Fig. 16–5). In addition, much detritus exported eastward from the mountain belt built a thick clastic wedge of fine sandstones and mudstones across Nebraska and Kansas. These mid-Cenozoic sediments preserve a record of the evolution of the mammals that is discussed in the latter part of this chapter.

Late Cenozoic Rejuvenation

In late Pliocene time the pattern of deposition and erosion in the Rockies and adjacent plains changed, and the relief that is characteristic of the Rockies of the present day developed (Fig. 16–9). Some geologists believe that the change can be accounted for by climatic factors; others ascribe it to regional uplift. Whatever the cause, the streams that once deposited sediment in the basins and on the clastic wedge in the plains began to excavate these poorly consolidated sedimentary rocks, and so brought the mountains into greater relief. Most of the mid-Cenozoic

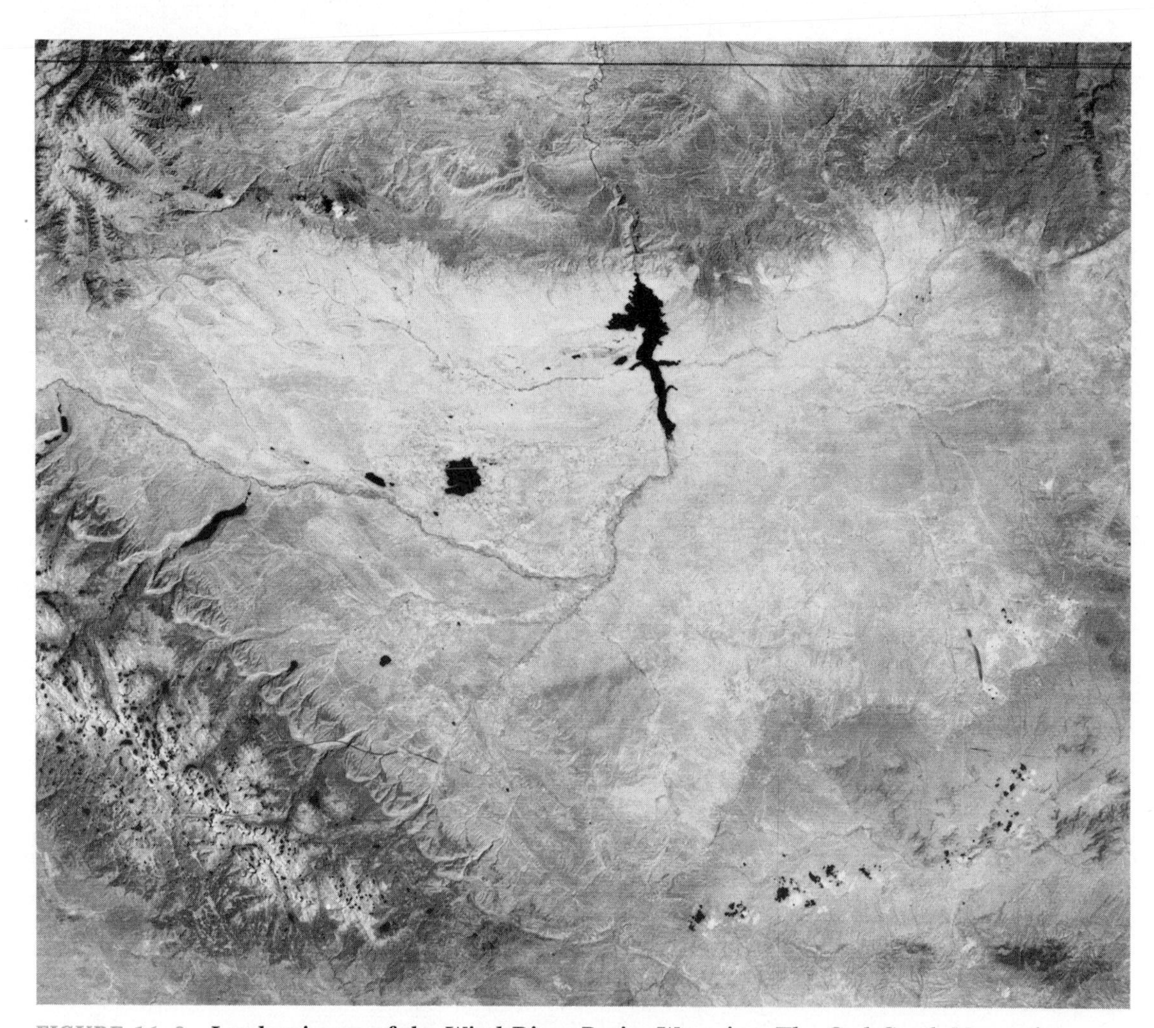

FIGURE 16–9 **Landsat image of the Wind River Basin, Wyoming. The Owl Creek Mountains across the top of the photograph, with the Washakie Needles (3809 m) in the upper left corner. In the lower left sedimentary beds dip away from the uplifted core of the Wind River Range. (Courtesy of NASA.)**

basin fillings were removed, leaving the pediments of the Subsummit surface high above the valleys and the basins floored with Eocene and Paleocene rocks. Only small remnants of the Oligocene, Miocene, and Pliocene sediments that once filled the basins are now found high on the mountainsides (Fig. 16–10).

East of the Rockies the quickened streams dissected the wedge of sediments formed in middle Cenozoic time and eroded most of the plains down to Cretaceous and Paleocene strata. The remnants of the Oligocene, Miocene, and Pliocene beds now constitute the "High Plains." If the surface of the High Plains is projected westward on a section toward the mountains, it will coincide with the Sherman erosion surface of the Laramie Range, one of the Subsummit surfaces (Fig. 16–7). At the "Gangplank," where the Union Pacific Railroad crosses the Laramie Range west of Cheyenne, a salient of the High Plains lies against the mountains and the continuity of erosional and depositional surfaces has been preserved.

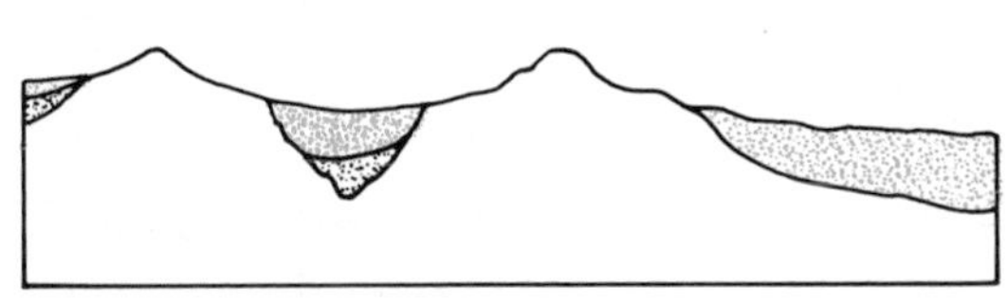

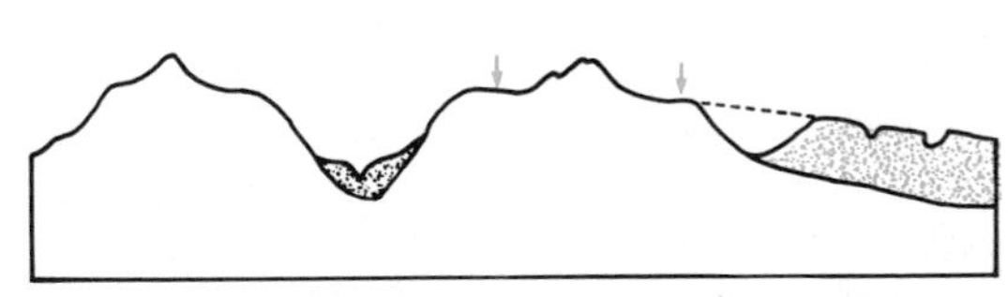

FIGURE 16–10 **Stages in the development of the central Rocky Mountains. The black pattern indicates early Cenozoic sediments; the colored pattern, Middle Cenozoic sediments. The arrows in the bottom figure indicate the Subsummit surfaces. The High Plains, remnants of the Middle Cenozoic detritus eroded from the mountains, is shown at the right. The vertical scale is greatly exaggerated.**

The Canyons of Colorado and Wyoming

Many of the rivers of Wyoming and Colorado follow courses totally out of adjustment with structure and topography. For example, the Bighorn River, instead of flowing eastward through a prominent gap in the Front Ranges of central Wyoming, turns northward and crosses the Owl Creek Mountains in a canyon 750 m deep; it then crosses the Bighorn Mountains before reaching the plains (Fig. 16–11). The Royal Gorge, cut by the Arkansas River across the Colorado Front Range, is one of the most spectacular of these canyons. At Devil's Gate, west of Alcova, Wyoming, the Sweetwater River cuts a canyon through a hard rock ridge, which could have been avoided by a course less than a kilometer away.

Rivers come to cut through resistant bedrock ridges in three ways. They may persist in their course as a ridge rises slowly across their path and progressively cut through it like a stationary saw cutting into a log raised against it. Such rivers are said to be **antecedent.** They may have established a course on an alluvial plain beneath which there is a buried transverse ridge. During uplift and erosion of the alluvium, the stream will encounter the ridge and, if competent, will maintain its course and grade, cutting a slot as the countryside around is lowered by other streams. Such rivers are said to be **superposed.** The river may be temporarily dammed behind a barrier rising across its flow, and spilling over may cut a canyon through to maintain its course. Such rivers have been called **anteconsequent.** All three of these explanations have been applied to the canyons of the central Rockies. The most popular explanation of the strange drainage system is that

FIGURE 16–11 **Aerial photograph of the Owl Creek Range, Wyoming, looking west. The canyon cut by the Bighorn River across the resistant core of the range during the canyon-cutting cycle is shown in the foreground. (Photograph by John S. Shelton.)**

the rivers were superposed across the ranges in late Pliocene time, a few million years ago, when the mid-Cenozoic basin fillings were excavated. Although such an explanation implies that a river may cut through 700 m of granitic rock in about 5 million years, we shall see below that a much greater rate of erosion is required for the development of other canyons in the Cordillera, such as the Grand Canyon.

BASIN AND RANGE PROVINCE

In the Basin and Range province (Figs. 16–12, 16–13) high ranges of complex structure are isolated from neighboring ranges by valleys filled with Cenozoic continental deposits. It is an area of low rainfall, a desert in the rain shadow of the Sierra Nevada. None of the streams or rivers in the northern sector (called the Great Basin) now reaches the sea. Infrequent heavy rains feed torrental streams that carry detritus and dissolved salts into closed basins (playas) where the water evaporates (Fig. 16–14).

The absence of early Cenozoic sediment in the Basin and Range province suggests that it was a highland at this time. Drainage of the area at the close of the Cretaceous, and during most of Cenozoic time, was westward to the Pacific across the site of the Sierra Nevada, which was a region of low relief until the end of the era.

Mid-Cenozoic Volcanism

About 40 million years ago in the Miocene Epoch the Basin and Range province was the site of extensive, explosive volcanic

FIGURE 16–12 **Basin and Range province near Elko, Nevada. Note the dark sculptured ranges and the light playas between them with interior drainage. Landsat satellite imagery. (Courtesy of NASA.)**

eruptions of dominantly silicic lavas. The rocks characteristic of these eruptions are of the type associated with island arcs (calc-alkaline, as defined in Chapter 2). The volcanic rocks are largely ash beds erupted from volcanoes of the Peléean type, in which hot gases, steam, and ash are discharged laterally from vents with great force. The clouds, called"nuées ardentes," are charged with pumice and ash and move

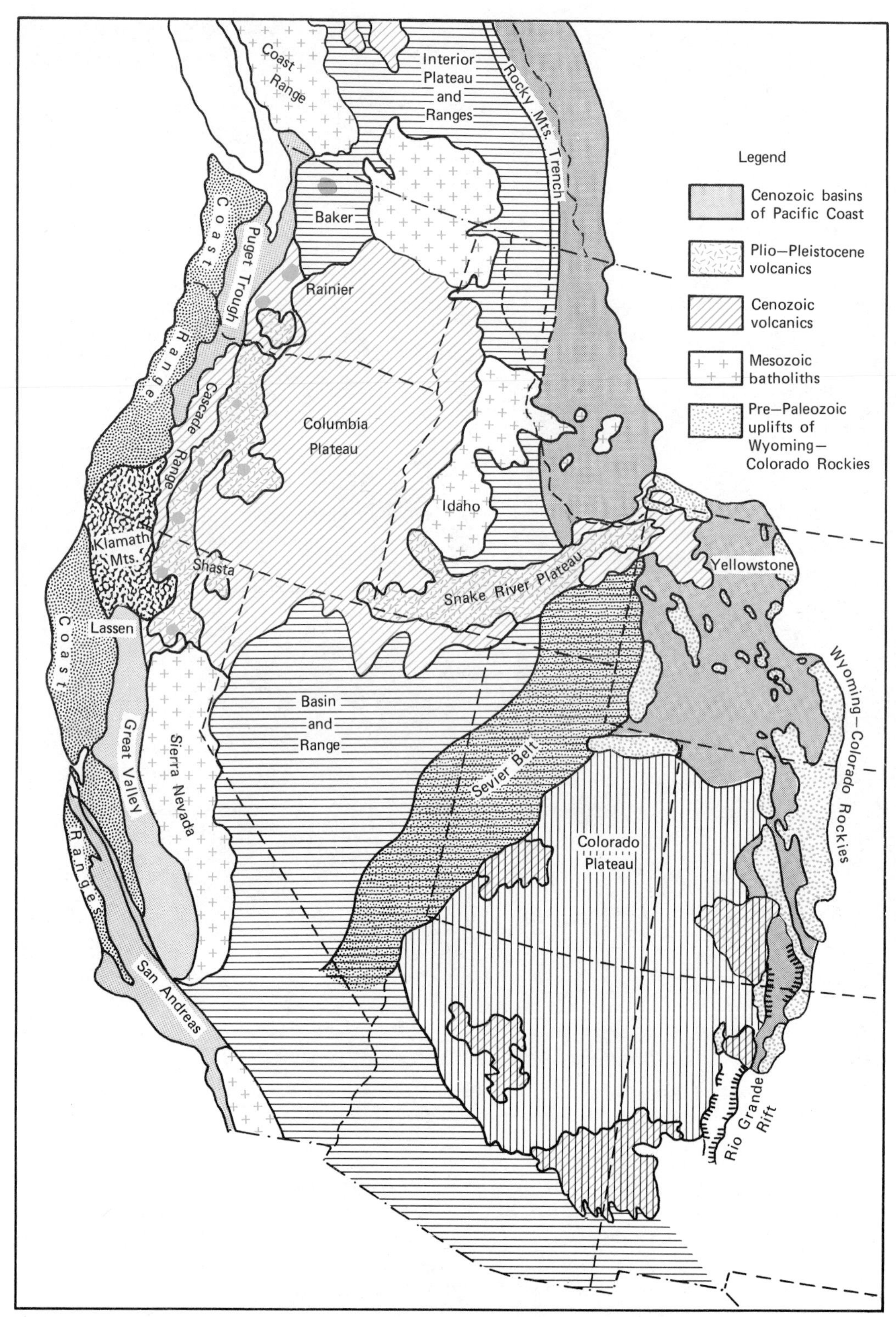

FIGURE 16–13 **Map of the central Cordillera showing the main structural provinces.**

FIGURE 16–14 **Oblique aerial photograph of Basin and Range topography showing flat basins floored with alluvium and the sharply defined ranges, many of which are bounded by fault scarps. (Courtesy of Fairchild Aerial Surveys.)**

rapidly downhill to spread out into flat country for many miles. The rocks so deposited are called **welded tuffs** (or ignimbrites), because the individual glassy particles kept hot in a cloud of gas are still plastic when they come to rest and are welded to their neighbors to form a solid rock.

Extensive lava flows built a plateau on the present site of the Sierra Nevada, but in the middle Cenozoic the altitude of the plateau was less than 900 m and drainage was maintained westward through it to the Pacific. This estimate of the altitude in the Miocene Epoch is made from the evidence of plants in sediments interbedded with the flows. Pliocene plants from these interbeds indicate that the altitude of this area remained moderate until about 5 million years ago.

Starting in middle Cenozoic time in the southern part of the province and moving progressively northward, the type of volcanism changed. Instead of viscous, explosive calc-alkaline type lavas, the volcanoes began to emit more liquid basaltic lavas. This change in petrology and style of volcanism has been attributed to the change in the nature of the plate boundary west of the Basin and Range province in later Cenozoic time. The calc-alkaline suite was erupted when the plate boundary west of the Basin and Range province was a subduction zone, but when the East Pacific spreading center moved against the zone and convergent motion along the plate boundary stopped, basaltic volcanism replaced explosive, calc-alkaline volcanism. Like the onset of the new volcanic style, the change in the plate boundary progressed from south to north (Fig. 16–2).

Late Cenozoic Extension and Uplift

The ranges of Nevada are tilted blocks of complexly deformed Paleozoic and Mesozoic rocks bounded on one or both sides by normal faults generally aligned in a north-south direction (Fig. 16–15). In this fault pattern the first geologists exploring the re-

FIGURE 16–15 **Mosaic of Landsat imagery of the northwestern states. The following are identified by numbers. 1. San Francisco Bay area. 2. Great Valley of California. 3. Sierra Nevada. 4. Basin and Range province (note the orientation of the dark colored ranges). 5. Lavas of the Columbia Plateau. 6. Idaho batholith. (Courtesy of NASA.)**

gion recognized the effects of uplift and the extension of the crust in an east–west direction. That vertical uplift is continuing to the present is shown by the frequency of earthquakes in the area and the fault scarps that traverse recent alluvial fans along the edges of the ranges. Like the last elevation of the Wyoming–Colorado Rockies, the widespread normal faulting of the Basin and Range province seems to have taken place in late Pliocene and Pleistocene time, that is, within the last 10 million years.

Mechanisms proposed to explain the Basin and Range province must account not only for its structure but also for the thinness of its crust and the abnormally high heat flow through the crust. The crust under the province is about 30 km thick and thickens abruptly to over 40 km under the Sierra Nevada and the Colorado Plateau on either side. Three mechanisms have been proposed to account for these features:

1. The westward movement of the Americas plate has caused it to over-

ride the East Pacific spreading center which, maintaining its nature beneath the continental plate, is pushing upward and outward on the crust below the Basin and Range area.

2. The faulting is a product of the horizontal shearing motion of the Americas and Pacific plates set up in late Cenozoic time, when the Cordilleran subduction zone was eliminated in this sector of the plate boundary.
3. The remnants of the descending Farallon plate, cut off when the Cordilleran subduction zone was eliminated in this sector, form a partially melted, upward-spreading mass of lighter rocks exerting upward and outward pressure on the Basin and Range area. As long as the convergent motion at the subduction zone kept the west side of the Americas plate under compression, this spreading force was ineffective, but when subduction stopped in the late Cenozoic, the upward and outward motion of the rising magma extended the crust into the present structure and changed the style of volcanic eruption.

The persistence of the East Pacific spreading center beneath the Basin and Range province now seems unlikely (see Introduction). Both the rise of subcrustal material and the shearing motion at the plate boundaries are likely to have influenced Basin and Range structure.

Sierra Nevada

Late Pliocene plants in the basin sediments of Nevada indicate a change from the warm, moist, subtropical climate of early Cenozoic time to a dry, savannah climate characteristic of regions with less than 40 cm of rainfall per year. On the west the Sierra Nevada began to rise along normal faults bounding its eastern side, a movement that in Pleistocene time carried it like a giant wedge 3000 m above the floor of the adjacent Owens Valley. The rising barrier placed Nevada in a rain shadow and initiated the aridity and interior drainage that now characterize the region. The history of the uplift is recorded in the valley profiles of the rivers that drain the gently sloping western flank of the range. The lower valley of the Merced River is a good example of a composite valley. The river now flows in a deep, V-shaped valley cut into a wider valley whose remnant walls appearing high above the river. Traces of an even wider valley that was eroded into an upland surface remain above the other two (Fig. 16–16). From this profile

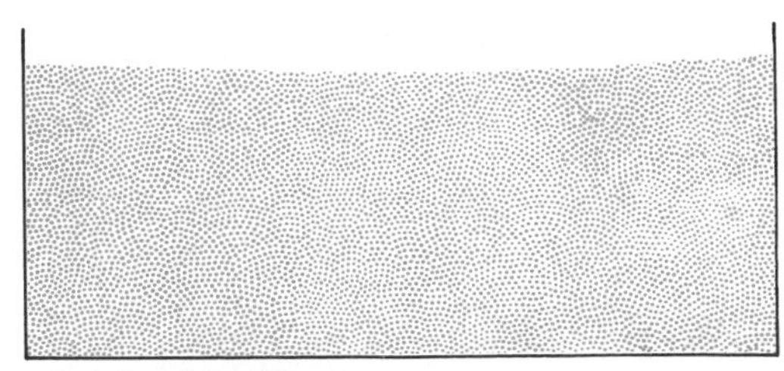

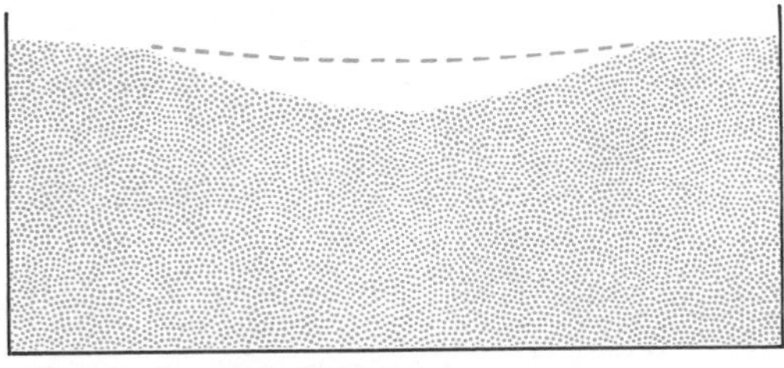

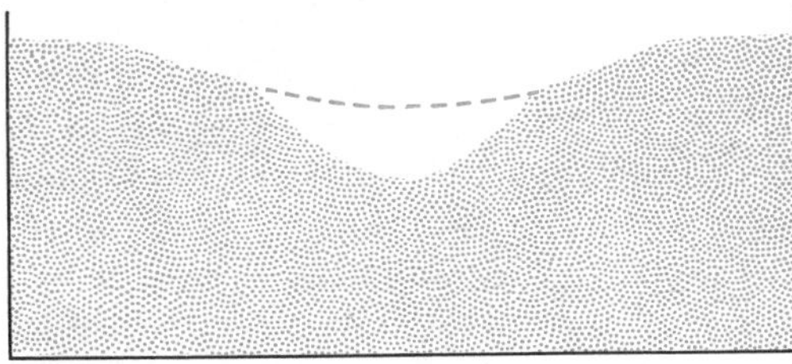

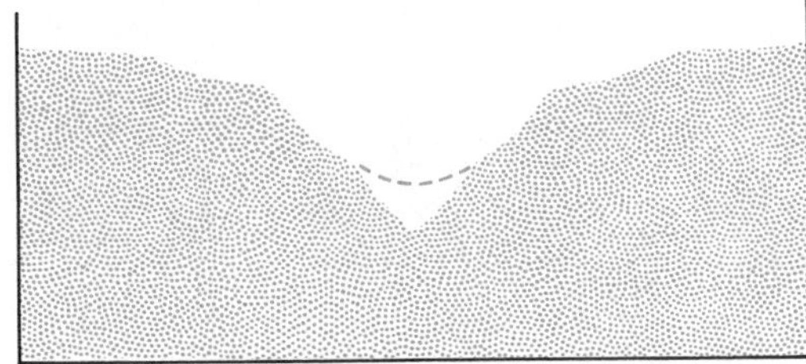

FIGURE 16–16 **Diagrammatic representation of an interpretation of the stages in the development of the transverse valley profiles in the Sierra Nevada.**

and a study of the sediments that accumulated to the west, in the Great Valley of California, geologists have read the following history. During most of Cenozoic time when the Sierra Nevada was less than 600 m above sea level, the oldest, and now highest erosion surface, the Boreal surface, was formed. The first uplift of the Sierra Nevada in Pliocene time started a new cycle of erosion and caused the Merced to cut a wide valley into the Boreal surface, forming the Broad Valley surface. The Pleistocene uplift of the range to its present height took place in two steps: in the first, or Mountain Valley stage, a valley was cut into the Broad Valley surface; in the second, or Canyon stage, the gorge in which the river now flows was formed. The erosion surfaces are preserved only in the lower parts of the Merced Valley because the headwaters were deeply eroded by glaciers during the Ice Age, forming the picturesque scenery of the Yosemite Valley. The dating of these surfaces and others within the Sierra Nevada has been a problem of reconciling contradictory geomorphological, paleobotanical, stratigraphic, and geochronological evidence (Fig. 16–17). The stepped topography of the mountains, ascribed by most geologists to a succession of erosion surfaces, has also been explained as a phenomenon of differential weathering and erosion.

COLUMBIA AND SNAKE RIVER PLATEAUS

The combined Columbia and Snake River plateaus represent one of the largest accumulations of lava in the world. One hundred

FIGURE 16–17 **Aerial view northward from the crest of the Sierra Nevada across the upper Kern Basin to the Great Western Divide. Some geologists believe that an ancient erosion surface is preserved on the flat summit of Mt. Whitney (4418 m) in the foreground; others attribute its flatness to high-altitude erosion processes. The surface of the Kern Basin is another controversial erosion surface. (Courtesy of F. E. Matthes, U. S. Geological Survey.)**

thousand cubic kilometers of lava covered large areas of Washington, Oregon, and southern Idaho to form the Columbia Plateau (Fig. 16–12). The basaltic lava must have been a rapidly moving liquid of low viscosty when it emerged from the conduits, which are believed to have been fissures in the crust rather than volcanoes, because single flows a hundred meters thick spread for over 150 km before they cooled enough to solidify. The first lava flowed out over a mountainous countryside with a relief of over 750 m. Flow after flow of the molten basalt filled the valleys until they were entirely buried in lava. More lava welled up, ultimately covering an area of 500,000 square kilometers (Fig. 16–18). The successive flows that built this great plateau were 30 to 150 m thick. The total thickness of the flows is difficult to measure because basement rocks are rarely exposed beneath the lava covering, but a well has penetrated over 3000 m of lava.

Continental sediments and tuffs interbedded with the lavas contain fossils which date most of the flows of the Columbia Pla-

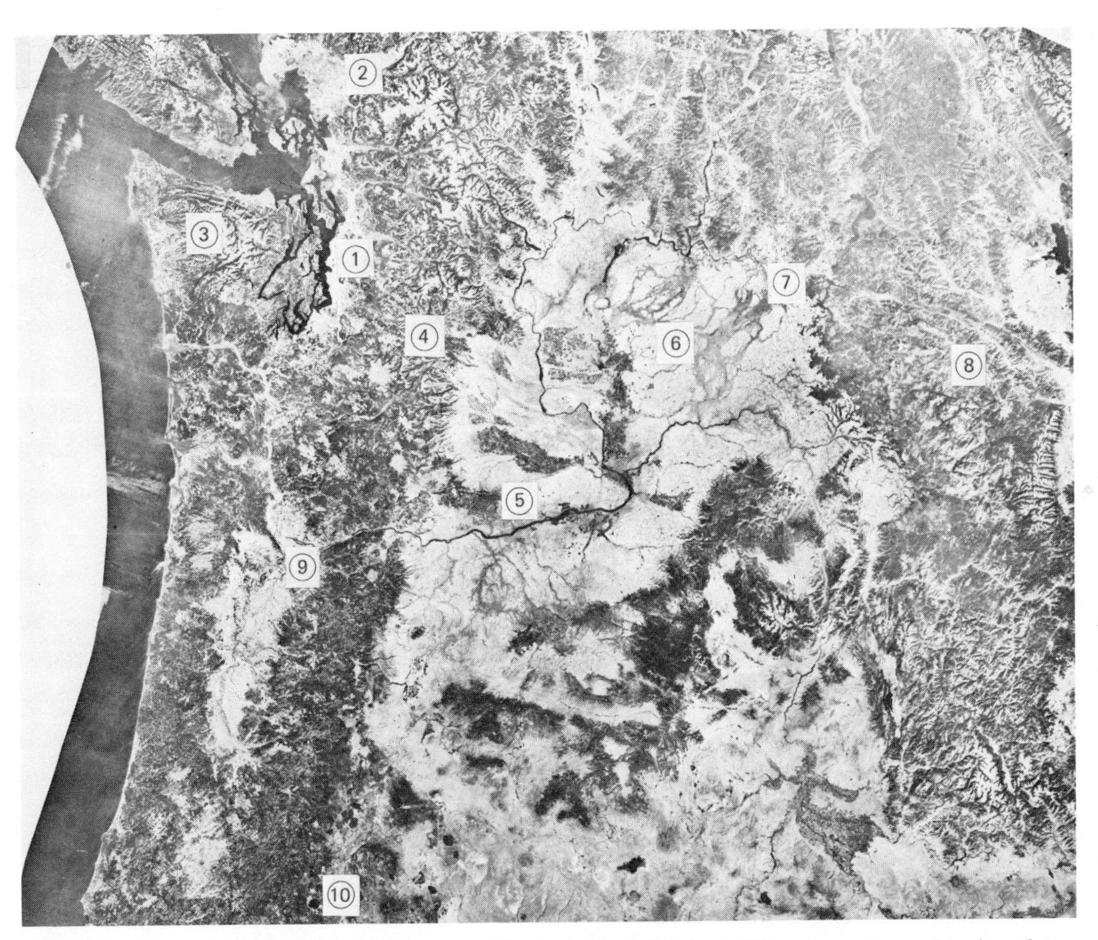

FIGURE 16–18 **Mosaic from Landsat satellite imagery of the northwestern states showing the setting of the Columbia Plateau. The following features are identified by numbers. 1. Seattle, in the Puget trough. 2. Vancouver, on the delta of the Frazer River. 3. The Olympic Mountains. 4. Mount Rainier, in the Cascade Range. 5. The Columbia River crossing the lava plateau. 6. The Channeled Scabland of the Columbia Plateau. 7. Spokane. 8. Idaho batholith. 9. Portland. 10. Crater Lake, in the Cascade Range. (Courtesy of NASA.)**

teau as of Miocene age. A large area of similar plateau basalts of Miocene and Pliocene age underlies the Neychako Basin of south-central British Columbia. The Snake River Plateau of southern Idaho is composed of flows that seem to have been erupted from volcanic centers rather than fissures into a large graben structure (Fig. 16–19). They are composed of basalt of the type associated with spreading centers in the oceans and are considerably younger than the Columbia Plateau flows. Extrusion in this area occurred in Pliocene and Pleistocene time, and the most recent flows may be only a few thousand years old.

At the eastern end of the Snake River Plateau, in the northwestern corner of Wyoming, is Yellowstone National Park, with its world-famous geysers, more than 100 in number, that intermittently spray hot, mineralized water and steam into the air (Fig. 16–20). "Old Faithful," the most famous of these erupting springs, sprays 50,000 liters of boiling water skyward 22 times a day, year in, year out. The heat is supplied from below where water seeping through cracks mixes with superheated steam from cooling igneous rocks. The heat may be derived

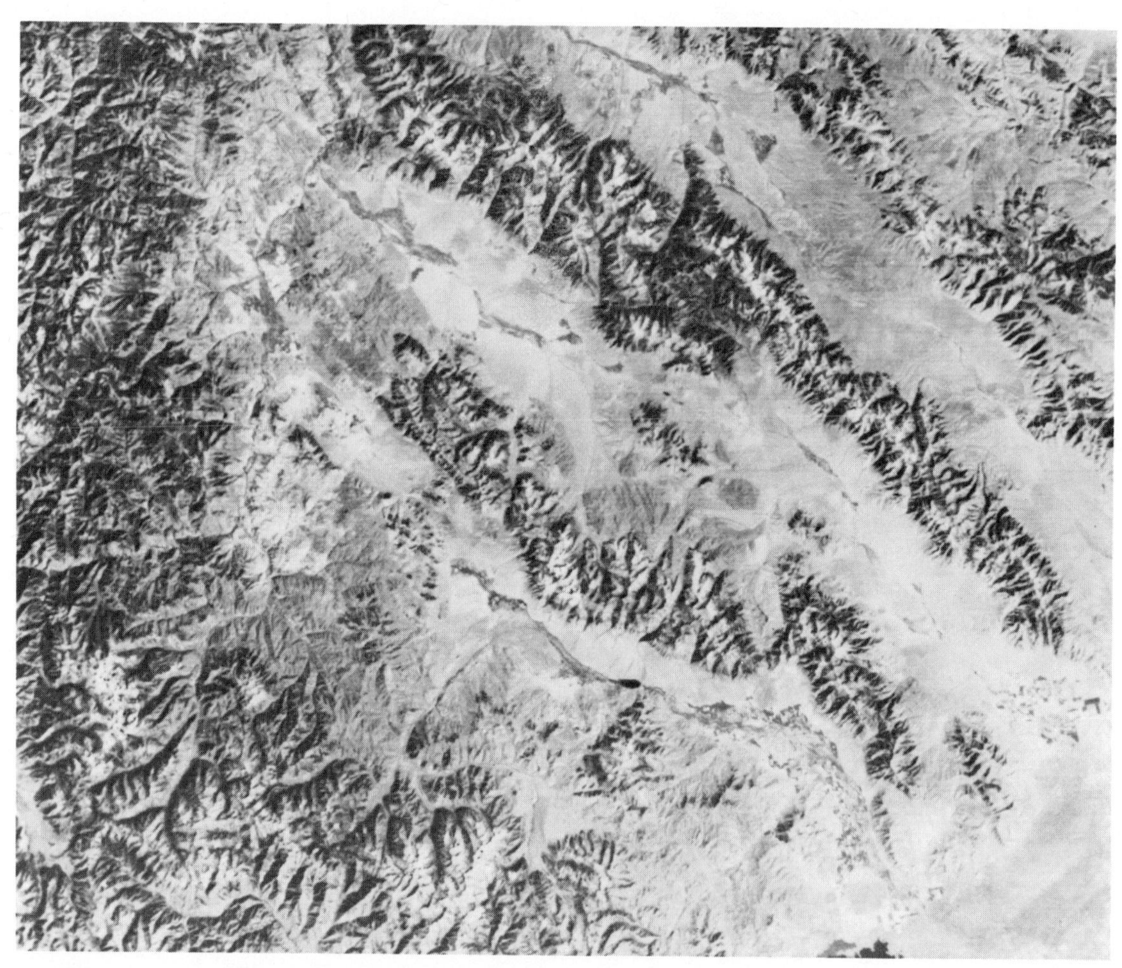

FIGURE 16–19 **Northern boundary of the Snake River Plateau from Landsat satellite imagery. On the left of the frame the highlands of the Idaho batholith. In the middle, the northwest-trending ranges of the Rocky Mountains cut off in the southeastern corner by the smooth Cenozoic lava plain along the Snake River. (Courtesy of NASA.)**

FIGURE 16–20 **Mosaic of Landsat images of the Idaho–Wyoming border area showing the relationship of the Snake River plain to the Yellowstone Plateau. 1. Snake River lava plain. 2. Yellowstone National Park and Yellowstone Lake. 3. Absaroka Range. 4. Teton Range and Jackson Lake. 5. Idaho–Wyoming thrust belt of the central Rockies. (Courtesy of NASA.)**

from the cooling of buried lava flows or from a body of intrusive igneous rock just below the surface.

The eastward progression of the center of igneous activity along the Snake River Plateau from western Idaho to western Wyoming during late Cenozoic time at a rate of about 4 cm per year has suggested to Jason Morgan that the great outpouring of lavas resulted from a mantle plume (see Chapter 1) over which the Americas plate was moving during late Cenozoic time, and which now lies near the site of Yellowstone Park. Other geologists have suggested that the plume is responsible for the general elevation and extension of the central part of the Cordillera in late Cenozoic time. They postulate that the North American plate is breaking up in this sector and that the Snake River Plateau is underlain by a zone of rifting where new oceanic crust is beginning to form. Similar lava fields elsewhere in the world, such as the North Atlantic basaltic province and the Deccan Plateau of India, have also been attributed to the effects of mantle plumes (the Iceland and Reunion plumes, respectively).

The Cascades

The Cascade Range of Washington, Oregon, and northern California (Fig. 16–21) in-

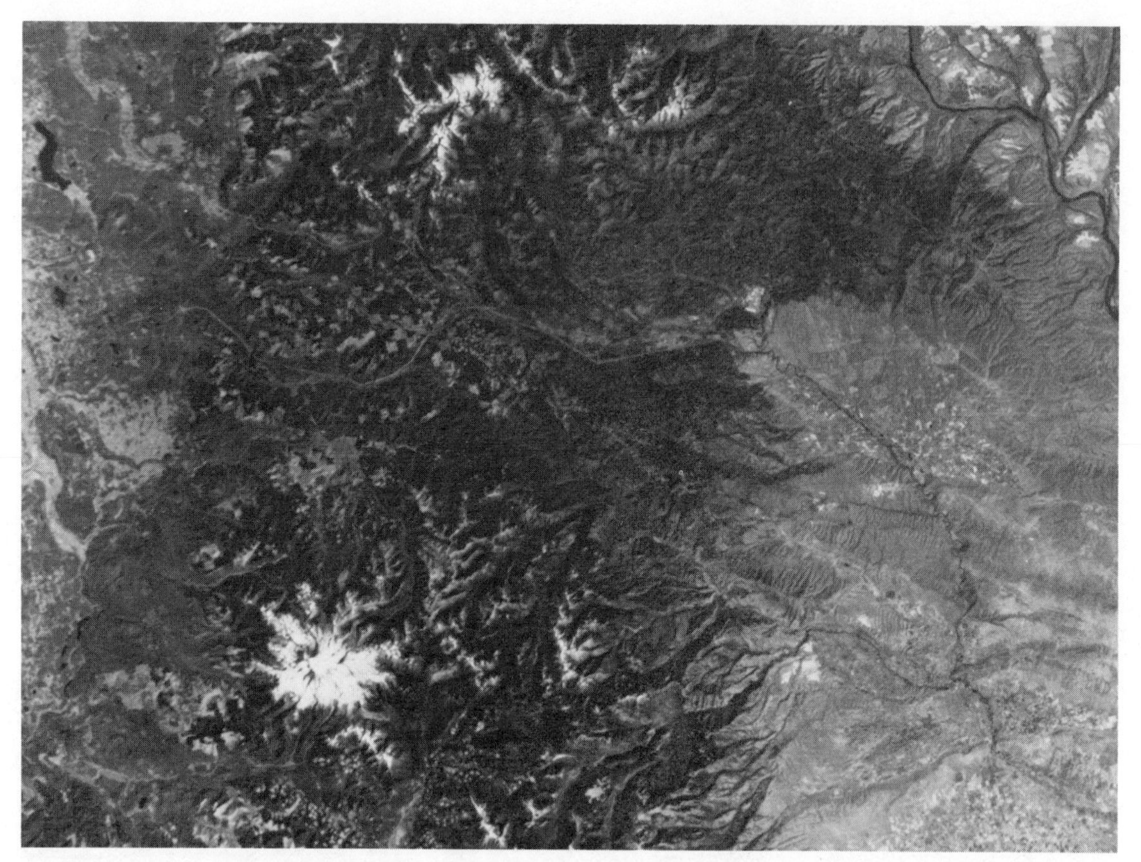

FIGURE 16–21 **Cascade Range of Washington. Landsat imagery showing Mount Rainier, the snow covered patch in the left corner, the Puget trough at the left, and the Columbia Plateau and the Columbia River on the right. Seattle is just off the frame in the upper left corner. (Courtesy of NASA.)**

cludes a group of volcanoes built on the western side of the Columbia lava plateau in Pleistocene time. The major peaks, such as Shasta (4316 m), Rainier (4413 m), Hood (3427 m), and Baker (3276 m) combine massive symmetry and graceful beauty with altitudes approaching the highest in the Cordillera (Fig. 16–22).

The lavas of the Cascades belt are as old as Eocene, but the major cones accumulated during Pleistocene time by the extrusion of andesite flows. At the southernmost of these cones, Lassen Peak (3186 m), a Peléean-type eruption took place in 1914 and 1915. In the mid-seventies fumarolic activity at Mount Baker melted large quantities of snow and ejected some ash, showing that this, and adjacent volcanoes of the chain, may only be dormant and not extinct. The internal structure of one of the Cascades volcanoes, the summit of which collapsed amid violent explosive eruptions, is exposed at Crater Lake, Oregon. The zone of recent volcanism extends northward into southern British Columbia, where undissected lava flows and cinder cones are evidence of activity within the last few thousand years.

The coincidence of the Cascade belt of modern volcanism with the subduction zone that exists between northern California and Vancouver Island suggests that these volcanoes were, and are, associated with the descent of the subducted ocean crust and

FIGURE 16–22 **Mount Rainier, Washington, an aerial photograph looking south toward the other Cascades peaks: Mount Adams (4391 m) at left, Mount Hood (3427 m) in the center distance, and Mount St. Helens (2848 m) on the right. (Courtesy of H. Miller Cowling, 116th Photo Section, Washington National Guard.)**

marginal sedimentary prism along the remaining segment of the Cordilleran subduction zone. The other belts of active volcanism in North America—the Aleutian Island arc and its continuation into the Alaska Range, and the central Mexican volcanic belt—are both associated with active subduction zones, the Aleutian and Middle America trenches. Within the Mexican belt are the highest peaks in Mexico, including the famous Popocatepetl (5452 m), which overlooks Mexico City.

COLORADO PLATEAU

The Colorado Plateau was little affected by the compressional forces of Cordilleran orogeny that deformed the Sevier thrust belt along its western boundary, but during late Cenozoic time it was the site of the intrusion of many small, igneous bodies and the eruption of volcanoes.

From his study of the Henry Mountains of Utah located near the center of the plateau (Fig. 16–23), G. K. Gilbert, in 1877, coined the term "laccolite" to describe a type of sill that domes the overlying beds, making room for molten rock. Later work has shown that the structure of the Henry Mountains is not as simple as Gilbert had conceived it, and that each mountain is composed of many small, tongue-shaped laccoliths (as laccolites are now called) surrounding a central stock. Simple, domed laccoliths, similar to the intrusions described in Gilbert's classic interpretation of the Henry Mountains, occur in the La Sal and Abajo mountains to the east.

During Pliocene and Pleistocene times a group of volcanoes known as the San Fran-

FIGURE 16–23 **Basins and ranges of the central Rocky Mountains and the Colorado Plateau.**

cisco Mountains was built by successive eruptions in the neighborhood of Flagstaff, Arizona (Fig. 16–24). Most of the volcanoes of the Hopi Buttes and Navajo volcanic fields have been eroded and only the resistant necks remain as buttes, surrounded by radiating dikes such as those at Ship Rock, New Mexico.

In addition to the minor intrusions mentioned above, a broad belt in central and southeastern Arizona, western New Mexico, and central Utah was intruded by a group of stocks ranging in composition from granite to diorite. Ore-forming fluids accompanying these intrusions deposited valuable copper ores in the host rocks of Paleozoic and pre-

FIGURE 16–24 **San Francisco volcanic field, north of Flagstaff, Arizona, looking southeast. An aerial photograph showing a recent, undissected lava flow that has issued from a small volcanic cone. (Photograph by John S. Shelton.)**

Paleozoic ages. Some of these ores are identical to those of the Colorado porphyry belt. The deposits at Bisbee, Arizona, have yielded 2.5 billion tons of copper and large quantities of silver, gold, and other metals. The colossal open pit at Bingham, Utah, is the largest copper mine in the United States and produces about 250,000 tons per year of this metal. A host of smaller mines are based on ore deposits formed during this Cenozoic episode of intrusion.

The great depth and spectacular scenery of the Grand Canyon, which incises the southwestern corner of the Colorado Plateau (Fig. 3–17), have stimulated much study of the time of uplift of this sector of the crust and the rate at which the Colorado River has cut this 1700 m deep canyon. Potassium–argon dating of lava pebbles in gravels deposited before and after the downcutting bracket the time of uplift as between 5 and 10 million years ago in early to middle Pliocene time. The uplift of the Colorado Plateau was another expression of the general Pliocene rejuvenation of the Cordillera.

CENOZOIC BASINS OF THE PACIFIC COAST

The Sierra Nevada of eastern California is separated from the highlands of the Coast Ranges by the Great Valley of California, a basin floored with Cenozoic rocks. The Coast Ranges are composed of metamorphosed and highly deformed late Mesozoic sedimentary rocks of the Franciscan and Knoxville groups. A system of faults, along which movement has been largely in a right-lateral sense, slice through the Coast Ranges. These faults join and split, forming a braided zone about 100 km wide (Fig. 16–25). Within this zone the San Andreas fault is the most persistent and can be followed from Los Angeles to the coast just north of San Francisco (Fig. 16–26). Just north of Los Angeles the fault zone swings eastward about 100 km where a group of east–west faults cut the Mesozoic highlands into blocks which define the Transverse Ranges. The longest of these east–west faults, the Garlock fault, forms the boundary between the Sierra Nevada and the Mohave Desert (Fig. 16–27). The fault system has broken the unity of the Coast Ranges, form-

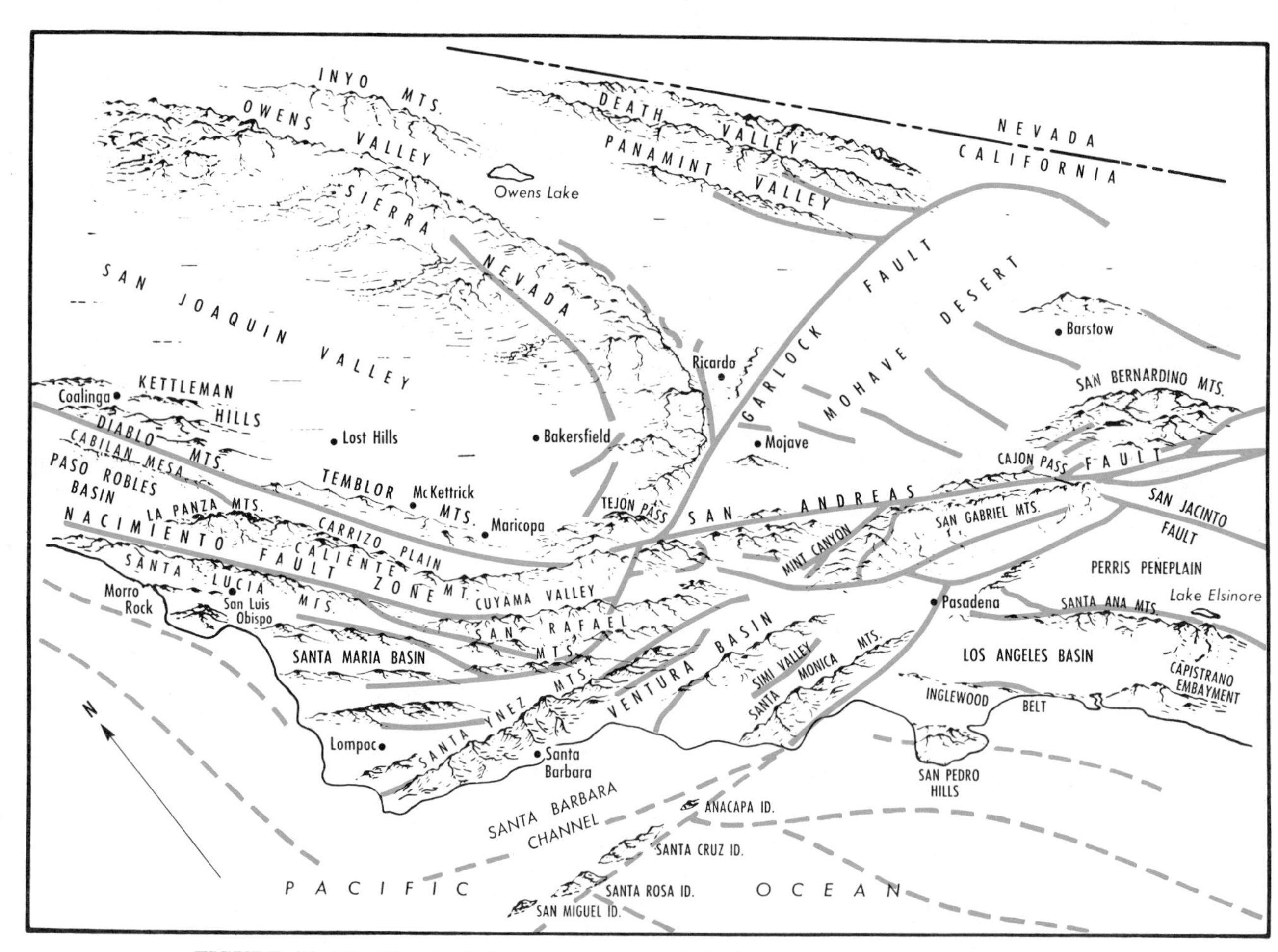

FIGURE 16–25 **Sketch of the topography and fault patterns of southern California. (After a diagram by L. D. D. Reed and J. S. Hollister; courtesy of the American Association of Petroleum Geologists; modified with the assistance of J. C. Crowell.)**

FIGURE 16–26 **Trace of the San Andreas fault near the northern end of the Temblor Range, California, looking north. Note the disruption of the drainage pattern caused by recent faulting. (Photograph by John S. Shelton.)**

ing basins in which Cenozoic sediments accumulated to great thicknesses.

Basins of Southern California

In the Los Angeles area the stratigraphy of the Cenozoic basins is known in great detail because their sedimentary fillings have been extensively drilled for oil. At times the basins were open to the sea to the west; at other times they were filled with continental deposits. Sedimentation in the basins was rapid and neighboring uplifts were available as suppliers of sediments. In the Ventura Basin 14,000 m of Cenozoic beds accumulated, of which 6000 m are of Eocene and Paleocene age. In the middle of the basins most of the beds are marine, but at the margins they interfinger with clastic, non-marine sediments that spread out as deltaic deposits from the highlands. Rapid lateral and vertical facies changes are characteristic of the sedimentary records of these basins.

By comparing foraminifers collected from these Cenozoic sediments with those now found off the California coast, geologists can deduce the depth of water in the basins millions of years ago. The similarity of foraminifers in Miocene beds to those that now live at depths of 1200 m in the Pacific Ocean is taken to indicate that the depth of water in some of the Miocene basins was similar. Only with relatively recent fossils are such determinations of

FIGURE 16–27 **Aerial photograph looking southeast from the southern end of the Great Valley of California across the Tejon Pass to Los Angeles beneath its smog blanket. Compare with Fig. 16–25 and note the Mojave desert on the upper left, the end of the Sierra Nevada on the middle left, the trace of the San Andreas fault crossing the photo from lower right to upper left and displacing the Garlock fault in the center of the photograph. (U.S.G.S. aerial photograph taken by the U. S. Air Force.)**

depth of deposition accurate, because the older the fossils are the less they resemble living species, and the less certain we are that they lived in similar environments.

The Cenozoic sediments of the basin fillings are folded, faulted, and interrupted by many unconformities. Episodes of deformation are distributed throughout the Cenozoic Era, and different basins have different tectonic histories. Numerous recent earthquakes indicate that the faults are still active, and deformation of the basin sediments continues to this day. Repeated land surveys that involve the determination of precise elevations show that parts of the crust are rising at appreciable rates. For instance, the Palmdale district north of Los Angeles rose about 25 cm in the early sixties. Accurate strain gauges placed near major faults like the San Andreas also reveal the slow deformation of the bedrock under stresses in the crust (Fig. 16–28).

The folding and faulting of the basin sediments in California have produced many traps for petroleum; the porous sandstone beds make good reservoirs. The reserves of the California oilfields are estimated at 4 billion barrels. It is one of the major oil-producing regions of the continent. Present exploration in this area has been concentrated on the extension of the Cenozoic basins under the continental shelf, as most of the onshore oil was found in the early part of this century.

FIGURE 16–28 **The Los Angeles area from Landsat satellite. Note the prominence of the San Andreas fault along the top of the photograph and the closeness with which the topography reflects the fault pattern shown in Fig. 16–25. The city occupies the lowland area in the lower right sector of the photograph. (Courtesy of NASA.)**

Fault Movements and Basin Deformation

The right-lateral movement along most of the faults bounding the basins of California reflects the northward movement of the Pacific plate relative to the Americas plate. The continuing motion of the lithosphere plates causes many minor and some major earthquakes along these faults in southern California. In the 1906 San Francisco earthquake the maximum horizontal displacement along the San Andreas fault was 6.2 m. Displacement of drainage lines and other topographic features indicate that this direction of motion has continued for at least the last few thousand years, and displacement of stratigraphic and structural markers suggests that hundreds of kilometers of displacement has taken place since mid-Cenozoic time. The closure of the Gulf of California, which has been opened by the right lateral movement between the Pacific and Americas plates requires 580 km of displacement. Tanya Atwater has calculated that the reconstruction of the Pacific plate boundaries, through interpretation of the patterns of magnetic anomalies, requires a relative movement of the two plates of 1400 km in the last 23 million years, but this displacement could have been partly taken up by other forms of deformation of the continental margin.

Many geologists believe that the formation, deformation, and sedimentation of the Cenozoic basins in California is a result of this motion. The basins have formed where blocks bounded by faults have been tilted or dropped, or where a sinuous change in the direction of a major fault has caused the crust to pull apart into a depression (Fig. 16–29). The repeated deformation of these basins during the Cenozoic Era probably is caused by the periodic adjustments along bounding faults as the lithosphere plates shear past each other.

Further evidence of the right-lateral movement along the Pacific–Americas plate boundary is found in the northern Cordillera. The displacement along the Denali fault, which traces an arc through southern Alaska, has been estimated at 400 km by the matching on either side of its trace of geological features separated by the movement.

Central Pacific Borderlands

Between northern California and Vancouver Island the Juan da Fuca plate (a remnant of the Farallon plate) is still being subducted beneath the Americas plate. Here, the motion of the plates is still convergent and the effects of the right-lateral shear so evident farther south and north are not present.

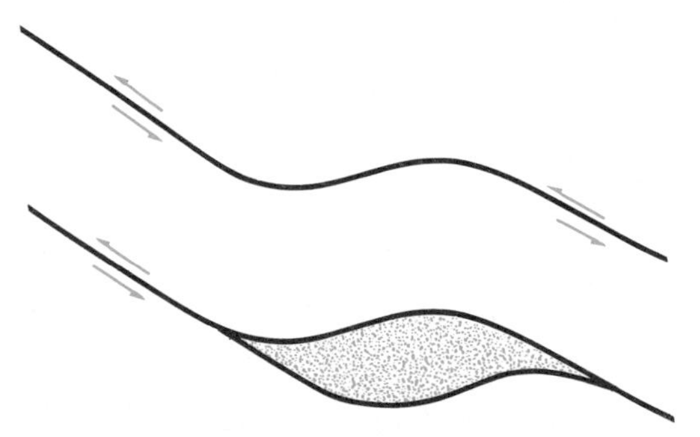

FIGURE 16–29 **Diagrammatic map of the formation of a "pull-apart" basin at the sinuous curve of a left-lateral fault.**

The Coast and Olympic mountains of Oregon and Washington are underlain by a very thick section of Cenozoic geosynclinal rocks, including turbidites and volcanic units several kilometers thick. These deposits show the effects of repeated deformation in many angular unconformities. This disturbed Cenozoic sequence continues beneath the narrow continental shelf off the coastline. At the base of the continental slope a wedge of sediment $2\frac{1}{2}$ km thick has built up within the last few million years. The movement of the Juan da Fuca plate is deforming this sedimentary wedge into anticlines and synclines that can be detected by geophysical surveys. This section of the Cordilleran coastline shows many features of sedimentation, tectonism, and volcanism that are typical of ancient geosynclines. That motion along the subduction zone during the past few thousand years has been small is shown by the absence of a topographic trench along the zone of accumulation and the absence of a Benioff zone (see Chapter 1) of earthquake loci indicative of the motion of the descending slab.

Although along some of the Pacific margin the conditions that resulted in the major deformation in the Cordillera have ceased to exist and transverse movements have replaced compressive ones, the Cordilleran subduction zone remains against Alaska, the central Pacific borderlands, and Mexico and, in a sense, the Cordilleran orogeny continues to this day.

CENOZOIC MAMMALS

The Cenozoic succession of terrestrial deposits in the intermontane basins of the Cordillera and the Great Plains preserves a record of the early Cenozoic diversification of the mammals and their evolution throughout the era. North America and Eurasia, the Holarctic realm, appear to be the areas of major evolutionary activity in the history of the mammals, areas from which many groups dispersed to the southern continents.

Primitive Mammals

Primitive mammals, both placentals and marsupials, were relatively small, and probably nocturnal and arboreal in habit. Their teeth were sharp-cusped, suitable for stabbing, shearing, and crushing small prey, such as insects and other terrestrial arthropods. This way of life has been further intensified in a living group of placental mammals, the order INSECTIVORA, represented by the shrews, moles, and hedgehogs. The living opossum is very similar to the early representatives of the marsupial stock. This modern animal is larger and eats a greater variety of foods, but the teeth and most skeletal features are, nevertheless, very close to the pattern of Late Cretaceous mammals.

Much of the mammalian radiation may be discussed in terms of dietary specialization from the primitive insectivorous pattern. In many mammalian groups the specialization of dentition is related to the available plant food. Within the broad herbivore adaptive zone many variants can be recognized. Several mammalian groups of small size have specialized the incisors as sharp chisels to gnaw plant food. Modern rodents are the most successful of gnawing forms, but many other mammalian groups of the Mesozoic and Cenozoic evolved similar adaptations. Larger herbivores may be classified as browsers or grazers, depending on whether they feed on relatively soft vegetable matter or grass. This is reflected in the configuration of the tooth cusps and the pattern of tooth growth. Large herbivores may either so increase their size that their bulk protects them from most predators (for example, the elephant or hippopotamus), or they become fast-running forms, such as the many members of the horse and cow groups. The latter strategy usually results in the evolution of long, slim limbs terminating in hooves. Such animals are collectively termed "ungulates."

The various adaptive patterns, as exemplified by diet, are not limited to individual taxonomic groups, but recur in various groups at different times and places. As the placentals and marsupials both radiated into different continental blocks fragmented by the break-up of Pangaea (Chapter 19), gnawing, browsing, and grazing herbivores developed in each area from primitive groups of a more or less insectivore level of development. Similarly, the carnivore pattern developed in all regions, although in South America and Australia the carnivores evolved from marsupials, and in the holarctic realm from placental stocks. Such attainment of similar specialized adaptive patterns with similar specialized anatomy is termed "convergence." It is a common feature of evolution.

In the northern continents the initial diversification of the placental mammals began late in the Mesozoic Era. Three orders can be identified in uppermost Cretaceous beds: the insectivores, condylarths (archaic herbivores), and primitive primates. The primates will be considered in more detail in Chapter 21.

Carnivores

A natural extension of the insectivore way of life was toward the role of the larger carnivores. Two orders of placentals specialized in this way of life. The creodonts of the Paleocene and Eocene (with stragglers in the late Cenozoic) evolved a number of dog-, hyena-, and bear-like lineages, which were successful in preying on primitive herbivores. At the same time a second group of small carnivores with larger brains evolved separately from the insectivores. The group developed a particularly effective combination of slashing teeth, termed "carnassials," involving the elongation into a sharp-edged ridge of the last upper premolar and the first lower molar (Fig. 16–30). This group, called the miacids, is ancestral to all modern carnivores. The miacids soon diversified into several types of terrestrial carnivores, the fissipeds, of which one line led to the civets, hyenas, and cats; the other led to the dogs, raccoons, bears, and weasels. A distinct group of living carnivores, known as pinnipeds, includes the seals and walruses. These evolved from early fissipeds, which were similar to the ancestors of the otter.

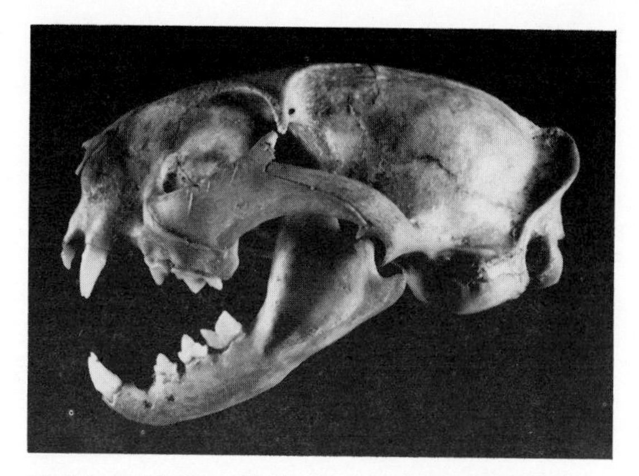

FIGURE 16–30 **Skull of a modern carnivore, the Lynx. The last teeth visible in the upper and lower jaws, termed carnassial, have specialized shearing surfaces. (Courtesy of the Redpath Museum.)**

The pinnipeds had lost most of their tail before becoming adapted to aquatic life, and now propel themselves with limbs that are modified into paddles.

The dog family (Canidae) is one of the oldest in the carnivore group, and is one of the few mammalian lineages in which increase in size to gigantism has not occurred. Within the cat family (Felidae) there developed the saber-toothed "tigers," whose upper canine teeth enlarged to dagger-like proportions and projected several centimeters below the lower jaw when the mouth was closed. The lower jaw opened the mouth on a special hinge to clear the stabbing teeth for killing large, thick-skinned prey (Fig. 16–31). The last and largest of the saber-toothed cats was the late Pleistocene genus *Smilodon.* A great many skeletons of this tiger-sized cat have been excavated from an asphalt seep at Rancho la Brea, now within the city of Los Angeles. The last saber-toothed cat died a few thousand years ago when large Pleistocene mammals, such as mastodons and ground sloths, on which they presumably preyed, became extinct, events to which the earliest human invaders of North America probably contributed.

Rodents

Among the herbivores, true rodents do not appear until the end of the Paleocene, although very rodent-like primates had appeared earlier in this epoch. The gnawing habit of the rodents has been extremely successful, leading to an extensive radiation of small to medium-sized herbivores of worldwide distribution. The rabbits (order LAGOMORPHA) are not closely related to the rodents, although they evolved very similar gnawing incisors. The ancestry of the rabbits has not yet been firmly established. The oldest representatives of the group have been found in late Paleocene beds of Mongolia. Unlike the rodents, the rabbits have never diversified, and modern forms generally resemble their early Cenozoic antecedents.

Bats presumably evolved from primitive insectivores, but the earliest (Eocene) bats are very close to modern bats and no intermediates are known.

FIGURE 16–31 **The saber-toothed cat, *Smilodon*, from the Pleistocene of North America. The skull is about one-sixth natural size, the animal about one-twelfth. (From Carroll Lane Fenton and Mildred Adams Fenton, *The Fossil Book.* Copyright © 1958 by Carrol Lane Fenton and Mildred Adams Fenton. Reprinted by permission of Doubleday & Co., Inc.)**

Condylarths

The condylarths, first known in the Late Cretaceous, are the basic ungulate stock. The early members are barely distinguishable from insectivores; the most advanced approach the modern hoofed orders. The best known genus *Phenacodus* (Fig. 16–32), has the dimensions of a tapir, but appeared as much like a carnivore as a herbivore in its proportions. The toes were hoofed, however, and the blunt cusps of the teeth suggest a diet of soft plant food.

Perissodactyls

Two groups of large herbivores dominate the modern fauna: perissodactyls, including the horse, that have an odd number of toes, and the artiodactyls, cows, pigs, etc., with an even number of toes. The perissodactyls evolved from the primitive herbivore stock, the condylarths, and appear at the very end of the Paleocene. The origin of the artiodactyls cannot yet be specified. They appear at the beginning of the Eocene.

In the perissodactyls the axis of the foot along which the weight of the body was supported lay through the third or middle toe. The mammals that developed hoofs were herbivores of the forest and plain, and initially depended on running to evade their enemies. The most primitive perissodactyl, *Hyracotherium* (Fig. 16–33), is closely related to the ancestry of the whole order, but

FIGURE 16–32 **Reconstruction of the Paleocene condylarth *Phenacodus*. (By Charles Knight, courtesy of the American Museum of Natural History.)**

FIGURE 16–33 **The earliest member of the horse family, *Hyracotherium*, sometimes called *Eohippus*, an animal about the size of a small dog. (Reproduced from J. Augusta and Z. Burian, *Prehistoric Animals*, London: Spring Books, 1956.)**

is particularly thought of as the earliest member of the horse family. The evolution of this family is one of the best-documented evolutionary sequences (Fig. 16–34). Between the upper Paleocene and the present, this lineage evolved from small, running, forest-dwelling animals having four toes on the front feet and three on the hind feet, with a dentition suitable only for chewing soft plant material, to large, galloping animals of the open plain, capable of chewing and digesting one of the toughest foods, grass. The bones of the feet became elongated, the teeth increased in complexity, the digits were reduced to a single unit, and the brain increased in size and complexity. These changes did not, however, occur at a regular rate, nor did they all occur in the many separate lineages. During the Eocene the horses increased little in size from the terrier-like *Hyracotherium,* but throughout the remainder of the Cenozoic their size increased gradually. The spread of grasslands in the middle Cenozoic provided an alternative diet, and the profound dental changes needed to accommodate it were initiated in a new lineage. The ancestors of the modern horse became grazers in Miocene time and evolved high-crowned teeth. The first of this advanced group that was adapted to life on the grassy plains, where the ability to run rapidly was a distinct advantage, was *Merychippus,* which walked on one toe and held the two shorter side toes off the ground. By the end of Miocene time one group of grazing horses (*Pliohippus*) had evolved to the single-toed stage. The modern horse, *Equus,* arose from *Pliohippus* in the Pleistocene Epoch, by which time all the side branches of the family had become extinct, except *Hipparion,* which retained three toes into early Pleistocene time.

The earliest horses were widespread in the Northern Hemisphere during the Eocene Epoch. Through the later Cenozoic, major

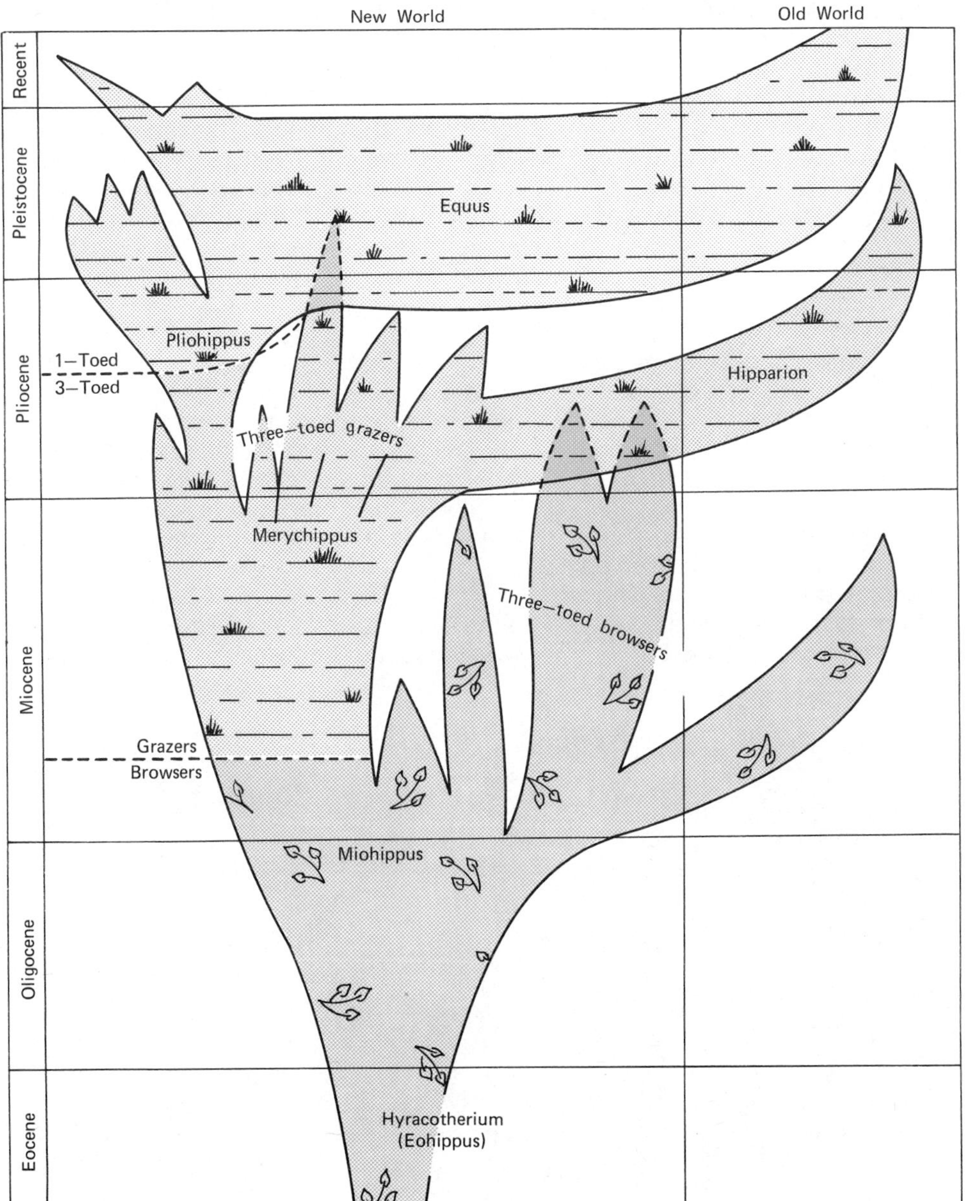

FIGURE 16–34 **The evolution of the horse family, the Equidae. The leaf and grass designs indicate browsing and grazing habits, respectively. (After G. G. Simpson.)**

advances in horse evolution appear to have been localized in North America, but migration to and from Europe and Asia occured repeatedly. In the Pliocene horses invaded South America. Despite their dominance in North America during most of the Cenozoic, horses became extinct here during the Pleistocene Epoch. The wild horses of this continent are all descendants of recent imports from Europe.

The modern African, Indian, and Indonesian rhinoceroses are remnants of a large, diverse group of mammals that originated in Eocene time and, for the rest of the Cenozoic, were widespread across the northern continents. Many of the early forms were swift animals of the plains, and did not have the horns that are so prominent on modern rhinoceroses. At the end of Oligocene time this stock produced *Baluchitherium,* the largest land mammal that ever lived—7 m in length and $5\frac{1}{2}$ m high at the shoulders. Its long front legs and neck allowed it to browse on foliage that smaller beasts could not reach.

During the Miocene Epoch a small rhinoceros, called *Diceratherium,* roamed the western plains in enormous herds. Their bones make up the bulk of the Agate Springs bone bed (Fig. 16–35), an accumulation of thousands of skeletons of animals probably drowned in a sudden flood. During the Ice Age, wooly rhinoceroses roamed much of Europe and Asia. The nature of the fur of these extinct animals is known through a study of two specimens preserved in a Polish oil seep, and also from drawings of the animals by prehistoric cavemen.

The titanotheres were heavy, hoofed mammals with primitive teeth (Fig. 16–36). This dominantly North American group appeared in Eocene time as contemporaries of the first horses, and lasted until early Oligocene time. The group increased in size rapidly and developed various patterns of horns on the head. The braincase remained small and primitive. The largest of the titanotheres, *Brontotherium,* had a thick-set body supported by pillar-like legs and stood $2\frac{1}{2}$ m high at the shoulder.

The chalicotheres (Fig. 16–37), a long-ranging (Eocene–Pleistocene), superficially horse-like group of perissodactyls, retained claws inherited from their condylarth ancestors, but are classed with the hoofed mammals on the basis of other features of their skeletons. Their clawed feet appear to have been adapted to digging roots.

Artiodactyls

In the artiodactyls, the weight of the animals was distributed between the third and fourth

FIGURE 16–35 **A portion of the Agate Springs bone bed (Miocene of Nebraska). Most of the bones are of the rhinoceros *Diceratherium,* but a few are of *Dinohyus* and *Moropus* (see Fig. 16–37). (Courtesy of the Smithsonian Institution.)**

FIGURE 16–36 **Reconstruction of a group of titanotheres belonging to the genus *Megacerops* (Oligocene beds of Saskatchewan). (Painting by P. R. Haldorsen; courtesy of the Geological Survey of Canada.)**

toes. When they rose up on their toes, achieving greater speed, they first became four-toed and then two-toed. These mammals make up a much larger group than the odd-toed hoofed mammals. They are divided into the SUINA, including pigs and hippopotamuses, and the RUMINANTIA, including camels, deer, giraffes, and cattle. Pigs first appeared in Oligocene time and became one of the most successful of the mammalian groups. They are highly intelligent animals, adapted to life in the forest and to an omnivorous diet. They tend to retain four digits in the feet. The hippopotamuses, adapted to a semi-aquatic life, did not appear until the Pliocene Epoch. An early offshoot comprised the entelodonts, sometimes referred to as the giant pigs. The skull of *Dinohyus* (Fig. 16–37), the largest and last, was nearly a meter in length.

The course of the evolution of the camels is almost as well known as that of the horses and is, in many respects, similar to it. The first camels were contemporaries of the first horses in the Eocene Epoch. By Oligocene time they had risen off their lateral toes and walked only on the third and fourth ones. In mid-Cenozoic time two branches of the camel family existed, the running camels and the giraffe camels. Typical of the latter group was *Alticamelus,* whose long neck supported its head 3 m above the ground. North America was the home of the camel through most of the Cenozoic Era, but during the Pleistocene Epoch camels mysteriously died out here and are now found only in South America (llama) and in the Old World.

The oreodonts were middle and late Cenozoic ruminants closely related to the camels but sheep-like in form and of small size. Great numbers of their skeletons have been found in North America, indicating that they roamed the plains in large herds. They retained the four-toed foot.

The first members of the deer family ap-

FIGURE 16–37 **In the foreground of this Miocene scene three entelodonts (*Dinohyus*) dig for roots. On the right are two chalicotheres (*Moropus*) with horse-like bodies but clawed feet. A few three-toed horses can be seen on the left (*Parahippus*). (Mural by Charles Knight; courtesy of the Field Museum of Natural History.)**

peared in Miocene time and have been predominantly an Old World group; today, this group includes a great variety of mammals, such as elk, moose, wapiti, and caribou. One of the most significant steps in the evolution of this family has been the acquisition of antlers that are discarded and regrown each year. The largest set of antlers unearthed was found in an Irish peat bog and was grown by the extinct "Irish elk" (actually a deer) of late Pleistocene time. Giraffes, close relatives of the deer, began later in the Miocene but are unknown in North America.

Cattle, or bovoids, are characterized by long legs adapted to running, and horns, which are not shed, on both the male and female. Unlike the horse, camel, and rhinoceros, the cattle are an Old World family which evolved in Europe and Asia late in Miocene time. Only a few members of this group, such as the bison, mountain sheep, and mountain goat, migrated to North America, and none reached South America.

Proboscidians

The first proboscidians (the group to which the elephants belong) were the size of a pig and had few distinct elephantine features, but by middle Cenozoic time several lines had evolved. The group appears to have originated in Africa, but by Miocene time appeared in North America. In some, both upper and lower incisor teeth were elongated, giving the animal four tusks; in others, the lower tusks were flattened, directed forward, and almost joined, forming a shovel-like projection in front of the mouth. Yet another branch, the deinotheres (Miocene–Pleistocene), had lower tusks that curved sharply backward toward the body. How the animal used such tusks is problematical, but it has been suggested that they were used in uprooting trees.

The jaws of early proboscidians were relatively long, compared with those of modern elephants, and contained many teeth bearing blunt cusps. These animals are mastodons (Fig. 16–38). Their teeth indicate that they were browsers. They lived in North America until Pleistocene time and then became extinct. One group of proboscidians, the elephants, branched from the mastodon stock in Pliocene time and evolved a striking dental apparatus for chewing abrasive grasses. Their teeth became large, with resistant enamel intricately

FIGURE 16–38 **Mastodons, skeletons, and a reconstruction from New York State. The background represents the Catskill Mountains as they may have looked late in the Ice Age. (Courtesy of the New York State Museum.)**

infolded (Fig. 16–39), but only one at a time was used on each side of the short jaws. The largest of the elephants, the woolly mammoth that inhabited North America, Europe, and Asia during the Ice Age, was about 4 m high at the shoulders and was covered with shaggy, reddish hair (Fig. 16–40). Almost as much is known about these extinct animals as if they still lived, because several specimens have been found frozen in the Siberian tundra in such good condition that the flesh was still preserved. One of the major sources of ivory has been Siberian mammoths' tusks, which reach a length of 4 m. These lumbering elephants were hunted by early European man, who painted their pictures on the walls of his caves. Both mammoths and mastodons were common in North America up to the close of the Wisconsin glaciation.

SUMMARY

By early Cenozoic time non-marine sedimentation in isolated successor basins had largely replaced marine sedimentation throughout the Cordillera. During the Cenozoic Era the East Pacific Rise progressively encountered the Cordilleran subduction zone, and the sections of the boundary between the oceanic plates and the Americas plate characterized by transverse motion increased in length at the expense of those along which subduction was taking place.

The relief of the Wyoming-Colorado Rockies in Eocene time was great, but the erosion products of the ranges were filling the basins with alluvial and lake sediments such as the Green River Formation. By mid-Cenozoic time pediments cut into the

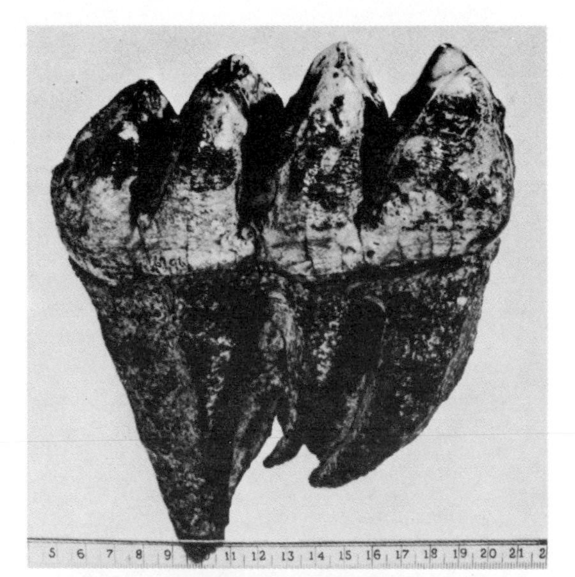

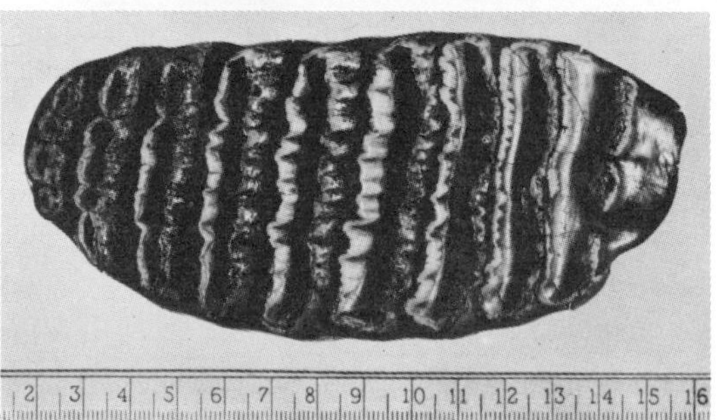

FIGURE 16–39 **Teeth of mastodon and mammoth. *Top:* A side view of a mastodon's tooth showing the paired cusps for crushing foliage, typical of a browsing animal. *Bottom:* A top view of a tooth of a mammoth or elephant showing the infolding of the enamel that strengthened the tooth for grazing. The scales are in centimeters. (Courtesy of the Redpath Museum.)**

ranges were continuous with the thick early Cenozoic sediments that partly filled the basins and considerably reduced the relief. At this time detritus from the mountains built a vast alluvial fan eastward across the plains. In late Pliocene time the present relief of the Rockies was produced by regional uplift or climatic change that caused the rivers to excavate the basin sediments and cut canyons through the emerging basement rocks as they were superposed across them.

The Basin and Range province was swept by nuées ardentes depositing ignimbrites in mid-Cenozoic time. Change to a less explosive volcanic activity coincided with the end of subduction movement along the Americas' plate boundary in this latitude. In late Cenozoic time the area of the Basin and Range was extended, probably by upward spreading of material beneath the crust and by the right lateral motion taking place along the Americas-Pacific boundary. The elevation of the Sierra Nevada along a fault on its east side in late Pliocene time cut off the supply of rain to the Basin and Range province, forming the desert of today.

In Miocene and Pliocene times highly fluid basaltic lavas poured out over much of Washington and Oregon and adjacent states to form the Columbia and Snake River plateaus. The Colorado Plateau was the site of numerous small volcanic intrusions and extrusions in late Cenozoic time. Its uplift resulted in the cutting of the Grand Canyon in the interval between 5 and 10 million years ago.

The major volcanic peaks of the Cascade Range built mostly in Pleistocene time are localized east of the remnant of the Cordilleran subduction zone.

The right lateral movement between the Americas and Pacific plates was responsible for the formation and deformation of a series of basins in southern California that filled with thick successions of Cenozoic sediments. Cumulative movement along the transverse faults along the Pacific margin in Cenozoic time may have amounted to over 1000 km.

Primitive mammals of the late Mesozoic were primarily small insectivores with sharp-cusped teeth for small prey. The major mammalian groups diversified in late Cretaceous and early Cenozoic time. Two groups of meat-eating mammals arose from insectivores in Paleocene time: a primitive order characteristic of the early Cenozoic, the creodonts, and the modern order Carniv-

FIGURE 16–40 **A herd of woolly mammoths in a Pleistocene landscape. This animal was common to North America and northern Europe. (Mural by G. A. Reid; courtesy of the Royal Ontario Museum.)**

ora, which includes dogs, cats, bears, seals, etc. Rodents developed large incisors for gnawing. Ungulates (hoofed mammals) arose from a primitive order, the condylarths, in late Cretaceous time. The horses evolved from condylarth ancestors in Eocene time and evolved along a well-documented course involving changes in toes, leg length, overall size, brain size, teeth, etc. Rhinoceros and camels were common mid-Cenozoic ungulates. Extinct groups of mid-Cenozoic ungulates include the titanotheres, chalicotheres, entelodonts, and oreodonts. Proboscidians of the mid-Cenozoic include types with shovel-like, recurved, and two pairs of tusks. Elephants are proboscidians that specialized for chewing grass by the development of large resistant teeth.

QUESTIONS

1. The canyons in the Wyoming Ranges are described as superposed. In what other ways can rivers cut valleys across resistant ridges?
2. Describe the conditions of topography, drainage, and climate under which the Green River Formation was laid down in western Wyoming. What fossils are found in these beds?
3. Make a tabular summary of the events since the beginning of the Paleozoic that have affected the Basin and Range province of Nevada. Include episodes of structural disturbance, erosion, and deposition.
4. What kinds of evidence can the historical geologist use to determine the relief and altitude of a mountain range of the distant past? Refer to the sections in this chapter on the Sierra Nevada and the Rockies.
5. Trace the relationship between subduction and volcanism in the Paleozoic, Mesozoic, and Cenozoic history of the Cordillera.
6. Both Paleocene and modern mammalian faunas include insectivores, small gnawing animals, large carnivores,

and herbivores. Indicate what orders are represented by these adaptive types in the two faunas. Suggest reasons to account for the persistence of some orders while others have become extinct.

SUGGESTIONS FOR FURTHER READING

CROWELL, J. C. "Origin of late Cenozoic basins in southern California." *Society of Economic Paleontologists and Mineralogists Special Publication 22,* 190–204, 1974.

EISBACHER, G. H. "Evolution of successor basins in the Canadian Cordillera." *Society of Economic Paleontologists and Mineralogists Special Publication 22,* 274–291, 1974.

EUGSTER, H. P., and SURDAM, R. C. "Depositional environment of the Green River Formation of Wyoming: a preliminary report." *Geological Society of America Bulletin,* 84: 1115–1120, 1973.

GINGERICH, P. D. "Patterns of evolution in the mammalian fossil record." In Hallam, A. ed. *Patterns of Evolution.* New York: Elsevier, Chapter 15, pp. 459–500, 1977.

LIPMAN, P., PROSLKA, H. J., and CHRISTIANSEN, R. L. "Cenozoic volcanism and plate tectonic evolution of the western United States, pt. 1—Early and Middle Cenozoic." *Royal Society, London, Philosophical Transactions,* 271A: 217–248, 1971.

ROBINSON, P. "Tertiary History," *in* "Geologic Atlas of the Rocky Mountain Region." *Rocky Mountain Association of Geologists Denver,* 233–242, 1972.

SCHOLZ, C. H., BARRAZANGEI, M., and SBAR, M. L. "Late Cenozoic evolution of the Great Basin, western United States as an ensialic interarc basin." *Geological Society of America Bulletin,* 82: 2979–2990, 1971.

STEWART, J. H., "Basin and range structure, a system of horst and graben produced by deep-seated extension." *Geological Society of America Bulletin,* 82: 1019–1044, 1971.

PART FIVE

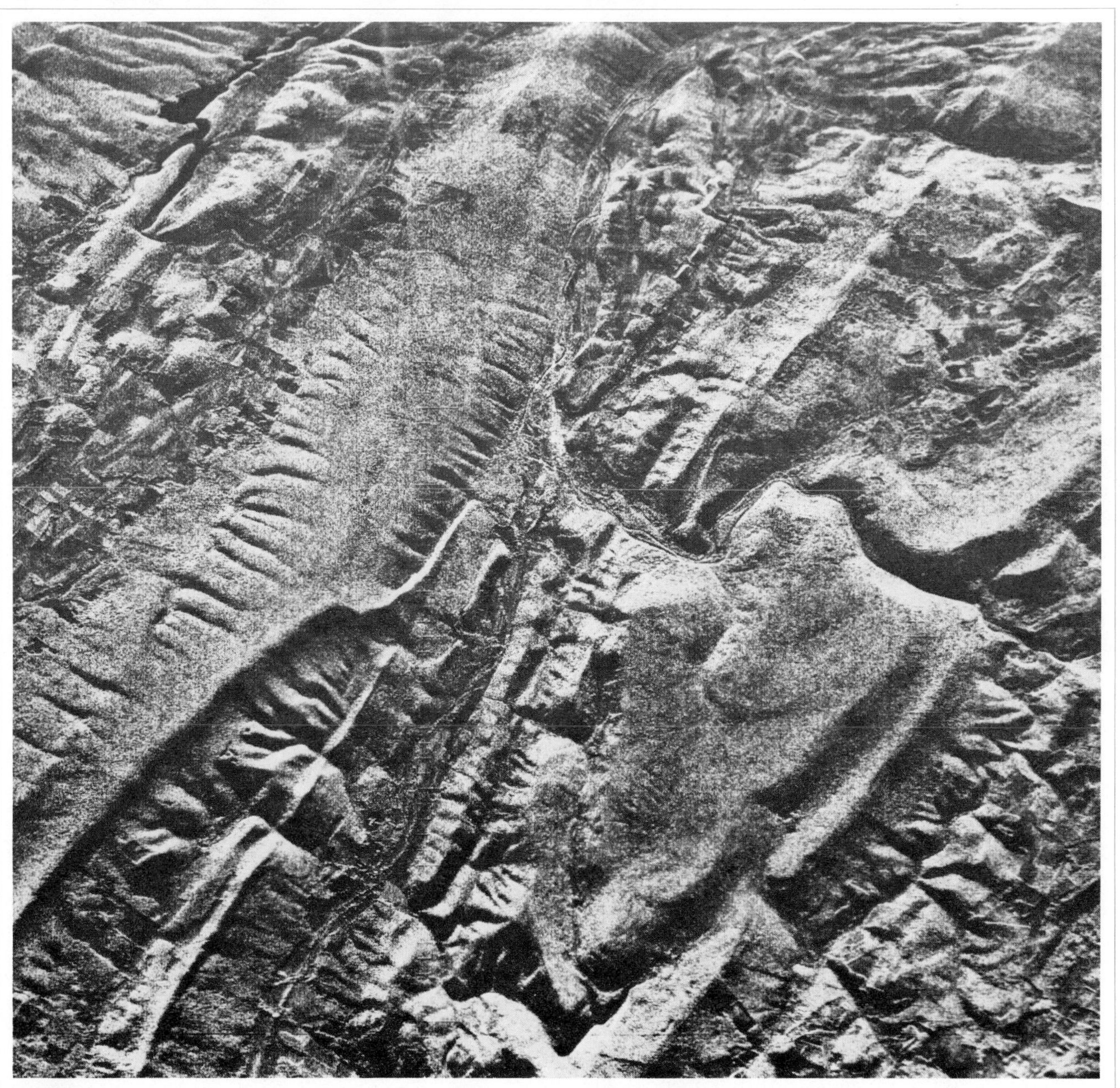

the appalachians

CHAPTER SEVENTEEN

REGIONAL SETTING

The Appalachian Mountains stretch 3000 km down the eastern side of the North American continent, from Newfoundland to Alabama (Fig. 17–1). Between Newfoundland and New York this mountain system forms the coastline of the continent, but from New York to Alabama a southward-widening coastal plain, built of Cenozoic and Mesozoic sediments, separates the deformed rocks of the Appalachian chain from the Atlantic Ocean. The northern end of the Appalachians in Newfoundland is covered by the waters of the Atlantic and by Mesozoic and Cenozoic beds that underlie the continental shelf. In Alabama Appalachian structures plunge beneath Cretaceous and younger sediments, forming the Mississippi embayment of the Gulf Coastal Plain. Structurally and lithologically similar rocks emerge on the western side of the Mississippi embayment in the Ouachita Mountains of Arkansas and Oklahoma, are covered southward by younger sediments in northern Texas, and emerge again in the Marathon region of West Texas. Although not strictly part of the geographic area called the Appalachians, the Ouachita and Marathon regions appear to be continuations of the Appalachian Mountain system. The geology of these two areas, and their subsurface connection to the Appalachians proper, is considered in the next chapter.

The strikes of the fold axes and faults in the Appalachian Mountain system are remarkably consistent compared to those of the Cordillera, where structures that cross the general northwest strike are common. Their trends define four arcs convex to the northwest (Fig. 17–2), separated by nodes at Roanoke, Virginia, the Hudson River of New York, and the Gulf of the St. Lawrence. The arcs are called the Southern, Central, Northern, and Newfoundland arcs. The Ouachita and Marathon sectors of the Appalachians can be considered to be additional arcs. The occurrence of the Hudson River node where the Canadian Shield projects into the Adirondak Mountains, suggests that the irregular cratonic margin of the continent may have held back the westward movement of the Appalachian folds in this sector, but the nature of the other nodes is obscure, hidden by covering water or sediments.

The Appalachians were named by the Spanish explorers of the 16th century after an Indian tribe, the Appalachees. These Indians lived in northern Florida and southern Georgia, far south of the end of the mountain chain.

Divisions of the Southern and Central Appalachians

South of the Hudson River the Appalachian chain can be divided naturally into four longitudinal belts: the Piedmont, Blue Ridge,

EARLY PALEOZOIC HISTORY

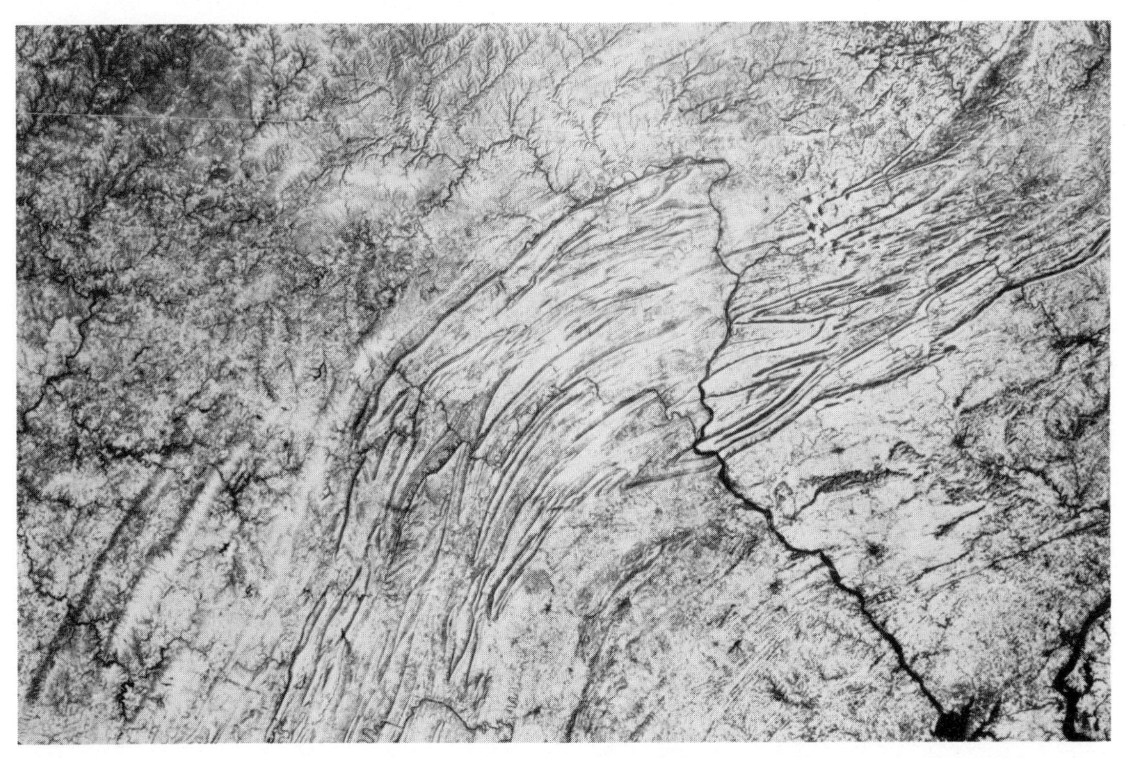

FIGURE 17–1 **The northern section of the Appalachian Mountain System as seem from the Landsat satellite. In this mosaic the folds of the Valley and Ridge province define the arcs of the central Appalachians. (Courtesy of NASA.)**

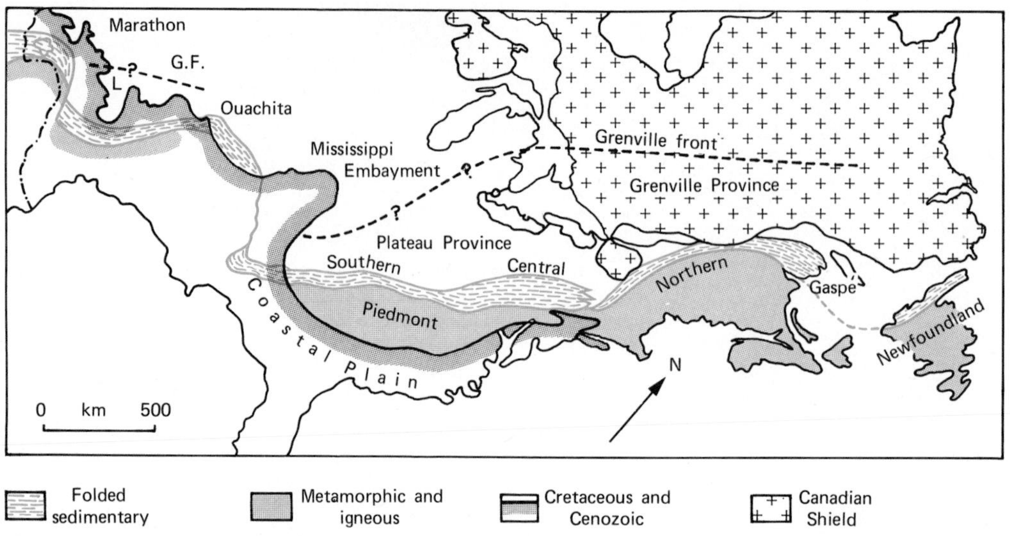

FIGURE 17–2 **The six arcs of the Appalachian Mountain system and the major structural provinces. G.F. = Grenville front, L = Llano dome.**

Valley and Ridge, and Appalachian Plateaus provinces (Fig. 17–3).

The rocks of the Piedmont are in contact with the sediments of the Coastal Plain at the southeastern side of the Appalachians, along a line known as the "Fall Line." The contact of resistant and easily eroded rock localizes waterfalls that form the upstream limit of navigation on the coastal rivers and, hence, the location of many towns such as Washington, Baltimore, Philadelphia, Richmond, etc. The Piedmont is a belt of low relief where slates, schists, gneisses, and plutonic rocks are weathered to depths of several hundred feet and outcrops are scarce. The rocks are believed to be the metamorphic equivalents of early Paleozoic sediments, but fossils that would prove this hypothesis are practically unknown, except in a few areas where the metamorphic grade is lower than usual.

On the northwestern side of the Piedmont the topography rises in an escarpment about 1000 m high along the border of the Blue Ridge province. The rocks underlying the Blue Ridge are complexly deformed and metamorphosed Hadrynian to Lower Cambrian sediments. In the central Appalachians the Blue Ridge belt is a single ridge, but in the southern segment it branches into several ridges and widens to form the Great Smoky Mountains, which include the highest point in the Appalachians, Mt. Mitchell (2037 m), in North Carolina.

The Valley and Ridge province is characterized by long ridges formed by resistant sandstones and valleys floored with shale and limestone (Fig. 17–4). These rocks are Paleozoic in age and little changed by metamorphism, although folded and cut by thrust faults. The "Great Appalachian Valley" forms the northwestern side of the Valley and Ridge province along much of its length.

The Appalachian Plateaus province, northwest of the Valley and Ridge, is underlain by very gently flexed upper Paleozoic sedimentary rocks. Although the rocks at the surface are generally little disturbed, they are locally folded and cut by faults, where décollement thrusts that underlie the eastern side of the plateaus come to the surface. The deep dissection of this region by streams has produced a mountainous topography.

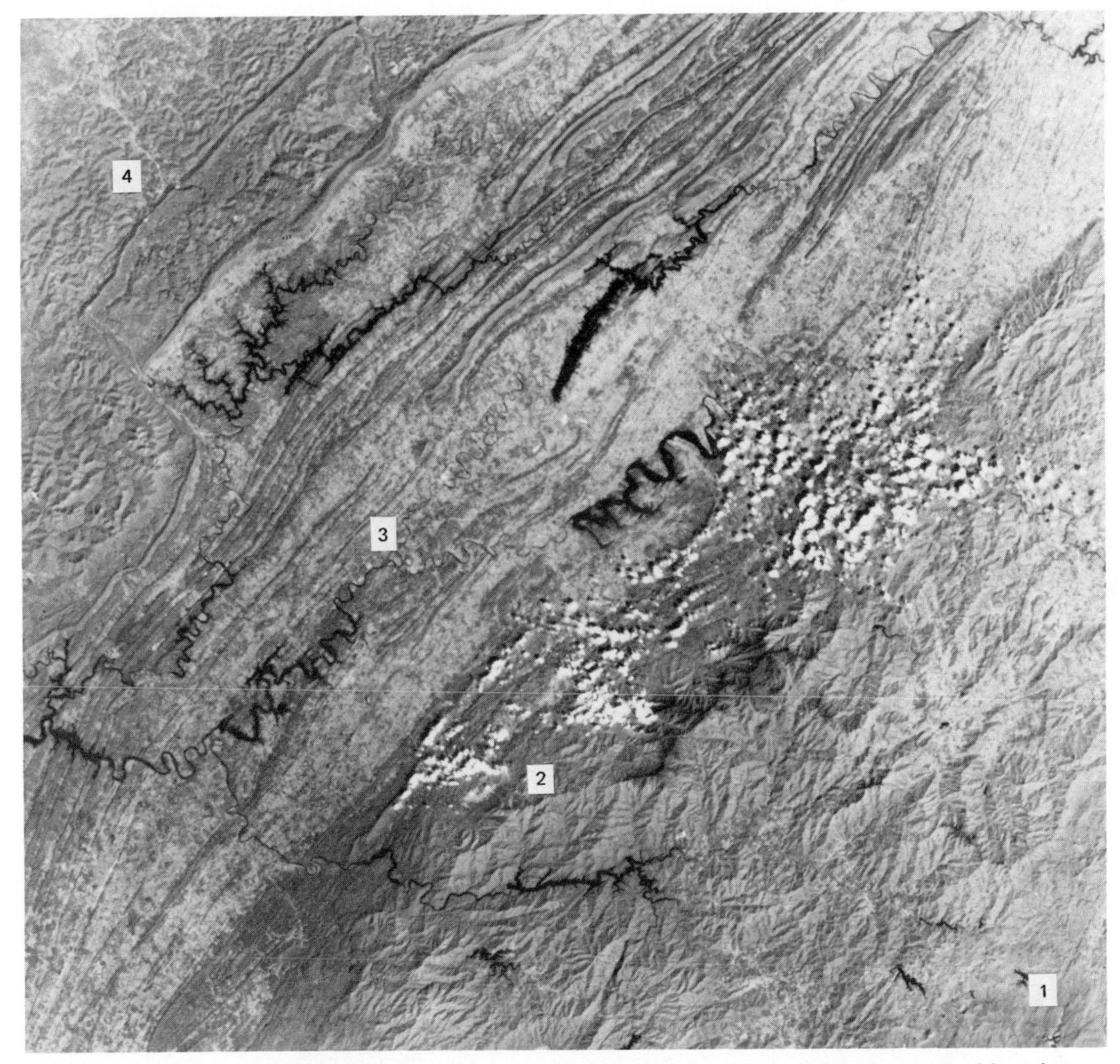

FIGURE 17–3 **Landsat photograph of the structural provinces of the southern Appalachians in eastern Tennessee and western North Carolina. 1. Piedmont. 2. Great Smoky Mountains of the Blue Ridge province. 3. Knoxville, Tennessee, in the Valley and Ridge province. 4. Plateau province. At number 4 the Pine Mountain thrust comes to the surface. (Courtesy of NASA.)**

Northern and Newfoundland Appalachians

The four-fold division of the Appalachians is difficult to recognize north of the Hudson Valley. In places, the intensely folded rocks of the mountain chain lie directly against the Canadian Shield, and in others they are separated from it by a narrow lowland of little deformed early Paleozoic rocks or by the St. Lawrence estuary. The Plateau province is not present. An anticlinal belt in which Precambrian rocks are locally exposed can be followed through the Green Mountains of Vermont, the Notre Dame and Shickshock Mountains of Quebec, and the Long Range of Newfoundland. Although its position within the mountain belt corresponds to that

FIGURE 17–4 **Topography of the Valley and Ridge province. The contrast between the resistant rock of the ridges, and the shales and carbonate rocks of the valleys, is emphasized by the trees growing on the former and the fields on the latter. Aerial photograph looking southwest, about 40 km west of Harrisburg, Pennsylvania. (Photograph by John S. Shelton.)**

of the Blue Ridge, its geology is quite different. In metamorphic grade and abundance of igneous intrusions the New England Upland and the Appalachians of the Canadian Maritime Provinces resemble the Piedmont, but fossils are more common in these northern areas and the history of deposition and deformation is better known. No equivalent of the Valley and Ridge province exists in the Canadian sector of the Appalachians.

The gross distribution of rocks in the Appalachian Mountain system is much like that in the Cordillera. The oceanward side is occupied by metamorphic and igneous rocks, both volcanic and intrusive, that were formed along the edge of the continent in a mobile zone of rifting, subduction, and orogeny that constitutes the Appalachian geosyncline. These now underlie the Piedmont and most of the New England and Maritime Appalachians. The continentward side of the chain is occupied by sedimentary rocks of cratonic facies, deposited on the miogeocline and deformed into open folds cut by low-angle thrust faults, but essentially untouched by metamorphism. These now underlie the Valley and Ridge and Plateau provinces. In the northern Appalachians the proximity of the belt of deformation to the Canadian Shield may have shortened or eliminated this belt of miogeoclinal deformation.

Unlike the Cordillera, the Appalachians are an old mountain chain worn low by erosion. The last compressive deformation took place in the late Paleozoic. We can study in the Appalachians the events that follow orogeny in a mobile belt; in the Cordillera the orogeny is not over. In addition, the Appalachians border an ocean which, as we saw in the initial chapters of this book, is relatively young, having been formed by continental rifting in Mesozoic time. The rifting did not follow the trend of the Appalachian belt everywhere, but cut across the northern segment, separating a once continuous mountain chain into European and North American segments. Many geologists now believe that Appalachian history reveals two episodes of continental rifting and one of continental collision.

BEGINNING OF THE APPALACHIAN RECORD

The oldest rocks in the deformed Appalachian belt are gneisses and schists exposed in the Blue Ridge province, along the Green Mountains of Vermont, and in Newfoundland, which are dated isotopically as about 1 billion years old. These rocks appear to be part of the Grenville structural province of the craton which subsided beneath the edge of the continent during the early Paleozoic deposition and was elevated in later mountain building. Hadrynian rocks, which overlie the Grenville basement, are the initial deposits of a sedimentary prism that was deformed to produce the Appalachian Mountain chain.

Basal Sediments

The Ocoee Group of the Great Smoky Mountains of Tennessee and North Carolina consists of 10,000 m of sandstones of high feldspar content, indicative of derivation from the weathering of a granitic terrane. The Hadrynian Ocoee Group is overlain by a succession of quartzites 2000 m thick, that appears to be largely of Ediacaran age and is called the Chilhowee Group. The appearance of trilobites in the upper part of the Chilhowee Group marks the base of the Cambrian System. The Hadrynian–Ediacaran succession of quartz-rich clastic sediments of great thickness can be followed northward discontinuously along the belt of the Blue Ridge. In Virginia and Maryland hundreds of meters of volcanic rocks appear at the base of the Chilhowee equivalents. This association of thick quartzites of Hadrynian to Early Cambrian age with volcanic flows can be followed still farther northward into Quebec, where these beds form the Oak Hill Group.

The lithology, provenance, and position along the edge of the platform suggest that these rocks once formed a continental shelf built eastward by streams draining the Proterozoic rocks of the Grenville mountain belt. These late Proterozoic and early Paleozoic continental shelf deposits on the Appalachian flank of the continent are analogous to the Windermere Group sediments (Fig. 14–5) that were building a shelf on the Cordilleran flank at the same time.

Cambrian and Ordovician Rocks of the Miogeocline

Events along the Appalachian edge of the continent in Cambrian time are recorded in sediments now folded into the Valley and Ridge province. The sandstones of the Lower Cambrian Series pass upward into mixed shales and limestones of the Middle Cambrian Series and into a thick succession of Late Cambrian and Early Ordovician carbonate rocks. This upward change of facies reflects the transgression of the Sauk Sea across the North American platform. At the height of the Sauk transgression little clastic sediment reached the miogeocline.

The eastern edge of the platform from Middle Cambrian to Middle Ordovician time has been compared with the present margin of the Bahama Bank. The shallow shelf sea in which carbonates were deposited was bordered by a steep slope offshore, leading to deeper water where shale accumulated. Along the crest of the slope carbonate blocks broke off the edge of the bank and slid into the deeper water shale environment. The resulting breccias are composed of limestone blocks that range in size from that of a fist to that of a house, and in age from Middle Cambrian to Middle Ordovician (Fig. 17–5). The blocks could have been broken from the bank edge by faulting or by earthquake shocks. By this slumping along a submarine escarpment the rocks and fossils of two very different environments of deposition have been mixed into a composite deposit. The breccias are distributed

FIGURE 17–5 **The Cow Head Breccia on the west coast of Newfoundland. The variety and irregularity of the limestone clasts are shown in this photograph of a bedding plane surface. (Courtesy of David Baird.)**

along the boundary between the Appalachians and the platform from Newfoundland to New York.

Northern Appalachians and Piedmont

In the Piedmont province rocks younger than Ordovician have not been identified, and Ordovician slates occur only in small synclines downfolded into the undated schists and gneisses of the belt in Virginia. That some of the unfossiliferous crystalline rocks of the central Appalachians may be equivalent to unmetamorphosed lower Paleozoic rocks of the Valley and Ridge province is suggested by comparisons of sequences of lithologies. For example, the succession of formations underlying New York City at the junction between the Piedmont and the northern Appalachians, is as follows:

Manhatten Schist
Inwood Marble
Lowerre Quartzite
unconformity
Fordham Gneiss

The gneiss is of Grenville age and, although the other formations have no fossils, their lithologic sequence is so similar to that in the Valley and Ridge that the Lowerre is considered to be Lower Cambrian, the Inwood is considered later Cambrian to early Ordovician, and the Manhattan is correlated with Middle Ordovidian shales. That such comparisons do not always give unequivocal results is shown by the decades of controversy over the Glenarm Series of Maryland, which is considered by some to be the metamorphosed equivalents of lower Paleozoic rocks.

In New England and the Maritime Provinces, Ordovician rocks are widespread and consist of fine-grained clastic sediments interbedded with great thicknesses of volcanics. In central Newfoundland these volcanic rocks reach thicknesses of 6000 m. The distribution of the sediments and volcanics suggests that in Ordovician time the Appalachian belt included volcanic island arcs that were the sources of the sediments, ash, and flows.

Interpretation of the Early Paleozoic Record

The Hadrynian and earliest Paleozoic sediments show that the North American continent was bordered by a continental shelf beyond which lay deeper water (Fig. 17–6). The ocean that bordered the North American plate on the Appalachian side in the early Paleozoic was not the Atlantic Ocean, for that ocean basin was not created until Pangaea rifted and the segments drifted apart in the early part of the Mesozoic Era (Chapter 2). This early ocean between Europe and North America has been referred to as the Protoatlantic, or better, as Iapetus (Eye-ăp-ĕt-us), a name suggested by W. B. Harland and R. A. Gayer. Iapetus, in Greek mythology, was the father of Atlas, from whose name the word "Atlantic" is derived.

Do we have any evidence of the time of formation of Iapetus? Some geologists have suggested that the Hadrynian sediments of the Appalachians are typical of coastlines

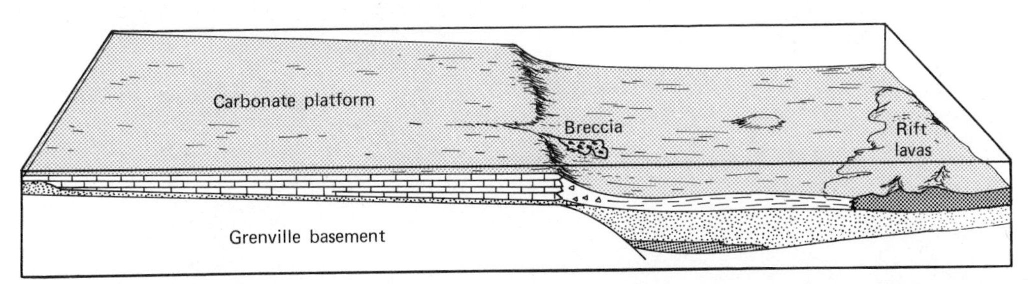

FIGURE 17–6 **Block diagram of the border of the North American plate in Middle to Late Cambrian time showing the environment of the slump breccias and the lavas released by the rifting of Iapetus. (After John Dewey and John Bird; courtesy of the Geological Society of America.)**

formed by continental rifting. They also suggest that the lavas associated with Hadrynian rocks along the continental margin could have been extruded through normal faults, tapping deep reservoirs during the rifting process. If these suggestions are acceptable, the Iapetus rift took place 700–800 million years ago.

The wide distribution and great thickness of volcanics in the Ordovician rocks of the northern and Newfoundland Appalachians indicate that subduction zones and island arcs were established along the edge of the North American plate in Ordovician time, and perhaps earlier in the Cambrian Period (Fig. 17–7).

The Avalon Belt

On the northeastern side of the Appalachians, in a zone that includes southern New Brunswick, Cape Breton Island, and the Avalon Peninsula of eastern Newfoundland (Fig. 17–8), Hadrynian rocks are as thick as those in the Great Smoky Mountains, but they are composed largely of lavas and redbed sandstones, both apparently of nonmarine origin. The geology of the Avalon belt is different from that of the rest of the Appalachians in several ways. Firstly, in the Avalon Peninsula rocks of Grenville age come to the surface, yet the region is separated from the edge of the Grenville prov-

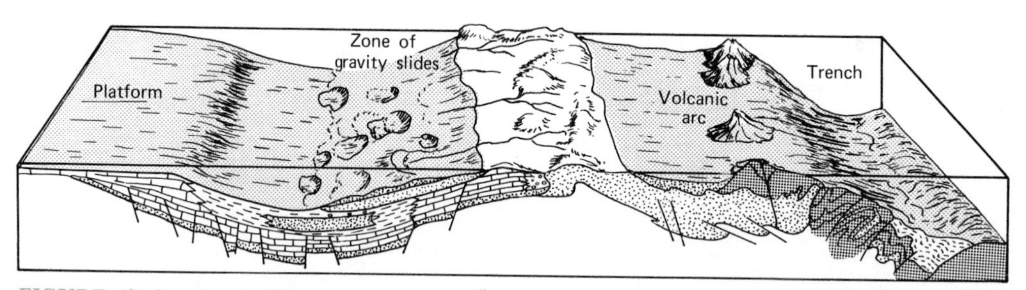

FIGURE 17–7 **Block diagram of the border of the North American plate in Middle Ordovician time at the start of the Taconic orogeny showing the zone where the gravity slides developed and a hypothetical reconstruction of the subduction zone to the east. (After John Dewey and John Bird; courtesy of the Geological Society of America.)**

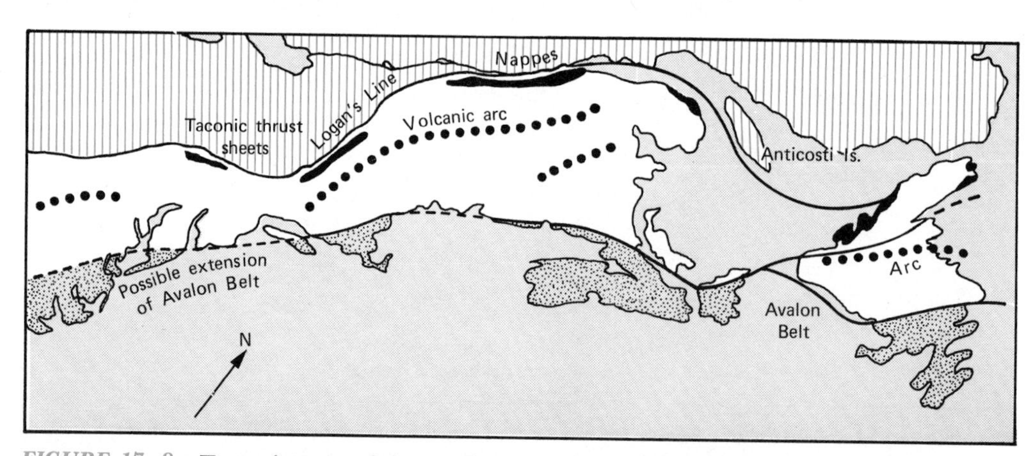

FIGURE 17–8 **Tectonic map of the northern part of the Appalachian belt in Ordovician time showing the Avalon belt, the Taconic thrust structures (black), and the postulated positions of the volcanic island arcs. (Modified from John Dewey and John Bird.)**

ince, exposed in the Long Range of western Newfoundland, by sediments and igneous rocks typical of deposition in a deep ocean basin. Secondly, the belt was intruded by granites and deformed at the beginning of the Cambrian Period, a time of orogeny in North Africa, but not one of widespread deformation in the Appalachians. Thirdly, the belt is bordered on the oceanward side, at present, by a great thickness of sparsely fossiliferous, folded, Ordovician, and possibly Cambrian sediments of the Meguma Group. One unit of the group is composed of laminated mudstones containing "dropstone" cobbles, the type of marine deposit formed at the margin of a glaciated area (see p. 156). This is the only deposit in North America so far attributed to the Ordovician–Silurian episode of continental glaciation, the center of which was North Africa (Fig. 17–9).

All these lines of evidence suggest that the Avalon belt may not be part of the sediments and volcanics that were deposited along the eastern border of the North American plate and deformed to produce the Appalachian fold-belt. Some geologists have suggested that before the late Paleozoic closure of Iapetus, the Avalon belt was part of the opposite side of the ocean or an isolated microcontinent plate. In the late Paleozoic, when the rift that became the North Atlantic began to form, it did not follow the old boundaries of Iapetus, but left a section of the far side, or the Avalon microcontinent, welded to North America.

Although the uniqueness of the Avalon belt is most easily identifiable in the northeastern Appalachians, certain of its features can be found in eastern New England, the eastern zones of the Piedmont, and a block of early Paleozoic rocks that lies buried beneath the Coastal Plain of northern Florida. In this block a thick succession of Hadrynian volcanic rocks is overlain by an unfolded, clastic sequence containing fossils of European aspect, ranging in age from Lower Ordovician to Middle Devonian. These rocks also are believed to be more closely related to the African side of Iapetus. In the southern Appalachians a major fault belt called the Brevard zone may mark the line along which the African segment of the Piedmont abuts the American rocks.

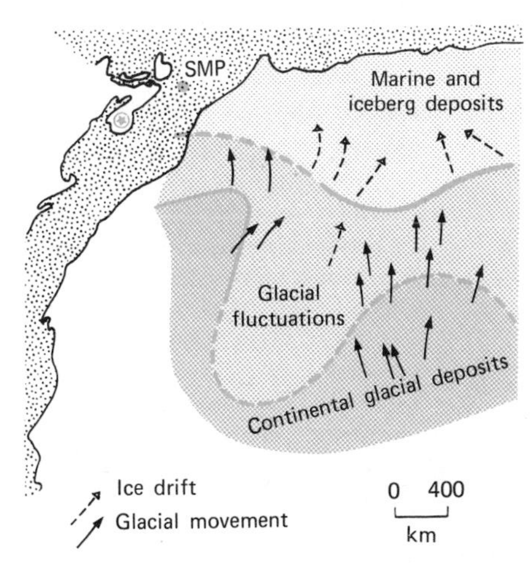

FIGURE 17–9 **Ordovician–Silurian glacial deposits of North Africa and their relationship to possible glacial-marine deposits in the Avalon belt of southern Nova Scotia (located at the star symbol). The continents have been fitted together as they might have been in mid-Paleozoic time. SMP = South Magnetic Pole of this time interval. (After Paul Schenk; courtesy of the Canadian Journal of Earth Sciences.)**

ORDOVICIAN SEDIMENTATION

From Hadrynian to Cambrian time there is little evidence of orogenic movements in the Appalachians outside the Avalon belt, and none of these events is of more than local importance. The tectonic quiescence of this first 350 million years of the Paleozoic was brought to a close at the beginning of Middle Ordovician time, when the first effects of major orogenic movements that are grouped as the Taconic orogeny are reflected in the sediments accumulating along the continental margin.

The Middle Ordovician Sediment Supply Reversal

In Jurassic time sedimentation in the western part of the North American platform showed a reversal of sediment supply due to the uplift of the Cordillera in orogeny. A similar reversal took place in Middle Ordovician time on the eastern side of the platform. Limestones of Middle Ordovician age at the edge of the platform and in the miogeocline become more argillaceous upward, and interfinger eastward with shales derived from source areas in the geosyncline. In the course of Ordovician time the area of shale, and later sandstone deposition, spread westward, steadily displacing the area of limestone deposition (Fig. 17–10). These same facies changes occur along the whole length of the miogeocline from Tennessee to Quebec, and must reflect a growth of a source of supply along the margin of the continent. In New York and Quebec the argillaceous limestones constitute the highly fossiliferous Trenton and Black River groups, and the shales constitute the Utica Group (Fig. 17–11). At the beginning of late Ordovician time the Utica facies spread over the platform as far as Lake Huron, and later in the epoch fine clastic sediment extended 1500 km from the Appalachians as far as Iowa, to form the Maquoketa Shale.

Thin, highly persistent layers of clay occur within Middle and Upper Ordovician limestones and shales along the miogeocline and adjacent platform. Although these beds are only a few centimeters thick, they can be traced for hundreds of kilometers from the shale into the limestone facies (Fig. 17–10). Because they are independent of facies, they make excellent key beds for establishing correlations. These clay beds, like the Late Cretaceous water-absorbing bentonites of the western plains, are believed to be the products of volcanic eruptions that spread ash into the surrounding seas. Because this clay has lost the ability to absorb water and swell, it is called **metabentonite**.

The Queenston Clastic Wedge

The spread of coarse, clastic sediments westward across the miogeocline and platform in Late Ordovician time is evidence of the increase in size and relief of a highland

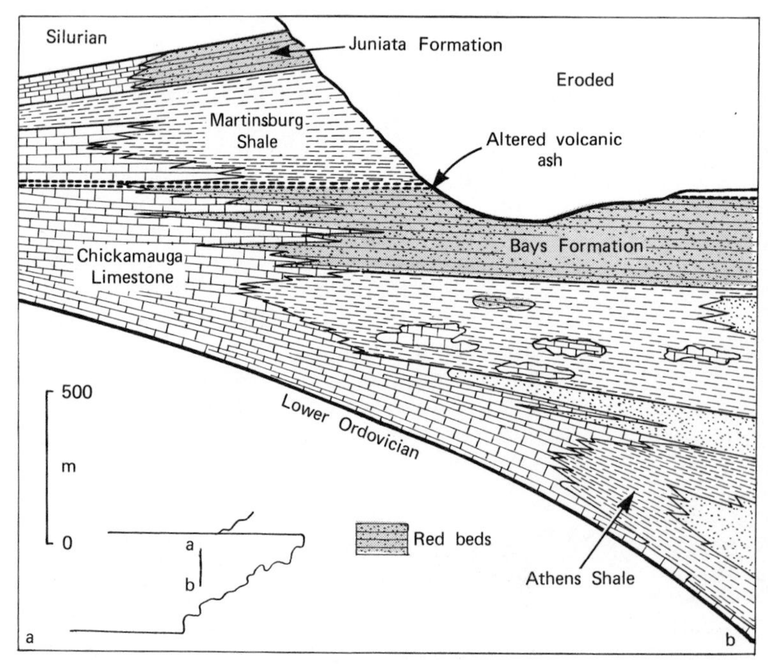

FIGURE 17–10 **Facies changes in the Middle and Upper Ordovician Series of Tennessee. Note the interfingering of the limestone facies deposited on the shelf with the shale deposited along the miogeocline; the shale facies spread progressively northward while the limestone facies retreated. (After John Rodgers.)**

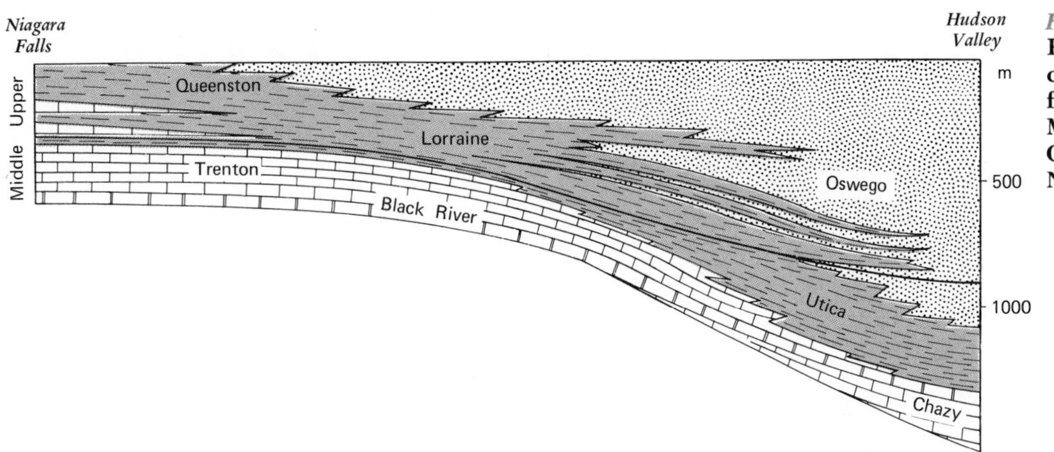

FIGURE 17–11
Restored and simplified cross-section of the facies changes in the Middle and Upper Ordovocian Series of New York.

produced by Taconic orogeny in the Appalachian geosyncline. These sandstones and shales form a clastic wedge named after one of its constituent formations, the Queenston red shales of the Niagara Escarpment (Chapter 3). Middle Ordovician clastic sediments are thickest in the southern Appalachians, but the Queenston wedge (Late Ordovician and Silurian clastic sediments) reaches its greatest thickness of 2000 m in eastern Pennsylvania. The rising highlands in the geosyncline built a compound delta, pushing the shoreline westward with an increasing supply of sand and silt and spreading red sediments as far as the present site of Lake Huron. The facies map of Fig. 11–14 shows the maximum extent of the clastic sediments of the Queenston clastic wedge and the distribution of sedimentary facies in Late Ordovician time. Over the broad alluvial plain formed by the rivers draining from the Taconic highlands, the landscape was a red color, barren despite the presence of an excess of water, for plants had not yet evolved the capability of growing on land. Only here and there a patch of yellow or green lichen on a boulder might have relieved the the monotonous dull red of the sediments deposited by the flooding rivers.

Each interfingering facies belt of the Queenston clastic wedge is characterized by a particular community of fossils, and each community contains different animals having trophic roles (see Chapter 4 for the description of trophic roles) appropriate to that environment. In the Lorraine Group of northern New York State (Fig. 17–11) Peter Bretsky distinguishes three Ordovician communities now represented by groups of fossils that occur together in beds of a particular lithology. The *Ambonychia–Modiolopsis* community is dominated by large, suspension-feeding bivalves, living on the sediment surface (that is, epifaunal), that give their names to the community (Fig. 17–12), and a lesser number of detritus-feeding gastropods. The community inhabited areas of irregular sedimentation with mobile bars near the shoreline of the Queenston delta. In the *Nuculites–Colpomya* community small detritus- and suspension-feeding bivalves, living beneath the sediment surface (that is, infaunal), predominate, and the containing sediments are silty shales and siltstones. This community probably lived in offshore waters of a normal marine environment. The *Dalmanella–Sowerbyella* community is characterized by epifaunal suspension-feeding brachiopods and bryozoans (Fig. 17–12), and occurs in calcareous siltstones and shales. It is believed to have lived in clearer marine waters, away from the influence of deltaic muds. An analysis of the differing roles played by each fossil organism in its community produces a deeper understanding of the environments of deposition in the Queenston clastic wedge (Fig. 17–13). This technique of community analysis by assignment of trophic roles can

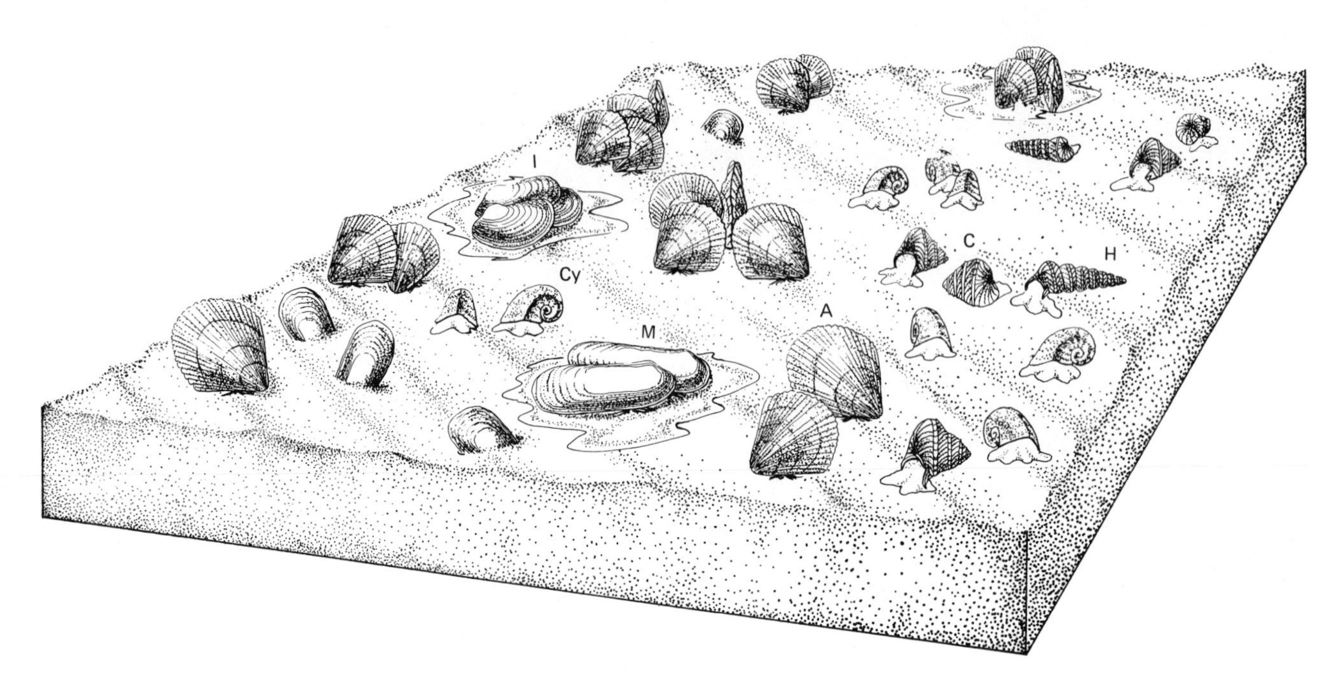

Ambonychia—Modiolopsis community

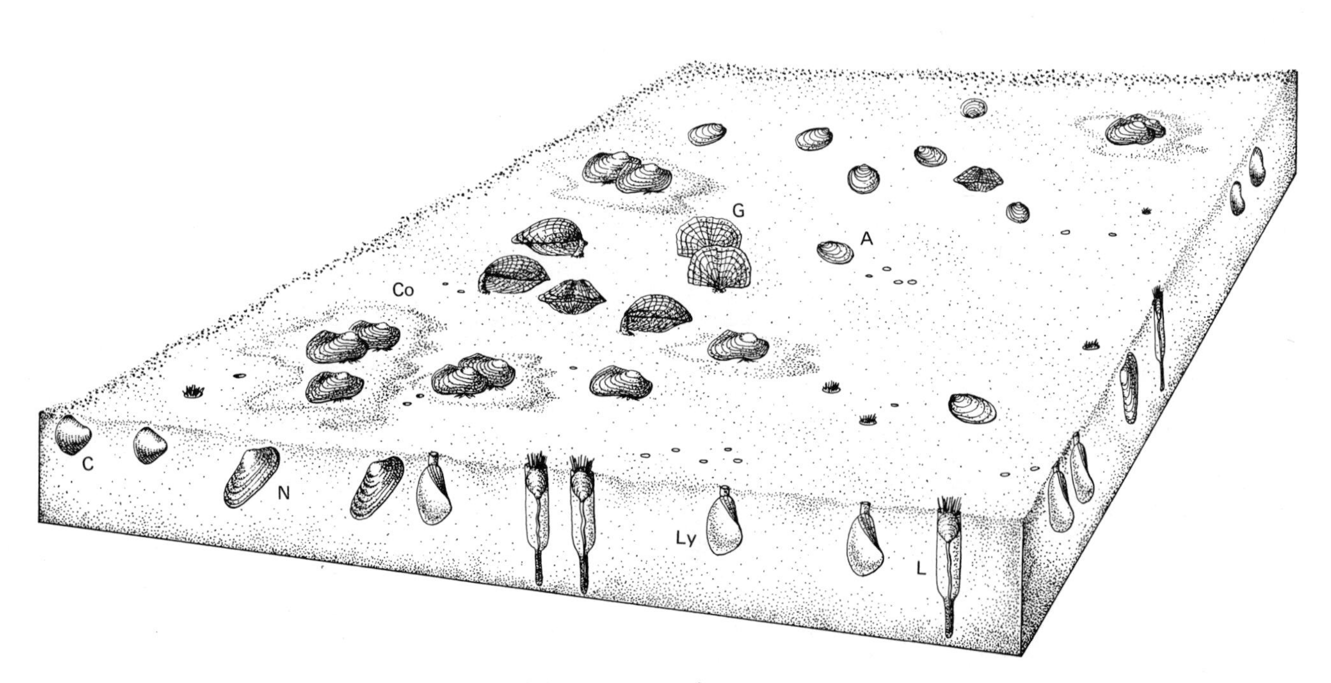

Nuculites—Colpomya community

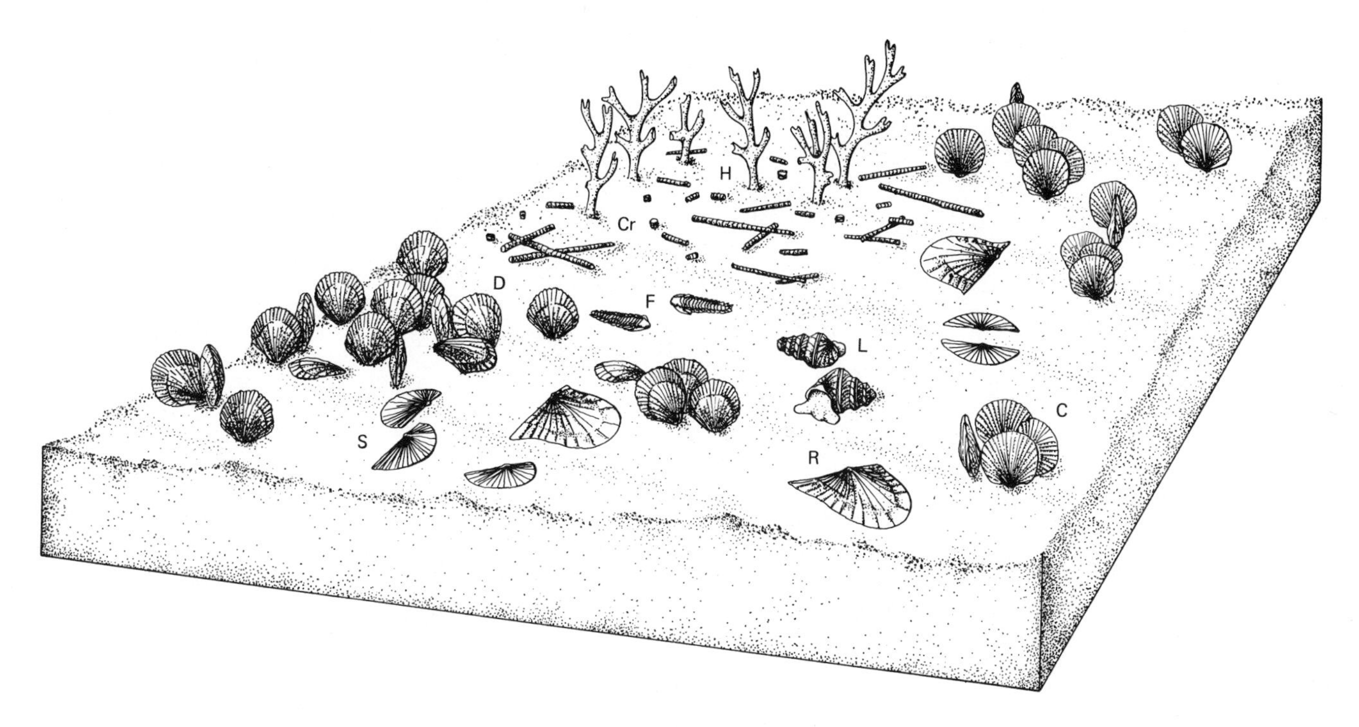

FIGURE 17–12 **Reconstructions of the Late Ordovician sea floor showing the habitats of three fossil communities of the Lorraine Group of New York. (From Peter Bretsky; courtesy of the New York State Museum.)**

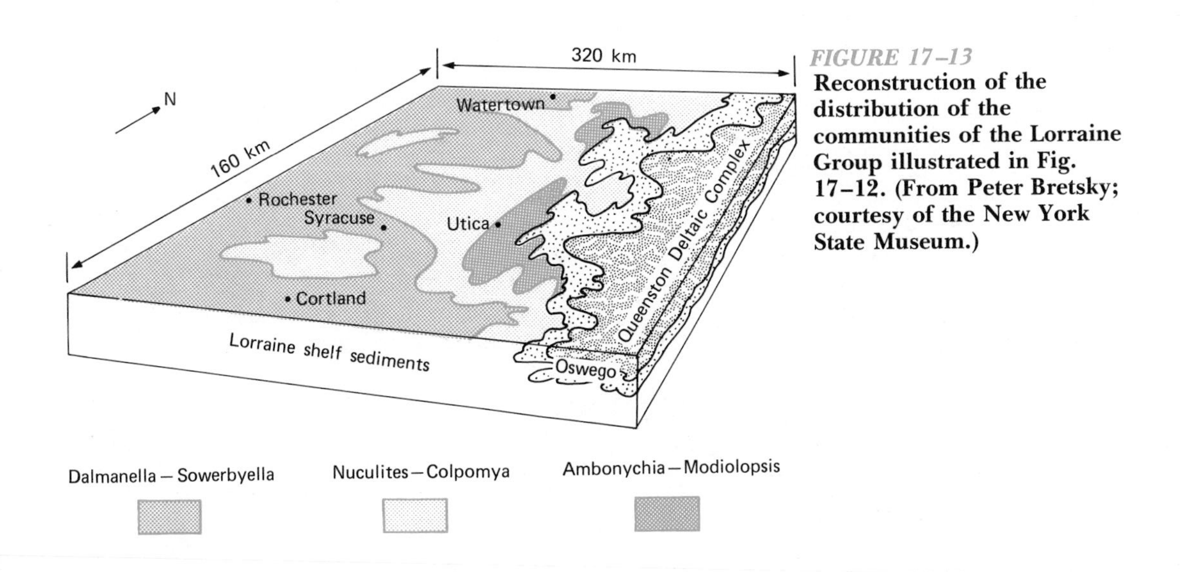

FIGURE 17–13 **Reconstruction of the distribution of the communities of the Lorraine Group illustrated in Fig. 17–12. (From Peter Bretsky; courtesy of the New York State Museum.)**

be applied to any sedimentary unit containing a number of environments.

TACONIC OROGENY

Structures of the Taconic Orogenic Belt

The stratigraphic record of the miogeocline and adjacent platform indicates the presence of highlands along the boundary of the North American plate from mid-Ordovician to the end of Silurian time. The nature of the deformation that accompanied this uplift is recorded in the rocks of the Appalachian Mountains. However, most of these rocks were deformed again at the end of the Devonian Period, and some a third time before the close of the Paleozoic Era. The effects of repeated orogenies have obscured those produced by the first one, and structural geologists may find it impossible to separate the mineral textures and structures caused by the Taconic deformation from those developed later.

The orogeny derives its name from the Taconic Range, which straddles the border of New York State with Massachusetts and Connecticut. The complex structure of these hills is now generally recognized as formed of a series of deformed plates, several tens of kilometers across, of graywackes and slates that were moved westward along low-angle thrust planes onto the limestones and sandstones that accumulated along the miogeocline. The sediments of the thrust plates were deposited in the deeper water of the geosyncline, now located on the east side of the Green Mountain anticline, and slid down a plane of low slope when that area was uplifted in Middle Ordovician time (Fig. 17–7). The time of emplacement of the lower thrust slices can be determined as Middle Ordovician, because they slid into the sea where graptolitic shales were accumulating. Some of the upper thrust sheets may have been emplaced in late Ordovician time. The movement of such large masses of rock was accompanied by extensive breakage, and the slide zone is characterized by spectacular breccias.

In the northern Appalachians the contact of the rocks deformed in the Taconic orogeny and the relatively undeformed miogeoclinal rocks is a zone of thrust faults and breccias that was first recognized as of regional significance by Sir William Logan, first director of the Geological Survey of Canada, and is named "Logan's Line." The structure of the Appalachians in Quebec, southeast of Logan's Line, has been interpreted as a series of nappes. A **nappe** is a recumbent anticline of regional scale, thrust for many kilometers along a low-angle fault within its lower limb. Such folds were first recognized in the Alps, but are now known to be a common feature of many mountain belts where there is much shortening and the rocks have behaved plastically under stress. In Quebec the deforming force from the southeast has piled the thrust sheets up against the craton (Fig. 17–14). In Newfoundland the structure of the western edge of the Appalachian belt is much like that of the Taconic Range (Fig. 17–15). Large masses of geosynclinal rocks have been transported by gravity sliding across an anticlinal axis onto the miogeoclinal facies.

The extent of Taconic deformation in the Piedmont province has been problematical, and in many places the regional structure of Piedmont rocks is still obscure. An interpretation of the structure of the Blue Ridge and Piedmont in South Carolina and Tennessee is shown in Fig. 17–16. Potassium–argon ages on metamorphic rocks and isotopic dating of igneous intrusions, discussed further below, suggest that part of this compressional deformation is of Ordovician age and part of Devonian age.

Dating the Taconic Orogeny

The existence of the Taconic orogeny was first recognized along the Hudson Valley, where Middle Ordovician shale is highly contorted and overlain unconformably by Upper Silurian limestone that has been gently folded by later movements. An unconformity like this indicates the time of folding only within wide limits, from the Middle Ordovician to the Late Silurian epochs. Farther south, the stratigraphic break between the beds bounding the unconformity diminishes. At Shawangunk Mountain, New York, the upper beds are Lower Silurian,

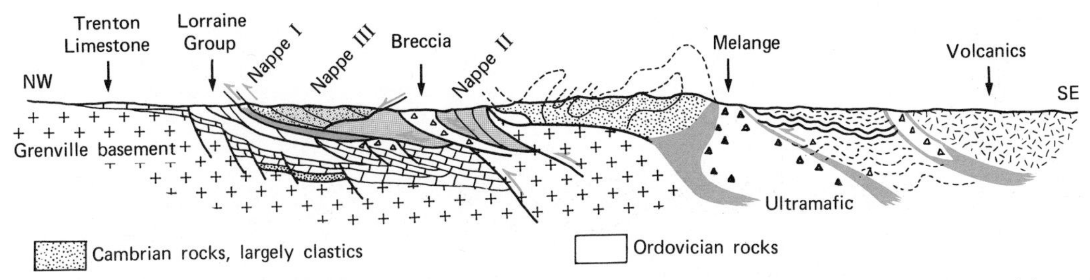

FIGURE 17–14 **Cross-section of the Taconic deformed belt in Quebec showing the position of the nappes and ultramafic rocks. (After Pierre St. Julien and Claude Hubert; courtesy of the American Journal of Science.)**

FIGURE 17–15 **Contorted Ordovician shale near the base of a thrust sheet, Port au Port Peninsula, western Newfoundland. (Geological Survey of Canada, Ottawa, Photograph 153816.)**

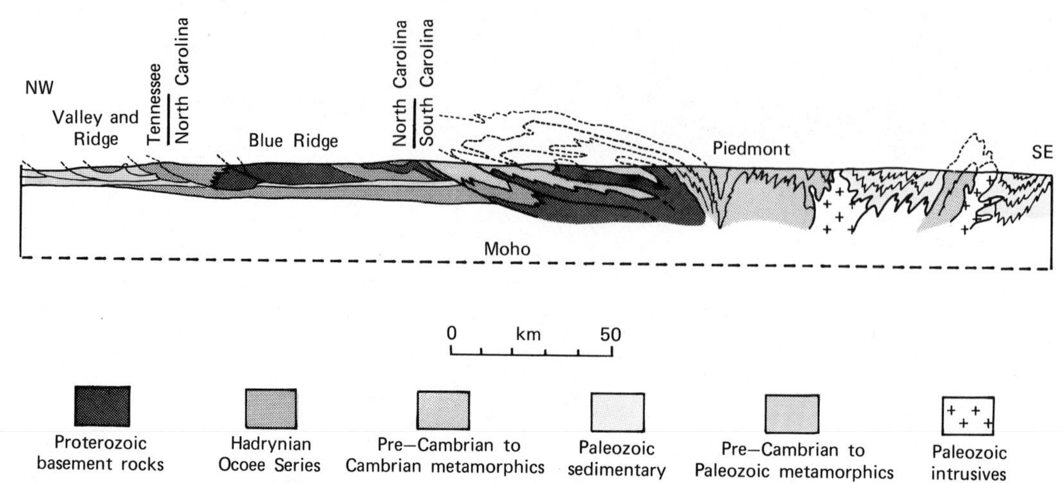

FIGURE 17–16 **Interpretive cross-section of the Blue Ridge and Piedmont between Knoxville, Tennessee (NW), and Aiken, South Carolina (SE), at the Coastal Plain overlap. (After R. D. Hatcher; courtesy of the Geological Society of America.)**

and in central Pennsylvania they are Upper Ordovician. Here some of the deformation is dated as within the Late Ordovician Epoch. The Taconic unconformity in the Appalachians of New England and eastern Canada is typically formed between Middle Silurian and Middle Ordovician rocks and, therefore, does not precisely date the time of deformation. In places, evidence of two periods of deformation within Ordovician time is suggested by a second unconformity.

Isotopic ages of about 440 million years of metamorphic rocks associated with Taconic structures in the northern Appalachians suggest that the major period of metamorphism was over by the end of the Ordovician Period. The last regional metamorphic event in the southern Piedmont has been dated as between 420 and 380 million years ago (Early Silurian to Early Devonian). The ages of plutonic rocks in the southern Piedmont also fall approximately into this range. In the central Appalachians ages of metamorphism in the range of 430 to 460 million years (Late Ordovician) have been reported. Intrusive rocks associated with Taconic orogeny have been dated as 462 million years in New Brunswick, at 432 million years in Nova Scotia, and at 440 million years in Newfoundland.

The wide spread of ages obtained by isotopic methods may reflect a series of events that we group somewhat arbitrarily into the Taconic orogeny, spread over a long period of time, or it may be the result of modification of the isotopic content of the rocks by later events. Because later periods of metamorphic reheating remove accumulated argon from older rocks, making them appear younger, some rocks dated by the potassium-argon method as of Ordovician age may be Precambrian rocks modified by Late Devonian metamorphism, and some dated as late Devonian may be Ordovician rocks modified by late Paleozoic reheating.

The stratigraphic, structural, and isotopic evidence shows that the Taconic orogeny was a complex sequence of events that reached a climax in Middle and Late Ordovician time, but extended over a longer interval. Different sections of the Appalachian geosyncline were subject to different styles of deformation at different times during this interval.

Intrusive Rocks

Both silicic and ultramafic bodies of igneous rocks were emplaced during the Taconic orogeny. In New England and the Maritime Provinces, igneous rocks of the Highlandcroft and Olivierian Series are associated with the Taconic deformation. That igneous intrusion took place in the Piedmont is shown by the isotopic dating mentioned above. Small intrusions of ultramafic rocks similar to those in the Cordillera are found in a relatively narrow belt along the length of the Piedmont, in the northern Appalachians, and in Newfoundland. In the north these bodies are almost always associated with Ordovician volcanic and sedimentary rocks, and have isotopic ages suggesting Middle Ordovician emplacement. Some geologists have ascribed the exposure of these rocks to the process of obduction, the overriding of oceanic lithosphere on to other oceanic or continental plates. In central Newfoundland, a section of igneous rock that grades from basal ultramafics through a layer completely composed of gabbro dikes, to pillowed oceanic basalts mixed with sediments, has been interpreted as a portion of the lithosphere of the Iapetus Ocean brought up by obduction and deformed in subsequent orogenic episodes.

The ultramafic rocks of the Appalachians are commercially important in southern Quebec, where they have been altered to serpentine that includes veins of the fibrous, fireproof mineral, asbestos. About 40 per cent of the asbestos mined in the world is derived from mines in this area.

The Taconic Orogeny and Plate Tectonics

John Bird and John Dewey have suggested that the events of the Taconic orogeny can be attributed to the development of a subduction zone along the Appalachian edge of the North American lithosphere plate in Early Ordovician time. The axis of the oceanic trench and the location of the postulated volcanic arc that accompanied it are shown in Fig. 17–8. The position of the arc is now marked by the thick Ordovician volcanic successions of central New England and central Newfoundland. The deformation of sediments in the trench and along the edge of the continental plate by the subduction of oceanic crust probably continued through the Ordovician and Silurian periods. The elevation of a ridge on the continent side of the volcanic arc resulted in the sliding westward of the slices of geosynclinal facies sediments that characterize the front of the Taconic deformed belt from New York to Newfoundland.

THE APPALACHIANS IN THE SILURIAN

Over most of the northern Appalachians angular unconformities exist between Silurian and Ordovician rocks, and Lower Silurian rocks are missing. In these areas (Fig. 17–17) the topography was mountainous at the close of Ordovician time. The extensive highland belt along the inner edge of the Appalachians acted as the main source of sediment supply to the Queenston clastic wedge. However, in a depression that traversed the present site of eastern Maine, northern New Brunswick, and the Gaspé Peninsula, deposition of Lower Silurian sediments followed those of the Ordovician without break. Another belt undisturbed by Taconic deformation crossed northern Nova Scotia and central Newfoundland. The Taconic orogeny raised welts in the Appalachian geosynclinal complex, but troughs between these welts continued to accumulate sediment into Silurian time.

Silurian Rocks of the Appalachian Geosyncline

During the course of the Silurian Period the highland areas steadily shrank in extent and relief, until by the beginning of the Devonian Period the sea was able to cover nearly the whole of the northern Appalachian geosyncline. Some of the early Silurian deposits in the troughs are coarse conglomerates derived from the neighboring Taconic highlands, but as erosion reduced the highlands the dominant sediments became siltstones and shales. In Middle Silurian time an extensive belt of carbonates containing stro-

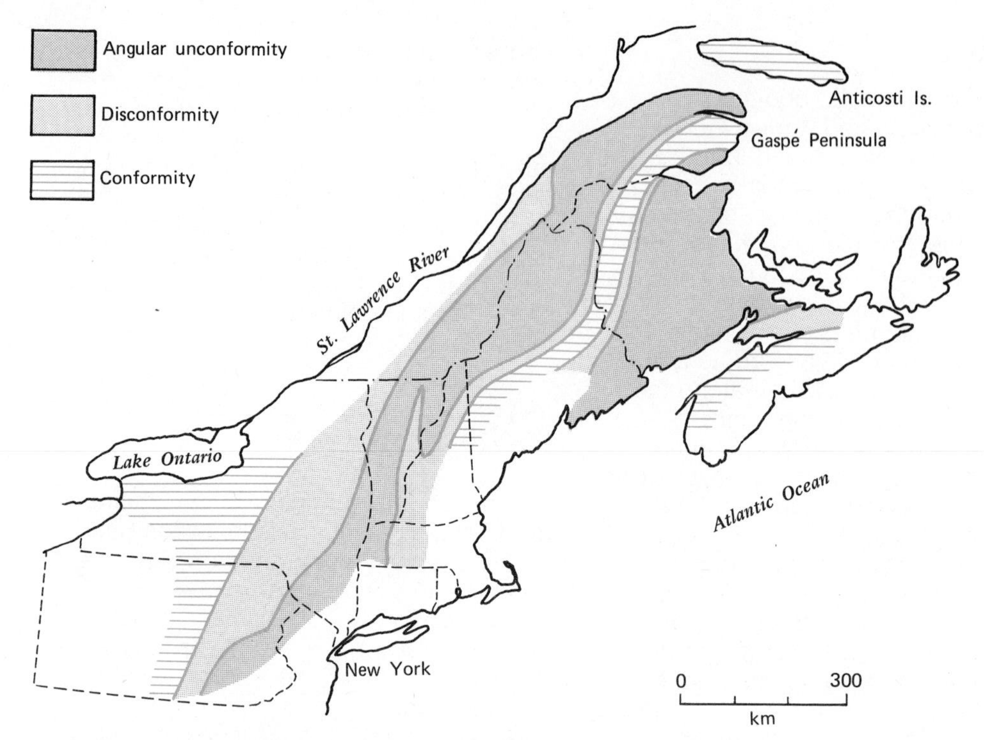

FIGURE 17–17 **Distribution of unconformities ascribed to the Taconic orogeny and their relationship to areas of uninterrupted deposition during the orogeny. (After L. Pavlides, A. Boucot, and B. Skidmore; courtesy of John Wiley & Sons, Inc., New York.)**

matoporoid–coral reefs was deposited in the siltstone facies of Maine, New Brunswick, and the Gaspé Peninsula of Quebec.

However, volcanic rocks continued to form an important part of the record in the Appalachian area, but the belt of volcanism (perhaps reflecting a change in the subduction zone) shifted to the position of the present coast of Maine and New Brunswick. At Eastport, Maine, the Silurian strata, 5000 m thick, are mostly water-laid ash with lavas of various kinds. On the northern shore of Baie des Chaleurs, Quebec, Silurian siltstones, nearly 2500 m thick, are capped by about 600 m of lavas erupted from a local volcanic center on the ocean floor (Fig. 17–18). In central Newfoundland Silurian volcanics are interbedded with marine sediments, which grade upward into nonmarine redbed sandstones and coarse conglomerates.

In summary, the Appalachian geosyncline in Silurian time was an area of decreasing relief. A subduction zone continued to consume oceanic lithosphere, narrowing the width of Iapetus, and maintaining volcanism along the continental margin.

The Shrinking Clastic Wedge

During Early Silurian time the supply of sediment from the Taconic highlands was sufficient to spread red sandstone and shale westward across New York State into Ontario, forming the redbeds of the Cataract Group, to lay down coarse conglomerates of the Shawangunk Formation near the mountain front in southern New York, and to deposit the extensive Tuscarora Sandstone in the Valley and Ridge province of the central Appalachians. Westward, away from the

FIGURE 17–18 **Silurian rocks on the south shore of the Gaspé Peninsula at Black Cape. The vertically dipping strata expose a section about 3000 m thick, one of the thickest Silurian sections on the continent. The volcanic rocks in this section are not shown in this photograph.**

supplying streams, the Cataract sandstones pass laterally into red shales of the Cabot Head Formation and finally, on the margins of the Michigan Basin, into the limestones characteristic of the shelf environment (see Chapter 4).

Middle Silurian rocks of the eastern side of the platform and the miogeocline record the transgression of the epeiric sea eastward toward the Taconic Mountains, as the supply of sediment from the erosion of the Taconic highlands dwindled. Middle Silurian sediments are divided here into a lower Clinton Group and an upper Albemarle Group. Near the mountains the Shawangunk Conglomerate was still accumulating in Clinton time, and a tongue of it, known as the Oneida conglomerate, spread westward as far as Lake Oneida, New York (Fig. 17–19). The Clinton Group of western New York is largely shale, but the proportion of limestone increases as the group is traced westward, and the proportion of sandstone increases eastward. In eastern Tennessee and Alabama the Clinton Group consists largely of interbedded sandstone and shale, which pass into limestones a short distance to the west. The narrowness of the Silurian belt of clastic rocks in the southern Appalachians is evidence of the lower relief of the Taconic highlands along the site of the Piedmont.

Beds of sedimentary iron ore are widespread in the Clinton Group. The ore consists of hematite (Fe_2O_3), in the form of sand-like pellets or oölites. Possibly the iron oxide precipitated directly from solution in the sea water on to nuclei, where it built up layers to form an oölith or, possibly, the original mineral to form the oöliths was chamosite, an iron silicate, which subsequently weathered on the ocean floor to hematite. The iron must have been derived from deep weathering of rocks in the Appalachian belt to the east, but its manner of transportation and precipitation remain a matter of contro-

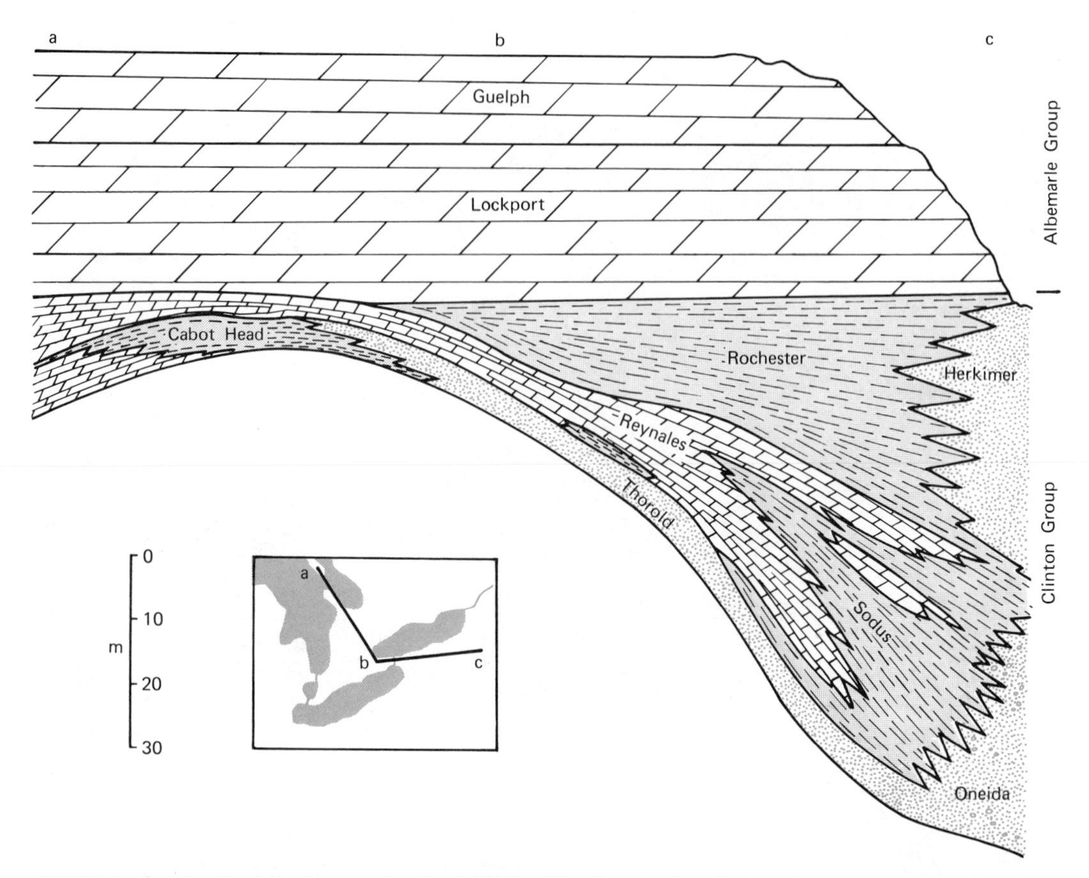

FIGURE 17–19 **Facies changes in the Middle Silurian rocks of the Niagara Escarpment from central New York to Manitoulin Island, along the section shown in the index map.**

versy. The ore beds thicken from about 2 m in New York to about 8 m near Birmingham, Alabama. In the north the ores were capable of supplying local forges only and have not been worked for many years, but at Birmingham they are the basis of a thriving steel-making industry.

Through the latter part of Middle Silurian and all of Late Silurian time, only fine gray mud was supplied to the miogeocline from the remains of the old Taconic highlands. On the adjacent platform Albemarle beds are largely limestones. During Late Silurian time evaporites accumulated in New York State, as described in Chapter 11. These salt deposits grade eastward into the red Bloomsburg mudstones, the last clastic sediments supplied by the Taconic generation of Appalachian Mountains. By the end of the Silurian Period the supply of clastics was so low that limestone was deposited along the site of the present Hudson Valley, where in the preceding period the Taconic Mountains had stood. The unconformity between the limestones and the contorted Ordovician shales represents 40 million years of erosion, the life span of this geosynclinal highland.

The growth of the highlands resulting from the Taconic orogeny is recorded in the sediments of the Middle and Late Ordovician Series; their reduction, in the sediments of the Silurian System. Together these strata of the Queenston clastic wedge record the changing relief of the Taconic Mountains.

SUMMARY

The southern and central Appalachians can be divided longitudinally into an igneous-metamorphic belt (Piedmont), a belt of complexly deformed late Proterozoic and early Paleozoic sediments (Blue Ridge), a belt of folded and faulted Paleozoic rocks (Valley and Ridge), and gently folded plateaus of late Paleozoic rocks. The northern Appalachians are largely composed of metamorphic and igneous rocks that cannot be divided into these regions.

The oldest sedimentary rocks of the Appalachian belt of Hadrynian age overlie basement rocks metamorphosed about 1 billion years ago. The Hadrynian to Lower Cambrian succession is composed largely of quartzites derived from the weathering of the Canadian Shield with some interbedded lavas. From Middle Cambrian to Middle Ordovician time the edge of the platform was the steep margin of a carbonate bank from which limestone blocks slid eastward into the deeper sea where shale was accumulating. The continental margin on which the late Proterozoic and early Paleozoic sediments accumulated was probably formed by the rifting that opened the Protoatlantic or Iapetus Ocean in Hadrynian time. The Avalon belt may be a part of the African side of the Iapetus Ocean left behind on the western side when the Atlantic Ocean formed.

In Middle Ordovician time the supply of sediment to the eastern margin of the continent reversed, and first shale and then coarse clastics spread west across the miogeocline to form the Queenston clastic wedge from highlands raised in the Appalachian geosyncline in the Taconic orogeny. This mid- to late Ordovician orogeny involved regional metamorphism, the intrusion of granitic and ultramafic rocks, and the thrusting and sliding westward of large plates of geosynclinal rocks on to the miogeocline. The orogenic episode may have been related to the plate interactions along a subduction zone formed at the west margin of the Iapetus Ocean.

In Silurian time the Appalachian area was one of decreasing relief, but extensive volcanism continued along the subduction zone at the continental margin. Along the miogeocline, Silurian rocks recorded the shrinking of the clastic wedge as the Taconic highlands were reduced by erosion.

QUESTIONS

1. Compare the Proterozoic and early Paleozoic history of the Appalachian and Cordilleran chains, citing specific formations that were deposited in similar tectonic settings.
2. Discuss the erosion, transport, and deposition of sediments in the Queenston clastic wedge, considering the probable effects of the absence of vegetation on these processes.
3. On thin tracing paper draw a map showing the main features of the Appalachian system. Turn the paper over, and beside the map of the Appalachians, draw a similar map of the Cordilleran mountain system, making the maps about the same length by changing the scale. Describe the geologic and geographic similarities and differences between the two systems when the mirror image of one is compared to the true image of the other.
4. Discuss the uncertainties inherent in using radioactive methods to find the age of an episode of deformation and intrusion in a complex mountain system like the Appalachians.

SUGGESTIONS FOR FURTHER READING

BIRD, J. M., and DEWEY, J. F. "Lithosphere Plate–Continental Margin Tectonics and the Evolution of the Appalachian Orogen." *Geological Society of America Bulletin,* 81: 1031–1060, 1970.

FISHER, G. W., PETTIJOHN, F. J., REED, J. C., and WEAVER, K. N. ed. *Studies of Appalachian Geology: Central and*

Southern. New York: Interscience Publishers, 460 p., 1970.

HATCHER, R. D. "Developmental Model for the Southern Appalachians." *Geological Society of America Bulletin,* 83: 2735–2760, 1972.

RODGERS, J. *The Tectonics of the Appalachians.* New York: Wiley–Interscience, 272 p., 1970.

______. "The Taconic Orogeny." *Geological Society of America Bulletin,* 82: 1141–1178, 1971.

ST. JULIEN, P., and HUBERT, C. "Evolution of the Taconian orogen in the Quebec Appalachians." *American Journal of Science,* 275-A: 337–362, 1975.

SCHENK, P. E. "Southeastern Atlantic Canada, Northwestern Africa and Continental Drift." *Canadian Journal of Earth Sciences,* 8: 1218–1251, 1971.

ZEN, E-AN, WHITE, W. S., and HADLEY, J. B. ed. *Studies of Appalachian Geology: Northern and Maritime.* New York: Interscience Publishers, 475 p., 1968.

CHAPTER EIGHTEEN

INTRODUCTION

During the second half of the Paleozoic Era the structural framework of the Appalachian Mountain chain reached its present form. The continental margin that was once the site of volcanic eruptions, extensive deformation, and rapid sedimentation lost its mobile nature and was stabilized. Through late Paleozoic orogeny a segment of the crust that was once oceanic had been added to the eastern margin of the North American plate.

At the end of the Paleozoic Era the continental masses of the world were united into a supercontinent—Pangaea. During late Paleozoic time the continent was assembled by the closing of the early Paleozoic ocean, Iapetus. The last orogeny in the Appalachian geosyncline has been attributed to this assembling process, that is, the collision of Europe and Africa with the Appalachian side of North America. Two late Paleozoic orogenies are distinguished: the Acadian orogeny of Middle and Late Devonian time; and the Alleghany, or Appalachian, orogeny of Late Pennsylvanian and Permian time. In order to understand the events taking place in late Paleozoic time on the Appalachian side of the present Atlantic Ocean, we must briefly consider the geology of the lands that now border its northern and eastern coasts, but were then in collision with North America.

At the beginning of the Devonian Period the highlands raised in the Taconic orogeny were submerged, or of low relief. Much of the northern Appalachian geosyncline was covered with limestones, shaly limestones, and siltstones during Early Devonian time, although volcanism continued in scattered localities. The Silurian belt of volcanism along the present Maine–New Brunswick coastline appears to have been land during Early Devonian time.

ACADIAN OROGENY

The northeastern corner of North America was called "Acadia" by the early settlers, and this name has been applied to the Devonian deformation in the Appalachian Mountain chain. The rocks of the Appalachian geosyncline were thrust northward and westward, and crumpled against the craton from Newfoundland to Pennsylvania. A narrow belt southeast of Logan's Line may have escaped the effects of Acadian deformation, but otherwise the full width of the Newfoundland and northern Appalachians was affected. Isotopic dating shows that the episodes of metamorphism and intrusion included in the Acadian orogeny occurred between 350 and 400 million years ago (the

LATE PALEOZOIC CONTINENTAL COLLISION

end of the Silurian to the end of the Devonian Period).

Stratigraphic evidence suggests that the Acadian orogeny had little effect on the southern, and most of the central, Appalachians. Unconformities in the anthracite belt show that the inner edge of the Valley and Ridge province at its northern end in Pennsylvania was caught up in some of the folding. Isotopic ages indicating metamorphism of Devonian age from the Blue Ridge and Piedmont can be interpreted as indicating widespread Devonian deformation or the resetting of radioactive series by late Paleozoic metamorphism.

Throughout the northern Appalachians the youngest marine sedimentary rocks are of Early Devonian age. In central New England these beds were folded and overlain unconformably by the Mississippian Moat volcanics. In the Gaspé Peninsula of Quebec the limestones pass up into non-marine, Middle Devonian, red sandstones indicative of uplift in the geosyncline. In the Gaspé a prominent Acadian unconformity is formed between these Middle Devonian beds and poorly dated Mississippian red conglomerates (Fig. 18–1). Here, the major period of deformation appears to have been in Middle Devonian time, but deformation apparently continued throughout the Devonian Period. In Newfoundland the whole central belt from the Avalon Peninsula to the Proterozoic rocks of the Long Range were folded and metamorphosed. Most of the deformation took place in Middle and early Late Devonian time, but isotopic ages of some intrusive rocks suggest that intrusion started as early as the Late Silurian Epoch.

Intrusion of granitic rocks took place on a grand scale throughout New England, the Maritime Provinces, and Newfoundland during the Devonian and into the Mississippian Period (Fig. 18–2). Base-metal mineralization associated with these granites supports several mines in New Brunswick and Newfoundland, but in New England no deposits of commercial value accompanied the Acadian intrusions.

CATSKILL CLASTIC WEDGE

Correlation of Deltaic Deposits

The Queenston clastic wedge reflected the rise of highlands of Taconic age in the northern part of the Appalachian geosyncline. The Catskill clastic wedge is evidence of the rise of the Acadian mountains in the same segment. The wedge is named from the hills that border the western side of the Hudson Valley in southeastern New York State. A stratigraphic section through the Devonian rocks of the Catskills shows upward changes that are the result of the ris-

FIGURE 18–1 **The Acadian unconformity at Baie des Chaleurs, Quebec. Conglomerates of the Bonaventure Formation (Mississippian) overlie the fine-grained sandstone of the Gascons Formation (Silurian), dipping steeply to the left at Pierre-Loiselle Cove. (Courtesy of Pierre-Andre Bourque.)**

ing highlands (Fig. 18–3). At the base of the Middle Devonian Series the Onondaga coralline limestone is evidence of the low relief in the Appalachians and clear water on the adjacent shelf. The limestone is overlain successively by units of black shale, gray shale, interbedded siltstone and sandstone, and at the top of the Devonian succession by about 1000 m of red and green sandstones and shales. The succession records a change of environments from clear water marine conditions, through shoreline and estuarine conditions, to an alluvial apron spread out in front of the growing Acadian Mountains. If stratigraphers were to measure two other sections, one at the Finger Lakes (Fig. 18–3) in the center of the state, and one at Lake Erie at its western side, they would record similar sequences of lithologies in each. The natural tendency of the early stratigraphers was to correlate shale with shale, sandstone with sandstone, and redbeds with redbeds.

The detailed paleontological and stratigraphic studies of George W. Chadwick and G. Arthur Cooper made these Devonian rocks a classic example of the divergence between facies boundaries and time horizons in a deltaic deposit. They established that in the Catskill clastic wedge time surfaces slope westward, facies boundaries slope eastward (Fig. 18–4). The establishment of time lines, which were the regional depositional surfaces in Devonian time, across these facies changes was largely by tracing key limestone beds. Four limestones beds a few meters thick can be followed from the nearshore facies into the black, offshore shales of the Middle Devonian Hamilton Group. These beds of distinctive lithol-

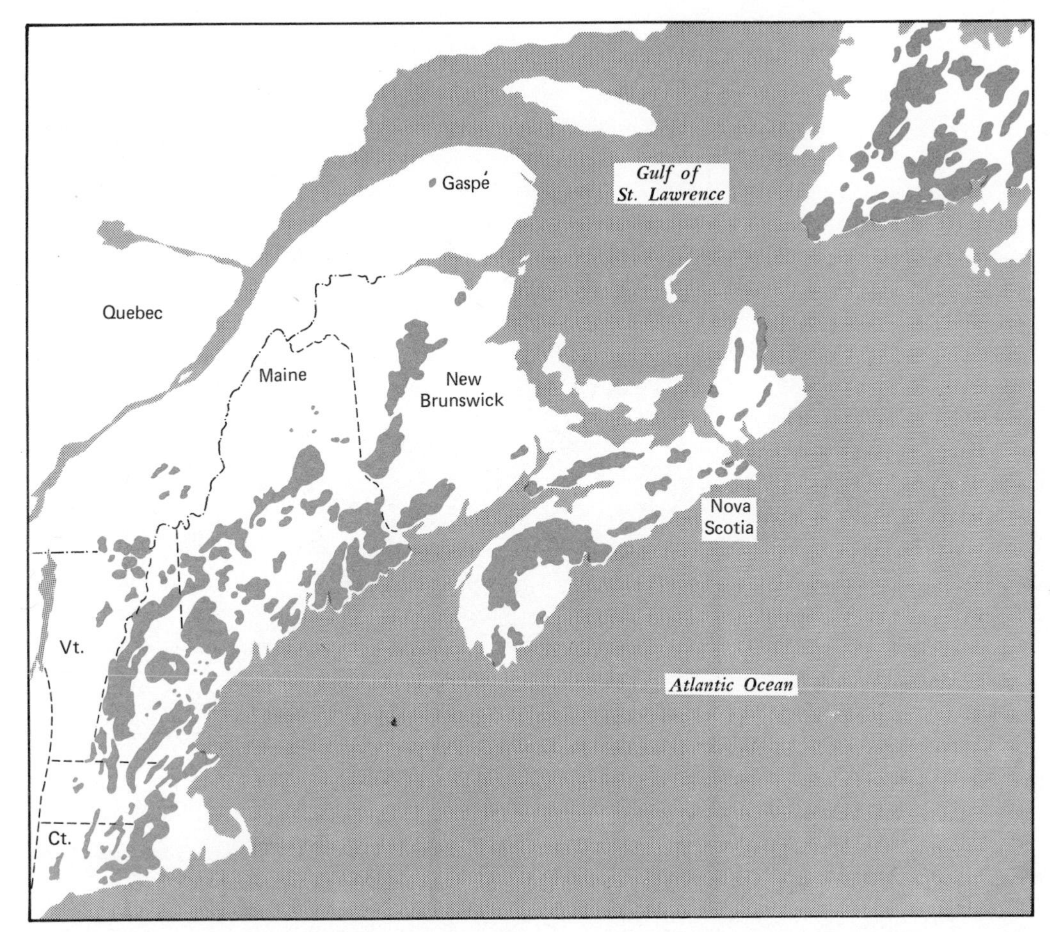

FIGURE 18–2 **Devonian intrusive rocks of the northern sector of the Appalachians. (Compiled from various sources.)**

ogy and fauna apparently represent brief periods in the history of the deltaic complex, when conditions for limestone deposition were widespread (Fig. 18–5). Such a widespread change from clastic deposition might have been connected with changes in sea level or in regional supply of sediment to the sea. During the deposition of two of these limestones the non-marine facies extended basinward, suggesting that lowering of sea level may have been the ultimate cause of the sudden change to limestone deposition throughout the basin of deposition. Whatever the cause of the interruption in the facies pattern marked by the limestone beds, they provide the basis of reconstructing the slope of the depositional surfaces which crossed the facies boundaries during the deposition of the Catskill wedge.

As the supply of detritus from the Acadian Mountains in the geosyncline grew, the surface of the alluvial apron pushed westward; the red sandstones were swept by the flooding streams over the estuarine sandstones with their fauna of bivalves, and the black shales with their fauna of brachiopods and fish, until at the end of the Devonian Period redbeds were deposited 500 km from the mountains in the region that is now Lake Erie.

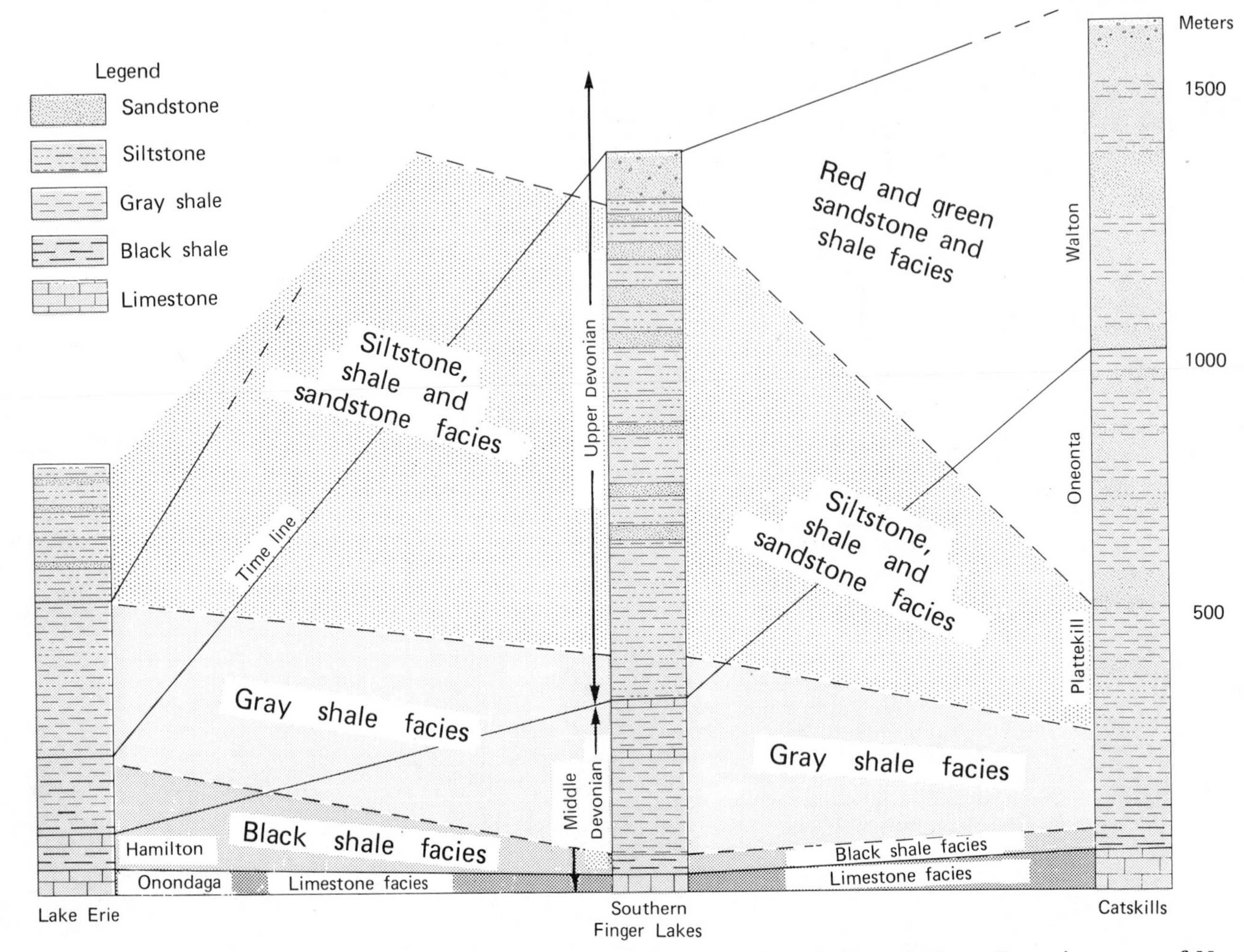

FIGURE 18–3 **Three stratigraphic sections through the Middle and Upper Devonian strata of New York at Lake Erie, the Finger Lakes, and the Catskill Mountains. Only the major facies are represented.**

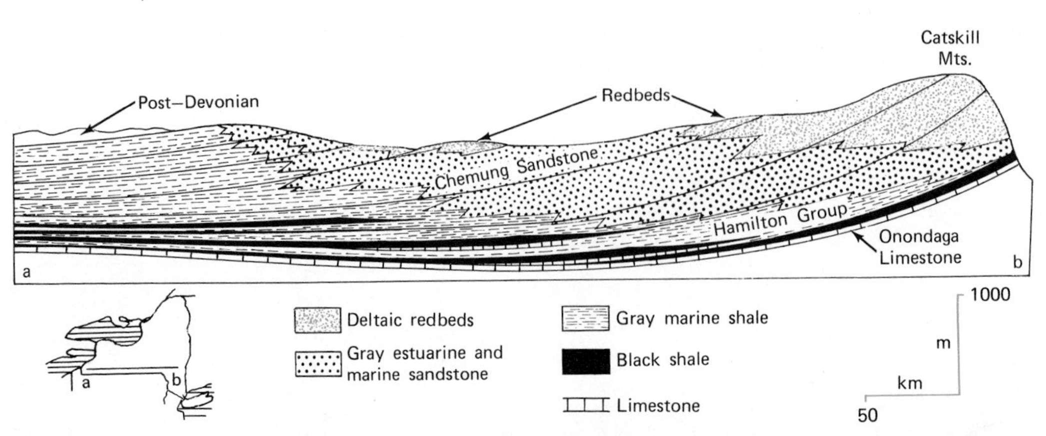

FIGURE 18–4 **Facies changes in the Middle and Upper Devonian strata of southern New York State. (Modified from the work of Chadwick and Cooper, after C. O. Dunbar and J. Rodgers, *Principles of Stratigraphy*. New York: John Wiley & Sons, Inc., 1957).**

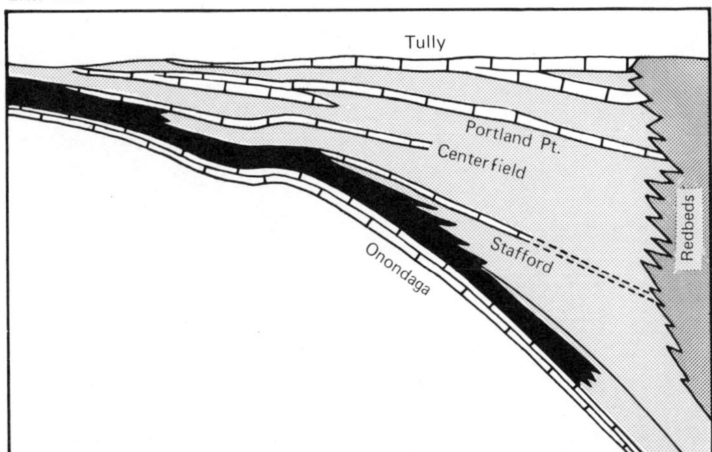

FIGURE 18–5 Correlation of the Middle Devonian Hamilton Group across the state of New York by means of key limestone beds. (Modified from G. A. Cooper and others.)

Facies Changes

The argillaceous limestones and shales of the Middle Devonian Hamilton Group contain some of the most beautifully preserved and abundant fossil invertebrates on this continent (Fig. 18–6). On the eastern side of the clastic wedge the beds equivalent to the Hamilton Group are red and gray sandstones of continental origin, containing freshwater fish and plant remains. During the excavation of a Catskill reservoir for New York City, a buried forest of the progymnosperm, *Aneurophyton,* and the scale tree, *Protolepidodendron,* was discovered in these sandstones. The stumps of the trees were preserved as they were buried in sand washed in by streams more than 360 million years ago.

The geometry and major facies changes in the clastic wedge can be illustrated with a facies map (Fig. 18–7) on which isopachs have been drawn. The thickest and coarsest parts correspond closely with those of the Queenston wedge; both thin southward along the valley and Ridge province. In Tennessee Late Devonian sedimentary rocks are thin and largely limestone, and still farther south no Devonian rocks are present in the Appalachians. Acadian uplift does not appear to have affected this area (Fig. 18–8).

Like many late Paleozoic clastic successions, the Catskill redbeds show cyclical alterations of fine, muddy members and coarse, sandy members. The sands have been interpreted as channel deposits of the rivers that formed braided courses on the alluvial apron; the muds, as interchannel deposits, formed as the rivers flooded over their banks. The cyclic repetition of these beds indicates that conditions that caused the cyclothemic pattern of deposition in Carboniferous time had already begun to develop in late Devonian time.

The significance of the Catskill beds was summed up by Joseph Barrell in the language of a more elegant era (1914): "The Devonian Mountains are gone, and where they once rose in defiant height their very foundations are broken and buried, but in the remnants of formations born of destruction, we may read the epitaph that records their greatness".

MID-PALEOZOIC FOLD BELTS OF THE NORTH ATLANTIC

On the northern coast of Newfoundland the folds and faults formed in the Taconic and Acadian orogenies strike out into the Atlantic Ocean, and beneath the waters are covered by a blanket of younger sedimentary rocks. Were they once continuous with a mountain chain on the other side of the At-

FIGURE 18–6 **A group of Devonian trilobites, *Phacops rana*, from the Hamilton Group of New York. (Courtesy of the Smithsonian Institution.)**

lantic, and have they since been separated from it by Mesozoic rifting? To answer this question we must turn our attention briefly to the other continents that border the North Atlantic.

Caledonian Mountain Belt

In the northwestern corner of Europe two mountain belts come together at the Atlantic Coast of Ireland, and diverge eastward around a platform area (Fig. 18–9). The northern mountain system is called the Caledonian, after the Roman name for Scotland; the southern is called the Hercynian, after the Roman name for the Hartz Mountains of central Germany. The Caledonian Mountain system extends northeastward through Norway. A thin belt of Caledonian deformation occupies the northeastern coast of Greenland and crosses the island of Spitzbergen at the entrance to the Arctic Ocean. Even when the continents are reassembled around the North Atlantic, these fragments of the Caledonian chain do not fit tightly with each other, nor with the end of the Appalachians in Newfoundland, because the continental shelves, beneath which the belt is presumed to be buried, intervene.

Although the hypothesis that the Appalachian and Caledonian structures were once continuous across the closed Atlantic is attractive, the time of deformation on the two sides of the Atlantic is not the same. The major orogeny in the Caledonian Mountains was of Late Silurian age. Granitic intrusions associated with it have isotopic ages of about 400 million years. There is little evidence of Middle and Late Ordovician orog-

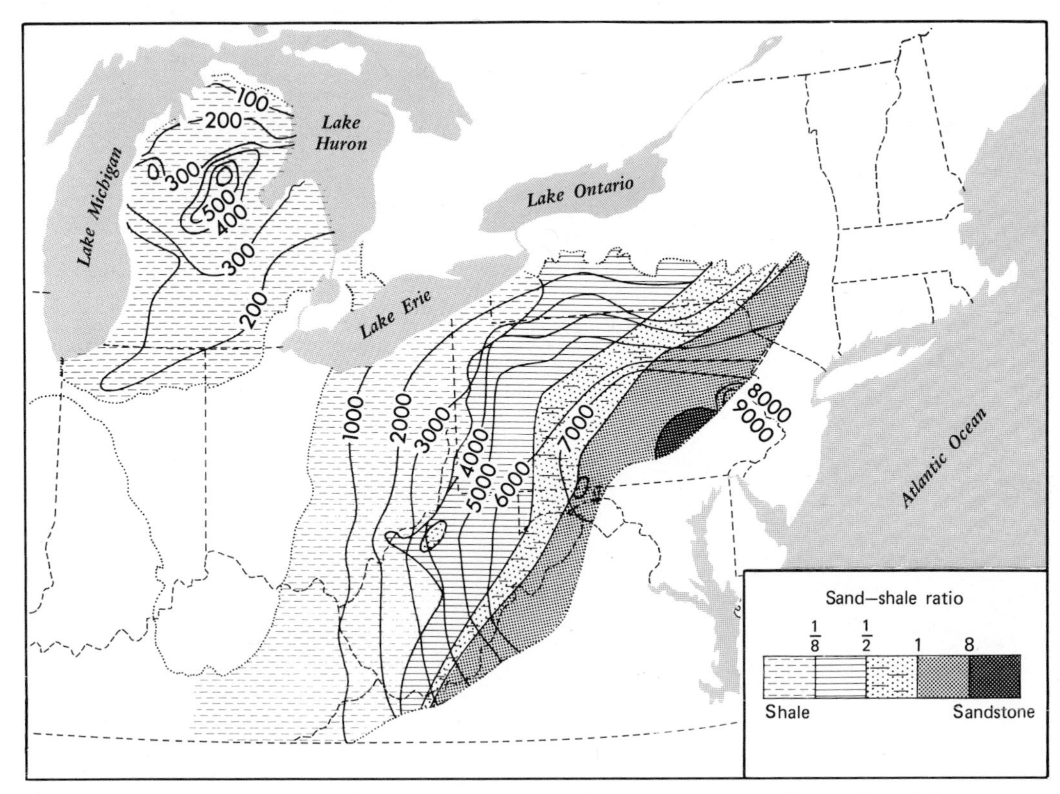

FIGURE 18–7 **Isopach and sand–shale ratio map of the Upper Devonian Series of the eastern United States, showing the geometry of the upper part of the Catskill clastic wedge. Isopachs are in feet, not meters. (After W. Ayrton; courtesy of the Pennsylvania Geological Survey.)**

eny in Europe, and only scant evidence of Late Silurian movements in North America. If the Caledonian–Acadian orogeny was caused by the closing of the Iapetus Ocean welding Europe and North America together, then these differences in timing could be caused by the collision having taken place, first in the north where the ocean was narrowest, and progressively later farther south. In addition, if local salients of the two continents collided first as the ocean closed, local variations in timing would be accounted for. If the collision of the North Atlantic continents caused the Acadian orogeny and the Avalon belt was once on the eastern side of Iapetus, then all that remains of that ocean is the deformed sediments and lavas of the Appalachians west of the Avalon belt. The rest of Iapetus has been subducted along the trench that was active along the Appalachian geosyncline in Ordovician to Devonian time, and along a similar subduction zone that may have existed on the European–African side of Iapetus in early Paleozoic time.

The closure of the northern end of Iapetus is reflected in the increasing similarity from Ordovician into Silurian time of the fossil faunas on either side of the ocean. In Early Ordovician time only the pelagic graptolites were common to the two sides, but by late Ordovician time brachiopods and trilobites that had free swimming larval stages were able to cross the closing gap. Not until Late Silurian time were brackish and fresh water animals able to mix as the northern end of Iapetus closed in the Caledonian orogeny.

The Caledonian orogeny immobilized much of northern Europe, forming a stable

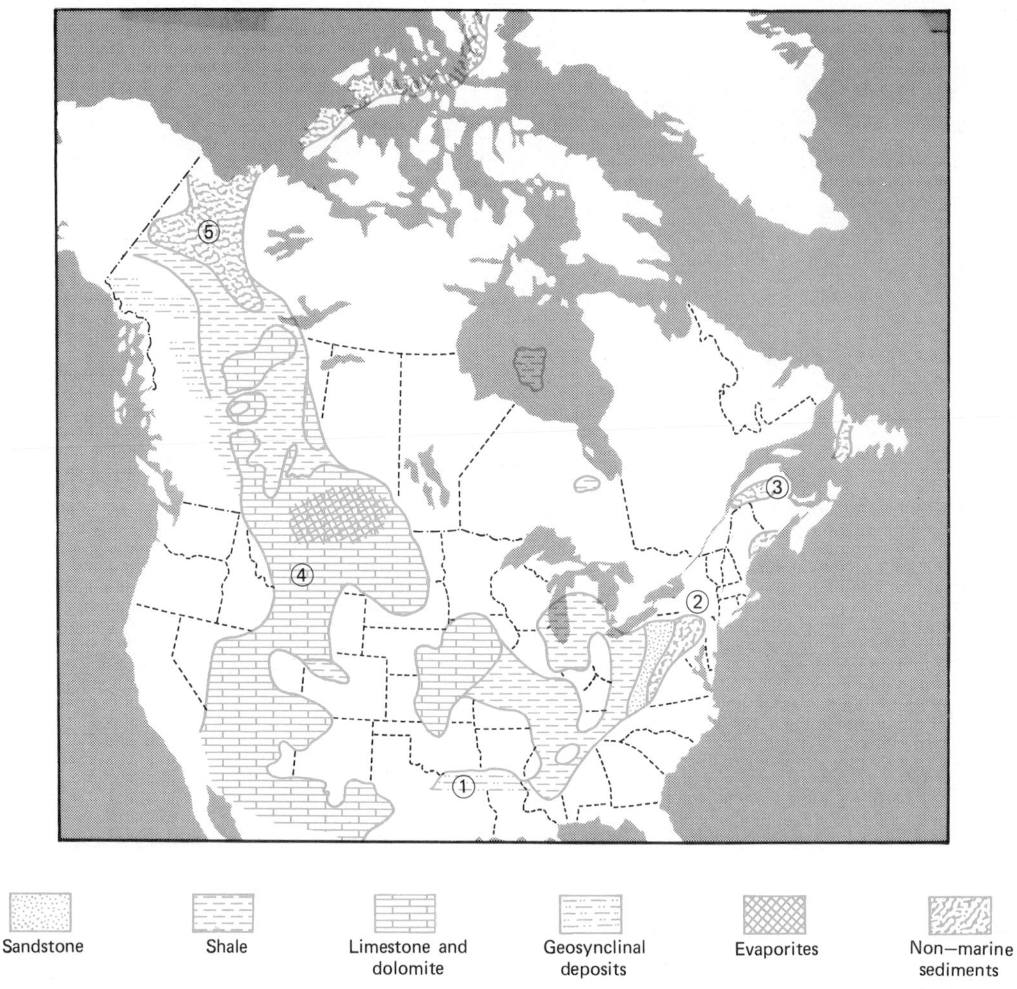

FIGURE 18–8 **Facies of the Upper Devonian Frasnian Stage in North America. 1. Arkansas Novaculite, 2. Catskill redbeds. 3. Escuminac beds. 4. Jefferson Limestone. 5. Imperial Formation redbeds.**

mass on which red sandstones of Devonian age accumulated in successor basins and on the flanks of the eroded fold belt. These sandstones are called the Old Red Sandstone in the British Isles. This formation is similar in many respects to the Catskill redbeds of New York State and the red Devonian sandstones deposited within the Appalachian fold belt, but it is unlikely that all these deposits were ever continuous; rather, they represent similar environments within or on the flanks of a developing mountain range.

Hercynian Belts

The late Paleozoic belt of deformation crosses central Europe and strikes out into the Atlantic across France and the southern tip of England and Ireland. Although the major deformation in the Caledonian belt was of Late Silurian age and that in the Hercynian belt was of Carboniferous and Permian age, the history of deformation in each of these belts spanned a much longer interval and, in fact, overlapped. Episodes of Devonian deformation took place in the Ca-

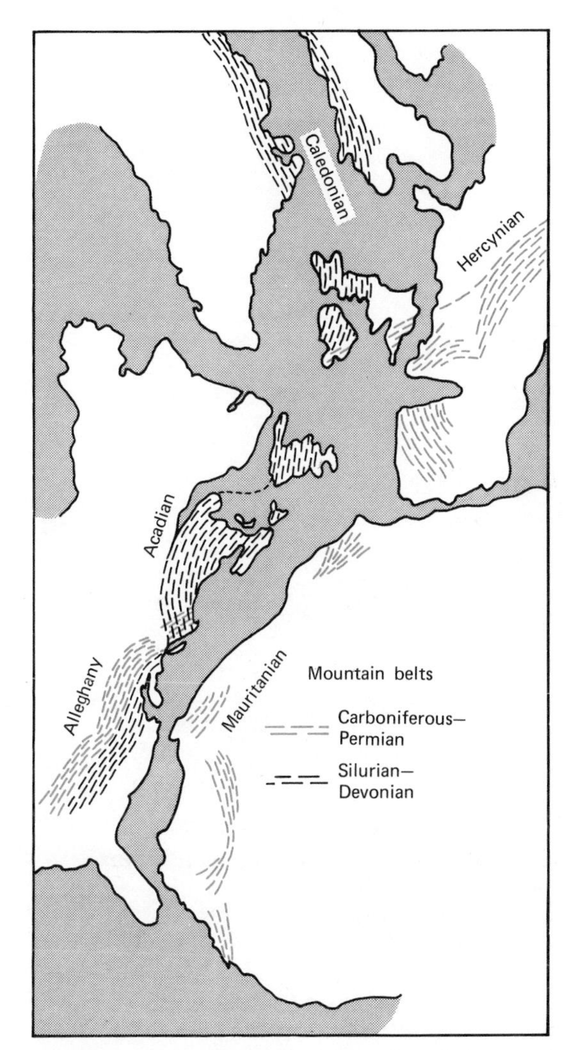

FIGURE 18–9 **Paleozoic fold belts of the North Atlantic region, plotted on the reconstruction of the continents in the late Paleozoic proposed by Bullard (Fig. 1–16). These were not necessarily the geographic relationships when the folds were formed as the effects of transform faulting have not been taken into account.**

ledonian belt of East Greenland and England, as well as in the Hercynian belt of central France.

The Paleozoic mountains of West Africa have been called the Mauritanians. They parallel the Appalachians on the eastern side of the Atlantic and, like them, were intensely deformed late in the Paleozoic Era. While the Appalachians were thrust westward, the Mauritanians were thrust eastward toward the cratonic interior of Africa. Earlier phases of deformation have been recognized in Ordovician–Silurian time in the south, and in Late Devonian time throughout the chain, a history remarkably similar to that of its Appalachian counterpart.

The Silurian–Devonian phases of orogeny in the northern Appalachians and Caledonian mountains have been interpreted as caused by collision between Europe and North America; the Pennsylvanian–Permian phases have been interpreted as caused by the collision of Africa with Europe and North America. The deformation caused by the closure of Iapetus appears to have progressed from north to south over an interval of 250 million years, but in detail the history of deformation is much more complex.

The detailed sequence of events and the exact fit of the continents and oceans into a picture of middle and late Paleozoic deformation and continental collision in the North Atlantic area have not yet been determined. Possibly, small microcontinents, like the Avalon belt, acted independently of Africa and America during the Paleozoic history of Iapetus. However, similarities of geology on either side of the Atlantic leave little doubt that the segments that are now separated then formed a major Paleozoic mountain system.

LATE PALEOZOIC BASINS OF THE NORTHERN APPALACHIANS

In Mississippian time the metamorphic and igneous terrane of the northern Appalachians was cut by a series of regional faults, which separated highland areas of the crystalline basement from depositional basins where erosion products accumulated (Fig. 18–10). During Mississippian time these basins were above sea level at times, and received deposits of red and gray, nonmarine sandstones and shales; at other times they were depressed to allow the entry of the sea and the deposition of limestone. When the connection to the sea was

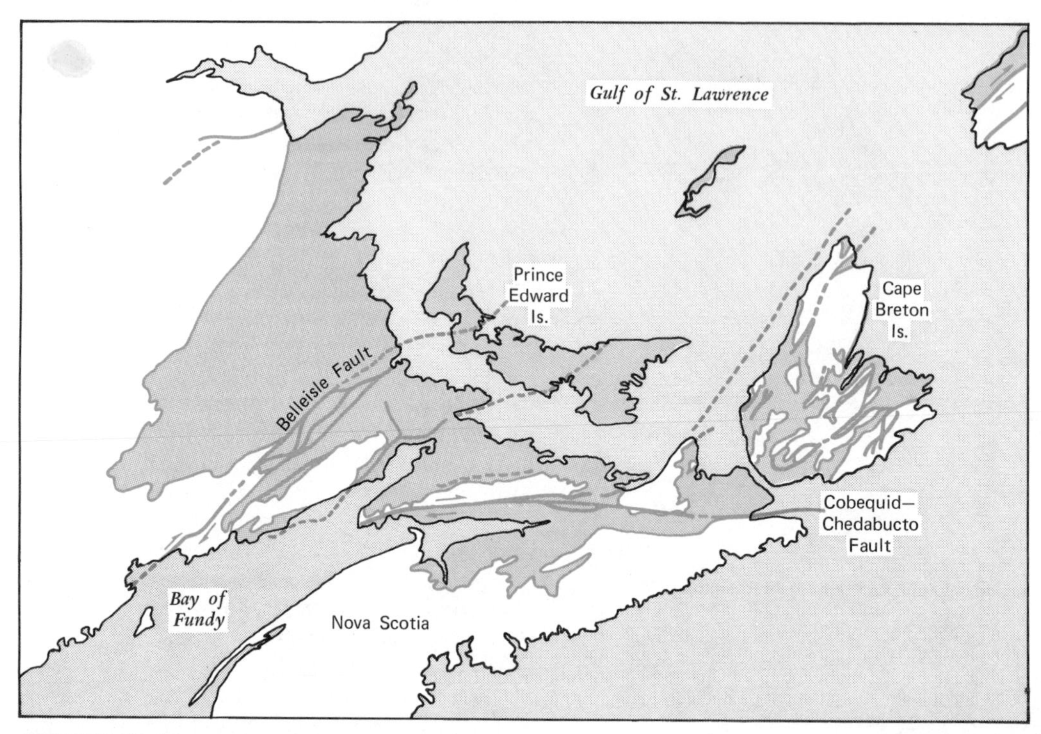

FIGURE 18–10 **Basins containing late Paleozoic sedimentary rocks in the Maritime Provinces of Canada and the major faults that bound and traverse the basins. (After Gregory Webb; courtesy of American Association of Petroleum Geologists.)**

restricted, salt, anhydrite, and gypsum accumulated. Evaporites of Mississippian age underlie much of the Gulf of the St. Lawrence. Under the weight of sediments accumulating above, the salt beds have acted plastically, flowing upward and piercing the overlying sediments in structures called salt domes. The formation of similar salt domes in the Gulf Coastal Plain is considered further in the next chapter.

The Pennsylvanian and Early Permian rocks in these basins are a non-marine, coal-bearing succession, interrupted by numerous unconformities that reflect the crustal instability of the fault-bounded basins. The rapid deposition of certain units is indicated by the 4000-m section of Pennsylvanian strata at Joggins, Nova Scotia, where early reptiles are found trapped in the stumps of fossil trees buried upright in the rapidly accumulating alluvial sandstones (see Chapter 13). Coal is mined from these basinal sediments at Sydney, Nova Scotia.

Two isolated basins containing late Paleozoic sedimentary rocks occur in southern New England. In the Boston Basin, the succession is 2000 m thick and consists of conglomerates and argillaceous sediments. The only evidence for dating these strata are two poorly preserved fossil tree trunks that may be as old as Devonian or as young as Permian. In this non-marine succession are lenses of the Squantum Formation, a coarse, unsorted conglomerate, consisting of boulders and cobbles of various lithologies set in an argillaceous matrix. Some investigators have identified the conglomerate as a till deposited by valley glaciers from the Acadian Mountains. The surficial characteristics of sand grains from the formation, seen in the scanning electron microscope, have been interpreted as characteristic of glacial deposits. Other geologists have suggested that the lenses are mudflow deposits and not indicative of either cold climate or extreme elevation. The Pennsylvanian sedimentary rocks of the Narraganssett Basin in Rhode Island consist of 4000 m of coal measures, conglomerate, sandstone and shale deposited in an intermontane environment. Metamorphism at the end of Pennsylvanian time raised the coal beds beyond the anthracite grade, forming graphite and making them unsuitable for use as fuel.

The late Paleozoic basins of the Northern and Newfoundland Appalachians are successor basins. Some of the movement along the faults that border them must have been vertical because the basins subsided to receive great thicknesses of sediment, and the adjacent blocks were raised to supply detritus. Many geologists now believe that horizontal movement parallel to the strike of the faults also took place on a large scale. The movement along the Cobequid–Chedabucto fault is believed to have shifted northwestern Nova Scotia about 250 km in a right-lateral sense. Similar movement of 200 km has been estimated for the faults running along the southern coast of New Brunswick (Fig. 18–10).

As these faults strike out into the Atlantic, we should find some evidence of their continuation in northern Europe. The movements on two of the major faults that cross Scotland, the Great Glen fault and the Highland Boundary fault, are believed to have major strike-slip components. Controversy exists as to which direction of motion has been dominant along these faults, but some evidence suggests a pre-Devonian phase in which motion was left-lateral, and a late Paleozoic phase in which it was right-lateral.

The late Paleozoic successor basins of the Appalachians can be compared to the Cenozoic basins of California. The tectonic settings of the two basin systems were different, but in both the structural behavior appears to have been controlled by lateral fault movements.

LATE PALEOZOIC CLASTIC WEDGE

During the Mississippian Period highlands in the Appalachian Piedmont continued to supply sediments westward to an alluvial apron at the sites of the present Valley and Ridge and Plateau provinces. In Pennsylvania the Pocono Formation at the base of the Mississippian System consists of gray sandstones with interbeds of conglomerate which, because of their resistance to erosion, underlie many of the ridges in the Valley and Ridge province. The systematic decrease in the maximum size of the pebbles in the Pocono conglomerates northwestward across Maryland and Pennsylvania shows that the source area of this formation was the Appalachian highlands in the present position of New Jersey. The Mauch Chunk Formation, which overlies the Pocono, is composed of less resistant red siltstones and shales. Southwestward, the clastic content of the system decreases and in Tennessee, Mississippian rocks are mostly limestones, like those of the Mississippi Valley (see Fig. 12–21). These facies patterns suggest that in the southern Appalachians, the Piedmont was not supplying clastic sediments during Mississippian time.

During the Pennsylvanian Period detritus from the Appalachian highlands continued to build the alluvial apron west of the Blue Ridge. The pattern of distributary channels on the delta shows clearly that the major source of sediment was northeast of the clastic wedge, in the area of the present northern Piedmont and Blue Ridge, but minor contributions were made from the Canadian Shield. This was a time of crustal, or sea-level, instability, when cyclothems were formed and the shoreline moved back and forth hundreds of kilometers across the platform. When the shoreline was withdrawn to the southwest, vast coal swamps stretching thousands of kilometers from the mountains were formed. The youngest beds in the late Paleozoic clastic wedge are of Early Permian age and occur in the Plateau province straddling the border between West Virginia and Pennsylvania. The dominance of nonmarine sandstones in these cyclothems shows that the supplying highlands must have been repeatedly uplifted from Middle Devonian time until the end of the era.

THE ALLEGHANY OROGENY

The term "Alleghany" was suggested by H. P. Woodward to designate the Pennsylvanian and Permian deformation in the Appalachian Mountain chain. The different spelling of the orogeny (Alleghany) and the plateau and sedimentary group (Allegheny) should be noted. The term "Appalachian orogeny" has also been applied to the late Paleozoic movements, but is now used to designate the whole set of orogenic episodes (Taconic, Acadian, Alleghany) that contributed to the formation of the Appalachian structures. The youngest rocks deformed in the Valley and Ridge belt in the Alleghany orogeny are of Late Pennsylvanian age, but in the Plateau country of western Pennsylvania and West Virginia Permian strata have been gently folded. The oldest beds overlying Appalachian structures unconformably are of Late Triassic age.

In the Taconic and Acadian orogenies geosynclinal rocks in the northern and Newfoundland Appalachians were thrust northwestward, deforming miogeoclinal rocks along the border of the North American plate, but in the southern Appalachians the miogeocline was not extensively deformed. In the Alleghany orogeny the miogeoclinal sedimentary rocks now underlying the Valley and Ridge and Plateau provinces were thrown into folds and cut by extensive thrust faults as they were compressed against the continent. The full length of the central and southern Appalachians appears to have been involved in the deformation, but only the southern part of New England was affected. In the Maritime Provinces and Newfoundland the rocks of the successor fault basins were tilted and broken, but show little sign of regional compression.

Structure of the Valley and Ridge Province

In Pennsylvania the ridges are composed of resistant sandstones of the Pottsville (Pennsylvanian), Pocono (Mississippian), and Tuscarora (Silurian) formations. From the air these ridges trace a series of broad, open, remarkably parallel folds that are typical of the central sector of the Valley and Ridge province (Fig. 18–11). Northward, the amplitude of the folds decreases and they die out into flat-lying beds in New York State; southward, into Tennessee, the province narrows and the shortening of the crust by deformation appears to be greater. In the Valley and Ridge belt the Paleozoic rocks are essentially unmetamorphosed, except along the eastern edge in Pennsylvania, where shales have been turned to slates and coals have been transformed into anthracite. Thrust faults are not conspicuous in the central sector of the Valley and Ridge province, and where they have been mapped at the surface they appear to be parallel to the bedding planes and located in the weak formations of the miogeoclinal assemblage, notably in the Cambrian shales. Their location has suggested to many geologists that the whole of the folded belt is underlain by a bedding plane, or décollement, thrust at this horizon, along which the overlying sedimentary rocks have sheared off the older rocks below (Fig. 18–12).

In the southern Appalachians the open folds of the central Appalachians are not common. The Paleozoic miogeoclinal succession appears to have been cut by numerous thrust faults into slices and piled against the platform (Fig. 18–13). Displacements of tens of kilometers have been mapped along décollement faults following incompetent layers in the succession. The contact of the Valley and Ridge with the Blue Ridge province in the southern Appalachians is a zone of intense overthrusting and folding (Fig. 17–16), whereas in the central Appalachians the Proterozoic rocks at the edge of the Blue Ridge are folded sharply downward beneath the Paleozoic rocks to the west.

The role of the Blue Ridge in the deformation of the Valley and Ridge rocks has been a matter of controversy. Some geologists have postulated that the Blue Ridge rocks acted as a piston, pushing against one side of the miogeoclinal sediments until they sheared from the basements rocks and were folded and faulted against the platform. Others, who doubt that stresses capable of causing intense deformation can be transmitted tens of kilometers from the "plunger" through sedimentary rock layers,

FIGURE 18–11 **Landsat imagery of the Harrisburg, Pennsylvania, region. The Susquehanna River crosses the area from northwest to southeast and empties into Chesapeake Bay. The upper left third of the area shows the ridges of the Valley and Ridge province; the middle half of the area represents the Piedmont and Blue Ridge crystalline rocks; and the extreme southeast corner is part of the Coastal Plain. (Courtesy of NASA.)**

have suggested that the décollement thrusts developed when slabs of the sedimentary succession tens of kilometers wide and one kilometer thick slid toward the platform as the area of the Blue Ridge and Piedmont was elevated to produce the necessary grade for the movement. These two hypotheses have also been applied to explain the

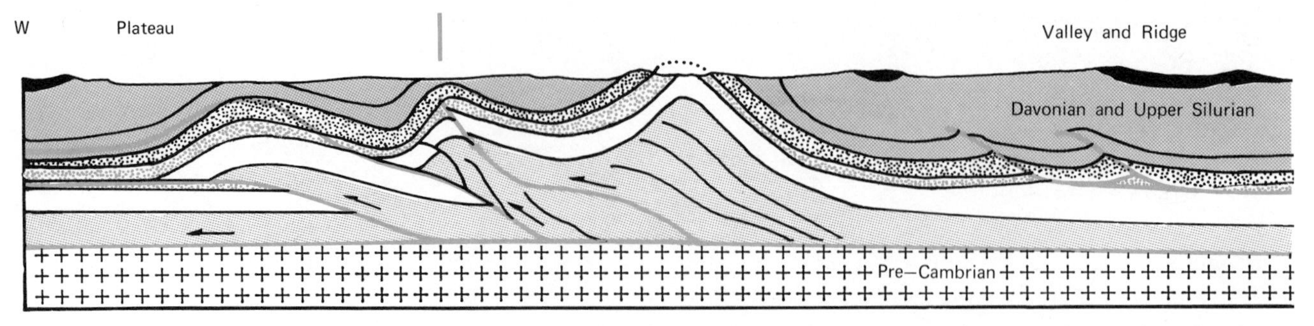

FIGURE 18–12 **Interpretative cross-section of the Valley and Ridge province near the southern border of Pennsylvania. (After V. E. Gwinn, courtesy of John Wiley & Sons, Inc., New York.)**

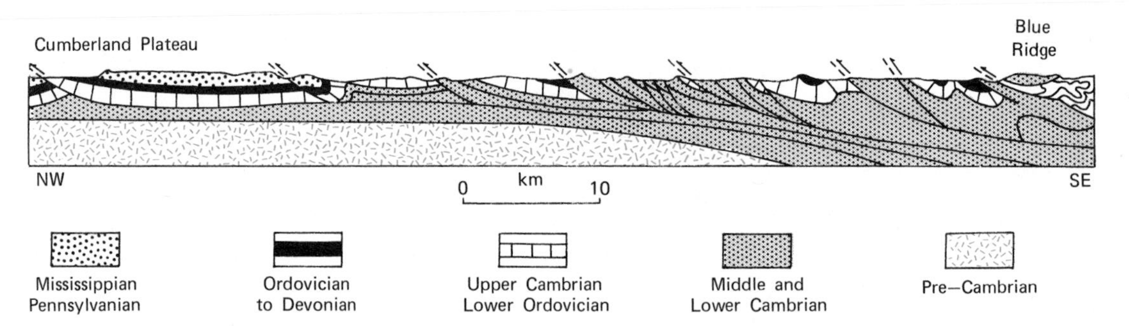

FIGURE 18–13 **Cross-section of the Valley and Ridge province in Tennessee showing the abundance of thrust faults. (After J. Rodgers; courtesy of the Kentucky Geological Survey.)**

deformation of miogeoclinal thrust belts in the Cordillera.

Most of the geological and geophysical evidence supports the interpretation of the Valley and Ridge structure in terms of décollement. This hypothesis is sometimes called the "thin-skinned" hypothesis. Some evidence in the Valley and Ridge province has been interpreted as indicating that the folds extend down into basement rocks which were involved in the Paleozoic deformation. Those supporting this "thick-skinned" hypothesis point out that stratigraphic sections measured in the axes of synclines are thicker than correlative sections in neighboring anticlines, and that the anticlinal rocks contain facies suggestive of shallow-water deposition while deeper water sediments accumulated in the synclines. They believe that this evidence shows that the folds were not all formed in late Paleozoic deformation, but were steadily increasing in amplitude during the whole Paleozoic while the sediments were being deposited. This evidence suggests a long history of deformation in the miogeocline, but does not prove basement participation. The major part of Valley and Ridge deformation must have taken place after Pennsylvanian beds were laid down, because the intensity of the deformation is similar throughout the Paleozoic succession, that is, it is not more intense in the older strata as would be expected if they had been involved in deformation throughout the era. However, early and middle Paleozoic disturbances of the continental border could have produced some hard-to-detect folding that was reflected in sedimentation patterns.

John Rodgers has used the analogy of a pile of crumpled rugs (the sedimentary strata) overlying floor boards (the basement) to illustrate the "thick-skinned" and "thin-skinned" hypotheses. The "thick-skinned"

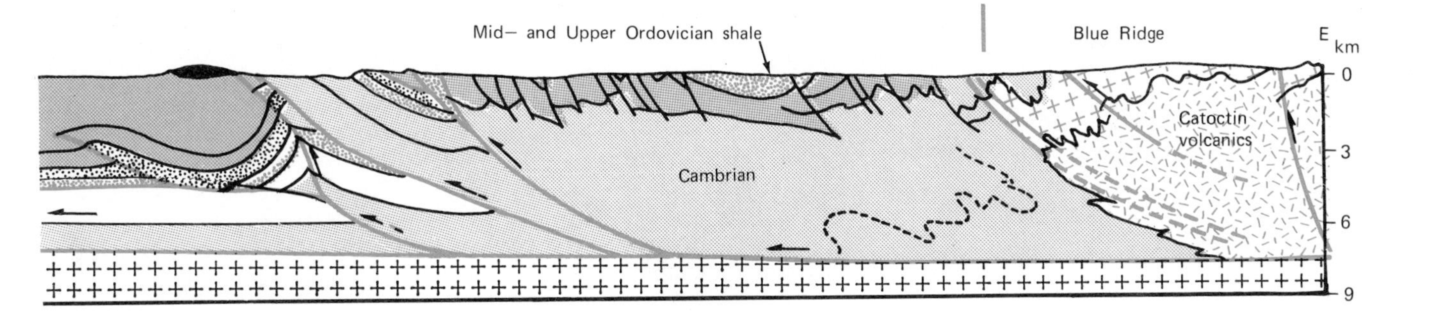

geologists believe that the folds in the rugs were caused by differential movement between the floor boards below, but the "thin-skinned" geologists believe that they were caused by the pushing of the pile of rugs from the edge, or the sliding of the rugs down a low slope.

Structure of the Appalachian Plateaus

The plateaus are underlain by Pennsylvanian rocks arranged in gentle, open folds, with the limbs dipping at only a few degrees. Although the structure of the plateau rocks seems simple at the surface, drilling has shown that more complex structures lie beneath. Wells drilled on structures that at the surface are simple anticlines have encountered complex structures at the level of the Lower Devonian, with thrust faults dipping away from the anticlinal axes. The disturbances at depth occur along décollement thrust faults that follow the level of late Silurian evaporites under the western plateaus. Under the influence of pressure from the east, or gravity acting down a west-facing slope, plates of sedimentary rocks about one kilometer thick have been released from the basement and moved westward for distances of several kilometers. The best documented example of this structure is found in the Cumberland Plateau of Tennessee where the Pine Mountain thrust block has moved 9 km northwestward along the late Silurian décollement surface (Fig. 17–3). The different forms of deformation of the Valley and Ridge, eastern Plateau, and western Plateau, may be the result of the westward stepping up of the plane of décollement from Middle Cambrian shales, to Upper Ordovician shales, to Upper Silurian evaporites.

Blue Ridge, Piedmont, and New England

The complex structure along the western side of the Blue Ridge (Fig. 18–12) shows that the rocks of this province were thrust strongly westward at the end of the Paleozoic. The extent of the deformation in the Blue Ridge and Piedmont in the Alleghany orogeny is recorded in isotopic changes in the metamorphic rocks. A belt along the eastern side of the Piedmont has yielded potassium–argon ages of about 250 million years, suggesting a period of widespread metamorphism and/or reheating in the late Paleozoic. However, the nature of the Piedmont's participation in the Alleghany orogeny remains obscure. The stratigraphic evidence indicates that it was uplifted and eroded and the structural evidence suggests that it was thrust westward, but whether regional metamorphism and intrusion took place at this time is still a matter of active investigation.

The presence of late Paleozoic rocks in the Boston and Narragansett basins permits the tracing of Alleghany deformation eastward across New England. In addition, potassium–argon ages on metamorphic rocks in southern New England give evidence that these rocks were reheated and their isotopic clocks reset in late Paleozoic time. The metamorphic Pennsylvanian rocks of Rhode Island are intruded by granites, and other small intrusions in southern New

England have also been identified as of late Paleozoic age.

THE OUACHITA AND MARATHON ARCS

Although the Ouachita and Marathon sectors of the Appalachian system have been traced for 1300 km along the strike, only 450 km of their folded rocks are exposed. The outlines of the rest of the belt have been followed by extensive drilling for oil through the post-Paleozoic strata that cover them in Texas, Arkansas, and Mississippi. Only where the deformed rocks are exposed at the surface—in the Ouachita Mountains of southern Oklahoma (Fig. 18–14) and Arkansas, in West Texas, and in small scattered areas of northern Mexico—can detailed structural and stratigraphic studies be made. Even in the Ouachita Mountains the bedrock is not exposed extensively, and the

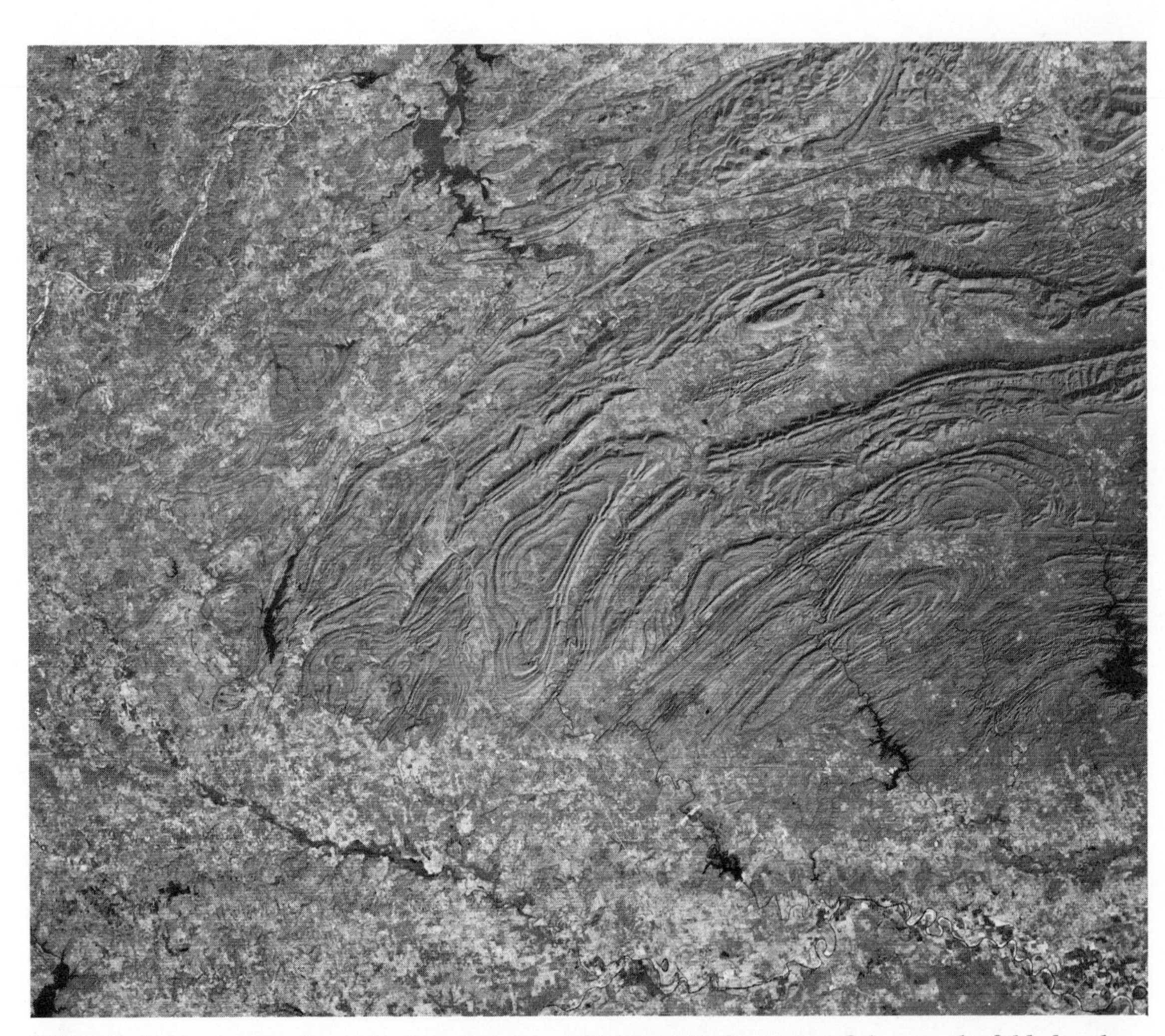

FIGURE 18–14 **Landsat imagery of southeastern Oklahoma. In the center of the area the folded rocks of the Ouachita Mountains are covered by the southward-dipping beds of the Coastal Plain. The Red River along the south margin of the photograph marks the boundary of Texas and Oklahoma. Compare with Fig. 18–15. (Courtesy of NASA.)**

limited nature of the outcrop has led to controversy concerning the structure.

Aulacogens

The Arbuckle trough extended through southern Oklahoma in early Paleozoic time, almost at right angles to the trend of the Ouachita Mountains whose thrust slices override its southern end (Fig. 18–15). It contains a great thickness of lower Paleozoic sedimentary rocks that overlie felsic lavas of late Proterozoic or earliest Paleozoic age. A second trough containing a thick section of early Paleozoic sedimentary rocks, the Reelfoot Basin, has been discovered beneath the northern end of the Mississippi embayment of the Gulf Coastal Plain. The reactivation of subsiding movements along the early Paleozoic faults bounding this basin may account for the thickening of the coastal plain sediments into the embayment in Cenozoic time. Recent movement along the same faults could have been responsible for the New Madrid, Missouri, earthquakes of 1811–12, which are thought to be the most powerful shocks in North America in historic time. The Ottawa River (Ontario and Quebec) flows in a similar trough bounded by nearly vertical faults trending westward at a high angle to the front of the northern Appalachians. Some of the faults are known to have been active in Hadrynian time, and some of the intrusions in the area also date from this time of Atlantic rifting. These fault-bounded basins, that diverge from folded mountain chains at a high angle, have been called **aulacogens.** According to plate tectonic theory, they form when continental separation takes place.

At the end of the Helikian Era the Appalachian margin of the North American plate was formed by the rifting that gave rise to Iapetus, the Paleozoic Atlantic. In the last chapter we saw that thick Hadrynian clastic sediments built a continental shelf along this margin, similar to the present shelf of the Atlantic seaboard. In addition, deep-reaching faults formed by the break-up brought to the surface lavas that occur in the top of the Proterozoic succession. The rift was located along the site of a spreading-ridge, where the oceanic crust of Iapetus was generated. Spreading along such a rift is postulated to have started at plumes where mantle material was welling up. Above such a plume, spreading would tend to be radial rather than linear, resulting in a triple junction with three spreading-ridges meeting in a single point (Fig. 18–16). Three is the minimum number of rifts that will satisfy such a spreading force. Such triple junc-

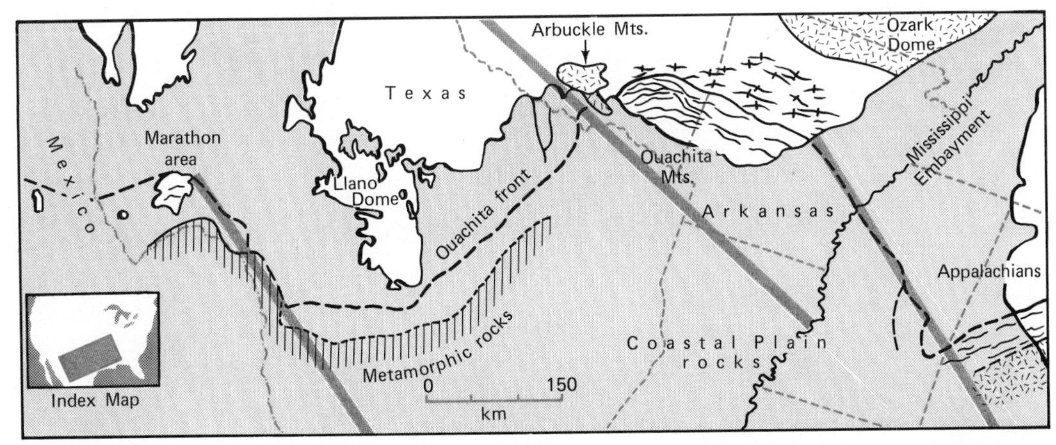

FIGURE 18–15 **Map of the Ouachita System. The straight, heavy colored lines are hypothetical transform faults postulated to account for the offsets of the structural trends that form the Ouachita and Marathon arcs.**

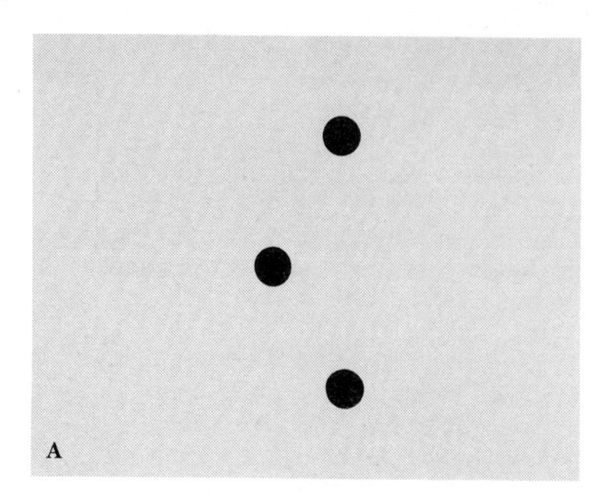

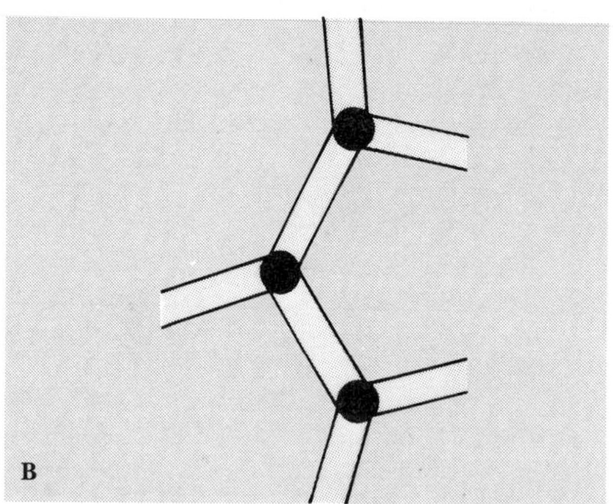

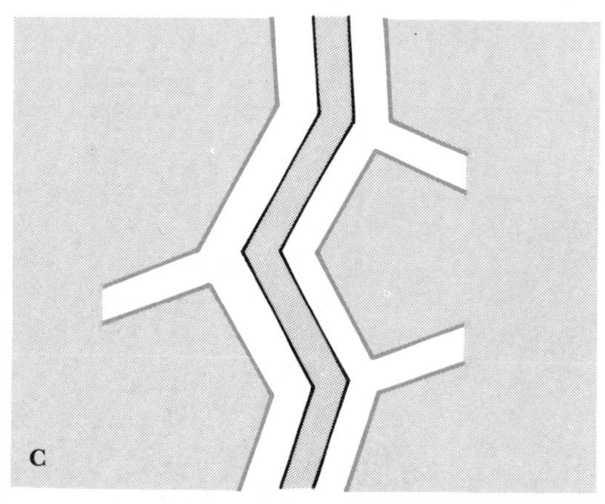

FIGURE 18–16 **Schematic diagram to illustrate the formation of aulacogens along the position of continental rifting. In A the rifting process is started by the establishment of 3 mantle plumes; in B each plume has caused rifting around itself in 3 directions; in C the continental masses begin to separate along two of the rift directions from each plume, leaving the third as a "failed arm," or aulacogen.**

tions exist in the present oceans at the intersection of two ridges (for example, the East Pacific and Galapagos ridges in Fig. 2–6). As the rift forms, spreading on one of the three arms may stop and the other two form part of a linear rift between two lithosphere plates. The "failed arm" becomes a zone of separation in the crust, extending at a high angle to the main rift, like the Arbuckle trough, the Reelfoot Basin, and the Ottawa–Bonnechere graben (Fig. 18–17).

Early Paleozoic Sedimentation in the Ouachitas

The pre-Carboniferous rocks in the Ouachita Mountains are a strange collection of bedded cherts, black shales, and sandstones (Fig. 18–18). On the platform to the north, correlative rocks are limestones continuous with those that built the carbonate platform of Cambrian and Ordovician time along the Appalachian continental margin.

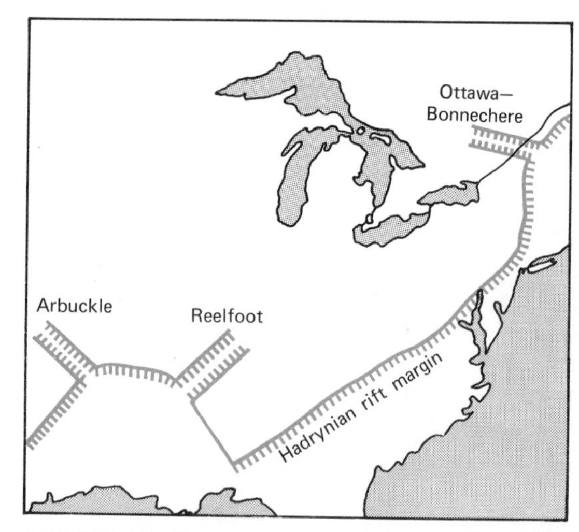

FIGURE 18–17 **Reconstruction, as aulacogens, of the basins marginal to the Hadrynian rift margin that started the formation of Iapetus.**

The early Paleozoic siliceous rocks of the Ouachita and Marathon mountains contain poorly preserved fossils of organisms that secerete siliceous hard parts, such as radiolarians and sponges, but these organisms do not seem to be present in sufficient numbers to account for the amount of silica in this succession. The thickest of the sili-

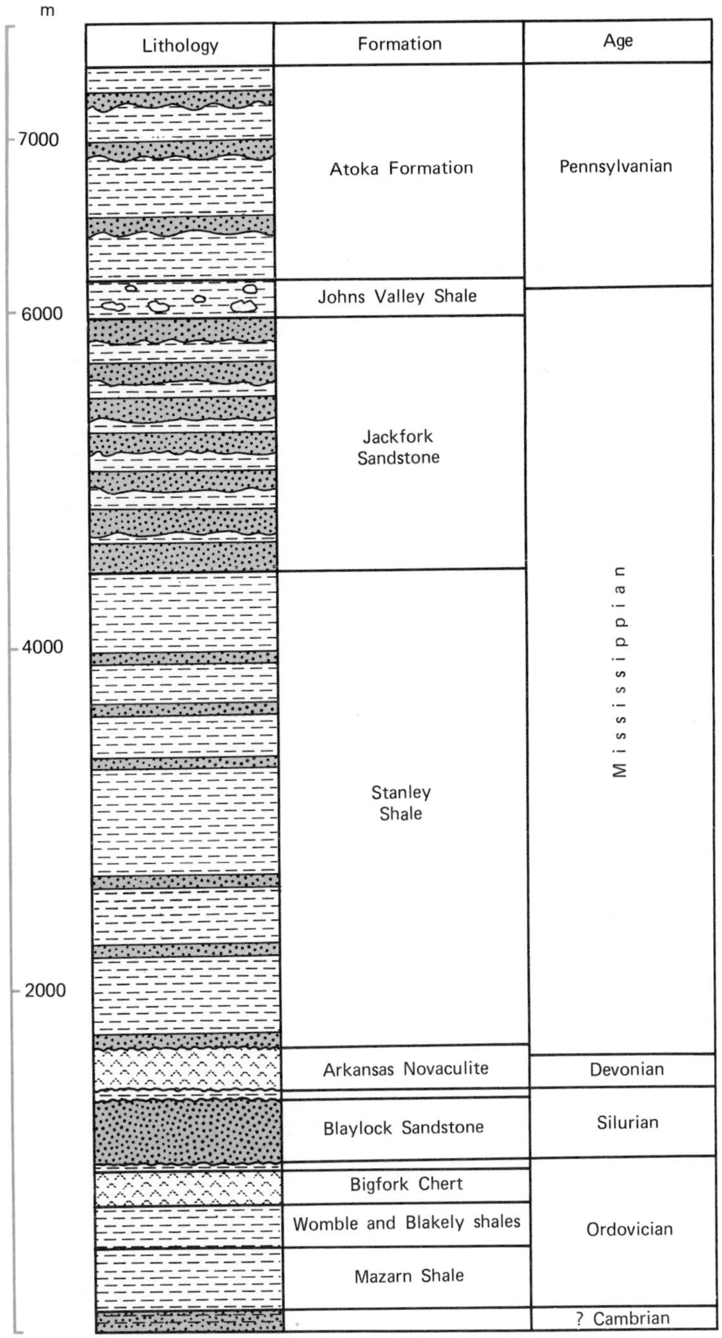

m

7000

6000

4000

2000

Lithology	Formation	Age
	Atoka Formation	Pennsylvanian
	Johns Valley Shale	Mississippian
	Jackfork Sandstone	
	Stanley Shale	
	Arkansas Novaculite	Devonian
	Blaylock Sandstone	Silurian
	Bigfork Chert	Ordovician
	Womble and Blakely shales	
	Mazarn Shale	
		? Cambrian

FIGURE 18–18 **Diagrammatic stratigraphic section of the sedimentary sequence in the Ouachita Mountains. Thicknesses are only approximate and change from place to place. (From information in P. T. Flawn and others.)**

ceous units (up to 290 m) is the Arkansas Novaculite of Early Devonian to Early Mississippian age. **Novaculite** is an almost white, bedded sedimentary rock like chert, but slightly coarser in grain. Its very fine and even grain, abrasive qualities, and great hardness make it suitable for the manufacture of grindstones and whetstones. Unlike the bedded cherts of the Cordilleran trough, those of the Ouachita facies are relatively thin and not interbedded with lava flows, although bentonites are common. Nevertheless, the processes of explosive volcanism and submarine weathering of volcanic ash on the ocean floor are the most satisfactory explanations so far proposed to account for these rocks. The volcanoes that supplied this ash must have been on an island arc south of the present outcrops, where Paleozoic rocks are now deeply buried beneath the coastal plain. The black shales with abundant organic matter and a graptolite fauna suggest that a relatively deep basin existed between the carbonate platform and the volcanic arc. In this basin the rate of accumulation of sediments was remarkably slow, for only 2000 m accumulated from Ordovician to Mississippian time. The pattern of sedimentation and volcanism indicates that the subduction zone that developed in Ordovician time along the Appalachian margin of the continent continued around to the Ouachita–Marathon sector.

The Late Paleozoic Clastic Wedge

At the beginning of the Mississippian Period the tempo of sedimentation in the Ouachita trough changed abruptly. Great thicknesses of rapidly deposited shales and sandstones overlie the early Paleozoic shales and cherts in the Ouachita Mountains (Fig. 18–18). The Stanley shale may be as thick as 3500 m, but its complex folding makes stratigraphic measurement difficult. The Jackfork Sandstone is up to 2000 m thick and was deposited in the rapidly filling trough from highlands south of the present exposures. Sandstone beds in these Ouachita formations have irregular lower surfaces, which are molds of irregularities in the top surface of the underlying shale layers. These irregularities are believed to have been eroded by turbidity currents that carried the sand down the flanks of the depositional basin into deeper water. The orientation of these "sole markings," as they are called, indicate that the turbidity currents flowed parallel as well as transverse to the length of the trough. The trends of the markings show that the currents flowed down the side of the trough and then turned along its axis.

The Johns Valley Shale is a relatively thin formation of peculiar lithology. The notable feature of this shale is the presence of large blocks surrounded by a fine-grained argillaceous matrix. The blocks are as large as 110 m across and consist of sedimentary rocks ranging in age from Late Cambrian to Early Pennsylvanian. Limestones of Late Cambrian and Early Ordovician age, deposited in the shelf environment north of the trough, form the majority of the blocks. One hypothesis accounts for their emplacement in the shale by uplift of the shelf along faults at the platform margin and the slipping and slumping of the blocks eroded from the fault escarpment down the shaly slope into the trough.

Sedimentation ended in Middle Pennsylvanian time in the Ouachita sector, but continued in the Marathon sector until the end of the period (Fig. 18–19). Thereafter, the Paleozoic sediments were thrust northward against the platform and the mountain belt was lifted into the zone of erosion. The orogeny in the Marathon sector appears to have been delayed until early Permian time, but the structural effects were much the same as those in the Ouachita belt.

Structure and Interpretation

The rocks of the Ouachita Mountains are cut by a few major and many minor thrust faults, along which they have been transported northward. They have been intensely folded locally, but in the exposed part of the mountain system metamorphism is weak to nonexistent. Wells drilled through the Coastal Plain sediments penetrate a belt of metamorphic rocks—phyllites, schists, and slates—south and east of the deformed sed-

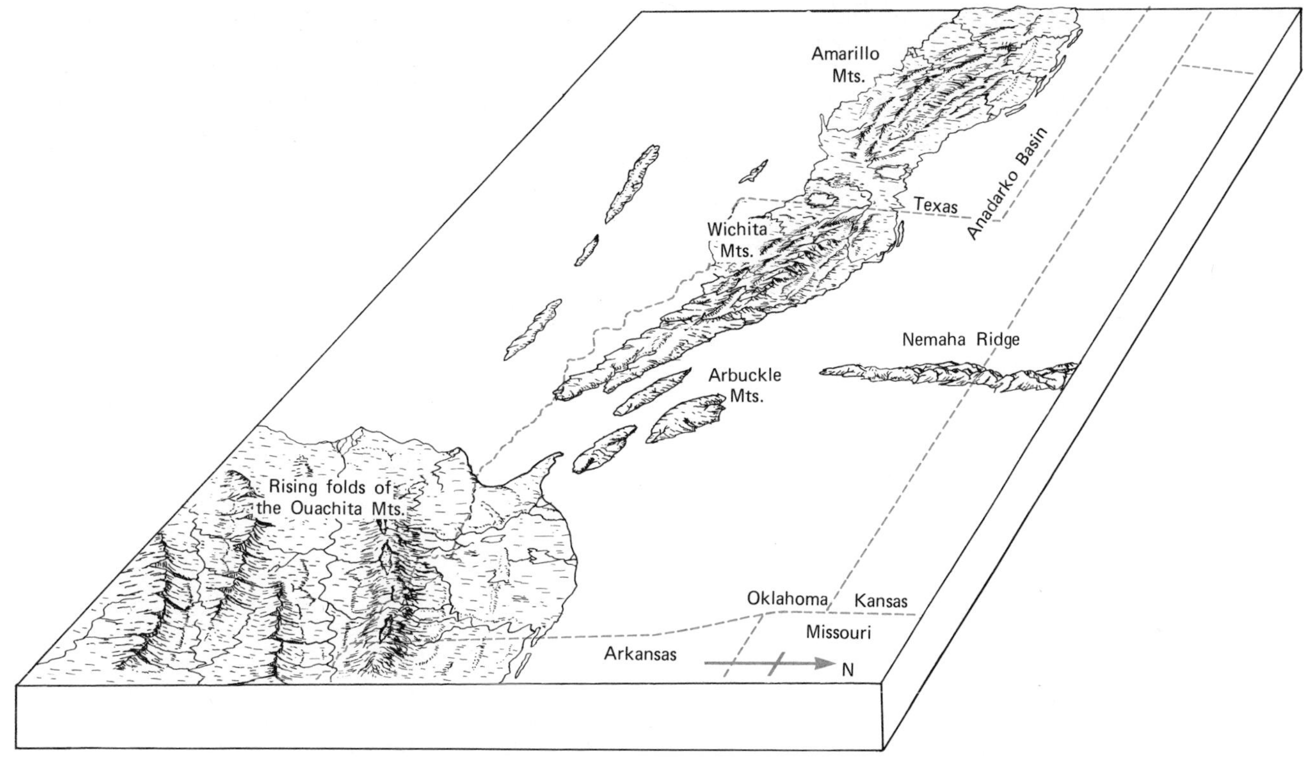

FIGURE 18–19 **Reconstruction of the geography of Oklahoma and adjacent areas in Middle Pennsylvanian (Des Moines) time. Clastic sediments derived from the domed uplifts of the Oklahoma Mountains were deposited in the Anadarko Basin. Most of the Ouachita geosyncline had been deformed into the zone of erosion, and growing folds contributed more sediment to the basin. (After C. W. Tomlinson and W. McBee.)**

imentary rocks, in a position similar to that of the Blue Ridge and Piedmont provinces (Fig. 18–15).

The history of the Ouachita–Marathon sector differs considerably from that of the rest of the Appalachian Mountain chain in that only one episode of orogeny is recorded in the rocks. The history of this southern margin of the continent can be interpreted in terms of continental rifting at the beginning of Hadrynian time; the establishment of a subduction zone, volcanic arc, and miogeocline during the early Paleozoic; and elevation, metamorphism, and thrusting of the arc–trench complex against the basin in Pennsylvanian and Mississippian time. The orogeny may have been related to the collision of South America with North America, but little evidence can be assembled to support or negate this hypothesis.

The Hidden Connections

The nature of the connection between the Ouachita Mountains and the southern end of the Appalachians under the Mississippi embayment has been a topic of intense research and speculation. The connection between the Marathon and Ouachita arcs is also buried by the Coastal Plain sediments (Fig. 18–15). Information on the buried structures is available from geophysical surveys and from wells drilled through the cover. Beneath the Coastal Plain of Alabama, the Valley and Ridge and Piedmont belts can be traced southwestward to eastern Mississippi, where the former appears to narrow and turn northward. Wells probing the continuation of the Ouachita belt eastward encounter black slate with many quartz veins, which is in abrupt, and possibly faulted, contact with undeformed Carboniferous rocks to the northeast. The absence of a folded belt analogous to the Valley and Ridge province along this junction, and a similar junction on the northeast side of the continuation of the Marathon belt in the subsurface, has suggested to some geologists that large-scale transform faulting cuts off both the Marathon and Ouachita arcs, displacing their continuations southeastward. Such transform faults could have been inherited from the Iapetus spreading-center that affected the Hadrynian rifting of the continental margin. However, much more drilling and geophysical work will be required before the nature of these hidden connections is firmly established.

By the end of the Paleozoic Era, with the formation of Pangaea, compressional deformation in the Appalachians was finished. The mountain belt formed a complex, major highland crossing the Laurasian segment of Pangaea. At the beginning of the Mesozoic the cycle of rifting and closure started again with the fragmentation of Pangaea. The events accompanying the break-up of the world continent are the subject of the next chapter.

SUMMARY

The Acadian orogeny affected the Northern and Newfoundland Appalachians between 400 and 350 million years ago (i.e., during the Devonian Period). Most of the belt was extensively intruded by granites and regionally metamorphosed. The highlands raised along the Appalachian trend in this orogeny were eroded to form the sediments of the Catskill clastic wedge. Correlation within this set of transgressing facies is provided by key limestone beds whose deposition was probably initiated by changing sea levels.

During Carboniferous time the area of Acadian deformation was cut by faults into successor basins in which limestone, evaporites, and later coal were deposited. Although some of the movement on the faults was vertical, much of it appears to have been lateral.

At the northern end the Appalachian fold belt strikes into the Atlantic Ocean and appears to be continued on its eastern side in the Caledonian and Hercynian orogenic belts of Europe. Progressive closure of the Iapetus Ocean from north to south and continental collision of North America with Europe and North Africa could have caused the deformation of the Appalachians in the Acadian and Alleghany orogenies.

Erosion of the highlands of the Piedmont area shed coarse clastic sediments westward on to the platform in Carboniferous time and contributed to the building of the deltaic swamps on which coal forests grew.

In the late Paleozoic Alleghany orogeny the miogeocline and shelf sediments of the southern and central Appalachains were deformed against the craton. In the Valley and Ridge province the folded rocks have sheared in décollement thrusts from the basement and along weak horizons localized by shales and evaporites in the Paleozoic succession. These thrusts extend under the surficially little deformed late Paleozoic beds of the Appalachain plateaus. The role of the Blue Ridge and Piedmont in the orogeny is unclear; some of its metamorphism and intrusions may date from early and middle Paleozoic orogenic episodes, and some may be of late Paleozoic age.

The opening of the Iapetus Ocean in Hadrynian time resulted in troughs called aulacogens extending at high angles from the rift margin into the platform. The Arbuckle trough, which is filled with thick, late Proterozoic and early Paleozoic sediments and lavas, was apparently formed in this way.

The Ouachita and Marathon arcs, which are the southern continuation of the Appalachian system, are largely covered by Cenozoic sediments. The early Paleozoic rocks in the Ouachita mountains are a thin, deep-water, siliceous succession apparently deposited in a subduction zone. In late Paleozoic time the southern continental border was the site of mountain building, and a clastic wedge was formed by detritus shed northward on to the platform. The breaking of the southern part of the Appalachians into Marathon and Ouachita sectors may be the result of large displacements along transform faults.

QUESTIONS

1. Compare the geometry and facies changes of the Catskill and Queenston clastic wedges.
2. Formulate alternative hypotheses to account for the absence of clastic sediments of the Catskill clastic wedge in the southern Appalachians, although isotopic ages suggest that orogeny in the Piedmont occurred at this time.
3. Summarize the evidence that middle and late Paleozoic orogeny in the Appalachians was the result of the collision of continental lithosphere plates rather than the interaction of oceanic with continental plates as in the Cordilleran orogeny.
4. Draw three maps representing late Ordovician, late Devonian, and Permian time, showing reconstructions of the areas of mountain building, on the eastern and southern margins of the North American plate, the positions of the Iapetus Ocean, the European, African, and South American plates, subduction zones around the ocean, volcanic arcs, and spreading centers. Locate the supposed position of the Avalon belt.
5. If the Squantum conglomerates are indicative of late Paleozoic glaciation in the Boston Basin, where were the nearest similar deposits at this time? From reconstructions of the continents in Figure 1–18, can you estimate their distance in kilometers?

SUGGESTIONS FOR FURTHER READING

BURKE, K., and DEWEY, J. F. "Plume-generated Triple Junctions: key indicators in applying Plate Tectonics to Older Rocks." *Journal of Geology,* 81: 406–433, 1973.

KAY, M. ed. "North Atlantic—Geology and Continental Drift." *American Association of Petroleum Geologists Memoir 12,* 1082 p., 1969.

KING, P. B. "Ancient southern Margin of North America." *Geology,* 3: 32–34, 1975.

KLEIN, G. de V., ed. "Late Paleozoic and Mesozoic continental sedimentation, northeastern North America." *Geological Society of America Special Paper 106,* 309 p., 1968.

MILICI, R. C. "Structural Patterns in the Southern Appalachians: Evidence for Gravity Slide Mechanism for Alleghanian Deformation." *Geological Society of America Bulletin,* 86: 1316–1320, 1975.

THOMAS, W. A. "Southwestern Appalachian Structural Systems beneath the Gulf Coastal Plain." *American Journal of Science,* 273-A: 372–390, 1973.

VIELE, G. W. "Structure and Tectonics of the Ouachita Mountains, Arkansas," in *Gravity and Tectonics,* DeJong, K. A., and Scholten, R., ed. New York: Wiley–Interscience, pp. 361–377, 1973.

CHAPTER NINETEEN

INTRODUCTION

At the end of the Paleozoic Era the continents of the world were united into a single landmass, Pangaea. The closing of Iapetus had deformed the borderlands of this early Paleozoic ocean, forming the Appalachian Mountains of North America, the Hercynian Mountains of Europe, and the Mauritanians of West Africa. The history of the fragmentation of this northwestern sector of Pangaea and the growth of the Atlantic Ocean was recorded in the initial sediments that were laid down along the southern and eastern margins of North America in middle Mesozoic time as this new basin of deposition formed.

The general pattern of the break-up of Pangaea was described in Chapter 1. At first the rifting and formation of a linear spreading-zone separated North from South America (Fig. 1–19). Gradually, the rift progressed northward, separating North America from Africa and later from Europe. Not until the later part of the Mesozoic Era did South America begin to separate from Africa as the South Atlantic spreading-center formed. The progressive expansion of the North Atlantic can be reconstructed by analysis of the magnetic stripes that have been impressed on the ocean floor by the reversing magnetic field of the earth, but definition of magnetic anomalies on the floor of the Gulf of Mexico and the Caribbean Sea is not as clear as that of the Atlantic, and the histories of these basins are, therefore, less certain.

The deeper sections of the Gulf of Mexico are underlain by oceanic crust, but no active or "fossil" spreading-center can be identified by magnetic anomalies, nor by the topography of the sea floor. The history of the Caribbea Sea and the islands of the Greater and Lesser Antilles along its northern and eastern margins is not directly related to the initial separation of the two Americas. The Caribbean plate (Figs. 1–12, 1–13, 19–1) appears to be an eastward extension between the two continents of the Pacific lithosphere, and its transit during late Mesozoic and Cenozoic time through the gap between them has resulted in the subduction of much of the oceanic lithosphere that was formed in the initial early Mesozoic spreading of North America from South America. One interpretation of this sequence of events is shown in Fig. 19–1.

A system of basins bounded by normal faults that were formed close to the present coastline in Late Triassic time constitute the first evidence of tension along the eastern and southern borders of North America. The sediments and lavas deposited in these basins mark the beginning of a new major interval of deposition along the Atlantic margin of the continent.

THE BREAK-UP OF PANGAEA

TRIASSIC RIFT BASINS

During Late Triassic time, the basins formed along the eastern side of the Appalachian Mountain chain were the sites of deposition of continental sediments. Most of these basins are bounded by normal faults, but some have been interpreted as downwarps in the pre-Triassic surface. In the Appalachian belt the Triassic basins extend from Nova Scotia to Georgia. By drilling through younger sediments and by geophysical measurements, additional basins have been discovered east of the Appalachians, below the coastal plains and the continental shelf (Fig. 19–2). Similar poorly dated rocks have been discovered by drilling at the base of the Mesozoic succession in the Gulf coastal region.

The nature of the unconformity surface beneath the Triassic sediments suggests that the late Paleozoic Appalachians had been reduced to rolling hills before deposition started. The pattern of normal faults that bound many of the basins shows that the crust was uplifted and subjected to tensional forces. The coarseness of the conglomerates along the sides of the basin is evidence that mountainous relief between the metamorphic and igneous rocks of the basin margins and the intermontane valleys was established and maintained by subsidence and uplift, despite erosion and fill, throughout the depositional episode. The sedimentary filling of the Connecticut Valley area, one of the best exposed and most extensively studied, is wedge-shaped in cross-section, bounded on one side by a normal fault and on the other by the tilted, pre-Triassic erosion surface (Fig. 19–3). At the time of deposition the areas of the Triassic sediments were considerably greater than at present, for the sedimentary fillings were tilted and reduced by erosion, probably in Early Jurassic time. The Triassic sediments reached thicknesses of 3000 m in Connecticut, and double that amount in New Jersey.

Newark Group Sedimentary Rocks

The sedimentary rocks that were deposited in the Late Triassic intermontane basins are known as the Newark Group. The group consists of three sedimentary facies: arkose and the red shales associated with it, black shales, and coarse conglomerates. Arkose is a sandstone in which more than a quarter of the grains are feldspar. Usually, but not always, it is colored deep red by the reddish feldspars, orthoclase and microcline, and by iron oxides staining the clayey matrix. Some of the shales contain the impressions of land plants and freshwater fish, and a few reptile fossils have been found in the arkose. In the southern basins thin beds of

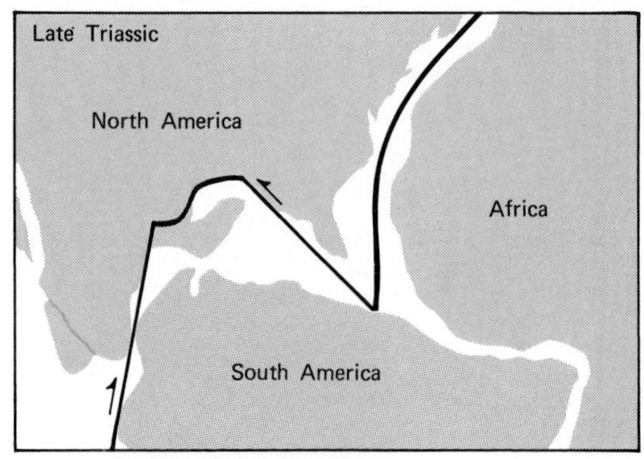

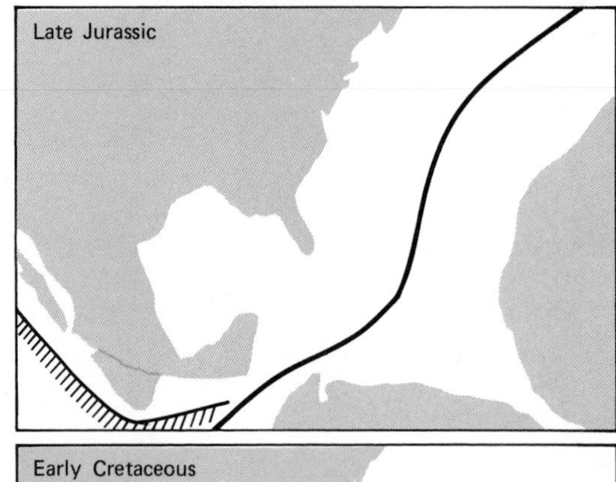

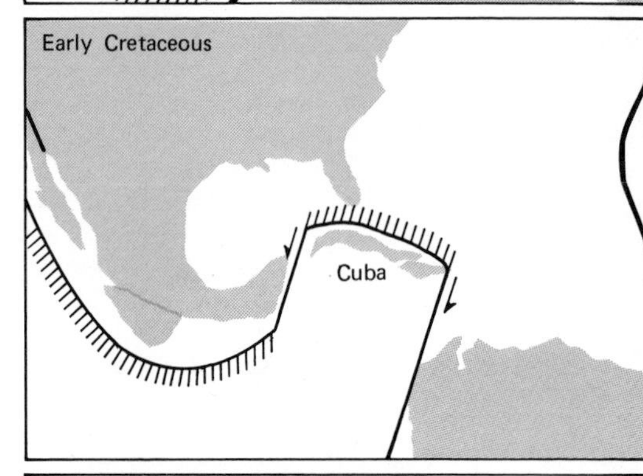

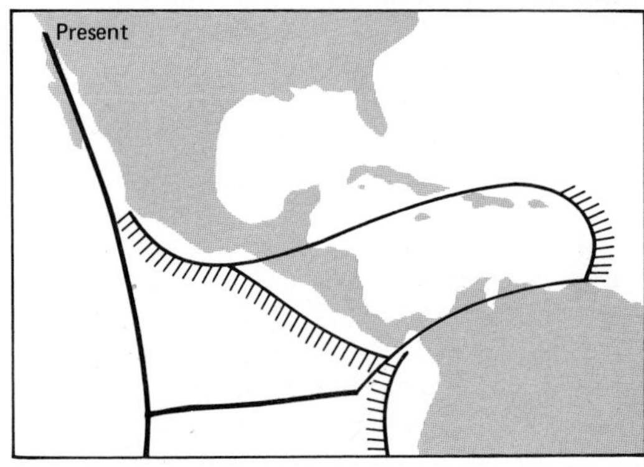

FIGURE 19–1 **An interpretation of the history of the Gulf of Mexico and Caribbean Sea in terms of plate tectonics. The heavy black lines are spreading-zones; the hatchured lines are subduction zones. (After G. W. Moore and L. Del Castillo; courtesy of the Geological Society of America.)**

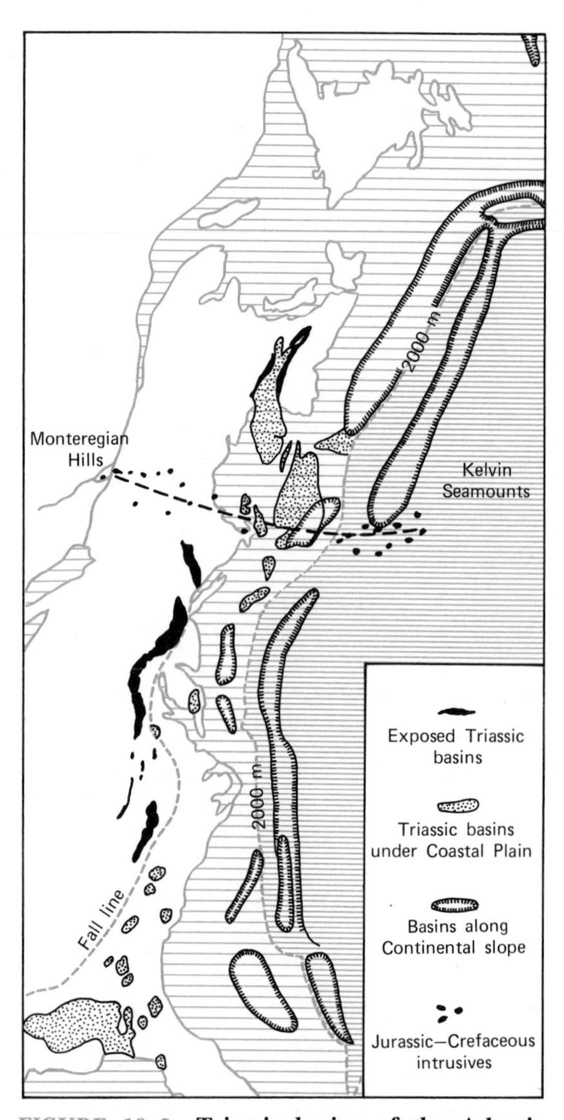

FIGURE 19–2 **Triassic basins of the Atlantic coastal region, and deep basins established by geophysical surveys along the continental slope. The curved line passing through New England is the axis of the Jurassic and Cretaceous intrusions. (Compiled from various sources.)**

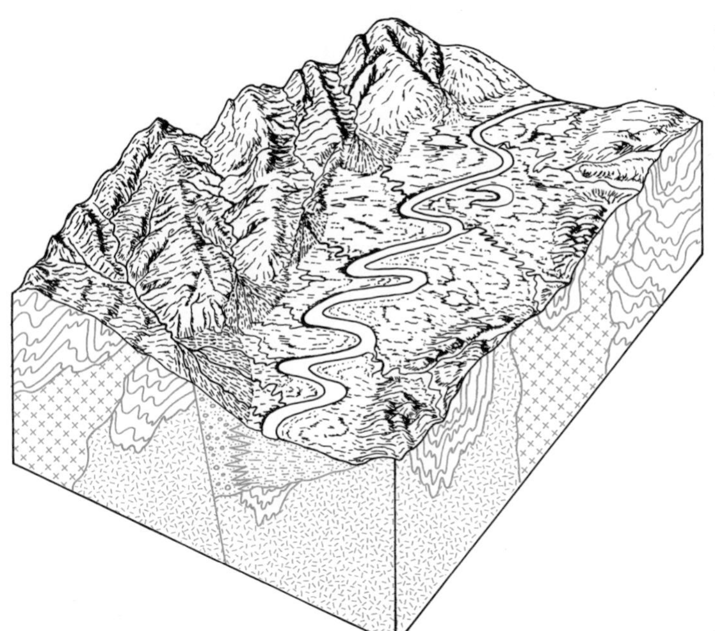

FIGURE 19–3 **Block diagram of a typical Triassic basin of eastern North America. Note the contrast between the side of the basin that was uplifted along the bounding normal fault and the low hills on the opposite side. Conglomerates forming alluvial fans against the mountains interfinger with sandstones accumulating in the valley.**

coal are found in the Newark Group. The coarse conglomerates along the fault margin of several of the basins represent indurated alluvial fan sediments dropped where fast-flowing streams from the highlands were discharged on to the valley floor.

The fresh feldspar and mica in the arkose were derived from granites and gneisses of the crystalline areas of the Appalachians, which formed the surrounding highlands. Only under conditions of considerable relief, rapid stream erosion in the mountains, and rapid deposition where the streams discharged into the plains could unstable minerals, such as plagioclase feldspar, be preserved to enter the accumulating arkose. The red iron oxides that color the matrix indicate accumulation under oxidizing conditions on land. Where drainage was interrupted in the valleys, lakes were formed and black clays were laid down with interbeds of coal recording the luxuriant vegetation. Dinosaurs walked and ran over the mud flats, leaving footprints behind in the drying sediment. These fossil footprints, found covering the bedding planes in some localities, suggest that dinosaurs were abundant in this environment. They also suggest that dinosaurs were social animals, for most of the prints are similarly oriented, as if formed by the passing of a dinosaur herd. Although tracks are locally abundant, few reptile skeletons have been found in Newark Group sediments.

Igneous Rocks of the Newark Basins

The tensional forces in the lithosphere opened deep-reaching channels along which basaltic and intermediate igneous rocks approached the surface to form dikes, sills, and pipes, and poured out onto the surface in lava flows. These intrusions and extrusions now form hills and ridges in the Triassic basins because the igneous rock is more resistant to erosion than the surrounding arkose and shale. Most of the flows welled up quietly from depth, but some evidence of explosive igneous activity has been found in the Connecticut Valley. The most famous of the intrusions is the Palisade Sill, which is grandly exposed on the bank of the Hudson River west of New York

City. The resemblance of this cliff to a palisade comes from its conspicuous columnar jointing.

Small intrusions of Triassic age are not confined to the North American coastal area, but occur along the Atlantic margin of West Africa and South America. If the orientation of the dikes is plotted on a map of Pangaea, they appear to be radial to a center located near the present position of the Bahamas. Radial tensional fractures develop in surface rocks above a rising column of material. The orientation of the dikes suggests that they occupied radial tensional fractures about a Late Triassic mantle plume that was an early part of the North Atlantic spreading-center. The Triassic basins do not mark the location of the Atlantic rift, but are a reflection of the tensional forces that were separating the lithosphere plates along the edges of the present continental platforms. The intrusive episode appears to have continued from Triassic into Early Jurassic time.

GEOMETRY OF THE COASTAL PLAINS

The eastern and southern coastal areas of North America are formed by a wedge of Mesozoic and Cenozoic sedimentary rocks that dip gently toward the Atlantic Ocean and the Gulf of Mexico. The land surface of this wedge constitutes the coastal plains; its seaward extension forms the continental shelves. The position of the shoreline, and consequently the relative widths of the coastal plains and continental shelves, has changed continually since the middle of the Mesozoic Era. The gentle seaward slope of the surface of the coastal plains is broken only by low escarpments that mark the outcropping of resistant strata, or by terraces that mark the former position of the sea or the abandoned banks of a river. The plain is usually divided into Atlantic and Gulf segments at Florida.

The beds of sedimentary rock beneath the coastal plains and shelves dip seaward at less than 1°, and consist largely of sands, silts, and clays eroded from the Appalachian highlands and the Rocky Mountains. Although the sedimentary strata are poorly exposed, the stratigraphy of the Gulf Coastal Plain rocks is known in great detail, for they have been penetrated to depths of over 7500 m by tens of thousands of holes drilled in the search for oil. Each sedimentary unit thickens, slowly at first, then more rapidly, as its distance from the continent increases. Information from drilling, supplemented by geophysical measurements, indicates that the sedimentary prism reaches a thickness of 12,000 m beneath the shelf of the Gulf of Mexico. This thickness has suggested to some geologists that the gulf is the site of a developing geosyncline, referred to as the Gulf Coast geosyncline. The sedimentary wedge has little in common with the Cordilleran or Appalachian geosynclinal sediments of the Paleozoic, but instead is analogous to the Hadrynian and early Paleozoic sediments that accumulated along the shores of the opening Iapetus as the Ocoee and Chilhowee groups.

Although the gross structure of the coastal plains and continental shelves is that of a simple wedge of strata dipping and thickening seaward, arches, domes, and basins of regional extent modify this simple picture. Where the sedimentary rocks are depressed into regional synclines, the belt of coastal plain rocks widens into the Rio Grande, East Texas, and Mississippi embayments (Fig. 19–4). Between them it narrows at the San Marcos, Sabine, Munroe, and Decatur arches. The embayments are located behind the salients in the Appalachian Mountain system, and the arches are behind the nodes between the arcs. A prominent feature of the Atlantic Coastal Plain is the Cape Fear Arch on the border of the two Carolinas, in line with the node between the southern and central Appalachians. A broad syncline, known as the Salisbury embayment, occurs in the neighborhood of Chesapeake Bay.

A zone of normal faulting 1600 km long can be followed from the Rio Grande to the northern boundary of Florida (Fig. 19–5). Most of these faults are downthrown toward the gulf, but some have the opposite displacement. A second major zone of normal

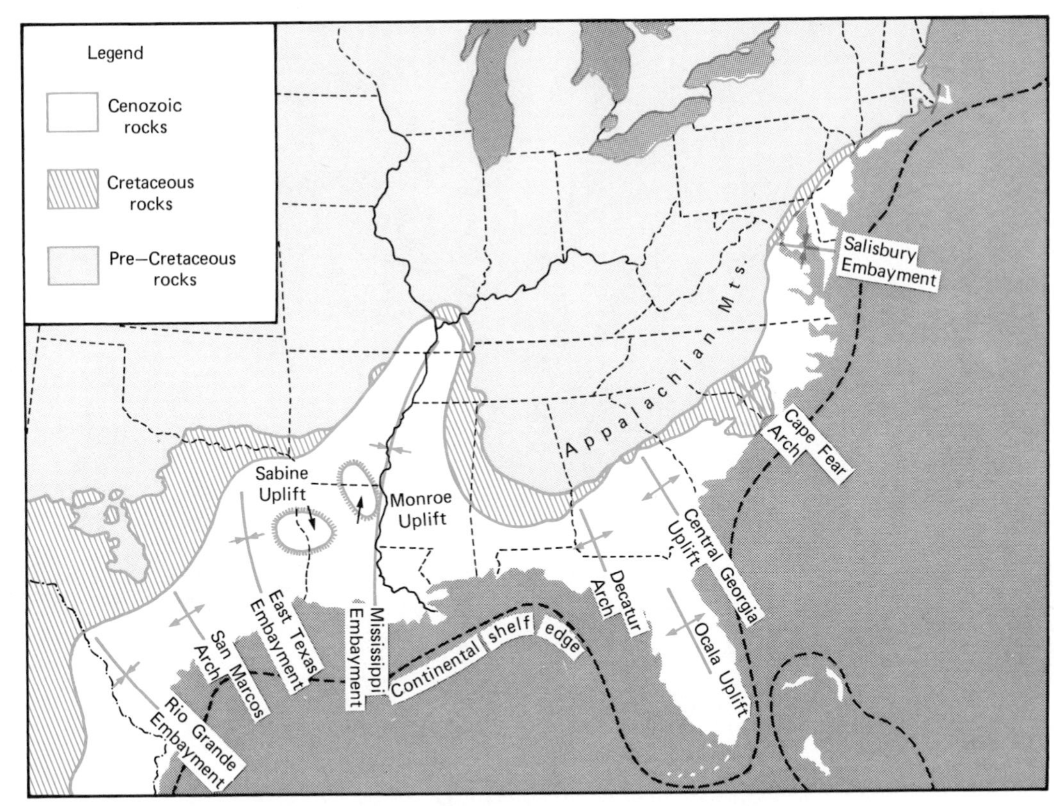

FIGURE 19–4 **Regional structures of the Atlantic and Gulf coastal plains.**

faulting occurs along the coastline in the Rio Grande and East Texas embayments. Many of these faults are known as "growth faults," because movement along them was repeated over an interval of millions of years. Since in a growth fault the downthrown side repeatedly subsides with respect to the upthrown side, greater thicknesses of sediment accumulate on this side of the fault and this sedimentary filling tends to decrease the relief between the two blocks. As a result the displacement of stratigraphic horizons along the fault increases downward (Fig. 19–6). The part of the Gulf Coastal Plain cut by the faults corresponds closely to that underlain by Mesozoic salt beds, and major fault trends follow the edge of the salt deposits. The faulting has been ascribed to the subsidence of the central part of the gulf basin, and the fracturing and slipping of segments of the sedimentary wedge toward the center of the basin. Ernst Cloos has suggested that the fault blocks are lubricated by the salt layer in their movement down the dip of the beds.

MESOZOIC STRATIGRAPHY OF THE COASTAL PLAINS

Detailed knowledge of the stratigraphy of the continental shelves is confined to the southern (Gulf of Mexico) and northern (Canadian continental shelf) sectors, for drilling in the Atlantic continental shelf of the eastern United States has only recently begun. The initial marine invasion of the coastal plains regions was of Early Jurassic age. By this time the rifting of the continents had advanced to the stage where the sea could enter the narrow embryonic Atlantic Ocean

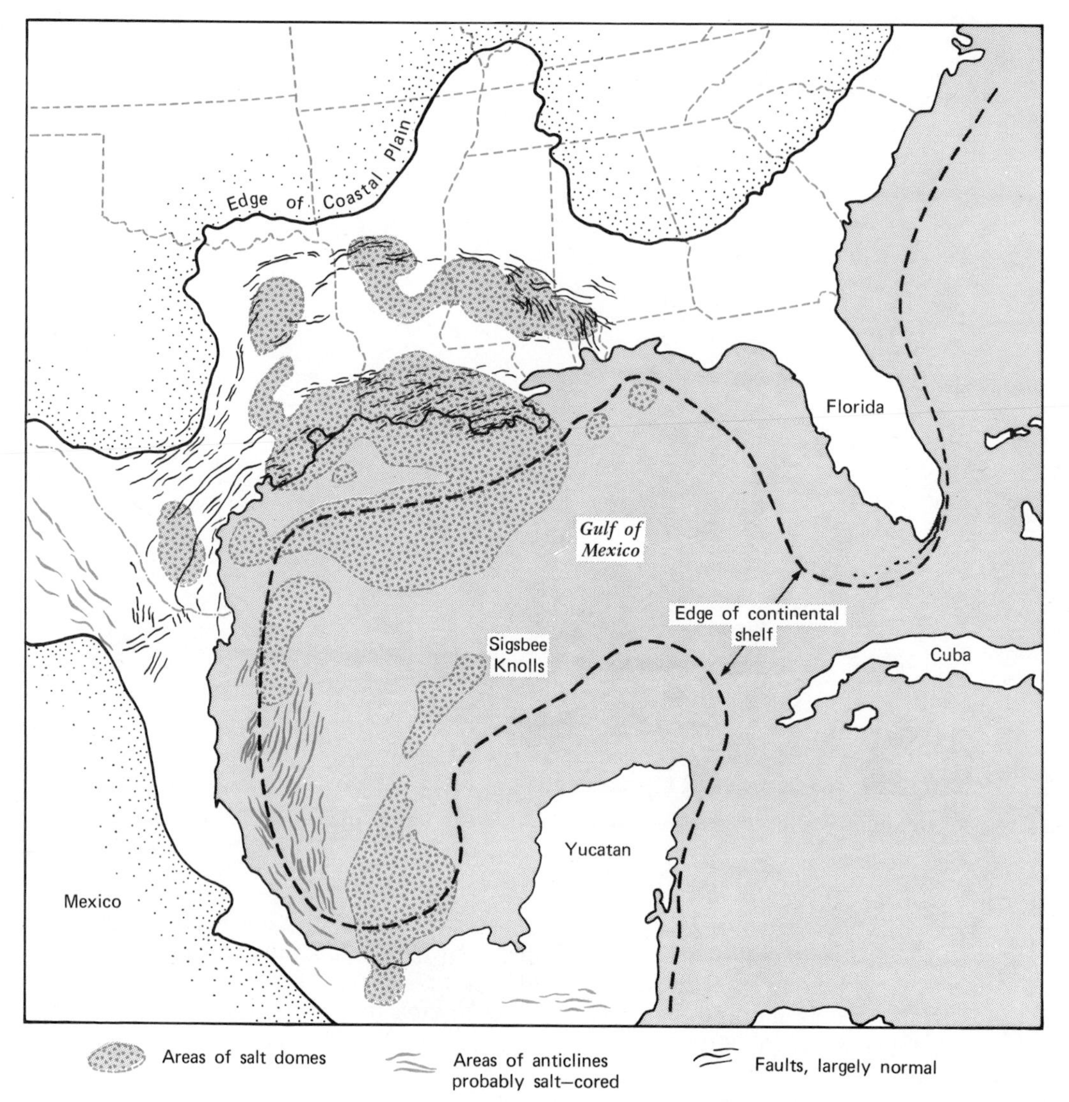

FIGURE 19–5 **The major features of the Gulf of Mexico that are related to the movement of salt.**

from Newfoundland to the Gulf. The zone of rifting was close to the Jurassic equator, and in the hot climate evaporation was high. In the Gulf region the Louann Salt was laid down over practically the whole area of the present basin, and east of Nova Scotia the Argo Salt was deposited during the same time interval. There is little evidence that the present area of the Atlantic continental shelf of the United States received any salt deposition. The mobility of the salt, to be discussed in more detail below, precludes any estimate of its original thickness.

Upper Jurassic deposits of both the Gulf and Canadian areas indicate that evaporitic conditions had ended and normal marine deposition of limestones and shales began. In southern Arkansas a group of oilfields produces from the oölitic Smackover Limestone of Late Jurassic age. Beneath the margin of the Canadian shelf this interval contains the Abenaki Limestone,

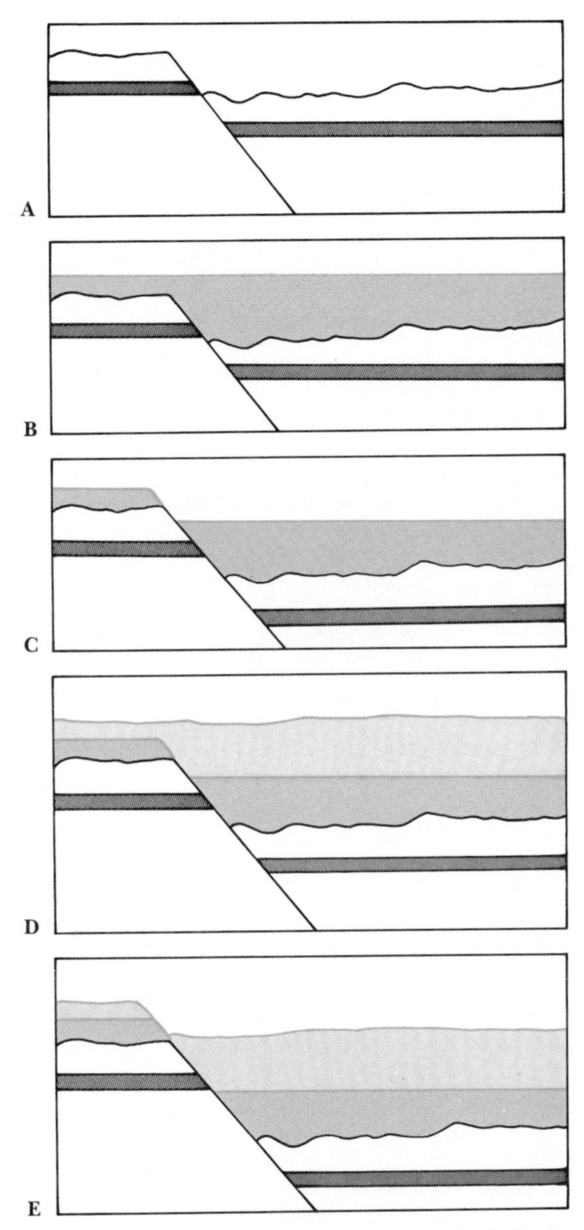

FIGURE 19–6 **Development of a growth fault. A. initial movement displacing a stratigraphic marker. B. deposition over the fault with thicker accumulation on the downthrown side. C. repeated movement. D and E. more deposition and repeated movement. Note that the stratigraphic separation of beds along the fault plane increases downward.**

a reef facies in which corals, sponges, and stromatoporoids were the principal framebuilders (Fig. 19–7). This Jurassic reefal facies is common in the Mediterranean region of Europe, but unknown elsewhere in North America.

The general pattern of facies changes along the Gulf Coast shoreline was established in Mesozoic time and continues to the present day. Inland, alluvial sands and clays were laid down in the flood plains of rivers or lakes. Red and gray silts and muds were deposited in estuaries near the shoreline, and sands and muds accumulated in deltas. Offshore marine clays accumulated and consolidated into shale; farther from the land and the source of clastics, marls, limestones, and chalks were deposited. These facies shifted back and forth across the coastal plain as the shoreline oscillated.

Cretaceous Limestones

In late Mesozoic time reefs were built in the region of the Gulf of Mexico and the Mediterranean Sea by a combination of organisms that had not previously been important in reefs. A group of bivalves began to live with one valve resting on, or cemented to, the ocean floor. The lower valve became enlarged, and the upper valve in some forms was eventually reduced to a cap on the lower one. During the Jurassic, one group of bivalves evolved a massive, lower valve that became coiled at the beak in a plane perpendicular to that between the valves. These shells, known as *Gryphaea,* are abundant in argillaceous rocks (Fig. 19–8B). Some Cretaceous bivalves became coiled in a plane parallel to that between the valves; these are grouped in the genus *Exogyra* (Fig. 19–8A). In others, one or both valves became so thickened that they were horn-shaped, like the corallum of a rugose coral. Such forms have been commonly called rudists (Fig. 19–8D) and lived in great reef-like banks in Cretaceous seas. They resembled the Permian brachiopods that developed similar coral-like form in response to a similar sedentary way of life. In forming Cretaceous reefs the rudists were joined by the stromatoporoids and the scleractinian corals.

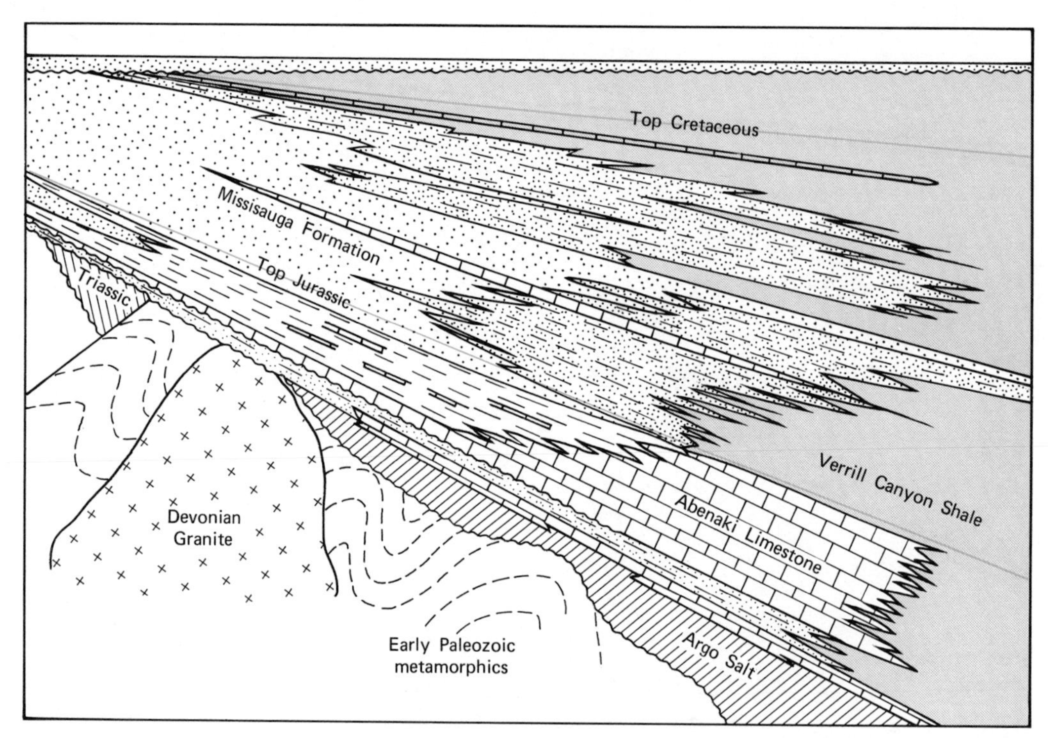

FIGURE 19–7 **A schematic cross-section of the Canadian continental shelf off Nova Scotia to show the facies changes. Not to scale. (After N. L. McIver; courtesy of the Canadian Journal of Earth Sciences.)**

The Paleozoic classes of corals, the TABULATA and RUGOSA, died out in the Permian Period. No corals are known from Early Triassic rocks, but in Middle Triassic time a new group appeared in the marine fauna, which has septa with sixfold symmetry and are, therefore, sometimes referred to as hexacorals. This group, formally designated the class SCLERACTINIA, soon came to have the dominant role in the reefal environment that they maintain to the present day. Their relationship to the RUGOSA is obscure, but they probably evolved from that stock during the worldwide restriction of the seas at the beginning of the Mesozoic Era. The combination of scleractinians and rudist bivalves formed the massive reef limestones of Early Cretaceous age that are the reservoirs for the amazingly productive Golden Lane oil fields of eastern Mexico.

Two other evolutionary events affected the deposition of sedimentary rocks in Cretaceous time. Until this period foraminifers lived on the ocean floor. During Cretaceous time planktonic forms evolved and became very successful in this new environment, so that their shells began raining down on the ocean floor and accumulating there as limestone oozes. At present oozes of the common planktonic foram *Globigerina* (Fig. 4–16) cover 130 million square kilometers of the ocean floor. Also during Cretaceous time a group of the yellow-green algae became planktonic in habit. They secreted minute rosettes of calcareous plates, called coccoliths, around their spherical bodies (Fig. 19–9). The coccolith-bearing algae were also highly successful in warm, Late Cretaceous seas and made important contributions to the lime muds accumulating on the ocean floors. These two organisms were major contributors to the formation of a sedimentary rock that is characteristic of the Cretaceous system, chalk.

FIGURE 19–8 **Mesozoic bivalves. A. *Exogyra*, view of one of the vales coiled at the beak in the plane of the valve. *B. Gryphaea*, side view of both valves. The lower valve is much larger than the nearly flat, upper valve and is coiled in a plane at right angles to that between the valves. *C. Inoceramus*, a group of specimens of this concentrically ribbed genus. D. *Hippurites*, a rudist with a lower valve of conical form like a rugose coral. (Courtesy of the Redpath Museum.)**

Chalk is a soft, white, very fine-grained, remarkably pure limestone composed primarily of the whole shells and the fragments of shells of micro-organisms, largely coccoliths and forams. The Upper Cretaceous chalk cliffs of the English Channel

FIGURE 19–9 **Electron photomicrograph of a coccolith from the Cretaceous Taylor Chalk of the Gulf Coast area (approximately 20,000×). (Courtesy of S. Gartner, Jr. and William W. Hay, University of Illinois.)**

suggested the name for the system. The lack of sedimentary structures attributable to waves or currents, and the abundance of planktonic fossils suggest that chalk was deposited below the zone agitated by the waves, but the presence of fossils of organisms that live in water of moderate depth indicates that deposition was not abyssal. The Upper Cretaceous marine sediments of the Gulf Coastal Plain contain several formations composed of chalk. (The material now used to mark "chalk-boards" is not chalk but a form of plaster made from gypsum.)

Greensand is another unusual sediment that is common in Cretaceous rocks. Much of the Lower Cretaceous Series of the type section of the system in England comprises the formation first called "The Greensand." In the Atlantic Coastal Plain, greensand has been quarried for use as a fertilizer and in water softeners. Greensand is composed largely of the dark green pellets of glauconite, a complex silicate mineral rich in iron and potassium and related to the micas and clays. The mineral is apparently formed on the ocean floor by processes not yet entirely understood.

Cretaceous Igneous Rocks

A suite of small intrusions was emplaced during late Mesozoic time in the Atlantic coastal belt of North America. The intrusions belong to the rock suite referred to as alkaline, because of its richness in sodium and potassium. In the Ouachita Mountains, stocks and dikes were intruded and some of the syenites have been weathered to produce bauxite deposits. A Late Cretaceous ultramafic stock in Arkansas has produced the only diamonds ever mined on this continent. Volcanic ash beds, dikes, and sills of this suite have been penetrated by many wells in the central Gulf region, and Pilot Knob, south of Austin, Texas, is believed to be the relic of an eroded volcano which was active at this time.

The largest group of late Mesozoic intrusions is located along a broad arc that passes through Montreal and Boston and extends into a line of sea mounts east of the continental shelf at the latitude of New York City (Fig. 19–3). These sea mounts have been called the Kelvin or New England chain. In Quebec these alkaline intrusives form a line of stocks, or possibly the feeders of now eroded volcanoes, that constitute the Monteregian Hills. In New England, the group of small intrusive bodies constitute the White Mountain magma series. Along this arc many recent, small earthquakes have taken place. The alignment of the intrusions, earthquakes, and sea mounts has been explained by localization along a transform fault associated with the formation of the Atlantic spreading zone, or as the trace of a mantle plume over which the continent has moved westward.

CENOZOIC SEDIMENTATION

The Cenozoic sediments of the Gulf and Atlantic coastal plains consist of sands, marls, clays, and shales; many of these are still unconsolidated and crumble in one's hands. Downward along the dip, each stratigraphic unit grades, with much interfingering, from a near-shore or continental facies to a marine facies. In addition to small-scale oscillations of the shoreline, a broad, cyclical change in its position can be recognized. Each cycle of sedimentation started as the advancing sea deposited calcareous marls, glauconitic sands, or dark clays. As the recessional phase of the cycle began, marine deposits were overlain by gray sands and, as the shoreline moved still farther outward, by non-marine and estuarine sands and lignite. These recessional deposits are deltaic in nature and were formed as the rivers from the interior of the continent built out the shore. Subsidence was apparently more rapid during the recessional phases of the cycle, for these strata are much thicker than are the strata of the transgressional, marine phase. Figure 19–10 illustrates two such cycles from the Eocene Series of the Gulf Coast. The thick, non-marine sand wedges of the recessional phase are represented by the Sparta and Yegua formations. The Crockett and Cane River shales belong to the thinner, transgressional phase. Throughout the Cenozoic, the shoreline moved generally seaward as each transgression fell short of the northward extent of its predecessor, and each delta was built farther out into the Gulf.

The first Cenozoic transgression deposited beds of Paleocene age, known as the Midway Group. The recessive phase of the Midway cycle is formed by 1500 m of deltaic Wilcox sandstones, which are among the most important reservoirs for oil and gas on the Gulf Coast. The next two cycles constitute the Claiborne Group.

The sea advanced and retreated again before the close of Eocene time, depositing the Jackson Group. The recessive phase of an Oligocene cycle is represented by the thick, oil-bearing Frio Sand. The cyclical movement of the shoreline was poorly defined in Miocene and Pliocene times, but the largely deltaic sediments of these epochs reach a thickness of more than 4500 m in the region of the Mississippi delta. This sedimentary record shows that the rapid sedimentation and concomitant subsidence, which are features of the delta region today, have been in progress since the beginning of Miocene time. Pleistocene

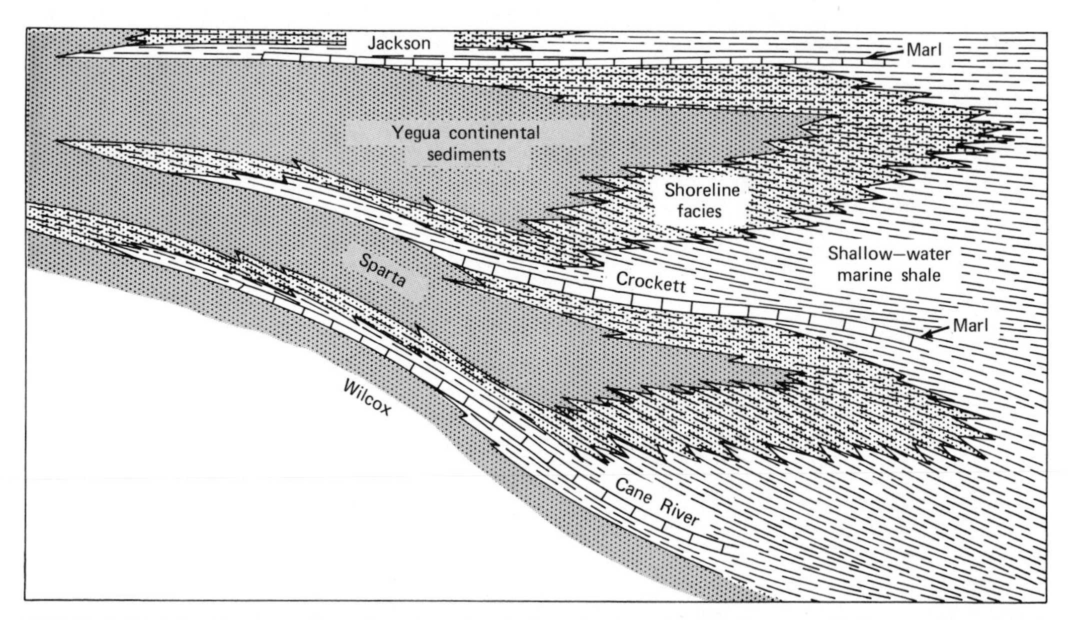

FIGURE 19–10 **Cross-section showing the facies changes in the Eocene Claiborne Group of central Louisiana. The Sparta and Yegua sands are regressive deposits separated by marine shales that were laid down during the transgression of the sea. (After S. W. Lowman; courtesy of the American Association of Petroleum Geologists.)**

deposits of this area are alluvial silts, gravels, and sands.

Rapidly changing facies make the Cenozoic and Cretaceous sediments difficult to correlate. This problem became acute when many oil wells were drilled in which accurate location of stratigraphic horizons was required. In the drilling process, minute shells of Foraminifera are brought up with the waste rock cuttings by the circulating mud that lubricates the bit. The use of these minute fossils for dating and correlating the rocks of the Gulf Coastal Plain was refined during the 1920's and 1930's by J. A. Cushman. As a result of his research, and that of other micropaleontologists, the Cenozoic and Mesozoic beds of this province are divided into many foraminiferal zones that can be recognized by a field paleontologist at the drilling site as the cuttings are collected.

The stratigraphy of the Atlantic Coastal Plain and continental shelf cannot be resolved into cycles as easily as can that of the Gulf Plain but in general, the times of marine transgression were the same in both areas and the stratigraphic units consist of similar continental, shoreline, and marine facies.

Two extensive areas of carbonate rocks in the dominantly clastic deposits of the coastal plains form the "arms" that embrace the Gulf of Mexico. In both Florida and the Yucatan Peninsula of Mexico, limestone began accumulating in Late Cretaceous time and has continued to build these platform areas to the present day. The Bahama Islands, although separated from Florida by a deep channel, are part of this carbonate province.

SALT-CORED STRUCTURES

While drilling for salt on a low hill in the Mississippi delta, Anthony Lucas noticed that the drill penetrated many pockets of gas. In 1901 he moved his drill over the border from Louisiana into Texas to test for oil in the rock below a low mound called Spindletop, similar to the mound he had drilled on in Louisiana. His first well was a "gusher," and 64 others drilled on the same hill each yielded 75,000 barrels a day. The hill at Spindletop was pushed up by the intrusion of salt into the sedimentary strata about 300 m below the surface. After Lucas' success, the search was begun for similar structures; in 75 years this search has resulted in the discovery of more than 300 other domes. The salt domes of the Gulf of Mexico are stocks of salt up to 10 km in diameter, which have penetrated as much as 10 km of sedimentary rock. The salt has flowed plastically upward through the strata from the Louann evaporites near the base of the coastal plain succession, squeezed by confining pressure produced by the weight of the overlying sediments (Fig. 19–11). The upward movement of the salt column was also aided by its lesser density with respect to the sediments around it. The columns of salt punctured the beds through which they moved, dragging them upward along their flanks. Higher beds were domed above the advancing salt. The movement of the salt into domes probably started as soon as about 3000 m of sediment had accumulated on top of the bed, and progressed as the Gulf subsided and additional sediments were loaded on to the salt layer.

The largest area of domes occurs along the coastline of eastern Texas and Louisiana, and extends out to the edge of the continental shelf (Fig. 19–5). That the source beds of the salt once extended to the center of the gulf is shown by the presence in deep water there of the Sigsbee Knolls, a series of small sea mounts which have been shown to be salt domes by the Deep Sea Drilling Project. On the western side of the Gulf the salt has moved into broad folds rather than into columns.

Like the Gulf Coast, the continental shelf of eastern Canada is underlain by many salt domes where the Argo Salt has moved upward through the overlying Jurassic and Cretaceous sediments.

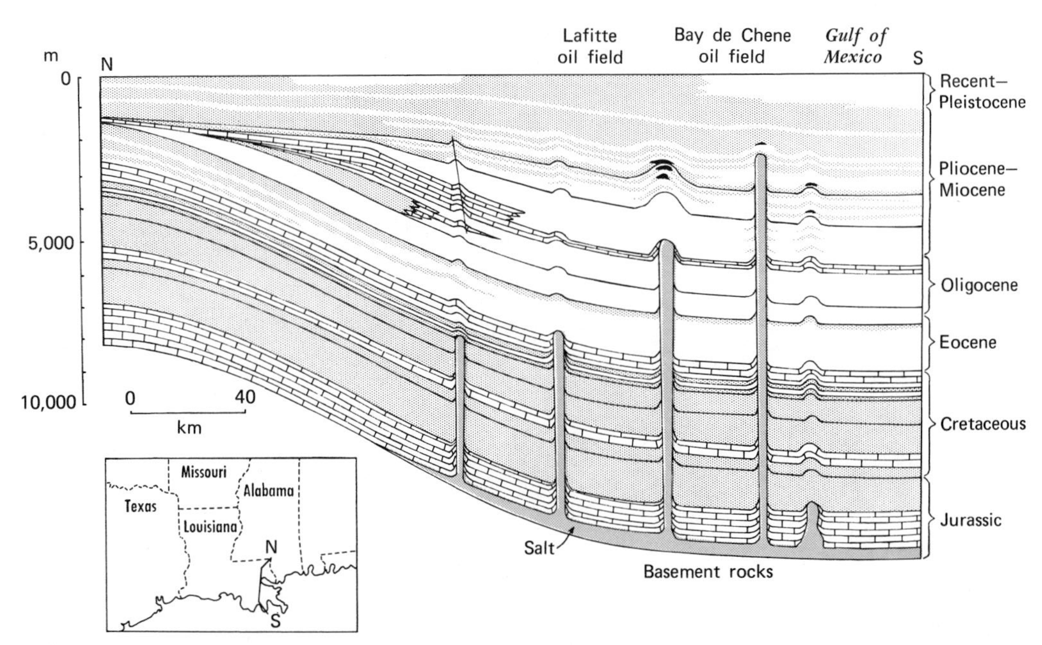

FIGURE 19–11 **Generalized cross-section of the Gulf Coastal Plain in southern Louisiana showing the relationship of the salt domes to the sediments. (Modified from J. B. Carsey; courtesy of the American Association of Petroleum Geologists.)**

PETROLEUM IN THE COASTAL PLAINS

The accumulation of oil and gas underground depends on the availability of (1) a source rock that can supply the hydrocarbons, (2) a migration route, (3) a porous reservoir rock, and (4) a trap. The alternating sands and shales of the Gulf area form ideal source and reservoir beds, respectively; the structures associated with salt intrusion form many traps, some of which are illustrated in Fig. 19–12. Porous beds domed into anticlines, the porous cap rock, beds truncated against the upthrust salt, and porous beds sealed off by faults or unconformities on the flanks of the domes, may all form reservoirs in the Coastal Plain oil fields.

The largest trap in the Gulf Region, at the East Texas Field, is formed by the truncation of the gently dipping Upper Cretaceous Woodbine Sandstone, and is sealed by the impervious Austin Chalk. Petroleum may also be trapped by facies changes as where beds of sand in such formations as the Frio or Yegua wedge out or become less porous upward along the dip.

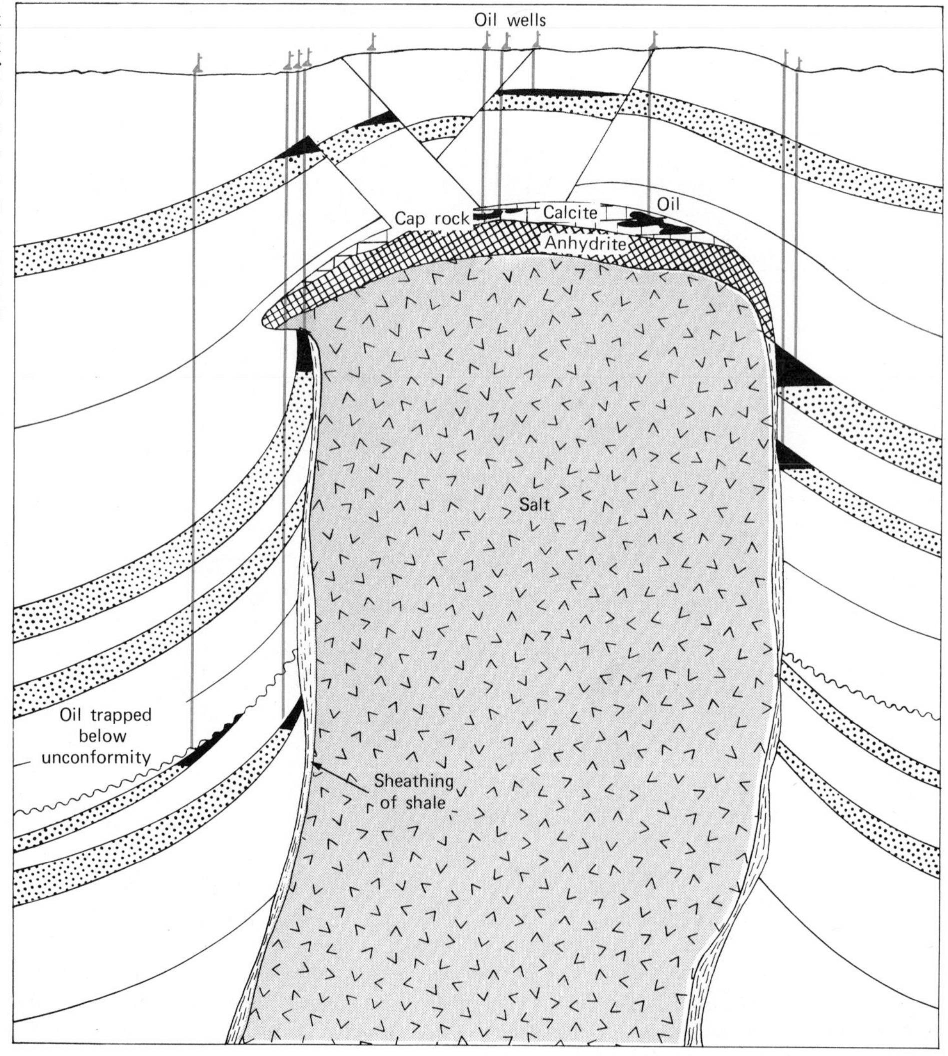

FIGURE 19–12 **Oil traps that may develop around a salt dome. In this cross-section of an idealized dome, the various positions in which petroleum can accumulate have been indicated. The cap rock of anhydrite and calcite is believed to be an accumulation of insoluble matter from the salt that was left behind as percolating water dissolved the rising salt column.**

To date no discoveries of petroleum have been made in the Atlantic Coastal plain or continental shelf of the United States. Despite an extensive exploration program, potential production of the Canadian shelf has been found only at the Sable Island Field on the edge of the shelf. Salt-cored structures have not been discovered as yet on the United States Atlantic shelf but geophysical surveys of the continental slope off the Carolinas indicate that domal uplifts cored by either plastic shales or salt have pushed through the sedimentary layers. The apparent lack of salt domes on the shelf decreases the probability of the discovery of a major new oil province like the Gulf of Mexico there.

EROSION OF THE APPALACHIAN MOUNTAINS

In Triassic time the original Appalachians, formed in the late Paleozoic, had been reduced to low relief, and a new generation of fault-block mountains was produced as a result of the rifting of Pangaea. The overlap of Triassic structures by Cretaceous sediments at the base of the Atlantic Coastal Plain succession shows that these Triassic mountains had little relief by the Cretaceous Period. The erosion surface, completed in Early Cretaceous time, that separates the Piedmont crystalline rocks and Triassic sediments below from the coastal plain succession above, is called the Fall Zone surface because it is exposed along the Fall Zone, or Fall Line, the contact of the Piedmont and Coastal Plain provinces (see Ch. 18).

The development of the Appalachian landscape through late Mesozoic and Cenozoic time to its present form has been interpreted in two different ways, which are examined below. Evidence has not as yet been assembled to justify the rejection of either hypothesis.

The Dissection of the Schooley Peneplain

The summits of the ridges of the Appalachians are remarkably smooth and level (Fig. 19–13), except where they are cut by major rivers (water gaps) or by valleys formerly occupied by rivers (wind gaps). The accordance of the summit levels has suggested to many geomorphologists—beginning with William Morris Davis, at the end of the 19th century—that the summits are remnants of a peneplain whose surface would be restored if the valleys between the ridges were filled in. This erosion surface has been commonly referred to as the Schooley peneplain. **Monadnocks** are hills that have escaped erosion in the process of peneplanation and rise conspicuously above the peneplain surface. Some mountains, such as the Great Smokies of Tennessee and the White Mountains of New Hampshire, rise above the Schooley surface and are considered to be monadnocks, a name derived from Mt. Monadnock in New Hampshire, which rises above the New England Upland surface. The Schooley surface could have been eroded only during late Mesozoic and early Cenozoic times, because it cuts the Fall Zone surface, which is of Early Cretaceous age. The modern topography of the Appalachians is believed to have resulted from the dissection of the surface in later Cenozoic time.

The concept of the Schooley peneplain offers a convincing explanation for the water gaps which the major rivers cut through the ridges of the Appalachian Mountains. Rivers such as the ancestral Potomac, Susquehanna, and Shenandoah are thought to have flowed in early Cenozoic time toward the Atlantic on alluvium that covered the peneplained bedrock surface, uninfluenced by the structure of the folded sediments beneath. The uplift of the Schooley surface cannot be dated exactly, but it could have occurred in mid-Cenozoic time and could have been responsible for the coarsening of coastal plain sediments in the Miocene Epoch. The rivers, quickened by uplift, started cutting into the alluvial covering and then into the bedrock beneath the Schooley plain. As the land was worn away, differential resistance to erosion caused the more resistant beds to be left as ridges and the valleys to be cut out of the more easily re-

FIGURE 19–13 **View of the smooth, level ridge tops of the mountains in the Valley and Ridge province. Looking southwest on a syncline at Trevorton toward the Susquehanna River. (Photograph by John S. Shelton.)**

moved shales and carbonates. The major streams that had flowed across belts of resistant rock on the plain continued to erode their courses, cutting water gaps through the ridges as they were uncovered, somewhat like a rotary saw cutting a log. The discontinuous nature of the uplift is believed to have allowed time for the cutting of such smaller and lower surfaces of planation as the Harrisburg and Somerville surfaces. The drainage was thus superposed across a topography where the relief first had been almost eliminated by peneplanation, and then re-established by uplift.

Dynamic Equilibrium

Some geologists do not believe that such erosion surfaces as the Schooley, Harrisburg, and Somerville are the remnants of peneplains or partial peneplains. They explain that the present landscape is the result of continuous erosion since Permian time, and believe that the different levels of topography are the results of the responses of different lithologies to the processes of weathering and erosion. The uniformity of altitude of the summits of the Appalachian ridges reflects the similar response of resistant beds of sandstone in a particular structural position to a particular set of weathering conditions. The lower erosion surfaces, such as the Harrisburg, are developed on softer rocks. The altitude of these surfaces can be no lower than the channels of the master streams; their gradients, in turn, are controlled by the grain size of the sediments they carry. Streams transporting coarse, resistant particles, such as sandstone derived from the Appalachian ridges, must maintain steep gradients. When the supply of this size particle is removed, the river adjusts to its new environment by cutting down through the non-resistant rocks, leaving them, for a time, as broad terraces (for example, the Harrisburg surface) to be dissected by later erosion. This theory has been called the theory of dynamic equilibrium, for it proposes that the landscape features are lowered by mass wasting and stream erosion at about the same rate; thus, the relief does not eventually decrease to an ultimate peneplain, but is maintained as the area is lowered by erosion, until the equilibrium is disturbed.

The Appalachian Mountains are the net result of the forces of deformation, uplift, and erosion at work for over half a billion years, since the formation of Iapetus. Deformation of the crust has shaped the structure; uplift has kept the rocks within the zone of erosion; and erosion and weathering have etched the weaker rocks from around the more resistant ones to produce the mountains that we see today.

ANIMAL DISTRIBUTION AND THE BREAK-UP OF PANGAEA

Study of the eastern side of what is now North America demonstrates two periods: one in the early Paleozoic and one toward the end of that era, when it was continuous with another large landmass to the east; and two other periods, in the middle Paleozoic and from the Mesozoic through the Recent, when it was bounded to the east by a large oceanic basin. The opening and closing of this basin has been the result of movements of continental plates. The separation and collision of the continents have important consequences for paleontology. The relative positions of the continents and oceanic basins strongly influenced the evolution and distribution of both marine and terrestrial organisms.

Late Paleozoic and Triassic

The early history of terrestrial vertebrates, from Late Devonian through Triassic times, reflects the unity of the major landmasses in the world continent—Pangaea. Most groups for which there is an adequate record show a cosmopolitain distribution. For most of the Carboniferous, the fossil record of terrestrial vertebrates is limited to Europe and North America. A specialized aquatic reptile, *Mesosaurus,* is known from numerous fossils in the very latest Carboniferous or earliest Permian deposits of western South Africa and eastern South America, but is not known from any of the northern continents. The distribution of this genus has long been cited as evidence that the southern continents were united in the late Paleozoic. By the beginning of the Mesozoic, the fossil record of terrestrial vertebrates becomes better known throughout the world. Although the record is not equally complete in all areas, and the distribution of some groups appears to be localized as a result of ecological requirements, there is little evidence of regionalism. The recent discovery of the Early Triassic mammal-like reptile, *Lystrosaurus,* in Antarctica extends the range of a genus earlier known from southern and eastern Asia and South Africa. The genus *Pro-*

terosuchus, or its close relatives, belonging to the group ancestral to dinosaurs, is even more widely distributed, appearing in Australia, China, India, South Africa, and Russia. Other members of the group are known from Antarctica. Several amphibian groups further demonstrate the cosmopolitan nature of the terrestrial fauna during the Permian and Triassic. Not only is there no evidence of important oceanic barriers, but there seem to have been few important climatological barriers. The evidence suggests a nearly uniform, equable climate throughout Pangaea.

In the later Mesozoic, the distribution of dinosaurs and other reptiles reflects the break-up of Pangaea, with the partial isolation of local faunas. This tendency becomes more conspicuous in the Cenozoic. The pattern of mammalian evolution is strongly influenced by the separation of the continents from one another by the oceans. None of the continents were completely isolated, but the amount of faunal interchange varied greatly from one area to another, and during different stages of the Cenozoic.

Early Mammals: Late Triassic–Early Cretaceous

Although technically still mammal-like reptiles, the late Triassic therapsids can be considered as initiating the mammalian radiation of the late Mesozoic and Cenozoic. They are known in western Europe, China, North and South America, and South Africa. The oldest currently known true mammals are found in the Late Triassic of Africa and the earliest Jurassic of Great Britain, but the transition between reptiles and mammals presumably occurred from a therapsid group of worldwide distribution. Most Jurassic fossil mammals are known from the northern continents, particularly Europe, North America and, to a lesser extent, eastern Asia. In the Jurassic, the fossil record is too incomplete to establish the significance of the early stages in the break-up of the continents to the evolution of mammals. By the time the major modern mammal groups, the placentals and marsupials, had evolved in the Late Cretaceous, the major continental plates had become isolated from one another. Despite the presence of advanced therapsids and early Mesozoic mammals in Africa and South America, they apparently did not evolve into marsupials or placentals in these areas. In fact, these continents may have been devoid of mammals in Middle Cretaceous time. For Australia, the subsequent history of mammalian evolution suggests that only the much more primitive, egg-laying mammals, the monotremes, the ancestors of the echidna and duck-billed platypus, were present on this continent in the Mesozoic.

Late Cretaceous and Early Cenozoic

The history of Late Cretaceous and Cenozoic mammals is one of repeated, but never continuous, interchange between the northern continents, intermittent invasion of the southern continents from the north, and occasional migration from the southern continents to the northern ones.

Australia was physically separated from the rest of the world throughout the Cenozoic. South America was isolated for much of that era. Africa was less effectively separated from Europe and Asia by physical barriers, but nevertheless, distinct mammalian groups developed there in isolation. North America and Eurasia, although in only limited and intermittent contact, constituted a single zoogeographic realm, the Holarctic, throughout most of the Cenozoic. Although some groups were limited to Asia or North America, they are exceptions in a fauna kept homogeneous by repeated migration across the North Atlantic, until that segment of the ocean opened in late Cenozoic time, and across the Bering Sea.

Origin of the Cenozoic Fauna

Marsupials and placentals presumably diverged from a single ancestral group at about the same time (Fig. 19–14). Their place of origin is uncertain. Within the Upper Cretaceous, placentals appear to be more common in Asia, marsupials more common in North America. With such a pattern, it is difficult to explain why marsupials, rather than placentals, are the first forms

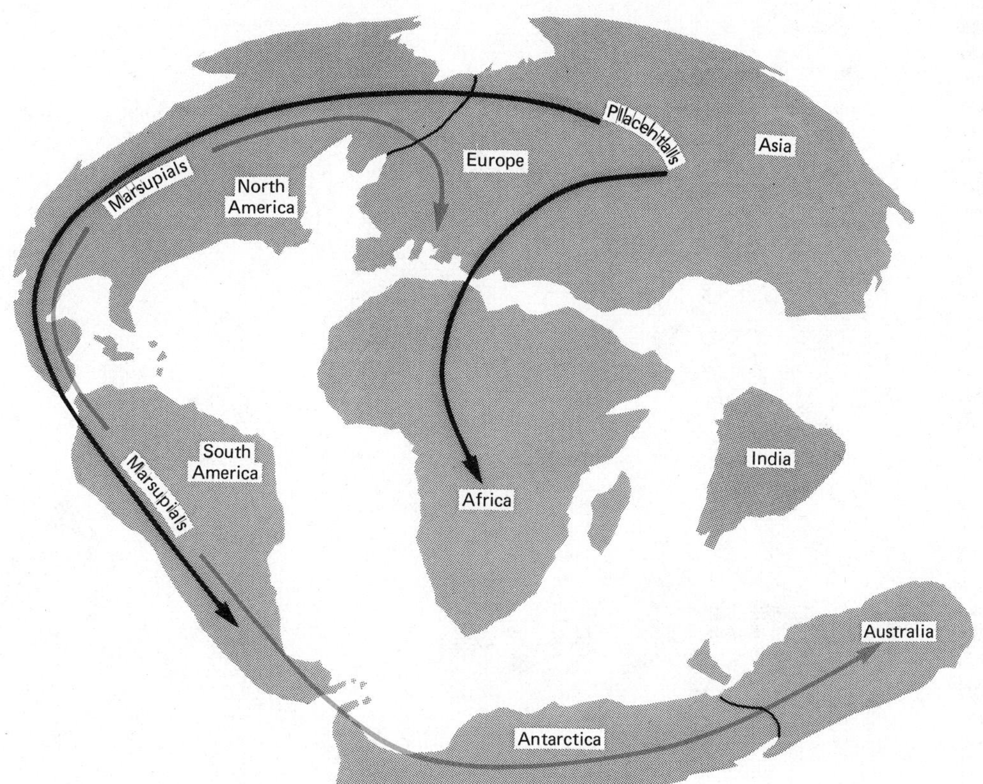

FIGURE 19–14 **Radiation of placental and marsupial mammals showing the probable distribution of the continents in the Late Cretaceous. (Data from R. Hoffstetter.)**

that invaded both South America and Australia, and formed an important element of the fauna of both continents in the early Cenozoic. There is no evidence of early marsupials in Africa. Most of the group vanished from North America in Late Cretaceous time, and from Europe and Asia (apparently) at about the same time. The opossum, one of the few marsupials that has done well in competition with placentals, reinvaded North America across the Panama land bridge in the late Cenozoic. Placentals are more effective in most environments than marsupials, and except for the opossum, Cenozoic marsupials have been able to multiply and diversify only in the absence of placentals.

Australia. For most of the Cenozoic, Australia remained separated from Asia and the rest of the world. The marsupials diversified in a manner parallel to placentals in other parts of the world, into marsupial "squirrels," "bears," "wolves," and a host of grazing herbivores, including the kangaroos (exceptional in being quite unlike any large placentals in their locomotor pattern) (Fig. 19–15). The only terrestrial placentals to invade Australia in the Cenozoic were rodents, coming across from Asia in Miocene and Pliocene times, but without becoming greatly different from the ancestral pattern.

South America. Mammalian evolution is more complex in South America. Early in the Cenozoic marsupials radiated into a host of niches occupied elsewhere by placentals. Smaller insectivore-like marsupials (e.g., a mole-analogue) and rodent-like forms (the Caenolestids) evolved here, as well as a form very similar to the saber-toothed cats (Fig. 19–16). The presence of a persistent chain of islands extending north toward Mexico made it possible for a few placentals to invade South America at various times during the Cenozoic, and the

FIGURE 19–15 **Australian marsupials. During the Cenozoic marsupials radiated widely, filling many of the adaptive zones occupied by placentals in the northern continents. The ancestral form may have resembled the living opossum. This general pattern is retained in the animal called Leadbeater's 'Possum (A). Squirrel-like phalangers (B) and rabbit-like bandicoot (C) parallel gnawing placentals in North America. A much larger animal with rodent-like teeth, *Diprotodon* (D), was present in the Pleistocene. The cat-like *Dasyurops* (E) and wolf-like *Thylacinus* (F) are becoming very rare in competition with recently invading placental carnivores. The koala (G) and kangaroo (H) are marsupial herbivores without close placental analogues. (Drawings by D. Alison; adapted from various sources.)**

FIGURE 19–16 **South American marsupials. Marsupials radiated widely to fill a continent largely devoid of placental insectivores, rodents, and carnivores. Subsequent invasion from the north by these groups is associated with the extinction of most of the specialized marsupials. The living oppossum (A) closely resembles the ancestral stock of all marsupials, and is now expanding its range in North America. The tiny shrew-like *Caenolestes* (B) also survives today. The rabbit-like *Argyrolagus* (C) survived into the Pleistocene, but is now extinct. The large carnivore *Prothylacinus* (D) and the false saber-tooth "tiger" *Thylacosmilus* (E) are known respectively from the Miocene and Pliocene. (Drawings by D. Alison; adapted from various sources.)**

marsupials never achieved as great a dominance of the fauna as in Australia.

In the late Eocene and Oligocene, primitive rodents invaded South America and radiated extensively, paralleling the pattern of rodent evolution in the northern continents and Africa. The most striking case of convergence is the evolution of a nearly exact replica of the African porcupine. So close is the similarity between specialized rodents in South America and Africa that biologists have repeatedly suggested a Cenozoic land bridge connecting the two continents, although there is no geological evidence for such a connection.

At approximately the same time as the invasion by northern rodents, South America received an influx of primates from North America. These were presumably relatively primitive forms, below the level of monkeys. Like the rodents, the primates became specialized in isolation from relatives in other parts of the world. Monkeys evolved there, as in Africa and Asia, but from a distinct, more primitive stock, and gave rise to the "spider" monkeys, with prehensile tails, that are typical of South America.

Another primitive placental stock that invaded South America, possibly early in the Paleocene, gave rise to the uniquely South American sloths, glyptodonts (Fig. 19–17), armadillos, and anteaters.

The most extensive radiation of uniquely South American mammals occurred among the large herbivores, derived from a primitive placental herbivore group that invaded South America in the Paleocene. The five derived orders show courses of evolution parallelling, in varying degrees, the elephants, horses, and camels from the rest of the world. Animals resembling the one-toed horse evolved in South America long before its analogue in the north.

At the end of the Pliocene, a continuous land connection with North America became established as the Cordilleran subduction zone was established across the westward side of the Caribbean Sea and built a volcanic isthmus. Mammalian groups that had been evolving in North America—placental carnivores and various orders of herbivores, including the elephant—extended their range south. Through competition and predation, the northern immigrants soon dominated the South American fauna, and nearly all of the larger native mammals became extinct. Surprisingly, few of the northern groups became permanently established in South America, and that continent remains strangely devoid of the larger herbivores; of the large horse and cow groups, only the camel's relative, the llama, remains. Some South American forms, notably the ground sloth and glyptodon, invaded North America across the Panama land bridge in Pleistocene time and ranged as far north as Alaska. These two species are now extinct, but two other South American immigrants, the opossum and armadillo, continue to extend their range into North America.

Africa. The pattern of mammalian evolution in Africa appears to have been quite different from that in either Australia or South America. Several distinct groups of animals evolved here and then radiated into the rest of the world. Early whales appear in the Eocene beds of Egypt. These forms apparently evolved along the margins of an extensive seaway (the Tethys) that separated Europe and Africa and extended eastward into southern Asia. Whales evolved from a group of terrestrial carnivores, not, apparently, from those that gave rise to the dogs, cats, and bears of the modern fauna, but from the primitive creodonts. When the whales first appear in the fossil record, the body skeleton is already highly specialized for swimming, for the hind limbs were lost and the forelimbs were modified into paddles. The skull retains the dentition and brain case typical of primitive creodonts. By Eocene time, the two modern groups of whales had evolved separate adaptations: toothed whales, including the dolphins, porpoises and killer whales; and the baleen, or whalebone whales. In the latter group, all the teeth are lost and a series of ridges of horny tissue developed, extending down from the upper jaw to strain small organisms, on which the animals feed, from the sea.

The specific ancestry of the Proboscidea (elephants, Figs. 16–40) is not established, but these animals are first known in Africa and then radiated throughout the world. In

FIGURE 19–17 **The ground sloths were accompanied into North America by an armored relative, *Glyptodon* (painting by Charles Knight, courtesy of the Field Museum of Natural History.)**

late Cenozoic time, they extended their range as far as South America and throughout the Holarctic landmass. Their current restriction to parts of Africa and India marks an extremely rapid diminution of their range in the Pleistocene.

Africa and part of Asia were also the ancestral homes of the higher primates—apes and man—to be discussed in Chapter 21.

Holarctic. As a result of limited connections with Africa, and intermittent opportunities for exchange with South America, the Holarctic area acted as a huge region for differentiation and distribution throughout the Cenozoic. Together with the more intensive study of fossil mammals in Europe and North America, one has the impression that most of the major events in the evolution of the group took place in this area. The main groups of mammals of the Holarctic realm have been discussed in Chapter 16.

SUMMARY

By Triassic time Appalachian relief had been much reduced, and the eastern margin of North America was divided into a series of fault-bounded basins in which coarse clastic sediments of the Newark Group accumulated. These basins and their associated igneous rocks may be related to the early phases of rifting of Pangaea.

The coastal plains and continental shelf are underlain by seaward-dipping and seaward-thickening Mesozoic and Cenozoic sediments.

The sedimentary prism thickens into regional basins and thins over arches whose axes are perpendicular to the coastline. Growth faults roughly concentric with the Gulf of Mexico have been formed by continued subsidence of its center and inward movement of the downthrown blocks along a lubricating salt layer.

As the Atlantic Ocean opened in Jurassic time extensive salt deposits were laid down in the restricted rift basin. Reefal limestones of Cretaceous age in the Gulf Coast region were formed by oyster-like bivalves and bivalves with conical attached valves called rudists combining with scleractinian corals. Chalk, a sediment particularly characteristic of Cretaceous seas, was formed largely of the tests of the newly evolved planktonic Foraminifera and the minute calcareous plates called coccoliths secreted by planktonic algae.

The Coastal Plain clastic wedge is composed of thick, regressive sandstones and thin, transgressive beds. The shoreline shifted back and forth across the area during the Cenozoic, but transgressions became progressively less extensive after the beginning of the Cenozoic Era.

Salt domes, structures characteristic of the Gulf Coastal Plain and Canadian continental shelf, have risen from Mesozoic salt beds, puncturing, doming, and fracturing the overlying strata and forming traps for oil. Many kinds of structural and stratigraphic traps for petroleum have been found in the Gulf Coastal Plain.

The evolution of the Appalachian landscape can be explained in terms of (1) a succession of peneplains and partial peneplains that record standstills in the process of uplift, or (2) a dynamic equilibrium in which the various topographic levels are determined by the interaction of lithology, structure, weathering, and erosion.

Little is known of Paleozoic terrestrial vertebrates outside of Europe and North America. What fossils are known suggest that a uniform fauna inhabited the world until the end of the Triassic. Local differences in ecology and depositional environment appear to have been more important than geographical distances in determining what fossil groups were preserved.

In the late Mesozoic and Cenozoic, regional differences became more significant, and the pattern of evolution was strongly influenced by the relative degree of separation of the continents. The Australian fauna was nearly completely isolated and consists largely of endemic species. South America was isolated for most of the Cenozoic and developed a unique fauna derived from occasional immigrants from North America. At the close of the Cenozoic, renewed contact with North America resulted in faunal interchange and, subsequently, the extinction of most of the specialized South American genera. Africa was partially isolated from Europe and Asia in the early Cenozoic, during which time the elephants and whales evolved. North America and Eurasia constitute a single faunal realm where climate rather than physical barriers has resulted in isolation of some orders, while others have migrated freely throughout most of the Cenozoic.

QUESTIONS

1. Under what conditions might a single kind of sediment be confined to a single time in the geological past? Cite examples other than those given in this chapter.
2. Contrast the structures caused by the tensional and vertical forces discussed in this chapter (basins, faults, salt intrusions, etc.) with those formed by compressional horizontal forces described in the preceding chapter.
3. Discuss the relationship between the plate tectonic events that formed the island chain of the Antilles and the Panama Isthmus and the development of the fauna of South America during the Cenozoic Era.
4. Compare the geomorphic evolution of the Appalachians and Rockies.

SUGGESTIONS FOR FURTHER READING

BALLARD, R. D., and UCHUPI, E. "Triassic Rift Structure in Gulf of Maine." *American Association of Petroleum Geologists Bulletin,* 59: 1041–1072, 1975.

BURK, C. A., and DRAKE, C. L. ed. *The Geology of Continental Margins.* New York: Springer Verlag, 1009 p., 1974.

COLBERT, E. H. *Wandering Lands and Animals,* New York: Dutton, 1973.

GARRISON, L. E., and MARTIN, R. G., Jr. "Geologic Structures in the Gulf of Mexico Basin." *U.S. Geological Survey Professional Paper 773,* 1973.

JANSA, L. F., and WADE, J. A. "Paleogeography and sedimentation in the Mesozoic and Cenozoic, southeastern Canada," in *Canada's Continental Margins. Canadian Society of Petroleum Geologists Memoir 4,* pp. 79–102, 1975.

McIVER, N. L. "Cenozoic and Mesozoic Stratigraphy of the Scotia Shelf. *Canadian Journal of Earth Sciences,* 9: 54–70, 1972.

MURRAY, G. E. *Geology of the Atlantic and Gulf Coastal Province of North America.* New York: Harper and Row, 1961.

PITMAN, W. C., and TALWANI, M. "Sea Floor spreading and the North Atlantic." *Geological Society of America Bulletin,* 83: 601–618, 1972.

POAG, C. W. "Stratigraphy of the Atlantic Continental Shelf and Slope of the United States." *Annual Review of Earth and Planetary Sciences,* 6, 1978.

SIMPSON, G. G. *The Geography of Evolution.* Philadelphia and New York: Chilton Books, 1965.

WILHELM, O., and EWING, M. "Geology and History of the Gulf of Mexico." *Geological Society of America Bulletin,* 83: 575–600, 1972.

PART SIX

the arctic and cenozoic ice age

CHAPTER TWENTY

INTRODUCTION

The northern margin of the North American continent with the frozen Arctic Ocean, is a complex of peninsulas and islands (Fig. 20–1). The third major folded mountain chain of the continent, the Innuitian Mountain system, crosses these islands. Its name is derived from the term used by the natives of this region to refer to themselves; Innuit means "the people" in the language of the Eskimos. The belt of Innuitian deformation, stretching from the northern Yukon to the north coast of Greenland, closes the ring of folded mountain chains around the continent and joins the Cordillera to the Caledonian segment of the Appalachian Mountain system in East Greenland.

The channels between the northern Arctic Islands have the appearance of a drowned river drainage system (Fig. 20–1). Straight coastlines suggest, and geophysical studies confirm, that some of the channels in the southern part of the archipelago are bounded by nearly vertical faults. The Arctic Islands are arranged in three rows or series roughly parallel to latitude lines, which reflect three major tectonic divisions (Fig. 20–2). The southern islands, from west to east—Banks, Victoria, Prince of Wales, King William, Somerset, and Baffin—are part of the craton. The middle set—Prince Patrick, Melville, Bathurst, Cornwallis, Devon, and the southern part of Ellesmere—represent mainly a miogeoclinal province. The northern tier—including Ellef Ringnes, Axel Heiberg, northern Ellesmere, and several smaller islands—are underlain mainly by late Paleozoic to Recent sediments of the Sverdrup Basin.

The Arctic Ocean has broad continental shelves north of Scandinavia and western Siberia, a narrow shelf off Alaska and the Arctic Islands of Canada, and a narrow, deep-water outlet to the Atlantic Ocean near the northeast corner of Greenland (Fig. 20–3). The deep basin of the Arctic Ocean contains the Canada Abyssal Plain north of Alaska and three prominent, subparallel ridges which cross from the Arctic Islands to the Siberian shelf. The Nansen Ridge (also called the Arctic Mid-Ocean Ridge and the Gakkel Ridge) is the continuation of the Mid-Atlantic Ridge, and is the site of current sea-floor spreading. The Lomonsov Ridge may be a slice of the continental shelf split off into the Arctic Ocean Basin by the rifting accompanying the formation of the Nansen Ridge. The Alpha Ridge (also called the Alpha Cordillera and, on the Russian side, the Mendeleyev Ridge) is of doubtful origin, but is believed by some geologists to be the site of sea-floor spreading between 40 and 60 million years ago. This interpretation is supported by the alignment of this ridge with the spreading center that

BORDERLANDS OF THE ARCTIC OCEAN

effected the separation of Greenland from Labrador in early Cenozoic time. Other geologists have interpreted the ridge as a submerged island arc overlying an ancient subduction zone.

The configuration of the Arctic Ocean Basin in Paleozoic time and the nature of its interaction with the northern boundary of the North American plate are subjects of much recent speculation but little agreement (see end of chapter).

TECTONIC DIVISIONS OF THE ARCTIC ISLANDS

The Platform

The pre-Paleozoic rocks of the Canadian Shield extend into the flat-lying platform rocks of the southern Arctic Islands in two promontories (Fig. 20–2). One crosses Victoria Island (Minto Arch) and the other extends up the Boothia Peninsula (Boothia Arch). These cratonic arches separate basins containing a relatively thin succession of lower Paleozoic sedimentary rocks. Typically, this succession consists of basal sandstones of late Proterozoic to Cambrian age, and overlying Ordovician, Silurian, and Devonian dolomites and limestones reaching a total thickness of about 1000 m. Small reefs are present at the top of this platform succession in the Devonian rocks of northern Banks Island. At the beginning of the Devonian Period the Boothia Arch was uplifted, probably along faults, and shed coarse, clastic sediments, forming a nonmarine alluvial fan and deltaic facies up to 3000 m thick in a deep trough on its flanks. Repeated subsequent uplifts of the arch are recorded by unconformities in the sediments on its sides. The youngest sedimentary rocks in the platform area are of Late Devonian age.

Parry Islands and Ellesmere Fold Belts

The early Paleozoic sedimentary rocks of this area are carbonates and shales, generally thicker than the platform rocks to the south, and are folded into open synclines and anticlines, and cut by steeply dipping thrust faults. These features are characteristic of miogeoclinal rocks. As vegetation is sparse at these latitudes, the anticlines and synclines stand out dramatically in an aerial view (Fig. 20–4). The continuity of the Parry Islands and Ellesmere fold belts is interrupted at Cornwallis Island, in line with the extension of the Boothia Arch, by a series of strike-slip faults and transverse folds that trend northwesterly (Fig. 20–1). These transverse structures of the Cornwallis fold belt are related to the Early Devonian uplift of the Boothia Arch, which extends under this

FIGURE 20–1 **Canadian Arctic Islands. Mosaic of images made by the Landsat satellite. Note the clarity of the folded patterns in the Parry Islands belt and the Sverdrup Basin. The islands can be identified with reference to Fig. 20–2. Most of the white on the photograph is sea ice. (Original photograph supplied by the Surveys and Mapping Branch, Department of Energy, Mines and Resources, Canada.)**

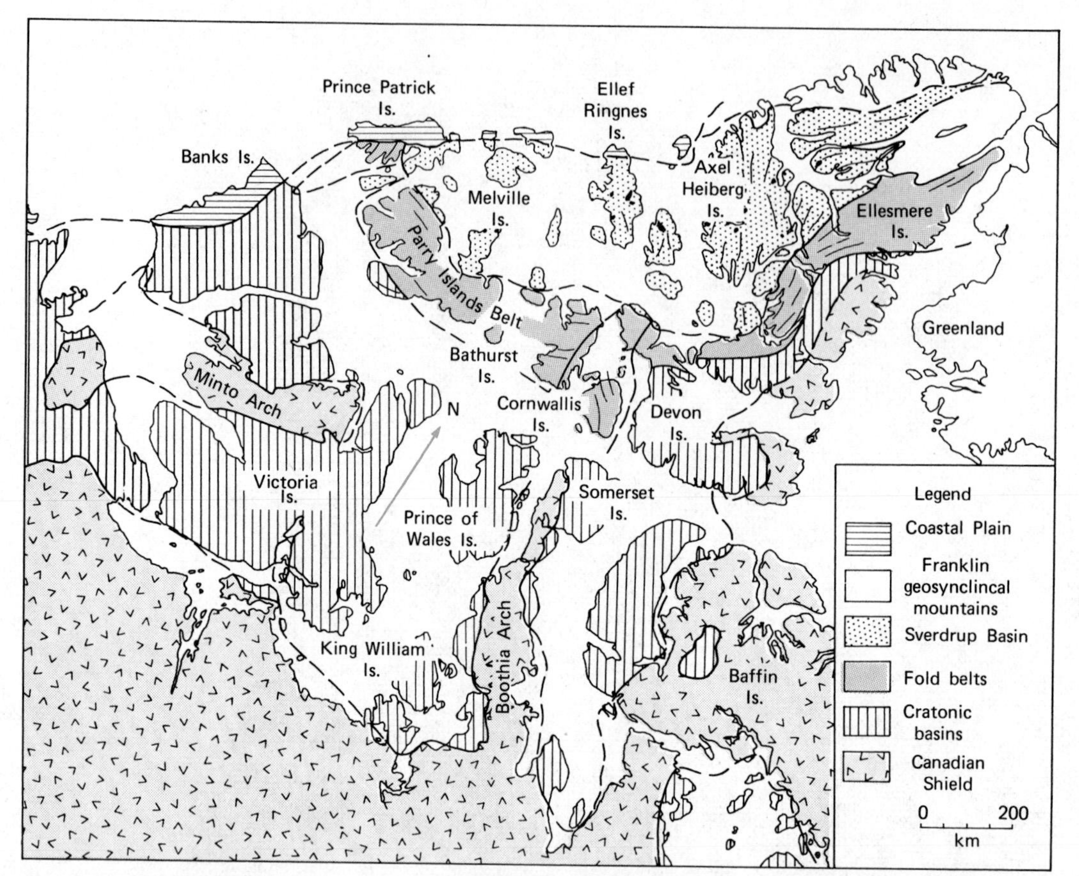

FIGURE 20–2 **Tectonic Map of the Arctic Islands. The structural trends in the fold belts are shown by short lines. The major evaporite intrusions in the Sverdrup Basin are represented by dark blobs.**

section of the miogeocline. The Ellesmere fold belt is much like that of the Parry Islands, but its folds strike north-northeast rather than east–west. At Cornwallis Island the trend of the miogeoclinal folding is bent northward through about 70° (Fig. 20–2).

Cambrian and Ordovician strata of the miogeocline (apart from the basal Paleozoic sandstones) are composed largely of limestone. In Early and Middle Ordovician time the access to the miogeocline of open water to the north was restricted by a barrier, now represented by a belt of dolomite. Behind this barrier evaporitic conditions developed and anhydrite and halite were deposited to a thickness of approximately 1000 m.

Northwest of a line that crosses the western side of Melville Island (Fig. 20–5), the dominantly carbonate facies passed abruptly into dark graptolitic shales during Ordovician and Silurian time. The Ibbett Bay Shale of this area is only about 1000 m thick, but represents Early Ordovician to Early Devonian time (approximately 100 million years). This shale facies, deposited in deep water, was separated from the shallower water carbonate–evaporite facies through much of this time interval by a belt of dolomite that has been interpreted as a barrier reef tract, now dolomitized so that the organisms that once built it have been obscured.

In Late Ordovician and Silurian time the shale facies moved southward over the car-

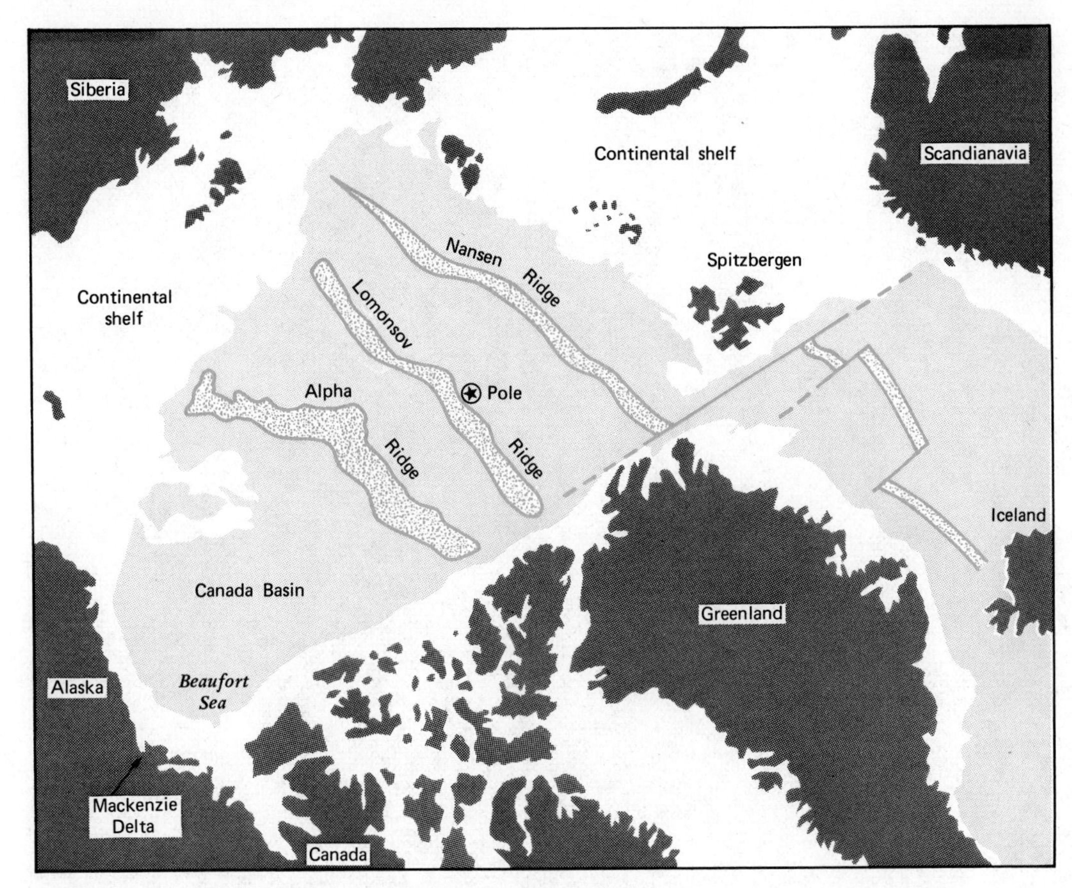

FIGURE 20–3 **Major features of the Arctic Ocean Basin.**

bonates of the miogeocline, forming a large embayment in Bathurst and eastern Melville islands, in which the Cape Phillips shale was deposited (Fig. 20–6). On the margins of the embayment almost 3000 m of porous, reef-like carbonates were deposited in Silurian time. Lead and zinc ores were deposited in porous Silurian carbonates on Cornwallis Island to form a deposit similar to that in the Middle Devonian reefs at Pine Point, described in Chapter 12.

In Early Devonian time the shale facies moved still farther southward and coarser sediments, derived from a northern source and from the Boothia Arch, were laid down above the shale. By Late Devonian time the flood of clastics from the north had converted the miogeocline into an alluvial plain, building a clastic wedge of non-marine sandstones. Late Devonian rocks are the youngest in the Parry Islands and Ellesmere belts. At the end of Devonian time the Paleozoic beds that had accumulated in the miogeocline were deformed and uplifted in an event called the Ellesmerian orogeny.

Franklin Geosyncline

The site of the source areas of most of the clastic sediments in the Parry Islands fold belt was north of the present outcrops of lower and middle Paleozoic rocks, but cannot be accurately located for it is now deeply covered by younger beds that occupy the Sverdrup Basin. Formations such as the Ibbett Bay and Cape Phillips indicate that in early Paleozoic time the miogeoclinal environment deepened northward, and the

FIGURE 20–4 **The open anticlines and synclines of the Parry Islands fold belt on Melville Island. Most of the rocks in the photograph are Upper Devonian sandstones and shales of the Hecla Bay and Gripper Bay formations. (Courtesy of the Department of Energy, Mines and Resources, Canada.)**

stratigraphic succession suggests that the source area north of this deep water became increasingly elevated during Silurian and Devonian time. At the northeastern end of the Sverdrup Basin in northern Ellesmere Island and Axel Heiberg Island, lower and middle Paleozoic sediments deposited on the southern flank of this source area reach thicknesses of thousands of meters.

The lower Paleozoic succession of northern Ellesmere Island contains flysch sequences, turbidites, volcanic flows, ash

FIGURE 20–5 **Aerial photograph of the Canrobert Hills, a section of the Parry Islands fold belt on the western side of Melville Island. The dark shales of the Lower Paleozoic Ibbett Bay Formation form the cores of the anticlines. In the center and background the lighter colored Pennsylvanian Canyon Fiord Formation overlies the shales unconformably. (Courtesy of the Department of Energy, Mines and Resources, Canada.)**

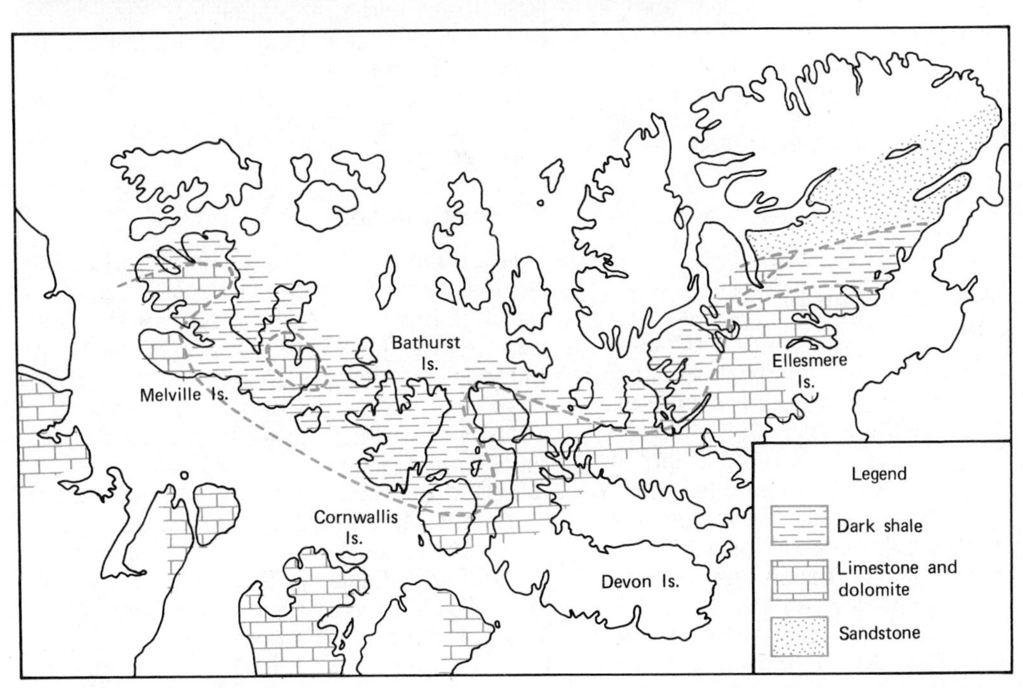

FIGURE 20–6 **Distribution of the major facies of Late Ordovician and Silurian rocks in the Canadian Arctic Islands. (After R. Thorsteinsson, H. D. Daae, and A. C. T. Rutgers.)**

beds, and chert that suggest that the source area to the north was an area of complex geology, considerable relief, and active volcanoes. The detrital particles in the sediments and isotopic evidence for late Proterozoic Grenville deformation along the northern Ellesmere Island coast indicate that the source area was not a simple volcanic island arc like those of the western Pacific. The trough in which the detrital lower and middle Paleozoic sediments of the Arctic Islands were deposited is called the Franklin (or Franklinian) geosyncline, after Sir John Franklin, an early explorer for the Northwest Passage. Most of the geosynclinal facies of the Franklin geosyncline is buried beneath the Sverdrup Basin and the waters of the Arctic Ocean.

In northern Ellesmere Island the Franklin geosynclinal rocks have been weakly metamorphosed, intensely deformed, and intruded by small masses of granitic rocks dated at about 360 million years (Late Devonian). Orogeny in the geosyncline may have started in Early or Middle Devonian time, but deformation spread southward and southeastward in late Devonian time, crumpling the miogeoclinal sediments and shearing them off the basement rocks in a zone of décollement along the Ordovician evaporites to produce the Parry Islands and Ellesmere fold belts. The Late Devonian clastic wedge in the Parry Islands was derived from the uplift and deformation of the Franklin geosyncline.

Extensions of the Innuitian Orogenic Belt

The Innuitian belt of mid-Paleozoic folding can be followed eastward from Ellesmere Island across the northern coastlands of Greenland. On the northeast point of that island the East Greenland and Innuitian zones of Paleozoic folding approach each other at right angles, but unfortunately for geologists, their area of intersection has been cut off by a major transform fault that shifts the mid-Atlantic spreading center across to the Nansen Ridge (Fig. 20–3), and the intersection is now covered by the Atlantic Ocean. The East Greenland Mountain system (Fig. 2–9) is the west side of the Caledonian Mountain system, deformed in Late Silurian time, and split into European and Greenland sections by the Mesozoic rifting of the Atlantic.

The westward continuation of the Innuitian belt is covered by younger sediments, but in the northern Yukon Territory and in northern Alaska a belt of middle Paleozoic deformation has been identified beneath the younger sediments, despite the obscuring effects of later orogenic movements that formed the Cordillera. The association in this area of facies changes in the early Paleozoic section much like that between the Ibbett Bay Shale and the carbonate facies in the Parry Islands belt, a Late Devonian clastic wedge, and small batholithic intrusions with Late Devonian isotopic ages, suggest that the Innuitian belt continued around the end of the Beaufort Sea. In early Paleozoic time the Cordilleran geosyncline in southern Alaska could have been separated from the early Paleozoic geosyncline in northern Alaska by a shelf area where carbonates were deposited (Fig. 20–7). Structurally and stratigraphically, the Alaska-Yukon area is one of the most complex on the continent, and further research will be necessary before the relationship of the Innuitian and Cordilleran belts is fully understood.

Sverdrup Basin

This roughly spoon-shaped basin which underlies the northern tier of the Arctic Islands contains late Paleozoic, Mesozoic, and Cenozoic rocks. In the deepest part of the "spoon," the succession reaches a thickness of 11,000 m. Structurally, the basin has been divided into a western section of simple synclinal structure, an eastern section folded in Cenozoic time, and a central section of simpler structure penetrated by many evaporite domes.

No Early Mississippian beds are known in the Sverdrup Basin but locally, Late Mississippian continental deposits overlie folded early Paleozoic rocks around its borders. The sea entered the basin in Early Pennsylvanian time, but was restricted by a

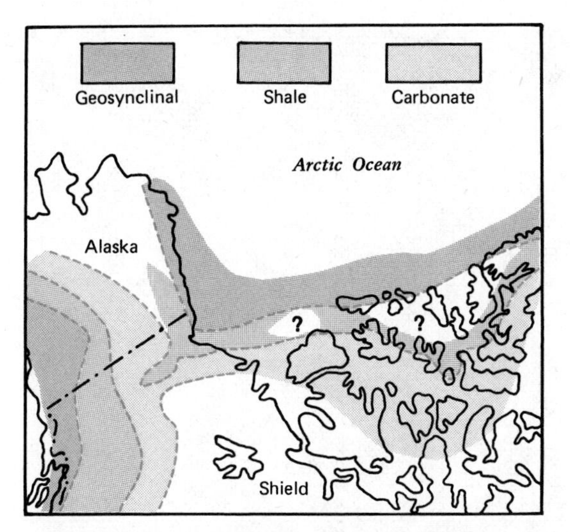

FIGURE 20–7 **An interpretation of the early and middle Paleozoic facies belts in Alaska and northern Canada showing the continuation of the Franklin geosyncline across the northern coast of Alaska. Post-Devonian deformation has probably distorted the original geographic relationships of these belts. (After maps by M. Churkin and M. Lerand.)**

sill, and thick evaporite deposits were laid down. Stratigraphic relationships in Ellesmere Island suggest that these evaporites were laid down in deep water. Pennsylvanian beds of the basin also contain porous limestone masses that were apparently formed as reefs.

The Mesozoic succession in the Sverdrup Basin consists largely of sandstones and shales. The clastic sediments were derived from source areas, possibly quite distant, to the south and west of the basin. Generally, a non-marine, deltaic facies occupies the southern and eastern sides of the basin and grades northwestward toward the center of the basin into marine shales. The youngest sediments in the Sverdrup Basin are the dominantly continental Eureka Sound sands and shales of Late Cretaceous and early Cenozoic age. A rich fauna of early Cenozoic vertebrates has been found in these sediments including mammals and reptiles indicative of a tropical climate.

Intrusions of evaporites are scattered throughout the basin. Over a hundred of these bodies have been mapped on the surfaces of the islands and detected geophysically beneath the sea between them. At the surface these intrusions are composed of anhydrite and gypsum, but wells passing into the domes and geophysical measurements indicate that these minerals form a cap on the top of a column of halite rising from Early Pennsylvanian source beds. In the eastern part of the basin the evaporites form the crests of open anticlines folded in late Cenozoic deformation (Fig. 20–8).

Various theories have been advanced to account for the folding in the eastern part of the Sverdrup Basin along dominantly north–south axes. The alignment of the structures with the northward extension of the Boothia Arch has suggested to some that they are related to renewed movements along that feature. Others believe that a late Cenozoic northward shift of about 500 km by Greenland and Ellesmere Island caused not only the folding in the Sverdrup Basin but also the northward bend of the whole Innuitian trend near Cornwallis Island.

Arctic Coastal Plain

Along the northwestern fringe of the Arctic Islands the unconsolidated Miocene sands of the Beaufort Formation form a narrow coastal plain and a wedge of sediments that thickens to several hundred meters below the narrow continental shelf. Farther south, around the margins of the Beaufort Sea (Fig. 20–2), the Arctic Coastal Plain and continental shelf are underlain by thousands of meters of Mesozoic and Cenozoic sediments. Geologically, these areas share many features with that of the Gulf Coastal Plain.

1. The sandstones and shales that have filled the edge of the Arctic Ocean Basin were derived from the uplift of the Cordillera in late Mesozoic time.
2. The seaward thickening wedge is cut by many nearly vertical faults, many of which show down-to-basin movement.
3. Many of these faults are growth faults.
4. Beneath the Beaufort Sea, dome-like structures rising from depth have

FIGURE 20–8 **Gypsum intrusive in the folded rocks of the Sverdrup Basin on Axel Heiberg Island (light mass in foreground). The syncline on the left is in Mesozoic strata. (Courtesy of the Department of Energy, Mines and Resources, Canada.)**

pierced the sediments. Whether these are cored with salt or plastic shales has not been determined.

5. The Mackenzie River has built a delta where the sediments are thickest, in much the same way that the Mississippi River has deposited a thick wedge of detritus in the Gulf of Mexico. The sedimentary succession north of the mouth of the Mackenzie reaches a thickness of 10,000 m.

The Arctic Coastal Plain can be followed along the south shore of the Beaufort Sea, between the Brooks Range of Alaska and the Arctic Ocean.

PETROLEUM RESOURCES OF THE ARCTIC BORDERLANDS

Most of the oil discovered so far has come from basinal areas of the earth underlain by great thicknesses of sedimentary rocks. Potential reservoirs must be covered with an adequate thickness of impermeable rocks or the oil will leak to the surface and be dissipated. Thus, in regions where the sedimentary covering of the basement rocks is less than 1000 m thick, oil and gas are missing or present only in small quantities. The sedimentary filling must be unmetamorphosed, for oil is not found in strata that have been affected by regional metamorphism.

Although regional metamorphism generally precludes the accumulation of oil in quantity, deformation of the strata by faulting and folding often forms structural traps in which petroleum accumulates. The continental side of miogeoclinal provinces, where sediments have been slightly disturbed by the movements in the geosyncline or by the instability of the cratonic margin, are among the most favorable places for the accumulation of oil. Examples of states with large production from such accumulations are Pennsylvania, Oklahoma, and Wyoming. Petroleum may be trapped in both flat-lying and folded sedimentary rocks by facies changes, such as the sealing of porous sand lenses by shale or the transition of limestone reefs into shales. Study of recent sediments shows that they contain only infinitesimal amounts of hydrocarbons, which can be flushed out with a great quantity of trapped water into adjacent porous beds as the sediments are compacted. If petroleum is to accumulate in quantity, the sedimentary basin must have a great volume of the source beds (probably dark shales) to supply the necessary hydrocarbons, because hydrocarbons are in such low concentrations.

The first well drilled for oil in the Canadian Arctic Islands was started in the winter of 1961–1962. Since then about 100 wells have been drilled in exploring this tectonic province, one of the last to be tested for petroleum resources. Although the conditions for the accumulation of petroleum appear to be appropriate, so far the results of exploration in the islands have been disappointing. The Silurian reef belts of the Parry Islands that appeared to be promising reservoirs have been found to be dry in most wells. However, on a small island off the northwest coast of Bathurst Island wells have encountered significant accumulations of oil in Devonian rocks. So far oil, in quantities large enough to be of commercial interest in such a remote area, has not been discovered in the Sverdrup Basin, although source beds, structural and stratigraphic traps, and appropriate reservoirs are all present. However, drilling has located large accumulations of natural gas, particularly along the southern boundary of the Sverdrup Basin in Melville Island. At present, sufficient gas has been proven to justify planning of a pipeline to bring it to southern markets, but the search continues for an oil field large enough to justify the great costs of transporting the oil to distant markets.

At the inner margin of the Arctic Coastal Plain in the northeastern corner of Alaska, the largest accumulation of petroleum yet found in North America was discovered in 1968 at Prudhoe Bay. The recoverable reserves of the Prudhoe Bay field have been estimated at 9.6 billion barrels of oil and 26 trillion cubic feet of gas. To tap this vast energy deposit, the Trans-Alaska pipeline has been built across some of the most formidable mountains in the world. The field is underlain by the Barrow Arch, a buried ridge of early Paleozoic rocks deformed in the Ellesmerian orogeny. Late Paleozoic and early Mesozoic beds arch over the crest of the ridge, are truncated by an unconformity, and their edges are sealed by a Cretaceous shale (Fig. 20–9). Petroleum migrating up the flank of the arch has been trapped under the unconformity in beds ranging from Mississippian to Jurassic age.

In the early 1970's several oil and gas fields were discovered on the Arctic Coastal Plain in the vicinity of the Mackenzie delta. The fields in this area are in the thick Cretaceous and Cenozoic clastic sequences, and produce from traps associated with the fault systems that traverse the seaward-dipping wedge. The successes south of the Beaufort Sea lend encouragement to geologists exploring the similar beds of the Sverdrup Basin.

PLATE TECTONIC INTERPRETATIONS OF ARCTIC GEOLOGY

The belt of Ellesmerian deformation can be followed from eastern Greenland around the edge of the Arctic Ocean Basin, at least as far as northern Alaska, and perhaps across the Bering Strait into Siberia. Sediments supplied to the Franklin geosyncline, the trough from which these middle Paleozoic mountains were formed, came from the present site of the Arctic Ocean. If the landmass supplying these sediments is reconstructed as a volcanic island arc, then a major subduction zone must have bordered the northern end of the North American lithosphere plate in early Paleozoic time and the Ellesmerian orogeny could be attributed to events associated with the subduction of oceanic crust along this zone. Some have suggested that the landmass north of the Franklin geosyncline in early Paleozoic time belonged to the Siberian lithosphere plate,

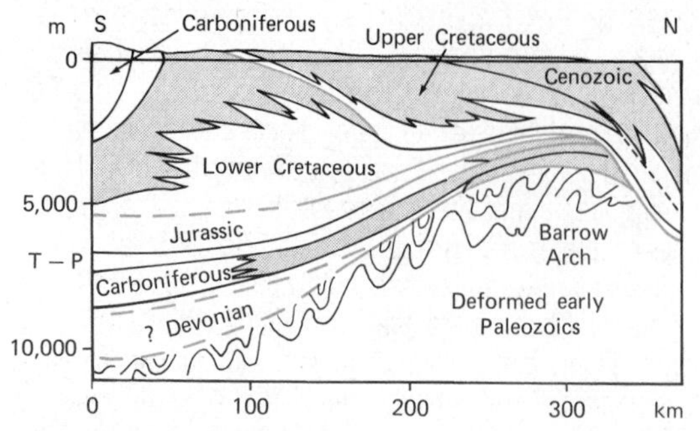

FIGURE 20–9 **Diagrammatic cross-section of the Prudhoe Bay structure. Petroleum is trapped below the unconformity in beds of Mississippian to Jurassic age. The colored pattern represents non-marine beds. (After D. L. Morgridge and W. B. Smith, Jr.; courtesy of the American Association of Petroleum Geologists.)**

and that the Ellesmerian orogeny was the result of the closing of the Arctic Ocean Basin and continental collision. They draw the analogy with the Acadian–Alleghany orogenies and the closure of Iapetus.

Much geological and geophysical work remains to be done around the margins and across the frozen surface of the Arctic Ocean Basin before we can be confident in our interpretations of its pre-Cenozoic history. The recent discovery of large quantities of petroleum in this, one of the last frontiers of exploration geology, will stimulate the search for a fuller understanding of the history of the lithosphere beneath the top of the world.

SUMMARY

The Innuitian Mountain System borders a thin band of platform rocks north of the Canadian Shield and can be traced from Alaska to the north coast of Greenland. The platform area is divided by the Boothia arch, an extension of the shield that was uplifted and eroded several times during the Devonian Period. The Parry Islands and Ellesmere fold belts are underlain by lower Paleozoic miogeoclinal rocks folded and thrust southward in the Ellesmerian orogeny of Late Devonian time. Most of the geosynclinal graywackes and lavas deposited in the Franklin geosyncline north of these fold belts is now hidden beneath younger rocks, but the geosynclinal facies forms the mountainous terrane along the north coast of Ellesmere Island. The deformation of these geosynclinal rocks was also the result of the Ellesmerian orogeny.

The Sverdrup Basin filled with late Paleozoic to Cenozoic sediments overlies and obscures much of the Franklin geosynclinal belt. Early Pennsylvanian salt beds at the base of the section have flowed into salt domes mantled with gypsum and penetrated the stratigraphic section. Late Cretaceous to Recent sediments around the southern borders of the Arctic Ocean form a coastal plain and continental shelf with many features in common with the Gulf Coastal Plain.

Most petroleum is found on the continentward side of miogeoclinal areas where structural and stratigraphic traps are common and thick sections of organic-rich shales are available to act as source beds. One of the largest accumulations of petroleum in North America has been trapped in late Paleozoic and early Mesozoic beds against an arch of early Paleozoic deformed rocks at Prudhoe Bay in northern Alaska. Substantial gas reserves have been proved in the Sverdrup Basin and Mackenzie delta regions.

QUESTIONS

1. Why did geologists expect to find oil in the Arctic Islands region?
2. Compare the Innuitian Mountain system as it is exposed in Bathurst, Melville, and adjacent islands with the Ouachita Mountains of Oklahoma and Arkansas in terms of exposure of the various zones of deformation, time of deformation, type of deformation, stratigraphic record, metamorphism, etc.
3. What climatic indicators are found in

the sediments of the Arctic islands to indicate that the present climate there is not typical of its geological history?

SUGGESTIONS FOR FURTHER READING

AITKEN, J. D., and GLASS, D. J., ed. *Canadian Arctic Geology.* Geological Association of Canada–Canadian Society of Petroleum Geologists, 368 p., 1973.

CHURKIN, M., JR. "Basement Rocks of the Barrow Arch, Alaska, and the Circum-Arctic Paleozoic Mobile Belt." *American Association of Petroleum Geologists Bulletin,* 59: 451–456, 1975.

DAAE, H. D., and RUTGERS, A. C. T. "Geological History of the Northwest Passage." *Bulletin of Canadian Petroleum Geology,* 23: 84–108, 1975.

LERAND, M. "Beaufort Sea, *in* Future Petroleum Provinces of Canada," R. G. McCrossan, ed. *Canadian Society of Petroleum Geologists Memoir 1,* 315–386, 1973.

MORGRIDGE, D. L., and SMITH, W. B. "Geology and Discovery of the Prudhoe Bay Field, eastern Arctic Slope, Alaska." *American Association of Petroleum Geologists Memoir 16,* 489–501, 1972.

PITCHER, M. G. ed. *Arctic Geology.* American Association of Petroleum Geologists Memoir 19, 747 p., 1973.

YORATH, C. J., PARKER, E. R., and GLASS, D. J., ed. "Canada's Continental Margins and Offshore Petroleum Exploration." *Canadian Society of Petroleum Geologists Memoir 4,* 501–700 (Arctic), 1975.

CHAPTER TWENTY-ONE

INTRODUCTION

Several times in the history of the earth large parts of its surface have been covered with ice. Episodes of glaciation in Hadrynian (Chapter 10), Late Ordovician (Chapter 17), and late Paleozoic (Chapter 13) times have already been briefly discussed. In late Cenozoic time the earth entered another glacial period and the 10 per cent of the land surface of the planet that is still covered with ice is evidence that we have not as yet emerged from this latest glaciation. This period of glaciation has been referred to as the "Great Ice Age."

The late Cenozoic glaciation is "great" because the effects of the growth and retreat of the ice sheets have such a large influence on human cultures. In northern latitudes the spread of the ice reshaped the landscape, modified the drainage, and deposited the soils of the land on which we live. In lower latitudes, not covered with ice, the climatic changes of this glacial interval left their impression on the landscape. Changes in sea level as the ice caps grew and shrank have modified the coastlines of the world. The present surface of the earth has been shaped by the erosional and depositional forces of the last half of the Cenozoic Era. Particularly in the northern latitudes, contours of the earth's surface are the products of the multitude of changes in the environment related to the Great Ice Age.

For many years the beginning of this glacial interval was equated with the beginning of the Pleistocene Epoch and was thought to have started about 1 million years ago. The Pleistocene Epoch was not, however, originally defined on the basis of glacial phenomena, but was established, like other Cenozoic epochs, on the basis of a fauna found in a marine section. Lyell's original definition of the Pleistocene on the basis of percentage of living fossils (see Chapter 5) is not practical. Stratigraphers have, therefore, tried to find a marine succession within which the base of the Pleistocene Series could be placed by agreement, so that a type section could be set up as a standard for correlation, as was done for other stratigraphic units. Apparently, the most promising of such sections is in southern Italy, where the change in marine fauna at the base of the Calabrian Stage has been proposed as a basis for defining the base of the Pleistocene Series. Correlation of continental glacial successions, which are generally lacking in marine fossils, with such a type section is difficult and can be based only on climatic interpretation of sedimentary successions or the careful study of intermediate sections where non-marine and marine facies interfinger. However, the use of the term "Pleistocene" for all late Cenozoic glacial deposits in the Northern

THE LATE CENOZOIC ICE AGE

Hemisphere is widespread and likely to persist for many years, until such correlations have been improved.

Recent evidence from Antarctica and mountainous areas has shown that late Cenozoic glacial expansions took place as long as 10 million years ago, during the Miocene Epoch. Although continental ice caps probably did not form in the Northern Hemisphere until millions of years later, the equating of the beginning of the Pleistocene Epoch with the onset of glacial conditions can no longer be maintained, and the Great Ice Age is now considered to have had its origins in mid-Cenozoic climatic cooling.

GROWTH OF GLACIERS

Trends of Late Cenozoic Temperatures

The temperature at which a sedimentary rock was deposited, or a fossil shell secreted, cannot be directly measured. Oxygen isotope measurements (briefly described in Chapter 7) have been extensively used within the last decade to study fossils in sediment cores taken from the deep ocean basins, but this method depends on certain assumptions regarding the isotopic composition of the oceans in past times that have been attacked as unjustified. Faunal, isotopic, and paleomagnetic analysis of such cores help to establish the details of climatic fluctuations in late Cenozoic time. Unfortunately, most of the cores are difficult to relate to the standard polarity reversal scale, and appear to give information about only the last few hundreds of thousands of years of the Cenozoic Era. Climatic information has been obtained from these cores by determining:

1. ratios of species of forams known to be warm or cold water species;
2. the direction of coiling of certain species of forams (*Globigerina truncatulinoides* and *G. pachyderma*) that are sensitive to water temperatures. Specimens living in warm waters are known to coil to the right; those in cold, to the left;
3. changes in the abundance of microorganisms in the core that would reflect changes in productivity of the waters and, hence, shifting climatic zones;
4. the $^{18}O/^{16}O$ ratio of planktonic and/or benthonic forams.

Most cores show many, almost rhythmic, oscillations of ocean temperatures for the last several hundred thousand years, with amplitudes of about 5°C (Fig. 21–1). Correlation between the temperature fluctuations and the late Cenozoic shallow marine and continental–glacial sedimentary successions

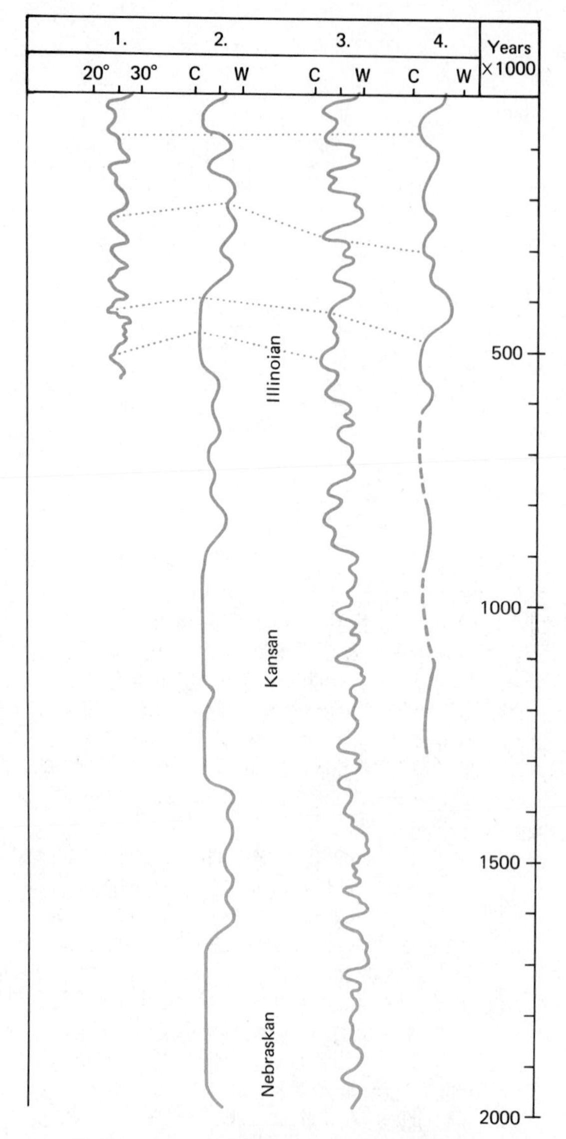

FIGURE 21–1 · **Climatic interpretations of the record of deep-sea cores. The interpretations from four areas have been placed in a time framework, and correlations between the interpretations have been indicated by dotted lines. A possible assignment to the Mississippi Valley glacial succession is also indicated in one curve. 1. Carribean Sea, 2. Atlantic Ocean, 3. Equatorial Atlantic Ocean, 4. sub-Antarctic region. (After H. B. S. Cooke; courtesy of Quaternary Research.)**

continues to be a problem. Similar temperature fluctuations during the last 100,000 years have been deduced from oxygen isotope studies of a core 1390 m long, which penetrated the Greenland ice cap.

Cenozoic plants are so similar to modern plants that estimates can be made of the ecological conditions under which a given fossil flora grew (Fig. 21–2). On the West Coast of the United States a subtropical Eocene flora gave way to a warm temperate one in Oligocene to Miocene time, and a cool temperate one in Pliocene time. The succession of post-Eocene floras in several parts of the world has been interpreted as showing a general slow cooling of the climate, with accelerated periods of cooling in early Miocene and early Pliocene time. These, and other measurements based on organisms, give qualitative measures only of the extent of cooling in late Cenozoic time.

Laurentide Ice Sheet

Snowline is the altitude above which more snow accumulates in the winter than melts in summer. At present this line is near sea level at the poles, and rises to above 5000 to 6000 m at the equator. The decrease in the mean annual temperature and the climatic changes that took place near the end of the Cenozoic Era in the Northern Hemisphere lowered the snowline over a wide area. When the snowline reached an altitude of about 900 m in central Quebec, the broad Laurentide Plateau in this area began to receive more snow in winter than melted in the summer (Fig. 21–3). Permanent snowfields appeared and grew steadily in thickness. Under the weight of the accumulating snow the base of the snowfield recrystallized into ice. When the ice reached a thickness of about 60 m and flowed radially outward under its own weight, a continental ice cap began to spread over the northeastern part of North America. On the north and east outlet glaciers flowing in valleys through the Torngat Mountains soon reached the sea and calved off into icebergs, but in the south the ice front advanced steadily toward the central lowlands of the continent.

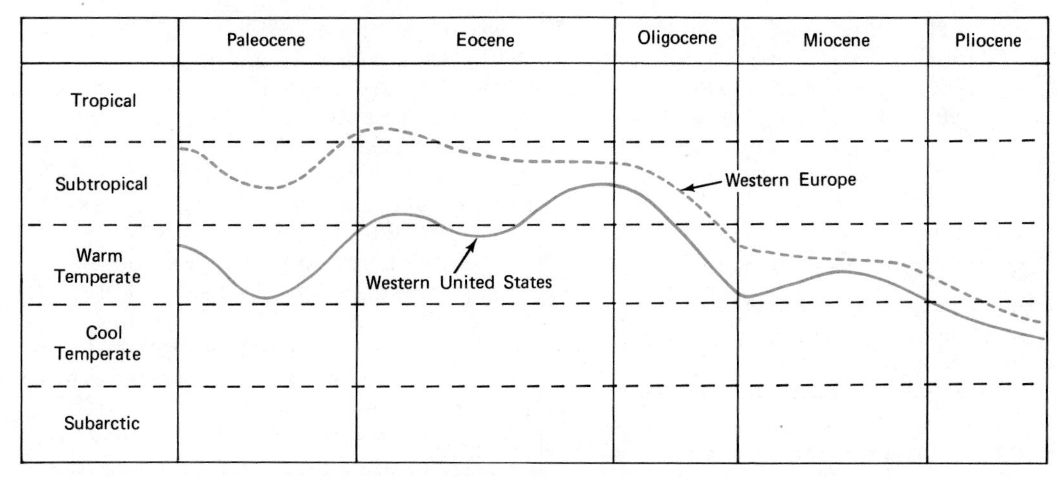

FIGURE 21–2 **Climatic fluctuations during the Cenozoic Era inferred from changes in fossil floras in the United States and Europe. (After E. Dorf; courtesy of John Wiley & Sons.)**

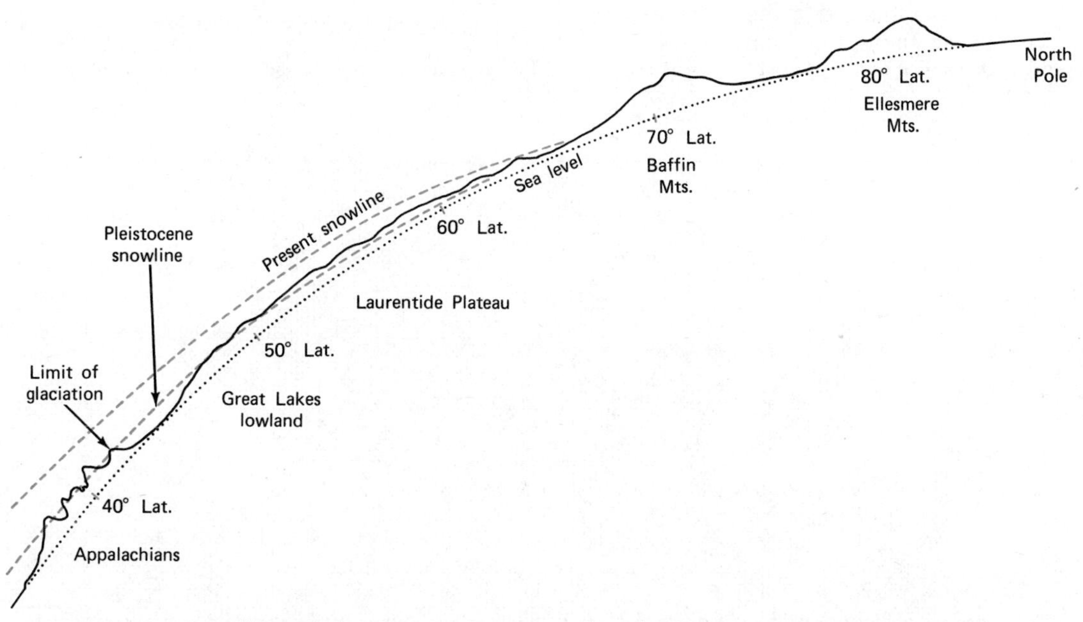

FIGURE 21–3 **Snowline relationships at present and at the formation of the Laurentide ice sheet. A moderate lowering of the snowline placed large areas of the Laurentide Plateau in the zone of snow accumulation and initiated the growth of the continental ice sheet. Relief of the mountains is greatly exaggerated.**

Several factors favor a rapid increase in the size of an ice cap from an embryonic stage. The smooth, white, snow-and-ice surface reflects more of the sun's heat than does darker soil and rock. The reflection of solar energy lowers the temperature of the air above the ice and favors greater expansion and further increase in reflection. By chilling the masses of humid air along its advancing front, the ice sheet induces more storms and increased snowfall along its margins. As more snow is supplied, the ice cap expands, chilling more air and inducing more snowfall. Both atmospheric phenomena act together, creating a rapid increase in the rate of expansion of the ice cap.

Cordilleran Glacier Complex

As the snowline lowered, valley glaciers formed and thickened in the northern Cordillera. In Canada, the glaciers moved down valleys, incorporating tributaries and eventually flowing out of the mountains to calve into the Pacific Ocean in the west and to push out into the plains in the east. Moist air masses from the Pacific that were carried eastward over the icy mountains by westerly winds continued to feed the glaciers with snow, filling the valleys until only a few peaks projected above the gently convex dome of ice, like the nunataks that rim the Greenland ice cap today.

The absence of glaciers in large areas of Alaska and a small part of the Yukon (Fig. 21–4) was apparently the result of a lack of precipitation in these areas. Conditions in which the temperature is low enough for glaciers to form, but the amount of snow falling is not great enough to support their growth, are found today in the Arctic Islands and seem to have existed in the northwestern corner of the continent during late Cenozoic time. Because the glaciers did not extend into the lowlands of the western Yukon, the rich gold placers of Bonanza Creek, the discovery of which led to the famous Klondike Gold Rush in 1898, were preserved from glacial erosion.

The Advance of the Ice Sheet

The advance of the first ice sheet across

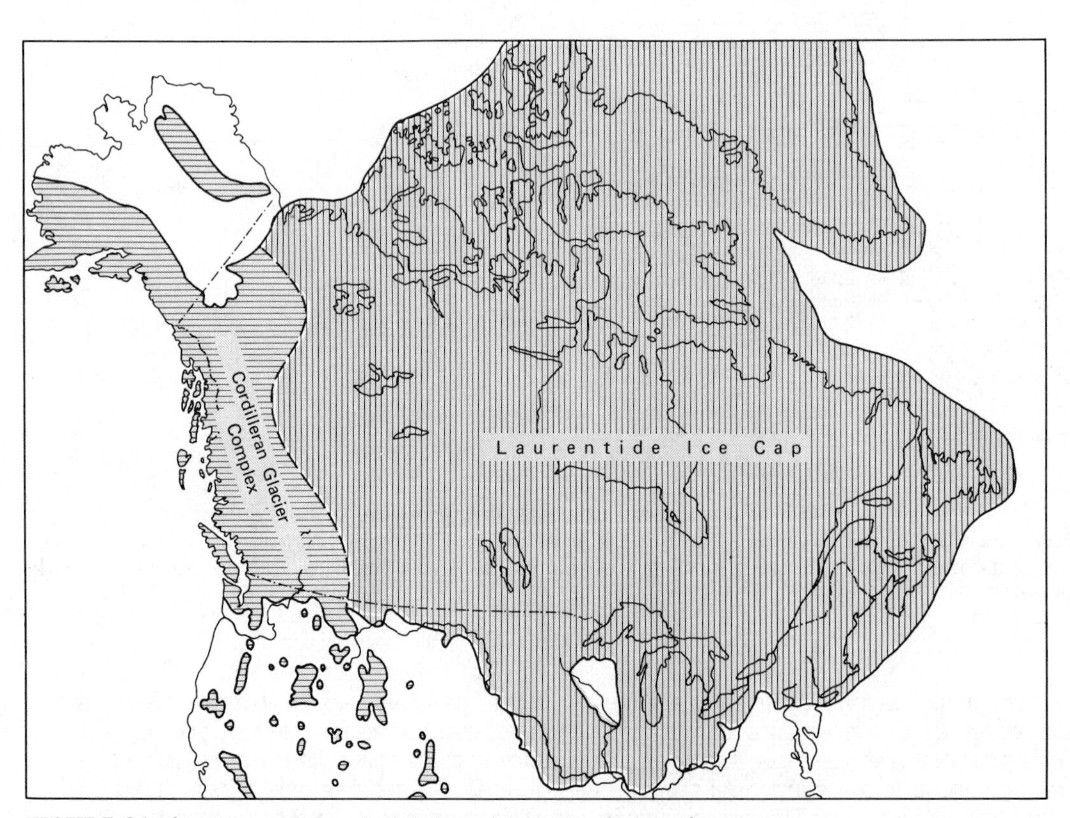

FIGURE 21–4 **Extent of the glacial deposits in North America.**

North America was probably slow and halting. The rate of the first advance is difficult to determine because the record of earliest glacial times has been obliterated or covered by later events. However, radiocarbon determinations (see p. 84) on a group of pinewood samples from a sheet of glacial till in Illinois indicate that the late Pleistocene ice front of 20,000 years ago advanced at about 100 m per year. Although the effects of the advancing ice and the changing climate on the plants and animals of North America were profound, they were slow. Occasionally, the ice advanced rapidly enough to overwhelm forests, and trees were smashed down and incorporated in the till beneath the glacier. The animals did not literally flee before the advancing ice, but followed the environment to which they were adapted as the climatic zones shifted slowly southward. An Arctic fauna moved into the northern United States, and the temperate fauna moved toward the tropics.

Beneath the glaciers, the ice eroded and carried away the soil and scoured the surface of the bedrock. When the ice moved over hills and mountains, it smoothed and rounded the peaks, leaving a subdued topography. Although the erosion of the bedrock of the Canadian Shield was not deep, the impression left by the passage of the ice can be seen in the streamlined form of hills and knobs of resistant rock.

CONDITIONS AT THE GLACIAL MAXIMUM

The Laurentide ice cap expanded until it covered one third of the continent. The position of a glacial terminus is governed by rates of wastage (largely melting and calving into water) and flow. When flow exceeds wastage, the ice front advances; when wastage exceeds flow, the ice front retreats. The front is stationary when wastage and flow are balanced; but like a conveyor belt, the moving ice brings rock debris ("drift"), ranging in size from clay to boulders, with it and deposits it at the terminus as moraines (Fig. 21–5).

When the terminus of the continental glacier in its advance southward reached a climate in which the average rate of wastage at the front exceeded the average rate at which more ice was accumulated by glacial flow, the glacier began to retreat. At every stage of the growth and shrinkage of the ice cap, minor local variations in climate resulted in limited retreats and advances of the lobes. Although an overall picture of glacial advance and retreat appears simple, in detail it was extremely complex.

Extent of Glaciation

At its greatest stage of expansion, the Laurentide ice cap calved into the Atlantic Ocean from the Baffin Sea to Long Island (Fig. 21–4). The ice pushed out along the broad continental shelf east of the Maritime Provinces and New England. Westward from Long Island, the ice margin extended northwestward through central Pennsylvania. The ice flowed as a broad lobe into the lowlands formed by the drainage basin of the Mississippi River and reached St. Louis, as far south as latitude 40°N. West of this lowland, the ice did not push as far south, for the land surface rises slowly toward the mountains. The front of the Laurentide ice sheet had an irregular margin about 150 km south of the 49th parallel, through western North Dakota and Montana.

The Cordilleran glaciers, flowing eastward from the Rockies, met the Laurentide ice cap along the foothills, from the 49th parallel to the Arctic Circle. The margin of the Cordilleran ice in Washington and Idaho was deeply lobate, for the glaciers flowed south from the main ice cap in valleys and lowlands. South of the main ice front, valley glaciers in the major Cordilleran ranges, such as the Sierra Nevada and the Colorado Front Range, expanded into lower altitudes but did not reach the plains.

At the same time, another ice cap covered Scandinavia, northern European Russia, and most of northern Germany. England was also glaciated as far south as the Thames. Northeastern Asia was largely free of ice because, although the climate was cold, snowfall was not adequate to form continental ice sheets. The expanded gla-

FIGURE 21–5 **Moraine on Victoria Island, Arctic Islands. The ice sheet advanced from right to left spreading a thin sheet of till over the area on the right. The moraine marking a temporary position of the ice terminus is the low ridge of irregular topography dotted with kettle lakes which stretches away into the distance on the left side of the photograph. (Courtesy of the Department of Energy, Mines and Resources, Canada.)**

ciers of the Alps came within a few hundred kilometers of joining the Scandinavian ice sheet, advancing from the north, at its maximum extent.

Nature of the Ice Margin

The terminal position of the ice margin in North America is marked by festoons of moraines. These were deposited from the melting glacier at times when a balance between wastage and flow stabilized the ice margin. The moraine system farthest from the glacial center is not necessarily the largest, for the ice may have occupied this advanced position for only a short time.

The front of the continental ice sheet was not a towering cliff; like that of modern glaciers, it rose at an angle of 7 to 9° at the margin and decreased in slope to 2 to 4° within 5 to 10 km. In the central part of the ice cap, the slope of the surface was probably a fraction of a degree. Although the thickness of the Laurentide ice cap at its center is difficult to determine, the heights of mountains overridden near its margins provide some evidence of the thickness of the peripheral ice. In the Appalachian sector the ice must have been more than 2000 m thick to cover such mountains as the Catskills (1280 m), the Adirondacks (1630 m), and the White Mountains (2000 m), all of which show evidence of the passage of the ice over their summits. At its southwestern margin on the plains, the ice cap was probably thinner, but estimates are difficult to make because the relief is less than a few hundred meters. The ice is estimated to have been 450 to 750 m thick along the 49th parallel during the climax of glaciation. Comparisons with modern ice caps suggest that the maximum thickness of the ice at the center of the Laurentide ice cap was about 3 km.

Pluvial Lakes

The worldwide change in climate brought surges of cold, polar air 1500 km farther south than they now reach. The zone of westerly winds, in which the polar and low-latitude air masses interact to produce cyclonic storms, moved nearer the equator. The shift of climatic belts brought more rain to the mid-latitude deserts of the world, such as the Sahara and the Basin and Range province. In these areas, existing lakes expanded during glaciation, and dry, closed basins were occupied by new lakes (Fig. 21–6). Although the late Cenozoic was a time of glaciation in high latitudes, it was a stormy time of increased rainfall in the low and middle latitudes. This period is said to be a pluvial epoch (that is, characterized by rain) in these latitudes, and the lakes that were formed or enlarged are known as **pluvial lakes.**

The largest of the pluvial lakes in North America was Lake Bonneville, which occupied several of the coalescing basins around the present site of Great Salt Lake in Utah. Its maximum extent can be traced in shoreline features, especially the wave-cut terraces that were left high on the basins when the water level dropped. At its highest level, the lake had an area of 50,000 square

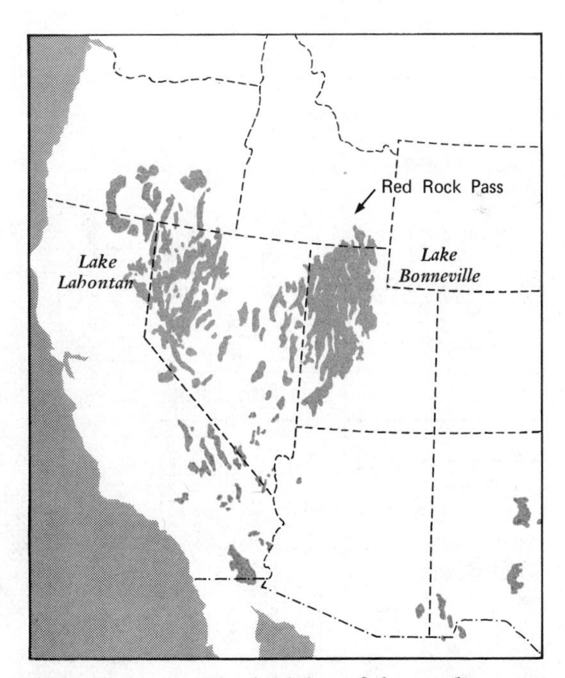

FIGURE 21–6 **Pluvial lakes of the southwestern states. (After W. D. Thornbury, courtesy of John Wiley & Sons, New York.)**

kilometers (similar to Lake Michigan) and a depth of 340 m. The lake level fluctuated with the rainfall, rising during pluvial intervals when the glaciers expanded, and falling when they retreated. At times the lake had no outlet and became saline as the water evaporated. About 30,000 years ago, the water was high enough to overflow through Red Rock Pass in northern Utah into the Snake River Valley and to the Pacific. Huge boulders strewn along the abandoned spillway suggest that a great volume of rapidly moving water once passed through, and geologists deduce that initially the water flowed over an alluvial dam which rapidly gave way, causing a catastrophic flood which moved the boulders down the Snake River Valley. The bedrock sill that now floors the pass is of the same elevation as a prominent shoreline in the lake basin, showing that once the alluvium was swept away and the resistant bedrock exposed in the spillway, the level of the lake was stabilized for some time while the shoreline features were formed. At the close of the last pluvial period Lake Bonneville shrank by evaporation to its present stage, now called the Great Salt Lake.

In northwestern Nevada, the expansion and coalescence of lakes in several intermontane valleys formed Lake Lahontan at the peak of the pluvial epochs. The history of this lake's fluctuations is as complex as that of Lake Bonneville.

STRATIGRAPHY OF THE DRIFT

The debris deposited directly from glacial ice is known as **till.** If meltwater has moved and sorted the debris before deposition, it is called **stratified drift.** Meltwater is almost always present in places where till is being deposited, so that the distinction between these two types of drift is not everywhere sharp. Tills form a highly varied group of

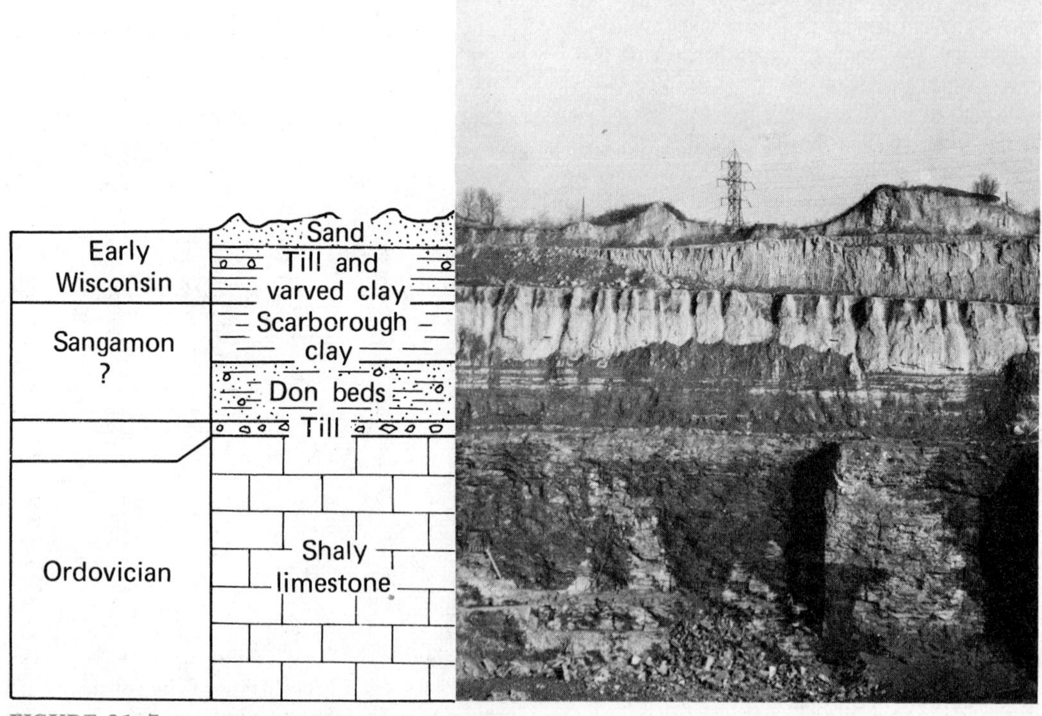

FIGURE 21–7 **Section of the glacial and interglacial sediments at Toronto Ontario. (Courtesy of A. K. Watt, Ontario Department of Mines.)**

sediments. They may be rich in boulders set in a clay matrix, or almost entirely clay with a few pebbles, or composed entirely of silt, sand, or boulders.

Soon after geologists recognized the drift of northern Europe and North America to be of glacial origin, they discovered that in many exposures layers of till are separated by sediment deposited in a warm climate. Such exposures could only be interpreted as indicating that the glaciers advanced more than once and that non-glacial sediments were deposited during intervals between these advances.

The Toronto Interglacial Section

The nature of interglacial sediments can be illustrated by exposures at Toronto, Ontario. The beds of this succession are shown in Fig. 21–7. Two main till horizons occur, one above the Ordovician bedrock and the other at the top of the section. Between these tills are a series of clays, sands, gravels, and peat beds overlain by a thick series of banded clays with till zones at three horizons. The lower beds of the interglacial sequence (the Don beds) contain a remarkable assemblage of interglacial fossils of trees, freshwater snails, clams, and mammals. The vertebrate fossils include the bones of woodchuck, deer, bison, bear, and an extinct giant beaver. Study of these fossils has led paleontologists to the conclusion that the plants and animals lived in a climate 3°C warmer than the present climate of Toronto. The Scarborough beds, above this lower assemblage, contain 14 species of trees and 72 species of beetles, which indicate a climate slightly cooler than the present one.

These fossils provide a key to the problem of the extent of withdrawal of the ice sheet during interglacial times. Because the fossils of the Don beds indicate a climate warmer than the present one at Toronto, the climate all over the continent was probably warmer and, therefore, less permanent ice was present in North America during the interglacial age represented by the Don beds than there is today. The interglacial succession at Toronto is proof that the ice cap melted and was reconstituted at least once in Pleistocene time, and other sections of similar stratigraphy record several other interglacial episodes and additional ice advances.

Interglacial Sediments

The interglacial sediments at Toronto were laid down and buried in a lacustrine basin that protected them from the erosion of the next ice advance. They consist of lake clays and sands, peat, and layered or varved clays. Trees fell into the lake from the margins, or were carried to it by streams, and sank to the bottom when they became waterlogged. The shells of freshwater clams and snails accumulated in the mud and sand, and insects that fell into or lived in the water were covered by layers of mud and preserved.

Peat is a common interglacial and postglacial deposit found over wide areas in the higher latitudes. It is formed in bogs, under cool climatic conditions, by the the partial decomposition of vegetable matter, such as moss, sedges, trees, and leaves. Although the larger parts of plants are difficult to identify botanically in the tangled mass of decayed vegetation, pollen grains can be identified in peat deposits. A record of the peat strata in a bog can be obtained by forcing a tube through the dead vegetation and retrieving the core in the tube. The pollen grains are extracted from the core and identified under the microscope. Because we know the conditions in which many of the plant species grow today, changes in climate during the accumulation of the peat bog can be estimated from the changing concentrations of pollen in the peat and the proportions of the various types of pollen along the length of the core (Figs. 21–8 and 21–9).

The laminated clays found in interglacial deposits consist of alternating dark and light layers. The light layers are coarse-grained and silty or sandy; the dark layers are fine-grained and contain more clay. Each pair is the deposit of a single year, and is called a **varve.** The varved clays were laid down in lakes receiving glacial

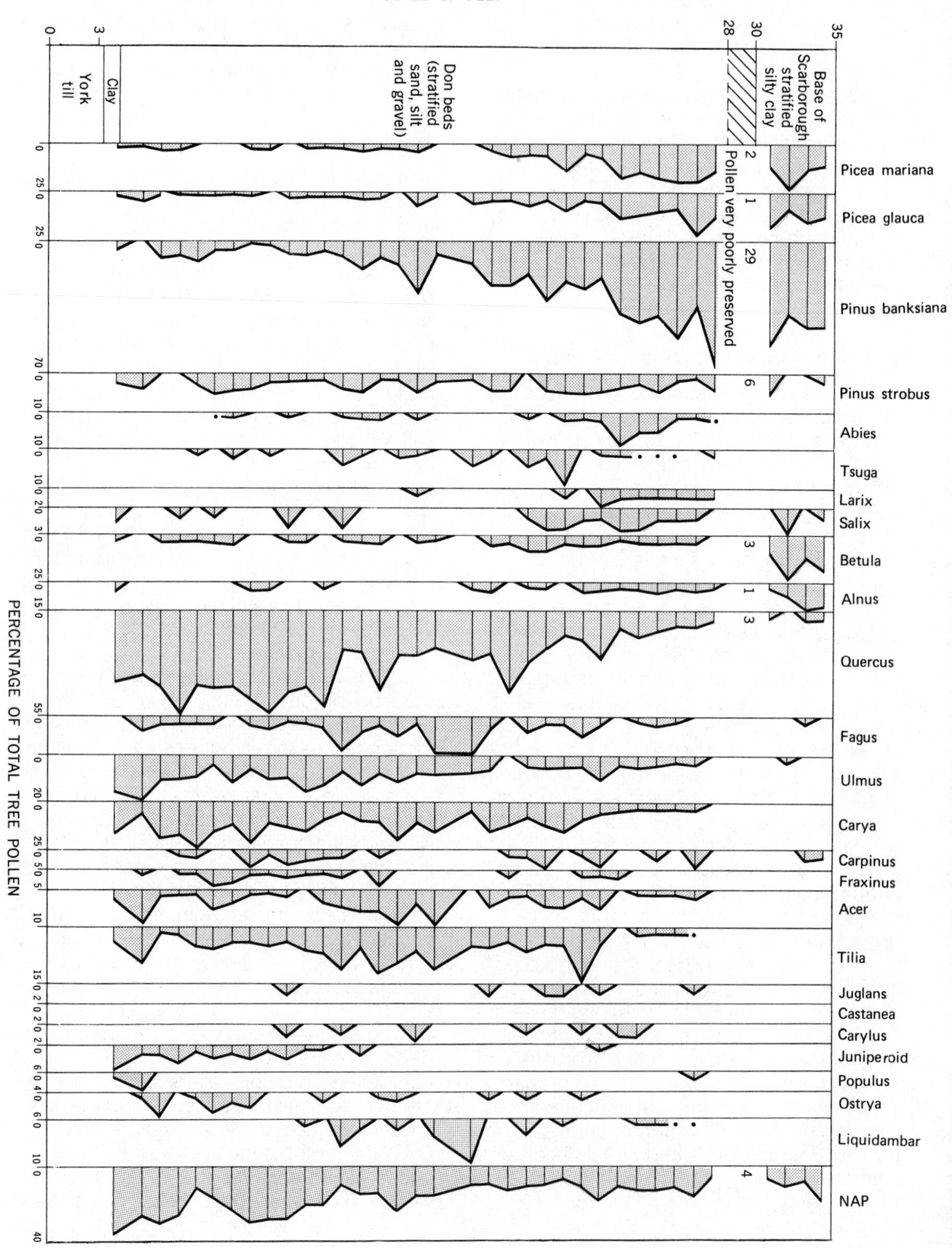

SCALE OF FEET
0
3
28
30
35
York till
Clay
Don beds (stratified sand, silt and gravel)
Base of Scarborough stratified silty clay
Pollen very poorly preserved
PERCENTAGE OF TOTAL TREE POLLEN
Picea mariana
Picea glauca
Pinus banksiana
Pinus strobus
Abies
Tsuga
Larix
Salix
Betula
Alnus
Quercus
Fagus
Ulmus
Carya
Carpinus
Fraxinus
Acer
Tilia
Juglans
Castanea
Carylus
Juniperoid
Populus
Ostrya
Liquidambar
NAP

FIGURE 21–8 **Pollen diagram of the Don Valley Brickyard section at Toronto. The changing proportion of the various kinds of tree pollen through the Don beds and into the Scarborough beds is shown by the graphs. NAP is the non-arboreal (not tree) pollen. (From J. Terasmae; courtesy of the Geological Survey of Canada.)**

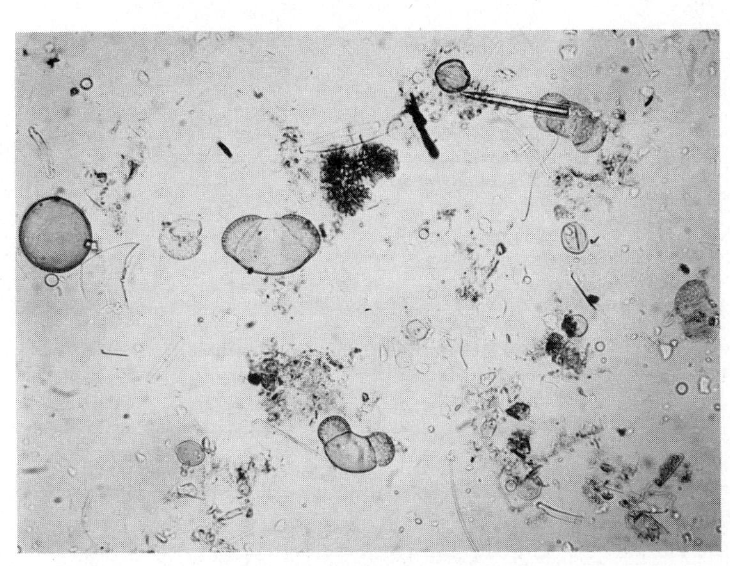

FIGURE 21–9 **Photomicrograph of pollen grains from Pleistocene lake sediments, ×200. (From J. Terasmae; courtesy of the Geological Survey of Canada.)**

FIGURE 21–10 **Varved clays of the interglacial section at Toronto. (Courtesy of P. F. Karrow, Ontario Department of Mines.)**

meltwater. In the spring thaw and in the early summer, the streams brought much silt and clay to the lakes as turbid flows. The coarse silt settled quickly to form a light layer, while the fine clay settled more slowly, largely in the autumn or in the winter, when the lake surface was frozen, to form a dark layer (Fig. 21–10). The intercalation of thin till layers with the varved clays at the top of the Toronto succession indicates that the ice advanced several times over the glacial lake in which the clay was being deposited.

Since each varve represents a yearly deposit, the rate of sedimentation in these glacial lakes can be determined by counting the layers and measuring their thickness. In Sweden, the elapsed time since the ice withdrew from certain moraines has been estimated by counting the number of varves in the lake deposits formed during the withdrawal of the ice. In North America this method had not been as useful, for the sedimentary record preserved by glacial lakes is less complete. A silty unconsolidated sediment called **loess** occurs between and above till sheets in the Mississippi Valley. It is characterized by silt-sized particles, lack of stratification, and blocky, vertical fracture that allows it to stand in almost vertical banks. Loess is thought to be deposited primarily by the wind from the fine sediment blown out of outwash plains. The loess of North America was derived from silt spread out by meltwater streams in the floodplains of rivers, draining the ice margin directly, and dried by the sun. Loess from the last glaciation now mantles much of the central United States from the Rocky Mountains to Pennsylvania. Loess is not indicative of a specific climate, because only moderately strong winds and a continually renewed supply of dry silt are necessary for its formation. Land snails, small vertebrates, and other fossils contained in loess provide data on past environments.

Weathered Zones and Soils

Conditions during interglacial times are revealed by the soils developed by weathering of till, interglacial loess, or clay after the ice withdrew. The weathering of this drift proceeds in a regular manner. First, the colorless iron compounds are oxidized to red and brown ferric compounds. Because oxidation is the most rapid weathering process, it penetrates deepest into the drift; therefore, the oxidized zone is the thickest in the soil profile (Fig. 21–11). The next process is the dissolution, or leaching, by groundwater of the finely divided carbonate grains or limestone pebbles in the drift. Hence, leached and oxidized drift form an intermediate zone in the soil profile. In the last phase of weathering the resistant silicate minerals are decomposed chemically to clay minerals.

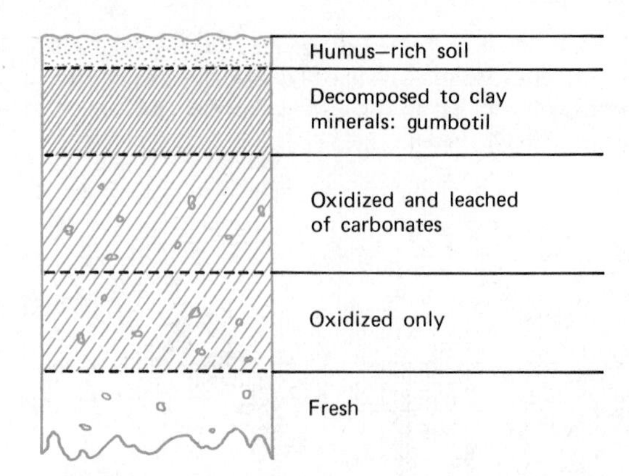

FIGURE 21–11 **Diagram of soil profile developed in till.**

In the standard soil profile shown in Fig. 21–11, the upper humus-rich layer is the "A" horizon, the decomposed zone is the "B" horizon, and the oxidized and leached zones are the "C" horizons. This soil profile may be completely or only partially developed and the soil zones may be thick or thin, depending on the length of time the drift was exposed to weathering agents, the climate under which the weathering took place, and the composition and texture of the drift from which the soil developed. The extent of weathering in interglacial soils has been used to estimate the relative lengths of the interglacial intervals.

Glacial and Interglacial Ages

In only a few localities can more than two till sheets be identified, but a composite

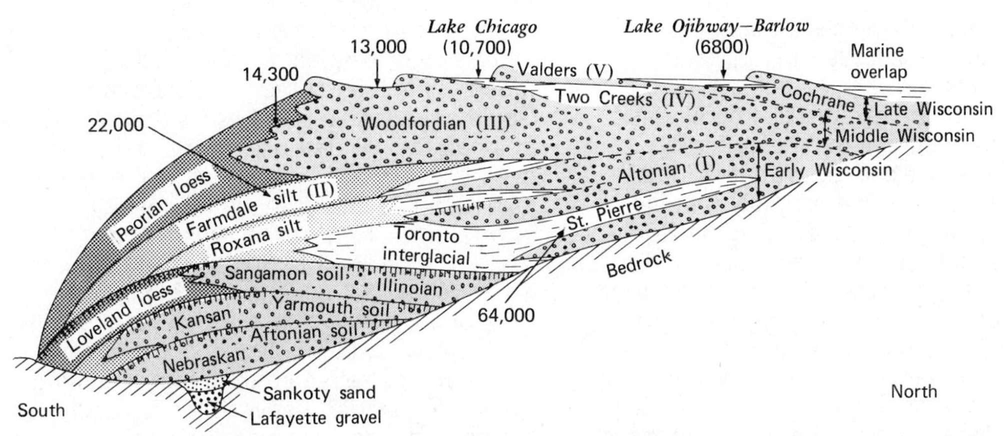

FIGURE 21–12 **Schematic cross-section showing the glacial sequence in the Great Lakes region of North America. Note the interbedding of glacial and interglacial deposits and the position of the loess formations. The section is about 1600 km long and thicknesses are immensely exaggerated. (Prepared by J. A. Elson.)**

section, such as that of Fig. 21–12, can be built up by correlating separate outcrops. From such correlations geologists have discovered that at least four major advances of the ice occurred in central North America in the late Cenozoic, and that three interglacial periods intervened during which the continent was largely free of ice. The advances took place during the Nebraskan, Kansan, Illinoian, and Wisconsin glacial ages, and were separated by the Aftonian, Yarmouth, and Sangamon interglacial ages, respectively.

The Wisconsin drift, representing the last major ice advance, covers most of the glaciated parts of North America, but in the Mississippi Valley the Wisconsin advance was less extensive than that of earlier glaciations, leaving Kansan and Illinoian tills exposed. The Nebraskan, Kansan, and Illinoian till can be studied in detail only in the Mississippi lowlands, for farther north the fourth advance of the ice (Wisconsin) eroded and incorporated the older drift into its own deposits. Because the Wisconsin drift is extensively exposed and has not been disturbed by the advance of later ice sheets, regional study of its stratigraphy and morphology has been possible. In the upper Mississippi Valley, five divisions of the Wisconsin drift have been recognized (Fig. 21–12). The persistent Farmdale Silt (II) separates a lower group of several Altonian (I) tills from the upper Woodfordian (III) Group. Woodfordian tills are separated from the Valders (V) Till by an intensely studied deposit, the Two Creeks bed (IV). During the Wisconsin glaciation, retreats were local, covering perhaps a few hundred kilometers; they do not involve the disappearance of the ice from the continent, as the interglacial deposits indicate.

Although the Mississippi Valley glacial sequence can be divided naturally into four major episodes of continental glaciation and three intervening interglacial episodes, evidence from the Cordillera and from Europe suggests that at least eight major advances took place in the late Cenozoic glacial interval. Although the early glaciations are generally poorly dated, the best information available now suggests that they began about 3 million years ago. Continental glacial deposits of these earlier glacial episodes in North America could have been destroyed by the advance of the Nebraskan glaciers, just as the Wisconsin advance wiped out much of the evidence of pre-Wisconsin glaciation, except where it was less extensive than its predecessors. The climatic fluctuations indicated by the analysis of deep-sea cores has also been interpreted as representing many more glacial advances than are recorded in glacial tills on the continents.

The sum of the evidence of continental glacial and deep-sea marine stratigraphies indicates that the cooling climate of late Cenozoic time was first expressed in the advance of mountain glaciers in the Northern Hemisphere and the expansion of the Antarctic ice in the Southern Hemisphere in late Miocene time, about 10 million years ago. By 2.5 to 3 million years ago the temperature had reached the point where large ice caps were forming in Canada and Scandinavia. Since that time the growth of these ice caps has repeatedly extended glacial ice into the middle latitudes. The intervening retreats and disappearances of the ice sheets were responses to climates as warm as, or warmer than, today's. Intuitively, we believe that the major retreats and advances in Europe were synchronous with those in North America, but their correlation is only well documented for the last, or Wisconsin, glaciation in North America and the Wurm glaciation of Europe.

DEGLACIATION

Although the continental ice sheet retreated, thinned, and eventually melted away several times in the last 3 million years, very little is known about the early episodes of deglaciation because the succeeding glacial advance redistributed the deposits laid down by the retreating ice. Much is known about the retreat of the Wisconsin ice sheet because the record of this retreat is written in till and stratified drift spread across the United States and Canada.

About 18,000 years ago the mean annual temperature in the northern latitudes began to rise. The melting of the ice sheet quickened, and the area covered by ice diminished. The regional snowline rose. The retreat of the ice front was an irregular process interrupted by many minor readvances.

Ice-Margin Lakes

Deglaciation in North America involved the melting of a mass of ice 15 million square kilometers in area and possibly 3 km thick at the center. The amount of water released (about 33 million cubic kilometers) certainly would have resulted in worldwide catastrophe had the melting occurred rapidly. However, it took place over an interval of more than 10,000 years. Although the melting lasted for a long time, the rivers flowing south still carried prodigious quantities of water to the sea, because the ice blocked most of the northerly drainage systems until deglaciation was practically complete.

As the ice retreated from higher ground, water was ponded between the ice and the drainage divide, forming giant ice-margin or proglacial lakes. Most proglacial lakes discharged southward, but some discharged along the margin of the ice front. As the glacier retreated from these ice-margin lakes, it uncovered successively lower outlets, and the lake levels dropped to the altitude of the new thresholds. Some outlets were abandoned when the isostatic rebound of the crust caused by the ice recession (see later discussion) raised the threshold, and with it the water level, until the lake overflowed through a formerly abandoned spillway. In other lakes erosion of the outlet changed the level of the water. Each ice-margin lake had many changes in level, and at each level shoreline features, such as cliffs, beaches, and deltas, were formed.

Some of these lakes were drained completely when the retreating ice uncovered an outlet lower than the floor of the lake basin; others still exist as remnants of the ice-margin bodies of water, even though the ice no longer forms their northern shore. We can recognize the past extent of these lakes by their abandoned beaches and lake-bottom deposits of sand, silt, and clay.

Lake Agassiz was the largest of the proglacial lakes in North America. Meltwater began to collect when the Laurentide ice sheet uncovered the basin of the Red River of the North. The deposits of Lake Agassiz extend from north-central Manitoba to northern South Dakota, and from near Lake Nipigon to central Saskatchewan (Fig. 21–13). The "first" Lake Agassiz was formed at the end of the Woodfordian Stage of the Wisconsin glaciation, and drained southward

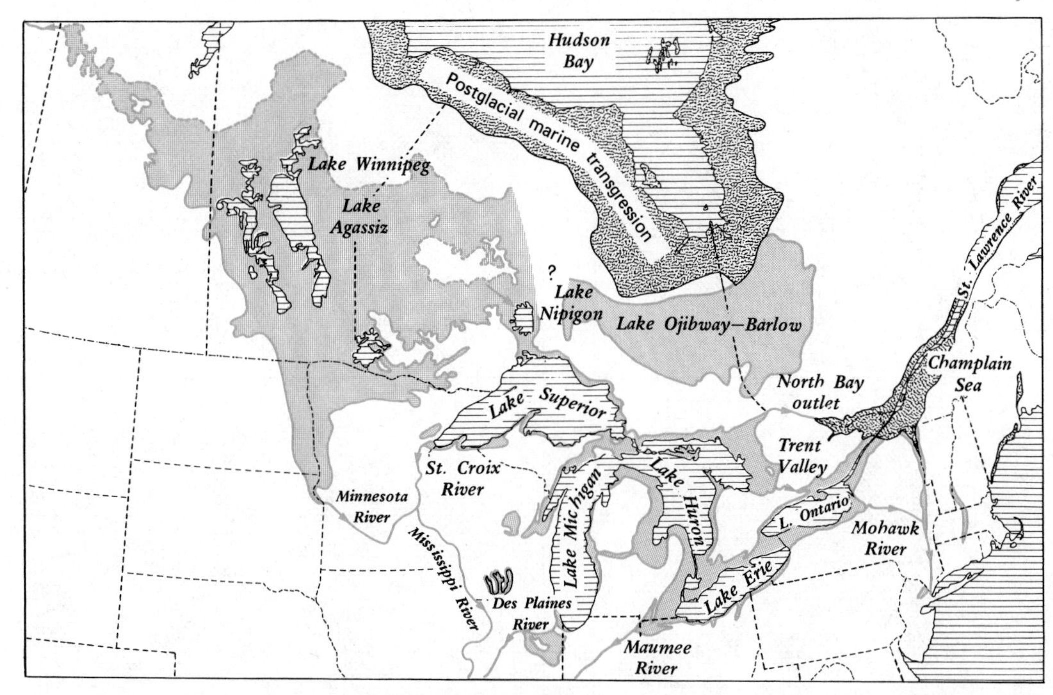

FIGURE 21–13 **Major glacial lakes of the Great Lakes area showing the major spillways through which the lakes drained. The shaded areas along the shores of the Great Lakes represent the areas flooded by the predecessors of the lakes. (Outline of Lake Agassiz after J. A. Elson.)**

through the Mississippi River system. As the ice retreated north of Lake Superior, the glacial lake drained eastward. The "second" lake was initiated by a readvance that blocked the eastern outlets. This lake originally discharged through the Minnesota River to the Mississippi Valley, but as the ice again retreated, the water was again discharged eastward into the Superior Basin. A third phase began when the ice readvanced to the Nipigon moraine. Even lower outlets were subsequently opened at the retreating ice margin and Lake Agassiz was drained,leaving only the present Lakes Winnipeg, Winnipegosis, and Manitoba.

The Great Lakes

Before the first ice sheet advanced over the central lowlands of the continent, the Great Lakes region was the drainage basin of a large river. The Laurentide glacier deepened this lowland by erosion, scoured the five lake basins, and depressed this segment of the crust under several kilometers of ice.

As soon as the ice of late Wisconsin time had melted north of the Mississippi watershed, water was ponded in the Lake Michigan and Lake Erie basins. The two resulting lakes, Lake Chicago and Lake Maumee, drained southward into the Mississippi River through the valleys of the Maumee and Wabash rivers, and through the valley of the Des Plaines River south of Chicago (Fig. 21–13). Early in this period of deglaciation, water was ponded in the Finger Lakes basins of central New York and drained southward through the valley of the Susquehanna River. Another spillway used by the ancestor of Lake Erie led eastward along the margin of the ice past Syracuse to the Mohawk Valley, the Hudson River, and the sea.

When the water first accumulated in the Superior Basin, it was separated from the rest of the lakes by ice extending south into the north end of Lake Michigan, and

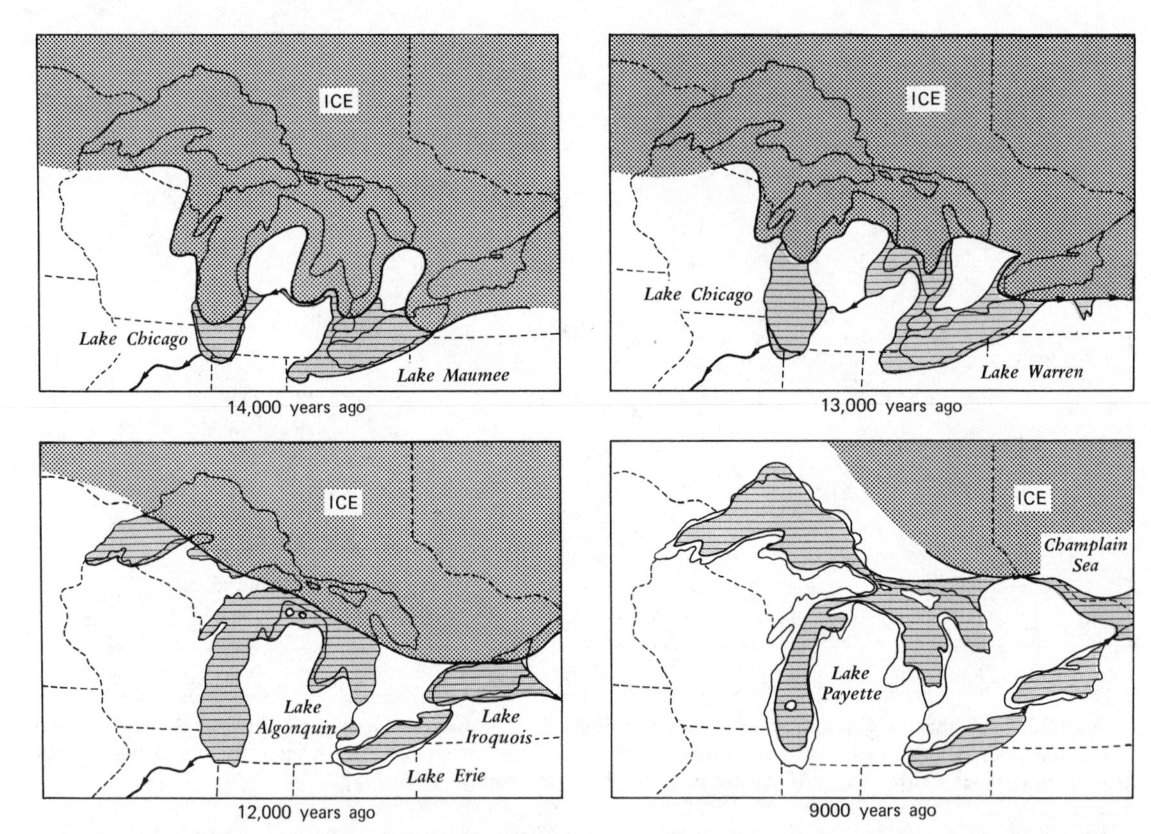

FIGURE 21–14 **Four stages in the complex development of the Great Lakes during deglaciation. Note that when the outlet was higher, the lakes were larger than at present; when the outlet was lower, as in the last map, they were smaller than at present. (After J. L. Hough; courtesy of the *American Scientist*.)**

drained southward through the St. Croix Valley to the Mississippi system. The ancestor of Lake Ontario, Lake Iroquois, drained into the Mohawk Valley at Rome, New York. At one time the Great Lakes drained from Georgian Bay across the Trent Valley outlet into Lake Ontario. As the ice margin retreated the channel moved to a position through North Bay and the Ottawa River (Fig. 21–14). The threshold of the North Bay Channel was so low when it was uncovered that nearly all the water in the upper lakes drained through it to the sea. The threshold was gradually raised to its present altitude of about 200 m by the isostatic rebound of the crust, and the lake basins again were filled and discharged southward through the Mississippi Valley. As the ice withdrew from the St. Lawrence Valley, the area was first occupied by a glacial lake and then by an embayment of the ocean called the Champlain Sea, which reached almost as far as Lake Ontario. The present discharge of the Great Lakes down the St. Lawrence River began only a few thousand years ago as the Champlain Sea retreated.

Meltwater Channels

Many of the valleys that extend southward from the glaciated part of the continent are much too large to have been cut by the small streams that now occupy them. These streams are said to be "underfit," because they do not fit the valleys they occupy. Other great channels that once led from moraines or from the abandoned shorelines of glacial lakes are now dry. From the gradients and cross-profiles of these valleys, geologists have deduced that they were chan-

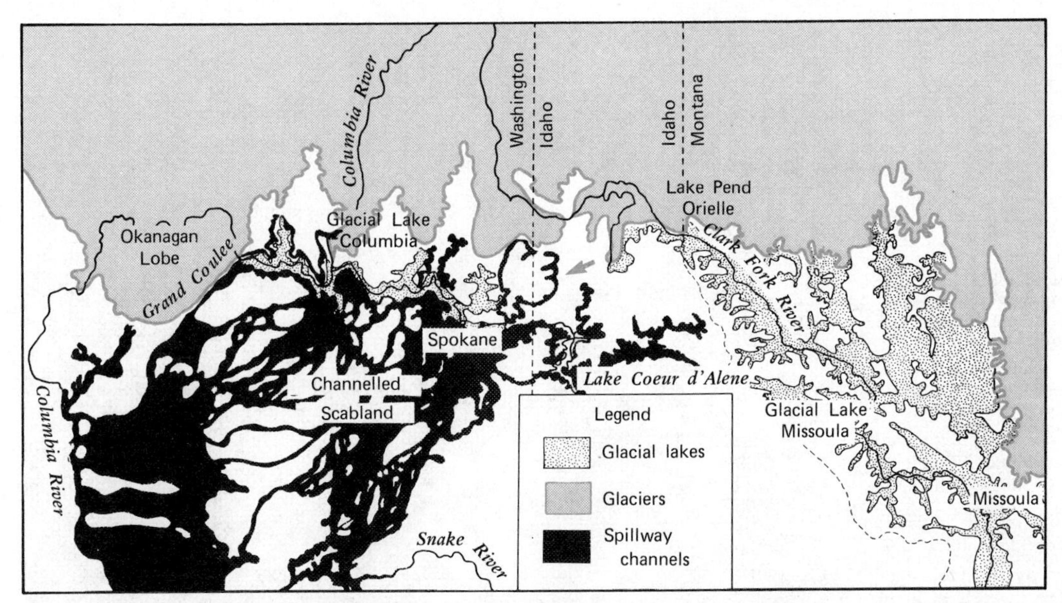

FIGURE 21–15 **Map of Glacial Lakes Missoula and Columbia and the Channeled Scabland of Washington. The arrow in the center represents the direction of the main discharge when the ice dam broke at Lake Pend Oreille. (Modified from G. M. Richmond and others.)**

nels along which great volumes of meltwater reached the sea. Near the ice margin where the water was loaded with debris, some streams built up the valley floors with thick deposits of sand and gravel. As the ice retreated and the sediment supply was trapped upstream in glacial lakes, the streams cut into the valley fillings and left terraces which, in many places, are now hundreds of feet above the valley floors.

The most striking meltwater channels are found in the middle of Washington State, in an area known as the Channeled Scabland (Fig. 16–18). A series of ice lobes extended southward along major north–south valleys at the edge of the Cordilleran glacier complex in northern Washington and Idaho, and dammed the drainage to the Pacific of such rivers as the Clark Fork, Spokane, and Columbia. Behind these lobes great volumes of meltwater were impounded in Glacial Lakes Columbia, Spokane, Coeur d'Alene, and Missoula (Fig. 21–15). Glacial Lake Missoula, in the valley of the Clark Fork River, was the largest of these, and held 2000 cubic kilometers of water. Some evidence suggests that the ice dam which held back glacial Lake Missoula formed and collapsed several times before the last catastrophic event about 13,500 years ago. With the disintegration of the ice dam, water rushed down the valleys of the Spokane and Columbia rivers, spreading coarse debris that marked its passage high on the valley walls. From the height of the flood deposits, the size of the boulders in them, and a cross-section of the valley, the maximum flow of the flood has been estimated at 40 cubic kilometers an hour. Where the path of the flood was blocked across the Columbia Valley by the Okanagan ice lobe, the water spread southward across the Columbia Plateau and cut a complex of anastomosing, broad channels, called coulees, through the loess soil and into the resistant basalt. The discharge along the margin of the Okanagan lobe eroded the Grand Coulee, spilling over the famous Dry Falls in a vast cataract (Fig. 21–16). Through these channels the water found its way back into the valleys of the lower Columbia and Snake Rivers, and eventually flowed out to the sea. In spite of an attachment to the principle of uniformitarianism, the geologist should not overlook

FIGURE 21–16 **Dry Falls in the Grand Coulee area, Washington. The aerial view is toward the north and shows the flat-lying lava flows of the Columbia Plateau, the lip of the Dry Falls, the plunge pool now occupied by Perch Lake, and the Grand Coulee, down which the diverted Columbia River flowed in the background. (Official U.S. Bureau of Reclamation photograph.)**

the occurrence of such "catastrophic" events in the history of the earth. The similarity of the topography of the Channeled Scabland to that of parts of the planet Mars supports the hypothesis that catastrophic waterfloods occurred in the history of that planet.

Sea Level Changes

When temperatures fell and the ice age began, precipitation falling on the land in the northern latitudes was not returned to the sea by streams, because it accumulated on the ground as snow and ice. As more and more water that had evaporated from the oceans was trapped in ice and snow on the continents, the amount of water in the ocean basins decreased. Sea level continued to fall until about 50 million cubic kilometers of ice had built up on the continents. When glacial ice reached its greatest extent and thickness, sea level had fallen about 105 m below its present position, and much of the continental shelf was exposed. Shoreline features—cliffs, beaches, river valleys, and sand dunes—were formed at this low water stage, but have since been inundated. The fall of sea level affected not only the northern latitudes but all the coastlines of the world. Sediments deposited in deeper water before the fall of sea level were churned up when the zone of wave motion reached to the top of the continental slope. Land bridges emerged from the receding waters, and animals from formerly isolated lands were able to cross and mix. For North America, the most important of these land bridges crossed what is now the Bering Sea. Siberian and Alaskan mammals were interchanged by means of this corridor, and human migrants reached the New World by this route.

As temperatures rose and the ice melted, water was returned to the sea, and its level rose. This fall and rise of the oceans must have taken place many times during the late Cenozoic Era. Fossils show that the climate of some of the interglacial ages was warmer than the present climate, and we infer from this that less ice existed on earth then than now. During warm interglacial ages, sea level was higher than at present, and deltaic deposits were laid down by rivers along these higher shorelines. Along the Gulf of Mexico these Pleistocene deltaic formations have been mapped and correlated with terraces cut in river valleys farther inland. In southwestern Louisiana where these deposits are best developed, four formations have been recognized in the Pleistocene sequence. Along the Atlantic Coastal Plain south of New Jersey, cliffs cut by the waves during interglacial times (when sea level was higher) can be traced parallel to the coast for 1300 km.

Climates and glaciers have fluctuated many times since the last Wisconsin glaciation 10,000 years ago, and the earth has been both warmer and cooler than at present. If all the ice in Greenland, the Arctic, and the Antarctic were to melt, sea level would rise almost 30 m, and the great seaport cities of the world would face serious flooding problems.

Isostatic Rebound

Geophysical measurements and comparisons with modern ice caps indicate that the ice was 3 km thick beneath the center of the Laurentide ice cap during the glacial maxima. How far the weight of such a large thickness of ice depressed the crust can be easily calculated from the relative densities of ice and of the subcrustal material displaced, assuming that isostatic adjustment to the load was perfect. Since the ratio of the two densities is 1:3, the crust was probably depressed about 1 km beneath the thickest ice. Geological evidence to confirm this, however, is hard to obtain. The plastic substratum below the crust must have flowed laterally to make way for the downbulging crust. As the ice melted and the load was removed, this material flowed back slowly at depth, and the crust rebounded upward toward its former position, restoring isostatic equilibrium. Although the ice melted relatively quickly the response of the earth was much slower, and rebound continues to this day.

This postglacial recovery is recorded in the raising and warping of glacial and postglacial marine and lacustrine shoreline features. Marine beaches in the Arctic 300 m above sea level could not have reached their position by a fall of sea level, for evidence from the unglaciated parts of the world shows that the sea level has risen during the last 20,000 years (Fig. 21–17). The beaches, therefore, must have been elevated by the isostatic rebound of the land as the ice load melted. They record the uplift only since the ice disappeared from the Arctic region and the sea came in. They do not record how much rebound accompanied the thinning of the ice sheet. Radiocarbon dating of these beaches shows that the rate of uplift was rapid immediately after the load was removed, but has been steadily decreasing since that time. Tide-gauge reading that have recorded the position of the land with respect to mean sea level over many years can also be used to prove the slow rise of the land within the glaciated areas.

Geologists have found that shoreline features formed in proglacial lakes, such as cliffs and beach ridges, slope downward away from the glaciated area, but are horizontal farther from the old glacial margin (Fig. 21–18). In addition, they have found that the beaches formed in the older lakes were tilted more than those of the younger lakes. The tilted shoreline features must have been horizontal when they were formed at the surface of a proglacial lake, and their tilting can be explained by the rebound of the land from its ice load. The line along which this tilting begins is called the hinge line, and its position is related to the margin of the ice at the time the shoreline features are formed. As the ice recedes, the

FIGURE 21–17 **Raised beaches on the shore of the Arctic Ocean, at Victoria Island. (Courtesy of J. G. Fyles, Geological Survey of Canada, Ottawa.)**

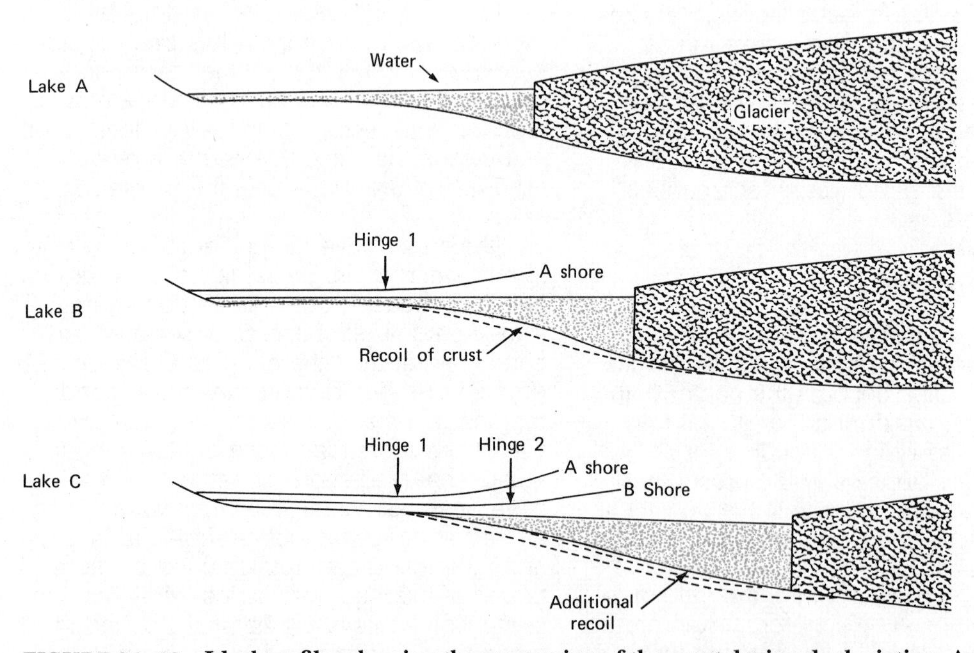

FIGURE 21–18 **Ideal profiles showing the upwarping of the crust during deglaciation. As lakes form at lower levels in successive stages, the beaches of former lakes are abandoned and bent upward by the rebound of the crust as the load is removed. (Courtesy of R. F. Flint.)**

hinge lines of the beaches of successively younger lakes are formed nearer to the glacial center (Fig. 21–18).

The Ice Age has supplied us with an illuminating experiment in the rate of response of the crust to large stresses and the workings of isostasy.

CAUSES OF GLACIATION

Many theories have been proposed to explain major climatic changes in earth history, some of which have been considered briefly in Chapter 7. Most have been specifically designed to account for the late Cenozoic Ice Age, but any acceptable theory should also be applicable to pre-Cenozoic glaciations. In addition to explaining why glacial ice covered large parts of the land area of the earth, such a theory should also account for the climatic fluctuations that caused the growth and retreat of the glaciers many times. Most, but not all, of the theories attribute intervals of glaciation to a general decline in world temperatures. Theories that involve only an increase in precipitation have been shown to be inadequate to explain the world pattern of late Cenozoic glaciation.

The major theories fall into six classes:

1. Changes in the emission of radiation by the sun.
2. Blockage of the radiation of the sun by cosmic dust.
3. Blockage of the radiation of the sun by changes in the earth's atmosphere (carbon dioxide, ozone, volcanic ash, etc.)
4. Variation in the distribution of solar radiation over the surface of the earth because of periodic changes in the astronomic position of the earth with respect to the sun.
5. Changes in the circulation pattern of oceans and atmosphere.
6. Lateral movements of the continents with respect to the poles, or general elevation of the continents.

Recent interest in the plate tectonics approach to world structure and history has focused attention on the last of these hypotheses. Paleomagnetic evidence has been used to suggest that glaciation may start when the polar regions reach positions with respect to the configuration of the continents and oceans that are thermally isolated. This theory maintains that when the polar regions are located in large oceanic areas, like the Pacific Ocean, extreme climatic zonation cannot develop, and world climates are equitable and mild. When the poles are located over large continental areas, such as Antarctica, or over largely enclosed marine basins, such as the Arctic Ocean, climatic zonation is intensified and glacial conditions spread from the polar regions to the mid-latitudes of the world. The location of the South Pole over the world continent, Pangaea, in late Paleozoic time is believed to be responsible for the glaciation of that interval.

The theory does not explain the fluctuating nature of the glacial advances. Maurice Ewing and William Donn suggested the repeated freezing and thawing of the Arctic Ocean to account for these. They believe that the late Wisconsin retreat of the glaciers was caused by the freezing of the Arctic Ocean, which prevented it from supplying Arctic Canada with the moisture necessary to nurture a continental ice sheet. If the ocean were free of ice, as they postulate it was at the start of the ice age, its water would mix with that of the North Atlantic Ocean over the threshold between Greenland and Scandinavia. The Atlantic would be cooled and the Arctic warmed. Evaporation from the Arctic Ocean would then so increase snowfall in northern Canada that the accumulation and advance of a continental ice sheet would begin. As water is withdrawn from the sea and accumulates as glacial ice, sea level would drop to a point at which the interchange of Arctic and Atlantic water across the narrow connection would be restricted and the Arctic Ocean would freeze. The removal of the source of moisture by this freezing would start the retreat of the glaciers and the rising of sea level to a point where the cycle could start again.

Many objections have been raised to this theory, and discussion of its merit continues. Perhaps the most damaging evidence is the lack of glacial features in areas that immediately adjoin the Arctic Ocean, for the effect of increased precipitation should have been most marked and the ice sheet should have been thickest there.

No general theory of the lowering of world temperatures in late Cenozoic time is widely accepted today, and these climatic changes remain as much a mystery as many other climatic changes of earth's history.

PRIMATES

Although primates were present throughout the Cenozoic Era, the ancestry of humans is considered in this chapter because they came to worldwide dominance during the late Cenozoic ice ages. Primates were one of the first advanced orders of mammals to become recognizably distinct from their insectivore ancestors. A single primate tooth has been found in Upper Cretaceous beds associated with the dinosaur, *Triceratops*.

In most skeletal features early primates retain a primitive pattern; insofar as they specialized, it was for an arboreal way of life. The primates became adapted to a life in the trees through the development of the following characteristics:

1. Forward-looking eyes, stereoscopic vision, and depth perception. The eyes became the principal sense organ, and the sense of smell was reduced.
2. A complex, large brain that facilitated coordinating reactions, particularly in the cerebral hemispheres, where the integration of sensory information takes place.
3. A thumb separable from and opposable to the rest of the digits, equipping the hand for grasping branches and, in connection with the improved brain, for the delicate and intelligent handling of objects; modification of laterally compressed claws to flattened nails.
4. Molar teeth with blunt cusps adapted to a varied diet, and skeletal elements that allowed a varied way of life (Fig. 21–19).

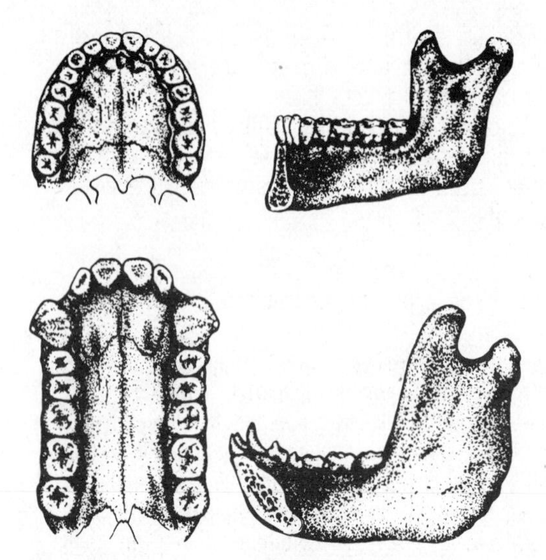

FIGURE 21–19 **Lower jaw of man (*top*) and gorilla (*bottom*). On the left is the dental arch as seen from below; on the right, the inside of the right half of the jaw. (From *General College Geology*; courtesy of A. J. Eardley and Harper & Row.)**

Prosimians: Tree Shrews, Lemurs and Tarsioids

Although we are principally interested in the anthropoid, or man-like, branch of the primates, the order includes many less advanced animals. Among these are the tree shrews, sometimes included among the insectivores but considered by many authorities to be the most primitive primates. They are small arboreal creatures, relatively large-brained, with an opposable thumb and big toe. They alone among the primates possess claws instead of nails. The lemurs—small, long-faced and long-tailed arboreal animals—are found in Paleocene and Eocene beds of North America and Europe. One of the best known is *Notharctus* (Eocene of North America), less than 30 cm in length. Thereafter, lemurs were apparently confined almost entirely to the island of Madagascar. In the comparative isolation of that island, numerous large lemurs evolved in the late Cenozoic, but without showing much advance over their Eocene ancestors.

Various small forms persist today. Slightly more advanced genera include the lorisids and bush-babies. The wide-eyed, hopping tarsioids are another group of primates that appeared early in the Cenozoic (Lower Eocene); they are now restricted to a single genus living in the jungles of eastern Asia. Like the lemurs, their remains are extremely rare in the middle and late Cenozoic rocks. Among their interesting features are their eyes, which are turned completely forward, their shortened nose, and their rounded face, which approximates that of the hominids.

Monkeys and Apes

All other primates are included in the Anthropoidea, or man-like group, which contains Old World monkeys, New World monkeys, apes, and man. In this group the four adaptations listed above are brought to various degrees of perfection. Fossils which may represent the oldest members of this assemblage are known from the Eocene of Burma.

In the Paleocene and Eocene, primitive primates were common and widespread in both Europe and North America. These continents were still close enough to permit migration, and three genera are common to both areas. Climatic changes in the middle Cenozoic restricted the previously extensive tropical regions, and the number of species became much reduced in what was to become the north temperate zone. With restriction of primates to equatorial areas, and the concomitant widening of the South Atlantic, the primates of the New and Old World became totally isolated from each other.

Early in the Oligocene Epoch, more advanced primates, the monkeys, appeared in both South America and northern Africa. Although sharing some advanced features, monkeys from these two areas are not closely related, but evolved separately from more primitive types of early Cenozoic primates.

Apes also first appeared in early Oligocene time and evolved through a late Miocene–early Pliocene form, *Dryopithecus* (Fig. 21–20) to the living great apes, the orangutan, chimpanzee, and gorilla. The fossil record of this evolution is exceedingly meager. The brain capacity and body size of the apes are greater than those of the early primates, but their teeth remained primitive, with prominent canines. The apes have lost their tail, and travel among the trees by swinging from branch to branch by their long arms. The heavier apes, the chimpanzee and gorilla, spend much of the time on the ground.

HOMINIDS

At about the Miocene–Pliocene boundary, one branch of the apes followed an evolutionary course that led to freedom from a life in the trees, perfection of an erect posture, and a two-legged walk. The brain of this family developed gradually to greater and greater complexity, and ultimately to a size three times that of the most advanced ape. With an erect posture, the forelimbs were free to handle objects, to make implements and weapons, and to investigate the world of nature. Because postnatal training of offspring was extended over a period of years, information and skill learned by one generation could be passed on to the next, and a body of knowledge could be accumulated in the rapidly developing cerebrum. This branch of the primates developed a social organization into families, clans, tribes, and nations, as had never been approached by any animal before. They were so successful in their exploitation of many different environments that they extended their range to all parts of the world and became the most abundant of the species of higher vertebrates. These advanced primates were men.

In zoological classifications, the family to which the human race belongs is known as the Hominidae. The differences separating members of this family from those of the Pongidae, or ape family, are few but significant. Certain differences in brain structure are important. Modern man has an average cranial capacity of approximately 1400 cubic centimeters, whereas the largest of the apes has a capacity rarely exceeding 500 cubic centimeters. The frontal part of

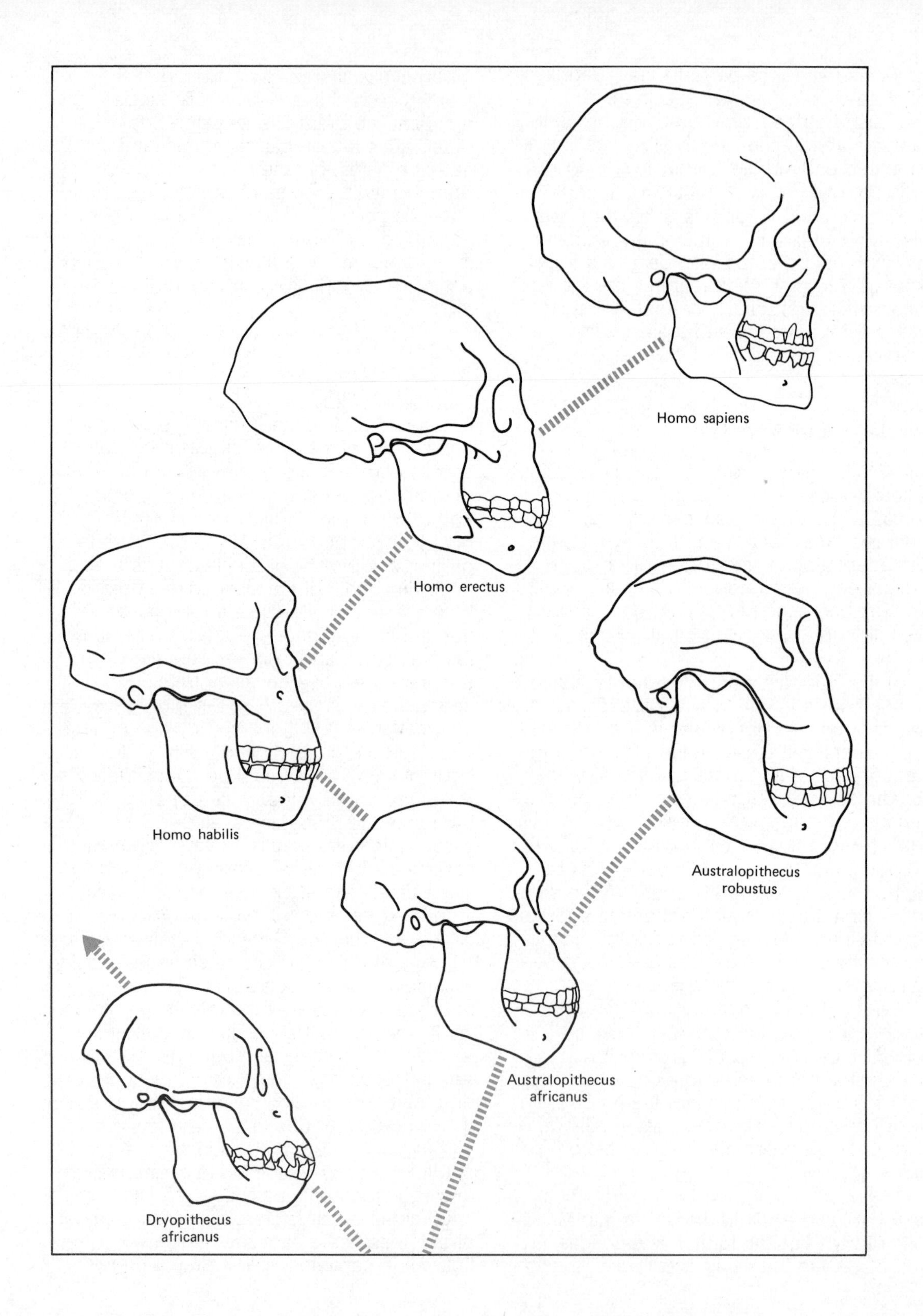
Homo sapiens
Homo erectus
Homo habilis
Australopithecus robustus
Australopithecus africanus
Dryopithecus africanus

FIGURE 21–20 **Generalized phylogeny of the Hominidae. *Dryopithecus*, from the late Miocene and early Pliocene, is a member of the Pongidae, related to the ancestry of the modern gorilla and chimpanzee. A contemporary genus, *Ramapithecus*, known only from jaw fragments, is considered to be an ancestral hominid. The species of *Australopithecus* are known from numerous remains in South and East Africa. *Homo habilis* is known from scattered remains in East Africa in beds dated at more than 2 million years, where it was a contemporary of both species of *Australopithecus*. *Homo habilis* has a significantly greater cranial capacity which suggests close relationship with modern humans. Intermediate forms make it difficult to distinguish clearly between the three species of the genus *Homo*.**

man's brain is larger than that of the ape's. The projection of the skull forward to accommodate this enlargement gives man a forehead, and the reduction in teeth and jaws results in a nearly vertical face. The line of the ape's teeth is a U-shape; man's is a parabola with no parallelism between the molar regions (Fig. 21–19). The ape's canines are always prominent, especially in the male; human canines barely project beyond the plane of the rest of the teeth. Differences in the mounting of the skull, the pelvis, and the leg and foot bones reflect the upright position of man's body and the stooped posture of the apes. These and many minor characteristics serve to set man apart from the ape morphologically. Socially, the difference between man and the apes is one of degree. Culturally, there is a vast difference. Man is the only animal to make complex tools and to make use of fire. Although vocalization is important for communication in other primates, man is the only animal who has developed conceptual language.

Ramapithecus

Ramapithecus, the earliest fossil hominid, is known from jaws found in the Upper Miocene and Lower Pliocene of Kenya and the Indian subcontinent. The diagnostic divergent lines of the molars, and the relatively smaller canines, but larger premolars, separate these jaws from those of fossil, or modern, apes (Fig. 21–19). The apparent small size of the front teeth suggests an inability to use them for stripping leaves from branches, as do the apes today, and suggests, therefore, a greater dependence on the use of the hands. Unfortunately, only jaws have been found, and no associated artifacts.

Australopithecus

Knowledge of Plio-Pleistocene hominids has increased enormously in recent years as a result of studies by the Leakeys and others in East Africa and Ethiopia. Their research, added to earlier information from South Africa, spans the last 6 million years of primate evolution.

Much of the fossil material represents a stage in evolution between that of *Ramapithecus* and the modern genus *Homo.* The specimens, assigned to several species, are placed in the genus *Australopithecus* (Fig. 21–20). On the basis of the many individuals found, most of the skeleton can be reconstructed. The structure of the vertebrae, pelvis, and feet indicate that *Australopithecus* was bipedal, although apparently not as efficiently as modern man. *Australopithecus* was small by modern standards, no more than 1.4 m in height and 30–45 kg in weight. The cranial capacity in some individuals exceeded 500 cc relative to body size, much greater than that of the apes. Within the genus *Australopithecus,* three species are recognized: *A. africanus, A. robustus,* and *A. boisei. Australopithecus africanus* was the smallest in stature and cranial capacity, but the most human in appearance. *Australopithecus robustus* had enormous jaws and teeth and accommodation for expanded jaw muscles; a high sagittal crest and laterally extended cheek bones gave an ape-like appearance to the skull. *A. boisei* shows these features expressed to an even greater degree.

There has long been debate over the degree of distinction between *Australopithecus africanus* and the more robust forms. That two closely related species of the same genus would live in one area at a given time is unusual, because they would compete intensively for the same food and space. *Australopithecus africanus* lived in the same general region in East and South Africa as *A. robustus* or *A. boisei* over a period of 1 to 2 million years. Some paleontologists believe that the three forms are extreme variants, based on sex, age, and individual differences, within a single species; others, that two or three distinct genera should be recognized. J. T. Robinson, for example, includes *A. africanus* in our own genus, *Homo,* and the larger species in the genus *Paranthropus.* He suggests that the large teeth of the *robustus* pattern were associated with a primarily herbivorous diet, associated with life in the forest. The smaller teeth of *africanus* suggest to him meat-eating and a proto-human hunting pattern, associated with life in the open grasslands. D. Pilbeam and S. J. Gould, however, suggest that all the described differences between *A. africanus, A. robustus,* and *A. boisei* can be attributed to allometric adjustments related to the larger size of the latter forms. *A. africanus* is suggested as the ancestral stock which gave rise to the other species, with whom it subsequently coexisted for 1 or 2 million years.

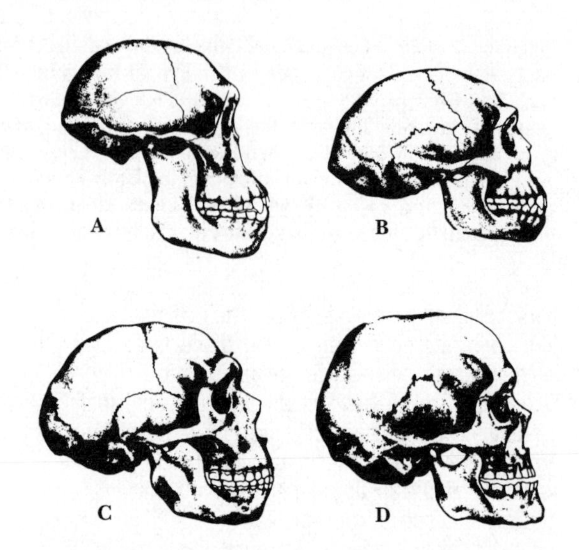

FIGURE 21–21 **Hominid skulls. A. *Australopithecus africanus.* B. Peking man, *Homo erectus.* C. Neanderthal man, *Homo sapiens neanderthalensis.* D. Cro-Magnon man, *Homo sapiens sapiens.* (From *Vertebrate Paleontology;* courtesy of A. S. Romer and the University of Chicago Press.)**

Homo

The most startling hominid discoveries in recent years are specimens of another species, contemporary with *Australopithecus,* 3 million years ago, with a cranial capacity of over 700 cc. As its body size was no greater than that of the larger specimens of *Australopithecus,* the relative size of the brain of this hominid was much larger than that of the species assigned to that genus. The differences are sufficiently great that the new discoveries were placed by L. Leakey in the genus *Homo,* as a new species, *Homo habilis* (Fig. 21–20). The skeleton also shows a higher degree of perfection of the bipedal stance than that of *Australopithecus.* In addition to the original material from Olduvai Gorge, a nearly complete skull attributed to *Homo habilis* has recently been discovered by R. Leakey near Lake Rudolf in Kenya, in beds dated at 2.8 million years.

Primitive stone tools are found with the remains of both *Australopithecus* and *Homo habilis* in East Africa, but which genus was responsible for their use is unknown.

Three species are currently included in the genus *Homo.* The name *Homo erectus* is applied to fossils about 200,000 years old from China and Java, and to others as old as about 3,000,000 years from East Africa (Fig. 21–21). Well-known members of this species (earlier called *Pithecanthropus* and *Sinanthropus*) have a cranial capacity slightly exceeding 1000 cc. The skull is low, the brow ridges prominent. The stature is intermediate between that of *H. habilis* and modern man. Gradual changes in the last 200,000 to 300,000 years culminated in the emergence of modern man, *Homo sapiens.* The cranial capacity of *Homo sapiens* is typically 400 cubic centimeters larger than that of *Homo erectus,* the skull is higher, the occiput less sharply angled, and the brow ridges less pronounced. Within the genus *Homo,* none of the three species can be readily categorized, for each shows extensive variability. Particular anatomical patterns may persist much longer in some

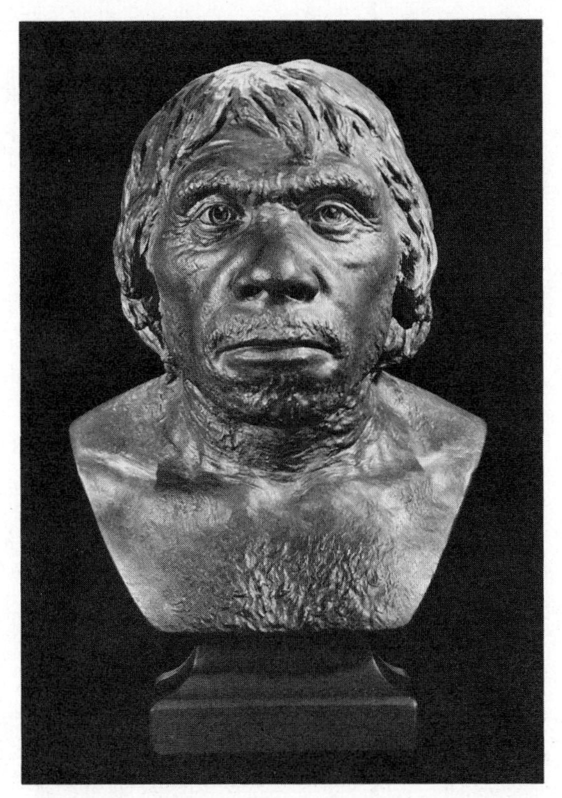

FIGURE 21–22 **Reconstructions of the heads of early humans.** ***Left: Homo erectus*** **(*Pithecanthropus*).** ***Right: Homo sapiens neanderthalensis.*** **(By J. H. McGregor; courtesy of the American Museum of Natural History.)**

areas than in others, with considerable overlap in duration. Some fully capable modern humans have a cranial capacity, for example, no larger than the average for the species *Homo erectus.*

Considerable variation in details of cranial anatomy within our own species has led to the recognition of various subspecies, such as those occupying different geographical regions today, and also those living at progressively earlier periods during the Pleistocene, such as the Neanderthal people of western Europe (Fig. 21–22). The Neanderthals were succeeded in Europe approximately 40,000 years ago by individuals of the modern subspecies *Homo sapiens sapiens.* In the Middle East and in North and East Africa, remains of fully modern humans are known from deposits approaching 100,000 years in age (Fig. 21–23).

EARLY MAN IN NORTH AMERICA

The primates became extinct in North America in the middle of the Cenozoic Era, soon after they evolved elsewhere to the monkey stage. Prehistoric man did not appear in the Americas until late in the Pleistocene, near the end of the time marked by the late Paleolithic culture (Old Stone Age) in Europe. Paleolithic people crossed to North America from Siberia over the Bering Strait which, during the last stages of glaciation, had emerged from beneath the sea. The record of man on this continent (Fig. 21–24) shows none of the early evolutionary and cultural steps that can be traced in Africa, Asia, and Europe.

The first human migrants into North America are difficult to date. The possibility of finding either skeletal remains or artifacts

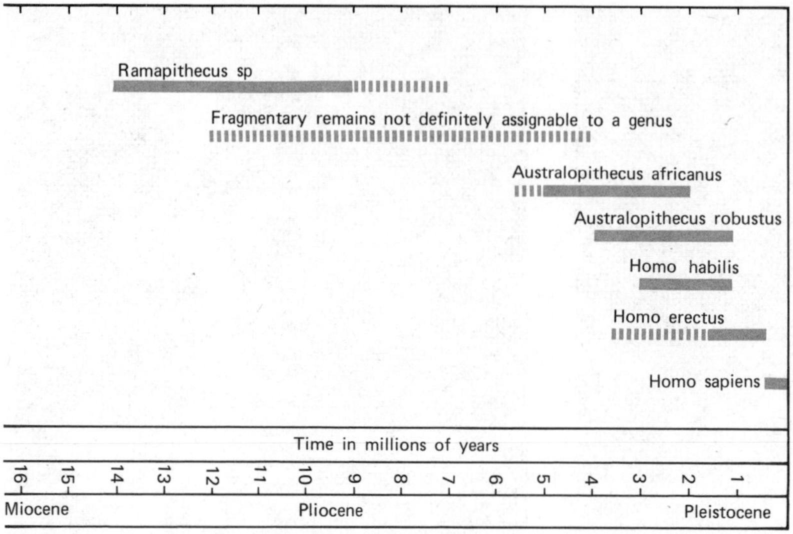

Ramapithecus sp
Fragmentary remains not definitely assignable to a genus
Australopithecus africanus
Australopithecus robustus
Homo habilis
Homo erectus
Homo sapiens
Time in millions of years
16 15 14 13 12 11 10 9 8 7 6 5 4 3 2 1
Miocene
Pliocene
Pleistocene

FIGURE 21–23 **Approximate time ranges of hominid species. Ranges of *Australopithecus boisei* and *A. robustus* not distinguished.**

left by members of initially very small populations is slight. The problem is compounded by the difficulty of dating the few remains that have been discovered. Radiocarbon dating is a very important method for determining the age of human remains and woody artifacts, but the rate of decay of carbon 14 is so rapid that it is very difficult to measure ages greater than about 50,000 years. Potassium–argon dating, so useful in the early Pleistocene and Cenozoic, is effective for dating rock in excess of 1 million years, but is of little use for younger rocks.

FIGURE 21–24 **Fluted points (Folsom type) from the Lindenmeier site, Colorado. (Courtesy of the Smithsonian Institution, Office of Anthropology.)**

The initial entrance of man into North America unfortunately falls between the limits of these two methods of dating. Other methods have been attempted but their usefulness has not yet been fully evaluated.

Radiocarbon dates of more than 40,000 years have been obtained for charcoal and burned bone associated with primitive human artifacts from Texas. Material from Peru has been dated at approximately 20,000 years ago. Other less reliably dated material suggests that early man may have entered North America as early as 100,000 years ago, not long after the emergence of the modern species, *Homo sapiens.*

According to R. S. MacNeish, several stages of cultural development can be discerned in North and South America prior to the emergence of the cultural complexes of the various Indian groups recognized by the early European explorers. The first three cultural levels, recognizable by distinctive artifact assemblages, apparently arose in eastern Asia, and then spread from north to south in the Americas. These are equivalent to time and degree of development of the late Old Stone Age cultures in Europe. The more sophisticated, early Neolithic cultural level apparently developed separately within tribes already spread throughout both North and South America; at least the specific types of artifacts cannot be readily derived from those developing at the same time in eastern Asia. The Bering Strait was certainly not a complete barrier to human movement at any time during the Pleistocene, and the Eskimos in the North American Arctic continue to move freely between Asia and North America.

Although people have inhabited North America only since the late Pleistocene, and had already achieved an essentially modern form by that time, early human remains and artifacts are associated with a fauna including a large number of extinct mammals. A striking feature of mammals associated with early man in North America is their large size. Early man certainly hunted the now-extinct mastodons, glyptodons, ground sloths, horses, and camels. Hunting patterns seem to have been specifically developed to kill very large numbers of these large herbivores by driving whole herds over cliffs or into swamps. There is now considerable controversy over the role of primitive people in bringing about the extinction of these animals. Primitive man certainly lived in North America for tens of thousands of years in association with numerous species which did become extinct, and must have affected, directly or indirectly, their chance of survival.

With the beginning of the New Stone Age cultures in both the New and Old World, man had achieved modern form in his skeleton, and with the beginning of agriculture, the foundations of what we consider as civilization were established. Documenting the history of civilization is the task of the historian, not the geologist. The time when man changed from a primitive hunter, at the mercy of his environment, to a civilized creature able to mold his environment to his advantage seems a good time to close the account of the historical geology of North America, and to let the archeologists and historians continue the tale.

SUMMARY

Late Cenozoic glaciation appears to be a response to Miocene cooling and does not correspond in time span to the Pleistocene Epoch. Evidence of the lowering of world temperatures can be obtained from deep-sea cores by analysis of fossil faunas and oxygen isotope ratios. About 3 million years ago lowering of the snowline caused by this climatic change caused large areas of the Laurentide Plateau to accumulate permanent snowfields and initiated the continental glaciation of North America.

Growth of an ice cap, once started, is speeded by the chilling of the moist air above it, which induces storms and increases precipitation on glacial margins.

The Laurentide ice cap advanced from central Quebec as far south as St. Louis, Missouri, and abutted on the west a complex of valley glaciers in the Cordillera thick enough to bury most of the mountains north of the 49th parallel. One third of the continent was covered with ice at the glacial maximum.

Festoons of moraines were left along the successive ice margins because of the conveyor-belt action of the retreating glaciers.

A comparison of the dimensions of modern ice caps with those of the Laurentide ice cap together with evidence of glaciated mountains near its southeastern margin suggest that the ice cap was 3000 m thick in the center and about 2000 m thick near the margin.

The increase in rainfall in the middle latitudes of the world during glacial advances caused the formation in desert regions of pluvial lakes, whose presence is recorded in abandoned shoreline features and lacustrine sediments separated by evaporites.

In the upper Mississippi Valley, four groups of till sheets (Nebraskan, Kansan, Illinoian, Wisconsin) represent four major advances of the glaciers. They are separated by interglacial sediments such as peat, clay, sand, varved clay, and loess, whose fossil faunas and pollen content indicate that climates as warm or warmer than the present one intervened between glaciations. Interglacial weathering of tills to form a soil profile is also evidence of the periodic retreat of the glaciers. As many as eight major advances of the ice may have taken place but later advances largely destroyed the record of earlier ones.

During deglaciation ice-margin lakes temporarily trapped meltwater and underwent complex changes in level as the ice withdrew and uncovered lower outlets and the land rose, closing other outlets. Occasionally, when ice or alluvial dams broke in catastrophic floods, the meltwater discharged down enlarged river valleys carrying extremely coarse sediment.

The storage of precipitation in the ice caps resulted in a drop in sea level of about 105 m during glacial advances and a rise of over 105 m during interglacial intervals. These changes are recorded in shoreline and submarine features all over the world.

The weight of the ice cap depressed the crust beneath its center about 1 km. The slow recovery of the crust can be studied in the raised marine beaches and tilted shorelines of glacial lakes.

Many theories have been proposed to explain the lowering of temperature that started the late Cenozoic ice age. Recently attention has been focused on the movement of the poles into thermally isolated positions with respect to the continents such as in the Antarctic continent and the Arctic Ocean.

The branch of the primates to which man belongs diverged from the apes in late Miocene time and specialized for erect posture and a two-legged walk. *Ramapithecus* is generally recognized as the first fossil hominid.

The genus *Australopithecus* comprises three species of Pliocene and Pleistocene hominids from Africa with cranial capacities of about 500 cc. The first species of the genus *Homo* (*H. habilis*) was a contemporary of the australopithecines about 3 million years ago. The modern subspecies of man (*H. sapiens sapiens*) appeared in Africa about 100,000 years ago and displaced Neanderthal man (*H. sapiens neanderthalensis*) from Europe about 40,000 years ago.

Prehistoric man did not appear in North America until late Pleistocene time and brought with him from Asia a culture equivalent to that of the Old Stone Age in Europe.

QUESTIONS

1. Using an atlas and other sources of geographical information, compare modern conditions of Antarctica with respect to climate, thickness and size of the ice cap, topography, etc., with northern North America during the glacial maximum.
2. List all the landscape features that are derived from continental glaciation and indicate in one or two sentences for each how they are formed.
3. What geological events can be considered as catastrophic? What evidence would such events leave in the stratigraphic record? Give examples from the historical record you have studied in this book.
4. Summarize the characteristics of the three species of *Australopithecus* and the three species of *Homo* discussed in this chapter and place them in their relative positions in a time scale.

SUGGESTIONS FOR FURTHER READING

Man and Primates

LEAKEY, R. E. "Skull 1470." *National Geographic.* 143 (6): 819–829, 1973.

______ "Hominids in Africa." *American Scientist,* 64 (2): 174–178, 1976.

MACNEISH, R. S. "Early man in the New World." *American Scientist,* 64 (3): 316–327, 1976.

PILBEAM, D. *Ascent of Man.* New York: Macmillan Publishing Co., Inc. 207 p., 1972.

———, and GOULD, S. J. "Size and scaling in human evolution." *Science,* 186: 892–901, 1974.

ANON. "Filling the gap," *Scientific American,* 234 (5): 56–57, 1976.

SIMONS, E. L. *Primate Evolution.* New York: Macmillan Publishing Co., Inc. 322 p., 1972.

Glaciation

FLINT, R. F. *Glacial and Quaternary Geology.* New York: John Wiley & Sons. 892 p., 1971.

MATSCH, C. L. *North America and the Great Ice Age.* New York: McGraw Hill, 131 p., 1976.

TUREKIAN, K. ed. *Late Cenozoic Glacial Ages.* New Haven: Yale University Press, 1971.

WRIGHT, A. E., and MOSELEY, F., ed. "Ice Ages: Ancient and Modern." *Geological Journal Special Issue 6,* 320 p., 1975.

APPENDIX

THE NAMES OF PLANTS AND ANIMALS

Before the sciences of botany and zoology were established, plants and animals were given names such as buttercup, dog, rose, or cat, which differed, of course, from language to language. As more of the world of nature was investigated, names were invented to designate types of organisms that were generally unknown to ordinary people. Had this scheme of naming continued, each language would have been burdened with many names for the types of animals and plants that modern scientists have distinguished, and it would have been almost impossible for scholars of one language to have assimilated the names proposed by scholars of other languages.

Fortunately, in 1735, before this confusion was far advanced, a Swedish naturalist named Karl von Linné (today the latinized form, Linnaeus, is widely used) proposed that animals and plants be named in Latin, the scholarly language of the 18th century. He also proposed that each organism be given two names: the first to indicate the general group of animals or plants to which the organism belonged, and the second to distinguish the particular type of animal within that group. These two names correspond in a general way to our family name (Smith) and given name (William). Our family names are written last, and our given names (which modify our family names, like adjectives) are written first: William Smith. Because modifying adjectives follow nouns in Latin, however, the group names of organisms are written first and the specific names last. Thus cats of all kinds were called *Felis* (Latin for cat) by Linnaeus. Within this large group he designated several kinds of cats, such as *Felis domestica* (the domestic cat), *Felis leo* (the lion cat), and *Felis tigris* (the tiger cat). The first name of these cats is known as their generic name, and all cats are said to belong to the genus (plural, genera) *Felis*. The second name is their specific name, and serves to distinguish the various types or species (singular, species) of cats within the genus *Felis*. Because of its simplicity and freedom from language barriers, Linnaeus' binomial system of nomenclature has been universally accepted. As a result, the names of animals and plants are the same in all languages, and to a large extent the relationships between organisms can be expressed by these names.

In this book the authors rarely refer to specific names, but the student should realize that every genus must have at least one species, and some have hundreds of species. As scientific investigation advances, a genus is sometimes found to be too inclusive, and the species in it are often regrouped within several newly defined genera. The original name is retained for one of the new and restricted genera. Thus, the

A SUMMARY OF THE PLANT AND ANIMAL KINGDOMS

genus *Spirifer,* named in the early days of paleontology, has in modern times been subdivided into scores of new genera, and the name *Spirifer* has been retained for one of them. The new genera, which were once included in the old genus, may collectively be referred to as *Spirifer sensu lato* (in the wide sense), and the new restricted genus becomes *Spirifer sensu stricto* (in the restricted sense). These are usually abbreviated to *Spirifer s.l.* and *Spirifer s.s.*

The relationship between organisms is further elaborated in a system of classification in which the genera, such as *Felis,* are grouped together into families, the families into orders, the orders into classes, the classes into phyla, and the phyla into kingdoms. The family of cats, the FELIDAE, belongs to the order CARNIVORA, which also includes most other meat-eating mammals and some extinct groups. This order belongs to the class MAMMALIA, which includes all the four-footed, warm-blooded animals with hair. The mammals are a division of the phylum CHORDATA, which comprises all the animals with backbones; this phylum, together with all the invertebrate phyla, is included in the ANIMALIA, or animal kingdom. The relationships between organisms is so complex that many more categories than those mentioned above are necessary to express these relationships. Detailed classifications contain such categories as superfamilies, suborders, superorders, and infraclasses.

The following systematic description of plants and animals is not intended to be complete but emphasizes groups that are important in the fossil record.

THE PLANT KINGDOM

Phylum Thallophyta

The thallophytes are a diverse group of primitive plants that are divided into several distinct phyla in modern detailed classifications, and include such forms as bacteria, slime molds, fungi, and seaweeds. Most of them have a negligible fossil record because they lack hard parts and disintegrate rapidly on death.

Bacteria. Bacteria, the smallest forms of living matter, occupy every environment in the modern world in astronomical numbers. Their paleontological record extends back to the pre-Paleozoic. Although bacteria are invisible to the naked eye, the chemical transformations they promote in their life processes are important in the formation of certain sedimentary rocks.

Diatoms. The diatoms (Fig. A–1) are microscopic plants that live in marine and freshwater environments. Their tests, or shells, composed of opaline silica, may accumulate in great numbers to form a layer of sili-

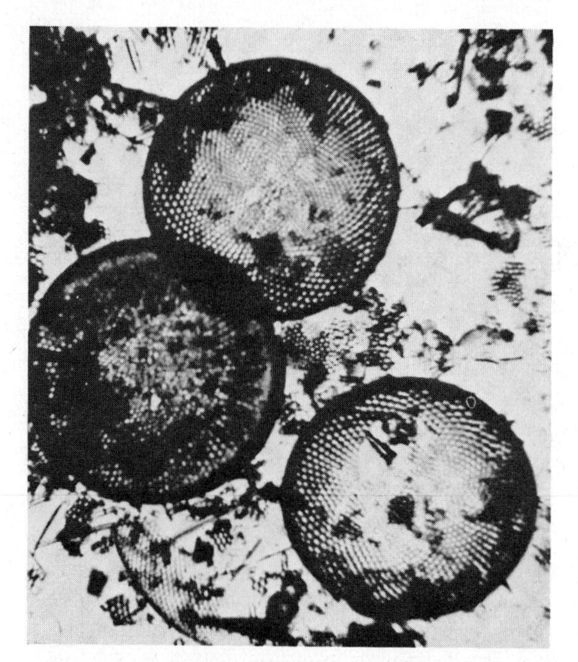

FIGURE A–1 **Recent diatom ooze from the Bering Sea (magnification ×1800). (Courtesy of Taro Kanaya.)**

ceous sedimentary rock known as diatomite. Because the colloidal silica of their tests is easily dissolved by water percolating through sedimentary rocks, diatoms are generally not well enough preserved in older sedimentary series to be positively identified.

Algae. The most important thallophytes, as far as the paleontologist is concerned, are the algae or seaweeds. The algae that secrete calcium carbonate contribute extensively to the formation of limestones, particularly reef limestones. The blue-green algae form fine filaments of cells which may secrete calcium carbonate on their outer surfaces. These primitive plants are responsible for the formation of such simple fossils as *Collenia* of the Belt Group, the cryptozoans of Paleozoic strata, and the stromatolites of all periods. The green algae form mats of floating scum on bodies of fresh water; some of them also secrete lime and have been important contributors to freshwater limestones. The red algae secrete a rigid, calcareous structure consisting of numerous regularly placed cells, and are the most important modern reef builders, with a paleontological record extending back to the early part of the Paleozoic Era. The brown algae include the common seaweed washed up along the seashore by storm waves; most of these do not secrete hard parts and are represented in the fossil record only by swirling markings and impressions called fucoids.

Phylum Bryophyta

The mosses and liverworts are grouped among the bryophytes. Although they are better adapted for life on land than the multicellular thallophytes, they are still in a very primitive stage of evolution and have not developed special vascular tissue for conducting fluids. Their fossil record is insignificant.

Phylum Rhyniophytina

Included in the phylum are the first land plants of Late Silurian time (Fig. A–2). These primitive plants lack true roots or leaves.

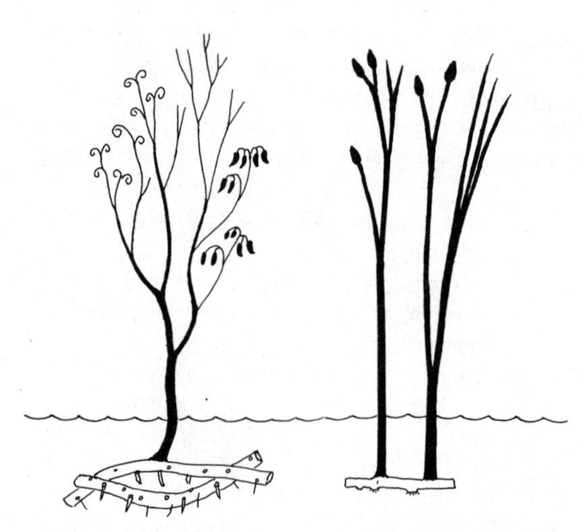

FIGURE A–2 **Primitive Devonian land plants. *A. Psilophyton princeps. B. Rhynia major.* The stems and spore cases are shown in solid black, the underwater creeping rootstocks in outline (approximately one-tenth actual size). (*A.* Modified from Dawson. *B.* Modified from Kidston and Lang.)**

From a horizontally creeping stalk, or rhizome, upright stems bearing small bracts branch off. Spore cases are carried at the tips of the stems. The rhyniophytes may be represented today by a few remote relatives of the Paleozoic plants, chiefly *Psilotum,* that live in tropical and subtropical environments. For the most part these plants are small, but some reached heights of several feet.

FIGURE A–3 ***Lycopodium*, a living lycopod with the spore cases in the axils of the leaves. (Courtesy of R. D. Gibbs.)**

Phylum Lycopsida

The scale trees, or lycopods, are another group of plants represented in the modern flora by a few small genera descended from their larger and more numerous Paleozoic ancestors. The modern lycopsids are commonly referred to as the club mosses. The most familiar of such plants is the ground pine, *Lycopodium,* a ground creeper of shady woodlands (Fig. A–3). The simple, vertical, blade-like leaves of the lycopsids are arranged in spiral rows on the stems. Reproduction is carried on by spores in small cases at the base of certain fertile leaves. These leaves usually grow together at the tips of the branches and form organs resembling the cones of pine trees.

The lycopods grew to heights of 30 m in the coal forests of late Paleozoic time. When the leaves that emerged directly from the trunks were shed, they left scars on the bark that gave the lycopods a scaly appearance. Some of the larger representatives of this group are described and illustrated in Chapter 13.

Phylum Sphenopsida (Arthrophyta)

Sphenopsids include the modern scouring rush and "horsetail fern" (Fig. A–4). They have a simple, longitudinally ribbed stem divided transversely by a series of joints. At these joints whorls of branches or leaves emerge. The spores are carried in cases at the tips of the stems. The late Paleozoic representatives of this phylum, *Calamites,* grew to tree-like size in the coal swamps. *Calamites* had a hollow stem whose stump was commonly filled with sand or silt when the plant died. Many of the fossils of this plant are interior molds left when the surrounding parts of the plant decayed.

Phylum Pteropsida (Ferns)

Ferns reproduce by means of spores contained in small, dark cases that appear periodically, usually on the undersides of certain leaves. They have a well-developed vascular system of cells specialized for conducting sap through the stem. The ferns probably were derived from a rhyniophytinian ancestor in Devonian time and have been abundant to the present day. During the late Paleozoic, many were tall enough to be called "tree ferns".

FIGURE A–4 **The modern arthrophyte *Equisetum*. (Courtesy of R. D. Gibbs.)**

Phylum Gymnospermae

Seed ferns (Pteridospermales). The foliage of the pteridosperms was similar to that of the ferns, but the plant produced seeds instead of spores. A spore is a single cell that is cast off by the plant and grows as it can, but a seed is fertilized and grows in the ovary for some time until, equipped with food, it is released to form a new plant. During Pennsylvanian time seed ferns grew to tree size, but in the Jurassic they became extinct. However, they gave rise to all the other seed-bearing plants.

Cycads and Cycadeoids. The cycads and cycadeoids bore naked seeds. They had long, unbranched trunks with a spread of palm-like leaves forming a crown. The cycads, which persist in the present as the sago palm, do not possess flowers (Fig. A–5). The cycadeoids, however, bore primitive flowers at the top of the trunk, and are believed to have given rise to the true flowering plants in the early Mesozoic Era.

Conifers. The conifers are a group of plants that bear unprotected seeds in cones and generally have blade-like or needle-like leaves, evergreen foliage, and a tree-like form (Fig. A–6). Cordaitids and ginkgoes are primitive conifers, but each is assigned to a separate class by some paleobotanists. Cordaitids were late Paleozoic trees that had long, swordlike leaves up to 1 m in length at the top of a high trunk. The ginkgoes are primitive conifers that are represented in the modern flora by a single species, the maidenhair tree. The peculiar fan-shaped leaves of these trees can be

FIGURE A–5 **Macrozamia, a modern cycad about 15 ft high (Australia). The fruit is borne at the top of the trunk and is hidden in this photograph by the long leaf branches. (Courtesy of R. D. Gibbs.)**

FIGURE A–6 **A modern North American conifer, *Tsuga*, the hemlock fir, showing the characteristic foliage and cones (½ actual size). (Courtesy of R. D. Gibbs.)**

found in rocks as old as Permian, and are abundant in Mesozoic strata. True conifers appeared late in the Paleozoic Era and reached their evolutionary peak in the Mesozoic Era. Representatives of this group, such as the pine, hemlock, spruce, yew, and redwood, are common elements of the modern coniferous flora.

Phylum Angiospermae

The angiosperms, or true flowering plants, have been immensely successful since their first appearance in the Cretaceous Period, and are represented by a quarter of a million species in the modern world. The seeds of these plants are not naked like those of the more primitive seed-bearing plants, but are protected by a capsule. Their leaves are common in the sedimentary record from Cretaceous time on, and many of the modern genera, such as the willow, birch, and magnolia, can be recognized easily in Cretaceous deposits. The angiosperms apparently developed from the cycadeoids and soon displaced them in the flora of late Mesozoic time. Today they include such diverse plants as the grasses, hardwood trees, most of our table vegetables, and all of our garden flowers.

THE ANIMAL KINGDOM

Phylum Protozoa

The simplest and most primitive animals include the protozoans. The phylum includes animals all of whose body functions—respiration, feeding, ejection of waste products, growth, and reproduction—are carried on within a single cell. Within this group is a host of largely microscopic animals, some of which are parasitic, some plant-like in their functions, some marine, some freshwater, some colonial, some single, some with calcareous or siliceous shells, and some naked. Nearly all can reproduce asexually by splitting in two, but most turn to sexual reproduction at certain stages of their life histories. The study of this phylum is important in the search for an understanding of the origin of life, because the simplest of these creatures must have been among the first eukaryotes. For the paleontologist, only those protozoans that secrete a preservable shell (generally called a test in this phylum) are of primary concern. Of these only two orders, the RADIOLARIA and the FORAMINIFERA, are represented by an adequate fossil record.

Most of the foraminifers, or forams, secrete a microscopic calcareous test composed of one or a series of chambers in a variety of patterns (Fig. A–7). Some forams secrete a test composed of sedimentary particles, largely sand grains, cemented together to form spheres, tubes, or groups of chambers. The first forams appear in the fossil record in Ordovician rocks and are of this type.

The other group of protozoans that secrete a test are the radiolarians. These microscopic animals construct a delicate framework of opaline silica in a variety of radiating patterns (Fig. A–8). However, such a fragile, open test composed of a relatively soluble substance, colloidal silica, is soon crushed or dissolved after it has been entombed in sediments. For this reason, most pre-Cenozoic radiolarian deposits have been converted into structureless chert or flint by solution and redeposition in percolating water.

Some chert nodules of Mesozoic and Paleozoic ages preserve vague radiolarian impressions, but in general the paleontological record of this group is poor.

Phylum Porifera

The lowest grade of multicellular life includes animals grouped in this phylum, the PORIFERA, or sponges. In these and all higher animals the cells that make up the organism are specialized for certain purposes and cannot live independently like protozoan cells. In its simplest form the sponge is like a vase with porous walls (Fig. A–9). Water is drawn in through the pores in the wall of the "vase" by the action of cells equipped with small, whip-like appendages whose movement creates a current in the water. Microscopic plants and animals in the circulating water are caught on cells inside the body cavity and provide food for the whole animal. In more advanced sponges, the interior surface lined with these food-absorbing cells is intricately folded into the thickened wall, thus enlarging the flagellate food-gathering surfaces.

The hard parts of the PORIFERA consist of calcareous or siliceous elements called spicules, which may be fused to form a continuous framework, or may be free in the soft body of the sponge. If the spicules are unconnected, they are released when the organic tissues of the animal decay and are scattered by currents on the ocean floor. Some sponges have a rigid calcareous skeleton composed of irregular, interlocking

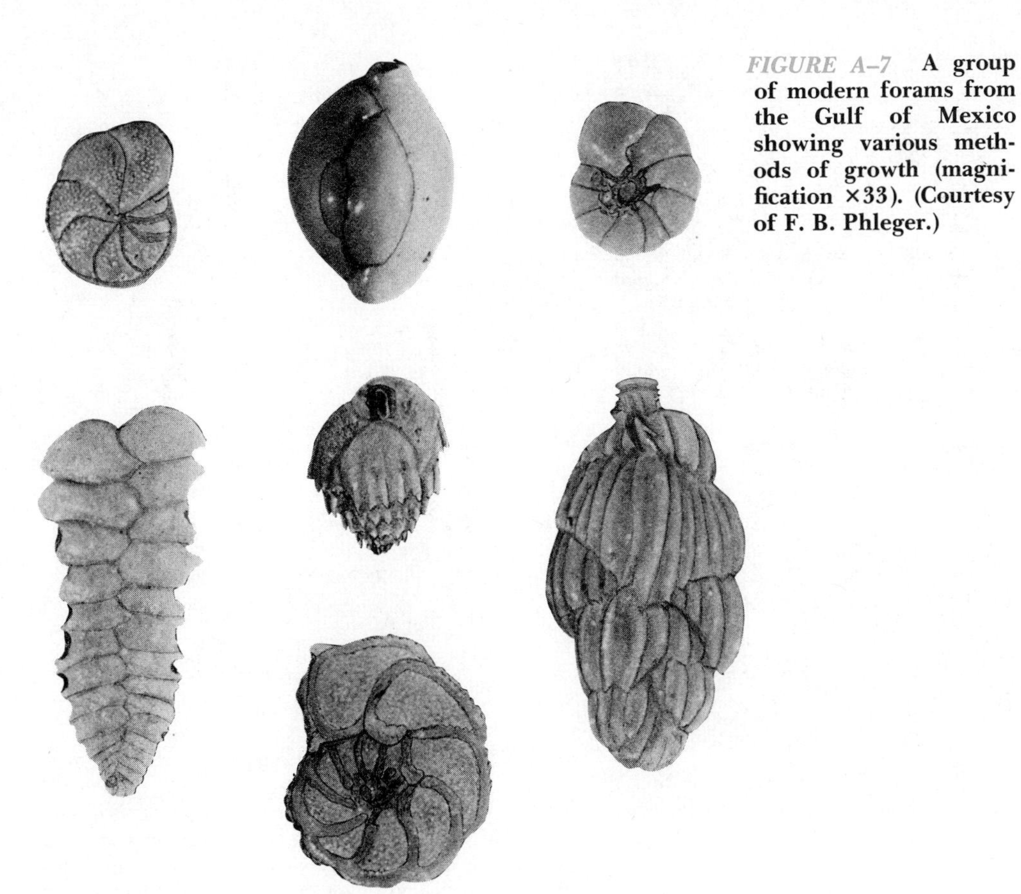

FIGURE A–7 **A group of modern forams from the Gulf of Mexico showing various methods of growth (magnification ×33). (Courtesy of F. B. Phleger.)**

spicules. Others, such as the glass sponges, have a skeleton composed of spicules of silica fused into a regular network. Still others are strengthened with spongin, a protein compound chemically similar to hair.

The STROMATOPOROIDEA were reef-building sponges that constructed a calcareous skeleton of laminae spaced about 5 per mm, separated by short pillars generally positioned at right angles to the laminae. From this simple plan the stromatoporoids evolved into encrusting and cabbage-like forms of great internal structural complexity (Fig. A–10). They thrived in the reefs of Silurian and Devonian times and helped to build the resistant framework that withstood the waves.

Phylum Archaeocyatha

The archaeocyathids were animals that flourished in the Early Cambrian Epoch and lingered on into the Middle Cambrian. Their hard parts were vase-shaped, and resembled a sponge in general, but were composed of two solid calcareous walls separated by a set of radial partitions, or septa. The soft parts of the animal were apparently confined to the space between the two walls. They presumably fed by absorbing micro-organisms through their walls from the circulating sea water, in much the same way as the sponges.

Phylum Coelenterata

This large and diverse group of animals is characterized by sac-like bodies with tenta-

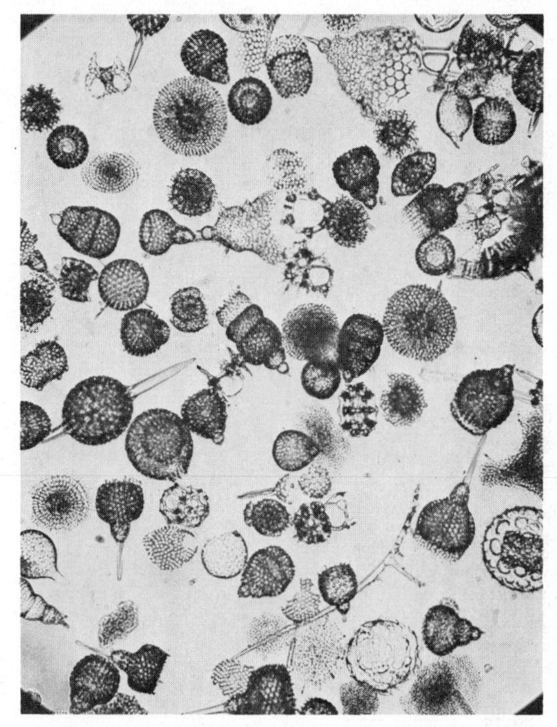

FIGURE A–8 **Radiolaria from Miocene beds, Philippine Islands (magnification ×70). (Courtesy of W. R. Riedel.)**

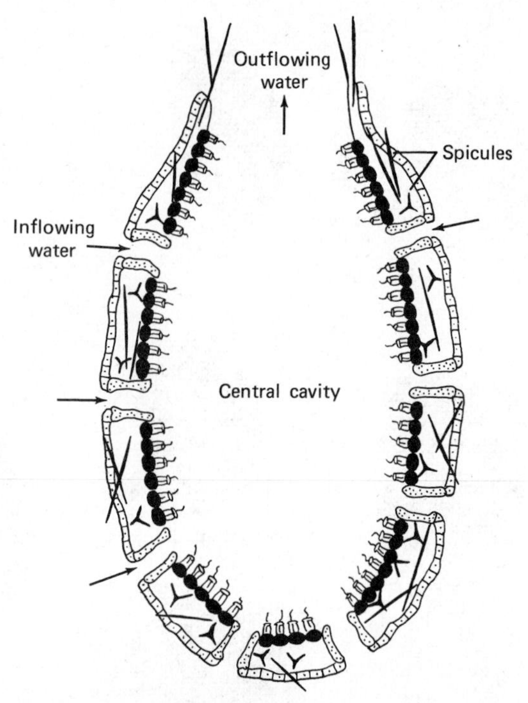

FIGURE A–9 **Diagrammatic cross-section of a simple sponge showing the direction of flow of water through the pores and central cavity. The dark objects embedded in the tissue are spicules. The three layers of the body are shown: the ectoderm, or epidermis, on the outside; the endoderm lining the interior and, in some cases, the canals; and mesogloea filling the space between.**

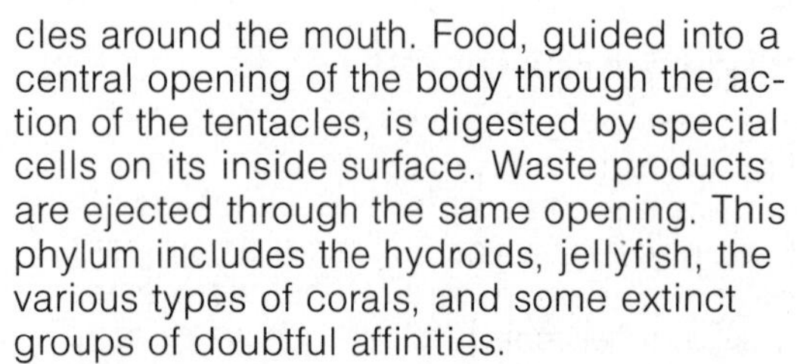

cles around the mouth. Food, guided into a central opening of the body through the action of the tentacles, is digested by special cells on its inside surface. Waste products are ejected through the same opening. This phylum includes the hydroids, jellyfish, the various types of corals, and some extinct groups of doubtful affinities.

The HYDROZOA, simplest class of the coelenterates, have an unmodified, sac-like disgestive system. Most of the members of this group are colonial animals in which each individual, or polyp, inhabits a small cup in a tree-like colony. Some polyps have special beak-like organs that clean the colony; others are specialized for stinging, and others for reproduction. The reproductive polyps give off small jellyfish at certain times of the year, which take part in a sexual reproductive process that results in a new colony of polyps. The hydrozoans are poorly represented in the fossil record.

A group of coelenterates, the jellyfish, forms the class SCYPHOMEDUSAE. Since they have no hard parts, jellyfish are represented in the fossil record by only rare impressions of their lower surfaces on the soft mud of the ocean floor.

Geologically, the corals, or ANTHOZOA, are the most important subdivision of the coelenterates. These animals can increase the area of the digestive surface on the interior of the body sac by folding the wall inward to form a series of radial partitions called mesenteries. The basal disc of the body is also folded radially so that the major ridges of the base alternate in position with the mesenteries (Fig. A–11). In

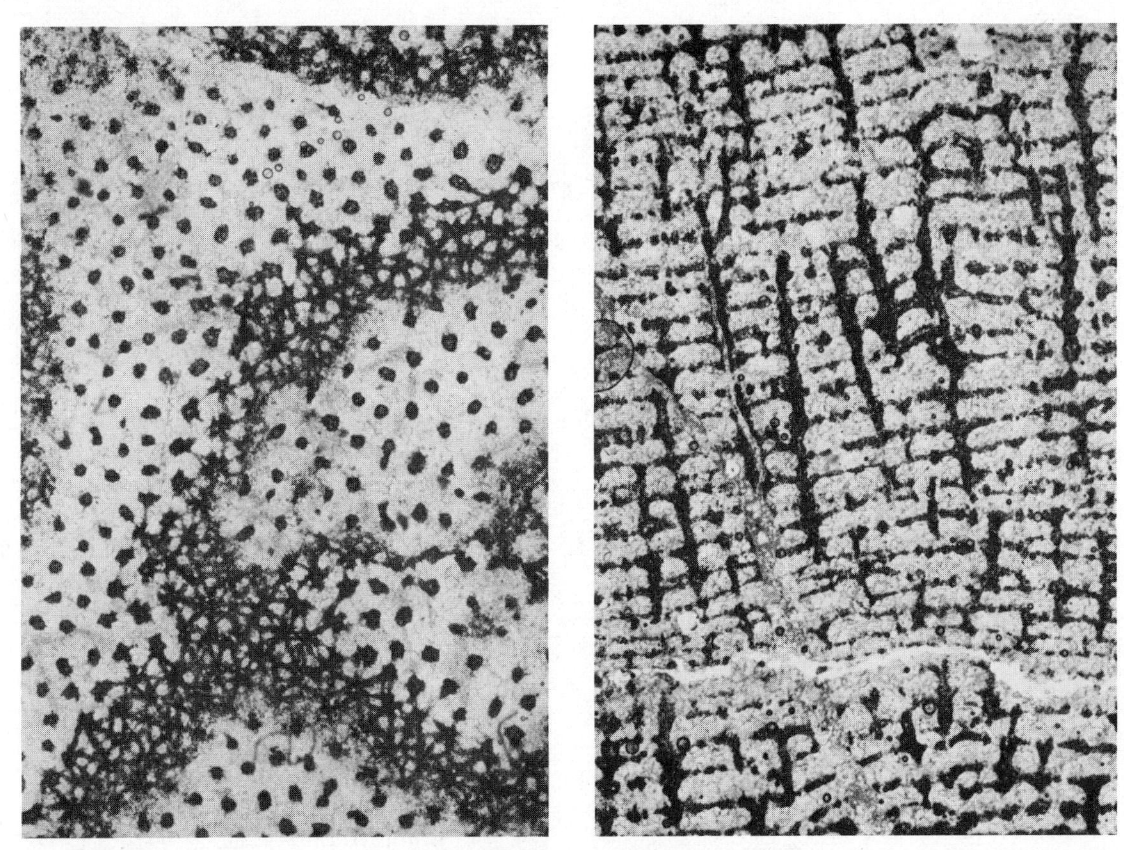

FIGURE A–10 ***Actinostroma*, a Devonian stromatoporoid. *Left:* A tangential section. *Right:* A longitudinal section showing pillars and laminae. From Evie Lake Reef, British Columbia (magnification ×40).**

most corals the cells on the basal exterior surfaces secrete calcium carbonate in the form of a plate or shallow cup surrounding the lower part of the body. Where the lime-secreting cells are folded inward at the base of the cup, they secrete a set of radial plates called septa. A cup, cone, or cylindrical structure, divided by radial partitions, is the typical fossil coral.

Some corals are colonial; others are solitary. Solitary corals secrete a cup which increases in diameter as the animal grows. In the process of growth the polyp periodically lifts itself upward in the cup, and in many cases partitions off the abandoned part with a calcareous plate called a tabula. The early smaller parts of the skeleton are left behind as the animal builds itself a horn-shaped support, which it occupies at the top in a depression called the calyx. Many individuals unite to form the large structures built by the colonial corals. The individuals occupy parallel tubes packed closely together to form a hemispherical, encrusting, or branching colony.

Corals are classified on the basis of the nature and symmetry of their septa. The most common Paleozoic corals had septa arranged in fourfold symmetry and are called RUGOSA, or sometimes the tetracorals. Most post-Paleozoic corals have septa arranged in sixfold symmetry and are called SCLERACTINIA. In certain colonial Paleozoic corals the septa were reduced to

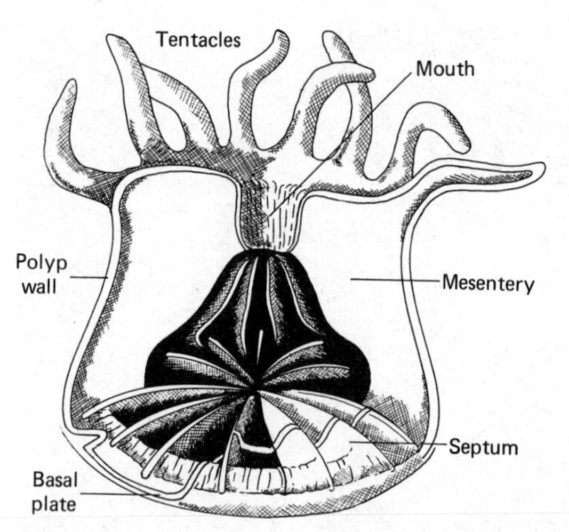

FIGURE A–11 **Cutaway drawing of a simple coral polyp, showing the digestive cavity divided radially by mesenteries, and the alternation of the mesenteries with septa in the basal part of the polyp.**

spines and low ridges. Because the individual tubes were relatively free of septa but crossed by prominent tabulae, these corals are called the tabulate corals (TABULATA).

Phylum Bryozoa

Although the bryozoan polyp superficially resembles that of the coelenterates, it is more complex. Instead of a simple sac for digestion, the bryozoan has a complete gut with separate mouth and anus. The individual is much smaller than the coral polyp and, with the exception of a single genus, always lives in colonies. The polyp of these colonies secretes a calcareous or chitinous "house" in the form of a box or a tube. The colony as a whole is made up of many such tubes or boxes and may be encrusting, branching, net-like, lacy, or hemispherical—or almost any shape (Fig. A–12). The internal structure of these colonies is often so complex that microscopic examination of thin-sections cut from them is required before they can be identified (Fig. A–13). Bryozoan colonies can generally be distinguished from colonial corals by the smallness of the cell and the absence of tabulae and septa. Since their beginning in the Ordovician Period, they have been exceedingly abundant and occasionally rock builders.

Phylum Brachiopoda

The soft parts of the brachiopods resemble those of the bryozoans, but the hard parts consist of two shells (or valves) secreted above and below the body. Although brachiopods can be gregarious, they never build colonial structures as do bryozoans. The brachiopod is attached to the sea floor by a muscular stalk or pedicle which emerges either between the valves or, more commonly, from a hole in the one known as the pedicle valve (Fig. A–14). The animal has either a continuous or a blind gut, a complex muscular system for the control of the valves, and delicate "arms" called brachia, which control respiration and the guiding of food toward the mouth. The brachia of many of these animals are supported by calcareous ribbons in the form of loops or spirals, or by short structures projecting from the brachial valve.

Most brachiopods secrete a shell of calcite, but primitive ones (class INARTICULATA) secreted chitin and calcium phosphate, which are commonly carbonized during preservation and appear as dark films on the bedding planes of rocks. The shells of advanced brachiopods (class ARTICULATA) are hinged, and interlock by tooth-and-socket structures. The hinged valves open to allow the animal to feed and breathe. The beaked part of the shell in the middle of the hinge marks the position from which growth started. The valves of most brachiopods show concentric lines of growth and are either folded or ridged in a radial pattern, presumably to give strength. In some brachiopods the hinge line is long and straight; in others it is short and slightly curved. In many brachiopods, the hinge line is interrupted in the pedicle valve by the passage of the pedicle to the exterior. The interior surfaces of both valves are scarred where the muscles that drew the valves together or spread them apart were attached.

Both brachiopod shells may be convex,

FIGURE A–12 **Small, twiglike bryozoans in Ordovician limestone from Montreal.**

or one may be convex and the other flat or concave. In all advanced and in most primitive brachiopods the valves are dissimilar because the larger one houses the pedicle opening; but each valve is symmetrical across an imaginary plane passing through the beaks perpendicular to one between the valves (Fig. A–14). Brachiopod shells may be distinguished from clam shells, which are mirror images of each other although neither is symmetrical. The brachiopods appeared in the Lower Cambrian Series, reached their acme in the middle and late part of the Paleozoic Era, and survive to the present day.

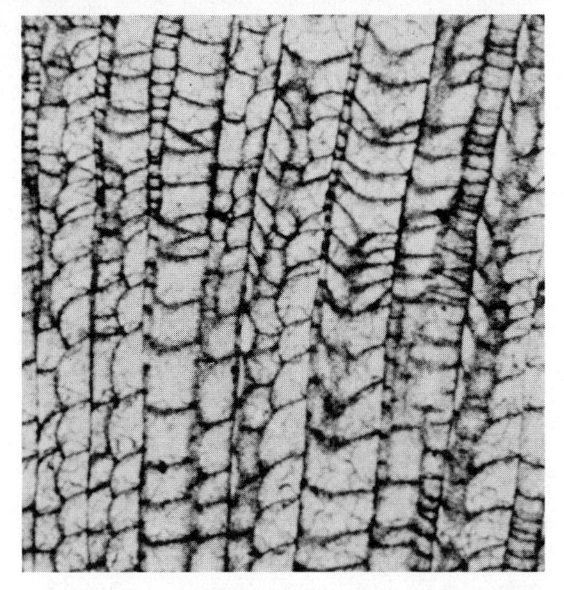

FIGURE A–13 **Transverse (*top*) and longitudinal (*bottom*) sections of *Prasopora,* a massive bryozoan colony, showing the complex internal structures of the fine tubes (magnification ×18). (Courtesy of M. A. Fritz.)**

Phylum Mollusca

The MOLLUSCA are a diverse group of animals, most of which secrete an external shell or shells of calcium carbonate. The phylum is divided into six classes, all of which have some fossil record; only three, however, the BIVALVIA, GASTROPODA, and CEPHALOPODA, are of primary importance to the paleontologist.

The bivalves are commonly called the clams. They have highly organized soft

  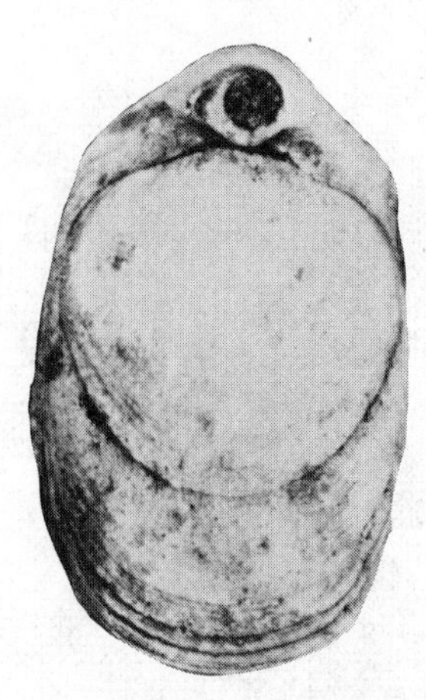

FIGURE A–14 **Brachiopods. *Left:* View of the pedicle valve. *Center:* Side view of the same specimen of the living brachiopod *Magellania*. *Right:* Brachial valve of the Miocene brachiopod *Olenothyris*, with the beak of the pedicle valve showing the pedicle opening. (*Olenothyris* courtesy of Ward's Natural Science Establishment, Inc., Rochester, N.Y.)**

parts which include a continuous gut, well-developed lamellar gills (hence their name LAMELLIBRANCHIA in British literature), and two tubes called siphons, which are essential for the feeding and respiration of the organism. Water filled with oxygen and minute food particles enters one of the siphons and leaves by the other, bearing waste products. Many clams can move slowly by thrusting out a muscular extension of their body (called the "foot"), but many are anchored to rocks or foreign objects by cementation or by organic threads.

The body of the bivalve is surrounded by a fleshy mantle that secretes the two valves of the shell. These are located on either side of the body, not in an over-and-under position like those of the brachiopod. The two valves are mirror images of each other, but in general are not bilaterally symmetrical. Each valve projects backward into a beak and is crossed by growth lines concentric with the beak (Fig. A–15). The valves fit together with teeth and sockets along the hinge line, which prevents lateral slippage between the beaks. In most shells the hinge is equipped with an elastic, organic ligament that is tensed when the valves are closed and relaxed when they are open. Two large muscles passing from valve to valve control this movement.

FIGURE A–15 **Interior and exterior of the two valves of the modern bivalve *Venus*, showing the lack of symmetry of the individual valves, the teeth and sockets, and the hinge line.**

The clam shell is composed of three layers. The outer layer is organic and prevents the inner layers of calcium carbonate from being dissolved in the surrounding sea or lake water. In freshwater clams this layer is thick, but in marine clams it is thin and easily worn away. The main body of the shell is composed of calcite, but the inner part is lined with aragonite (another form of calcium carbonate) in the form of mother-of-pearl. Since aragonite is considerably more soluble than calcite, bivalve shells in older rocks are generally less well preserved than those of the brachiopods, which secrete a shell entirely of calcite. The bivalve shell may be ornamented with nodes, ridges, spines, and projecting growth lines. Bivalves are first recorded in Cambrian rocks, and in general have been increasing in number and variety ever since.

The gastropods, or snails, have a single shell which is usually coiled like a circular staircase or corkscrew, although a few secrete a shell coiled in a single plane like a watch spring (Fig. A–16). The snails have a distinct head equipped with tentacles, eyes, and a mouth. The animal creeps on a muscular lobe of the body called the foot and carries its twisted shell on its back. In times of danger the snail withdraws into its shell. Some have a horny or calcareous plate that acts as a trap door to close the shell aperture once the animal is inside. The shell is a simple coiled cone and is not divided into chambers. Gastropod shells may be ornamented in various ways with spines, nodes, and keels (Fig. A–16). In older sedimentary rocks they are poorly preserved because the shell substance is largely aragonite, and many of the earlier fossils are only molds or fillings of the interior. The gastropods appeared in the Cambrian Period and have been abundant in marine environments since that time. They have also been successful in invading freshwater and terrestrial environments.

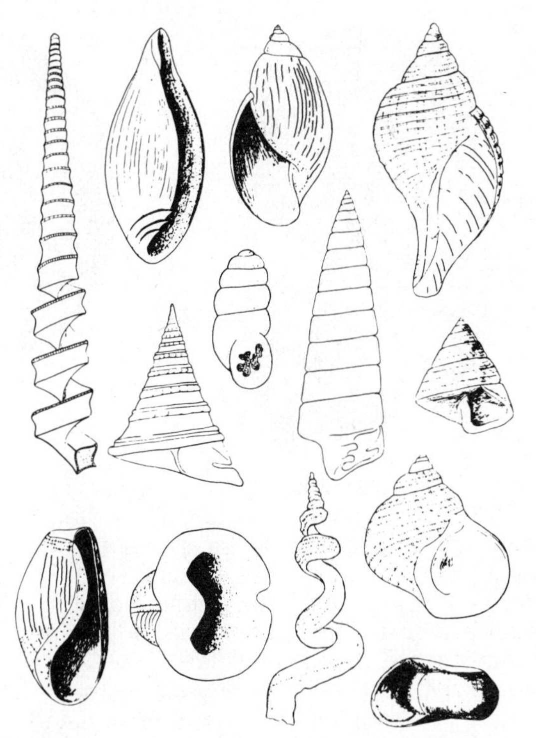

FIGURE A–16 **Variety in gastropod shell form. In all cases the spire, sometimes difficult to identify, points upward. Note that the mouth is usually on the right when facing the observer. (From *Treatise on Invertebrate Paleontology;* courtesy of the Geological Society of America and the University of Kansas Press.)**

The cephalopods include such diverse animals of the modern oceans as the squid, nautilus, octopus, and cuttlefish. As the name cephalopod (Greek: head-foot) suggests, the head of these animals is surrounded by tentacles (a modified foot) which capture food. The cephalopod shell is fundamentally an external, chambered cone, but some of the members of this class secrete their shells within the soft body tissues. The modern pearly *Nautilus,* the last of the great tribe of Paleozoic nautiloids, illustrates the main features of the shell structure (Fig. A–17). The shell is coiled in a single plane, and the outer whorl, in which the animal lives, covers all the earlier ones. The part of the shell not occupied by the animal is progressively partitioned off by concave, calcareous septa and the spaces so formed are filled with gas which buoys the animal up in the water. The septa are pierced by a calcareous tube (the siphuncle) that carries an extension of the soft parts of the animal (the siphon) through the abandoned cham-

FIGURE A–17 **Exterior and longitudinal section of the modern pearly *Nautilus*. Notice the septa and their extension backward to form a discontinuous tube for the siphon (1/4 actual size). (From *Treatise on Invertebrate Paleontology;* courtesy of the Geological Society of America and the University of Kansas Press.)**

bers. The line of contact between a septum and the inside surface of the exterior wall of the shell is called the suture. The earliest cephalopods had straight, conical shells in which the suture lines were simple and nearly circular. The cephalopods with simple septa are called nautiloids. They flourished from Late Cambrian to Silurian time, but thereafter dwindled in numbers to the single genus of the modern oceans, *Nautilus*.

In Devonian time another type of cephalopod appeared in which the edges of the septa were crenulated, forming an undulating suture line. All the Paleozoic members of this advanced group, called the ammonoids, were coiled in a single plane, but different patterns of coiling appeared frequently in Mesozoic groups. The ammonoids evolved through a progressive complication of the suture line from the simplest type, with broad flowing curves (goniatites, through a type in which one side of the suture inflections was crinkled (ceratites), to a type in which the whole of the suture was intensely folded (ammonites).

The belemnites were squid-like cephalopods with an internal shell similar in many ways to that of a straight nautiloid shell. The phragmocone was small, conical, and chambered and was encased posteriorly in a solid structure of calcite, called a guard, of approximately the size and shape of a cigar (Fig. A–18). This guard is generally all that is preserved of the belemnite. From the front end of the phragmocone extended a rarely preserved, horny support for the front part of the body, called a pen. The guards of belemnites are common fossils in Mesozoic sedimentary rocks and are used in correlation.

Phyla Annelida

The annelids, or segmented worms, include the earthworms, sandworms, and leeches. They are poorly represented in the fossil record, but their hard siliceous mouth parts, known as scolecodonts, were occasionally preserved. The burrows of annelid worms are not uncommon in sandstone formations.

Phylum Arthropoda

The arthropods are complex invertebrates whose body is enclosed in a tough, chitinous, segmented exoskeleton. The limbs are jointed, allowing the animal to move

FIGURE A–18 **Belemnite guard and phragmocone. Half the guard is cut away to show the chambered phragmocone within.**

freely, and from this feature the name ARTHROPODA (jointed legs) was derived. The arthropods have been among the most successful of all animals since their appearance in Cambrian time. They have diversified into marine, freshwater, terrestrial, and aerial environments and include such different animals as the crabs, barnacles, trilobites, insects, spiders, and scorpions. Although nearly every order has some representation in the fossil record, only four have been preserved in any numbers: the TRILOBITA, INSECTA, OSTRACODA, and EURYPTERIDA.

The trilobites were a group of exclusively Paleozoic arthropods that were divided by two longitudinal grooves into an axial lobe and two lateral lobes (Fig. A–19). The upper surface of the trilobite was protected by a thickened chitinous carapace. The body was divided into a head (cephalon), thorax, and tail (pygidium). The thorax was distinctly segmented, and each segment bore on the underside a pair of doubly branched limbs used in crawling or swimming. The segmentation of the head and tail was less obvious. The axial lobe of the head (glabella) was raised and bordered by lower areas called cheeks, on which multifaceted eyes were located. The grooves indenting the axial lobe marked the boundaries between the segments that fused to form the head. The limbs of trilobites were apparently unsuitable for preservation, for they are among the rarest of fossils. Trilobites, like most other arthropods, grew by discarding or molting their exoskeleton and secreting a larger one. Many trilobite fossils are undoubtedly these discarded carapaces.

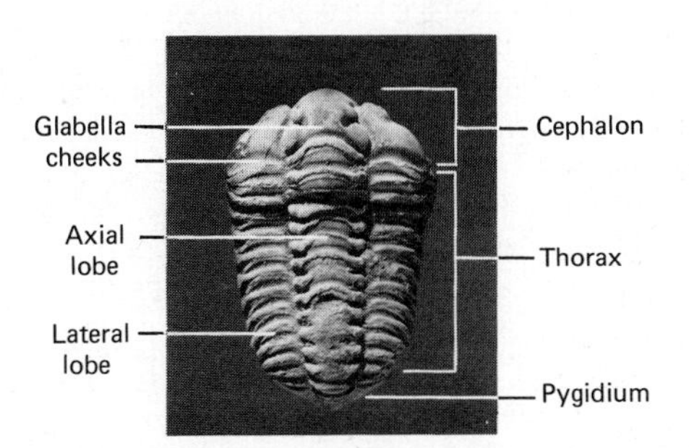

FIGURE A–19 **A typical trilobite, *Calymene*, showing the subdivision into cephalon, thorax, and pygidium, and other features.**

The insects are the most numerous and diverse group of anthropods. Like most animals living in terrestrial and aerial environments, however, their fossil record is unsatisfactory. Apart from a few exceptional sedimentary deposits which contain them in great numbers (Fig. A–20), the insects are practically unrepresented in the sedimentary sequence.

The ostracods are members of the crustacean arthropods, and are related to the crabs and lobsters. They are small—in most cases minute—aquatic animals that secrete a double-valved shell carried on either side of the body and hinged at the top. Appendages projecting between the edges of the shell are used by the animal to swim. Only the shells of these animals are preserved as fossils, but they occur in great numbers in some beds. Because they are small enough to be brought intact to the surface in the cuttings from oil wells, these shells have been extensively used for correlation. They may be smooth or complexly ornamented with ridges, spines, nodes, frills, and pits. Ostracods are abundant in marine and freshwater sediments from Ordovician time on. Crabs and lobsters begin to be common at the end of the Mesozoic.

The eurypterids were a group of Paleozoic arthropods that resembled, and were closely related to, the scorpions. Unlike the scorpions, however, they were aquatic. The eurypterid body consisted of an apparently unsegmented head (on the underside of which the appendages were attached), a segmented abdomen, and a spike-like or button-like tail. In some eurypterids the appendages were modified for walking, swimming, or catching food.

The spiders are close relatives of the eurypterids. Like the insects, they are poorly represented as fossils, but very primitive forms are known from rocks as old as mid-Devonian.

Phylum Echinodermata

Although echinoderms do not appear to be the most advanced of the invertebrates,

FIGURE A–20 **Senoprosopis, a fly from Oligocene beds at Florissant, Colorado. Note the preservation of the fine setae on the legs. (Courtesy of F. M. Carpenter.)**

most paleontologists believe that chordates and echinoderms sprang from a common ancestor. Echinoderms have a complex system of tubes for carrying water through the body, and a test composed of calcite plates transversed by multitudes of fine canals. Each of the plates is a single calcite crystal. Most echinoderms have a fivefold symmetry superposed on a bilateral symmetry. Some are sedentary, rooted by a stem to the ocean floor throughout their adult life. The mouths of these animals open upward, and the stems that connect them to the substratum are on the opposite side. Free echinoderms, such as the sea stars and sea urchins, have no stems and the mouths open downward.

The body of the anchored echinoderms is housed in a small sac of plates called the calyx. The calyx is attached to a column of centrally perforated, disc-like plates resembling beads on a necklace. They are held together by a strand of muscular tissue that passes through the openings. The mouth is generally surrounded by branched arms arranged in fivefold symmetry. Food particles trapped by the arms pass along food grooves to the mouth.

The carpoids form what is probably the most primitive group of echinoderms. Their bilateral symmetry is supposedly characteristic of the ancestral echinoderm, and they lack a water-vascular system (Fig. A–21).

The cystoids are primitive anchored echinoderms, with fivefold symmetry generally poorly developed (Fig. A–22). In some cystoids the food-gathering arms are absent and the food grooves are confined to the surface of the calyx; in others, one or two arms may be present. Their plates are

FIGURE A–21 **The carpoid *Enoploura* showing upper and lower surfaces. One "arm" is missing, one is broken. The resemblance to cystoids (Fig. A–22) is marked, but the two groups are separated by important morphological differences. (Courtesy of K. E. Caster.)**

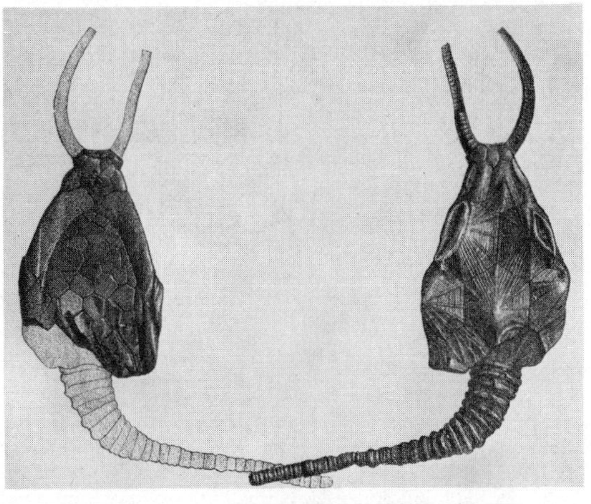

FIGURE A–22 ***Pleurocystites,* an Ordovician cystoid somewhat resembling the carpoids in the flattened form, arrangement of plates, arms, and stem. However, many cystoids lacked one or more of these features. (Courtesy of the Geological Survey of Canada.)**

perforated by pores arranged in a variety of patterns.

The blastoids were a group of Paleozoic echinoderms that had calices composed of 13 plates (Fig. A–23). Five food grooves extended downward over the sides of the calyx and were bordered by rows of small platelets. Each groove was lined with small arms which guided the food into the mouth.

The crinoids have been the most successful of the attached echinoderms since their appearance in Early Ordovician time. The crinoid calyx is composed of symmetrically arranged plates. The numerous arms occur in multiples of five and are branched many times. The plates of crinoids, like those of most attached echinoderms, tend to fall apart after the animal dies and are deposited by the waves as crinoidal limestones. Some crinoids shed their stems when they are mature and float free in the open ocean.

The only abundant free echinoderms are the asteroids (sea stars) and the echinoids (sea urchins). The asteroids are usually shaped like a short-armed, five-pointed star. The food grooves along the undersides of the arms converge on the central mouth. The groove of each arm is lined with minute tube feet which, controlled by the interior water-circulatory system, move the animal over the ocean floor.

The soft parts of the echinoids are enclosed in globular to discoidal shells that are covered with movable spines. Each spine projects from a protuberance on the plates (Fig. A–24), which are arranged in ten compound zones around the animal from "pole to pole." Alternate zones are perforated with double pores for the extension of the tube feet. The animal progresses by the movement of the spines and tube feet on its lower surface. In primitive, radially symmetrical echinoids, the mouth is at the center of this lower surface and the anus is diametrically opposite at the top. In the advanced types, the mouth evolved to a forward position—in the direction of movement—and the anus moved backward, destroying the radial symmetry and leading to

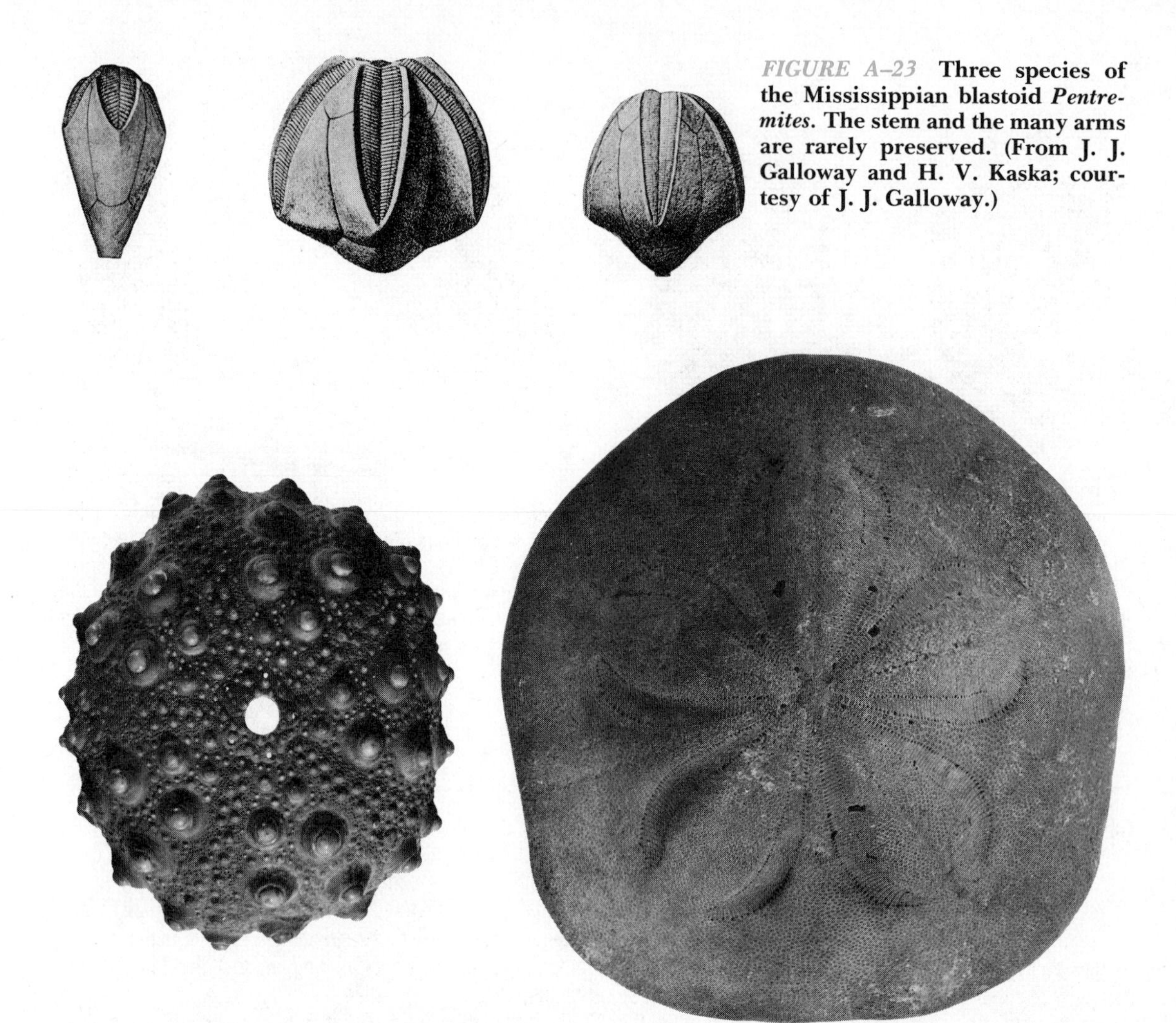

FIGURE A–23 **Three species of the Mississippian blastoid *Pentremites*. The stem and the many arms are rarely preserved. (From J. J. Galloway and H. V. Kaska; courtesy of J. J. Galloway.)**

FIGURE A–24 ***Left:* The regular echinoid *Heterocentrotus*. The large tubercles bear spines as long as the diameter of the test; the anal opening is at the top. *Right:* The irregular echinoid *Clypeaster* with five petalloid areas; the anal opening on the underside is not visible. (Courtesy of the Redpath Museum.)**

heart-shaped echinoids with mouth and anus at opposite sides of the lower surface. In such asymmetrical echinoids the plates bearing the pores are commonly arranged in a petal-shaped pattern on the upper surface.

Phylum Hemichordata

The living acorn worms are worm-like filter feeders with numerous gill slits, and the pterobranchs are small colonial tentacle-feeders. Larval development resembles that of the chordates.

Class Graptolithina. The graptolites are an extinct group of colonial organisms apparently, though obscurely, related to the hemichordates. All graptolites secreted colonial structures of chitin that are generally preserved as carbonized films on the bedding planes of black shales. The colonies are composed of a number of minute cups, each of which housed a polyp-like individual. The first graptolites built net-like structures in which the individuals occupied cups on branches of the net. More advanced graptolite colonies consisted of one or a few branches with cups arranged in a single or double row. The graptolites apparently floated, attached either to seaweed or to floats of their own making. They are excellent index fossils in rocks of Ordovician and Silurian ages.

Phylum Chordata, Subphylum Vertebrata

The vertebrates are chordates possessing a backbone, and are grouped into five familiar divisions: fish, amphibians, reptiles, birds, and mammals. The fish include such a diversity of primitive vertebrates that in most classifications they are represented by four classes: AGNATHA, PLACODERMI, CHONDRICHTHYES, and OSTEICHTHYES.

Agnatha. The most primitive of the vertebrates, the agnaths, lack jaws. Their mouths are merely holes on the undersides of their heads. This group is represented today by the lampreys, which are parasites of larger fish. Lampreys have a circular, rasping mouth by which they attach themselves to other fish and eat their flesh and suck their blood. During Ordovician and most of Silurian time, the only fish of which we have a record belong to this jawless class, but they were much different from their modern parasitic descendants.

Most of the early agnaths were characterized by a covering of bony plates and, hence, were called ostracoderms (armored skin) by paleontologists. The head was encased by a solid shield of bone above, and by a mosaic of smaller plates below in which the mouth and gill openings were located. Behind this rigid head the flexible body was covered with smaller plates. The body and the tail propelled the fish through the water. A few of the agnaths had pectoral fins for balance and guidance, but, some had spines which probably served the same purpose, although inefficiently. With their heavy armor they must have been poor swimmers, and probably lived much of their lives groveling on the sea floor in search of organic debris.

The fossil record of the agnaths begins in the Late Cambrian Series, but only in rocks of Late Silurian and Devonian age was a record of the diversity and number of this race preserved. None of these later ostracoderms was ancestral to the higher fishes. The more advanced vertebrates must have branched off from an ancestor of the AGNATHA, perhaps as early as the Cambrian Period.

Placodermi. The placoderms were the first of the jawed vertebrates. They flourished during Late Silurian and Devonian times, dwindled to insignificance by late Paleozoic time (when the bony fish replaced them), and became extinct before the close of the era. They are a diverse class whose subdivisions have little in common.

Chondrichthyes. The CHONDRICHTHYES, or sharks, have an almost entirely cartilaginous skeleton, and as a consequence have a poor fossil record. They lack an air bladder, or lung, which is common in other fish. The sharks have been a successful branch of the vertebrates ever since they first appeared in the Devonian Period.

Osteichthyes. Nearly all the fish of the modern fauna belong to the class OSTEICHTHYES, commonly referred to as the bony fish. These animals have a strong internal skeleton, a rigid skull composed of many bones, and two pairs of efficient paired fins that guide and balance the fish in the water. The divisions of the osteichthyans are particularly significant for the paleontologist: one comprises the ray-finned fish (ACTINOPTERYGII), and the other, lobe-finned fish (SARCOPTERYGII).

The first of the ray-finned fish appeared in the Devonian Period; they were apparently very well adapted to fast swimming, for they soon replaced the competing placoderms as masters of the sea. Early members of this lineage had thick, shiny, diamond-shaped scales, and lungs that served as a swim-bladder that controlled the buoyancy.

From the standpoint of evolution, the sarcopterygians are the more important group of the bony fish. The fins of these fish, unlike those of the rayfins, are supported by a framework of bone with elements homologous to those of the limbs of land vertebrates. Some of the sarcopterygians retained functional lungs; in other fish these were converted into a swim-bladder. Two branches of this group diverged early in the Devonian Period: the lungfish and the crossopterygians, which eventually gave rise to the land vertebrates. Three species of lungfish have survived to the present day, but they are only remnants of an order that was more abundant in the Mesozoic Era. A branch of the crossopterygians, not ancestral to the higher vertebrates, persists to the present day and is represented by a few specimens named *Latimeria,* first caught in 1939 off the coast of Madagascar. Skeletal comparisons of the Devonian crossopterygians with latest Devonian amphibians leave little doubt that one evolved from the other.

Amphibia. The amphibians are the most primitive of the land vertebrates. Their eggs are typically laid in water and their young commonly have a larval aquatic stage. The amphibians have not been entirely freed from dependence on water, and few of them can live far from it. The frog is the most familiar of modern amphibians.

The most primitive amphibians yet found come from beds of latest Devonian age. Nearly all the Paleozoic amphibians were low, squat creatures who walked with their elbows and knees held outward in a slow, sprawling, wallowing gait. During most of Carboniferous time, the amphibians were unchallenged as rulers of the land. When the reptiles developed late in the Paleozoic Era, however, the amphibians were relegated gradually to the insignificant role they play in the fauna of the modern world.

One of the most important groups of amphibians of late Paleozoic time was the labyrinthodonts. These animals are so named because the enamel of their conical teeth is infolded to make a labyrinthine pattern in cross-section. In many respects these animals were similar to their crossopterygian ancestors.

Reptilia. The reptiles achieved freedom from the water because their eggs were encased in membranes which protected the embryo as it developed. This great step forward in adaptation to land life is difficult to place in time, for eggs are rarely preserved in accumulating sediments. The oldest known egg (Early Permian) postdates the oldest reptilian skeleton (Early Pennsylvanian). Reptiles quite probably evolved from amphibians during the Mississippian Period.

The most primitive order of reptiles, from which all others appear to have arisen, is called the COTYLOSAURIA, or the stem reptiles. The diversification of the reptiles into many different environments hitherto unoccupied by vertebrates gave rise to an evolutionary "explosion" in Pennsylvanian time. By Late Permian time all subclasses of reptiles were represented in the fossil record: the stem reptiles and turtles (ANAPSIDA), the mammal-like reptiles (SYNAPSIDA), and a large group (DIAPSIDA) that includes the dinosaurs and their ancestors the thecodonts, and the snakes, lizards, and crocodiles.

Aves. The birds are warm-blooded, flying animals characterized by various structural adaptations to flight, such as wings, feathers, and hollow bones. The first birds, found in Upper Jurassic rocks, show many of the characteristics of the reptilian skeleton, and they obviously were descended from the branch of the reptiles that included the dinosaurs. The birds evolved rapidly through late Mesozoic and Cenozoic times, and their ability to fly was perfected by a reduction in the length of tail, an enlargement of the breastbone (anchoring the flight muscles), the fusing of the bones of the hand (giving support to the wing), and many other structural changes. The fossil record of the birds is meager, for animals living in the aerial environment have little chance of being preserved in the sedimentary record.

Mammalia. Some time prior to the end of the Triassic Period, the first mammals emerged from the stock of the mammal-like reptiles. Most of the mammalian skeletal characteristics, such as differentiated teeth, single jaw bone, double ball-and-socket arrangement at the back of the head, and secondary palate, had been attained or approached by various members of the mammal-like reptiles. The basic mammalian features of reproduction—live birth, and suckling—cannot be found in the fossil record; nor normally can the hairy covering, which is another feature of nearly all mammals.

The primitive Mesozoic mammals have been divided into several orders. The pantotheres are of the greatest evolutionary importance, for this group gave rise to the higher mammals at the beginning of the Cretaceous Period. The advanced mammals constitute two numerically unequal groups: the marsupials, whose young are born in an underdeveloped state and must live for some time in their mother's pouch before becoming independent, and the placentals, which retain their young inside the body until they are much more completely developed.

INDEX

Page numbers in *italic* type refer to pages showing pertinent illustrations. Where illustrations appear within a section of several pages, they have not been noted. Page numbers followed by an asterisk refer to the page on which the term is defined and on which it appears in **boldface** type.